Lecture Notes in Computer Science 9156

Commenced Publication in 1973
Founding and Former Series Editors:
Gerhard Goos, Juris Hartmanis, and Jan van Leeuwen

More information about this series at http://www.springer.com/series/7407

Osvaldo Gervasi · Beniamino Murgante
Sanjay Misra · Marina L. Gavrilova
Ana Maria Alves Coutinho Rocha · Carmelo Torre
David Taniar · Bernady O. Apduhan (Eds.)

Computational Science and Its Applications – ICCSA 2015

15th International Conference
Banff, AB, Canada, June 22–25, 2015
Proceedings, Part II

 Springer

Editors
Osvaldo Gervasi
University of Perugia
Perugia
Italy

Ana Maria Alves Coutinho Rocha
University of Minho
Braga
Portugal

Beniamino Murgante
University of Basilicata
Potenza
Italy

Carmelo Torre
Polytechnic University
Bari
Italy

Sanjay Misra
Covenant University
Canaanland
Nigeria

David Taniar
Monash University
Clayton, VIC
Australia

Marina L. Gavrilova
University of Calgary
Calgary, AB
Canada

Bernady O. Apduhan
Kyushu Sangyo University
Fukuoka
Japan

ISSN 0302-9743   ISSN 1611-3349   (electronic)
Lecture Notes in Computer Science
ISBN 978-3-319-21406-1   ISBN 978-3-319-21407-8   (eBook)
DOI 10.1007/978-3-319-21407-8

Library of Congress Control Number: 2015943360

LNCS Sublibrary: SL1 – Theoretical Computer Science and General Issues

Springer Cham Heidelberg New York Dordrecht London
© Springer International Publishing Switzerland 2015

Printed on acid-free paper

Springer International Publishing AG Switzerland is part of Springer Science+Business Media
(www.springer.com)

Preface

The year 2015 is a memorable year for the International Conference on Computational Science and Its Applications. In 2003, the First International Conference on Computational Science and Its Applications (chaired by C.J.K. Tan and M. Gavrilova) took place in Montreal, Canada (2003), and the following year it was hosted by A. Laganà and O. Gervasi in Assisi, Italy (2004). It then moved to Singapore (2005), Glasgow, UK (2006), Kuala-Lumpur, Malaysia (2007), Perugia, Italy (2008), Seoul, Korea (2009), Fukuoka, Japan (2010), Santander, Spain (2011), Salvador de Bahia, Brazil (2012), Ho Chi Minh City, Vietnam (2013), and Guimarães, Portugal (2014). The current installment of ICCSA 2015 took place in majestic Banff National Park, Banff, Alberta, Canada, during June 22–25, 2015.

The event received approximately 780 submissions from over 45 countries, evaluated by over 600 reviewers worldwide.

Its main track acceptance rate was approximately 29.7 % for full papers. In addition to full papers, published by Springer, the event accepted short papers, poster papers, and PhD student showcase works that are published in the IEEE CPS proceedings.

It also runs a number of parallel workshops, some for over 10 years, with new ones appearing for the first time this year. The success of ICCSA is largely contributed to the continuous support of the computational sciences community as well as researchers working in the applied relevant fields, such as graphics, image processing, biometrics, optimization, computer modeling, information systems, geographical sciences, physics, biology, astronomy, biometrics, virtual reality, and robotics, to name a few.

Over the past decade, the vibrant and promising area focusing on performance-driven computing and big data has became one of the key points of research enhancing the performance of information systems and supported processes. In addition to high-quality research at the frontier of these fields, consistently presented at ICCSA, a number of special journal issues are being planned following ICCSA 2015, including TCS Springer (*Transactions on Computational Sciences,* LNCS).

The contribution of the International Steering Committee and the International Program Committee are invaluable in the conference success. The dedication of members of these committees, the majority of whom have fulfilled this difficult role for the last 10 years, is astounding. Our warm appreciation also goes to the invited speakers, all event sponsors, supporting organizations, and volunteers. Finally, we thank all the authors for their submissions making the ICCSA conference series a well recognized and a highly successful event year after year.

June 2015

Marina L. Gavrilova
Osvaldo Gervasi
Bernady O. Apduhan

Organization

ICCSA 2015 was organized by the University of Calgary (Canada), the University of Perugia (Italy), the University of Basilicata (Italy), Monash University (Australia), Kyushu Sangyo University (Japan), and the University of Minho, (Portugal)

Honorary General Chairs

Antonio Laganà	University of Perugia, Italy
Norio Shiratori	Tohoku University, Japan
Kenneth C.J. Tan	Sardina Systems, Estonia

General Chairs

Marina L. Gavrilova	University of Calgary, Canada
Osvaldo Gervasi	University of Perugia, Italy
Bernady O. Apduhan	Kyushu Sangyo University, Japan

Program Committee Chairs

Beniamino Murgante	University of Basilicata, Italy
Ana Maria A.C. Rocha	University of Minho, Portugal
David Taniar	Monash University, Australia

International Advisory Committee

Jemal Abawajy	Deakin University, Australia
Dharma P. Agrawal	University of Cincinnati, USA
Claudia Bauzer Medeiros	University of Campinas, Brazil
Manfred M. Fisher	Vienna University of Economics and Business, Austria
Yee Leung	Chinese University of Hong Kong, SAR China

International Liaison Chairs

Ana Carla P. Bitencourt	Universidade Federal do Reconcavo da Bahia, Brazil
Alfredo Cuzzocrea	ICAR-CNR and University of Calabria, Italy
Maria Irene Falcão	University of Minho, Portugal
Marina L. Gavrilova	University of Calgary, Canada
Robert C.H. Hsu	Chung Hua University, Taiwan
Andrés Iglesias	University of Cantabria, Spain
Tai-Hoon Kim	Hannam University, Korea
Sanjay Misra	University of Minna, Nigeria
Takashi Naka	Kyushu Sangyo University, Japan

Rafael D.C. Santos   Brazilian National Institute for Space Research, Brazil
Maribel Yasmina Santos   University of Minho, Portugal

Workshop and Session Organizing Chairs

Beniamino Murgante   University of Basilicata, Italy
Jorge Gustavo Rocha   University of Minho, Portugal

Local Arrangement Chairs

Marina Gavrilova   University of Calgary, Canada (Chair)
Madeena Sultana   University of Calgary, Canada
Padma Polash Paul   University of Calgary, Canada
Faisal Ahmed   University of Calgary, Canada
Hossein Talebi   University of Calgary, Canada
Camille Sinanan   University of Calgary, Canada

Venue

ICCSA 2015 took place in the Banff Park Lodge Conference Center, Alberta (Canada).

Workshop Organizers

Agricultural and Environment Information and Decision Support Systems (AEIDSS 2015)

Sandro Bimonte   IRSTEA, France
André Miralles   IRSTEA, France
Frederic Hubert   University of Laval, Canada
François Pinet   IRSTEA, France

Approaches or Methods of Security Engineering (AMSE 2015)

TaiHoon Kim   Sungshin W. University, Korea

Advances in Information Systems and Technologies for Emergency Preparedness and Risk Assessment (ASTER 2015)

Maurizio Pollino   ENEA, Italy
Marco Vona   University of Basilicata, Italy
Beniamino Murgante   University of Basilicata, Italy

Advances in Web-Based Learning (AWBL 2015)

Mustafa Murat Inceoglu   Ege University, Turkey

Bio-inspired Computing and Applications (BIOCA 2015)

Nadia Nedjah   State University of Rio de Janeiro, Brazil
Luiza de Macedo   State University of Rio de Janeiro, Brazil
  Mourell

Computer-Aided Modeling, Simulation, and Analysis (CAMSA 2015)

Jie Shen   University of Michigan, USA, and Jilin University, China
Hao Chen   Shanghai University of Engineering Science, China
Xiaoqiang Liun   Donghua University, China
Weichun Shi   Shanghai Maritime University, China

Computational and Applied Statistics (CAS 2015)

Ana Cristina Braga   University of Minho, Portugal
Ana Paula Costa   University of Minho, Portugal
  Conceicao Amorim

Computational Geometry and Security Applications (CGSA 2015)

Marina L. Gavrilova   University of Calgary, Canada

Computational Algorithms and Sustainable Assessment (CLASS 2015)

Antonino Marvuglia   Public Research Centre Henri Tudor, Luxembourg
Beniamino Murgante   University of Basilicata, Italy

Chemistry and Materials Sciences and Technologies (CMST 2015)

Antonio Laganà   University of Perugia, Italy
Alessandro Costantini   INFN, Italy
Noelia Faginas Lago   University of Perugia, Italy
Leonardo Pacifici   University of Perugia, Italy

Computational Optimization and Applications (COA 2015)

Ana Maria Rocha   University of Minho, Portugal
Humberto Rocha   University of Coimbra, Portugal

Cities, Technologies and Planning (CTP 2015)

Giuseppe Borruso   University of Trieste, Italy
Beniamino Murgante   University of Basilicata, Italy

Econometrics and Multidimensional Evaluation in the Urban Environment (EMEUE 2015)

Carmelo M. Torre   Polytechnic of Bari, Italy
Maria Cerreta   University of Naples Federico II, Italy
Paola Perchinunno   University of Bari, Italy

Simona Panaro	University of Naples Federico II, Italy
Raffaele Attardi	University of Naples Federico II, Italy
Claudia Ceppi	Polytechnic of Bari, Italy

Future Computing Systems, Technologies, and Applications (FISTA 2015)

Bernady O. Apduhan	Kyushu Sangyo University, Japan
Rafael Santos	Brazilian National Institute for Space Research, Brazil
Jianhua Ma	Hosei University, Japan
Qun Jin	Waseda University, Japan

Geographical Analysis, Urban Modeling, Spatial Statistics (GEOGAN-MOD 2015)

Giuseppe Borruso	University of Trieste, Italy
Beniamino Murgante	University of Basilicata, Italy
Hartmut Asche	University of Potsdam, Germany

Land Use Monitoring for Soil Consumption Reduction (LUMS 2015)

Carmelo M. Torre	Polytechnic of Bari, Italy
Alessandro Bonifazi	Polytechnic of Bari, Italy
Valentina Sannicandro	University Federico II of Naples, Italy
Massimiliano Bencardino	University of Salerno, Italy
Gianluca di Cugno	Polytechnic of Bari, Italy
Beniamino Murgante	University of Basilicata, Italy

Mobile Communications (MC 2015)

| Hyunseung Choo | Sungkyunkwan University, Korea |

Mobile Computing, Sensing, and Actuation for Cyber Physical Systems (MSA4CPS 2015)

| Saad Qaisar | NUST School of Electrical Engineering and Computer Science, Pakistan |
| Moonseong Kim | Korean Intellectual Property Office, Korea |

Quantum Mechanics: Computational Strategies and Applications (QMCSA 2015)

Mirco Ragni	Universidad Federal de Bahia, Brazil
Ana Carla Peixoto Bitencourt	Universidade Estadual de Feira de Santana, Brazil
Roger Anderson	University of California, USA
Vincenzo Aquilanti	University of Perugia, Italy
Frederico Vasconcellos Prudente	Universidad Federal de Bahia, Brazil

Remote Sensing Data Analysis, Modeling, Interpretation and Applications: From a Global View to a Local Analysis (RS2015)

Rosa Lasaponara	Institute of Methodologies for Environmental Analysis, National Research Council, Italy

Scientific Computing Infrastructure (SCI 2015)

Alexander Bodganov	St. Petersburg State University, Russia
Elena Stankova	St. Petersburg State University, Russia

Software Engineering Processes and Applications (SEPA 2015)

Sanjay Misra	Covenant University, Nigeria

Software Quality (SQ 2015)

Sanjay Misra	Covenant University, Nigeria

Advances in Spatio-Temporal Analytics (ST-Analytics 2015)

Joao Moura Pires	New University of Lisbon, Portugal
Maribel Yasmina Santos	New University of Lisbon, Portugal

Tools and Techniques in Software Development Processes (TTSDP 2015)

Sanjay Misra	Covenant University, Nigeria

Virtual Reality and Its Applications (VRA 2015)

Osvaldo Gervasi	University of Perugia, Italy
Lucio Depaolis	University of Salento, Italy

Program Committee

Jemal Abawajy	Deakin University, Australia
Kenny Adamson	University of Ulster, UK
Filipe Alvelos	University of Minho, Portugal
Paula Amaral	Universidade Nova de Lisboa, Portugal
Hartmut Asche	University of Potsdam, Germany
Md. Abul Kalam Azad	University of Minho, Portugal
Michela Bertolotto	University College Dublin, Ireland
Sandro Bimonte	CEMAGREF, TSCF, France
Rod Blais	University of Calgary, Canada
Ivan Blecic	University of Sassari, Italy
Giuseppe Borruso	University of Trieste, Italy
Yves Caniou	Lyon University, France
José A. Cardoso e Cunha	Universidade Nova de Lisboa, Portugal
Leocadio G. Casado	University of Almeria, Spain

Carlo Cattani	University of Salerno, Italy
Mete Celik	Erciyes University, Turkey
Alexander Chemeris	National Technical University of Ukraine KPI, Ukraine
Min Young Chung	Sungkyunkwan University, Korea
Gilberto Corso Pereira	Federal University of Bahia, Brazil
M. Fernanda Costa	University of Minho, Portugal
Gaspar Cunha	University of Minho, Portugal
Alfredo Cuzzocrea	ICAR-CNR and University of Calabria, Italy
Carla Dal Sasso Freitas	Universidade Federal do Rio Grande do Sul, Brazil
Pradesh Debba	The Council for Scientific and Industrial Research (CSIR), South Africa
Hendrik Decker	Instituto Tecnológico de Informática, Spain
Frank Devai	London South Bank University, UK
Rodolphe Devillers	Memorial University of Newfoundland, Canada
Prabu Dorairaj	NetApp, India/USA
M. Irene Falcao	University of Minho, Portugal
Cherry Liu Fang	U.S. DOE Ames Laboratory, USA
Edite M.G.P. Fernandes	University of Minho, Portugal
Jose-Jesus Fernandez	National Centre for Biotechnology, CSIS, Spain
Maria Antonia Forjaz	University of Minho, Portugal
Maria Celia Furtado Rocha	PRODEB/UFBA, Brazil
Akemi Galvez	University of Cantabria, Spain
Paulino Jose Garcia Nieto	University of Oviedo, Spain
Marina Gavrilova	University of Calgary, Canada
Jerome Gensel	LSR-IMAG, France
Maria Giaoutzi	National Technical University, Athens, Greece
Andrzej M. Goscinski	Deakin University, Australia
Alex Hagen-Zanker	University of Cambridge, UK
Malgorzata Hanzl	Technical University of Lodz, Poland
Shanmugasundaram Hariharan	B.S. Abdur Rahman University, India
Eligius M.T. Hendrix	University of Malaga/Wageningen University, Spain/The Netherlands
Tutut Herawan	Universitas Teknologi Yogyakarta, Indonesia
Hisamoto Hiyoshi	Gunma University, Japan
Fermin Huarte	University of Barcelona, Spain
Andres Iglesias	University of Cantabria, Spain
Mustafa Inceoglu	EGE University, Turkey
Peter Jimack	University of Leeds, UK
Qun Jin	Waseda University, Japan
Farid Karimipour	Vienna University of Technology, Austria
Baris Kazar	Oracle Corp., USA
DongSeong Kim	University of Canterbury, New Zealand
Taihoon Kim	Hannam University, Korea

Ivana Kolingerova	University of West Bohemia, Czech Republic
Dieter Kranzlmueller	LMU and LRZ Munich, Germany
Antonio Laganà	University of Perugia, Italy
Rosa Lasaponara	National Research Council, Italy
Maurizio Lazzari	National Research Council, Italy
Cheng Siong Lee	Monash University, Australia
Sangyoun Lee	Yonsei University, Korea
Jongchan Lee	Kunsan National University, Korea
Clement Leung	Hong Kong Baptist University, Hong Kong, SAR China
Chendong Li	University of Connecticut, USA
Gang Li	Deakin University, Australia
Ming Li	East China Normal University, China
Fang Liu	AMES Laboratories, USA
Xin Liu	University of Calgary, Canada
Savino Longo	University of Bari, Italy
Tinghuai Ma	NanJing University of Information Science and Technology, China
Sergio Maffioletti	University of Zurich, Switzerland
Ernesto Marcheggiani	Katholieke Universiteit Leuven, Belgium
Antonino Marvuglia	Research Centre Henri Tudor, Luxembourg
Nicola Masini	National Research Council, Italy
Nirvana Meratnia	University of Twente, The Netherlands
Alfredo Milani	University of Perugia, Italy
Sanjay Misra	Federal University of Technology Minna, Nigeria
Giuseppe Modica	University of Reggio Calabria, Italy
José Luis Montaña	University of Cantabria, Spain
Beniamino Murgante	University of Basilicata, Italy
Jiri Nedoma	Academy of Sciences of the Czech Republic, Czech Republic
Laszlo Neumann	University of Girona, Spain
Kok-Leong Ong	Deakin University, Australia
Belen Palop	Universidad de Valladolid, Spain
Marcin Paprzycki	Polish Academy of Sciences, Poland
Eric Pardede	La Trobe University, Australia
Kwangjin Park	Wonkwang University, Korea
Ana Isabel Pereira	Polytechnic Institute of Braganca, Portugal
Maurizio Pollino	Italian National Agency for New Technologies, Energy and Sustainable Economic Development, Italy
Alenka Poplin	University of Hamburg, Germany
Vidyasagar Potdar	Curtin University of Technology, Australia
David C. Prosperi	Florida Atlantic University, USA
Wenny Rahayu	La Trobe University, Australia
Jerzy Respondek	Silesian University of Technology Poland
Ana Maria A.C. Rocha	University of Minho, Portugal

Humberto Rocha	INESC-Coimbra, Portugal
Alexey Rodionov	Institute of Computational Mathematics and Mathematical Geophysics, Russia
Cristina S. Rodrigues	University of Minho, Portugal
Octavio Roncero	CSIC, Spain
Maytham Safar	Kuwait University, Kuwait
Chiara Saracino	A.O. Ospedale Niguarda Ca' Granda - Milano, Italy
Haiduke Sarafian	The Pennsylvania State University, USA
Jie Shen	University of Michigan, USA
Qi Shi	Liverpool John Moores University, UK
Dale Shires	U.S. Army Research Laboratory, USA
Takuo Suganuma	Tohoku University, Japan
Sergio Tasso	University of Perugia, Italy
Ana Paula Teixeira	University of Tras-os-Montes and Alto Douro, Portugal
Senhorinha Teixeira	University of Minho, Portugal
Parimala Thulasiraman	University of Manitoba, Canada
Carmelo Torre	Polytechnic of Bari, Italy
Javier Martinez Torres	Centro Universitario de la Defensa Zaragoza, Spain
Giuseppe A. Trunfio	University of Sassari, Italy
Unal Ufuktepe	Izmir University of Economics, Turkey
Toshihiro Uchibayashi	Kyushu Sangyo University, Japan
Mario Valle	Swiss National Supercomputing Centre, Switzerland
Pablo Vanegas	University of Cuenca, Equador
Piero Giorgio Verdini	INFN Pisa and CERN, Italy
Marco Vizzari	University of Perugia, Italy
Koichi Wada	University of Tsukuba, Japan
Krzysztof Walkowiak	Wroclaw University of Technology, Poland
Robert Weibel	University of Zurich, Switzerland
Roland Wismüller	Universität Siegen, Germany
Mudasser Wyne	SOET National University, USA
Chung-Huang Yang	National Kaohsiung Normal University, Taiwan
Xin-She Yang	National Physical Laboratory, UK
Salim Zabir	France Telecom Japan Co., Japan
Haifeng Zhao	University of California, Davis, USA
Kewen Zhao	University of Qiongzhou, China
Albert Y. Zomaya	University of Sydney, Australia

Reviewers

Abawajy Jemal	Deakin University, Australia
Abdi Samane	University College Cork, Ireland
Aceto Lidia	University of Pisa, Italy
Acharjee Shukla	Dibrugarh University, India
Adriano Elias	Universidade Nova de Lisboa, Portugal
Afreixo Vera	University of Aveiro, Portugal
Aguiar Ademar	Universidade do Porto, Portugal

Aguilar Antonio	University of Barcelona, Spain
Aguilar José Alfonso	Universidad Autónoma de Sinaloa, Mexico
Ahmed Faisal	University of Calgary, Canada
Aktas Mehmet	Yildiz Technical University, Turkey
Al-Juboori AliAlwan	International Islamic University Malaysia, Malaysia
Alarcon Vladimir	Universidad Diego Portales, Chile
Alberti Margarita	University of Barcelona, Spain
Ali Salman	NUST, Pakistan
Alkazemi Basem Qassim	University, Saudi Arabia
Alvanides Seraphim	Northumbria University, UK
Alvelos Filipe	University of Minho, Portugal
Alves Cláudio	University of Minho, Portugal
Alves José Luis	University of Minho, Portugal
Alves Maria Joo	Universidade de Coimbra, Portugal
Amin Benatia Mohamed	Groupe Cesi, France
Amorim Ana Paula	University of Minho, Portugal
Amorim Paulo	Federal University of Rio de Janeiro, Brazil
Andrade Wilkerson	Federal University of Campina Grande, Brazil
Andrianov Serge	Yandex, Russia
Aniche Mauricio	University of São Paulo, Brazil
Andrienko Gennady	Fraunhofer Institute for Intelligent Analysis and Informations Systems, Germany
Apduhan Bernady	Kyushu Sangyo University, Japan
Aquilanti Vincenzo	University of Perugia, Italy
Aquino Gibeon	UFRN, Brazil
Argiolas Michele	University of Cagliari, Italy
Asche Hartmut	Potsdam University, Germany
Athayde Maria Emilia Feijão Queiroz	University of Minho, Portugal
Attardi Raffaele	University of Napoli Federico II, Italy
Azad Md. Abdul	Indian Institute of Technology Kanpur, India
Azad Md. Abul Kalam	University of Minho, Portugal
Bao Fernando	Universidade Nova de Lisboa, Portugal
Badard Thierry	Laval University, Canada
Bae Ihn-Han	Catholic University of Daegu, South Korea
Baioletti Marco	University of Perugia, Italy
Balena Pasquale	Polytechnic of Bari, Italy
Banerjee Mahua	Xavier Institute of Social Sciences, India
Barroca Filho Itamir	UFRN, Brazil
Bartoli Daniele	University of Perugia, Italy
Bastanfard Azam	Islamic Azad University, Iran
Belanzoni Paola	University of Perugia, Italy
Bencardino Massimiliano	University of Salerno, Italy
Benigni Gladys	University of Oriente, Venezuela

Bertolotto Michela	University College Dublin, Ireland
Bilancia Massimo	Università di Bari, Italy
Blanquer Ignacio	Universitat Politècnica de València, Spain
Bodini Olivier	Université Pierre et Marie Curie Paris and CNRS, France
Bogdanov Alexander	Saint-Petersburg State University, Russia
Bollini Letizia	University of Milano, Italy
Bonifazi Alessandro	Polytechnic of Bari, Italy
Borruso Giuseppe	University of Trieste, Italy
Bostenaru Maria	"Ion Mincu" University of Architecture and Urbanism, Romania
Boucelma Omar	University of Marseille, France
Braga Ana Cristina	University of Minho, Portugal
Branquinho Amilcar	University of Coimbra, Portugal
Brás Carmo	Universidade Nova de Lisboa, Portugal
Cacao Isabel	University of Aveiro, Portugal
Cadarso-Suárez Carmen	University of Santiago de Compostela, Spain
Caiaffa Emanuela	ENEA, Italy
Calamita Giuseppe	National Research Council, Italy
Campagna Michele	University of Cagliari, Italy
Campobasso Francesco	University of Bari, Italy
Campos José	University of Minho, Portugal
Caniato Renhe Marcelo	Universidade Federal de Juiz de Fora, Brazil
Cannatella Daniele	University of Napoli Federico II, Italy
Canora Filomena	University of Basilicata, Italy
Cannatella Daniele	University of Napoli Federico II, Italy
Canora Filomena	University of Basilicata, Italy
Carbonara Sebastiano	University of Chieti, Italy
Carlini Maurizio	University of Tuscia, Italy
Carneiro Claudio	École Polytechnique Fédérale de Lausanne, Switzerland
Ceppi Claudia	Polytechnic of Bari, Italy
Cerreta Maria	University Federico II of Naples, Italy
Chen Hao	Shanghai University of Engineering Science, China
Choi Joonsoo	Kookmin University, South Korea
Choo Hyunseung	Sungkyunkwan University, South Korea
Chung Min Young	Sungkyunkwan University, South Korea
Chung Myoungbeom	Sungkyunkwan University, South Korea
Chung Tai-Myoung	Sungkyunkwan University, South Korea
Cirrincione Maurizio	Université de Technologie Belfort-Montbeliard, France
Clementini Eliseo	University of L'Aquila, Italy
Coelho Leandro dos Santos	PUC-PR, Brazil
Coletti Cecilia	University of Chieti, Italy
Conceicao Ana	Universidade do Algarve, Portugal
Correia Elisete	University of Trás-Os-Montes e Alto Douro, Portugal
Correia Filipe	FEUP, Portugal

Correia Florbela Maria da Cruz Domingues	Instituto Politécnico de Viana do Castelo, Portugal
Corso Pereira Gilberto	UFPA, Brazil
Cortés Ana	Universitat Autònoma de Barcelona, Spain
Cosido Oscar	Ayuntamiento de Santander, Spain
Costa Carlos	Faculdade Engenharia U. Porto, Portugal
Costa Fernanda	University of Minho, Portugal
Costantini Alessandro	INFN, Italy
Crasso Marco	National Scientific and Technical Research Council, Argentina
Crawford Broderick	Universidad Catolica de Valparaiso, Chile
Crestaz Ezio	GiScience, Italia
Cristia Maximiliano	CIFASIS and UNR, Argentina
Cunha Gaspar	University of Minho, Portugal
Cutini Valerio	University of Pisa, Italy
Danese Maria	IBAM, CNR, Italy
Daneshpajouh Shervin	University of Western Ontario, Canada
De Almeida Regina	University of Trás-os-Montes e Alto Douro, Portugal
de Doncker Elise	University of Michgan, USA
De Fino Mariella	Polytechnic of Bari, Italy
De Paolis Lucio Tommaso	University of Salento, Italy
de Rezende Pedro J.	Universidade Estadual de Campinas, Brazil
De Rosa Fortuna	University of Napoli Federico II, Italy
De Toro Pasquale	University of Napoli Federico II, Italy
Decker Hendrik	Instituto Tecnológico de Informática, Spain
Degtyarev Alexander	Saint-Petersburg State University, Russia
Deiana Andrea	Geoinfolab, Italia
Deniz Berkhan	Aselsan Electronics Inc., Turkey
Desjardin Eric	University of Reims, France
Devai Frank	London South Bank University, UK
Dwivedi Sanjay Kumar	Babasaheb Bhimrao Ambedkar University, India
Dhawale Chitra	PR Pote College, Amravati, India
Di Cugno Gianluca	Polytechnic of Bari, Italy
Di Gangi Massimo	University of Messina, Italy
Di Leo Margherita	JRC, European Commission, Belgium
Dias Joana	University of Coimbra, Portugal
Dias d'Almeida Filomena	University of Porto, Portugal
Diez Teresa	Universidad de Alcalá, Spain
Dilo Arta	University of Twente, The Netherlands
Dixit Veersain	Delhi University, India
Doan Anh Vu	Université Libre de Bruxelles, Belgium
Durrieu Sylvie	Maison de la Teledetection Montpellier, France
Dutra Inês	University of Porto, Portugal
Dyskin Arcady	The University of Western Australia, Australia

Eichelberger Hanno	University of Tübingen, Germany
El-Zawawy Mohamed A.	Cairo University, Egypt
Escalona Maria-Jose	University of Seville, Spain
Falcão M. Irene	University of Minho, Portugal
Farantos Stavros	University of Crete and FORTH, Greece
Faria Susana	University of Minho, Portugal
Fernandes Edite	University of Minho, Portugal
Fernandes Rosário	University of Minho, Portugal
Fernandez Joao P.	Universidade da Beira Interior, Portugal
Ferrão Maria	University of Beira Interior and CEMAPRE, Portugal
Ferreira Fátima	University of Trás-Os-Montes e Alto Douro, Portugal
Figueiredo Manuel Carlos	University of Minho, Portugal
Filipe Ana	University of Minho, Portugal
Flouvat Frederic	University New Caledonia, New Caledonia
Forjaz Maria Antónia	University of Minho, Portugal
Formosa Saviour	University of Malta, Malta
Fort Marta	University of Girona, Spain
Franciosa Alfredo	University of Napoli Federico II, Italy
Freitas Adelaide de Fátima Baptista Valente	University of Aveiro, Portugal
Frydman Claudia	Laboratoire des Sciences de l'Information et des Systèmes, France
Fusco Giovanni	CNRS - UMR ESPACE, France
Gabrani Goldie	University of Delhi, India Galleguillos Cristian, Pontificia Universidad Catlica de Valparaso, Chile
Gao Shang	Zhongnan University of Economics and Law, China
Garau Chiara	University of Cagliari, Italy
Garcia Ernesto	University of the Basque Country, Spain
Garca Omar Vicente	Universidad Autònoma de Sinaloa, Mexico
Garcia Tobio Javier	Centro de Supercomputación de Galicia, CESGA, Spain
Gavrilova Marina	University of Calgary, Canada
Gazzea Nicoletta	ISPRA, Italy
Gensel Jerome	IMAG, France
Geraldi Edoardo	National Research Council, Italy
Gervasi Osvaldo	University of Perugia, Italy
Giaoutzi Maria	National Technical University Athens, Greece
Gil Artur	University of the Azores, Portugal
Gizzi Fabrizio	National Research Council, Italy
Gomes Abel	Universidad de Beira Interior, Portugal
Gomes Maria Cecilia	Universidade Nova de Lisboa, Portugal
Gomes dos Anjos Eudisley	Federal University of Paraba, Brazil
Gonçalves Alexandre	Instituto Superior Tecnico Lisboa, Portugal

Gonçalves Arminda Manuela	University of Minho, Portugal
Gonzaga de Oliveira Sanderson Lincohn	Universidade Do Estado De Santa Catarina, Brazil
Gonzalez-Aguilera Diego	Universidad de Salamanca, Spain
Gorbachev Yuriy	Geolink Technologies, Russia
Govani Kishan	Darshan Institute of Engineering Technology, India
Grandison Tyrone	Proficiency Labs International, USA
Gravagnuolo Antonia	University of Napoli Federico II, Italy
Grilli Luca	University of Perugia, Italy
Guerra Eduardo	National Institute for Space Research, Brazil
Guo Hua	Carleton University, Canada
Hanazumi Simone	University of São Paulo, Brazil
Hanif Mohammad Abu	Chonbuk National University, South Korea
Hansen Henning Sten	Aalborg University, Denmark
Hanzl Malgorzata	University of Lodz, Poland
Hegedus Peter	University of Szeged, Hungary
Heijungs Reinout	VU University Amsterdam, The Netherlands
Hendrix Eligius M.T.	University of Malaga/Wageningen University, Spain/The Netherlands
Henriques Carla	Escola Superior de Tecnologia e Gestão, Portugal
Herawan Tutut	University of Malaya, Malaysia
Hiyoshi Hisamoto	Gunma University, Japan
Hodorog Madalina	Austria Academy of Science, Austria
Hong Choong Seon	Kyung Hee University, South Korea
Hsu Ching-Hsien	Chung Hua University, Taiwan
Hsu Hui-Huang	Tamkang University, Taiwan
Hu Hong	The Honk Kong Polytechnic University, China
Huang Jen-Fa	National Cheng Kung University, Taiwan
Hubert Frederic	Université Laval, Canada
Iglesias Andres	University of Cantabria, Spain
Jamal Amna	National University of Singapore, Singapore
Jank Gerhard	Aachen University, Germany
Jeong Jongpil	Sungkyunkwan University, South Korea
Jiang Bin	University of Gävle, Sweden
Johnson Franklin	Universidad de Playa Ancha, Chile
Kalogirou Stamatis	Harokopio University of Athens, Greece
Kamoun Farouk	Université de la Manouba, Tunisia
Kanchi Saroja	Kettering University, USA
Kanevski Mikhail	University of Lausanne, Switzerland
Kang Myoung-Ah	ISIMA Blaise Pascal University, France
Karandikar Varsha	Devi Ahilya University, Indore, India
Karimipour Farid	Vienna University of Technology, Austria
Kavouras Marinos	University of Lausanne, Switzerland
Kazar Baris	Oracle Corp., USA

Keramat Alireza	Jundi-Shapur Univ. of Technology, Iran
Khan Murtaza	NUST, Pakistan
Khattak Asad Masood	Kyung Hee University, Korea
Khazaei Hamzeh	Ryerson University, Canada
Khurshid Khawar	NUST, Pakistan
Kim Dongsoo	Indiana University-Purdue University Indianapolis, USA
Kim Mihui	Hankyong National University, South Korea
Koo Bonhyun	Samsung, South Korea
Korkhov Vladimir	St. Petersburg State University, Russia
Kotzinos Dimitrios	Université de Cergy-Pontoise, France
Kumar Dileep	SR Engineering College, India
Kurdia Anastasia	Buknell University, USA
Lachance-Bernard Nicolas	École Polytechnique Fédérale de Lausanne, Switzerland
Laganà Antonio	University of Perugia, Italy
Lai Sabrina	University of Cagliari, Italy
Lanorte Antonio	CNR-IMAA, Italy
Lanza Viviana	Lombardy Regional Institute for Research, Italy
Lasaponara Rosa	National Research Council, Italy
Lassoued Yassine	University College Cork, Ireland
Lazzari Maurizio	CNR IBAM, Italy
Le Duc Tai	Sungkyunkwan University, South Korea
Le Duc Thang	Sungkyunkwan University, South Korea
Le-Thi Kim-Tuyen	Sungkyunkwan University, South Korea
Ledoux Hugo	Delft University of Technology, The Netherlands
Lee Dong-Wook	INHA University, South Korea
Lee Hongseok	Sungkyunkwan University, South Korea
Lee Ickjai	James Cook University, Australia
Lee Junghoon	Jeju National University, South Korea
Lee KangWoo	Sungkyunkwan University, South Korea
Legatiuk Dmitrii	Bauhaus University Weimar, Germany
Lendvay Gyorgy	Hungarian Academy of Science, Hungary
Leonard Kathryn	California State University, USA
Li Ming	East China Normal University, China
Libourel Thrse	LIRMM, France
Lin Calvin	University of Texas at Austin, USA
Liu Xin	University of Calgary, Canada
Loconte Pierangela	Technical University of Bari, Italy
Lombardi Andrea	University of Perugia, Italy
Longo Savino	University of Bari, Italy
Lopes Cristina	University of California Irvine, USA
Lopez Cabido Ignacio	Centro de Supercomputación de Galicia, CESGA
Lourenço Vanda Marisa	University Nova de Lisboa, Portugal
Luaces Miguel	University of A Coruña, Spain
Lucertini Giulia	IUAV, Italy
Luna Esteban Robles	Universidad Nacional de la Plata, Argentina

M.M.H. Gregori Rodrigo	Universidade Tecnológica Federal do Paraná, Brazil
Machado Gaspar	University of Minho, Portugal
Machado Jose	University of Minho, Portugal
Mahinderjit Singh Manmeet	University Sains Malaysia, Malaysia
Malonek Helmuth	University of Aveiro, Portugal
Manfreda Salvatore	University of Basilicata, Italy
Manns Mary Lynn	University of North Carolina Asheville, USA
Manso Callejo Miguel Angel	Universidad Politécnica de Madrid, Spain
Marechal Bernard	Universidade Federal de Rio de Janeiro, Brazil
Marechal Franois	École Polytechnique Fédérale de Lausanne, Switzerland
Margalef Tomas	Universitat Autònoma de Barcelona, Spain
Marghany Maged	Universiti Teknologi Malaysia, Malaysia
Marsal-Llacuna Maria-Llusa	Universitat de Girona, Spain
Marsh Steven	University of Ontario, Canada
Martins Ana Mafalda	Universidade de Aveiro, Portugal
Martins Pedro	Universidade do Minho, Portugal
Marvuglia Antonino	Public Research Centre Henri Tudor, Luxembourg
Mateos Cristian	Universidad Nacional del Centro, Argentina
Matos Inés	Universidade de Aveiro, Portugal
Matos Jose	Instituto Politecnico do Porto, Portugal
Matos João	ISEP, Portugal
Mauro Giovanni	University of Trieste, Italy
Mauw Sjouke	University of Luxembourg, Luxembourg
Medeiros Pedro	Universidade Nova de Lisboa, Portugal
Melle Franco Manuel	University of Minho, Portugal
Melo Ana	Universidade de São Paulo, Brazil
Michikawa Takashi	University of Tokio, Japan
Milani Alfredo	University of Perugia, Italy
Millo Giovanni	Generali Assicurazioni, Italy
Min-Woo Park	SungKyunKwan University, South Korea
Miranda Fernando	University of Minho, Portugal
Misra Sanjay	Covenant University, Nigeria
Mo Otilia	Universidad Autonoma de Madrid, Spain
Modica Giuseppe	Università Mediterranea di Reggio Calabria, Italy
Mohd Nawi Nazri	Universiti Tun Hussein Onn Malaysia, Malaysia
Morais João	University of Aveiro, Portugal
Moreira Adriano	University of Minho, Portugal
Moerig Marc	University of Magdeburg, Germany
Morzy Mikolaj	University of Poznan, Poland
Mota Alexandre	Universidade Federal de Pernambuco, Brazil
Moura Pires João	Universidade Nova de Lisboa - FCT, Portugal
Mourão Maria	Polytechnic Institute of Viana do Castelo, Portugal

Mourelle Luiza de Macedo	UERJ, Brazil
Mukhopadhyay Asish	University of Windsor, Canada
Mulay Preeti	Bharti Vidyapeeth University, India
Murgante Beniamino	University of Basilicata, Italy
Naghizadeh Majid Reza	Qazvin Islamic Azad University, Iran
Nagy Csaba	University of Szeged, Hungary
Nandy Subhas	Indian Statistical Institute, India
Nash Andrew	Vienna Transport Strategies, Austria
Natário Isabel Cristina Maciel	University Nova de Lisboa, Portugal
Navarrete Gutierrez Tomas	Luxembourg Institute of Science and Technology, Luxembourg
Nedjah Nadia	State University of Rio de Janeiro, Brazil
Nguyen Hong-Quang	Ho Chi Minh City University, Vietnam
Nguyen Tien Dzung	Sungkyunkwan University, South Korea
Nickerson Bradford	University of New Brunswick, Canada
Nielsen Frank	Université Paris Saclay CNRS, France
NM Tuan	Ho Chi Minh City University of Technology, Vietnam
Nogueira Fernando	University of Coimbra, Portugal
Nole Gabriele	IRMAA National Research Council, Italy
Nourollah Ali	Amirkabir University of Technology, Iran
Olivares Rodrigo	UCV, Chile
Oliveira Irene	University of Trás-Os-Montes e Alto Douro, Portugal
Oliveira José A.	University of Minho, Portugal
Oliveira e Silva Luis	University of Lisboa, Portugal
Osaragi Toshihiro	Tokyo Institute of Technology, Japan
Ottomanelli Michele	Polytechnic of Bari, Italy
Ozturk Savas	TUBITAK, Turkey
Pagliara Francesca	University of Naples, Italy
Painho Marco	New University of Lisbon, Portugal
Pantazis Dimos	Technological Educational Institute of Athens, Greece
Paolotti Luisa	University of Perugia, Italy
Papa Enrica	University of Amsterdam, The Netherlands
Papathanasiou Jason	University of Macedonia, Greece
Pardede Eric	La Trobe University, Australia
Parissis Ioannis	Grenoble INP - LCIS, France
Park Gyung-Leen	Jeju National University, South Korea
Park Sooyeon	Korea Polytechnic University, South Korea
Pascale Stefania	University of Basilicata, Italy
Parker Gregory	University of Oklahoma, USA
Parvin Hamid	Iran University of Science and Technology, Iran
Passaro Pierluigi	University of Bari Aldo Moro, Italy
Pathan Al-Sakib Khan	International Islamic University Malaysia, Malaysia
Paul Padma Polash	University of Calgary, Canada

Peixoto Bitencourt Ana Carla	Universidade Estadual de Feira de Santana, Brazil
Peraza Juan Francisco	Autonomous University of Sinaloa, Mexico
Perchinutino Paola	University of Bari, Italy
Pereira Ana	Polytechnic Institute of Bragança, Portugal
Pereira Francisco	Instituto Superior de Engenharia, Portugal
Pereira Paulo	University of Minho, Portugal
Pereira Javier	Diego Portales University, Chile
Pereira Oscar	Universidade de Aveiro, Portugal
Pereira Ricardo	Portugal Telecom Inovacao, Portugal
Perez Gregorio	Universidad de Murcia, Spain
Pesantes Mery	CIMAT, Mexico
Pham Quoc Trung	HCMC University of Technology, Vietnam
Pietrantuono Roberto	University of Napoli "Federico II", Italy
Pimentel Carina	University of Aveiro, Portugal
Pina Antonio	University of Minho, Portugal
Piñar Miguel	Universidad de Granada, Spain
Pinciu Val	Southern Connecticut State University, USA
Pinet Francois	IRSTEA, France
Piscitelli Claudia	Polytechnic University of Bari, Italy
Pollino Maurizio	ENEA, Italy
Poplin Alenka	University of Hamburg, Germany
Porschen Stefan	University of Köln, Germany
Potena Pasqualina	University of Bergamo, Italy
Prata Paula	University of Beira Interior, Portugal
Previtali Mattia	Polytechnic of Milan, Italy
Prosperi David	Florida Atlantic University, USA
Protheroe Dave	London South Bank University, UK
Pusatli Tolga	Cankaya University, Turkey
Qaisar Saad	NURST, Pakistan
Qi Yu	Mesh Capital LLC, USA
Quan Tho	Ho Chi Minh City University of Technology, Vietnam
Raffaeta Alessandra	University of Venice, Italy
Ragni Mirco	Universidade Estadual de Feira de Santana, Brazil
Rahayu Wenny	La Trobe University, Australia
Rautenberg Carlos	University of Graz, Austria
Ravat Franck	IRIT, France
Raza Syed Muhammad	Sungkyunkwan University, South Korea
Rinaldi Antonio	DIETI - UNINA, Italy
Rinzivillo Salvatore	University of Pisa, Italy
Rios Gordon	University College Dublin, Ireland
Riva Sanseverino Eleonora	University of Palermo, Italy
Roanes-Lozano Eugenio	Universidad Complutense de Madrid, Spain
Rocca Lorena	University of Padova, Italy
Roccatello Eduard	3DGIS, Italy

Rocha Ana Maria	University of Minho, Portugal
Rocha Humberto	University of Coimbra, Portugal
Rocha Jorge	University of Minho, Portugal
Rocha Maria Clara	ESTES Coimbra, Portugal
Rocha Miguel	University of Minho, Portugal
Rodrigues Armanda	Universidade Nova de Lisboa, Portugal
Rodrigues Cristina	DPS, University of Minho, Portugal
Rodrigues Joel	University of Minho, Portugal
Rodriguez Daniel	University of Alcala, Spain
Rodrguez Gonzlez Alejandro	Universidad Carlos III Madrid, Spain
Roh Yongwan	Korean IP, South Korea
Romano Bernardino	University of l'Aquila, Italy
Roncaratti Luiz	Instituto de Física, University of Brasilia, Brazil
Roshannejad Ali	University of Calgary, Canada
Rosi Marzio	University of Perugia, Italy
Rossi Gianfranco	University of Parma, Italy
Rotondo Francesco	Polytechnic of Bari, Italy
Roussey Catherine	IRSTEA, France
Ruj Sushmita	Indian Statistical Institute, India
S. Esteves Jorge	University of Aveiro, Portugal
Saeed Husnain	NUST, Pakistan
Sahore Mani	Lovely Professional University, India
Saini Jatinder Singh	Baba Banda Singh Bahadur Engineering College, India
Salzer Reiner	Technical University Dresden, Germany
Sameh Ahmed	The American University in Cairo, Egypt
Sampaio Alcinia Zita	Instituto Superior Tecnico Lisboa, Portugal
Sannicandro Valentina	Polytechnic of Bari, Italy
Santiago Jnior Valdivino	Instituto Nacional de Pesquisas Espaciais, Brazil
Santos Josué	UFABC, Brazil
Santos Rafael	INPE, Brazil
Santos Viviane	Universidade de São Paulo, Brazil
Santucci Valentino	University of Perugia, Italy
Saracino Gloria	University of Milano-Bicocca, Italy
Sarafian Haiduke	Pennsylvania State University, USA
Saraiva João	University of Minho, Portugal
Sarrazin Renaud	Université Libre de Bruxelles, Belgium
Schirone Dario Antonio	University of Bari, Italy
Schneider Michel	ISIMA, France
Schoier Gabriella	University of Trieste, Italy
Schuhmacher Marta	Universitat Rovira i Virgili, Spain
Scorza Francesco	University of Basilicata, Italy
Seara Carlos	Universitat Politècnica de Catalunya, Spain
Sellares J. Antoni	Universitat de Girona, Spain
Selmaoui Nazha	University of New Caledonia, New Caledonia
Severino Ricardo Jose	University of Minho, Portugal

Shaik Mahaboob Hussain	JNTUK Vizianagaram, A.P., India
Shoikho Kamel	KACST, Saudi Arabia
Shen Jie	University of Michigan, USA
Shi Xuefei	University of Science Technology Beijing, China
Shin Dong Hee	Sungkyunkwan University, South Korea
Shojaeipour Shahed	Universiti Kebangsaan Malaysia, Malaysia
Shon Minhan	Sungkyunkwan University, South Korea
Shukla Ruchi	University of Johannesburg, South Africa
Silva Carlos	University of Minho, Portugal
Silva J.C.	IPCA, Portugal
Silva de Souza Laudson	Federal University of Rio Grande do Norte, Brazil
Silva-Fortes Carina	ESTeSL-IPL, Portugal
Simão Adenilso	Universidade de São Paulo, Brazil
Singh R.K.	Delhi University, India
Singh V.B.	University of Delhi, India
Singhal Shweta	GGSIPU, India
Sipos Gergely	European Grid Infrastructure, The Netherlands
Smolik Michal	University of West Bohemia, Czech Republic
Soares Inês	INESC Porto, Portugal
Soares Michel	Federal University of Sergipe, Brazil
Sobral Joao	University of Minho, Portugal
Son Changhwan	Sungkyunkwan University, South Korea
Song Kexing	Henan University of Science and Technology, China
Sosnin Petr	Ulyanovsk State Technical University, Russia
Souza Eric	Universidade Nova de Lisboa, Portugal
Sproessig Wolfgang	Technical University Bergakademie Freiberg, Germany
Sreenan Cormac	University College Cork, Ireland
Stankova Elena	Saint-Petersburg State University, Russia
Starczewski Janusz	Institute of Computational Intelligence, Poland
Stehn Fabian	University of Bayreuth, Germany
Sultana Madeena	University of Calgary, Canada
Swarup Das	Ananda Kalinga Institute of Industrial Technology, India
Tahar Sofiène	Concordia University, Canada
Takato Setsuo	Toho University, Japan
Talebi Hossein	University of Calgary, Canada
Tanaka Kazuaki	Kyushu Institute of Technology, Japan
Taniar David	Monash University, Australia
Taramelli Andrea	Columbia University, USA
Tarantino Eufemia	Polytechnic of Bari, Italy
Tariq Haroon	Connekt Lab, Pakistan
Tasso Sergio	University of Perugia, Italy
Teixeira Ana Paula	University of Trás-Os-Montes e Alto Douro, Portugal
Tesseire Maguelonne	IRSTEA, France
Thi Thanh Huyen Phan	Japan Advanced Institute of Science and Technology, Japan

Thorat Pankaj	Sungkyunkwan University, South Korea
Tilio Lucia	University of Basilicata, Italy
Tiwari Rupa	University of Minnesota, USA
Toma Cristian	Polytechnic University of Bucarest, Romania
Tomaz Graça	Polytechnic Institute of Guarda, Portugal
Tortosa Leandro	University of Alicante, Spain
Tran Nguyen	Kyung Hee University, South Korea
Tripp Barba, Carolina	Universidad Autnoma de Sinaloa, Mexico
Trunfio Giuseppe A.	University of Sassari, Italy
Uchibayashi Toshihiro	Kyushu Sangyo University, Japan
Ugalde Jesus	Universidad del Pais Vasco, Spain
Urbano Joana	LIACC University of Porto, Portugal
Van de Weghe Nico	Ghent University, Belgium
Varella Evangelia	Aristotle University of Thessaloniki, Greece
Vasconcelos Paulo	University of Porto, Portugal
Vella Flavio	University of Rome La Sapienza, Italy
Velloso Pedro	Universidade Federal Fluminense, Brazil
Viana Ana	INESC Porto, Portugal
Vidacs Laszlo	MTA-SZTE, Hungary
Vieira Ramadas Gisela	Polytechnic of Porto, Portugal
Vijay NLankalapalli	National Institute for Space Research, Brazil
Vijaykumar Nandamudi	INPE, Brazil
Viqueira José R.R.	University of Santiago de Compostela, Spain
Vitellio Ilaria	University of Naples, Italy
Vizzari Marco	University of Perugia, Italy
Wachowicz Monica	University of New Brunswick, Canada
Walentynski Ryszard	Silesian University of Technology, Poland
Walkowiak Krzysztof	Wroclav University of Technology, Poland
Wallace Richard J.	University College Cork, Ireland
Waluyo Agustinus Borgy	Monash University, Australia
Wanderley Fernando	FCT/UNL, Portugal
Wang Chao	University of Science and Technology of China, China
Wang Yanghui	Beijing Jiaotong University, China
Wei Hoo Chong	Motorola, USA
Won Dongho	Sungkyunkwan University, South Korea
Wu Jian-Da	National Changhua University of Education, Taiwan
Xin Liu	École Polytechnique Fédérale de Lausanne, Switzerland
Yadav Nikita	Delhi Universty, India
Yamauchi Toshihiro	Okayama University, Japan
Yao Fenghui	Tennessee State University, USA
Yatskevich Mikalai	Assioma, Italy
Yeoum Sanggil	Sungkyunkwan University, South Korea
Yoder Joseph	Refactory Inc., USA
Zalyubovskiy Vyacheslav	Russian Academy of Sciences, Russia

Sponsoring Organizations

ICCSA 2015 would not have been possible without the tremendous support of many organizations and institutions, for which all organizers and participants of ICCSA 2015 express their sincere gratitude:

University of Calgary, Canada (http://www.ucalgary.ca)

University of Perugia, Italy (http://www.unipg.it)

University of Basilicata, Italy (http://www.unibas.it)

Monash University, Australia (http://monash.edu)

Kyushu Sangyo University, Japan (www.kyusan-u.ac.jp)

Universidade do Minho, Portugal (http://www.uminho.pt)

Contents – Part II

**Workshop on Computational Geometry and Security Applications
(CGSA 2015)**

Workshop on Computational Algorithms for Sustainability Assessment (CLASS 2015)

Workshop on Chemistry and Materials Sciences and Technologies (CMST 2015)

Workshop on Computational Optimization and Applications (COA 2015)

Workshop on Cities, Technologies and Planning (CTP 2015)

Workshop on Bio-inspired Computing and Applications (BIOCA 2015)

Wave Algorithm for Recruitment in Swarm Robotics

Luneque Silva Jr.[1](✉) and Nadia Nedjah[2]

[1] Systems Engineering and Computer Science Program,
Federal University of Rio de Janeiro, Rio de Janeiro, Brazil
luneque@cos.ufrj.br
[2] Department of Electronics Engineering and Telecommunications,
State University of Rio de Janeiro, Rio de Janeiro, Brazil
nadia@eng.uerj.br

Abstract. In the last decades, studies on swarm robotics have grown significantly, with different new aspects becoming of interest in research. Although the primary interest in swarm robotics should be the application, which is the task that robots should perform, other aspects are also studied. For instance, in a robotic swarm, it may be necessary to select the robots that would perform a given task, aiming at a division of labor among the members. A similar problem is that of robot recruitment, required when a specific robot needs assistance to execute some tasks and thus needs to recruit other robots of the swarm to do so. In this paper, a distributed algorithm, termed the wave algorithm, is exploited to perform the recruitment based mainly on robot message passing to neighboring robots.

Keywords: Swarm robotics · Message passing · Recruitment · Wave algorithms · Distributed algorithms

1 Introduction

In a *robotic swarm*, the members of a group of mobile robots work together to achieve some required collective behavior [1]. In such systems, the agents are autonomous robots that follow simple rules and dispose of only local and limited capabilities of sensing and communication. The swarm has no central control or global information sharing. At the microscopic level, even individual robots are not aware of the complete behavior of the system.

The non-centralized approach of a robotic swarm is classically inspired by the observation of social insects and animals, such as ants, bees, birds and fishes. Such species exhibit some kind of *swarm intelligence* [2], where a successful collective behavior emerges from interactions between individuals. In general, the

L. Silva Jr.—The work of this author was supported by CNPq (National Council for Scientific and Technological Development) grants no. 141901/2013-6.

© Springer International Publishing Switzerland 2015
O. Gervasi et al. (Eds.): ICCSA 2015, Part II, LNCS 9156, pp. 3–13, 2015.
DOI: 10.1007/978-3-319-21407-8_1

emerged behavior is robust, flexible and scalable, which are all desired characteristics in multi-agent systems.

Swarm robots are devices with limited capabilities of computational processing, sensing, communication and movement. The task performed by the swarm is usually achieved by the execution of algorithms in each robot. This kind of system has many similarities with the execution of distributed algorithms by a multiprocessor system. In this paper, the behavior of an individual robot is defined by an algorithm, that takes advantage of basic concepts of distributed systems, such as the *direct communication* with explicit messages between processes. However, swarm robotics may also use *indirect communication*, sensing other robots or interacting with the environment [3].

In this work a distributed algorithm is designed to perform the recruitment of robots in a swarm system. Each robot runs the algorithm locally, sending messages to robots in its neighborhood. The reception of a message affects the local execution of the algorithm. This asynchronous communication allows the coordination between robots.

The paper is organized as follows. In Section 2, some relevant works covering recruitment are presented. Section 3 shows an overview of the swarm requirements, while Section 4 shows the basic wave algorithm. The Section 5 describes the task of recruitment. Section 6 presents some performance results of simulated behaviors. The conclusion of this work, in which we also point out some relevant future work, is presented Section 7.

2   Related Works

Numerous studies on swarm robotics have been conducted in the last decades. For a broad review, the surveys in [1], [4] and [5] present methods of design, modeling and a list of behaviors and problems that would be done by swarms.

The recruitment in a robotic swarm is a kind of task allocation, where the selection of robots is a reaction to events in the environment. In [6], the idea of recruitment is related with an *activation-threshold* model, where individuals react to stimuli associated to a specific task. Thus, individuals spatially closer to the work to be done have more chance to be recruited. It is based on *tandem running* of some ant species, where the recruiting ant periodically waits for the recruited ant which in turn touches the recruiter to indicate that it can continue. However, the technique allows for only one robot to be recruited at a time.

In [7], the authors demonstrate that a single unmanned aerial vehicle can recruit unmanned ground vehicles in a land mine detection task. The recruiter uses an *emotional model* to request and receive assistance from other members. In [8], the authors present the group formation in heterogeneous swarms. An aerial robot searches for tasks in the environment, then it recruits groups of wheeled robots to perform these tasks.

In [9], the robots recruitment is done using chemical substances, which is inspired by the *pheromones* of some species of social insects, such as ants and termites. Such robots are capable of sensing and releasing small portions of ethanol, which works like the pheromone trail of ants. In this case, the recruitment is used to perform the tasks of foraging and collective transport.

3   Robot Requirements and Simulation Platform

In this context, a robot is an agent capable of moving in the environment. In this work, a robot is defined as having two movement engines, allowing it to go forward and backward, but with no direct right-to-left displacements. Robots with this movement are called *non-holonomic*.

An individual robot is an agent having some computational capabilities. Furthermore, it may sense the environment via some built-in sensors. The main sensory capability allows distance measurement to neighboring robots in the swarm. This is a local knowledge, because a robot can only sense robots in a small neighborhood.

Robots may also have some kind of wireless communication, allowing message passing among them. This communication may be a *local broadcast* message, where a robot broadcasts some information to all robots in the neighborhood, or a *direct message*, identifying both the sender and the receiver.

The proof of the concept of the proposed solution is verified through simulations using V-Rep [10]. This is a virtual robot experimentation platform that includes several models of industrial and academic robots. It also allows the design of customized robots. V-Rep can also simulate real-world physics and object interactions.

4   General Distributed Systems

A distributed system is considered to be an interconnected collection of autonomous nodes. In general, the nodes can be processes, processors, computers, etc, that have their own private control. Nonetheless, at the same time, a node can be able to exchange information with other nodes [11]. Following this description, the swarm may be considered as a distributed system, with robots being the autonomous nodes and the interconnection as the message passing between them using wireless communication.

4.1   Wave Algorithms

Wave Algorithms are a class of distributed algorithms where a node, identified as the initiator, sends messages to neighbors, which in turn start to send messages to their neighbors. A wave algorithm must satisfy the following three requirements:

1. Each computation is finite.
2. Each computation contains at least one decide event.
3. In each computation every decide event is causally preceded by an event in each process.

An *initiator* is a process that starts its local algorithm spontaneously, upon a specific state of an internal trigger. A non-initiator starts its local algorithm upon the arrival of a message. The computation of this distributed algorithm is called a *wave*, which is a sequence of *events*, where each event is a message sent or received. There is also a special type of internal event called a *decide event*. A wave algorithm exchanges a finite number of messages, and then makes a decision. A centralized wave algorithm is that with only one initiator, while all other nodes are non-initiators.

4.2 PIF and SYN Algorithms

Wave algorithms are used as sub-tasks within other application algorithms. Examples of these sub-tasks includes the broadcast of information and the synchronization between processes.

Let *msg* be a message that must be broadcasted to all nodes in the system. An initiator must execute a special notify event when all nodes have received *msg*. It can be done with a wave algorithm, in this case called *Propagation Information with Feedback*, or PIF Algorithm [12]. Likewise, wave algorithms may also be used when a *Global Synchronization* between the nodes must be achieved. In this case the wave is called SYN Algorithm [13]. Consider a distributed system with two event types, E_a and E_b that can occur in different nodes. If all E_a events must occur before the first E_b event, then a wave algorithm should be employed to synchronize events with the decision event in initiator.

5   Robot Recruitment

In this work, the recruitment is started by one or more arbitrary leader. Each wave form a group of robots. However, before the recruiting messages start propagating through the swarm each robot must know its neighborhood. These two steps, the neighborhood discovering and the recruitment are discussed as follows.

5.1   Neighborhood Discovering

To allow the broadcast of messages that will create the group, each robot must know its neighborhood, which is the set of neighbors within the robot range of communication. We will call *neighborhood$_i$* the set of *neighboring robots* that can send/receive messages to/from *robot$_i$*. In Figure 1(a), robots A and B have four neighbors each one, with hatched area representing the neighborhood. Robot C is neighbor of both A and B. If each robot in swarm knows its neighbors, the swarm can be represented as a connectivity graph, as shown in Figure 1(b).

Two ways to make robots know their neighbors are considered. The first way is using a static neighborhood, wherein a robot has within its own control a list of all neighbors. The second way is to make all robots send and wait for a message that contains the individual identifier of the sender during an initial

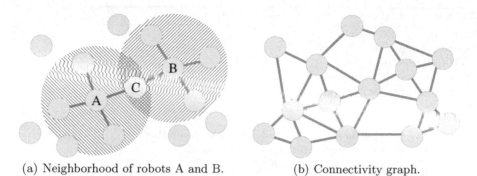

(a) Neighborhood of robots A and B. (b) Connectivity graph.

Fig. 1. Neighborhood in the swarm of robots

Algorithm 1. Neighbors Discover

Require: *time_limit*;
Ensure: *neighbors_list*;
 1: initialize *counter*, $number_{neighbors}$;
 2: start broadcasting (locally) msg_0;
 3: **while** *counter* < *time_limit* **do**
 4:   receive msg_0 from neighbors;
 5:   **if** msg_0 is new **then**
 6:     save neighbor identifier in *neighbors_list*;
 7:     increment $number_{neighbors}$;
 8:   **end if**
 9:   increment *counter*;
10: **end while**;
11: stop broadcasting msg_0

---

phase of the computation. In this work, this *dynamic neighborhood* is used. After some time, a robot receives the identifiers of all the robots in its proximity, thus creating a list of neighbors. This task is shown in Algorithm 1, which is executed simultaneously in all robots. The size of *time_limit* is the time threshold to receive the message msg_0 from neighbors.

5.2 Recruitment Wave

The proposed recruitment in a robotic swarm is based on the execution of a wave algorithm. The main idea is shown in Algorithm 2. An initiator starts broadcasting a specific message, that is spread through the swarm in a neighborhood basis. This means that when a non-initiator node receives a message, it also forwards it to its neighbors. Note that the initiator message creates a father-child dependency, *i.e.* if the first message that arrives at node B came from node A, then it is said that A is the father node of B. Whenever a node receives the initiator message, it will be recruited, becoming a child node. When a node, which has received the initiator message, has no child nodes, it starts

Algorithm 2. Wave
***
Require: *neighbors_list*;
Ensure: father node, child nodes;
 1: **if** initiator **then**
 2:     start broadcast message;
 3:     receive the feedback messages from neighbors;
 4:     make a decision;
 5: **else**
 6:     receive the first message (the sender becomes the father);
 7:     start broadcast message (the receiver become child);
 8:     receive messages from child nodes;
 9:     send the feedback message to father node;
10: **end if**
***

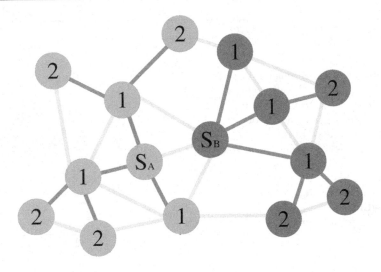

Fig. 2. Two spanning trees resulted from waves

broadcasting a feedback message, that is echoed back to its father node. All nodes, upon the reception of the initiator message, wait for feedback messages from its child nodes. Intermediate non-initiator nodes also broadcast this kind of feedback message as soon as they complete receiving the feedback messages from all their child nodes. This process is repeated at different nodes until the initiator receives the feedback messages from all its child nodes. At this point, the initiator knows that all nodes in the swarm have received his original message, and thus proceeds with making a decision. In this case, the decision will be the start of the task for which robots are being recruited.

The messages transmitted in the swarm contain the identifier of the robot and the identifier of its father node. This identifier allows the robots to know from which neighbor each message came from. During the execution of the recruitment, each wave creates a spanning tree, as seen in Figure 2. Two initiators, S_A and S_B, are represented. The number inside each node indicates the distance in hops to one of the initiators. This wave can be interpreted as a PIF algorithm: the initiator recruits neighborings with the spread of messages. It can also be interpreted as a SYN algorithm, because the wave can separate the execution of two sequential tasks.

6   Experiment Setup

(a) Real-world Kilobot.   (b) Simulated Kilobot in V-Rep.

Fig. 3. The real and simulated Kilobot robot

The evaluation of the proposed swarm behavior was done using simulations executed by the V-Rep Version 3.1.1 [10]. We used a simulation model based on Kilobot robot [14]. Figure 3 shows the real-world and simulated version of Kilobot. This robot have 3 legs that supports the circuit board above the surface and allows the robot to move by vibration due to the activation of the two step motors. The vibration allows the robot to move forward and/or make curves. The Kilobot also has an infrared communication system that allows it to send and receive messages to/from neighboring robots. A RGB LED placed in the circuit board can be used to visually signalize specific internal states of the robot. Besides all these Kilobot characteristics, it was chosen to be used in this work because of the similarities of the programming of the physical robot and the corresponding simulated model provided in V-Rep. The message waves used in recruitment algorithm can be implemented in real Kilobot with few minor changes.

To evaluate the proposed method, we analyzed the outcome of the recruitment under different conditions, varying the number of robots and the number of initiators. Figure 4 shows the results for the recruitment of two groups in swarms of 8, 16, 32 and 64 robots. The color of each robot represents its group. The two initiators were placed in the first and last line of the first column (upper-left and bottom-left corners). This positions allows the formation of symmetrical balanced groups, *i.e.* with the same number of robots. The balance may not occur when initiators are placed in other positions inside the swarm. One example

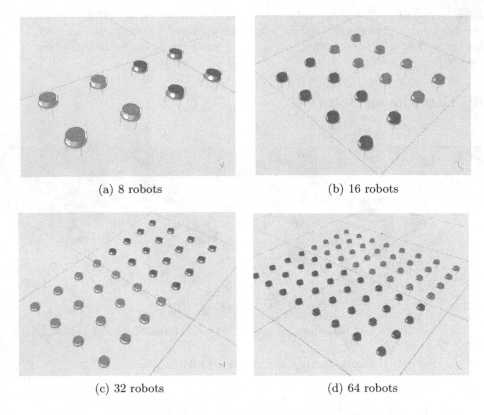

(a) 8 robots   (b) 16 robots

(c) 32 robots   (d) 64 robots

Fig. 4. Swarms with two initiators

is shown in Figure 5. The two initiators are positioned in the bottom-left and upper-right corners in Figure 5 (a). This position is changed to the other corners in Figures 5 (b)-(d). In these cases, only one initiator recruits the robots on the diagonal, forming unbalanced groups.

The results shown in Figure 6 report on the impact of the number of robots and the number of groups in the recruitment time. Each simulation was performed 10 times, and the average execution time is shown. This time is counted starting from the first message sent by an initiator until it receives the last message from the child nodes. There is a reduction in recruitment time when a smaller swarm is considered. This is expected as the wave needs less time to reach all robots. This reduction also occurs when more initiators are used. When a robot is recruited by an initiator, it cannot be recruited by other initiators. For instance, with two initiators, the propagating waves divide the swarm in two groups. In other words, the recruiting time depends on the number of hops of the messages from the initiator to the robots that have no child nodes.

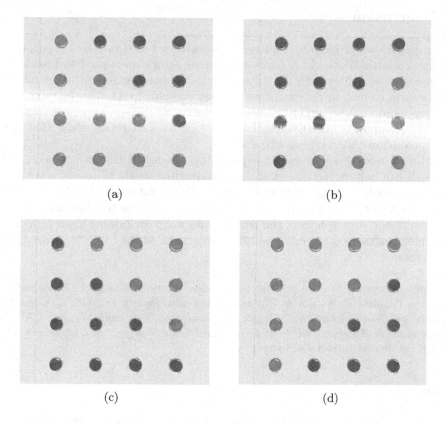

(a)   (b)

(c)   (d)

Fig. 5. Swarms with two initiators

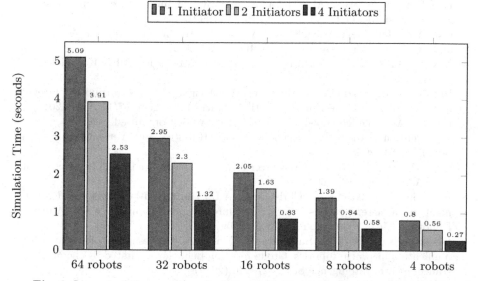

Fig. 6. Impact of the number of robots and initiators in the recruiting time

7   Conclusion

In this paper, we propose a recruitment algorithm based on the Propagating Information with Feedback. A wave distributed algorithm is used to broadcast messages through the swarm. Thus, an initiator robot can recruit other robots in the swarm to help in the execution of some complex tasks, like the transport of a huge object, that cannot be carried by a single robot. Tests performed in the V-Rep Simulator, using a virtual model of Kilobot robot, show that the distributed algorithm works as expected.

In future works, we intend to explore the control of the size of recruited groups. It can help in the case that the initiator just wants to recruit a specific number of robots, and not the entire swarm. Other extensions of this research include the use of this recruitment as a basis to the execution of other tasks. After the recruitment step, robots of the group could start performing aggregation or deployment tasks, or other more complex tasks, such as the collective movement or transport.

Acknowledgments. We are grateful to Brazilian agencies FAPERJ (*Fundação de Amparo à Pesquisa do Estado do Rio de Janeiro, www.faperj.br*), CNPq (*Conselho Nacional de Desenvolvimento Científico e Tecnológico, www.cnpq.br*) and CAPES (*Coordenação de Aperfeiçoamento de Pessoal de Ensino Superior, www.capes.gov.br*) for their continuous financial support.

References

1. Şahin, E.: Swarm Robotics: From Sources of Inspiration to Domains of Application. In: Şahin, E., Spears, W.M. (eds.) Swarm Robotics 2004. LNCS, vol. 3342, pp. 10–20. Springer, Heidelberg (2005)
2. Bonabeau, E., Dorigo, M., Theraulaz, G.: Swarm intelligence: from natural to artificial systems. Oxford University Press (1999)
3. Cao, Y.U., Fukunaga, A.S., Kahng, A.: Cooperative mobile robotics: Antecedents and directions. Autonomous Robots 4(1), 7–27 (1997)
4. Navarro, I., Matía, F.: An introduction to swarm robotics, ISRN Robotics (2012, 2013)
5. Brambilla, M., Ferrante, E., Birattari, M., Dorigo, M.: Swarm robotics: a review from the swarm engineering perspective. Swarm Intelligence 7(1), 1–41 (2013)
6. Krieger, M.J., Billeter, J.-B.: The call of duty: Self-organised task allocation in a population of up to twelve mobile robots. Robotics and Autonomous Systems 30(1), 65–84 (2000)
7. Gage, A., Murphy, R.R.: Affective recruitment of distributed heterogeneous agents. In: AAAI, pp. 14–19 (2004)
8. Pinciroli, C., O'Grady, R., Christensen, A.L., Dorigo, M.: Self-organised recruitment in a heteregeneous swarm. In: International Conference on Advanced Robotics, ICAR 2009, pp. 1–8. IEEE (2009)
9. Fujisawa, R., Dobata, S., Sugawara, K., Matsuno, F.: Designing pheromone communication in swarm robotics: Group foraging behavior mediated by chemical substance. Swarm Intelligence 8(3), 227–246 (2014)

10. Rohmer, E., Singh, S.P.N., Freese, M.: V-rep: a versatile and scalable robot simulation framework. In: Proc. of The International Conference on Intelligent Robots and Systems (IROS) (2013)
11. Tel, G.: Introduction to distributed algorithms. Cambridge University Press (2000)
12. Segall, A.: Distributed network protocols. IEEE Transactions on Information Theory **29**(1), 23–35 (1983)
13. Finn, S.: Resynch procedures and a fail-safe network protocol. IEEE Transactions on Communications **27**(6), 840–845 (1979)
14. Rubenstein, M., Ahler, C., Nagpal, R.: Kilobot: A low cost scalable robot system for collective behaviors. In: 2012 IEEE International Conference on Robotics and Automation (ICRA), pp. 3293–3298. IEEE (2012)

Efficient Spacial Clustering in Swarm Robotics

Nicolás Bulla Cruz[1], Nadia Nedjah[1($\boxtimes$)], and Luiza de Macedo Mourelle[2]

[1] Department of Electronics Engineering and Telecommunication
Engineering Faculty, State University of Rio de Janeiro, Rio de Janeiro, Brazil
{nicolas,nadia}@eng.uerj.br
[2] Department of System Engineering and Computation, Engineering Faculty,
State University of Rio de Janeiro, Rio de Janeiro, Brazil
ldmm@eng.uerj.br

Abstract. This paper proposes and evaluates a clustering technique of swarm robots into ζ classes. Based only on the local information coming from neighboring robots and the distribution of virtual tokens in the system, the robots of the swarm can be grouped into different classes. The proposed technique acts in a distributed manner and without any global knowledge or movement of the robots. Depending on the amount and weight of the tokens available in the system, robots exchange information to reach a token uniform distribution. The clustering technique is inspired by the settling process of liquids of different densities. Using information gathered from neighboring robots, a token density is computed. As a result, the tokens with higher weights form a cluster first, shifting those of lower weight, until they form differentiated bands for each group, thus completing the clustering of the robots.

1 Introduction

Multi-robots systems provide advantages over one individual robot when a task that requires higher execution speed, higher precision and fault tolerance is performed [7]. When there are two or more tasks to be performed and the set of robots is heterogeneous, it is possible to group them according to their functionalities. In case the set of robots is homogeneous, the grouping can be implemented according to the distance between the robots and the places where the different tasks must be performed.

In this work, an spacial clustering method, based on [2], is used. In [2], only two classes are allowed. Through message passing among neighbouring robots, this method allows clustering without robot movement. Based on token clustering method, the proposed method employs a virtual token, denominated as *load*. Exploring the characteristics of the *load*, it is possible to determine to which class a robot belongs. The aim of this work is to generalize the method proposed by [2] for $\zeta \geq 2$ classes. The proposed method also takes into account the problem of information loss, in order to minimize the impact of the original load losses.

The rest of this paper is organized in six sections. In Section 2, the problem of clustering is briefly presented, as well as the problem of robots clustering.

O. Gervasi et al. (Eds.): ICCSA 2015, Part II, LNCS 9156, pp. 14–27, 2015.
DOI: 10.1007/978-3-319-21407-8_2

In Section 3, related works are discussed. In Section 4, the main steps of the distributed algorithm, proposed herein, for robots clustering into $\zeta \geq 2$ classes is described. In Section 5, some aspects of the implementation are detailed. In Section 6, experimental results are analyzed. In Section 7, some conclusions based on the proposed algorithm are presented, along with some directions for future work.

2   Clustering

Clustering is the name given to a group of computational methods for classifying elements in groups, based on their characteristics or some degrees of similarity. The basic idea consists of putting individuals, that are similar according to some predefined criteria, in the same group [6]. Groups in a system must be described in terms of internal homogeneity and external separation. That is, the elements of the same group must be mutually similar yet distinct from those included in other groups. Clustering in swarm robotics has two main purposes, depending on the elements to be grouped: *(i)* the token clustering, which deals with passive elements, distributed throughout the environment; *(ii)* the clustering of the robots themselves.

2.1   Token Clustering

For token clustering, the behavior of swarm robots, that have the ability to move tokens from one point to another, is studied. The robots, programmed with simple rules, can gather homogeneous elements in only one cluster and, depending on the sensing characteristics of the robots, can gather heterogeneous elements in different clusters.

One aspect that has to be taken into account, in token clustering, is the movement physics of the passive element, as well as the physics of the robot movement itself. In [4], an approach called *non-physical system (aphysical system)* is described, where the behavior of token clustering, done by the robots, is considered, without taking into account the physical structure of the robots or that of the tokens. In other words, the robots are represented only by the manipulation of the tokens. As a result of this research, it was concluded that the increase in the number of clustered elements lead to a considerable increase in the clustering speed.

2.2   Robots Clustering

Multi-robots systems consist mainly of many simple robots, that generally have low computational capacity, due to cost restrictions. However, working together, robots can solve complex problems. In order to get the best efficiency in solving problems, the original problem needs to be divided into many subproblems, which will be distributed among individual robots or groups of robots. Robots clustering attempts to manage the division of a large group of robots into many smaller groups, in order to allocate the tasks.

3   Related Work

One of the pioneer works developed for token clustering was published in [1]. The employed robots were equipped with a claw that enables disc movement, aiming at gathering discs of similar characteristics.

In [4], the behavior of robots while performing token clustering, is studied. In order to understand this behavior, first it was simulated without taking into account the physical structure of the robots or that of the tokens. The simulation results lead to the conclusion that the increase in the amount of elements already clustered facilitates the clustering of the remaining elements.

Based on the tokens clustering existing work, other studies were carried out, aiming at robots clustering themselves. In [5], robots clustering behavior was performed. In this work, robots could perform two initial actions: stay still, waiting for another robot to arrive, in order to start a cluster, or keep in movement until finding still robots that already clustered. After starting the first clusters, robots could move in search for a better cluster. The evaluation of the found cluster is implemented through the robot vision, where the most dense cluster is considered the best.

Based on another perspective, in [3], a robots segregation method is proposed, based on the effect of *Brazilian Nuts* (*Castanhas do Brasil*) [8]. The simulation of this effect is reached through the random movement of the robots, with a common attraction point, and the repulsion between neighbors. This last parameter depends on a virtual radius defined by the communication range. Robots with the same virtual radius are grouped.

After collecting information from different clustering methods, in [2] a robots spatial clustering method is developed, where robots need a local knowledge of the system. Through virtual tokens exchange, robots can gather into clusters. This strategy provides an increase in clustering speed and a decrease in energy spent, due to the fact that robots need to communicate only for information exchange and do not need to move. In this work, only one kind of token is implemented and, hence, the algorithm works only when two clusters are required.

4   Clustering Algorithm

The proposed algorithm is based on the spatial clustering algorithm developed in [2]. This algorithm is distributed and allows the swarm robots to form clusters, exchanging information with neighboring robots. Here, the algorithm is generalized for $\zeta \geq 2$ classes.

The clustering algorithm is mainly based on tokens movements, called *loads*, by the robots. For ζ classes, there will be $\zeta - 1$ types of *loads*. These loads can be dynamic or static. A robot can have only one static load, but may receive any number of dynamic loads. The static load determines the robot's class, while the dynamic loads represent the movement of the token within the system. If a robot has a static load, independently of its class, it is called *loaded*. On the other hand, if a robot does not have any static load, it is called *unloaded*. A local variable u_i indicates that robot i has an static load, with $u_i \in \{0, 1, 2, \ldots, \zeta - 1\}$. Note that robot i is unloaded when $u_i = 0$. This variable is also used to indicate the class weight.

In order to guide the movement of a load, the local variable $v_i \in [0, 1]$ of robot i is used. It determines the local density of the stationary load, using a distributed consensus filter, in such a way that, depending on the clustered loads weight, it is possible to establish if the current load in robot i should keep static or will need to move. The aim of this variable is to form bands of density related with the load associated weight. The density update is performed according to Equation 1:

$$v_i^+ = v_i + \delta \left(\sum_{j \in \mathcal{N}_i} (v_j - v_i) + \sum_{j \in J_i} \left(\frac{u_j}{\zeta - 1} - v_i \right) \right), \tag{1}$$

where v_i^+ is the updated value of v_i, $\mathcal{N}_i$ is the set of neighboring robots of robot i, $J_i = \mathcal{N}_i \bigcup i$ and δ is a constant, such that $0 \leq \delta \leq \frac{1}{n}$, where n is the maximum number of robots in the swarm. The number of density bands is determined by the number of classes of the clustering process. In order to establish the bands, a maximum and a minimum value of density for each class is computed, according to Equation 2:

$$\frac{u_i}{\zeta} < v_i \leq \frac{u_i + 1}{\zeta}. \tag{2}$$

Note that ζ classes are defined by the density bands $]0, \frac{1}{\zeta}],]\frac{1}{\zeta}, \frac{2}{\zeta}], \ldots,]1 - \frac{1}{\zeta}, 1]$.

During the clustering process, the loads behavior is ruled by five phases, which make them circulate through the swarm, searching for a higher density region. Figure 1 shows the transition diagram of the load phase, wherein i is the robot's identifier and $j \in \mathcal{N}_i$ its neighboring robots identifiers.

Phase 1: Stationary beginning

This phase is executed only when the load is stationary, as indicated by $u_i \neq 0$. That is, when the robot is loaded. In this case, the load is retained and is only moved after a certain pre-defined period of time. This gives the loaded robot some time to share the acquired knowledge about the environment and probably leading to an increase of the density value of its region. Therefore, a counter C_i is initialized, according to the waiting time, and only decreased in case the density stays out of the band determined by Equation 2. When C_i reaches 0, the load is moved with a certain probability. This probability allows for a random increase in terms of waiting time, so as the robot has more time to share its knowledge. The load is no more static and, thus, it becomes dynamic. At the end of this stage, u_i is updated as 0.

Phase 2: Ascending movement

During this phase, a dynamic load is moved from one robot to another, always looking for a higher density region, *i.e.* a larger v_j among neighbors, $j \in \mathcal{N}_i$. The aim of this movement is to reach the center of the cluster

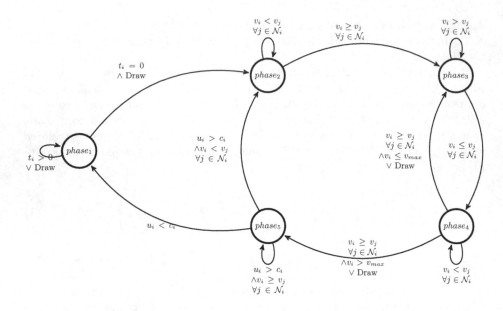

Fig. 1. Transition diagram of the load phase

formed by the robots that are already loaded. When a load ascends to a local maximum, and assuming that robot i is at the center of the cluster, i.e. $v_i \geq v_j$ for all neighbors $j \in \mathcal{N}_i$, the density v_i is stored and the load is sent to a random neighbor. Then, the load goes through Phase 3.

Phase 3: Descending movement

During this phase, a dynamic load is moved from one robot to another, always searching for a lower density region among neighbors. The aim of this movement is to restart the search for a more higher density region, avoiding that the load gets stuck into a local maximum. When the load descends until a local minimum, and assuming that robot i represents this local minimum, i.e. $v_i \leq v_j$ for all neighbors $j \in \mathcal{N}_i$, the load is sent to a random neighbor. Then, the load moves to Phase 4.

Phase 4: Slow ascending movement

Similarly to Phase 1, here the load is moved to an area with high density. When the load reaches a local maximum, its value is compared to the that obtained during Phase 2. If the new local maximum is greater than the stored one, then the load moves to Phase 5. Otherwise, it goes either to Phase 5 with a certain probability p or to Phase 3 with probability $1 - p$. This intends to find a new loaded region.

Phase 5: Slow descending movement

In this phase, the dynamic load is moved from robot to robot, searching the neighbor $j \in \mathcal{N}_i$ with high density v_j, such that $v_j < v_i$, until reaching a robot with $u_i = 0$ or a robot with $u_i < u_d$, where u_d identifies the dynamic load class. If the dynamic load finds an unloaded robot, it moves to Phase

0 and, thus, the robot stays loaded with $u_i \neq 0$. If the dynamic load finds a loaded robot, but with $u_i < u_d$, then the robots exchange their loads. In other terms, the dynamic load becomes static, changing to Phase 1, and the static load changes to dynamic, leaving the robot loaded with the highest weight load. This provides a higher priority in clustering to loads of higher class. The lowest weight load stays in Phase 5, keeping the clustering process. If the load reaches a local minimum and can not find an unloaded robot or another one loaded with $u_i < u_d$, then the load is sent to a randomly chosen neighbor, changing to Phase 2, thus iterating the clustering process. Note that when the load reaches Phase 1, counter C_i is reset.

5   Implementation Issues

The proposed clustering algorithm was implemented using a swarm of Kilobots [9]. This robot has $33mm$ of diameter and $34mm$ of height, equipped with an ATmega 328p processor (8bit@8MHz), a Flash memory of 32 KB and an EPROM of 1 KB, an RGB LED, an infrared transmitter and an infrared receiver. Each robot of the swarm has an unique identifier, which is required during robot-to-robot communication.

In this implementation, the different possible states of a robot, the kinds of load and the step of a dynamic load are visualized by a distinct LED color. The aim is to observe the initial load of each robot, the movement of the loads and the final load of each robot.

The infrared transmitter and receiver are located in the central part under the robot. The transmitter has an isotropic emission and the receiver has an standard reception, allowing the robots to receive messages uniformly from all directions in a $10cm$ radius. When the transmitter is active, any robot in the neighborhood can receive the reflected signal on the surface.

Each robot of the swarm implements two types of communication: a local broadcast communication, in such a way that the information is shared among all neighborhood, and a point-to-point communication, through which the load is transmitted from one robot to another. Note that the loss of information is not important during the local broadcast communication, since, in this case, the consensus filter compensates for that loss. Hence, a confirmation from the receiver is not needed. On the other hand, the transmission of a load from one robot to another is essential for the algorithm. Hence, two confirmations for load reception are required. This guarantees that no loss of information occurs. The first confirmation aims at avoiding the loss or duplication of a sent load. The second confirmation aims at preventing the loss of a second load sent consecutively by the same robot. The process of confirmation of load reception is organized in 6 transactions: *(i)* the emitter robot keeps transmitting the message with the information about the load; *(ii)* the receiver robot, upon receiving the load, starts to transmit the first confirmation; *(iii)* upon reception of the first confirmation, the emitter robot stops sending the first message and starts sending the second message, waiting for a second confirmation from the receiver robot, inhibiting it of sending the first confirmation; *(iv)* after receiving the second confirmation, the receiver robot starts sending the second confirmation; *(v)* once the second confirmation is received, the receiver robot sends a third message, finishing the confirmation process; *(vi)* the third message makes the receiver robot to stop transmitting the second confirmation, thus ending the communication.

6   Performance Results

The first results obtained in this work were compared with the work reported in [2]. The comparison does not aim at obtaining better results, but at assessing the system response, as well as obtaining a reference convergence time. The metrics used to evaluate the performance of the swarm clustering are the *linear separability* and the *class unbalance*.

The linear separability represents the possibility of visualizing the groups individually, *i.e.* the possibility of drawing a line that separates one group from another. This property is quantified according to Equation 3:

$$separability = \frac{R_e}{n},\tag{3}$$

where n is the total amount of robots in the swarm and R_e is the number of clustered robots in the wrong way. If the linear separability is very good, the result is 0, otherwise the result approaches 1.

The class unbalance evaluates if the number of robots is the same for all classes. The class unbalance is quantified according to Equation 4:

$$unbalance = \frac{R_m}{n},\tag{4}$$

where R_m is the number of robots of the class with the least number of members and, as before, n is the total number of robots of the swarm. For two classes with the same number of robots, the best result for unbalance is 0.5 and the worst result is 0. In order to visualize the results more clearly and generalize the result for more that two classes, the result of the unbalance is multiplied by the number of classes in the system, as described by Equation 5:

$$unbalance = \frac{R_m}{n}\zeta.\tag{5}$$

In the case study reported in [2], 15 robots of type Foot-bot were used, a two-class system was adopted, using 8 initial loads randomly distributed among the robots of the swarm. Counter C_i was initialized with 120 and the value of δ with $\frac{1}{15}$. The same parameters were used in our tests. The probabilities used in Phase 1 and Phase 5 were established as 20% and 10%, respectively. These values were chosen after several simulations.

At $t = 0s$, the linear separability is 0.4 and the unbalance is 0.933 in both systems. At $t = 75s$, the swarm of Foot-bots shows a linear separability of 0.594

and an unbalance of 0.466. On the other hand, the swarm of Kilobots shows a linear separability of 0.2 and an unbalance of 0.666. At $t = 125s$, the Foot-bots converge to a solution, with a linear separability of 0.066 and an unbalance of 0.933. Note that the linear separability is not equal to zero. This is because the loss of one load happened in this system and only 7 loads, from the initial 8, clustered. On the other hand, the swarm of Kilobots converge at $t = 130s$, but with a linear separability of 0 and an unbalance of 0.933. Note that in both cases the unbalance is not equal to 1. This is due to the fact that the number of robots in the swarm is odd. Then, the classes proportion will never be equal.

Figure 2 shows the temporal evolution of the linear separability and of the unbalance, as achieved by our implementation, for the described case study.

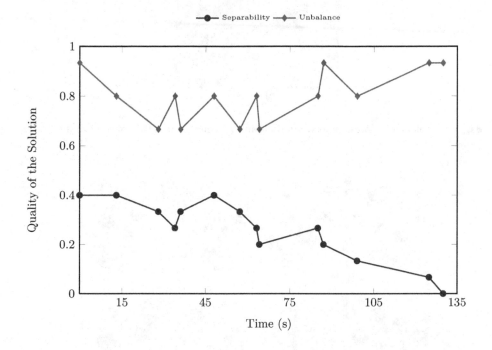

Fig. 2. Temporal evolution of the linear separability and of the unbalance for a system with 15 robots and 2 classes

Figure 3 shows the response evolution of the system, as presented in [2]. The convergence of the system happens in $125s$. Figure 4 shows the response evolution of the system implemented in the Kilobot swarm, where convergence happens only in $130s$.

Clustering results were also obtained increasing the number of classes in the system. The experiments were performed for 3 and 4 classes. In the experiment with 3 classes, the convergence time was of $360s$. In this case, 15 robots were used, with 10 initial loads, where half of them was of type 1 and the other half

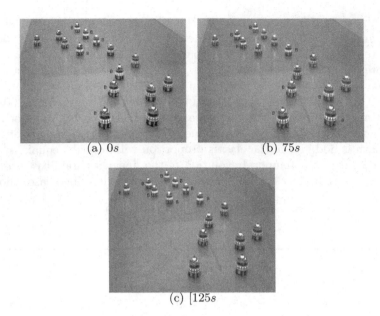

Fig. 3. Implementation in Foot-bots done in [2], where unloaded robots are indicated by letter B

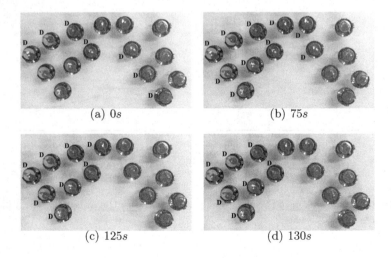

Fig. 4. Implementation in Kilobots, where unloaded robots are indicated by letter D

of type 2. At $t = 0s$, the linear separability is of 0.7 and the unbalance of 1. At $t = 180s$, the swarm presents a linear separability of 0.5625 and an unbalance of 1. At $360s$, the swarm converges to a solution, with a linear separability of 0

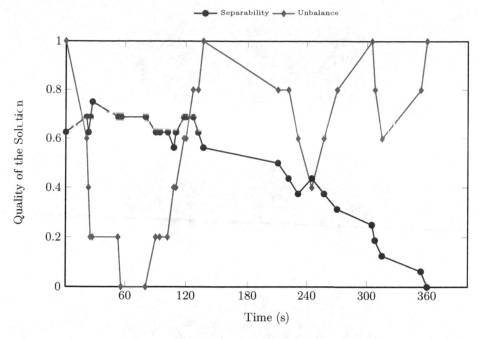

Fig. 5. Temporal evolution of the linear separability and of the unbalance for a system with 15 robots and 3 classes

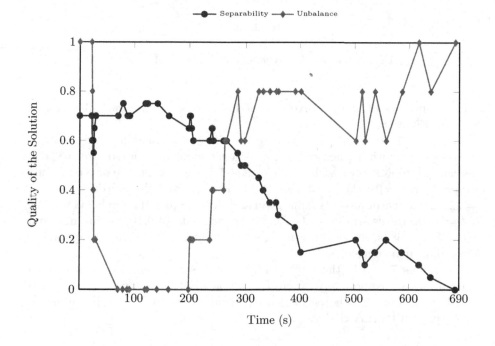

Fig. 6. Temporal evolution of the linear separability and the unbalance for a system with 20 robots and 4 classes

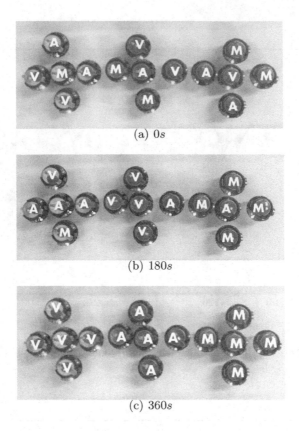

(a) 0s

(b) 180s

(c) 360s

Fig. 7. Swarm of 15 Kilobots, clustered in 3 classes

and an unbalance of 1. The curves for the linear separability and unbalance are shown in Figure 5.

In Figure 6, the curves for linear separability and unbalance are shown for an experiment with 4 classes. The convergence time is achieved after $683s$. In this case, 20 robots were used, with 15 loads: 5 loads of type 1, 5 of type 2 and 5 of type 3. At $t = 0s$, the linear separability is of 0.7 and the unbalance of 1. At $t = 227s$, the swarm presents a linear separability of 0.6 and an unbalance of 0.2. At $t = 455s$, the swarm presents a linear separability of 0.15 and an unbalance of 0.8. At $t = 683s$, the swarm converges to a solution, with a linear separability of 0 and an unbalance of 1.

In Figures 7 and 8, the implemented systems responses are presented for 3 and 4 classes, respectively, where the robot LED color must be interpreted according to Table 1. The performed experiments videos are available via the link http://bit.ly/1mVxBci.

(a) 0*s*

(b) 227*s*

(c) 455*s*

(d) 683*s*

Fig. 8. Swarm of 20 Kilobots, clustered in 4 classes

Table 1. Robot LED colors for examples with 3 and 4 classes

Class	With no dynamic loads	With dynamic loads
0	Green (V)	Turned off (D)
1	Blue (A)	Cyan (C)
2	Magenta (M)	White (B)
3	Red (R)	Yellow (Y)

7   Conclusions

In this work, an spacial clustering algorithm within a swarm of robots is proposed, using virtual tokens. In order to achieve the expected clustering results, the robots do not need to move, since they only need to exchange information using a wireless communication. The communication is only required between robots of the same neighborhood. The robots do not have a global knowledge about the state of the whole swarm.

Due to the lack of related work in the field of clustering applied to robot swarms, the results were only compared to a clustering of two classes, yielding good results. Although the convergence time was higher, the implemented system did not loose loads. Besides, the proposed algorithm is more general, as it allows clustering into two or more classes.

As for future work, an analysis of the impact of the number of robots on the clusters frontiers and clustering time will be evaluated, as well as the impact of the number of robots in each cluster. A comparison with results from stationary *ad hoc* network clustering methods and/or stationary wireless sensor networks is also recommended.

References

1. Beckers, R., Holland, O.E., Deneubourg, J.-L.: From local actions to global tasks: Stigmergy and collective robotics. In: Artificial life IV, vol. 181, p. 189 (1994)
2. Di Caro, G.A., Ducatelle, F., Gambardella, L.: A fully distributed communication-based approach for spatial clustering in robotic swarms. In: Proceedings of the 2nd Autonomous Robots and Multirobot Systems Workshop (ARMS), affiliated with the 11th International Conference on Autonomous Agents and Multiagent Systems (AAMAS), Valencia, Spain, pp. 153–171 (June 5, 2012)
3. Gross, R.., Magnenat, S., Mondada, F.: Segregation in swarms of mobile robots based on the brazil nut effect. In: IEEE/RSJ International Conference on Intelligent Robots and Systems, IROS 2009, pp. 4349–4356 (October 2009)
4. Kazadi, S., Abdul-Khaliq, A., Goodman, R.: On the convergence of puck clustering systems. Robotics and Autonomous Systems **38**(2), 93–117 (2002)
5. Lee, C., Kim, M., Kazadi, S.: Robot clustering. In: 2005 IEEE International Conference on Systems, Man and Cybernetics, vol. 2, pp. 1449–1454 (2005)
6. Linden, R.: Técnicas de agrupamento. Revista de Sistemas de Informação da FSMA **1**(4), 18–36 (2009)

7. Marjovi, A., Choobdar, S., Marques, L.: Robotic clusters: Multi-robot systems as computer clusters: A topological map merging demonstration. Robotics and Autonomous Systems **60**(9), 1191–1204 (2012)
8. Rosato, A.: Katherine J Strandburg, Friedrich Prinz, and Robert H Swendsen. Why the brazil nuts are on top: Size segregation of particulate matter by shaking. Physical Review Letters **58**(10), 1038 (1987)
9. Rubenstein, M., Ahler, C., Nagpal, R.: Kilobot: A low cost scalable robot system for collective behaviors. In: 2012 IEEE International Conference on Robotics and Automation (ICRA), pp. 3293–3298. IEEE (2012)

Simulating the Effect of Cell Migration Speed on Wound Healing Using a 3D Cellular Automata Model for Multicellular Tissue Growth

Belgacem Ben Youssef$^{(\boxtimes)}$

Department of Computer Engineering, College of Computer and Information
Sciences, King Saud University, Riyadh, Saudi Arabia
BBenyoussef@ksu.edu.sa

Abstract. We present the simulation of the effect of cell migration
speed on wound healing using a three-dimensional computational model
for multicellular tissue growth. The computational model uses a discrete approach based on cellular automata to simulate wound-healing
times and tissue growth rates of multiple populations of proliferating
and migrating cells. Each population of cells has its own division, motion,
collision, and aggregation characteristics resulting in a number of useful
system parameters that allow us to investigate their emergent effects. Our
sequential performance results point to the need of porting the model to
modern high performance machines to harness the computational power
available in multicore and GPU-based computers. Discrete systems of
this kind can be a valuable approach for studying many complex systems, including biological ones.

Keywords: Cellular automata · 3D model · Tissue growth · Wound
healing · Cell migration

1 Introduction

Cell motility is important for the proliferation of mammalian cells. In addition, it
is paramount in many physiological processes such as angiogenesis, wound healing, inflammation, and tumor cell metastasis [1]. Increased motility of cells significantly enhances their proliferation rates, and thus directly affects the population
dynamics of tissue growth. The structure of natural tissues is supported by an
extracellular matrix (ECM) that has the form of a three-dimensional network of
cross-linked protein strands (see Figure 1, for an example). The ECM plays many
important roles in tissue development. Biochemical and biophysical signals from
the ECM modulate fundamental cellular activities, including adhesion, migration,
proliferation, differentiation, and programmed cell death [3]. Scaffold properties,
cell activities like adhesion or migration, and external stimuli that modulate cellular functions are among the many factors that affect the growth rate of tissues [4].

© Springer International Publishing Switzerland 2015
O. Gervasi et al. (Eds.): ICCSA 2015, Part II, LNCS 9156, pp. 28–42, 2015.
DOI: 10.1007/978-3-319-21407-8_3

As a result, the development of bio-artificial tissue substitutes involves extensive and time-consuming experimentation. The development of computational models with predictive abilities could enhance progress in this area. In this context, the simulation of the effect of cell migration under different conditions is necessary to evaluate their charactersitics, screen many alternatives, and choose only the most promising ones for laboratory experimentation.

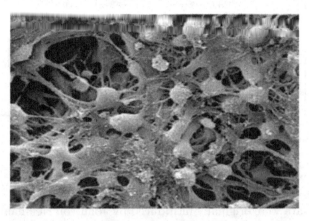

Fig. 1. A scanning electron micrograph displaying the three-dimensional structure of an extracellular matrix

This research describes a three-dimensional cellular automata (CA) model to simulate the growth of three-dimensional tissues consisting of more than one cell type in a wound-healing environment. The corresponding discrete model is an extension of a previously developed base model that accounted for only a single type of cells [5]. The model incorporates all the elementary features of cell division and locomotion including the complicated dynamic phenomena occurring when cells collide and aggregate. Each computational element is represented by a site within a cubic lattice. While the assumption of cubic living cells does not reflect the true morphology of migrating or confluent mammalian cells, it allows us to use data structures that minimize memory and computational time requirements. Here, each computational site interacts with its neighbors that are to its north, east, west, south, and immediately above it or below it. This is known as the von Neumann neighborhood in three dimensions [6]. Our objective is to evaluate the effects of cell migration speed on the tissue growth rate and wound-healing time in the context of a mixed wound-seeding topology employing two types of cell populations. In particular, we explore the following question:

– What are the effects of cell migration speed on the wound-healing time and tissue growth rate?

In the next section, we define the concept of cellular automata. This is followed by a concise review of related work and a short description of the development of the model. We then present the corresponding sequential algorithm and include its flowcharts. Before concluding, we give an overview of the important parameters and inputs of the model and discuss our performance and simulation results.

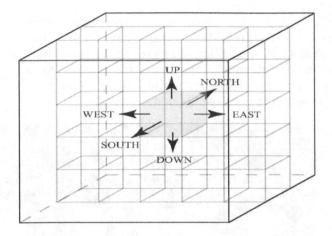

Fig. 2. The von Neumann neighborhood in three dimensions used in our CA model

2   Cellular Automata Concepts

Cellular automata were originally introduced by John von Neumann and Stanislaw Ulam as a possible idealization of biological systems with a particular purpose of modeling biological self-reproduction [7]. This approach has been used since then to study a wide variety of physical, chemical, biological, and other complex natural systems [7].

We consider d-dimensional cellular automata consisting of an array D of lattice cells covering a finite domain. Any cell c is uniquely identified by d integer coordinates $(i_1, i_2, \ldots, i_d)$, where $1 \leq i_1 \leq N_1$, $1 \leq i_2 \leq N_2$, ..., and $1 \leq i_d \leq N_d$. Let Ω be the set of all computational sites in the cellular space and N be the total number of such sites such that $N = N_1 \times N_2 \times \ldots \times N_d$. A cellular automaton satisfies the following properties:

1. Each cell c interacts only with its neighbor cells defined by a neighborhood relation that associates with the cell c a finite list of neighbor cells $c + \nu_1, c + \nu_2, \ldots, c + \nu_k$. In general, the neighborhood vector (or neighborhood index), $V = [\nu_1, \nu_2, \ldots, \nu_k]$, may vary from one cell to another.
2. Each cell can exist in one of a finite number of states. This finite list of states will be listed by Q. In the simplest case of two-state automata, $Q = \{0, 1\}$.
3. Each function $X : \Omega \longrightarrow Q$ defining an assignment of states to all cells in the cellular space Ω is called a configuration. Then, x_c is called the state of the cell c under configuration X.
4. For any cell c in the cellular space, there exists a local transition function (or rule) f_c, from Q^k to Q, specifying the state of the cell at time level $t + 1$ as a function of the states of its neighbors at time level t. That is, $x_c^{t+1} = q^{t+1}(c) = f_c(x_{c+\nu_1}^t, x_{c+\nu_2}^t, \ldots, x_{c+\nu_k}^t)$.
5. The simultaneous application of the local transition functions f_c to all the cells in a cellular space defines a global transition function F which acts on

the entire array transforming any configuration X^t to a new configuration X^{t+1} according to $X^{t+1} = F(X^t)$.

These properties imply that each cellular automaton is a discrete dynamical system. Starting from an initial configuration X^0, the cellular array follows a trajectory of configurations defined by the global transition function F. All possible configurations of the cellular automaton define a set Φ, whose cardinality can be quite large. For instance, using $N_1 = N_2 = N_3 = 5$ and $Q = [0, 1]$, the number of configurations in Φ would be equal to $2^{5 \wedge 5 \wedge 5} \approx 4.254 \times 10^{37}$ configurations.

We can now define parallel discrete iterations for a cellular automaton as follows:

$$\begin{cases} X^0 & \text{is given in } \Phi \\ X^{t+1} = F(X^t), \end{cases} \tag{1}$$

for $t = 0, 1, 2, \ldots$ or equivalently:

$$\begin{cases} X^0 & = (x_1^0, x_2^0, \ldots x_N^0) \text{ is given in } \Phi \\ X_f^{t+1} & = f_i(x_1^t, x_2^t, \ldots, x_N^t), \end{cases} \tag{2}$$

for $t = 0, 1, 2, \ldots$ and $i = 1, 2, 3, \ldots, N$. The preceding two equations, or rules, imply that the parallel discrete iterations update the states of all cells at the same time. It should be noted here that the transition functions of cellular automata need not be algebraic in form and may be rule-based. A potentially important feature of cellular automata is the capability for *self-reproduction* through which the evolution of a configuration yields several separated, yet identical copies of the configuration. Moreover, cellular automata rules may map several initial configurations into the same final configuration, thus leading to microscopically *irreversible* time evolution in which trajectories of different states may merge [8].

3   Related Work

Various modeling approaches have been used to simulate the population dynamics of proliferating cells. These models can be classified as: deterministic, stochastic, or based on cellular automata and agents. We briefly review a few of the recent cellular automata and agent-based lattice-free models to simulate tissue growth. Chang and his team developed a 3-D cellular automata based model to describe the growth of microbial cell units [9]. This model considered the effects of bacterial cell division and cell death. Other CA-based models have also been used to solve more specific biological modeling problems. For instance, Kansal et al. developed a model to simulate brain tumor growth dynamics [10]. Their model utilizes a few automaton cells to represent thousands of real cells, thus reducing the computational time requirements of the model while limiting its ability to track individual cells in the cellular space. Another CA model was used by Cickovski et al. as a framework to simulate morphogenesis [11]. This model used a hybrid approach to simulate the growth of an avian limb. The cellular automaton governed cell interactions while reaction-diffusion equation solvers were used to determine the concentration levels of surrounding chemicals.

Some of the agent-based models apply the dynamics of cell proliferation and death to describe tissue pattern formation and growth [12]. Other related models are suitable for describing the locomotion of a fixed number of cells where cells move relatively slowly with respect to other processes like the diffusion of soluble substances [13]. Additional models employ feedback mechanisms between cells and the substrate to model cells entering and leaving the tissue and to establish homeostasis in such systems [14]. Some of the agent-based models use regular triangulation to generate the neighborhood topology for the cells, thus allowing for a continuous representation of cell sizes and locations in contrast to grid-based models [15]. Others utilize multiscale approaches to model collective phenomena in multicellular assemblies, including inflammation and wound healing [16].

4   The Computational Model

The growth of tissues is a complex biological process. In this model, the migration and proliferation of mammalian cells are considered to be mainly characterized by the following four subprocesses: cell division, cell motion, cell collision, and cell aggregation. For a detailed account of the modeling steps of each of these subprocesses, we refer the reader to related reference [5].

4.1   States of the Cellular Automaton

The model is a discrete system operating in a cellular space containing $N = N_x \times N_y \times N_z$ computational sites. Cells in the cellular space interact with their neighbors at equally spaced time intervals $t_1, t_2, \ldots, t_r, t_{r+1}, \ldots$ where $t_{r+1} = t_r + \Delta t$ for all r. An occupied computational site must describe the current state of a given cell using a set of values. These values must describe the asynchronous proliferation and persistent random walks of multiple cell types. In building an adequate state definition, sufficient information must be provided about the history so that given the current state, the past is statistically irrelevant for predicting all future behavior pertinent to the application at hand [17]. Based on these specifications, the state x_i of an automaton containing a living cell must specify the following set of parameters:

1. The cell type.
2. The direction of cell motion.
3. The cell speed.
4. The time remaining until the next direction change.
5. The time remaining until the next cell division.

The average speed of migrating cells is controlled by varying the value of the time interval, . This is due to the fact that migrating cells cover a fixed distance in each step. Another means of regulating the speed of locomotion is the ability to adjust the transition probability for the stationary state. Therefore, a migrating cell of type j in automaton i must only specify the direction of locomotion and the times which remain until the next direction change and the next cell division in its state x_i. The state of an arbitrary automaton i, thus, takes values from the following set of eight-digit integer numbers $\Psi = \{klmnpqrs/k, l, m, n, p, q, r,$ and $s \in \mathbb{N}\}$, where k is the cell type. The direction of motion is identified by the direction index l. When l is equal to 0, the cell is in the collision stationary state. When the value of l is in the range of 1 to 6, it represents one of six directions the cell is currently moving in. When the value of l is 7, it enters an aggregation stationary state where it "sticks" to another cell of the same type potentially forming cellular aggregates. The digits mn denote the persistence counter. This counter represents the time remaining until the next change in the direction of cell movement. The cell phase counter is given by the remaining four digits $pqrs$ and holds the time remaining before the cell divides.

5 Sequential Algorithm

5.1 Initial Conditions

The initial parameters for the simulation are first read from the input data file. Then, the computational sites to be occupied by the cells at the start of this simulation run are selected based on the seeding mode of the initial cell distribution. For each occupied site, we assign a cell state based on the population characteristics of that cell type. The direction index is randomly selected, the persistence counter is assigned a properly chosen value, and the cell phase counter is set based on experimentally determined cell division data.

5.2 Iterative Operations

At each time step $t_{r+1} = t_r + \Delta t$, for $r = 0, 1, 2, \ldots$

1. Randomly select a computational site.
2. If this site is occupied by a cell c and the phase counter is zero then it is time for this cell to divide and the division routine is called.
3. If this site is occupied by a cell c and the persistence counter is zero, then it is time for this cell to change directions and the direction change routine is called.
4. If this site is occupied by a cell c and both the phase and persistence counters are not zero, attempt to move this cell to a neighboring site in the direction indicated by the direction index of its current state.
   (a) If this neighboring site is free, then mark it for cell c and decrement the phase and persistence counters by one.
   (b) If this neighboring site is occupied by a cell from a different type, then cell c remains in the current site and both cells enter the stationary state due to collision. Their persistence counters are set accordingly while their respective phase counters are decremented by one.

(c) If this neighboring site is occupied by a cell from the same type, then cell c remains in the current site and both cells enter the aggregation stationary state. The persistence counters for both cells are set to the appropriate waiting time and their phase counters are decremented by one.

5. Select another site (randomly) and repeat Steps 2-4 until all sites have been processed.
6. Update the states of all sites so that the new locations of all cells are computed.
7. If confluence has not been reached, proceed to the next time step.

The flowcharts of the main module of the sequential algorithm and those of the division and direction change routines are displayed in figures 3 through 6, respectively.

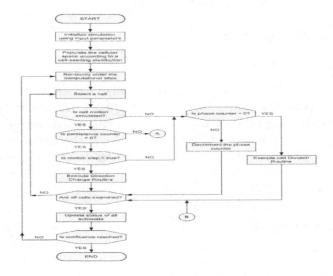

Fig. 3. Flowchart of the main module of the sequential algorithm (part 1 of 2)

6   Simulation Parameters for Wound Healing

6.1   Cell Seeding Distribution

In this study, we consider a wound-seeding topology where a wound in the shape of an empty cylinder is centered in the cellular grid with all surrounding sites occupied by two types of cells. This topology simulates the cell migration and proliferation phase of wound healing. This model does not attempt to describe all the steps of the complicated wound-healing process [18]. We associate one type of cell distributions with this seeding topology, known as the mixed distribution. All cell

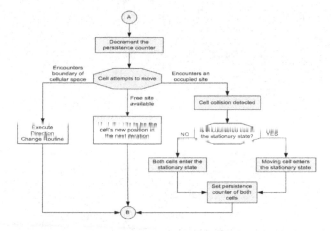

Fig. 4. Continuation of the flowchart of the main module (part 2 of 2)

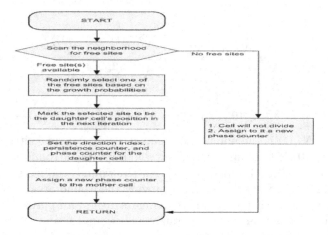

Fig. 5. Flowchart showing the cell division routine

types are seeded together in all areas surrounding the denuded area of the wound environment using a uniformly random placement of cells. During the simulation, cells of both types can migrate freely in the wound area. Figure 7 illustrates an example of this seeding distribution.

6.2   Cell Population Dynamics

Starting with a total number of seed cells equal to N_0, the CA rules transform the cellular array to simulate the dynamic process of tissue growth inside the wound environment. At some time t after the start of the simulation, $N_c(t)$ sites of the cellular automaton are occupied by cells. We define a measure to indicate the volume coverage at time t inside the wound area as the cell volume fraction $k(t)$, as follows:

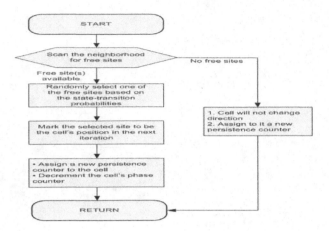

Fig. 6. Flowchart depicting the direction change routine

$$k(t) = \frac{N_c(t) - N_0}{N - N_0} = \frac{\sum\limits_{i=1}^{n} (N_{c_i}(t) - N_{c_i}(0))}{N - \sum\limits_{i=1}^{n} N_{c_i}(0)},$$

with $N_0 = N_c(0) = \sum\limits_{i=1}^{n} N_{c_i}(0)$ and where $N(= N_x \times N_y \times N_z)$ is the size of the cellular space, $N_{c_i}(t)$ is the number of occupied computational sites by cell type i at time t, $N_{c_i}(0)$ is the number of seed cells of type i surrounding the wound, and n is the number of cell types ($n \geq 1$). For the wound seeding, the cell volume fraction indicates the fraction of cells occupying the wound area at a given time. Thus, the time to reach full volume coverage can be an approximation to the wound-healing time, an important parameter in wound healing research.

The overall tissue growth rate represents the increase in volume coverage, within the wound area, with respect to time. To this end, the tissue growth rate measure is given by the following formula:

$$\frac{dk(t)}{dt} = \frac{\sum\limits_{i=1}^{n} (N_{c_i}(t) - N_{c_i}(t - \Delta t))}{\Delta t \times (N - N_0)} = \frac{\sum\limits_{i=1}^{n} (N_{c_i}(t) - N_{c_i}(t - \Delta t))}{\Delta t \times (N - \sum\limits_{i=1}^{n} N_{c_i}(0))}.$$

Here, $k(t)$ is the cell volume fraction at time t as given above and Δt is the time step in hours or days, depending on the resolution of the time scale utilized in the model. The simulation continues until all sites are occupied by cells, that is until $k(t)$ equals one. The movement of cells will slow down due to breaks in the persistent random walks, cell collisions, and cell aggregations. Thus, only a fraction of the total cells, $N_c(t)$, will move in the time interval $[t, \, t + \Delta t]$ and

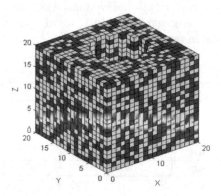

Fig. 7. An example of a mixed wound-seeding topology, comprised of two types of cell populations, is depicted. A wound of cylindrical shape, with a diameter of 10 and a height of 20, inside a $20 \times 20 \times 20$ cellular grid is exhibited.

the effective speed of migration, $S_e(t)$, in a wound-seeding environment can be computed by using the following equation:

$$S_e(t) = \frac{N_m(t)}{N_{c,w}(t)} \times S \,, \tag{3}$$

where $N_m(t)$ refers to the number of moving cells in the time interval $[t,\ t + \Delta t]$, $N_{c,w}(t)$ is the number of occupied sites inside the wound area at time t, and S is the individual cell "swimming" speed. We also define the cell heterogeneity measure H as the ratio of the initially seeded number of cells from population 1 to that from population 2. This is given by:

$$H = \frac{\text{initial number of (faster) cells from population 1}}{\text{initial number of (slower) cells from population 2}} \,.$$

That is, when $H = 9$ there are 9 cells from population 1 for every cell from population 2.

6.3   Additional Simulation Inputs

The simulation results of the proliferation of multiple cell types are obtained for a $200 \times 200 \times 200$ cellular array where two cell populations are used with a wound diameter of 100 and a height of 200. A confluence parameter of 100%, an average waiting time of 2 hours for the six directions of motion and 1 hour for the two stationary states are utilized. The cell speed of population 1 is assigned values of $1, 2, 5, 10$, and $50 \mu m$ per hour while the speed of migration of cells in population 2 is kept constant at $1 \mu m$ per hour. In addition, a cell heterogeneity ratio of $H = 1$ is maintained throughout the simulations so that an equal number of cells from both populations is used to populate the tissue surrounding the wound. Each cell is modeled as a cubic computational element whose sides are assumed to be equal to $10\mu m$ in length. As given in Table 1, we also employ different division time distributions for these two cell populations.

Table 1. Division time distributions for the two cell populations

	Cell Populations	
Division Times	Cell Population 1	Cell Population 2
12 - 18 hrs	64%	4%
18 - 24 hrs	32%	32%
24 - 30 hrs	4%	64%

6.4 Sequential Performance Results

We implemented the model sequentially on a single node of a high-performance cluster, made available to us by the InfoNetMedia Centre at Simon Fraser University, Canada. Each node uses an Intel P4 3.0-GHz processor with 1 GB of RAM and runs the Gentoo Linux operating system with a GCC compiler version 3.4.4. We compiled the sequential program using the following command: `g++ -O2 -march=pentium4 pro-gram.cpp`. The `-O2` option provides the highest optimization level in the g++ complier without introducing errors into the application while the `-march=pentium4` option instructs the compiler to generate Intel Pentium 4 processor-efficient code.

Due to a limited per-node memory capacity, the largest cellular array size for the sequential runs was $330 \times 330 \times 330$. The following performance results (see Table 2) were obtained for a wound-seeding distribution with a ratio $H = 1$ and cell migration speeds of $10\mu m/hr$ and $1\mu m/hr$ for cell population 1 and cell population 2, respectively. We note that running the model with a large array size is a computationally demanding task that requires small time steps to accurately describe the dynamics of multiple cell populations. In addition to the size of the cellular array, several input parameters affect the execution time needed to run a simulation, including the size of the cylindrical wound, cell migration speed, and cell division time. A simulation using a $200 \times 200 \times 200$ cellular array with a wound diameter of 100 takes $4,059$ seconds (or 1.13 hours) to run serially. Such grid represents a cubical tissue object whose side is only equal to 2 mm and containing a wound with 1 mm in diameter. These performance results point to the need to use parallel computing systems in order to simulate wound healing in the context of multicellular tissues of larger sizes.

7   Simulation Results and Discussion

We discuss our results that simulate the effect of varying the migration speed of cell population 1 on the wound-healing time and tissue growth rate, and then present results that display the temporal evolution of the average speed of all

Table 2. Sequential execution times of the model using different cellular array sizes and wound diameters. The height of the wound is set equal to one dimension of the cellular grid.

Cellular Array	Wound Diameter	Sequential Execution Time (secs)
150 × 150 × 150	75	1, 593
200 × 200 × 200	100	4, 059
250 × 250 × 250	125	8, 536
300 × 300 × 300	150	15, 908
330 × 330 × 330	165	21, 867

cells inside the wound area. Figure 8(a) shows the temporal evolutions of volume coverage as the cell speed of population 1 is varied from $1\mu m/hr$ to $50\mu m/hr$ while cells of population 2 move at a fixed speed of $1\mu m/hr$. A constant cell heterogeneity ratio of $H = 1$ is also maintained in a mixed wound-seeding distribution throughout these simulations. Broadly speaking, we note that volume coverage inside the wound increases with time until it reaches confluence for all values of cell-population-1 speeds. We also observe that as the motility of the cell increases, the proliferation rate increases and hence confluence is attained faster; thus, healing the wound much more quickly (from nearly 24 days to about 5 days). Higher motility of cells decreases the impact of contact inhibition on the proliferation rate as it reduces the formation of cell colonies. Cells moving at increased speeds are the first to enter the denuded wound area seeking to populate its empty sites. This delays the formation of cell colonies and leads to faster proliferation by mitigating the impact of contact inhibition. Furthermore, part (b) of the same figure illustrates the impact of varying the cell speed of population 1 on the overall tissue growth rate. The figure clearly depicts that increasing the cell migration speed leads to higher rates of tissue growth (from a low of about 0.07 to a high of nearly 0.52). Increasing cell migration speeds even to very large values continues to impact positively both the tissue growth rate and the time to reach confluence in the case of this wound-seeding topology. In all simulations, the tissue growth rate increases initially, reaches a maximum and then decreases as a result of contact inhibition brought about by cell-colony formation and merging events that eventually lead to the closure of the wound area and its healing.

Figure 9 depicts the temporal evolution of the effective migration speed, S_e, of all cells in the denuded area of the wound for different cell-population-1 speeds. At the beginning of the simulations, cells move into the wound at their peak speeds. Then, the overall cell speeds drop rapidly as the wound area becomes congested with new daughter cells; and collisions as well as aggregations become more frequent. The average speed decreases with time and shows a drastic decline as confluence is attained due to the formation of local cell clusters and their subsequent mergings.

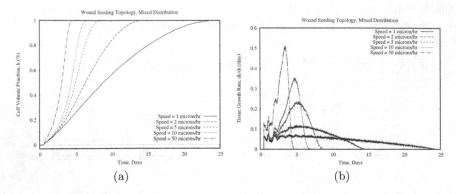

Fig. 8. The temporal evolution of the (a) cell volume fraction and (b) overall tissue growth rate as the cell speed of population 1 is varied from 1 to $50\mu m/hr$. Cells in population 2 move at a fixed speed of $1\mu m/hr$

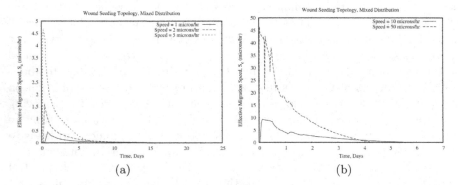

Fig. 9. Effect of cell motility on the overall migration speed, S_e, of all cells inside the wound area for cell-population-1 speeds ranging from (a) 1 to $5\mu m/hr$ and (b) 10 to $50\mu m/hr$

8   Conclusion and Future Work

We described herein a three-dimensional computational model for the growth of multicellular tissues based on the concept of cellular automata to simulate wound healing. The model incorporates many aspects of cell behavior involving cell migration, division, collision, and aggregation while including multiple cell types. We presented simulation results from the serial implementation of the model using a mixed wound-seeding distribution showing the effect of varying cell migration speed of one cell population on the tissue growth rate and wound-healing time. Our simulation results indicate that increasing cell migration speeds leads to a sharp decrease in wound-healing time and that this improvement continues unabated even for larger cell speeds reaching $50\mu m/hr$. Further, our performance results point to the need for using parallel systems such as shared-memory and heterogeneous architectures, including multicore CPU and GPU machines to run the model with much larger cellular grids in order to simulate tissue growth for more realsitic wound sizes [19]. As part of our future work, we will also consider the use of agent-based modeling for multicellular tissue growth and wound healing [20] [21].

Acknowledgments. The author would like to acknowledge the support for this research from the Research Centre in the College of Computer & Information Sciences and the Deanship of Scientific Research at King Saud University.

References

1. Safferling, K., et al.: Wound Healing Revised: A Novel Reepithelialization Mechanism Revealed by In Vitro and In Silico Models. J. Cell Biol. **203**(4), 691–709 (2013)
2. Palsson, B.O., Bhatia, S.N.: Tissue Engineering. Pearson Prentice Hall, Upper Saddle River (2004)
3. Soll, D., Wessels, D.: Motion Analysis of Living Cells: Techniques in Modern Biomedical Microscopy. Wiley-Liss, New York (1998)
4. Langer, R., Vacanti, J.P.: Tissue Engineering. Science **260**, 920–926 (1993)
5. Ben Youssef, B.: A Visualization Tool of 3-D Time-Varying Data for the Simulation of Tissue Growth. Multimed. Tools Appl. **73**(3), 1795–1817 (2014)
6. Tchuente, M.: Computation on Automata Networks. In: Soulie, F.G., Robert, Y., Tchuente, M. (eds.) Automata Networks in Computer Science: Theory and Applications, pp. 101–129. Princeton University Press, Princeton (1987)
7. Deutsch, A., Dormann, S.: Cellular Automaton Modeling of Biological Pattern Formation: Characterization, Applications, and Analysis. Birkhauser, Boston (2005)
8. Chopard, B., Droz, M.: Cellular Automata Modeling of Physical Systems. Cambridge University Press, Cambridge (1998)
9. Chang, L., Gilbert, E.S., Eliashberg, N., Keasling, J.D.: A Three-Dimensional, Stochastic Simulation of Biofilm Growth and Transport-Related Factors that Affect Structure. Micro-biology **149**(10), 2859–2871 (2003)
10. Kansal, A.R., Torquato, S., Harsh IV, G.R., Chiocca, E.A., Deisboeck, T.S.: Simulated Brain Tumor Growth Dynamics Using a Three-Dimensional Cellular Automaton. J. Theor. Biol. **203**(4), 367–382 (2000)
11. Cickovski, T.M., et al.: A Framework for Three-Dimensional Simulation of Morphogenesis. IEEE ACM T. Comput. Biol. Bioinformatics **2**(4), 273–288 (2005)
12. Schaller, G., Meyer-Hermann, M.: Multicellular Tumor Spheroid in an Off-Lattice Voronoi-Delaunay Cell Model. Phys. Rev. E **71**(5 pt. 1), 051910 (2005)
13. Beyer, T., Meyer-Hermann, M.: Delauny Object Dynamics for Tissues Involving Highly Motile Cells. In: Chauviere, A., Preziosi, L., Verdier, C. (eds.) Cell Mechanics: From Single Scale-Based Models to Multiscale Modeling, pp. 417–442. CRC Press (2010)
14. Fu, Y.X., Chaplin, D.D.: Development and Maturation of Secondary Lymphoid Tissues. Annu. Rev. Immunol. **17**, 399–433 (1999)
15. Beyer, T., Schaller, G., Deutsch, A., Meyer-Hermann, M.: Parallel Dynamic and Kinetic Regular Triangulation in Three Dimensions. Comput. Phys. Commun. **172**(2), 86–108 (2005)

16. Cordelia, Z., Mi, Q., An, G., Vodovotz, Y.: Computational Modeling of Inflammation and Wound Healing. Adv. Wound Care **2**(9), 527–537 (2013)
17. Bratley, P., Fox, B.L., Schrage, L.E.: A Guide to Simulation, 2nd edn. Springer, New York (1987)
18. Majno, G., Joris, I.: Cells, Tissues, and Disease: Principles of General Pathology. Oxford University Press, New York (2004)
19. Ben Youssef, B.: A Parallel Cellular Automata Algorithm for the Deterministic Simulation of 3-D Multicellular Tissue Growth. Cluster Comput. (2015). doi:10.1007/s10586-015-0455-7
20. An, G., Mi, Q., Dutta-Moscato, J., Vodovotz, Y.: Agent-Based Models in Translational Systems Biology. Wiley Interdiscip. Rev. **1**(2), 159–171 (2009)
21. Azuaje, F.: Computational Discrete Models of Tissue Growth and Regeneration. Brief. Bioinforma. **12**(1), 64–77 (2011)

Multiple Classifier System with Metaheuristic Algorithms

Maciej Quoos, Iwona Pozniak-Koszalka, Leszek Koszalka[✉], and Andrzej Kasprzak

Department of Systems and Computer Networks,
Wroclaw University of Technology, Wroclaw, Poland
maciej.quoos@gmail.com,
{iwona.pozniak-koszalka,leszek.koszalka,
andrzej.kasprzak}@pwr.edu.pl

Abstract. Ability of Multiple Classifier Systems (MCSs) to deliver correct prediction, even though all of the classifiers from its ensemble are wrong, confirms that MCS approach possess large potential in the field of pattern recognition. This work focuses on the problem of fuser design. While most MCSs use popular Voting combiner, authors of this work investigate prediction accuracy of MCS with either Evolutionary or Particle swarm algorithms applied as the fusers. Obtained results suggest, that aforementioned fusion algorithms increase accuracy performance on the evaluated datasets in comparison to the popular Voting combiner.

Keywords: Evolutionary algorithm · Particle swarm · Fuser design · Classifier ensemble · Optimization

1   Introduction

Impact of machine learning on our everyday life-routine can be noticed everywhere. From medical applications in which the machines support doctors in delivering diagnosis - to the video games which are able to respond to our gestures.

In order to deliver satisfactory predictions which would meet the aforementioned challenges, one would need a well-trained classifier which could provide satisfactory results for all possible input data. This task turned out to be very difficult, in some situations even impossible. Therefore alternative to a single classifier has been proposed – Multiple classifier system (MCS). The idea of MCS [1] is to combine set of experts in different areas together, so that they would contemplate each other. The typical architecture of MCSs is shown in Figure 1.

One of the issues encountered during MCS design is choosing the fusion method which would be able to exploit the best qualities of each classifier from the ensemble [2]. Very popular approach to this problem is using Voting combiner, where decision made by the biggest number of classifiers is chosen as an output [3]. Alternative to this method is weight selection [4]. Collective decision making methods exploit classifier fusion based on discriminant analysis [5]. In this research focus has been put on the analysis of what is the best way of assigning weights for a fuser based on discriminants.

© Springer International Publishing Switzerland 2015
O. Gervasi et al. (Eds.): ICCSA 2015, Part II, LNCS 9156, pp. 43–54, 2015.
DOI: 10.1007/978-3-319-21407-8_4

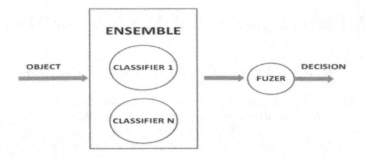

Fig. 1. Multiple classifier system architecture

For this purpose, the methods based on artificial intelligence [6] can be used. In the paper, both evolutionary and particle swarm algorithms are applied as a MCS fusion techniques. Obtained prediction accuracies on three different real life datasets are compared with the popular and simple Voting combiner performance.

The rest of the paper is organized as follows. Section 2 provides mathematical formulation of the fuser design problem. Section 3 describes shortly both evolutionary and particle swarm algorithms and its implementation details. Section 4 contains experimental set up, and Section 5, the experimental results. In final Section 6, work is summarized and final conclusions are drawn.

2   Problem Statement

The goal of the pattern recognition task, explained with details in [5] is to assign given object to one of the predefined classes $i \in C = \{1, ..., C\}$, based on its features X. This assumption leads to a classifier Ψ, defined as the following function:

$$\Psi : X \to C \tag{1}$$

which classifies the object to one of the predefined classes. In the following research we investigate set of n classifiers, each of which makes decision on the basis of values of discriminants. Let $F^{(l)}(i, x)$ means such a function of class i and given value of x. This function is used by the l-th classifier $\Psi^{(l)}$. Following classifier $\widehat{\Psi}(x)$ is used to obtain common decision of the classifier ensemble:

$$\widehat{\Psi}(x) = i \quad if \quad \hat{F}(i, x) = \max_{k \in C} \hat{F}(k, x) \tag{2}$$

In [5], an interesting experiment has been conducted, showing that there is a possibility of fuser based on discriminants producing correct decision, even though all individual classifiers are wrong. Therefore the focus of the optimization task in the conducted research was to find the matrix of weights:

$$W = \begin{bmatrix} W_1^1 & \cdots & W_1^C \\ \vdots & W_l^i & \vdots \\ W_n^1 & \cdots & W_n^C \end{bmatrix} \tag{3}$$

which consists of weights assigned to each classifier and a class number that modifies the discriminant of class i as follows:

$$F(i,x) = \sum_{l=1}^{n} W_l^i F^{(l)}(i,x), \tag{4}$$

such that assures the lowest frequency of misclassification of MCS:

$$\hat{\Psi}(W) = \min P_e(W), \tag{5}$$

where P_e is the misclassification frequency. Moreover W is subject to constraints

$$0 \le W_l^i \le 1 \tag{6}$$

3 Algorithms

The algorithms considered for solving the problem are Evolutionary Algorithm (EVO), and Particle Swarm Optimization algorithm (PSO), both of which are compared with the popular algorithm called Voting combiner.

3.1 Evolutionary Algorithm (EVO)

The proposed evolutionary algorithm implementation has been inspired by [7] and is as follows:

- For each member of the population, value of the fitness function is calculated according to (5).
- The parent's representation is obtained using one of the three approaches: random, k-best and roulette wheel.
- Mutation involves adding a vector of numbers randomly generated according to the normal density distribution (mean equal to 0 and the standard deviation set to 1).
- The crossover operator generates two offspring members on the basis of two parents according to the two-point crossover rule.
- A selection of individuals from the population is formed by merging the descendant population with a set of individuals created by mutation and crossover.
- The chance of selection particular individual to the next iteration is one of the following: every member survives, only the descendant population survives or |P| best members survive, where P is the size of the initial population.
- A validation set is used in order to calculate the fitness of the individuals in the same way as for the regular population assessment. The procedure breaks the optimization process if the number of iterations reaches the maximum value.

The weight matrix (3) is encoded as a column vector:

$$W = [W_1^1, ..., W_1^C, W_2^1, ..., W_2^C, ..., W_n^1, ..., W_n^C]^T \tag{7}$$

3.2   Particle Swarm Optimization Algorithm (PSO)

The proposed particle swarm algorithm implementation has been inspired by [8] and is as follows: The population called swarm consists of N particles moving around a D-dimensional search space, where D is the length of the vector W described by (7). The position of the i-th particle is represented by $x_i = (x_{i1}, x_{i2}, ..., x_{iD})$ with corresponding velocity written as $v_i = (v_{i1}, v_{i2}, ..., v_{iD})$. Both position and velocity are limited within $[X_{min}, X_{max}]^D$ and $[V_{min}, V_{max}]^D$, respectively. Additionally, knowledge from the neighboring particles has to be considered in order to find the best solution. The best encountered position of the i-th particle, denoted as *pbest_i*, is represented by $p_i = (p_{i1}, p_{i2}, ..., p_{iD})$. The best value of the all individual *pbest_i* values is denoted as *gbest* and is expressed as $g = (g_1, g_2, ..., g_D)$. At any iteration, position and velocity of each particle is updated as follows:

$$v_{id}^{new} = w \times v_{id}^{old} + c_1 \times r_1 \times (pbest_{id} - x_{id}^{old}) + c_2 \times r_2 \times (gbest - x_{id}^{old}) \tag{8}$$

$$x_{id}^{new} = x_{id}^{old} + v_{id}^{new} \tag{9}$$

where r_1 and r_2 are the random numbers between (0, 1) and c_1 and c_2 are the accelerations which control the distance a particle moves in a single iteration [9]. The parameters v_{id}^{new} and v_{id}^{old} denote the velocities of the new and old particles, respectively.

The w denotes inertia weight as proposed in [10], which linearly decreases from $w_{max} = 0.9$ to $w_{min} = 0.4$ throughout the search process and is expressed as

$$w = (w_{max} - w_{min}) \times \frac{iter_{max} - iter}{iter_{max}} + w_{min} \tag{10}$$

Encoding of the individual particle is as in (7).

4   Experimentation System

The proposed experimentation system is shown in Figure 2. The first step of the procedure presented in Figure 2 is to feed the ensemble with a training set from a given dataset. The input parameters at this stage called ensemble parameters are type and number of individual classifiers that are used in the ensemble. In the conducted experiment following pool of five classifiers was arbitrary chosen:

- k-Nearest Neighbor classifier,
- Quadratic classifier,
- Support Vector classifier,
- Parzen density classifier,
- Neural network trained with back-propagation.

Aforementioned pool ensures that the heterogeneous ensemble has enough diversity, so that each classifier brings new quality to the system.

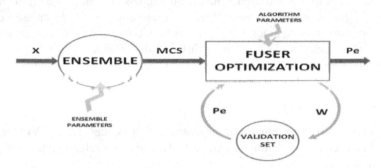

Fig. 2. Experimentation system

After training of the MCS on the training set, the matrix of weights (3) is being optimized using validation set. At this stage, number of algorithm parameters is delivered such as the size of the population and probabilities of crossing-over and mutation for the EVO algorithm and number of particles in the PSO. Finally the misclassification error is calculated on testing set. In the following research, stacked topology of the MCS has been proposed due to the pool of classifiers being trained over the same feature space of a training set [11]. All experiments have been conducted using Matlab R2013b and pattern recognition toolbox PR Tools.

5   Investigations

To perform comparative analysis of the aforementioned optimization methods, the preliminary tests were conducted in order to find out the best set of parameters for each method using grid search and 10-fold cross validation for each cell of the grid. In the following part, average misclassification frequencies and standard deviation of each experiment are presented for various input parameters and various datasets. Below, in Table 1 are presented chosen real-life datasets from UCI repository [12] used in the conducted research.

Table 1. Selected benchmark datasets

Dataset	No. of objects	No. of features	No. of classes	Training +validation size	Test size
Car_evaluation	1 728	6	4	1 038	690
Haberman	306	3	2	184	122
Breast_cancer	699	9	2	420	279

The datasets used in this study cover a variety of dimensions, data size and data complexity. To evaluate the performance of different methods with respect to various data dimensions, both low and high dimensional datasets are selected with the number of features ranging from three to nine. In terms of data complexity, the datasets contain different curvature on boundaries, such cases as both convex and concave are selected. Summarizing, the selected datasets cover different data characteristics. Therefore, the experimental results can give a rigorous evaluation of the optimization methods.

5.1 Haberman Dataset

Tables 2-4 presented below consist of misclassification frequencies of EVO algorithm with fixed number of population size equal to 10 and fixed selection and various reproduction methods.

Table 2. Error frequency of EVO algorithm for random reproduction method

Crossover probability	Mutation probability				
	0.01	0.1	0.2	0.3	0.4
0.60	0.28±0.01	0.22±0.01	0.20±0.01	0.21±0.01	0.22±0.01
0.80	0.24±0.02	0.25±0.02	**0.19±0.01**	0.21±0.02	0.23±0.02
0.99	0.22±0.00	0.21±0.00	0.23±0.02	0.24±0.01	0.24±0.01

Table 3. Error frequency of EVO algorithm for k-best reproduction method

Crossover probability	Mutation probability				
	0.01	0.1	0.2	0.3	0.4
0.60	0.24±0.01	0.21±0.01	0.24±0.01	0.22±0.02	0.21±0.01
0.80	0.23±0.00	0.26±0.02	0.20±0.01	**0.19±0.02**	0.25±0.01
0.99	0.21±0.00	0.22±0.01	**0.19±0.02**	0.24±0.01	0.22±0.02

Table 4. Error frequency of EVO algorithm for roulette wheel reproduction method

Crossover probability	Mutation probability				
	0.01	0.1	0.2	0.3	0.4
0.60	0.22±0.00	0.20±0.01	0.21±0.01	0.21±0.02	0.21±0.02
0.80	0.21±0.01	0.24±0.01	**0.18±0.00**	0.19±0.01	0.25±0.01
0.99	0.20±0.00	0.21±0.01	0.19±0.00	0.22±0.02	0.22±0.02

As can be noticed from Tables 2-4 and Figure 3, the best result for Haberman dataset using the EVO algorithm has been obtained using roulette wheel reproduction method with selection method being the |W| best members. The tuning parameters for which the misclassification frequency is the lowest (18%) are cross-over probability equal to 80% and mutation probability equal to 20%.

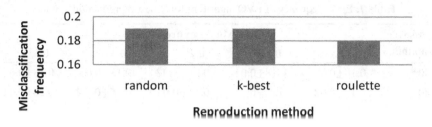

Fig. 3. Misclassification frequency of various reproduction methods

Table 5 presented below consists of misclassification frequencies of PSO algorithm for various numbers of particles.

Table 5. Error frequency of PSO algorithm

	Number of particles				
5	10	15	20	25	30
0.26 ±0.02	0.25±0.01	0.24±0.01	**0.22±0.00**	0.23±0.00	0.23±0.00

As can be noticed from Table 5, the lowest misclassification frequency (22%) for the PSO algorithm for Haberman dataset has been obtained using the number of particles in a swarm equal to 20.

5.2   Car Evaluation Dataset

Tables 6-8 presented below consist of misclassification frequencies of EVO algorithm with fixed number of population size equal to 10 and various reproduction methods.

Table 6. Error frequency of EVO algorithm for random reproduction method

Crossover probability	Mutation probability				
	0.01	0.1	0.2	0.3	0.4
0.60	0.18±0.02	0.17±0.00	0.21±0.02	0.18±0.01	0.18±0.02
0.80	0.17±0.01	**0.14±0.02**	0.19±0.01	0.17±0.02	0.20±0.02
0.99	0.19±0.01	0.16±0.01	0.17±0.00	0.17±0.01	0.21±0.01

As can be noticed from Tables 6-8 and Figure 4, the best result for Car dataset using the EVO algorithm has been once again obtained using roulette wheel reproduction method with selection method being the |W| best members. In this case both random and roulette wheel obtained similar accuracy, however the lowest frequency (14%) obtained with roulette wheel has smaller standard deviation.

Table 7. Error frequency of EVO algorithm for k-best reproduction method

Crossover probability	Mutation probability				
	0.01	**0.1**	**0.2**	**0.3**	**0.4**
0.60	0.16±0.01	0.18±0.01	0.21±0.02	0.18±0.01	0.18±0.02
0.80	0.17±0.02	0.16±0.01	0.17±0.01	**0.15±0.02**	0.20±0.01
0.99	0.19±0.02	0.18±0.01	0.16±0.01	0.17±0.02	0.21±0.01

Table 8. Error frequency of EVO algorithm for roulette wheel reproduction method

Crossover probability	Mutation probability				
	0.01	**0.1**	**0.2**	**0.3**	**0.4**
0.60	0.15±0.01	**0.14±0.01**	0.17±0.01	0.15±0.01	0.18±0.02
0.80	0.17±0.01	0.15±0.01	0.15±0.01	0.15±0.01	0.18±0.02
0.99	0.16±0.01	0.15±0.01	0.14±0.01	0.16±0.02	0.19±0.01

The tuning parameters for which the misclassification frequency is the lowest are crossover probability equal to 60% and mutation probability equal to 10%.

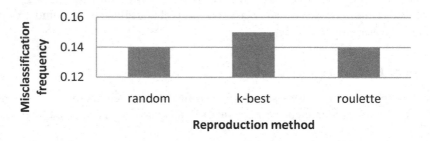

Fig. 4. Misclassification frequency of various reproduction methods for Car dataset

Table 9 consists of misclassification frequencies of PSO algorithm for various numbers of particles. It can be noticed from Table 9, that the lowest misclassification frequency (12%) for the PSO algorithm for Car dataset has been obtained using once again number of particles in swarm equal to 20.

Table 9. Error frequency of PSO algorithm

Number of particles					
5	**10**	**15**	**20**	**25**	**30**
0.15 ±0.01	0.13±0.02	0.13±0.01	**0.12±0.01**	0.16±0.02	0.14±0.01

PSO algorithm obtained better results for three different numbers of particles compared to the best result achieved with EVO algorithm.

5.3 Breast_cancer Dataset

Tables 10-12 presented below consist of misclassification frequencies of EVO algorithm with fixed number of population size equal to 10 and various reproduction methods.

Table 10. Error frequency of EVO algorithm for random reproduction method

Crossover probability	Mutation probability				
	0.01	0.1	0.2	0.3	0.4
0.60	0.05±0.01	0.05±0.01	0.05±0.01	0.05±0.01	0.06±0.01
0.80	0.04±0.00	0.05±0.01	0.04±0.01	**0.03±0.00**	0.03±0.01
0.99	0.04±0.01	0.06±0.02	0.04±0.01	0.04±0.01	0.04±0.01

Table 11. Error frequency of EVO algorithm for k-best reproduction method

Crossover probability	Mutation probability				
	0.01	0.1	0.2	0.3	0.4
0.60	0.04±0.01	0.04±0.01	0.05±0.02	0.05±0.01	0.06±0.01
0.80	0.05±0.01	0.05±0.01	**0.03±0.01**	0.04±0.00	0.04±0.01
0.99	0.04±0.01	0.04±0.00	0.04±0.01	0.03±0.01	0.05±0.01

Table 12. Error frequency of EVO algorithm for roulette wheel reproduction method

Crossover probability	Mutation probability				
	0.01	0.1	0.2	0.3	0.4
0.60	0.04±0.01	0.04±0.01	0.04±0.01	0.05±0.01	0.06±0.01
0.80	0.04±0.01	0.04±0.01	**0.03±0.00**	0.03±0.01	0.05±0.01
0.99	0.04±0.01	0.04±0.01	0.04±0.01	0.03±0.01	0.04±0.01

Conclusions drawn from Tables 10-12 and Figure 5 are that for Breast cancer dataset we obtained equally good results (3%) for all methods, but both random and roulette wheel reproduction methods have smaller standard deviation for 80%/30% and 80%/20% respectively. Moreover, the results obtained for the roulette wheel are more stable on a very good level for various parameters; therefore it seems that this method seems to be more immune to changes of the parameters.

The reason why the k-best reproduction method turned out to produce the worst results in all three datasets can be the elitism which allows surviving and reproducing only the strongest members, therefore after little iteration there is a very small diversity between particular members. Table 13 consists of misclassification frequencies of PSO algorithm for various numbers of particles.

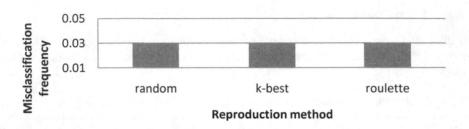

Fig. 5. Misclassification frequency of various reproduction methods for Breast_cancer dataset

Table 13. Error frequency of PSO algorithm

		Number of particles			
5	**10**	**15**	**20**	**25**	**30**
0.06 ±0.02	0.04±0.00	**0.03±0.00**	0.03±0.00	0.05±0.01	0.04±0.01

5.4   Comparative Analysis

Comparative analysis of two algorithms with popular Voting combiner, which simply chooses the class which has the biggest support from the individual classifiers, has been conducted, using the parameters for each algorithm that proved to be the best for each dataset. Performance of all three methods has been examined using exactly the same training and testing sets. Results are presented in Table 14.

It can be noticed from Table 14 that for Haberman dataset the best result has been obtained using Evolutionary algorithm with 18% misclassification frequency, followed by 21% of Particle Swarm Optimization algorithm.

Table 14. Comparative analysis of error frequency of algorithms with Voting combiner

Dataset	EVO	PSO	VOTING
Haberman	**0.18 ± 0.01**	0.21 ± 0.01	0.25 ± 0.01
Car_evaluation	0.14 ± 0.00	**0.13 ± 0.01**	0.18 ± 0.02
Breast_cancer	0.03 ± 0.00	0.03 ± 0.00	0.03 ± 0.00

For the Car_evaluation dataset PSO algorithm has achieved a little edge with 13% over 14% of the EVO. In both datasets, the simple voting combiner has turned out to give significantly worse results. For the third dataset – Breast cancer, all three algorithms have achieved very good accuracy of 97%. The reason might be a very small number of objects belonging to second class, therefore even with random shooting, we would be capable of delivering very good results.

6   Conclusion

In the paper, authors have proposed applying EVO or PSO algorithms as a combiner technique for heterogeneous classifier ensemble on real life datasets.

During performed experiments, it has been shown, that both examined meta-heuristic methods deliver better results than the simple and popular Voting combiner, therefore one can conclude that democracy, which is used by this combiner, is not always the best solution to our problems. According to the 'No free lunch' theorem [13], however, which says that there does not exist a single best solution to every problem, it may turn out that for different datasets, the aforementioned algorithms may not deliver such good results in comparison to the Voting combiner, therefore there is still a need for further research.

Another promising direction for further research might be optimization of the weights assigned to particular features of each object. Moreover, the experimentation system for testing the algorithms could be improved and extended to ensure designing multistage experiments [14] and processing output data [15] in the automatic way.

Acknowledgement. This work was supported by the statutory funds of the Department of Systems and Computer Networks, Faculty of Electronics, Wroclaw University of Technology under S40029/K0402.

References

1. Ho, T.K., Hull, J.J., Shrihari, S.N.: Decision combination in multiple classifiers systems. IEEE Trans. on Pattern Analysis and Machine Intelligence **16**(1), 66–75 (1994)
2. Budnik, M., Pozniak-Koszalka, I., Koszalka, L.: The usage of the k-nearest neighbor classifier with classifier ensemble. In: Proc. to ICCSA (2012)
3. Van Erp, L., Vuurpijl, L.G., Schomaker, L.R.B.: An overview and comparison of voting methods for pattern recognition. In: Proc. of IWFHR, Canada, pp. 195–200 (2002)
4. Wozniak, M., Jackowski, K.: Some remarks on chosen methods of classifier fusion based on weighted voting. In: Corchado, E., Wu, X., Oja, E., Herrero, Á., Baruque, B. (eds.) HAIS 2009. LNCS, vol. 5572, pp. 541–548. Springer, Heidelberg (2009)
5. Wozniak, M., Zmyslony, M.: Designing combining classifier with trained fuser – analytical and experimental evaluation. In: Proc. 10th International Conference on Intelligent Systems Design and Applications (2010)
6. Russell, S., Norvig, P.: Artificial Intelligence: A Modern Approach, 2nd ed., pp. 97–104. Prentice Hall (2003). ISBN 978-0137903955
7. Kuncheva, L.I., Jain, L.C.: Designing classifier fusion systems by genetic algorithm. IEEE Transactions on Evolutionary Computation **4** (2000)
8. Macaš, M., Gabrys, B., Ruta, D., Lhotská, L.: Particle swarm optimisation of multiple classifier systems. In: Sandoval, F., Prieto, A.G., Cabestany, J., Graña, M. (eds.) IWANN 2007. LNCS, vol. 4507, pp. 333–340. Springer, Heidelberg (2007)
9. Alam, S., Dobbie, G., Koh, Y.S., Riddle, P., Rehman, S.U.: Research on particle swarm optimization based clustering: A systematic review of literature and techniques. Swarm and Evolutionary Computation (2014)

10. Chen-Hong, Y., Chih-Jen, H., Li-Yeh, C.: Linearly decreasing weight particle swarm optimization with accelerated strategy for data clustering. IAENG **37**(3) (2010)
11. Wozniak, M., Grana, M., Corchado, E.: A survey of multiple classifier system as hybrid systems. Information Fusion **16**, 3–17 (2014)
12. Asuncion, A., Newman, D.J.: UCI ML repository. University of California, School of Information and Computer Science (2007). http://www.ics.uci.edu/~mlearn/MLRepository.html
13. Wolpert, D.H.: The supervised learning no-free-lunch theorems. In: Proc. 6th World Conference on Soft Computing in Industrial Applications (2001)
14. Ohia, D., Koszalka, L., Kasprzak, A.: Evolutionary algorithm for solving congestion problem in computer networks. In: Velásquez, J.D., Ríos, S.A., Howlett, R.J., Jain, L.C. (eds.) KES 2009, Part I. LNCS, vol. 5711, pp. 112–121. Springer, Heidelberg (2009)
15. Kaminski, R.T., Koszalka, L., Pozniak-Koszalka, I., Kasprzak, A.: Evaluation and comparison of task allocation algorithms for mesh networks. In: Proc. 9th International Conference on Networks, pp. 104–108. IEEE Computer Society Press (2010)

A Robust Clustering via Swarm Intelligence

Sadrollah Abbasi[2], Sajad Manteghi[1], Ali Heidarzadegan[2(✉)], Yasser Nemati[2],
and Hamid Parvin[1]

[1] Department of Computer Engineering, Mamasani Branch,
Islamic Azad University, Mamasani, Iran
s.manteghi@mail.sbu.ac.ir, parvin@iust.ac.ir
[2] Department of Computer Engineering, Beyza Branch, Islamic Azad University, Beyza, Iran
heidarzadegan@beyzaiau.ac.ir

Abstract. A very promising approach to reach a robust partitioning is to use ensemble-based learning. In this way, the classification/clustering task is more reliable, because the classifiers/clusterers in the ensemble cover the faults of each other. The common policy in clustering ensemble based learning is to generate a set of primary partitionings that are different from each other. These primary partitionings could be generated by a clustering algorithm with different initializations. It is popular to filter some of these primary partitionings, i.e. a subset of the produced partitionings is selected for the final ensemble. The selection phase is done to reach a diverse ensemble. A consensus function finally aggregates the ensemble into a final partitioning called also the consensus partitioning. Another alternative policy in the clustering ensemble based learning is to use the fusion of some primary partitionings that come from naturally different sources. On the other hand, swarm intelligence is also a new topic where the simple agents work in such a way that a complex behavior can be emerged. The necessary diversity for the ensemble can be achieved by the inherent randomness of swarm intelligence algorithms. In this paper we introduce a new clustering ensemble learning method based on the ant colony clustering algorithm. Indeed ensemble needs diversity vitally and swarm intelligence algorithms are inherently involved in randomness. Ant colony algorithms are powerful metaheuristics that use the concept of swarm intelligence. Different runnings of ant colony clustering on a dataset result in a number of diverse partitionings. Considering these results totally as a new space of the dataset we employ a final clustering by a simple partitioning algorithm to aggregate them into a consensus partitioning. From another perspective, ant colony clustering algorithms have many parameters. Effectiveness of the ant colony clustering methods is questionable because they depend on many parameters. On a test dataset, these parameters should be tuned to obtain a desirable result. But how to define them in a real task does not clear. The proposed clustering framework lets the parameters be free to be changed, and compensates non-optimality of the parameters by the ensemble power. Experimental results on some real-world datasets are presented to demonstrate the effectiveness of the proposed method in generating the final partitioning.

Keywords: Swarm intelligence · Ant colony · Data fusion · Clustering

© Springer International Publishing Switzerland 2015
O. Gervasi et al. (Eds.): ICCSA 2015, Part II, LNCS 9156, pp. 55–70, 2015.
DOI: 10.1007/978-3-319-21407-8_5

1   Introduction

Data clustering as an important unsupervised learning task is a very challenging problem. The objective function of clustering is to partition a set of unlabeled objects into homogeneous groups or clusters [1]. There are many applications that may use clustering techniques to discover hidden structures in data, such as data mining [1], information retrieval, image segmentation, and machine learning [2]. In real-world problems, the clusters may have different shapes, sizes, degrees of data sparseness, and degrees of data separation. To use any clustering technique, we need to define a similarity measure. The similarity measure should receive two patterns as input, and then give a value (often between [0; 1]) reflecting similarity of the two input patterns. When there is no prior knowledge about cluster shapes, choosing a specialized clustering method is not an easy task [3]. For extracting a proper partitioning out of a given dataset, one requires both clustering expertise and insight about dataset to choose a single appropriate clustering algorithm and also to set parameters of the selected clustering algorithm to appropriate values. Instead of running the risk of picking an inappropriate clustering algorithm, we can benefit from a different alternative solution. The alternative solution is based on the ensemble learning. It can be achieved by applying all possible clustering algorithms (and possibly with all of their initializations) to the dataset and then combining their output partitionings into a consensus partitioning. This is the basic idea behind cluster ensembles [4].

Studies in the field of data mining in the last few years have tended to ensemble methods (combinational methods). Clustering ensemble methods as a subfield of data mining also attempt to find a better, more robust, novel and accurate clustering solution by fusing information from several primary data partitionings [5]. Ensemble learning originates from wisdom of crowd in the society and humanity sciences. In the ensemble-based learning, it is believed that the clustering/classification task is more reliable, because the clusterers/classifiers in the ensemble can cover the faults of each other.

The common policy in clustering ensemble based learning is to generate a set of primary partitionings that are different from each other. These primary partitionings could be generated by a clustering algorithm with different initializations. Some of the produced partitionings may be selected for the final ensemble. The selection phase is done to reach a diverse ensemble. A consensus function finally aggregates the ensemble into a final partitioning named also a consensus partitioning. Another alternative policy in the clustering ensemble learning is to use the fusion of some primary partitionings that come from naturally different sources to ensure they are diverse. In any ensemble diversity between elements of the ensemble has a vital role. It means if an ensemble lacks diversity, it will be surly ineffective. It has been widely used as an inevitable element in a classifier ensemble. Expectedly diversity concept is also vital for a pool of partitionings to be considered as a successful ensemble. To ensure that a number of partitionings are diverse one can use the same strategies used to build a diverse classifier ensemble that include [7, 15, 16]:

1. Different subsets of features: Each partition available in the ensemble is obtained by employing a clustering algorithm on a different projection of dataset.

2. Different clustering algorithms: Each partition available in the ensemble is obtained using a different clustering algorithm.

3. Randomizing: Each partition available in the ensemble is obtained by employing a clustering algorithm with a different initialization. Note that some clustering algorithms are inherently sensitive to its different initialization.

4. Different datasets: Each partition available in the ensemble is obtained employing clustering algorithm on a different resampled sub-dataset of original dataset.

There are many types of consensus functions that solve this problem heuristically. Most of them require the number of clusters to be specified a priori, but in practice the number of clusters is usually unknown. A new consensus function is proposed for cluster ensemble based on swarm intelligence [8] that addresses this problem. In particular, given a set of partitions, we apply ant clustering to the co-association matrix computed from the ensemble to produce the final partition, and automatically determine the number of clusters.

The first ant colony clustering model was introduced by Deneubourg et al. [9]. His model possesses the swarm intelligence of real ants, and was inserted into a robot for the object collecting task. Lumer and Faieta [10] improved upon Deneubourg's model by adding the Euclidean distance formula to the similarity density function and giving ants three kinds of abilities: speed, shortterm memory, and behavior exchange.

It is inspired by how ants organize their food in their nests. Ant clustering typically involves two key operations: picking up an object from a cluster and dropping it off into another cluster [11]. At each step, some ants perform pick-up and drop-off based on some notions of similarity between an object and the clusters. Azimi et al. define a similarity measure based on the co-association matrix [12]. Their clustering process is completely decentralized and self-organized, allowing the clustering structure to emerge automatically from the data. As a result, we can accurately determine the number of clusters in the data. The experimental results show that the proposed consensus function is very effective in predicting the number of clusters and also achieves reliable clustering performance. In addition, by introducing some simple heuristics, we can detect the marginal and outlier samples in the data to improve our final clustering.

Liu et al. propose a method for incrementally constructing a knowledge model for a dynamically changing database, using an ant colony clustering. They use information-theoretic metrics to overcome some inherent problems of ant-based clustering. Entropy governs the pick-up and drop behaviors, while movement is guided by pheromones. They show that dynamic clustering can provide significant benefits over static clustering for a realistic problem scenario [13].

In this paper it is tried to bring two successful concepts in the field of clustering: (a) ensemble concept and (b) swarm concept. It is tried to bring two successful concepts in the field of clustering: (a) ensemble concept and (b) swarm concept. We introduce a new clustering ensemble learning method based on the ACC algorithm. Indeed ensemble needs diversity vitally and swarm intelligence is inherently involved in randomness. Different runnings of ACC algorithm with different initializations on a dataset result in a number of diverse partitionings. Considering these results totally as a new space of the dataset we employ a simple partitioning algorithm to aggregate them into a consensus partitioning. Our experimental results on some real-world datasets have been presented previously to demonstrate the effectiveness of the proposed method in generalization of the final partitioning [6]. However this paper explores the effectiveness of the proposed method by some new metrics to generalize the conclusion. To show how much the swarm intelligence has contribution in reaching better results is the question to be answered in the paper.

2   Ant Colony Clustering

General form of ant colony clustering algorithm is presented here. The algorithm includes a population of ants. Each ant is operating as an autonomous agent that reorganizes data patterns during exploration to reach an optimal clustering. Pseudo code of ant colony clustering algorithm is depicted in Fig. 1.

```
initializing parameter;
for each ant a
            place random a in a position not occupied by other ants;
end;
for each object o
            place random o in a position not occupied by other objects;
end;
for t=1:t_max
        for each ant a
                g=select a random number uniformly from range [0,1];
                r=position(a)
                if(loaded(a) and (is_empty(r)))
                    if(g<p_drop)
                                    o=drop(a);
                                    put(r,o);
                                    save(o,r,q);
                    end;
                end;
                elseif(not (loaded(a) or (is_empty(r))))
                    if(g<p_pic)
                                    o=remove(r);
                                    pick_up(a,o);
                                    search_and_jump(a,o);
                    end;
                end;
                else
                            wander(a,v,N_dir);
                end;
        end;
end;
```

Fig. 1. Pseudo code of original ant colony clustering algorithm

At the first step each object represented by a multi-dimensional vector in the original feature space is randomly scattered in a two-dimensional space. In each step each ant randomly searches the space. They use its short-term memory to jump into a location that is potentially near to an object. They can pick up or drop an object using a probability density obtained by equation 1.

$$f(o_i) = \max\left\{0, \frac{1}{s^2} \sum_{o_j \in Neigh_{s \times s}(r)} \left[1 - \frac{d(o_i, o_j)}{\alpha(1 + \frac{v-1}{v_{max}})}\right]\right\} \tag{1}$$

Neigh$_{sxs}$(r) is the observable local area (or the set of observable rooms) for an ant located at room *r*. *Neigh$_{sxs}$(r)* must be adjacent to the location *r*. It is worthy to mention that each room including *Neigh$_{sxs}$(r)* and *r* is a two-dimensional vector. The function *d(o$_i$,o$_j$)* is the distance between two objects *o$_i$* and *o$_j$* in the original feature space. It is calculated by equation 2. Threshold α is a parameter that scales the distance between each pair of objects and speed parameter *v* control the volume of feature space that an ant explores in each epoch of algorithm.

$$d(o_i,o_j) = \sqrt{\sum_{k=1}^{m}(o_{ik}-o_{jk})^2} \qquad (2)$$

where *m* is the number of original features and where *o$_{ik}$* is *k*-th feature of object *o$_i$*. Probability that an unloaded ant takes an object that is in the room occupied by the ant, obtained from the equation 3.

$$P_{pick}(o_i) = (\frac{k_1}{k_1+f(o_i)})^2 \qquad (3)$$

k$_1$ is a fixed threshold to control the probability of picking an object. The probability that a loaded ant lays down its object is obtained by equation 4.

$$P_{drop}(o_i) = \begin{cases} 2f(o_i) & if\ f(o_i) < k_2 \\ 1 & if\ f(o_i) \geq k_2 \end{cases} \qquad (4)$$

k$_2$ is a fixed threshold to control the probability of dropping an object. Similarity measure, speed parameter, local density and short-term memory are described in following.

2.1   Perception Area

Number of data objects observed by an ant in a two-dimensional area *s*. It is considered as one of the effective factors controlling the overall similarity measure and consequently the accuracy and the computational time of the algorithm. If *s* is large, it will cause the rapid formation of clusters and therefore generally fewer developed clusters. If *s* is small, it will cause the slower formation of clusters and therefore the number of clusters will be larger. It is worthy to mention that selecting this parameter is a very important factor. While selecting a large value can cause premature convergence of the algorithm, selecting a small value also causes late convergence of the algorithm.

2.2   Similarity Scaling Factor

Scaling parameter value α is defined in the interval (0, 1]. If α is large, then the similarities between objects will increase, so it is easier for the ants to lay down their objects and more difficult for them to lift the objects. So if α is large, fewer clusters are formed and it will be highly likely that well-ordered clusters will not form. If α is small, the similarities between objects will reduce, so it is easier for the ants to pick up objects and more difficult for them to remove their objects. So many clusters that can be well-shaped are created. On this basis, the appropriate setting of parameter α is very important and should not be data independent.

2.3   Speed Parameter

Speed parameter v can uniformly be selected form range $[1, v_{max}]$. Rate of removing an object or picking an object up can be affected by the speed parameter. If v is large, few rough clusters can irregularly be formed on a large scale view. If v is small, then many dense clusters can precisely be formed on a small scale view. The speed parameter is a critical factor for the speed of convergence. An appropriate setting of speed parameter v may cause faster convergence.

2.4   Short Term Memory

Each ant can remember the original real features and the virtual defined two-dimensional features of the last q objects it drops. Whenever ant takes an object it will search its short term memory to find out which object in the short term memory is similar to the current object. If an object in memory is similar enough to satisfy a threshold, it will jump to the position of the object, hoping the current object will be dropped near the location of the similar object, else if there is no object in memory similar, it will not jump and will hold the object and will wander. This prevents the objects originally belonging to a same cluster to be spitted in different clusters.

2.5   Drawbacks of Original Ant Colony Clustering Algorithm

The original ant colony clustering algorithm presented above suffers two major drawbacks. First many clusters are produced in the virtual two-dimensional space and it is hard and very time-consuming to merge them and this work is inappropriate.

The second drawback arises where the density detector is the sole measure based on that the clusters are formed in the local similar objects. But it fails to detect their dissimilarity properly. So a cluster without a significant between-object variance may not break into some smaller clusters. It may result in forming the wrong big clusters including some real smaller clusters provided the boundary objects of the smaller clusters are similar. It is because the probability of dropping or picking up an object is dependent only to density. So provided that the boundary objects of the smaller clusters are similar, they placed near to each other and the other objects also place near to them gradually. Finally those small clusters form a big cluster, and there is no mechanism to break it into smaller clusters. So there are some changes on the original algorithm to handle the mention drawbacks.

2.6   Entropy Measure of Local Area

Combining the information entropy and the mean similarity as a new metric to existing models in order to detect rough areas of spatial clusters, dense clusters and troubled borders of the clusters that are wrongly merged is employed.

Shannon entropy information has been widely used in many areas to measure the uncertainty of a specified event or the impurity of an arbitrary collection of samples. Consider a discrete random variable X, with N possible values $\{x_1, x_2, ..., x_N\}$ with probabilities $\{p(x_1), p(x_2), ..., p(x_N)\}$. Entropy of discrete random variable X is obtained using equation 5.

$$H(X) = -\sum_{i=1}^{N} p(x_i) \log p(x_i) \tag{5}$$

Similarity degree between each pair of objects can be expressed as a probability that the two belong to the same cluster. Based on Shannon information entropy, each ant can compute the impurity of the objects observed in a local area L to determine if the object o_i in the center of the local area L has a high entropy value with group of object o_j in the local area L. Each ant can compute the local area entropy using equation 6.

$$E(L \mid o_i) = - \sum_{o_j \in Neigh_{sxs}(r)} P_{i,j} \times \frac{\log_2(p_{i,j})}{\log_2 |Neigh_{sxs}(r)|} \tag{6}$$

where the probability $p_{i,j}$ indicates that we have a decisive opinion about central object o_i considering a local area object o_j in its local area L. The probability $p_{i,j}$ is obtained according to equation 7.

$$P_{i,j} = \frac{2 \times |D(o_i, o_j)|}{n} \tag{7}$$

where n ($n = |Neigh_{sxs}(r)|$) is the number of neighbors. Distance function $D(o_i, o_j)$ between each pair of objects is measured according to equation 8.

$$D(o_i, o_j) = \frac{d(o_i, o_j)}{norm(o_i)} - 0.5 \tag{8}$$

where $d(o_i, o_j)$ is Euclidian distance defined by equation 2, and $norm(o_i)$ is defined as maximum distance of object o_i with its neighbors. It is calculated according to equation 9.

$$norm(o_i) = \max_{o_j \in Neigh_{sxs}(r)} d(o_i, o_j) \tag{9}$$

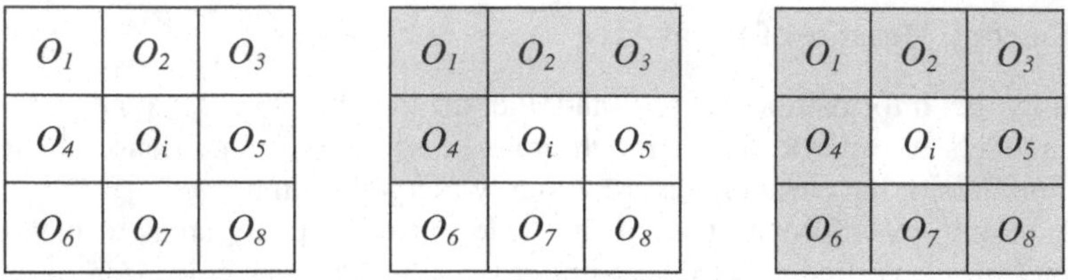

O_1	O_2	O_3
O_4	O_i	O_5
O_6	O_7	O_8

O_1	O_2	O_3
O_4	O_i	O_5
O_6	O_7	O_8

O_1	O_2	O_3
O_4	O_i	O_5
O_6	O_7	O_8

Fig. 2. Three examples of local area objects

Now the function $H(L|o_i)$ is defined as equation 10.

$$H(L|o_i) = 1 - E(L|o_i) \tag{10}$$

Three examples of local area objects on a 3×3 (=9) neighborhood depicted in the Fig. 2. Different classes with different colors are displayed.

When the data objects in the local area L and central object of the local area L exactly belong to a same cluster, i.e. their distances are almost uniform and low values, such as the shape or the form depicted by the left rectangle of Fig. 2, uncertainty is low and $H(L|o_i)$ is far from one and near to 0. When the data objects in the local area L and central object of the local area L belong to some completely different separate clusters, i.e. their distances are almost uniform and high values, such as the shape or the form depicted by the right rectangle of Fig. 2, uncertainty is again low and $H(L|o_i)$ is far from one and near to 0. But in the cases of the form depicted by the middle rectangle of Fig. 2 where some data objects in the local area L and central object of the local area L exactly belong to a same cluster and some others does not, i.e. the distances are not uniform, the uncertainty is high and $H(L|o_i)$ is far from 0 and close to 1. So the function $H(L|o_i)$ can provide ants with a metric that its high value indicates the current position is a boundary area and its low value indicates the current position is not a boundary area.

In ant-based clustering, two types of pheromone are employed: (a) cluster pheromone and (b) object pheromone. Cluster pheromone guides the loaded ants to valid clusters for a possible successful dropping. Object pheromone guides the unloaded ants to lose object for a possible successful picking-up.

Each loaded ant deposits some cluster pheromone on the current position and positions of its neighbors after a successful dropping of an object to guide other ants for a place to unload their objects. The cluster pheromone intensity deposited in location j, by m ants in the colony at time t is calculated by the equation 11.

$$rc_j(t) = \sum_{a=1}^{m} \left[\mu^{(t-t_a^1)} \times C \times E(L|o_j) \right] \tag{11}$$

where C is cluster pheromone constant, t_a^1 is the time step at that a-th cluster pheromone is deposited at position j, and μ is evaporation coefficient. On other hand, an unloaded ant deposits some object pheromone after a successful picking-up of an object to guide other agents for a place to take the objects. The object pheromone intensity deposited in location j, by m ants in the colony at time t is calculated by the equation 12.

```
Input:
      QD, itr, q, AntNum, Data, O, C, k₁, k₂, v_max, period, thr, st, distributions of v, α and μ
initializing parameter using distributions of v, α and μ;
for each ant a
            place random a in a position not occupied by other ants in a plane QD*QD;
end;
for each object o
            place random o in a position not occupied by other objects in the plane QD*QD;
end;
success(1:ant)=0;
failure(1:ant)=0;
for t = 1:itr
            for each ant a
                        g=select a random number uniformly from range [0,1];
                        r=position(a)
                        if(loaded(a) and (is_empty(r)))
                                    if(g<p_drop)
                                                o=drop(a);
                                                put(r,o);
                                                save(o,r,q);
                                    end;
                        elseif(not (loaded(a) or (is_empty(r))))
                                    if(g<p_pic)
                                                o=remove(r);
                                                pick_up(a,o);
                                                search_and_jump(a,o);
                                                success(a)=success(a)+1;
                                    else
                                                failure(a)=failure(a)+1;
                                    end;
                        end;
                        else
                                    wander(a,v,N_dir); // considering the defined pheromone
                        end;
            end;
            if( t mod period==0)
                        for each ant a
                                    if(success(a)/(failure(a)+success(a))>thr)
                                                α(a)=α(a)+st;
                                    else
                                                α(a)=α(a)-st;
                                    end;
                        end;
            end;
end;
```

Fig. 3. Pseudo code of modified ant colony clustering algorithm

$$ro_j(t) = \sum_{a=1}^{m} \left[\mu^{(t-t_a^2)} \times O \times H(L \mid o_j) \right] \tag{12}$$

where O is object pheromone constant, and t_a^2 is the time step at that a-th object pheromone is deposited at position j. Transmission probabilities of an unloaded ant based on that ant moves from the current location i to next location j from its neighborhood can be calculated according to equation 13.

$$P_j^u(t) = \begin{cases} 1/w & if \sum_{j=1}^{w} ro_j(t) = 0 \forall j \in N_{dir} \\ \dfrac{ro_j(t)}{\sum_{j=1}^{n} ro_j(t)} & otherwise \end{cases} \tag{13}$$

where N_{dir} is the set of possible w actions (possible w directions to move) from current position i. Transmission probabilities of a loaded ant based on that ant moves from the current location i to next location j from its neighborhood can be calculated according to equation 14.

$$P_j^l(t) = \begin{cases} 1/w & if \sum\limits_{j=1}^{w} rc_j(t) = 0 \forall j \in N_{dir} \\ \dfrac{rc_j(t)}{\sum\limits_{j=1}^{n} rc_j(t)} & otherwise \end{cases} \qquad (14)$$

2.7   Modified Ant Colony Clustering

Combining the information entropy and the mean similarity as a new metric to existing models in order to detect rough areas of spatial clusters, dense clusters and troubled borders of the clusters that are wrongly merged is employed.

After all the above mentioned modification, the pseudo code of ant colony clustering algorithm is presented in the Fig. 3. For showing an exemplary running of the modified ant colony algorithm, take a look at Fig. 4. In the Fig. 4 the final result of modified ant colony clustering algorithm over Iris dataset is presented.

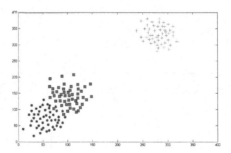

Fig. 4. Final result of modified ant colony clustering algorithm over Iris dataset

It is worthy to mention that the quantization degree parameter (QD), the algorithm maximum iteration t_{max}, queue size parameter (q), ant number parameter ($AntNum$), object pheromone parameter (O), cluster pheromone parameter (C), $k1$ parameter, $k2$ parameter, maximum speed parameter (v_{max}), period parameter, update parameter (thr) evaporation parameter μ and step of update for α parameter (st) are respectively set to 400, 5000000, 20, 240, 1, 1, 0.1, 0.3, 150, 2000, 0.9, 0.95 and 0.01 for reaching the result of Fig. 4. Parameter α for each ant is extracted from uniform distribution of range [0.1, 1]. Parameter v for each ant is extracted from uniform distribution of range [1, v_{max}].

It is important to note that the result shown in Figure 4 is a well separated running of algorithm. So it is considered as a successful running of algorithm. The algorithm may also converge to a set of overlapping clusters in an unsuccessful running.

3   Proposed New Space Defined by Ant Colony Algorithm

Both ACC algorithm and MACCA have many parameters. Their effectiveness is questionable because they depend on many parameters. On a test data, these parameters should be well-tuned to obtain a desirable result. But how to define them in a real task does not clear. The proposed method lets the parameters be free to be changed, and compensates non-optimality of their parameters by the ensemble power.

Indeed the main idea behind the proposed clustering framework is the usage of ensemble learning concept in the field of ACC. The MACCA is very sensitive to the proper initialization of its parameters. If one employs a proper initialization for parameters of the MACCA, the final discovered clusters can be desirable. On the other hand, an ensemble needs diversity to be successful. It can be inferred that by several runs of MACCA with different initializations for parameters, we can reach several partitions that are very diverse. So we use an ensemble approach to overcome the problem of well-tuning of its parameters. The main contribution of the paper is illustrated in Figure 5.

As it is depicted in Figure 5, a dataset is fed to as many as *max_run* different MACCAs with the different initializations. By running each MACCA, an output in a 2-dimension plane, denoted by virtual 2-dimension, is produced. Therefore, we obtain *max_run* virtual 2-dimensions, one per each run of MACCA. Then by considering all these virtual 2-dimensions as a new space with $2 \times max\_run$ dimensions, we reach a new data space which is named mapped dataset. It means that ith feature of mapped dataset will be the dimension x of the output of rth MACCA where $r = \frac{i}{b}$, if i is odd; otherwise, it will be the dimension y of the output of rth MACCA. We can finally employ a clustering algorithm on mapped dataset to extract the final partitioning. So the proposed method has two main contributions: (a) proposing a method that bypasses well-tuning of parameters of the original ACC algorithm and MACCA, (b) proposing a framework to construct a diverse and suitable clustering ensemble.

4   Experimental Study

This section evaluates the result of applying the proposed clustering framework on some real datasets available at UCI repository [14]. The proposed method is examined over 6 different standard datasets. Brief information about the used datasets is available in Table 1. More information is available in (Newman et al. 1998).

The main metrics based on which a partitioning is evaluated are discussed in the first subsection of this section. Then the settings of experimentations are given. Finally the experimental results are presented.

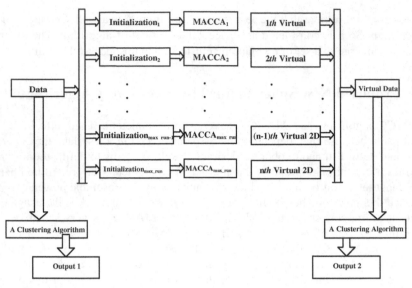

Fig. 5. Proposed framework to cluster a dataset using ant colony clustering algorithm

4.1 Evaluation Metric

After producing the final partitioning, the most important issue is the method of its evaluation. The evaluation of a partitioning is a very important and challenging task due to the lack of supervisor. Here the normalized mutual information (NMI) between the output partitioning and real labels of the dataset is considered as the main evaluation metric of a partitioning [4, 15]. The output partitioning is the one that is obtained as the final partitioning of applying any clustering method on the dataset. It means that after applying a clustering algorithm on the dataset, the partitioning obtained by applying a clustering algorithm is output partitioning of that clustering algorithm. The true labels of the dataset can be used after termination of a clustering algorithm to evaluate how good clustering algorithm has done the clustering task [4, 15]. The NMI between two partitionings, P^a and P^b, is calculated based on equation (15):

$$NMI\left(P^a, P^b\right) = \frac{-2 \displaystyle\sum_{i=1}^{k_a} \sum_{j=1}^{k_b} n_{ij}^{ab} \log\left(\dfrac{n_{ij}^{ab} . n}{n_i^a . n_j^b}\right)}{\displaystyle\sum_{i=1}^{k_a} n_i^a \log\left(\dfrac{n_i^a}{n}\right) + \sum_{j=1}^{k_b} n_j^b \log\left(\dfrac{n_j^b}{n}\right)} \qquad (15)$$

where n is the total number of samples and n_{ij}^{ab} denotes the number of shared patterns between clusters $C_i^a \in P^a$ and $C_j^b \in P^b$; n_i^a is the number of patterns in the cluster i of partition a; also n_j^b are the number of patterns in the cluster j of partition b.

Another alternative to evaluate a partition is the accuracy metric, provided that the number of clusters and their true assignments are known. To compute the final performance of a clustering algorithm in terms of accuracy, one can first re-label its output partitioning in such a way that has maximal matching with the ground true labels (the true labels of dataset) and then counting the percentage of the true classified samples. So the error rate can be determined after solving the correspondence problem between the labels of derived and known clusters [16]. The Hungarian algorithm is employed to solve the minimal weight bipartite matching problem. It has been shown that it can efficiently solve the label correspondence problem [17]

Table 1. Experimental results in terms of accuracy

Dataset Name	Fuzzy k-means output 1		Fuzzy k-means output 2	
	Accuracy	Normalized Mutual Information	Accuracy	Normalized Mutual Information
Image-Segmentation	52.27	38.83	54.39	40.28
Zoo	80.08	79.09	81.12	81.24
Thyroid	83.73	50.23	87.94	59.76
Soybean	90.10	69.50	94.34	80.30
Iris	90.11	65.67	93.13	75.22
Wine	74.71	33.12	76.47	35.96

4.2   Experimental Settings

The quantization degree parameter (QD) is a very important parameter in qualification of the final clustering. The high value QD directly increases the time burden. Time order of the proposed algorithm is related to QD quadratically. So if it is set very high, the algorithm may fail to find a solution. On the other hand, setting a low value to QD, results in limiting space for the cluster points to be appropriately shaped. The queue size parameter (q) as it has been mentioned before should be a value that covers some of the main dense locations in the space. Experimentally we found it that choosing a value 20 is always more than necessity. As it relates to the time order of the proposed algorithm linearly so increasing it may cause some problem, so we use a value no more than 20. The object pheromone parameter (O), the cluster pheromone parameter (C), $k1$ parameter, $k2$ parameter, maximum speed parameter (v_{max}), period parameter, update parameter (thr) evaporation parameter μ and step of update for α parameter (st) are set to the values from previous papers, because their concepts have not changed during their usage in this paper.

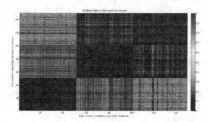

Fig. 6. Similarity matrix of data points in original Iris dataset

The quantization degree parameter (QD), the algorithm maximum iteration t_{max}, queue size parameter (q), ant number parameter ($AntNum$), object pheromone parameter (O), cluster pheromone parameter (C), $k1$ parameter, $k2$ parameter, maximum speed parameter (v_{max}), period parameter, update parameter (thr) evaporation parameter μ and step of update for α parameter (st) are respectively set to 400, 5000000, 20, 240, 1, 1, 0.1, 0.3, 150, 2000, 0.9, 0.95 and 0.01 in all experimentations as before. Parameter α for each ant is extracted from uniform distribution of range [0.1, 1]. Parameter v for each ant is extracted from uniform distribution of range [1, v_{max}]. Fuzzy k-means (c-means) is employed as the base clustering algorithm to perform final clustering over original dataset and new defined dataset. Parameter max_run is set to 30 in all experimentations. So the new defined space has 60 virtual features. Number of real cluster in each dataset is given to fuzzy k-means clustering algorithm in all experimentations.

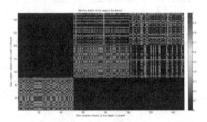

Fig. 7. Similarity Matrix of data points in mapped Iris dataset

As it is inferred from the Table 1, the new defined feature space is better clustered by a base clustering algorithm rather than the original space.

4.3   Experimental Results

Table 1 shows the performance of the fuzzy clustering in both original and defined spaces in terms of accuracy and normalized mutual information. All experiments are reported over means of 10 independent runs of algorithm. It means that experimentations are done by 10 different independent runs and the final results are averaged and reported in the Table 1.

To demonstrate the effectiveness of the clustering framework, consider Fig. 6 and Fig. 7. In Fig. 6 and Fig. 7, a more comprehensive example of similarity matrices of data points in the original and mapped (by ant colony clustering) Iris dataset is presented. Fig. 6 shows the similarity matrix of data points in original Iris dataset. It is worthy to mention that the order of data points in each presented similarity matrix is based on real labels of the dataset.

Fig. 7 shows the similarity matrix of data points in the mapped Iris dataset. As it is inferred from Fig. 6 and Fig. 7 comparatively, the similarity matrix in the mapped Iris dataset is more discriminative than the similarity matrix in the original Iris dataset.

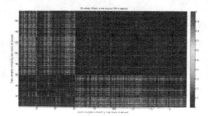

Fig. 8. Similarity matrix of data points in original Wine dataset

To generalize the conclusion the same experimentation is repeated and presented over Wine dataset. Fig. 8 shows the similarity matrix of data points in original Wine dataset. Fig. 9 shows the similarity matrix of data points in the mapped Wine dataset. As it is again obvious from Fig. 8 and Fig. 9 comparatively, the similarity matrix in the mapped Wine dataset is more discriminative than the similarity matrix in the original Wine dataset.

5   Conclusion and Future Works

In this paper it is tried to bring two successful concepts in the field of clustering: (a) ensemble concept and (b) swarm concept. Based on them a new clustering ensemble framework is proposed. Indeed different runnings of ant colony clustering algorithm result in a number of diverse partitionings. In the proposed framework we use a type of modified ant colony clustering algorithm and produce an intermediate space considering their outputs totally as a defined virtual space. After producing the virtual space we employ a base clustering algorithm to obtain the consensus partitioning.

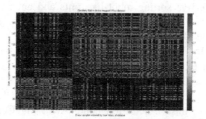

Fig. 9. Similarity Matrix of data points in mapped Wine dataset

The experiments show that the clustering task in the defined data space outperforms in comparison with the clustering in original data space. It is concluded that the new defined feature space is better clustered by any base clustering algorithm rather than the original space. As it is shown, the similarity matrix in the mapped dataset is more discriminative than the similarity matrix in the original dataset.

As a future work, a weighting mechanism to participate the output of each ant colony clustering based on its fitness can be examined. The effect of niching and elitism mechanisms can be studied. One can also turn to other types of swarm based clustering algorithms.

References

1. Alizadeh, H., Minaei, B., Parvin, H., Moshki, M.: An asymmetric criterion for cluster validation. In: Mehrotra, K.G., Mohan, C., Oh, J.C., Varshney, P.K., Ali, M. (eds.) Developing Concepts in Applied Intelligence. SCI, vol. 363, pp. 1–14. Springer, Heidelberg (2011)
2. Ayad, H.G., Kamel, M.S.: Cumulative Voting Consensus Method for Partitions with a Variable Number of Clusters. IEEE Trans. on Pattern Analysis and Machine Intelligence 30(1), 160–173 (2008)
3. Faceli, K., Marcilio, C.P., Souto, D.: Multi-objective clustering ensemble. In: Proceedings of the Sixth International Conference on Hybrid Intelligent Systems (2006)
4. Newman C.B.D.J., Hettich S., Merz C.: UCI repository of machine learning databases (1998). http://www.ics.uci.edu/~mlearn/MLSummary.html
5. Roth, V., Lange, T., Braun, M., Buhmann, J.: A Resampling Approach to Cluster Validation. Intl. Conf. on Computational Statistics, COMPSTAT (2002)
6. Strehl, A., Ghosh, J.: Cluster ensembles - a knowledge reuse framework for combining multiple partitions. Journal of Machine Learning Research 3(Dec), 583–617 (2002)
7. Kennedy, J., Russell, S.: Swarm Intelligence. Morgan Kaufmann, San Francisco (2001)
8. Kuncheva, L.I.: Combining Pattern Classifiers, Methods and Algorithms. Wiley, New York (2005)
9. Azimi, J., Cull, P., Fern, X.: Clustering ensembles using ants algorithm. In: Mira, J., Ferrández, J.M., Álvarez, J.R., de la Paz, F., Toledo, F. (eds.) IWINAC 2009, Part I. LNCS, vol. 5601, pp. 295–304. Springer, Heidelberg (2009)
10. Tsang, C.H., Kwong, S.: Ant Colony Clustering and Feature Extraction for Anomaly Intrusion Detection. Studies in Computational Intelligence (SCI) 34, 101–123 (2006)
11. Liu, B., Pan, J., McKay, R.I.(Bob): Incremental clustering based on swarm intelligence. In: Wang, T.-D., Li, X., Chen, S.-H., Wang, X., Abbass, H.A., Iba, H., Chen, G.-L., Yao, X. (eds.) SEAL 2006. LNCS, vol. 4247, pp. 189–196. Springer, Heidelberg (2006)
12. Deneubourg, J.L., Goss, S., Franks, N., Sendova-Franks, A., Detrain, C., Chretien, L.: The dynamics of collective sorting robot-like ants and ant-like robots. In: International Conference on Simulation of Adaptive Behavior: from Animals to Animates, pp. 356–363. MIT Press, Cambridge (1991)
13. Lumer, E.D., Faieta, B.: Diversity and adaptation in populations of clustering ants. In: International Conference on Simulation of Adaptive Behavior: from Animals to Animates, pp. 501–508. MIT Press, Cambridge (1994)
14. Maheshkumar S., Gursel S.: Application of machine learning algorithms to KDD intrusion detection dataset within misuse detection context. In: Int. Conf. on Machine Learning, Models, Technologies and Applications, pp. 209–215. CSREA Press, Las Vegas (2003)
15. Munkres, J.: Algorithms for the Assignment and Transportation Problems. Journal of the Society for Industrial and Applied Mathematics 5(1), 32–38 (1957)
16. Parvin, H., Beigi, A.: Clustering ensemble framework via ant colony. In: Batyrshin, I., Sidorov, G. (eds.) MICAI 2011, Part II. LNCS, vol. 7095, pp. 153–164. Springer, Heidelberg (2011)

PheroSLAM: A Collaborative and Bioinspired Multi-agent System Based on Monocular Vision

Evandro Luis S. Falleiros[1]([✉]), Rodrigo Calvo[2], and Renato P. Ishii[3]

[1] Instituto Federal de Mato Grosso Do Sul, Dourados, Mato Grosso Do Sul, Brazil
evandro.falleiros@ifms.edu.br
[2] Universidade Estadual de Maringá, Maringá, Paraná, Brazil
calvo.rodrigo@gmail.com
[3] Universidade Federal de Mato Grosso Do Sul,
Campo Grande, Mato Grosso Do Sul, Brazil
renato@facom.ufms.br

Abstract. Multi-robot applications have been extensively discussed and, recently, they are essential for solving problems in robotics field. Nevertheless, development multi-robot real-time applications is usually a complex task, in which it is necessary to design robust environments to support implementation scenarios. In order to deal with such scenarios, this paper proposes PheroSLAM, a bio-inspired multi-robot system based on monocular camera which adopt an extended version of Ant Colony Optimization approach to coordinate multiple-robot teams in the problem related to localization and mapping simultaneously (SLAM). Moreover, robots launch repulsive articial pheromone around themself, creating a repulsive trail in PheroSLAM system. This pheromone trail must be avoided by the other robots, since it denotes an area that have been recently explored. A vision-based SLAM mechanism is also used to provide visual odometry information and to build a 3D feature-based map, considering that every robot must be able to localize itself in the explored environment. Usually, the SLAM problem is solved by cameras or robots remotely controlled. Therefore, the relevance of the proposal is to extend an SLAM problem for many robots and promote the robots move autonomously in the environment according a bio-inspired coordination strategy. Experimental evidences indicated the dispersibility of the PheroSLAM system, increasing the covered area of an environment. Also, results showed that the coordination strategy is efficient and satisfactory to accomplish the exploration task.

Keywords: Multi-robot systems · Multi-robot framework · Ros applications

1 Introduction

In recent years, robots have been used to solve several problems, such as discovering and exploring uncertain environments. In those scenarios, the human

O. Gervasi et al. (Eds.): ICCSA 2015, Part II, LNCS 9156, pp. 71–85, 2015.
DOI: 10.1007/978-3-319-21407-8_6

life can be at risk due to inhospitable environments. For tasks not performed by humans, it is reasonable to employ mobile robots to solve them. In some cases, a multi-robot system is recommended to reduce the problem complexity solving it collaboratively.

Regarding to collaborative mobile robotic, the information about the robot's current location and orientation is important to know and they can be necessary to know the relative pose of a robot in relation to others. This ability is essential for tasks such as, surveillance, environment exploration, disaster management, industrial inspection, odor detection and environment conservation [1].

A representative map is needed for several tasks whose robots should percept the environment i.e., robots work together in a sensing problem which depend on a map to locate themselves in unknown environment. However, if there is not available a map of environment the robots will not recognize the environment previously. Therefore a map must be constructed at the same time wherein the robots localize themselves in the environment.

This problem is called Simultaneous Localization And Mapping (SLAM) and relevant works have proposed algorithms to solve and improve it [2–5,7,8], considering several different kinds of sensors to acquire data with statistically independent errors [10]. Such optical sensors may be one-dimensional, 2D laser, 3D High Definition, 2D or 3D sonar sensors and one or more 2D cameras. Since 2005, there has been intense research into visual SLAM (VSLAM) using primarily visual sensors (camera), according to the increasing ubiquity of cameras such as those in mobile devices [11].

Furthermore, this paper proposes PheroSLAM, a collaborative and bio-inspired Multi-Agent System (MAS) based on monocular vision to coordinate multi-robot team in exploration task. Considering that every robot on exploration task must to be able to localize itself in the explored environment, PheroSLAM uses an extended version of MonoSLAM algorithm, proposed by [12]. In our approach, MonoSLAM provides visual odometry data and to build a probabilistic 3D feature-based map. Here, our work contributes by extending MonoSLAM coupling it into multi-agents (robots) in order to navigate, explore and to map environments autonomously. Current MonoSLAM approach is focused in applications to real-time 3D localization and mapping for a high-performance full-size humanoid robot, and live augmented reality with a hand-held camera.

In PheroSLAM, every robot involved on exploration task runs MonoSLAM algorithm in order to estimates its own position. However, every robot also requires an efficient strategy to move around and explore an environment. In this sense, PheroSLAM proposes a modified version of the Ant Colony Optimization approach as an extension of the Inverse Ant System-Based Surveillance System (IAS-SS) presented in [13].

IAS-SS approach uses repulsive pheromone substances to coordinate multi-robot teams in exploration and surveillance tasks. The collaborative behavior of robots is repulsive to the deposited substance (pheromone) while they move. It denotes a repulsive behavior, instead of the tradicional Ant Colony attractive behavior. Briefly, the repulsion is addressed to avoid areas that have been earlier explored. Based in the IAS-SS approach, PheroSLAM proposes an autonomous exploration algorithm named LPCS (Low Pheromone Concentration Search). LPCS mainly consists on searching areas with low pheromone concentration to autonomously guide multi-robot teams in exploration tasks. On each exploration iteration, a robot receives a target area, provided by LPCS, that must be followed. The idea is that LPCS leads every robot to areas that have not been earlier visited.

Focusing on proposed system validation, simple exploration task were developed on the Robot Operating System (ROS). The experiments were guided by two main approaches. The first consists on real-time environment experimentation, considering real small indoor rooms and camera image data. These experiments were carried out in order to test MonoSLAM robustness. The second approach consists on simulation experiments in order to prove that multi-robot team can be used to navigate, explore and to map environments autonomously. The experiments showed that PheroSLAM is robust enough to autonomously guide multi-robot teams in simple exploration tasks, considering small indoor rooms.

This paper is organized as follows: Section 2 reviews related work; Section 3 presents the system overview, including the description of MonoSLAM and IAS-SS approaches; The experimental results and some discussion are shown in Section 4; and finally, the conclusions and future works are presented in Section 5.

2   Related Works

Robot navigation strategies are driven by the knowledge of environment where it is and of the its position heading to accomplish a task. On the one hand, a robot needs information about the environment map to find out its position. On the other hand, the robot needs its position to build the map. Hence, one of the challenges of the mobile robotics is to provide these data while the robot navigates. That problem is known as Simultaneous Localization And Mapping (SLAM). It was originally introduced through the papers [2,3]. After that, several approaches have proposed algorithms to solve and improve it. The main mechanisms for solving the SLAM problem are based on Kalman filter (KF) SLAM [4], particle filters [5,6] and Graph SLAM [14–16].

Independently of the mechanism adopted for solving the SLAM problem, the sensor parameter must be addressed. Different sensors can cause oscillations in the performance of the algorithm. One of the sensor models, sonar sensor is cheap but its resolution is limited. Also, it suffers from noisy data. Although laser sensors have limited view field, its resolution is high. Generally, sonar and laser sensors are used for 2D mapping. Cameras as sensors also have limited view field, but that sensor is able to collect large amount of data even in a small image. The use of cameras is the focus of this paper.

Cameras are used to construct 2D and 3D maps. Combining the sensor parameter with the mentioned mechanisms for the SLAM problem, an algorithm, named Monocular SLAM, is proposed in [7]. The algorithm that requires a single camera, named Monocular SLAM, generates and tracks sparse feature-based models of the world. The algorithm creates a dense 3D surface model and immediately uses it for dense camera tracking via whole image registration captured by a hand-held camera. Aerial vehicle is considered in [8] and equipped with cameras. In this case, the SLAM problem is solved using stereo vision or monocular camera. The algorithm uses the graph-based formulation where the solution is a configuration of the nodes which minimizes the error introduced by the constraints. To increase the SLAM execution speed, the approach in [9] uses extended KF and a stereo camera to detect two images of the environment simultaneously. For each pair of images, the distance between the landmark and the camera is computed using a revised algorithm.

The SLAM problem can be performed more efficiently if a group of robots is considered to construct cooperatively a map of an environment. The idea of multi-robots using camera is adopted in [17]. Each agent builds its own 2D and 3D feature maps. The maps are merged matching overlapping tendency of any two mates. In [18], several cameras are used to build a global 3D map of dynamic environments. The cameras share a common view work collaboratively for executing the mapping and localization tasks with more robustness. An algorithm for localization of multiple robots is described in [19]. The authors use the KF algorithm to localize each robot according to their speed command and pose information. The robots are equipped with camera sensor to identify landmarks and correct their pose. Besides of landmarks in the environment, there are landmarks attached to robots. When a robot finds a mobile landmark, it only estimates its pose with uncertainty. For the case of the high uncertainty, the robot requires collective localization of the robots near. Thus, all of them will similar high uncertainty of their pose. However, if one of them finds a immobile landmark, they can be localized.

For multi-robot SLAM problem, the environment map should be generated by fusion of the built local maps of each robot. In [20] the map fusion is divided into an alignment problem and a map merging problem. The alignment calculates the transformation among the individual maps with different coordinate systems. After that, aligned data should be merged to generate the global map. The maps are aligned by means of 3D landmark-based techniques. In [21], Micro Aerial Vehicles detect features from an environment to a central ground station that create individual maps for each vehicle. The maps are joined whenever the station detects overlaps. Thus, the vehicles can estimate their pose in a global map.

Although there are several approaches to solve the multi-robot SLAM problem, most of them does not present a mechanism for autonomous navigation of the robots. In this case, the SLAM problem is solved using hand-held cameras. In others approaches, the navigation is based solely on mathemetical formulation. Thus, the robots are very parameter dependent and the system performance suffer critical degradation due to robot failure. Bio-inspired and evolutionary theories provide fundamentals to design alternative strategies. Particularly, the artificial analog versions of biological mechanisms that define the social organization dynamics, observed in some swarm systems, are very appropriate in applications involving multiple agents, for example, decentralized control, communication and coordination. For this reason, this paper proposes an extension of SLAM algorithm for multi-robots where the robots are able to navigate autonomously according to a bio-inspired technique without prior knowledge of the environment.

3   System Overview

The PheroSLAM approach is basically composed by the MonoSLAM and IAS-SS approaches. The Figure 1 shows the architecture of the PheroSLAM as an aggregation of the both approaches for solving the SLAM problem autonomously using mult-robots. In that figure, the IAS-SS approach is broken down in two modules (Navigation Controller and Dispersion Module) to emphasize that one of them, Navigation Controller module, is modified to the PheroSLAM approach.

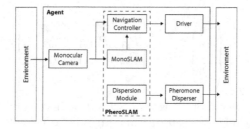

Fig. 1. PherosSLAM architectural diagram for a single robot

Since the PheroSLAM is supplied by only a monocular camera as sensor, some data are obtained from the MonoSLAM to enable the bio-inspired strategy, in particular the Navigation Controller module, such as: odometry, the camera pose estimation as along as a robot move, the robot estimated path in the environment, two 3D probabilistic feature-based maps where the first contains all identified features of the environment and the seconde one contains only the features that are closer of the position of the camera (considering an specific radius distance). The generated data can be virtually visualized in ROS RViz that is a package for 3D visualization tools.

3.1   Vision-Based SLAM

To build a map of an environment, a robot must be able to localize itself and estimate the distance to other robots. For this issue, it is adopted the ego-motion that consists on estimating the robot relative position in an environment. Indeed the robots have only one kind of sensor, the monocular camera. Thus, the robots must predict their position in the environment map using only vision information from the sensor. In this sense, in order to solve the ego-motion problem, we extended the MonoSLAM system proposed in [12].

MonoSLAM is a real-time vision-based SLAM system which can recover the 3D trajectory of a monocular camera through a previously unknown scene. In our proposal, every robot is able to run an instance of MonoSLAM alorithm and build a probabilistic 3D feature-based map. This map is a key concept at the MonoSLAM approach that represents, at any instant, a snapshot of the current estimates of the state of the camera and all features of interest. From the initialization of the MonoSLAM, the probabilistic map is continuously updated by the Extended Kalman Filter. During the camera motion, features are observed and the probabilistic state estimation of the camera is updated. At instants when new features are discovered, they are stored in the 3D probabilistic map with new states. In some cases, features can also be deleted (see [12]).

The probabilistic 3D map is mathematically represented by a state vector $\hat{x}$ and a covariance matrix M, where $\hat{x}$ is composed of state estimates of the features and camera; and M is a square matrix that can be partitioned into sub-matrix elements:

$$\hat{x} = \begin{pmatrix} \hat{x}_v \\ \hat{y}_1 \\ \hat{y}_2 \\ \vdots \end{pmatrix}, M = \begin{bmatrix} P_{xx} & P_{xy1} & P_{xy2} & \cdots \\ P_{y1x} & P_{y1y1} & P_{y1y2} & \cdots \\ P_{y2x} & P_{y2y1} & P_{y2y2} & \cdots \\ \vdots & \vdots & \vdots & \ddots \end{bmatrix} \tag{1}$$

The camera's state vector $\hat{x}_v$ is represented by a vector r^W, an orientation quaternation q^{RW}, a velocity vector v^w and an angular velocity vector ω^R, where W is a fixed world frame and R is a robot camera frame:

$$\hat{x}_v = \begin{pmatrix} \hat{r}^W \\ \hat{q}^{WR} \\ \hat{v}^W \\ \hat{\omega}^W \end{pmatrix} \tag{2}$$

In order to provide visual odometry information to PheroSLAM, an existing ROS MonoSLAM implementation [22] was modified and adapted to the PheroSLAM system. Figure 2 shows MonoSLAM execution over ROS in an indoor environment. The left side of the figure shows the proccessing of the camera image frame by MonoSLAM and the capturing of the all identified features. The estimated 3D position of every feature is plotted in the right side of the figure. The red spheres denote 3D estimated features positions in the probabilistic feature-based map. At every iteration of the MonoSLAM algorithm, new features of interest may be identified and the probabilistic map is updated.

3.2   A Bio-inspired Vision-Based SLAM System for Multi-robot Coordination

Multi-robot coordination has an important role in cooperative tasks. As regards the exploration task for unknown environments, it desirable that the robots

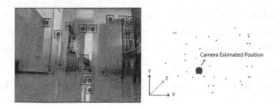

Fig. 2. MonoSLAM algorithm execution with real-time camera data

search for unvisited regions. However, coordinate robots in unknown environment can be a complex task. In this sense, coordination strategies are essential to robots solve that problem efficiently. In cooperative multi-robot exploration problems, the robots must coordinately explore each unknown area until all the environment is considered as completely visited. There are many possible approaches that solve this problem.

As long as the robots accomplish the environment exploration, a sparse feature-map can be built according to the MonoSLAM algorithm. However, as mentioned above, an effective strategy is required to coordinate multi-robots and achieve the performance improvements in such task. Considering biological mechanisms which define social organization of collective systems, PheroSLAM presents a mechanism for environment exploration based on an extension of the Inverse Ant System-Based Surveillance System (IAS-SS) presented in [13].

The IAS-SS is a coordination strategy for multi-robot in exploration and surveillance tasks. That strategy is motivated by artificial ant system theory proposed in [23]. Here, the robots (or artificial ants) deposit a substance, called pheromone, with repulsive property. While the robots navigate, they deposit pheromone to mark some regions of the environemnt as explored. The logic of the adopted strategy is opposite of the traditional ant system theory. That is, the robots are repelled from areas with high amount of pheromone, because it indicates that a robot was in these areas previously.

In the traditional ant system algorithm, the pheromone is deposited only in a specific position of the ant, signaling its exact path. Differently, in IAS-SS, the robots leave pheromone on a wide area in its front considering the range of sensor area. However, both the traditional ant system and the IAS-SS coordination strategy consider only 2D spaces. In this sense, the built pheromone-concentration maps are useful only for ground robots. There are scenarios that a 3D pheromone map is necessary, mainly, when aerial vehicles are considered. Although the PheroSLAM is evaluated according to experimentos using ground robots, the proposed system provides a 3D pheromone map to support aerial vehicles in future researches.

Indeed, the PheroSlam system is an integration of the MonoSLAM algorithm and the modified IAS-SS coordination. It allows the robots to build a map and localize theirself in an unknown environment while they explore the environment autonomously. Also, the PheroSLAM system provide 3D navigation capabilities by mean the feature-based maps obtained in the MonoSLAM algorithm.

The presented proposal here differs from others because the robots are able to move autonomously spreading out each other to distribute the exploration task while the SLAM problem is solved.

Pheromone Releasing and Evaporation Mechanisms. The main adaptions from the IAS-SS coordination strategy to the bio-inspired mechanism for navigation in the PheroSLAM system is the manner how the pheromone is released in the environment. Since the robots, in PheroSLAM, have only a camera as sensor, the pheromone is deposited in any position of a region of the environment that is detected by an image. Thus, the pheromone is not deposited only in a 2D planar space as in IAS-SS. Another issue highlighted here is the amount of pheromone deposited. In IAS-SS, the amount of pheromone in a position is dependent on the distance between it and the position of the robot that left such substance. For the PheroSLAM system, the deposited amount of the pheromone is the same in any position of the environment.

At every instant, the robots release a specific amount of pheromone around itself, according to the captured image. The pheromone is spread considering a spherical limitation area around the robot, as shown in the Figure refimg:dispersion. In each iteration t, an amount of pheromone is released around the robot estimated position. In practical terms, repulsive pheromone trails are created in the PheroSLAM problem. In this sense, the PheroSLAM manages a 3D pheromone-based map along the navigation.

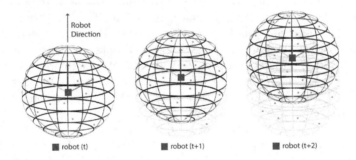

Fig. 3. Pheromone dispersion approach. The red, green and blue axis represents X, Y and Z axis respectively. The orange points represent the deposited pheromone.

Initially, PheroSLAM algorithm defines a spherical structure considering the robot pose estimation over the map. Differently from the IAS-SS approach, a predefined amount of 3D points are randomly generated in the pheromone map using normal (Gaussian) distribution. All the points are generated inside this spherical structure limits, see Figure 3. It is noteworthy that the spherical wireframe is invisible, since it is just used to limit pheromone dispersion around the robot. The amount of pheromone and the spread radius were empirically defined during PheroSLAM experimentatal phase.

The pheromone is not a stable substance. It evaporates according to a specific rate at every instant t. After while, the pheromone concentration tend to decrease under an evaporation rate. Thus, avoided areas previously become attractive to be visited again. That behavior is essential for spreading out the robots in the environment and executing the surveillance task. The pheromone evaporation is independent of any action performed by robots. That phenomena occurs in the Dispersion Module of the PheroSLAM architecture (Figure 1)

Similar to the IAS-SS, the total amount of pheromone that evaporates $\Phi_q(t)$ in a specific position q at time t is mathematically modeled by:

$$\Phi_q(t) = \rho\tau_q(t) \tag{3}$$

where ρ, $0 \leq \rho \leq 1$, is the evaporation rate and $\tau_q(t)$ is the total amount of pheromone at position q at time t.

In PheroSLAM, the pheromone is modeled as 3D points in a point cloud. Considering that PheroSLAM searches for areas with low pheromone concentration, it is accounted the incidence of position with pheromone rather than measuring concentration rate per deposited pheromone. Also a life time is defined for the deposited pheromone. Mathematically, the total amount of pheromone $\tau_q(t)$ in position q at time t in PheroSLAM is given by:

$$\tau_q(t) = (\tau_q(t-1) - \Phi_q(t-1)) + \sum_{i=1}^{N} \Delta_q^k(t) \tag{4}$$

where $\Delta_q^k(t)$ is the amount of pheromone deposited by $k-th$ robot in position q at time t.

In PheroSLAM, each robot randomly generates ten 3D points around itself at time t, considering a spherical limit of one meter. The number of the pheromone points deposited around the robots was defined empirically. Thus, $\Delta_q^k(t)$ is constant at every iteration.

Search Algorithm for Regions with Low Pheromone Concentrations.
In tradicional ant colony approaches, every robot deposits an amount of pheromone in their own covered path [23]. It highlights the past of each robot in the explored environment, creating a pheromone trail that consists on its own history along a task. This trail supports the robots in future situations for exploration and surveillance tasks.

Although exists distinct kinds of pheromone released by ants, in PheroSLAM each robot creates an artificial repulsive trail of pheromone. The absence of this kind of substance indicates areas with high potencial for possible monitoring and exploration. On the other hand, areas with high concentration of repulsive pheromone denotes areas that were recently visited. This repulsive behavior causes the robots to stay away from each other. It benefits surveillance tasks [13].

In PheroSLAM, simultaneously to the pheromone dispersion process, a robot also searches for areas that presents low pheromone concentration. The idea is that a robot must avoid areas with high pheromone concentration, since this areas have been earlier visited. For exploration task, it may be desired that an area do not be constantly revisited.

When a robot locally identifies the area with the lowest pheromone concentration, this area is considered as a goal, and the robot must follow this target. In order to find this target area, it is proposed a modification of the algorithm carried out in the Controller Navigation module of the IAS-SS approach. In the PheroSLAM approach, a simple algorithm, named LPCS (Low Pheromone Concentration Search algorithm), is designed for the Controlled Navigation module. LPCS subdivides, recursively, square cells into new four equal-sized subcells, as shown in Figure 4. The LPCS algorithm finishes when all the 4 subdivided quadrants presents no pheromone concentration.

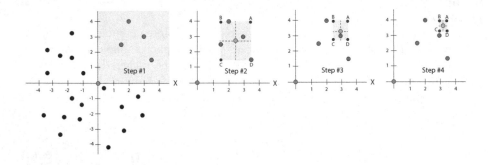

Fig. 4. LPCS algorithm execution. The red point represents the origin point, originally the robot pose. The green point represents the target pose.

Initially, the LPCS algorithm finds the number of the pheromone points occurrence in the every quadrant. The quadrant that presents the lowest number of pheromone points is selected to be divided recursively. It has been defined as stop criterion to find a quadrant with none pheromone point. Thus, this quarter provides more information gain, since it presents the lowest pheromone concentration inside the spherical limits. Also, this quarter is a potential area to be explored. This process can be better understood observin the in Algorithm 1.

After identifying a target using LPCS algorithm, the Controller Navigation send a command to the Driver. Then, the predicted goal target must be followed. However, a robot does not necessarily will reach its respective target. The target is used as a new promising area to be followed, however, another robots may be trying to achieve the same area. In those cases, every robot should be able to dynamically change its own path, searching for an area that provides greater information gain.

4   Experimental Results

The PheroSLAM system and all of its experiments were developed on the Robot Operating System (ROS)[1], in C++ under Ubuntu Linux 14.04. Experimental

---

[1] http://www.ros.org/

input : A cloud of points $detected_p$, containing only pheromone that is within robot sensing area, and an origin point $origin$

output: A quaternation $target$, that represents the area with lowest pheromone concentration

begin

$q_1 \longleftarrow \forall p(x, y, z) \in detected_p$ where $p.x > origin.x \wedge p.y > origin.y$;

$q_2 \longleftarrow \forall p(x, y, z) \in detected_p$ where $p.x < origin.x \wedge p.y > origin.y$;

$q_3 \longleftarrow \forall p(x, y, z) \in detected_p$ where $p.x < origin.x \wedge p.y < origin.y$;

$q_4 \longleftarrow \forall p(x, y, z) \in detected_p$ where $p.x > origin.x \wedge p.y < origin.y$;

Find the quadrant with the lowest amount of pheromone;

$lowest \longleftarrow findLowest(q_1, q_2, q_3, q_4);$

Determine a center point within the quadrant with the lowest amount of pheromone;

$target \longleftarrow findCenter(lowest);$

if $numberOfPheromone(lowest) = 0$ **then**

| *Just return the target;*

| **return** $target;$

else

| *Recursive call of LPCS;*

| $LPCS(target, lowest);$

end

end

Algorithm 1. LPCS

results were carried out to evaluate our assumption, that is, can MonoSLAM be extended to multi-agents through bio-inspired techniques in order to navigate, explore and and map environments autonomously? To answer this question, we using two different environments: a real-time development environment, with experiments using hand-held camera data stream; and a simulation environment, using Gazebo[2] simulator along with ROS.

Focusing on MonoSLAM implementation tests, we used the `Logitech HD Pro Webcam C920` camera to provide image frames. The image frames are broadcasted using the USB Video Class (UVC) ROS nodes. Two distinct approaches were used to guide the experimental real-time tests: a single-robot approach, using only one camera to explore a controlled environment; and a multi-robot approach, using two or more cameras to explore the same uncertain environment. These approaches were applied in order to verify MonoSLAM and PheroSLAM robustness considering real-time applications.

In the early tests of MonoSLAM implementation, a simple indoor environment was explored. A top view of the explored environment and the existing obstacles are shown in Figure 5. In this specific experiment, two robots were employed in a simple exploration task. The little points represent each deposited pheromone estimated position and the thicker represent each captured feature estimated position. Both robots started at the same point in the map. After the

---

[2] http://gazebosim.org/

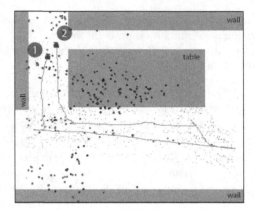

Fig. 5. A Top-view RViz snapshot of an indoor environment exploration. Transparent objects represent the environment obstacles. Thicker points represent features and the thinner corresponds to pheromone.

beginning of the exploration task, one of the cameras changed its path to the right, based on PheroSlam approach.

In order to establish a visual comparison between the single-robot and the multi-robot approaches, the resulting data from the traditional MonoSLAM and PheroSLAM approaches were displayed in ROS RViz. Figure 6 shows a sequence of RViz snapshots showing the probabilistic pheromone 3D map construction over the time (in seconds). The pheromone cloud was generated and plotted using a single camera along a simple exploration task. We consider that PheroSLAM produces a pheromone trail that can be virtually visualized by every robot in a mobile-robot team.

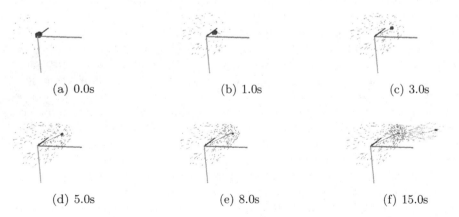

Fig. 6. A sequence of RViz snapshots showing the probabilistic 3D pheromone map construction. Red, green and blue axis represents X, Y and Z axis, respectively, in a 3D projection.

As previously said, PheroSLAM experiments were carried out using Gazebo simulator along with ROS. In Gazebo, a small simple indoor scene was simulate. In this simulated scene, three robots were positioned at known distances. In our approach, before PheroSLAM start running, all robots position must be informed. This can be done through simple configuration files. These initial positions are all considered in order to guarantee PheroSLAM approach correctness. It is noteworthy that PheroSLAM does not provide robot team formation capabilities yet. This kind of feature will be provided in future work.

PheroSLAM also provides a simple mechanism to avoid collisions based on feature detection. Figure 7 shows a situation in which collision avoidance is necessary. This mechanism simple consists on avoiding closer features. In this sense, when a robot predicts a set of closer features, it immediately turns to avoid this set.

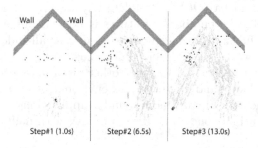

Fig. 7. A sequence of snapshots showing PheroSLAM collision avoidance system in action. Each of these steps was captured on RViz in a specific time t, given in seconds. The blue points represents every feature predicted position. The pheromone trail is illustrated by the gray points.

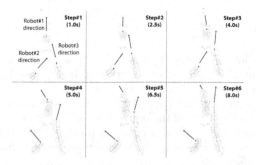

Fig. 8. PherosSLAM top-view snapshots showing the repulsive behavior. The yellow points represents the repulsive pheromone trail generated by each robot.

Finally, Figure 8 shows some top-view PheroSLAM repulsive behavior snapshots along the time (in seconds). This snapshots were taken in RViz during an exploration experiment. It can be noted that the robots clearly avoid each other, since they are able to identify repulsive trails using the LPCS algorithm. In this sense, PheroSLAM is able to guide every robot to continuously explore the environment searching for areas the have not been explored earlier.

5   Conclusions

In this paper we presented PheroSLAM, a bio-inspired and vision-based system that uses a modified version of the Ant Colony approach to coordinate multi-robot teams in exploration tasks. PheroSLAM provides an useful environment for collaborative applications with multi-robot teams that uses Visual-Based SLAM to predict and estimate robots positioning.

Some applications have been implemented in order to improve and test the robustness of the PheroSLAM approach. The experiments showed that the implemented solution is able to run real-time and it is robust enough to execute and to support multi-camera applications considering small indoor rooms.

Some limitations were also identified, mainly in MonoSLAM algorithm. MonoSLAM fits well considering about 100 features, since the complete MonoSLAM algorithm has $O(N^2)$ complexity, where N is the number of features. Unfortunately, the number of features is bonded but it turns out to be sufficient with careful map management to span a small indoor room. In this sense, PheroSLAM works fine only in small rooms.

Future works will be devoted to apply other vision-based SLAM algorithms in order to manage larger and more complex environments. Formation strategies will also be implemented. Finally, more complex applications are under consideration. These applications concern multi-robot MAV with multi-camera systems.

Acknowledgments. This paper is based upon work supported by Fundação de Apoio ao Desenvolvimento do Ensino, Ciência e Tecnologia do Estado de Mato Grosso do Sul-FUNDECT; Conselho Nacional de Desenvolvimento Científico e Tecnológico - CNPq; and Programa de Educa ção Tutorial - PET. Any opinions, findings, and conclusions or recommendations expressed in this material are those of the authors and do not necessarily reflect the views of FUNDECT, CNPq and PET.

References

1. Aulinas, J., Petillot, Y., Salvi, J., Lladó, X.: The slam problem: a survey. In: Proceedings of the 2008 Conference on Artificial Intelligence Research and Development: Proceedings of the 11th International Conference of the Catalan Association for Artificial Intelligence, pp. 363–371. IOS Press, Amsterdam (2008)
2. Smith, R.C., Cheeseman, P.: On the representation and estimation of spatial uncertainty. The International Journal of Robotics Research **5**(4) (1986)
3. Smith, R., Self, M., Cheeseman, P.: Autonomous robot vehicles. In: Cox, I.J., Wilfong, G.T. (eds.) Autonomous Robot Vehicles, pp. 167–193. Springer-Verlag New York Inc., New York (1990)
4. Myung, H., Lee, H.K., Choi, K., Bang, S.: Mobile robot localization with gyroscope and constrained kalman filter. International Journal of Control, Automation and Systems **8**(3), 667–676 (2010)
5. Murphy, K.: Bayesian map learning in dynamic environments. In: In Neural Info. Proc. Systems (NIPS), pp. 1015–1021. MIT Press (2010)
6. Montemerlo, M., Thrun, S., Koller, D., Wegbreit, B.: FastSLAM: a factored solution to the simultaneous localization and mapping problem. In: Proceedings of the AAAI National Conference on Artificial Intelligence. AAAI, Edmonton (2002)

7. Newcombe, R.A., Lovegrove, S., Davison, A.J.: DTAM: dense tracking and mapping in real-time. In: IEEE International Conference on Computer Vision, ICCV 2011, Barcelona, Spain, pp. 2320–2327, 6–13 November 2011
8. Steder, B., Grisetti, G., Stachniss, C., Burgard, W.: Visual slam for flying vehicles. IEEE Transactions on Robotics 24(5), 1088–1093 (2008)
9. Wang, T.Q., Chen, C.H.: Improved simultaneous localization and mapping by stereo camera and surf. In: 2013 CACS International Automatic Control Conference (CACS), pp. 204–209, December 2013
10. Magnabosco, M., Breckon, T.P.: Cross-spectral visual simultaneous localization and mapping (slam) with sensor handover. Robotics and Autonomous Systems 61(2), 195–208 (2013)
11. Karlsson, N., Di Bernardo, E., Ostrowski, J., Goncalves, L., Pirjanian, P., Munich, M.: The vslam algorithm for robust localization and mapping. In: Proceedings of the 2005 IEEE International Conference on Robotics and Automation, 2005. ICRA 2005, pp. 24–29, April 2005
12. Davison, A.J., Reid, I.D., Molton, N.D., Stasse, O.: Monoslam: Real-time single camera slam. IEEE Trans. Pattern Anal. Mach. Intell. 29(6), 1052–1067 (2007)
13. Calvo, R., de Oliveira, J.R., Figueiredo, M., Romero, R.A.F.: Inverse aco applied for exploration and surveillance in unknown environments. In: The Third International Conference on Advanced Cognitive Technologies and Applications, Rome, Italy, pp. 142–147 (2011)
14. Yin, J., Carlone, L., Rosa, S., Bona, B.: Graph-based robust localization and mapping for autonomous mobile robotic navigation. In: 2014 IEEE International Conference on Mechatronics and Automation (ICMA), pp. 1680–1685, August 2014
15. Thrun, S., Montemerlo, M.: The GraphSLAM algorithm with applications to large-scale mapping of urban structures. International Journal on Robotics Research 25(5/6), 403–430 (2005)
16. de la Puente, P., Rodriguez-Losada, D.: Feature based graph-slam in structured environments. Autonomous Robots 37(3), 243–260 (2014)
17. Riaz un Nabi Jafri, S., Ahmed, W., Ashraf, Z., Chellali, R.: Multi robot slam for features based environment modelling. In: 2014 IEEE International Conference on Mechatronics and Automation (ICMA), pp. 711–716, August 2014
18. Zou, D., Tan, P.: Coslam: Collaborative visual slam in dynamic environments. IEEE Trans. Pattern Anal. Mach. Intell. 35(2), 354–366 (2013)
19. Han, J.S., Ji, S.H., Kim, K.H., Lee, S.M., Choi, B.W.: Collective robot behavior controller for a security system using open sw platform for a robotic services. In: 2011 11th International Conference on Control, Automation and Systems (ICCAS), pp. 1402–1404, October 2011
20. Ballesta, M., Gil, A., Reinoso, O., Juliá, M., Jiménez, L.M.: Multi-robot map alignment in visual slam. WTOS 9(2), 213–222 (2010)
21. Forster, C., Lynen, S., Kneip, L., Scaramuzza, D.: Collaborative monocular slam with multiple micro aerial vehicles. In: IROS, pp. 3962–3970. IEEE (2013)
22. Russo, L., Rosa, S., Bona, B., Matteucci, M.: A ros implementation of the mono-slam algorithm. International Journal of Computer Science & Information Technology 6(1) (2014)
23. Dorigo, M.: Optimization, learning and natural algorithms. PhD thesis, Dipartimento di Elettronica, Polit. di Milano (1992)

Workshop on Computer Aided Modeling, Simulation, and Analysis (CAMSA 2015)

Ego-Motion Estimation by Using the Integration of Coplanarity and Collinearity Conditions

Yuan Kong Sih, Hone-Jay Chu[✉], and Yi-Hsing Tseng

Department of Geomatics, National Cheng Kung University, Tainan City, Taiwan
honejaychu@gmail.com

Abstract. Collinearity condition is generally used to establish relations between 3D objects and image points and calculate the absolute orientation parameters of image pairs in photogrammetry research. Given that the collinearity condition is a nonlinear system, the appropriate initial values of unknown absolute orientation parameters must be first established in the iterative least-squares solution. This research considers the coplanarity condition to solve relative orientation parameters as initial values and to solve absolute orientation parameters from the collinearity condition. The proposed method can provide a strategy for the motion estimation of a single camera. First, the algorithm automatically acquires conjugate image points between sequential images. Information on conjugate image points can provide information that can be used to solve relative orientation parameters from the coplanarity condition. Second, the absolute orientation parameters of a camera can be solved through the iterative least-squares method with the aid of ground control points. In addition to the absolute orientation parameters of a camera, the object coordinates of conjugate image points can be acquired. Finally, the camera trajectory can be obtained by repeating the procedure. This research conducted experiments in indoor and outdoor environments, and the results show that the proposed procedure is effective and feasible.

Keywords: Collinearity condition · Coplanarity condition · Relative orientation · Absolute orientation · Motion estimation · Photogrammetry

1 Introduction

Ego-motion estimation is an automatic estimation of camera motion from image sequences ([1], [2]). Many techniques for ego-motion estimation are based on different applications. Nistér et al. [3] presented an application system called visual odometry, which is used to estimate the motion of stereo or single moving cameras in vehicles. Visual odometry also allows for enhanced navigational accuracy in rovers or robots on the Mars [4]. Visual odometry solutions for navigating and tracking the movement of objects by using two or more synchronized and self-oriented cameras simplifies the identification of absolute object coordinates with respect to a camera coordinate system. Cheng et al. [5] presented a visual odometry system that uses a pair of stereo images to calculate the position of a rover, including the coordinates and attitudes of

© Springer International Publishing Switzerland 2015
O. Gervasi et al. (Eds.): ICCSA 2015, Part II, LNCS 9156, pp. 89–103, 2015.
DOI: 10.1007/978-3-319-21407-8_7

three axes, by tracking the motion of terrain features between images. Visual odometry covers many of the existing approaches. The relative orientations of current and previous camera frames were provided according to a five-point algorithm [6]. Kitt [7] estimated the relative displacement between two sequential camera positions using stereo sequences. Avidan & Shashua [8] considered fundamental matrices along a sequence of images based on the trifocal tensor. Trifocal tensor can be determined from image feature correspondences without knowledge of calibration [9]. Geiger et al. [10] proposed a sparse feature matcher approach with an efficient visual odometry algorithm that provides stereo matching to generate consistent 3D point clouds to build 3D maps in real time. Davison et al. [11] proposed SLAM, a simultaneous localization and mapping approach that uses a single moving camera on the basis of probabilistic mapping, motion modeling and active measurement of visual natural landmarks.

Constructing the relation among object points, image points, and perspective center of the camera is important to determine the position and attitude of the camera in photogrammetry. One of the methods used is bundle adjustment based on the collinearity condition. The collinearity condition which establishes the relation among the object point, image point, and camera position has been extensively applied in various photogrammetry applications. Bundle adjustment considers the information of objects and their corresponding image points and image coordinates to estimate the positions and attitudes of cameras in triangulation calculations [12]. Mouragnon et al. [13] used bundle adjustment to simultaneously refine the coordinates of feature points and camera positions. The solid solution of exterior orientation parameters can be obtained through an iterative least-squares solution with a set of fine initial values for unknown parameters. In addition, other solutions for absolute and relative orientation problems have also been proposed ([12], [14]). For example, the coplanarity condition and direct linear transformation were proposed to estimate the orientation parameters of cameras. Coplanarity condition that two perspective center, any object point and the corresponding image point on the pairs of images lie in a common plane. Applying the coplanarity condition to the matched feature points between two subsequent images solves relative orientation parameters.

This study aims to integrate the coplanarity condition with the collinearity condition equations. The coplanarity condition can provide the required initial values for collinearity condition equations. For an initial guess of collinearity condition equations, the coplanarity condition was proposed to calculate the relative orientation parameters between image pairs. This method identifies the exterior orientation parameters of sequential images covered by automatically detected conjugate points.

2   Methodology

Generally, image coordinates are linked to corresponding object points by perspective projection. An object point and its corresponding image point will be positioned on a collinear line with the perspective center of the camera. The collinearity condition is shown in Eq. (1), where m_{ij} denotes the nine elements in rotation matrix M, which are applied to transform the object coordinates in the mapping frame to image coordinates in the camera frame.

$$x_a - x_0 = -f \frac{m_{11}(X_A - X_L) + m_{12}(Y_A - Y_L) + m_{13}(Z_A - Z_L)}{m_{31}(X_A - X_L) + m_{32}(Y_A - Y_L) + m_{33}(Z_A - Z_L)}$$

$$y_a - y_0 = -f \frac{m_{21}(X_A - X_L) + m_{22}(Y_A - Y_L) + m_{23}(Z_A - Z_L)}{m_{31}(X_A - X_L) + m_{32}(Y_A - Y_L) + m_{33}(Z_A - Z_L)} , \tag{1}$$

where (X_A, Y_A, Z_A) : the position of object point A in the mapping frame; (X_L, Y_L, Z_L) : the position of the camera's perspective center in the mapping frame; (x_a, y_a) : the position of object point A in the camera frame; (x_0, y_0) : the image coordinates of the principal point; f : the focal length.

Furthermore, coplanarity is the condition in which two camera stations of stereo images, an object point, and its corresponding image points on two images lie simultaneously on a common plane (Fig. 1). This condition is illustrated in Eq. (2). The rotation matrix M is applied to produce image coordinates that are parallel to the first image coordinate system.

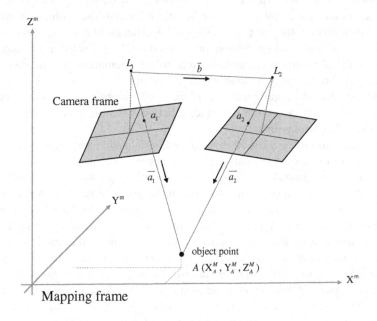

Fig. 1. Coplanarity condition

$$F = \begin{vmatrix} b_X & b_Y & b_Z \\ u_1 & v_1 & w_1 \\ u_2 & v_2 & w_2 \end{vmatrix} = 0 , \tag{2}$$

where

$$\bar{b} = \begin{bmatrix} b_X \\ b_Y \\ b_Z \end{bmatrix} = \begin{bmatrix} X_{L_2} - X_{L_1} \\ Y_{L_2} - Y_{L_1} \\ Z_{L_2} - Z_{L_1} \end{bmatrix}, \quad \bar{a_1} = \begin{bmatrix} u_1 \\ v_1 \\ w_1 \end{bmatrix} = \begin{bmatrix} x_{a_1} - x_0 \\ y_{a_1} - y_0 \\ -f \end{bmatrix}, \quad \bar{a_2} = \begin{bmatrix} u_2 \\ v_2 \\ w_2 \end{bmatrix} = M^T \begin{bmatrix} x_{a_2} - x_0 \\ y_{a_2} - y_0 \\ -f \end{bmatrix}.$$

The coplanarity condition expresses that the two perspective centers and two conjugate image points lie on a common plane, as described in the preceding part. All 12 exterior orientation parameters of the 2 images appear in Eq. (2). The model coordinate system in the dependent relative orientation is parallel to the first image's coordinate system, thus leaving two position parameters and three orientation parameters of the second image to be determined. Therefore, 7 parameters are fixed prior to solving the relative orientation. The remaining 5 parameters are the unknown relative orientation parameters, which are $\left(Y_{L_2}, Z_{L_2}, \omega_2, \varphi_2, \kappa_2 \right)$ of the second image.

The relative orientation parameters can be calculated by using the least-squares adjustment with the iterative solution. More observations should be obtained to achieve a more precise solution and to allow the detection of incorrect measurements. After the calculation of relative orientation, the absolute orientation parameters are calculated on the basis of the collinearity condition. Given that the problem is nonlinear, the approximations for the initial values must be generated before the process of the solution. The parameters of the relative orientation can be provided as the approximations for the initial values in the solution of absolute orientation. The relative orientation parameters are considered the approximations of the absolute orientation and can be used to calculate the approximate coordinates of feature points in the object coordinate. Last but not the least, the scale factor between two images can be determined by using ground control points.

The flowchart in Figure 2 can be divided into initialization and sequential processes. The information obtained from calculating the absolute orientation parameters of the first two images and ground coordinates of the conjugate points is provided for the automatic calculation of sequential images. When the absolute orientation parameters of all images are calculated, the image locations are connected to the moving trajectory. Figure 2(a) illustrates the initialization that requires matching feature points between the first two images and ground control points before calculating the absolute orientation parameters of the initial two images. The relative orientation parameters are determined by the coplanarity condition with observations of matching feature points. By using an iterative least-squares solution with a set of initial values for unknown parameters, the absolute orientation parameters of the initial two images can be calculated and the coordinates of feature points can be acquired based on collinearity condition.

Figure 2(b) illustrates the sequential process that requires information from initialization to calculate the absolute orientation parameters of the sequential images. After acquiring the absolute orientation parameters of the initial two images, the process begins with feature detection and matching for the following images. The relative orientation parameters are determined by the coplanarity condition. Moreover, the information of the absolute orientation in the previous image will be considered the initial values in the iterative solution of the relative orientation. The unknown

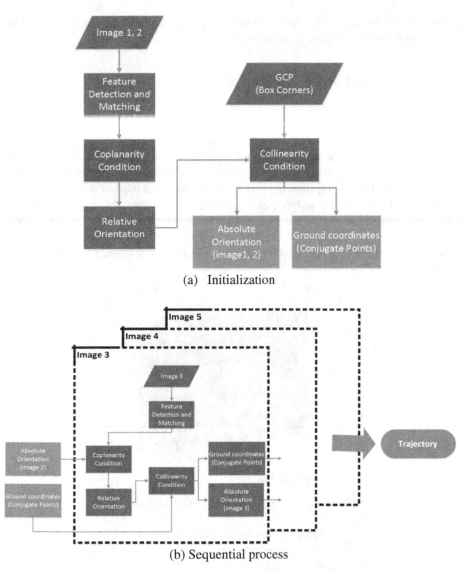

(a) Initialization

(b) Sequential process

Fig. 2. Flow chart of the localization and orientation of sequential images

parameters merely include the absolute orientation parameters of the following image and the object coordinates of the feature image points within each step. According to the previous result, the absolute orientation parameters of the subsequent image and newly generated ground control points provide effective observations for calculating the absolute orientation parameters of the following image. The result of the previous relative orientation parameters can be used as the initial values of the unknowns regarding the absolute orientation parameters to further calculate the object coordinates of feature points. The use of iterative least squares allows the calculation of the absolute orientation parameters of the following image. The object coordinates of feature points can also be acquired from the iteration. As repeatedly shown in this procedure, the trajectory of the camera can be calculated.

3   Experiment and Case Studies

3.1   Local Reference System

The localization and orientation of a single camera can be applied with one or more objects observed with respect to the local reference body. If the coordinates of the object and reference system are known, the absolute orientation can be derived from the result of the bundle adjustment [15]. A cuboid box was used to establish the local reference system. The cuboid has eight spots on its edges that can be considered control object points in the reference system after measuring their lengths. The real cuboid we made and its information are shown in Figure 3. The method can track the movement of a moving camera with respect to a stationary reference by using sequential images.

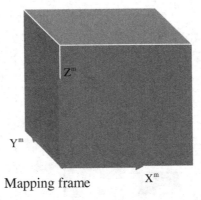

Fig. 3. Local reference system from the cuboid box

Table 1. Information of the cuboid box

Cuboid box	Size
Length	0.336 m
Width	0.230 m
Height	0.380 m

The camera was mounted on a tripod and captured images with a size of 4592 × 3056 for processing. The route was planned for each test, and the distance was measured between every two stations. Without any additional sensor, our scheme can be used directly for motion estimation with respect to the initial position of the camera. The ground control points in the local reference system can be used to determine the absolute scale factor among coordinate systems. The speed-up robust features (SURF) algorithm [16] was used to connect the sequential images and match the feature image points, through which the absolute orientation with respect to the local reference system of the camera can be determined. Three cases were designed for experiments to validate the feasibility of integrating the coplanarity and collinearity conditions to estimate the motion of a moving camera, as discussed in the following sections.

3.2   Case I: Straight Line

Case I was designed as a straight line in an indoor scene. The experimental route is located in the corridor of the Geomatics Department of the National Cheng Kung University (NCKU). Ground truth data were measured by the laser distance meter and total station. Case I contains a total of 30 designed stations. The direction from the first to the second station is perpendicular to the shooting direction of the camera to establish good geometry for determining the absolute scale parameter. The distance of the baseline is 1.315 m, whereas the distance between the following stations is 1 m each.

3.3   Case II: Loop

Case II was designed as an indoor rectangular loop. The experimental loop lies on an area in the first floor of the Geomatics Department of NCKU. Ground truth data were measured by laser distance meter and total station. A total of 25 stations were designed. The distance between each station is 0.7 m.

3.4   Case III: Outdoor

In Case III, the experimental route lies in front of the library of the NCKU campus. A total of 21 stations were designed. The distances between stations were measured by the laser distance meter and total station. The average distance between stations in a straight line and in corners is 1.16 and 0.83 m.

4   Results

Figure 4 shows the plots of ground truth data and the calculated trajectory in Case I. The results of Case I prove that our scheme is effective for estimating the motion of a moving camera. Table 2 compares the traveled distance of ground truth data with our results. The difference in total traveled distances is less than 50 cm, and the error is 1.93% for approximately 30 m. Table 3 shows the root mean square errors (RMSEs) and mean differences of the three axes. The mean differences in the X, Y, and Z directions are

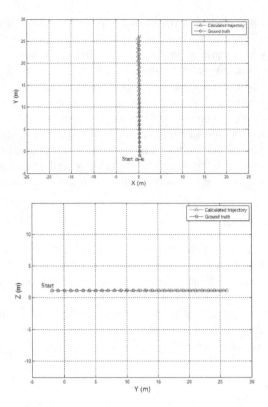

Fig. 4. Ground truth data and the calculated trajectory in Case I

Table 2. Length of true and calculated trajectory of Case I

Case	Stations	True Trajectory	Our Scheme	Relative Precision
Straight line	30	29.476 m	28.915 m	1.93 %

Table 3. Mean differences and RMSEs in Case I

	X	Y	Z
Mean diff. (m)	−0.181	−0.241	−0.005
RMSE (m)	±0.226	±0.309	±0.01

−0.181, −0.241, and −0.005 m, respectively. These results prove that planimetric positions and altitudes are accurate, and that the drift situation is tolerant.

Figure 5 shows the variations in distance between the ground truth data and the calculated trajectory in the three axes between stations. Variations in the Z axis are the smallest among the three axes, and variations in the X and Y axes are all within 5 cm. Figure 6 illustrates the error propagation in Case I. The accumulation errors in the X and Y directions of the terminal point are 0.39 and 0.53 m, respectively, which are within the acceptable range.

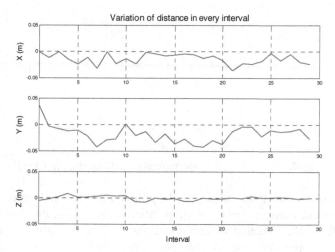

Fig. 5. Variation of the calculated trajectory and ground truth data in the three axes in Case I

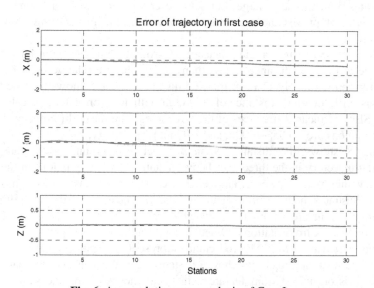

Fig. 6. Accumulative error analysis of Case I

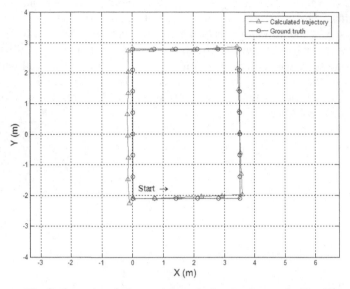

Fig. 7. Ground truth data and the calculated trajectory in Case II

Figure 7 displays the trajectories from the ground truth data and our results in Case II. As the camera moves along the route parallel to the X axis, the shooting angle is perpendicular to the direction of movement. As the camera moves along the route parallel to the Y axis, the shooting angle is parallel to the direction of movement. The results of Case II affirm that the shooting angle does not have to be fixed with a specific value. With this advantage, our scheme can be applied in arbitrary situations. The experiment proves the feasibility of different camera shooting angles during movement.

Table 4 compares the traveled distance of ground truth data with our results. The difference in traveled distance is less than 0.2 m, and the relative error can achieve 1.07% for approximately 17 m of traveled distance. Table 5 shows that the loop misclosure is 0.191 m after a distance of 16.983 m with an error of 1.12%. Table 6 shows the RMSEs and mean differences of the three axes. The mean differences in the X, Y, and Z directions are −0.05, −0.006, and 1.02 m, respectively.

Figure 8 shows the distance variations between ground truth data and the calculated trajectory in the three axes between stations. Most variations in the X and Y axes are within 5 cm. However, large errors occurred in the last stations in the Z axis. Figure 9 illustrates the error propagation in Case II. The errors in the three directions of the terminal point are within the acceptable range. Figure 10 displays a false-color overlay of images from Stations 20 and 21, with a color-coded plot of corresponding points connected by a line. The red circles represent the feature points of Station 20, whereas green circles represent those of Station 21. We can deduce that the errors in the Z axis originated from feature points that mostly lie on the same plane with

Table 4. Length of true and calculated trajectory of the Case II

Case	Stations	True Trajectory	Our Scheme	Relative Precision
Loop	25	16.8 m	16 983 m	1.07 %

Table 5. Misclosure of the Case II

Case	Misclosure	Relative Precision
Loop	0.191 m	1.12 %

Table 6. Mean differences and RMSEs in Case II

	X	Y	Z
Mean diff. (m)	−0.05	−0.006	−0.102
RMSE(m)	±0.09	±0.069	±0.160

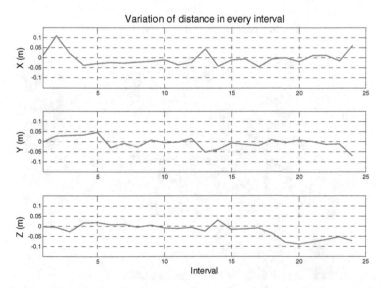

Fig. 8. Variation of the calculated trajectory and ground truth data in the three axes in the two stations in Case II

similar depth values. These feature points are unable to provide good geometry with respect to camera position and can lead to poor precision in the Z axis.

Figure 11 plots the ground truth data and the calculated trajectory in Case III. Figures 12 and 13 present the details of the trajectory errors in each station. The errors in the X direction tend to gradually accumulate with 1.05 m error after making a turn. However, the errors in the Y (0.16 m) and Z (−0.11 m) directions are relatively small and steady. Table 7 shows the RMSEs and mean differences of the three axes. The mean difference is 0.37 m in the X direction, 0.22 m in the Y direction, and −0.06 m in the Z direction. Table 8 compares the traveled distance of ground truth data with

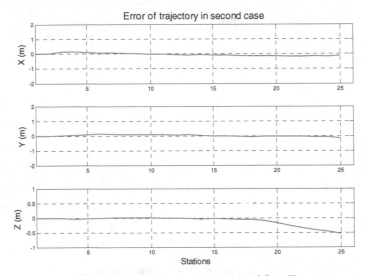

Fig. 9. Accumulative error analysis of Case II

Fig. 10. Corresponding feature points between two images

our results. The difference is approximately 60 cm, and the relative error is 3.04% in approximately 21 m. This proves that the results on the direction of forward motion are accurate. After all, single camera motion estimation can be applied in situations wherein the camera is mounted on a vehicle or is stably handheld.

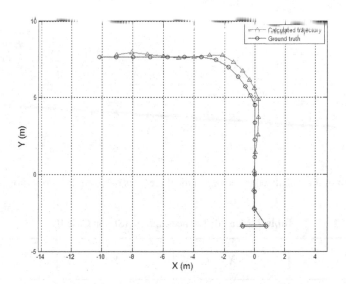

Fig. 11. Ground truth data and the calculated trajectory in Case III

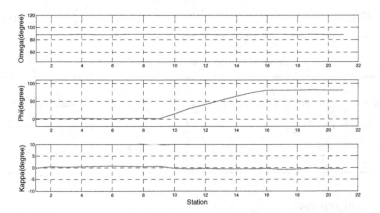

Fig. 12. Orientation variation of calculated trajectory in each station

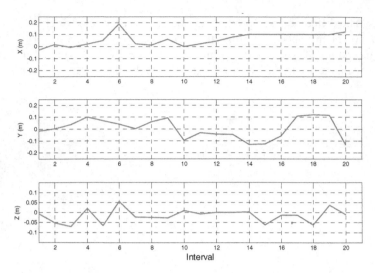

Fig. 13. Error of the calculated trajectory in the three axes between the two stations (interval) in Case III

Table 7. Mean differences and RMSEs in Case III

	X	Y	Z
Mean diff. (m)	0.373	0.224	-0.09
RMSE (m)	±0.375	±0.139	±0.09

Table 8. Metric precision of the Case III

Case	Stations	True Trajectory	Our Scheme	Relative Precision
III	21	21.332 m	20.702 m	3.04 %

5   Conclusions

A scheme is presented to estimate the motion of a single camera on the basis of photogrammetry. This study integrates the collinearity and coplanarity conditions to automatically calculate the absolute orientation parameters of a camera with the aid of the SURF algorithm to detect and match the feature points of sequential images. Despite being a nonlinear equation, applying the coplanarity condition to the matched feature points between two subsequent images can solve relative orientation parameters in a straightforward manner. Relative orientation parameters can be used as initial values for absolute orientation parameters determined on the basis of the collinearity condition by using observations for ground control points. This research establishes a known reference system, that is, a cuboid box, to rapidly measure ground points. The presented method can be applied for indoor navigation, unmanned aerial vehicles, autonomous under water vehicles, and planetary rovers.

References

1. Tian, T., Tomasi, C., Heeger, D.: Comparison of approaches to egomotion computation. In: IEEE Computer Society Conference on Computer Vision and Pattern Recognition: 315 (1996)
2. Luhmann, T.: Precision potential of photogrammetric 6DOF pose estimation with a single camera. Isprs Journal of Photogrammetry and Remote Sensing **64**(3), 275–284 (2009)
3. Nistér, D., Naroditsky, O., Bergen, J.: Visual odometry. In: Proceedings of the 2004 IEEE Computer Society Conference on Computer Vision and Pattern Recognition, CVPR 2004, vol. l(1), p. I-652 (2004)
4. Maimone, M., Cheng, Y., Matthies, L.: Two years of Visual Odometry on the Mars Exploration Rovers. Journal of Field Robotics **24**(3), 169–186 (2007). doi:10.1002/rob.20184
5. Cheng, Y., Maimone, M., Matthies, L.: Visual odometry on the mars exploration rovers. In: 2005 IEEE International Conference on Systems, Man and Cybernetics, vol. 1, pp. 903–910 (2005)
6. Nistér, D.: An efficient solution to the five-point relative pose problem. IEEE Transactions on Pattern Analysis and Machine Intelligence **26**(6), 756–770 (2004)
7. Kitt, B., Geiger, A., Lategahn H.: Visual odometry based on stereo image sequences with ransac-based outlier rejection scheme. In: Intelligent Vehicles Symposium (IV), 2010 IEEE (2010)
8. Avidan, S., Shashua, A.: Threading fundamental matrices. IEEE Transactions Pattern Analysis and Machine Intelligence **23**(1), 73–77 (2001)
9. Torr, P.H.S., Zisserman, A.: Robust parameterization and computation of the trifocal tensor. Image and Vision Computing **15**(8), 591–605 (1997)
10. Geiger, A., Ziegler, J., Stiller, C.: StereoScan: dense 3d reconstruction in real-time. In: Intelligent Vehicles Symposium (IV), 2011 IEEE, pp. 963–968 (2011)
11. Davison, A.J., Reid, I.D., Molton, N.D., Stasse, O.: MonoSLAM: Real-time single camera SLAM. IEEE Transactions Pattern Analysis and Machine Intelligence **29**(6), 1052–1067 (2007)
12. Mikhail, E.M., Bethel, J.S., McGlone, J.C.: Introduction to Modern Photogrammetry. John Wiley & Sons, New York (2001)
13. Mouragnon, E., Lhuillier, M., Dhome, M., Dekeyser, F., Sayd, P.: Generic and real-time structure from motion using local bundle adjustment. Image and Vision Computing **27**(8), 1178–1193 (2009)
14. Juea, L.: RESEARCH ON CLOSE-RANGE PHOTOGRAMMETRY WITH BIG ROTATION ANGLE. Astronomy and astrophysics (Berlin), 1, 1 (2008)
15. Haralick, R.M., Lee, D., Ottenburg, K., Nolle, M.: Analysis and solutions of the three point perspective pose estimation problem. In: IEEE Computer Society Conference Computer Vision and Pattern Recognition, Proceedings CVPR 1991, pp. 592–598. IEEE (1991)
16. Bay, H., Ess, A., Tuytelaars, T., Van Gool, L.: Speeded-Up Robust Features (SURF). Computer Vision and Image Understanding **110**(3), 346–359 (2008)

Semi-Lagrangian Semi-Implicit Vertically Splitting Scheme for Nonhydrostatic Atmospheric Model

Andrei Bourchtein[✉]

Institute of Physics and Mathematics, Pelotas State University, Pelotas, Brazil
bourchtein@gmail.com

Abstract. A semi-Lagrangian semi-implicit scheme with additional vertical splitting is developed for nonhydrostatic atmospheric model. The essential elements of the scheme are solution of the trajectory equations for advective part, separation of the vertical normal modes in fast and slow processes with respect to gravity wave propagation speed for each mode, explicit and simpler approximation of slower vertical modes and implicit time differencing for faster modes. This approach allows us to choose the time step based on accuracy considerations and substitute 3D elliptic problems inherent for semi-implicit time differencing by a set of 2D elliptic problems. The performed numerical experiments show computational efficiency of the proposed scheme and accuracy of the predicted atmospheric fields.

Keywords: Atmosphere modeling · Nonhydrostatic model · Numerical schemes

1 Introduction

The continuously increasing power of computers has reached the point when real-time simulations based on the fully compressible equations of the atmosphere (Euler or Navier-Stokes ones) are possible. Although the hydrostatic assumption is proved to be a good approximation for synoptic and global scale processes and many of the current numerical prediction and climate simulation models are still employ the so-called primitive equations (including the hydrostatic relation), there is a strong current trend to return to non-filtering governing equations. It seems that the two most important reasons behind this are: the aim to improve the forecasting skills for meso-scale phenomena where hydrostatic balance is not quite accurate, and the need to reduce the cost of development and maintenance of highly complex forecast systems, which can be achieved by using unified models suitable for both global and meso scales.

Analysis of the linearized Euler equations reveals the three principal types of atmospheric processes: acoustic, gravitational and inertial waves. The propagation speed of these oscillations is different: acoustic waves propagate in the atmosphere with velocities above 330 m/s, the propagation speed of gravity waves ranges from 0 up to 300 m/s, and inertial waves move at most at 50 m/s. Also the energy contribution of these oscilations is quite different in the Earth's atmosphere: acoustic waves are always negligible, gravity waves carry a small part of the energy for most

© Springer International Publishing Switzerland 2015
O. Gervasi et al. (Eds.): ICCSA 2015, Part II, LNCS 9156, pp. 104–118, 2015.
DOI: 10.1007/978-3-319-21407-8_8

phenomena, and inertial processes contain the main part of energy (especially for large scales) [13]. Nevertheless, opting to employ the full atmospheric equations for better description of the principal phenomena, one automatically involves the fastest acoustic and gravity waves and, therefore, should deal with a stiff system. This property of stiffness (together with the nonlinearity) strongly influences the choice of numerical method.

On the one hand, fully implicit time discretizations are not used due to the nonlinearity of the governing equations, because it leads to solution of nonlinear algebraic systems of high order at each time step that is very problematic and computationally expensive task. On the other hand, fully explicit schemes are not a practical choice for atmospheric models due to stability requirements resulting from the high stiffness of the fully compressible equations. Roughly speaking, the Courant-Friedrichs-Lewy (CFL) condition of numerical stability can be formulated in the following way: the time step of numerical schemes is proportional to the mesh size of spatial grid and inversely proportional to the maximum propagation speed of the processes approximated explicitly. It implies that fully explicit numerical schemes have very rigid restriction on the time step due to the presence of the fast acoustic waves. In fact, according to the accuracy requirements for modeling meso-scale processes, one can use the time steps about 10 min (for synoptic-scale phenomena the time step can be increased up to 30 min). However, the numerical stability of explicit schemes requires much smaller time steps. For example, the use of a vertical resolution of 1 km demands the restriction of 1 sec on time steps. Even if the vertical approximation is implicit, for horizontal grids with 20 km mesh size the maximum stable time step in horizontally explicit schemes should be less then 30 sec. Current models employ more than 30 vertical levels in the atmosphere layer of 20-30 km and horizontal resolution of meso-scale models varies from 1 to 20 km, which means that restrictions on time step may be even more severe [12,19,23].

There are two modern general approaches to time differencing in the fully compressible atmospheric models. The first one is split-explicit time differencing, which traces back to the model of Klemp and Wilhelmson [15] and allows the use of different time steps for fast and slow parts of general solution. Usually it includes implicit discretization in vertical for the fast waves to avoid super small time steps because of fine vertical resolution [19,20,23]. The second one is semi-Lagrangian semi-implicit (SLSI) time differencing originally developed for nonhydrostatic models by Tanguay et al. [24] and based on semi-Lagrangian description of advection joined with implicit approximation of linear terms responsible for the fast processes and explicit approximation of nonlinear terms [21,22,24]. The last method has been successfully used in a wide range of the hydrostatic models and apparently is the most popular approach for the hydrostatic-based simulations [3,5,14,21,25]. Both approaches are reviewed in the papers by Saito et al. [19] and Steppeler et al. [23], and their descriptions can be found in a number of papers (e.g. [11,16,20] for split-explicit models and [7,8,10,22,26] for SLSI models and references therein).

In this study we apply additional vertical decoupling in the context of the SLSI method in order to separate the fast and slow vertical modes and eliminate the necessity of implicit approximation for slower gravity waves. Technically it allows transformation of 3D elliptic problems, which arise at each time step due to implicit approximation of linear terms, to a set of 2D Helmholtz equations and makes possible to solve only a few of the last equations related to the fastest processes. In this way, coupling different numerical techniques we are able to construct computationally efficient and accurate SLSI model for the nonhydrostatic equations of the atmosphere.

2   Governing Nonhydrostatic Equations

Using time coordinate t, horizontal Cartesian coordinates x, y of a conformal mapping of the Earth's surface and geometric vertical coordinate z, the governing equations of the nonhydrostatic atmosphere (Euler equations) can be written as follows [13]:

$$\frac{du}{dt} = fv - RTP_x,$$ (1)

$$\frac{dv}{dt} = -fu - RTP_y,$$ (2)

$$\frac{dw}{dt} = -g - RTP_z,$$ (3)

$$\frac{dP}{dt} = -\frac{c_p}{c_v}\left(u_x + v_y + w_z\right),$$ (4)

$$\frac{dT}{dt} = \frac{RT}{c_p} \cdot \frac{dP}{dt}.$$ (5)

Here u, v, w, P, T are unknown functions, namely, u and v are the horizontal velocity components, w is the vertical velocity component, $P = \ln p$, p being the pressure, T is the temperature. The individual 3D derivative is

$$\frac{d\varphi}{dt} = \varphi_t + u\varphi_x + v\varphi_y + w\varphi_z \quad, \quad \varphi = u, v, w, P, T$$ (6)

and the following parameters are used: f is the Coriolis parameter, g is the gravitational acceleration, R is the gas constant of dry air, c_p and c_v are the specific heat at constant pressure and volume, respectively. Hereinafter the subscripts t, x, y, z denote the partial derivatives, while all other subscripts are used only for indexing.

For implementation of the SLSI scheme it is convenient to represent the thermodynamic variables and the Coriolis parameter using their mean values $\overline{T} = const$, $\overline{P}_z = -g/R\overline{T}$, $\overline{f} = const$ and deviations: $T = \overline{T} + T'$, $P = \overline{P} + P'$, $f = \overline{f} + f'$ and rewrite equations (1)-(5) in the following form (from now on, the full quantities are substituted by deviations, so the primes for deviations are omitted without ambiguity):

$$\frac{du}{dt} = \bar{f}v - R\bar{T}P_x + N_u \ , \tag{7}$$

$$\frac{dv}{dt} = -\bar{f}u - R\bar{T}P_y + N_v \ , \tag{8}$$

$$\frac{dw}{dt} = \frac{g}{\bar{T}}T - R\bar{T}P_z + N_w \ , \tag{9}$$

$$\frac{dP}{dt} = -\frac{c_p}{c_v}\left(u_x + v_y + w_z\right) + \frac{g}{R\bar{T}}w \ , \tag{10}$$

$$\frac{dT}{dt} = \frac{R\bar{T}}{c_p} \cdot \frac{dP}{dt} - \frac{g}{c_p}w + N_T \ . \tag{11}$$

Here N_u, N_v, N_w, N_T contain nonlinear terms (except for advection) and variable coefficient terms and can be expressed as follows

$$N_u = fv - RTP_x, N_v = -fu - RTP_y, N_w = -RTP_z, N_T = \frac{RT}{c_p}\left(\frac{dP}{dt} - \frac{g}{R\bar{T}}w\right). \tag{12}$$

3   Semi-Lagrangian Semi-implicit Vertically Splitting Scheme

The design of the developed algorithm is based on a general outline of the three-time-level SLSI method [3,4,7,8,24] and vertical decoupling previously tested in hydrostatic models [3,4,5,6]. First, let us write down the formulas of the standard SLSI time discretization for (7)-(11):

$$\frac{u^\tau - u^{-\tau}}{2\tau} = \bar{f}\frac{v^\tau + v^{-\tau}}{2} - R\bar{T}\frac{P_x^\tau + P_x^{-\tau}}{2} + N_u, \tag{13}$$

$$\frac{v^\tau - v^{-\tau}}{2\tau} = -\bar{f}\frac{u^\tau + u^{-\tau}}{2} - R\bar{T}\frac{P_y^\tau + P_y^{-\tau}}{2} + N_v, \tag{14}$$

$$\frac{w^\tau - w^{-\tau}}{2\tau} = \frac{g}{\bar{T}}\frac{T^\tau + T^{-\tau}}{2} - R\bar{T}\frac{P_z^\tau + P_z^{-\tau}}{2} + N_w, \tag{15}$$

$$\frac{P^\tau - P^{-\tau}}{2\tau} = -\frac{c_p}{c_v}\frac{D^\tau + D^{-\tau}}{2} - \frac{c_p}{c_v}\frac{w_z^\tau + w_z^{-\tau}}{2} + \frac{g}{R\bar{T}}\frac{w^\tau + w^{-\tau}}{2} \ , \tag{16}$$

$$\frac{T^\tau - T^{-\tau}}{2\tau} = \frac{R\bar{T}}{c_p}\frac{P^\tau - P^{-\tau}}{2\tau} - \frac{g}{c_p}\frac{w^\tau + w^{-\tau}}{2} + N_T \ . \tag{17}$$

Here, $D = u_x + v_y$ is the horizontal divergence, τ is the time step, the superscript "τ" denotes a value at the arrival point of 3D trajectory at the new time level $t_{n+1} = (n+1)\tau$, the superscript "$-\tau$" denotes a value at the departure point of the trajectory at the past time level $t_{n-1} = (n-1)\tau$, and the time dependent functions without superscripts are evaluated as the mean values between the values at the arrival and departure points at the current time level $t_n = n\tau$:

$$\varphi^\tau = \varphi(t_{n+1}, P_a), \varphi^{-\tau} = \varphi(t_{n-1}, P_d), \varphi = \frac{\varphi(t_n, P_a) + \varphi(t_n, P_d)}{2}, \quad \varphi = u, v, w, P, T \qquad (18)$$

P_a and P_d - arrival and departure points, respectively.

A simpler SL quasi-explicit (SLEX) time differencing can be written as

$$\frac{\hat{u}^\tau - u^{-\tau}}{2\tau} = \bar{f}\bar{v} - R\bar{T}P_x + N_u, \qquad (19)$$

$$\frac{\hat{v}^\tau - v^{-\tau}}{2\tau} = -\bar{f}\bar{u} - R\bar{T}P_y + N_v, \qquad (20)$$

$$\frac{\hat{w}^\tau - w^{-\tau}}{2\tau} = \frac{g}{\bar{\bar{T}}} \frac{\hat{T}^\tau + T^{-\tau}}{2} - R\bar{T} \frac{\hat{P}_z^\tau + P_z^{-\tau}}{2} + N_w, \qquad (21)$$

$$\frac{\hat{P}^\tau - P^{-\tau}}{2\tau} = -\frac{c_p}{c_v} D - \frac{c_p}{c_v} \frac{\hat{w}_z^\tau + w_z^{-\tau}}{2} + \frac{g}{R\bar{T}} \frac{\hat{w}^\tau + w^{-\tau}}{2}, \qquad (22)$$

$$\frac{\hat{T}^\tau - T^{-\tau}}{2\tau} = \frac{R\bar{T}}{c_p} \frac{\hat{P}^\tau - P^{-\tau}}{2\tau} - \frac{g}{c_p} \frac{\hat{w}^\tau + w^{-\tau}}{2} + N_T. \qquad (23)$$

In both SLSI and SLEX approximations the arrival points (x^τ, y^τ, z^τ) are given points of the computational grid and the departure points $(x^{-\tau}, y^{-\tau}, z^{-\tau})$ of 3D trajectories are found using the iterative algorithm proposed by Tanguay et al. [17,24]:

$$x^{-\tau}(s+1) = x^\tau - 2\tau u\left(\frac{x^\tau + x^{-\tau}(s)}{2}, \frac{y^\tau + y^{-\tau}(s)}{2}, \frac{z^\tau + z^{-\tau}(s)}{2}\right), \qquad (24)$$

$$y^{-\tau}(s+1) = y^\tau - 2\tau v\left(\frac{x^\tau + x^{-\tau}(s)}{2}, \frac{y^\tau + y^{-\tau}(s)}{2}, \frac{z^\tau + z^{-\tau}(s)}{2}\right), \qquad (25)$$

$$z^{-\tau}(s+1) = z^\tau - 2\tau w\left(\frac{x^\tau + x^{-\tau}(s)}{2}, \frac{y^\tau + y^{-\tau}(s)}{2}, \frac{z^\tau + z^{-\tau}(s)}{2}\right), \qquad (26)$$

where s is the iteration number. The condition for the convergence of these iterations is [4,18]:

$$\tau \le \frac{2}{3V_d}, \quad V_d = \max\left(|u_x|, |u_y|, |u_z|, |v_x|, |v_y|, |v_z|, |w_x|, |w_y|, |w_z|\right). \tag{27}$$

For the maximum values of the wind component variations $V_d \approx 5 \cdot 10^{-4} s^{-1}$, the maximum allowable time step is $\tau \approx 20$ min.

Notice that a solution to the SLEX approximation can be found in a simple almost explicit way. Indeed, equations (21)-(23) can be combined in one equation for the function $\hat{P}^\tau$:

$$-\hat{P}^\tau_{zz} + \frac{g}{R\overline{T}} \hat{P}^\tau_z + \frac{1}{\tau^2 c_a^2} \hat{P}^\tau = L_a, \tag{28}$$

where L_a includes only the functions at the past and current time levels and $c_a^2 = \frac{c_p}{c_v} R\overline{T}$ is the squared propagation speed of the horizontal acoustic waves. The last equation is completed by the boundary conditions at the lower and upper levels of the atmosphere

$$w = 0 \quad at \quad z = 0, \; z = z_{up} \tag{29}$$

expressed in the terms of the pressure function

$$-\hat{P}^\tau_z + \frac{g}{c_p\overline{T}} \hat{P}^\tau = L_b \quad at \quad z = 0, \; z = z_{up}, \tag{30}$$

where L_b is a given right-hand side. Solution of the vertically discretized one-dimensional boundary value problem (28), (30) can be found by a fast version of the Gauss elimination (sometimes called the Gelfand-Thomas algorithm) with the number of operations proportional to the number of the grid points [12]. This is the only implicit relation that should be solved in the quasi-explicit scheme, all other relations are explicit. Therefore, the SLEX approximation allows computationally efficient implementation at each time step, but the number of the time steps is large due to a strict stability condition

$$\tau \le \frac{h_g}{\sqrt{2}c_a^2}, \tag{31}$$

where h_g is the spatial mesh size used for an approximation of gravity waves. For the typical values $h_g \approx 20$ km, and the standard second order central difference approximation of the pressure gradient in (19),(20) and the horizontal divergence in (22), condition (31) requires the use of the time steps less than 1 min.

In order to increase the allowable time step a more implicit time differencing should be introduced. To this end, consider the difference between the SLSI and SLEX schemes, which can be represented in the form:

$$u^* - \tau \overline{f} v^* + \tau R \overline{T} P^*_{x} = L_u , \tag{32}$$

$$v^* + \tau \overline{f} u^* + \tau R \overline{T} P^*_{y} = L_v , \tag{33}$$

$$w^* - \tau \frac{g}{\overline{T}} T^* + \tau R \overline{T} P^*_{z} = L_w , \tag{34}$$

$$P^* + \tau \frac{c_p}{c_v} D^* + \tau \frac{c_p}{c_v} w^*_{z} - \tau \frac{g}{R \overline{T}} w^* = L_P , \tag{35}$$

$$T^* - \frac{R \overline{T}}{c_p} P^* + \tau \frac{g}{c_p} w^* = L_T , \tag{36}$$

where the differences between SLSI and SLEX prognostic values are denoted by $\varphi^* = \varphi^* - \hat{\varphi}^\tau$, $\varphi = u, v, w, T, P$ and the linear terms L_φ do not contain the SLSI prognostic values:

$$L_u = \tau \overline{f} \tilde{v} - \tau R \overline{T} \widetilde{P}_x , \ L_v = -\tau \overline{f} \tilde{u} - \tau R \overline{T} \widetilde{P}_y , \ L_w = 0 , \ L_P = -\tau \frac{c_p}{c_v} \tilde{D} , \ L_T = 0 ; \tag{37}$$

$\tilde{\varphi} = \hat{\varphi}^\tau - 2\varphi + \varphi^{-\tau}$, $\varphi = u, v, w, T, P$. Considering that the SLEX prognostic values are already computed, equations (32)-(36) represent the linear system for the corrections φ^*. Solving this system is equivalent to finding the SLSI prognostic values and, due to an implicit approximation of the fast waves, it requires solution of 3D boundary value problem for Helmholtz type equation. However, applying vertical splitting, it is possible to reduce this computationally expensive task to solution of a set of 2D elliptic problems without compromising the stability and accuracy of the numerical solution. In the remaining part of this section, we describe this splitting technique following the construction given for the hydrostatic models in [4,5].

The first step toward vertical splitting is the elimination of the unknown functions w^* and T^* from equations (34)-(36) that leads to the following equation for P^*:

$$-P^*_{zz} + \frac{g}{R \overline{T}} P^*_z + \frac{1}{\tau^2 c_a^2} P^* + \frac{1}{\tau R \overline{T}} \kappa^2 D^* = L_1 , \tag{38}$$

where

$$L_1 = \frac{1}{\tau^2 c_a^2} \kappa^2 L_P , \ \kappa^2 = 1 + \tau^2 \frac{g^2}{c_p \overline{T}} . \tag{39}$$

The boundary conditions (29) for the function w^* imply

$$P^*_z - \frac{g}{c_p \overline{T}} P^* = 0 \ \ at \ z = 0, \ z = z_{up} , \tag{40}$$

It can be shown that the spectrum of the corresponding Sturm-Liouville problem

$$-F_{zz} + \frac{g}{R\overline{T}}F_z + \frac{1}{\tau^2 c_a^2}F = \lambda F \tag{41}$$

$$F_u - \frac{g}{c_p\overline{T}}F = 0 \quad at \quad z = 0, \ z = z_{up} \tag{42}$$

is positive and simple

$$\lambda_0 = \alpha^2 \frac{c_v R}{c_p^2} + \beta, \quad \lambda_k = \left(\frac{k\pi}{z_{up}}\right)^2 + \frac{\alpha^2}{4} + \beta, \quad k = 1,2,\dots \ ; \tag{43}$$

$$\alpha = \frac{g}{R\overline{T}}, \quad \beta = \frac{1}{\tau^2 c_a^2} \tag{44}$$

with the corresponding eigenfunctions

$$F_0 = e^{\mu_0 z}, \quad F_k = e^{\mu_k z}\left(A_k \cos\eta_k z + \sin\eta_k z\right), \quad k = 1,2,\dots \ ; \tag{45}$$

$$\mu_0 = \frac{g}{c_p\overline{T}}, \quad \mu_k = \frac{g}{2R\overline{T}}, \quad \eta_k = \frac{k\pi}{z_{up}}, \quad A_k = \frac{k\pi}{z_{up}}\frac{2Rc_p\overline{T}}{g\left(2R - c_p\right)}, \quad k = 1,2,\dots \tag{46}$$

Using expansion in the eigenfunctions (45)

$$\varphi^*(t, x, y, z) = \sum_k \varphi_k^+(t, x, y)F_k(z), \quad \varphi = u, v, P \tag{47}$$

one can rewrite equations (32),(33),(38) for the amplitudes φ_k^+ (subscript k is omitted in amplitudes for simplicity of notation):

$$u^+ - \overline{\tilde{f}}v^+ + \tau Q^+_x = L_u^+, \tag{48}$$

$$v^+ + \overline{\tilde{f}}u^+ + \tau Q^+_y = L_v^+, \tag{49}$$

$$Q^+ + \tau\frac{1}{\tau^2\lambda_k}\kappa^2 D^+ = L_3^+, \tag{50}$$

where

$$Q^+ = R\overline{T}P^+, \quad L_3^+ = \frac{1}{\lambda_k}R\overline{T}L_2^+. \tag{51}$$

Each of the systems (48)-(50) is a set of vertically decoupled 2D equations (for different k) representing semi-implicit approximation of the linearized shallow water equations with the gravity-wave propagation speed

$$c_k = \sqrt{\frac{1}{\lambda_k\tau^2}\kappa^2}. \tag{52}$$

The characteristic values of c_k decrease fast with respect to k. For example, in the case $z_{up} = 20$ km and $\tau = 20$ min, the first twenty values of c_k are (in m/s):

$$325, \ 299, \ 111, \ 58.4, \ 39.4, \ 29.6, \ 23.8, \ 19.8, \ 17.0, \ 14.9,$$
$$13.2, \ 11.8, \ 10.8, \ 9.93, \ 9.16, \ 8.51, \ 7.94, \ 7.45, \ 7.01, \ 6.62. \tag{53}$$

Similar results are obtained for the vertically discretized problem.

If all the systems (48)-(50) are solved, then we return to the SLSI scheme and the only stability criterion is the condition of the convergence of iterative algorithm for finding the trajectories of air particles (27), which allows large time steps. If only the first I fastest vertical modes are corrected using (48)-(50), then the additional stability condition can be approximately expressed as

$$\tau \leq \frac{h_g}{\sqrt{2} c_{I+1}}, \tag{54}$$

where h_g is the mesh size of the horizontal grid used for approximation of gravity waves and c_{I+1} is the maximum gravity wave speed of the uncorrected modes (that is, the fastest vertical mode approximated by SLEX scheme). For the characteristic horizontal mesh size of 20 km, it is not advisable to use the time step larger then 20 min due to accuracy considerations. If only the first ten vertical modes are corrected, while the remaining modes are solved in SLEX scheme, then the time steps up to 20 min are admissible according to condition (54). In the models with about 40 vertical levels and more used in current practice [7,8,12,25], it means that less than one fourth of all vertical modes should be corrected, which essentially reduces the computational time required for implicit treatment of gravity oscillations.

Each of systems (48)-(50) is solved by using a reduction to 2D Helmholtz type equation for Q^+:

$$Q^+ - \frac{1}{\lambda_k} \kappa^2 \frac{1}{1+\tau^2 \bar{f}^2} \left(Q_{xx}^+ + Q_{yy}^+ \right) = L_Q^+ , \tag{55}$$

$$L_Q^+ = L_3^+ - \frac{1}{\tau \lambda_k} \kappa^2 \frac{1}{1+\tau^2 \bar{f}^2} \left(L_{ux}^+ + L_{vy}^+ + \bar{f} \left(L_{vx}^+ - L_{uy}^+ \right) \right) . \tag{56}$$

The last elliptic equation is solved using BOXMG software [2,9], which allows us to employ rectangular grids with an arbitrary number of points. Through numerical experiments, the optimal version of the multigrid algorithm has been chosen, which consists of usual V-cyclic method with two cycles for the first vertical mode ($k = 1$) and one cycle for other ones ($k = 2,...,10$). With one four-color Gauss-Seidel point relaxation sweep performed on each grid both before dropping down to next coarser grid and before interpolation to previous finer grid, this multigrid algorithm halves the computational time of the solution of the last equation as compared to the traditional SOR method for grids with 100x100 points.

For given amplitude corrections Q^+, the amplitudes u^+ and v^+ are found explicitly by the first two formulas (48),(49) in system (48)-(50). Computed amplitude corrections are added to the amplitudes of the fastest vertical modes and inverse vertical transformation returns the physical fields composed of faster modes evaluated by SLSI scheme and slower modes computed in SLEX scheme.

4   Numerical Experiments

The described SLSI numerical scheme for nonhydrostatic model (referred to as SLSIN in this section) was implemented on horizontal grid of 20 km mesh size and vertical resolution of 30 levels for short-range forecasting of the atmosphere phenomena on space domain with extension of 3000 km in both horizontal directions and about 15 km in altitude. First, extended-time integrations were run in order to verify the stability of the time differencing. As expected according to the linear stability results, the integrations are stable for time steps up to 20 min, although the forecast accuracy was lost after about 48 hours of integration. Notice that such behavior is not a disadvantage in regional modeling, since numerical solutions of regional atmospheric models are subject to strong influence of the horizontal boundary conditions and after 36-48 hours of integration lose their accuracy due to strong interference of the boundary conditions, whatever numerical scheme is used [1]. For this reason, the accuracy assessment in regional modeling is usually restricted to the time period of 36 hour integration.

For evaluation of accuracy and computational efficiency of SLSIN scheme, numerical forecasts starting from actual atmospheric data were computed. The integrations were carried out on the horizontal domain centered at Porto Alegre city ($30^0 S$, $52^0 W$) and the initial and boundary conditions were obtained from objective analysis and global forecasts of the National Centers for Environmental Prediction (NCEP). The scheme performance was compared with that of SLSIS scheme (similar SLSI scheme designed for hydrostatic equations and described in [4]) and Eulerian leapfrog scheme (LF) applied to hydrostatic model. The last is a simple popular explicit scheme tested in different models of the atmosphere. The SLSIS scheme was run with the 20 min time step and LF with 30 sec time step (the maximum time step allowable by CFL condition).

The averaged root-mean-square differences between 12-h, 24-h and 36-h forecasts and NCEP analysis are shown in Figs. 1,2 for two elements: geopotential height at the 500 hPa and 850 hPa pressure levels. These characteristic elements are traditionally used for verification of forecast skill in the numerical weather prediction systems: the first reflects the dynamics of the middle troposphere and the second is important for determination of humidity and cloud processes [1]. In both figures the solid line is used for LF scheme, the dashed line for SLSIS scheme and the dotted line for SLSIN scheme.

As seen, both SLSI schemes produce forecasts of slightly superior quality than LF scheme. The accuracy of both SLSI forecasts is virtually coincides and is within expected level of accuracy for adiabatic models [1]. The computational cost of one SLSIS forecast is about 10% of the LF computational time and SLSIN scheme requires about 20% additional computational time as compared with SLSIS scheme due to more complex governing equations.

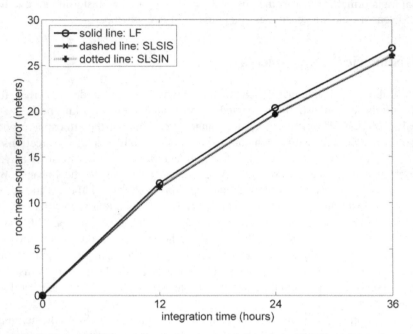

Fig. 1. Root-mean-square error of geopotential forecast at 500 hPa pressure level

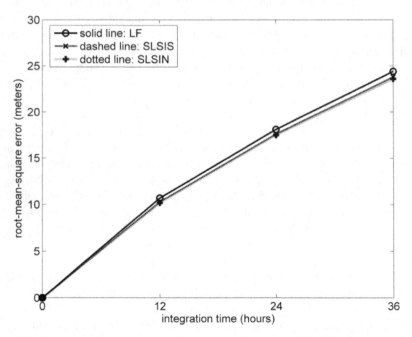

Fig. 2. Root-mean-square error of geopotential forecast at 850 hPa pressure level

Similar experiments were also performed using two more schemes for the nonhyd rostatic model: the standard fully explicit leap-frog scheme (LFN) and horizontally explicit vertically implicit scheme (HEVIN), which treats all the processes explicitly except for the acoustic waves approximated implicitly in the vertical direction [19,23]. The former was integrated with time step of 0.5 sec, while the latter - with 20 sec time step (according to the corresponding CFL conditions). The results of integration are shown in Figs. 3,4 together with the evaluations of SLSIN scheme (the last are the same as in Figs. 1,2).

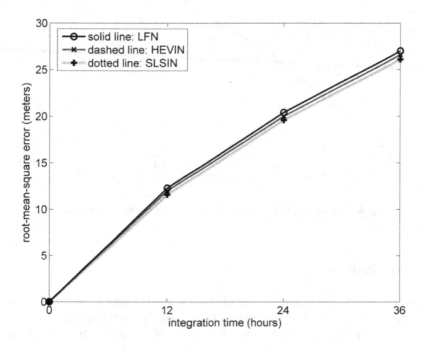

Fig. 3. The same as Fig.1 but for the three nonhydrostatic schemes

Again the semi-Lagrangian scheme (SLSIN) provides more accurate results in comparison with the Eulerian-based approximations (LFN and HEVIN). Probably the application of additional space filtering can improve the accuracy of LFN and HEVIN forecasts, but these options were not tested. As for the computational requirements, one SLSIN forecast takes less than 1% of the LFN computational time and less than 10% of the HEVIN forecast time. Thus, the performed evaluations show the validity of the applied vertical splitting and the efficiency of the developed scheme.

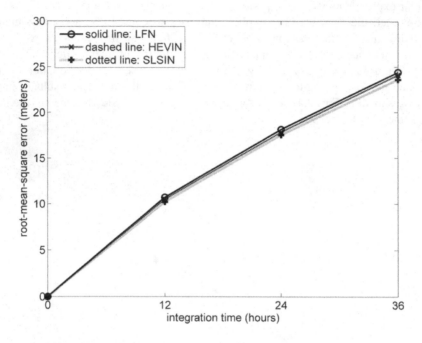

Fig. 4. The same as Fig.2 but for the three nonhydrostatic schemes

5   Conclusions

In this study semi-Lagrangian semi-implicit time discretization was applied to the fully compressible nonhydrostatic equations of the atmospheric dynamics. A possibility to use computationally efficient implementation of the algorithm and to keep increased time steps was achieved by applying the vertical decoupling technique that allows a separate treatment of the fast and slow gravity waves. For the typical horizontal mesh size of 20 km used in regional atmospheric models, the allowable time step is about 20 min, which is close to the time steps chosen in accordance with the accuracy requirements. When tested on actual atmospheric datasets, the proposed scheme produces forecasts of the meteorological fields with a good level of accuracy for the considered class of models and shows significant reduction of the computational cost as compared to the standard semi-implicit algorithms.

Acknowledgements. This research was supported by the Brazilian science foundations CNPq and FAPERGS.

References

1. Anthes, R.A., Kuo, Y.H., Hsie, E.Y., Low-Nam, S., Bettge, T.W.: Estimation of skill and uncertainty in regional numerical models. Q. J. R. Meteorol. Soc. **115**, 763–806 (1989)
2. Bandy, V., Sweet, R.: A set of three drivers for BOXMG: a black box multigrid solver. Comm. Appl. Num. Methods **8**, 563–571 (1992)
3. Bates, J.R., Moorthi, S., Higgins, R.W.: A global multilevel atmospheric model using a vector semi-Lagrangian finite-difference scheme. Part I: Adiabatic formulation, Mon. Wea. Rev. **121**, 244–263 (1993)
4. Bourchtein, A.: Semi-Lagrangian semi-implicit space splitting regional baroclinic atmospheric model. Appl. Numer. Math. **41**, 307–326 (2002)
5. Bourchtein, A., Bourchtein, L.: Semi-Lagrangian semi-implicit time-splitting scheme for a regional model of the atmosphere. J. Comput. Appl. Math. **227**, 115–125 (2009)
6. Burridge, D.M.: A split semi-implicit reformulation of the Bushby-Timpson 10 level model. Quart. J. Roy. Meteor. Soc. **101**, 777–792 (1975)
7. Côté, J., Gravel, S., Methot, A., Patoine, A., Roch, M., Staniforth, A.: The operational CMC-MRB global environmental multiscale (GEM) model. Part I: Design considerations and formulation, Mon. Wea. Rev. **126**, 1373–1395 (1998)
8. Davies, T., Cullen, M.J.P., Malcolm, A.J., Mawson, M.H., Staniforths, A., White, A.A., Wood, N.: A new dynamical core for the Met Office's global and regional modeling. Q. J. Roy. Met. Soc. **131**, 1759–1782 (2005)
9. Dendy, J.E.: Black box multigrid. J. Comp. Phys. **48**, 366–386 (1982)
10. Diamantakis, M., Davies, T., Wood, N.: An iterative time-stepping scheme for the Met Office's semi-implicit semi-Lagrangian nonhydrostatic model. Q. J. Roy. Met. Soc. **133**, 997–1011 (2007)
11. Dudhia, J., Bresch, J.F.: A global version of the PSU-NCAR mesoscale model. Mon. Wea. Rev. **130**, 2989–3007 (2002)
12. Durran, D.: Numerical Methods for Wave Equations in Geophysical Fluid Dynamics. Springer, New York (1999)
13. Holton, J.R.: An Introduction to Dynamic Meteorology. Academic Press, New York (2004)
14. Hortal, M.: The development and testing of a new two-time-level semi-Lagrangian scheme (SETTLS) in the ECMWF forecast model. Q. J. R. Met. Soc. **128**, 1671–1687 (2002)
15. Klemp, J., Wilhelmson, R.: The simulation of three-dimensional convective storm dynamics. J. Atm. Sci. **35**, 1070–1096 (1978)
16. Klemp, J., Skamarock, W.C., Dudhia, J.: Conservative split-explicit time integration methods for the compressible nonhydrostatic equations. Mon. Wea. Rev. **135**, 2897–2913 (2007)
17. McDonald, A.: Accuracy of multiply upstream, semi-Lagrangian advective schemes. Mon. Wea. Rev. **112**, 1267–1275 (1984)
18. Pudykiewicz, J., Benoit, R., Staniforth, A.: Preliminary results from a partial LRTAP model based on an existing meteorological forecast model. Atmos.-Ocean **23**, 267–303 (1985)
19. Saito, K., Ishida, J., Aranami, K., Hara, T., Segawa, T., Narita, M., Honda, Y.: Nonhydrostatic atmospheric models and operational development at JMA. J. Meteorol. Soc. Japan **85**, 271–304 (2007)
20. Skamarock, W.C., Klemp, J.B.: A time-split nonhydrostatic atmospheric model for weather research and forecasting applications. J. Comp. Phys. **227**, 3465–3485 (2008)

21. Staniforth, A., Côté, J.: Semi-Lagrangian integration schemes for atmospheric models - A review. Mon. Wea. Rev. **119**, 2206–2223 (1991)
22. Staniforth, A., Wood, N.: Aspects of the dynamical core of a nonhydrostatic, deep-atmosphere, unified weather and climate-prediction model. J. Comp. Phys. **227**, 3445–3464 (2008)
23. Steppeler, J., Hess, R., Schattler, U., Bonaventura, L.: Review of numerical methods for nonhydrostatic weather prediction models. Met. Atm Phys. **82**, 287–301 (2003)
24. Tanguay, M., Robert, A., Laprise, R.: A semi-implicit semi-Lagrangian fully compressible regional forecast model. Mon. Wea. Rev. **118**, 1970–1980 (1990)
25. Temperton, C., Hortal, M., Simmons, A.J.: A two-time-level semi-Lagrangian global spectral model. Q. J. R. Meteorol. Soc. **127**, 111–126 (2001)
26. Yeh, K.S., Cote, J., Gravel, S., Methot, A., Patoine, A., Roch, M., Staniforth, A.: The CMC-MRB global environmental multiscale (GEM) model. Part III: Nonhydrostatic formulation. Mon. Wea. Rev. **130**, 339–356 (2002)

Workshop on Computational and Applied Statistics (CAS 2015)

Methods of Speeding up of Diameter Constrained Network Reliability Calculation

Denis A. Migov[1][(⊠)] and Sergei N. Nesterov[2]

[1] Institute of Computational Mathematics
and Mathematical Geophysics SB RAS, Novosibirsk, Russia
mdinka@rav.sscc.ru
[2] Novosibirsk State University, Novosibirsk, Russia
cepera@inbox.ru

Abstract. We consider the problem of reliability calculation of networks with diameter constraint. It is assumed that the network has unreliable communication links and perfectly reliable nodes. The diameter constrained reliability (DCNR) for such network is defined as a probability that every pair of terminals of network is connected by operational paths with a number of included edges less or equal to a given integer. The problem of computing this characteristic is known to be NP-hard, just like the problem of computing the probability of network connectivity. We propose new methods of speeding up of DCNR calculation. These methods are the edge reduction and the edge selection strategy for DCNR computing by the well-known factoring method. Also, we propose the parallel method for DCNR calculation and the method of cumulative updating of lower and upper bounds of DCNR, which allows to decide the feasibility of a network with respect to given threshold without performing exhaustive calculation of reliability. Experiments show an applicability of proposed methods.

Keywords: Network reliability · Random graph · Diameter constraint · Series-parallel transformation · Factoring method · Parallel algorithm

1 Introduction

Networks where the edges are subject to random failures are studied in the present article. As a rule networks with unreliable elements are modeled by a probabilistic graph in which an operational probability is associated with every element (a node or an edge). The most common reliability measure of such networks is the probability that all the terminal nodes in a network can keep connected together, given the reliability of each network node and edge [1]. Despite the associated problems are NP-hard, we can conduct the exact calculation of

---

Supported by Russian Foundation for Basic Research under grants 14-07-31069, 13-07-00589.

O. Gervasi et al. (Eds.): ICCSA 2015, Part II, LNCS 9156, pp. 121–133, 2015.
DOI: 10.1007/978-3-319-21407-8_9

reliability for networks with dimension of a practical interest by taking into consideration some special features of real network structures and based on modern high-speed computers[2–7].

In practice, it is often not enough to have a path between chosen pair of nodes, but it is necessary to have a path passing via a limited number of communication links [8]. For example, if there is a constraint on the time it takes to transmit the data between two nodes, T, then the number of transit nodes participating in this transmission must not exceed T/t, where t is the time it takes to process the data on each network node. Restriction on the number of hops is also critical for data transmission through a wireless sensor networks. Thus, we arrive to a different reliability measure. The diameter constrained network reliability (DCNR) is a probability that every two nodes from a given set of terminals are connected with a path of length less or equal to a given integer. By the length of a path we understand the number of edges in this path.

This reliability measure was introduced in [9,10] and studied in more detail in [11–14]. The problem of computing this measure in general is known to be NP-hard, just like the problem of computing the probability of network connectivity. Now the complexity of DCNR calculation is completely studied for different diameter values and number of terminals [14]. The factoring method is commonly used for calculation of different reliability measures. In [10] the modification of this method is proposed for DCNR calculation. The proposed modification is based on prior generation of a set of all paths with limited length for each pair of terminals. This method is much faster than basic factoring method in the case with diameter constraint. Despite this method works correctly with arbitrarily chosen pivot edge, right choice of a pivot edge may significantly decrease the total number of recursive calls of the factoring procedure. In the present study we propose the edge selection strategy for DCNR computing. Furthermore, we've obtained analog of the well-known series-parallel transformation for DCNR calculation, which removes degree 2 vertex from graph [2] and replaces two incident edges with a single edge with an equivalent reliability. These methods allow decrease significantly the calculation time of DCNR in comparison with existent methods.

The new approach in area of network reliability analysis (without diameter constraint) was introduced in [15,16]: cumulative update of lower and upper bounds of network reliability for faster feasibility decision. This method allows to decide the feasibility of a given network without performing exhaustive calculation. For example, if lower bound becomes greater than given threshold during cumulative updating, we can stop calculation and decide that network is reliable enough. If given threshold becomes greater than upper bound, we decide that network is unreliable. The method was further developed with the help of network decomposition [17]. In [18] it is shown that such approach may be useful for network topology design and optimization. In the present study we propose similar method of cumulative updating of lower and upper bounds for DCNR.

2   Definitions and Notations

We represent a network with unreliable edges and perfectly reliable nodes by an undirected graph $G = (V, E)$. For each edge e the presence probability $0 \leq r_e \leq 1$ is given. Further on we refer to this probability as edge reliability. Also the set of terminals K is given. It is supposed that the network operates well when any pair of terminals can establish a connection via only the operational edges.

An elementary event is a special realization of the graph defined by existence or absence of each edge. Thus, total number of elementary events is $2^{|E|}$. The probability of an elementary event equals the product of probabilities of existence of operational edges times the product of probabilities of absence of faulty edges.

A reliability of G with diameter constraint d is defined as the sum of elementary events in which every pair of terminals can be connected by a path of length at most d. By a path we mean a path without loops, by a length of path we mean number of edges which belong to this path. We denote defined reliability measure by $R_K^d(G)$.

3   Factoring Method for Diameter Constrained Network Reliability Calculation

The definition of DCNR gives a method for computation of this parameter. However, such direct approach leads us to an exhaustive search of all graph realizations; thus, this method is ineffective even for small scale networks. That's why the other methods are used for calculation of different reliability measures. The most common method among them is the factoring method, which can be applied to any network reliability measure, including DCNR. The factoring method divides the probability space into two sets, based on the success or failure of one graph's particular unreliable element: a node or an edge. We consider networks with unreliable edges only, so we may choose any unreliable edge. The chosen edge is called pivot. So we obtain two subgraphs, in one of them the pivot edge is absolutely reliable and in the second one the pivot edge is absolutely unreliable that is, absent. The probability of the first event is equal to the reliability of pivot edge; the probability of the second event is equal to the failure probability of pivot edge. Thereafter obtained subgraphs are subjected to the same procedure. The total probability law gives expression for the network reliability. For DCNR we have the following formula [10]:

$$R_K^d(G) = r_e R_K^d(G/e) + (1 - r_e)R_K^d(G \backslash e), \tag{1}$$

Where $G \backslash e$ is graph G without edge e, G/e is graph G with absolutely reliable edge e. Recursions continue until a graph is obtained, in which at least one pair of terminals cannot be connected by path of limited length (returns 0), or all pairs of terminals are connected by absolutely reliable paths (returns 1).

A modified factoring method for DCNR calculation was proposed by Cancela and Petingi [10]. This method is much faster than basic factoring method in the

Input: $G = (V, E)$, d, P_d, $P(e)$, $np(s,t)$, $links(p)$, $feasible(p)$, $connected(s,t)$, $connectedPairs$

Output: $R_K^d(G)$

```
 1  Function FACTO( np(s,t), links(p), feasible(p), connected(s,t),
    connectedPairs)
 2  |   RContract ← 0
 3  |   RDelete ← 0 ?
 4  |   e ← arbitrary edge : 0 < r_e < 1
 5  |   "Contract" branch:
 6  |   foreach p = (s,...,t) in P(e) such that feasible(p) = true do
 7  |   |   links(p) ← links(p) − 1
 8  |   |   if connected(s,t) = false and links(p) = 0 then
 9  |   |   |   connected(s,t) ← true
10  |   |   |   connectedPairs ← connectedPairs + 1
11  |   |   |   if connectedPairs = (k×(k−1))/2 then
12  |   |   |   |   RContract ← 1
13  |   |   |   |   GoTo "Delete" branch
14  |   |   |   end
15  |   |   end
16  |   end
17  |   RContract ←FACTO(np(s,t), links(p), feasible(p), connected(s,t),
    connectedPairs)
18  |   "Delete" branch:
19  |   foreach p = (s,...,t) in P(e) such that feasible(p) = true do
20  |   |   feasible(p) ← false
21  |   |   np(s,t) ← np(s,t) − 1
22  |   |   if np(s,t) = 0 then
23  |   |   |   RDelete ← 0
24  |   |   |   GoTo Final computations
25  |   |   end
26  |   end
27  |   RDelete ←FACTO(np(s,t), links(p), feasible(p), connected(s,t),
    connectedPairs)
28  |   Final computations:
29  |   R_K^d(G) = r_e × RContract + (1 − r_e) × RDelete
30  |   return R_K^d(G)
31  end
```

1. Pseudocode of the CPFM

diameter constrained case (1). The main feature of the modified factoring method is operating with list of paths instead of operating with graphs. In the preliminary step for any pair of terminals s, t the list $P_{st}(d)$ of all paths with limited length between s, t is generated. It automatically removes all edges which don't belong to any such path from consideration. For example, all so called "attached trees" without terminals are no longer considered. By P_d the union of $P_{st}(d)$ for all pairs of terminals is denoted. By $P(e)$ the set of paths of P_d which include link e is denoted.

Parameters of the modified factoring procedure aren't graphs. Instead we use 6 parameters, which describe the corresponding graph from the viewpoint of P_d. Below we outline these parameters and the algorithm's pseudocode. Further on we refer to this method as CPFM.

- np_{st}: the number of paths of length at most d between s and t in the graph being considered.
- $links_p$: the number of non-perfect edges (edges e such that $r(e) < 1$) in path p, for every $p \in P_d$.
- $feasible_p$: this is a flag, which has value False when the path is no longer feasible, i.e. it includes an edge which failed; and True otherwise.
- $connected_{st}$: this is a flag, which has value True when s and t are connected by a perfect path of length at most d and False otherwise.
- $connectedPairs$: this is the number of connected pairs of terminals (those between which there is a perfect path of length at most d).

4 The Acceleration of Calculation of Diameter Constrained Network Reliability

One of the main reasons, which make the calculation of diameter constrained network reliability much more complicated in comparison with other network reliability measures, is the lack of methods for decreasing of recursions quantity. For example, for k-terminal network reliability calculation we may use a lot of network reduction and decomposition methods. We can also apply a specific edge selection procedure in order to choose the optimal element for factoring procedure, and some other techniques, which are not adopted yet or cannot be applied for DCNR calculation in general.

We propose to use the edges reduction (the series transformation) and the edge selection strategy for calculation by directed CPFM. The proposed edge selection strategy allows us to choose a next pivot edge during factoring.

4.1 Edges Reduction

One of the most effective methods for acceleration of k-terminal network reliability calculation is the series-parallel transformation. Series transformation removes degree 2 vertex from graph [2] and replaces two incident edges with a single edge with an equivalent reliability. In this case the reliability of the initial graph is equal to the reliability of resulting graph multiplied by special factor, dependent on reliabilities of removed edges. Note that two parallel edges may arise as a result of this transformation. The parallel transformation replaces these edges with a single edge with an equivalent reliability. The series-parallel transformations can be applied recursively until the network does not have any degree 2 vertices or multi-edges. Networks which can be reduced by series-parallel transformation to one edge are said to be series-parallel networks. Thus, one of our main objectives is to find out whether it is possible to apply analogous transformation for networks with diameter constraint.

In a case of diameter constraint series transformation is possible only if degree 2 node isn't a terminal (see Fig. 1). Any path which connects any two terminals and passes through the edge (s, v) must obligatorily pass through the edge (v, t), because v is not a terminal. Therefore, $P(s, v) = P(v, t)$ and edges (s, v) and (v, t) can be replaced with the edge (s, t) with reliability $r_{(s,v)} \times r_{(v,t)}$.

Fig. 1. Reduction of edges incident to a degree 2 node

Parallel transformation in this case is impossible. Let's assume that nodes s and t were connected by the edge f before union of edges (s, v) and (v, t) into edge $g = (s, v)$. Then equality $P(g) = P(f)$ isn't true in general because a larger number of paths with limited length between various pairs of terminals can pass through f, then through g.

We note that if the list of paths $P(e)$ is formed after the proposed series transformation, the newly formed edge increases the length of any path which is passed through this edge. If the transformation is accomplished after the list was formed, then we may leave one of two coincident sets in this list: $P(s, v)$ or $P(v, t)$.

In general we may perform the same transformation with any subgraph without terminals instead of single node (see Fig. 2). Then any path which is connecting any two terminals and passing through the edge (s, v) is obligatorily passing through the edge (u, t). Therefore, the equality $P(s, v) = P(u, t)$ remains correct and it is possible to accomplish the factoring of these two edges at once.

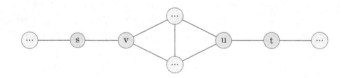

Fig. 2. Edges reduction in general case

Thus, on the preliminary step it is possible to reduce the chains with random length to one edge. It leads to significant decreasing of the recursions quantity and complexity of computations.

4.2   Pivot Edge Selection Strategy for the Factoring Method

The factoring method works correctly with arbitrarily chosen pivot edge. However, right choice of a pivot edge may significantly decrease the total number of recursive calls of the factoring procedure. The problem of pivot edge optimal choice is still unsolved [10].

We propose to regularize edges by $|P(e)|$ maximum criterion. As it was mentioned above in section 3, before DCNR calculation by CPFM for each edge e we should make $P(e)$ — the list of paths which belongs to P_d and include this edge. Thus, we use an edge e for factoring before an edge f if $|P(e)| \geq |P(f)|$. For the majority of graphs the proposed way of edge selection strategy shows speeding up of DCNR calculation in comparison with random choice of pivot edge. Nevertheless, this selection strategy is not the optimal one. For example, let us consider a graph in Fig. 3 with $K = \{1, 2, 3, 4\}$ and $d = 4$. The recursions quantity without any special strategy for edge selection is 602. Proposed edge selection strategy allows to reduce recursions quantity to 216. However, the exhaustive search of all possible variants gives us the optimal edge selection strategy, which further reduces recursion quantity to 128.

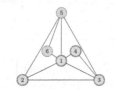

Fig. 3. Tested graph

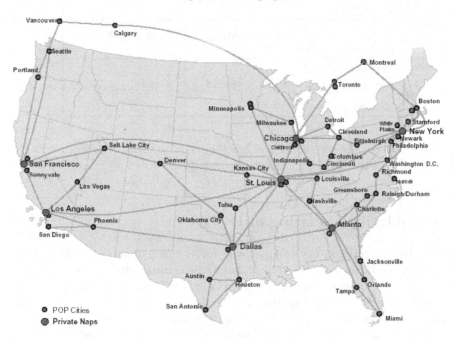

Fig. 4. Tested network

5   Case Studies

In order to find the accelerating effect of the proposed methods numerical experiments were performed on topology of Intellinet network (Fig. 4). Private nodes were chosen as terminals of the network. It was assumed that for each edge its reliability is equal to 0.7. The computations were done for different diameter values: 4, 5, 8, 10. The obtained DCNR values were 0.2992265977, 0.39167458235, 0.604070784917, and 0.650312593213 respectively.

The CPFM was chosen as the basic method, which has been improved by three ways: edge selection (ES) strategy, edges reduction, edge selection strategy, and edge reduction. Tables 1 and 2 show computational time and recursions count for each method. Experiments were performed on Intel Xeon E31240 3.3 GHz, 8 cores.

Table 1. Results of the experiments

Algorithm	$d=4$		$d=5$	
	Time	Recursions	Time	Recursions
Factoring method	87m	11721542454	>24 hours	–
Factoring method with ES strategy	7m 6s	1058428568	>24 hours	–
Factoring method with edges reduction	0.193s	401010	13.045s	27323118
Factoring method with ES strategy and edges reduction	0.041s	83648	1.826s	4275506

As we can see, the edge reduction has the most accelerating effect for DCNR calculation, and the pivot edge selection strategy allows to additionally speed it up.

6   Cumulative Updating of DCNR

Authors in [15] propose the original way for decision making about if a network is reliable in terms of network probabilistic connectivity (without diameter constraint). The main idea of the proposed method is to obtain exactly if a network is feasible without calculating exact value of network reliability. For this purpose some threshold R_0 is defined and the lower bound RL and the upper bound RU of $R(G)$ are calculated cumulatively in such a way that on l-th iteration $RL_i \geq RL_{l-1}$ and $RU_l \leq RU_{l-1}$. RL_0 and RU_0 are initialized as zero and one, respectively. Decision process stops when RL_l exceeds R_0 or R_0 exceeds RU_l. In the first case the network is reliable, and in the second one the network is unreliable.

Table 2. Results of the experiments

Algorithm	$d=8$		$d=10$	
	Time	Recursions	Time	Recursions
Factoring method	>24 hours	–	>24 hours	–
Factoring method with BM strategy	>24 hours	–	>24 hours	–
Factoring method with edges reduction	7m 10.179s	643915162	14m 49.793s	974464256
Factoring method with ES strategy and edges reduction	52.978s	101076210	2m 55.716s	253120356

For obtaining RL and RU exact values of reliabilities of some "subnetworks" must be calculated along with probabilities of their realizations. Suppose that during factoring procedure we obtain L final graphs: $G_1, G_2, \ldots, G_L$, for which the reliability can be easily calculated. Let P_l for $1 \leq l \leq L$ be the probability to have G_l. Then, $\sum_{l=1}^{L} P_l = 1$, and the reliability of the initial graph G can be formulated as

$$R(G) = \sum_{l=1}^{L} P_l R(G_l). \tag{2}$$

For any $1 \leq k \leq L$ we have the following inequality [15]:

$$\sum_{l=1}^{k} P_l R(G_l) \leq R(G) \leq 1 - \sum_{l=1}^{k} P_l (1 - R(G_l)). \tag{3}$$

This inequality gives the algorithm for cumulatively updating of the lower and upper bounds of $R(G)$. Whenever the ATR of G_l for $1 \leq l \leq L$ is resolved, we can update RL_l, and RU_l as

$$RL_l = RL_{l-1} + P_l R(G_l)$$
$$RU_l = RU_{l-1} - P_l (1 - R(G_l)). \tag{4}$$

As l increases, RL_l and RU_l approach $G(R)$. Once either RL_l or RU_l reaches R_0, the proposed algorithm concludes the feasibility of G: if RL_l reaches R_0, G is feasible; if RU_l passes R_0, G is infeasible.

We have applied this approach for DCNR. For this purpose we need to add parameter P_l, which is initially equal to 1. This parameter is being updated after line 5 of Algorithm 1: $P_l = P_l * r_e$ and after line 18: $P_l = P_l * (1 - r_e)$. Instead lines 12 and 23 we should insert lines with updating of values RU and RL resprctivly: $RL = RL + P_l$ and $RU = RU - P_l$.

Diagram (fig. 5) shows how RL, RU are changing during proposed procedure for topology of Intellinet network (fig. 4) for the diameter value 12. Edge reliability is equal 0.7 for each edge. R_0 value was equal to exact value of DCNR. Calculation time was about 12 minutes.

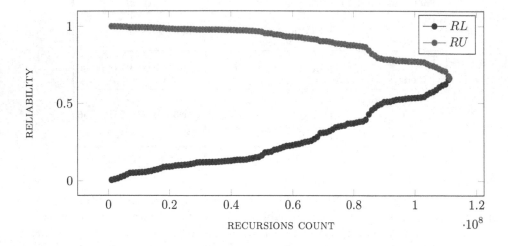

Fig. 5. D = 12, P = 0.7

7   Parallel Algorithm for DCNR Calculation

In this section we introduce an algorithm with use of MPI for DCNR calculation for supercomputers with distributed memory. The basic idea is based on CPFM method: during the factoring procedure one part of work (for example, "contracting" of an edge) stays on paternal process while another one is being sent to some idle process.

It was decided to send to a child process all required data for calculation with a probability of corresponding event. That allows us to store only a part of result reliability on every process; it should be just summarized at the end of factoring algorithm. A probability of every partial "sub-network" is a multiplicative factor which initialized by 1 for initial network. It changes during factoring procedure: the probability of $G \backslash e$ is the probability of G multiplied by r_e, and for G/e – by $(1 - r_e)$.

As in our previous work on parallel computing of network reliability [7] (without diameter constraint), we chose "Master-Slave" parallel programming model. The main idea of such approach is that one *master* process controls all other *guided* processes.

In [7] one important parameter of the algorithm was studied — the lower limit of a dimension of graph that could be assigned to another process. Here we study the analogue of this parameter in case of diameter constraint. In this case it turns to a upper limit of considered edges amount. After the amount of considered edges exceeds this limit, current process stops sending data to the master process and executes all procedures without any help. We define this parameter as N_{Edges}.

For numerical experiments we choose a grid 5x5 topology (fig. 6), it contains 25 vertexes and 4 edges. Number of terminals was equal to 5, diameter was equal to 9.

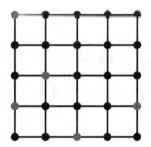

Fig. 6. Tested graph for the parallel algorithm

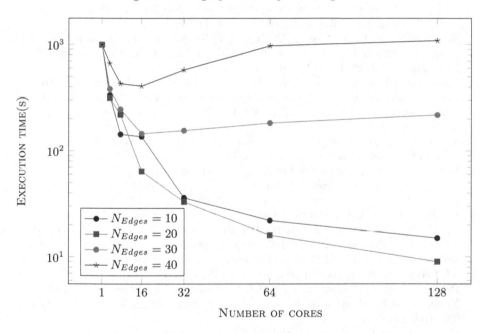

Fig. 7. Scalability of the parallel algorithm

Fig. 7 shows the scalability of the proposed algorithm for different values of N_{Edges}. As we can see the optimal value is between 10 and 20.

The results show that the CPFM works well in parallel implementation, as well as the factoring method in parallel implementation for all-terminal network reliability [7].

8   Conclusion

The analogue of the well-known series-parallel transformation for DCNR calculation is proposed in the present study. The problem of the optimal choice of pivot edge for DCNR calculation by CPFM is studied as well. As a result the edge selection strategy is proposed. As it has been shown, these methods allow to significantly reduce the number of recursive calls in CPFM and complexity of DCNR computation. However, the problem of the optimal choice of pivot edge is still unsolved. Also, the parallel implementation of the CPFM and the method of cumulative updating of lower and upper bounds of DCNR were proposed. Cumulative updating allows to decide the feasibility of a network with respect to given threshold without performing exhaustive calculation of reliability. The proposed methods may be useful for network topology design and optimization.

References

1. Ball, M.O.: Computational complexity of network reliability analysis: An overview. IEEE Transactions on Reliability **35**, 230–239 (1986)
2. Shooman, A.M., Kershenbaum, A.: Exact graph-reduction algorithms for network reliability analysis. In: IEEE Global Telecommunications Conference GLOBECOM 1991, pp. 1412–1420. IEEEP Press, New York (1991)
3. Rodionova, O.K., Rodionov, A.S., Choo, H.: Network probabilistic connectivity: exact calculation with use of chains. In: Bubak, M., van Albada, G.D., Sloot, P.M.A., Dongarra, J. (eds.) ICCSA 2004. LNCS, vol. 3036, pp. 565–568. Springer, Heidelberg (2004)
4. Migov, D.A., Rodionova, O.K., Rodionov, A.S., Choo, H.: Network probabilistic connectivity: using node cuts. In: Zhou, X., et al. (eds.) EUC Workshops 2006. LNCS, vol. 4097, pp. 702–709. Springer, Heidelberg (2006)
5. Xiao, Y.: An Efficient Network Reliability Computation Method Based on Isomorphism Judgment. Journal of Networks **9**(8), 2037–2044 (2014)
6. Forghani-elahabad, M., Mahdavi-Amiri, N.: A New Efficient Approach to Search for All Multi-State Minimal Cuts. IEEE Trans. on Reliability **63**(1), 154–166 (2014)
7. Migov, D.A., Rodionov, A.S.: Parallel implementation of the factoring method for network reliability calculation. In: Murgante, B., et al. (eds.) ICCSA 2014, Part VI. LNCS, vol. 8584, pp. 654–664. Springer, Heidelberg (2014)
8. Pandurangan, G., Raghavan, P., Upfal, E.: Building Low-Diameter Peer-to-Peer Networks. IEEE Journal on Selected Areas in Communications (JSAC) **21**(6), 995–1002 (2003)
9. Petingi, L., Rodriguez, J.: Reliability of Networks with Delay Constraints. Congressus Numerantium **152**, 117–123 (2001)
10. Cancela, H., Petingi, L.: Diameter constrained network reliability: exact evaluation by factorization and bounds. In: Int. Conf. on Industrial Logistics, Okinawa, Japan, pp. 359–356 (2001)
11. Cancela, H., Robledo, F., Rubino, G., Sartor, P.: Monte Carlo Estimation of Diameter Constrained Network Reliability Conditioned by Pathsets and Cutsets. Computer Communications **36**(6), 611–620 (2013)

12. Jin, R., Liu, L., Ding, B., Wang, H.: Distance-Constraint Reachability Computation in Uncertain Graphs. In: VLDB Endowment, vol. 4, pp. 551–562 (2011)
13. Migov, D.A.: Computing Diameter Constrained Reliability of a Network with Junction Points. Automation and Remote Control **72**(7), 1415–1419 (2011)
14. Canale, E., Cancela, H., Robledo, F., Rubino, G., Sartor, P.: On Computing the 2-Diameter-Constrained K-Reliability of Networks. International Transactions in Operational Research **20**(1), 49–58 (2013)
15. Won, J.-M., Karray, F.: Cumulative Update of All-Terminal Reliability for Faster Feasibility Decision. IEEE Trans. on Reliability **59**(3), 551–562 (2010)
16. Won, J.-M., Karray, F.: A Greedy Algorithm for Faster Feasibility Evaluation of All-Terminal-Reliable Networks. IEEE Trans. on Systems, Man, and Cybernetics, Part B Cybernetics **41**(6), 1600–1611 (2011)
17. Rodionov, A.S., Migov, D.A., Rodionova, O.K.: Improvements in the Efficiency of Cumulative Updating of All-Terminal Network Reliability. IEEE Trans. on Reliability **61**(2), 460–465 (2012)
18. Nechunaeva, K.A., Migov, D.A.: Speeding up of genetic algorithm for network topology optimization with use of cumulative updating of network reliability. In: ACM IMCOM 2015, article 42, ACM New York (2015)

Parallel Super-Resolution Reconstruction Based on Neighbor Embedding Technique

Marwa Moustafa[2(✉)], Hala M. Ebied[1], Ashraf Helmy[2], Taymoor M. Nazamy[1], and Mohamed F. Tolba[1]

[1] Faculty of Computer and Information Sciences, Ain Shams University, Cairo, Egypt
halam@fcis.asu.edu.eg, fahmytolba@cis.asu.edu.eg
[2] Data Reception, Analysis and Receiving Station Affairs, National Authority for Remote, Sensing and Space Science, Cairo, Egypt
{marwa,akhelmy}@narss.sci.eg

Abstract. Super Resolution (SR) is a technique to recover a high-resolution (HR) image from different noisy low resolution (LR) images. The missing high-frequency components in LR images should be restored correctly in HR image. Because of the extensive size of satellite images, the utilize to parallel algorithms can accomplish results more quickly with accurate results. This paper proposes an accelerated parallel implementation for an example based super-resolution algorithm, Neighbor Embedding (NE), using GPU. The NE trains the dictionary with patches obtained from a single image in the training phase. Euclidean distances are used to obtain the optimal weights that will be used in the construction of high-resolution images. Compute Device Unified Architecture (CUDA) by NVidia's has been used to implement the proposed parallel NE. Different experiments have been carried out on a synthetic test image and satellite test image. The proposed GPU implementation of the NE was benchmarked against the serial implementation. The experimental results show that the speed of the implementation depends on the image size. The speed of the GPU implementation compared to the serial one using CPU ranged from 20× for small images to more than 30× for large image size.

Keywords: Super-Resolution · Neighbor Embedding · CUDA · GPU

1 Introduction

High-resolution images are highly desired and often required in modern remote sensing applications such as marine and ocean management, bio-resources, environmental management, coastal zone study, water resources and urban planning. Super resolution is an economical technique used to provide the required high resolution images for scientific and commercial use. The problem of constructing a high resolution image from single or multiple low resolution images is well known as the super resolution approach.

Different modern imaging sensors have different limitations such as; the capacity of the transmission channel, environmental conditions and the length. Low resolution

© Springer International Publishing Switzerland 2015
O. Gervasi et al. (Eds.): ICCSA 2015, Part II, LNCS 9156, pp. 134–143, 2015.
DOI: 10.1007/978-3-319-21407-8_10

(LR) images are acquired due these limitations. Different observation models are used to map the relation between the HR and LR image. Super resolution could be broken up into two tasks. The first task is estimating the pixel shift between the different LR images while the second task is to recover the high resolution image from these LR images. SR is an inverse problem that is well known as ill-conditioned. Various optimization techniques are used to find an HR image that is a close approximation of the original scene.

Super resolution was proposed by Tsai and Huang [13] in their pioneer work. A super resolution reconstruction method has been introduced and has been applied to Landsat image. Super resolution techniques are categorized into a frequency and a spatial domain. In the frequency domain, the SR technique has been extended by using Fourier transform, discrete cosine transform (DCT) and wavelet transform based. The majority of recently developed techniques have been done within the spatial domain area. Spatial domain algorithms include iterative back projections, projected onto convex sets and probabilistic learning methods such as Maximum Likelihood, Maximum a Posteriori, Total Variance. etc

Learning based method is one of the promising solutions to produce a high resolution image. Prior knowledge defines the relationship between LR and HR patches and the relation between the LR and HR patches are defined in the training phase in order to construct a model that will be used to produce the required HR image.

This paper focuses on generating a high-resolution satellite image from a single low-resolution image by using the Neighbor Embedding algorithm. The paper studies the behavior of both the serial and parallel implementation of Neighbor Embedding towards the state-of-art of interpolation methods (bicubic algorithm). Landsat-8 scene image is used in the experiments and both serial and CUDA GPU implementations are evaluated and compared.

The paper is organized as follows. In section 2, we discuss the related work. In section 3, the Neighbor Embedding is illustrated. Section 4 describes the proposed parallel implementation of NE algorithm. Experimental results are shown and discussed in section 5. Finally, section 6 concludes the paper.

2   Related Work

Image enhancement is an active research area where different smoothing and interpolation techniques have been introduced. Limitations of these methods include blurred edge or artifacts. Recently, learning based method had been used to overcome these limitations. This section reviews the recent researchers work in this area.

In [1], a novel approach for single image super resolution (SR) had been proposed. The proposed method is based on a learning dictionary and it is of central importance in the image SR application. Kernel principal component analysis (KPCA) is adopted as high order statistics to learn the dictionary. Kernel ridge regression is used to map between LR image patches and the HR coding coefficients. The proposed solution reduced the time complexity of learning and testing. Experimental results show that the proposed method is effective and efficient in comparison with state-of-art algorithms.

In [3], a computationally efficient single image SR method has been introduced. The Multiple linear mappings (MLM) technique is used to transform the LR features subspace into HR sub-spaces directly. The high number of non-linear feature space of LR images is clustered into linear sub spaces to learn the multiple LR sub-directory then HR sub-dictionaries is learned based on the assumption that the LR–HR features share the same representation coefficients. Fast non-local means algorithm is applied to construct a simple, effective similarity based regularization term for SR enhancement. Experimental results indicate that the proposed approach is both quantitatively and qualitatively and outperforms other application oriented SR methods while maintaining relatively low time and space complexity [3].

Zhou and Liao [4] proposed a visual perceptual PCA (VP-PCA) technique by adopting structural similarity (SSIM) as the object function. Training data had been clusters VP-PCA had been applied to each cluster to calculate the coefficients. Feature similarity (FSIM) is used to combine traditional SR results and SR results using VP-PCA to form the final results. Experimental results showed the superiority of the proposed algorithm over the state-of-the-art methods in both quantitative and visual comparisons. By the increasing interest in super resolution reconstruction via sparse representation, Juan et al. [9] proposed a new method for image super resolution using a local sparse model on image patches. A new dictionary training formulation had been adopted in which the optimization problem had been solved using the stochastic gradient algorithm. The Experimental results had been compared with the recently proposed joint dictionary learning method for image super resolution, the proposed method showed visual, PSNR and SSIM improvements. [5], in order to enhance the resolution and overcome undesired artifacts, the sparse based system dictionary had been introduced based on non-local methods. Local similarity had been used on small spatial patches of the image to eliminate the artifacts. Quantitative results on several test datasets are promising.

Neighbors embedding (NE) technology is efficient and robust especially in a single image super resolution. In [6], a novel Dual-Geometric Neighbor Embedding (DGNE) approach has been adopted. Exploring the geometric structure of both the feature domain and spatial domain has been done. Multi-view features of image patches and their spatial neighbors are jointly sparsely coded, via a tensor-simultaneous orthogonal matching pursuit algorithm and different experiments had been done on some benchmark images. The experimental results indicated that DGNE is comparable to some state-of-the-art SISR approaches without additional regularizers.

Chang et al. [10] used the YIQ model in order to learn only the Y component while I and Q channels copied from LR to the reconstructed HR image. The image is converted from the RGB model of a color image to the YIQ model. In [2], Chan et al. also learned only the Y component and the interpolated I and Q channels are just added to the luminance to form the whole image in the YIQ space. However, they represented the RGB components of a color image to be the three imaginary parts of a quaternion, and directly used neighbor embedding to learn the target HR image with the quaternion matrix model [2].

Ramakanth and Babu [7] proposed an SR reconstruction technique where a single image had been used to train a dictionary. Different state-of-art methods had been used in teaching a pair of low and high-resolution dictionaries for multiple images. These trained dictionaries are used to replace the patch of LR images with appropriate matches of HR. In their work, approximate nearest neighbor fields (ANNF) had been used to perform super-resolution (SR). Experiments had been done to compare the proposed algorithm against various state-of-the-art algorithms, and the results showed that the proposed SR reconstruction technique is able to achieve better and faster recovery without any training.

In [8], a novel single image super resolution (SR) method is introduced. The proposed method had adopted the low-rank matrix recovery (LRMR) and neighbor embedding (NE) where the underlying structures of sub-spaces spanned by similar patches had been explored using LRMR. In the training phase, training patches are divided into groups before the LRMR technique is used to learn the latent structure of each group. NE algorithm is performed on the learned low-rank components of HR and LR patches to produce SR. Experimental results show that the proposed approach outperformed state-of-art algorithms in both quantitatively and perceptually.

3   Super Resolution Method

In this section, the example based algorithm SR reconstruction method is presented inspired by Chang et al. [10]. The presented method consists of two phases, 1) the Construction of the training Dictionary phase and 2) SR phase. In the training phase, learning dictionary is constructed by extracting the patch pairs from high and low resolution images. In the SR phase, the HR image is reconstructed by searching the stored patches in the dictionary for a best match for input images.

In order to construct a high-resolution image, NE uses manifold learning called locally linear embedding. The performance of the NE algorithm is affected by both neighbor size and feature representation. Let X_s, Y_s, X_t, Y_t represent set of single or multiple low resolution image in training, corresponding high resolution image, input image for testing and corresponding high resolution image respectively. $X_s = \{x_s^i\}_{i=1}^m, Y_s = \{y_s^i\}_{i=1}^m, X_t = \{x_t^j\}_{j=1}^m, Y_t = \{y_t^j\}_{j=1}^m$. Where $x_s^i, y_s^i, x_t^j, y_t^j$ represent training low resolution patches, training high resolution patches, low resolution images and estimated high resolution patches respectively.

Neighbor embedding for image super-resolution can be stated in three steps as shown in Fig. 1. The first step divides the input LR image into m patches and then calculates k-nearest neighbors to each patch. The second step computes the optimal weights w_{ij} by minimizing reconstruction errors with the k-nearest neighbors N_t^q using eq. 1. The last step computes an HR patch with optimal weights using eq. 2.

$$\varepsilon^i = \min \left\| x_t^i - \sum_{x_s^i \in N_t^i} w_{ij} x_i^i \right\|^2 \tag{1}$$

Where $\sum_{x_s^i \in N_t^i} w_{ij} = 1$ and $w_{ij} = 0$ if $x_i^i \notin N_t^i$

$$y_t^i = \sum_{x_s^i \in N_t^i} w_{ij} y_s^i \tag{2}$$

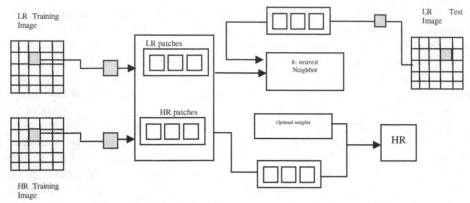

Fig. 1. Neighbor embedding super-resolution

4   GPU Implementation

Compute Unified Device Architecture (CUDA) is a framework for programming computing devices with a large amount of cores executing data-parallel codes. CUDA is implemented by NVIDIA to enable a dramatic increase in computing performance. C/C++ CUDA extensions are adopted for the proposed parallel solution. A single instruction multiple data (SIMD) model is used by the CUDA framework, in which a large number of paralled threads executed are used to increase the performance boosts. CUDA is organized into grids that consist of several blocks and each block has a fixed number of threads. Grids and clocks share the same dimension that could be one, two and three dimensions but two dimensions is commonly used in image processing.

The intensive computation part in the EN SR algorithm is the calculated distance. In this paper, two CUDA implementations for pairwise distances are presented. The first implementation is simple, while the second implementation used the shared memory. The data matrix is n×m of n number of patches and m features and both implementations calculate the n×n matrix of pairwise Euclidean distances. In the first implementation, the computation is divided into a number of thread blocks; each block is two dimensional of 16x16 and the grid is configured into two-dimensional. The first dimensional is calculated using (number of features /16) while the second is calculated using (number of patches/16).

In shared memory implementation, the GPU device has registers and a local memory for each thread, shared memory for each block and a global memory for all blocks in a grid. The shared memory for a block of threads is fast, yet it is limited in size. The needed data is loaded from the global memory into the device shared memory. Each thread loads the data needed using the indexed blockIdx.x, blockIdx.y, blockIdx.x, blockIdx.y. . __syncthreads() is used as a barrier, which enforces the threads to wait until each thread reaches the barrier. Fig. 2 illustrates the high performance of the CUDA code. Assume we want to calculate the entry (i,j), row 16x n in matrix A has to be copied to the shared memory. All the threads assigned to the entries (i, j+1), (i, j+2), ...(i, j+15) are responsible for copying their corresponding entries to the fast shared memory.

```
__global__ void gpudist2(float*out, float*in, int
n,int m){
    __shared__ float Ys[16][16];
    __shared__ float Xs[16][16];
    int bx = blockIdx.x, by = blockIdx.y;
    int tx = threadIdx.x, ty = threadIdx.y;
    int yBegin = by * 16* m;
    int xBegin = bx * 16* m;
    int yEnd  = yBegin + m - 1, y, x, k, o;
    float tmpx, ss = 0;
 for(y=yBegin,x=xBegin;y<=yEnd;y+=16,x+=16){
        Ys[ty][tx] = in[y + ty*m + tx];
        Xs[tx][ty] = in[x + ty*m + tx];
        __syncthreads();
     for(k=0;k<16;k++){
            tmpx = Ys[ty][k] - Xs[k][tx];
            ss+= tmpx * tmpx;
     }
        __syncthreads();
}
    oo=by*16*n+ty*n+bx*16 + tx;
 out[oo] = sqrtf(ss);
}
```

Fig. 2. Parallel GPU EN SR Implementation

5   Experimental Results

In this section, our serial and GPU implementations are evaluated on both synthetic test images and satellite test images. The peppers image from USC-SIPI image database [15] is used as the syntactic image. The real satellite image is a 1024x1024 window from Landsat-8 image [16] with a band combination of (4, 3, and 2) was used to obtain a natural color image. The satellite image scene is located in Egypt at latitude of 30°18'23.08"N and longitude of 31°56'32.50"E which acquired on the 25/10/2014. The adapted window is chosen with a diversity of artificial features such as urban, roads and agriculture areas.

In both cases, serial and GPU implementations produce HR images with fine details. The hardware configuration used to test our GPU parallel code is NVIDIA NVS 5200M. The GF117 architecture is based on the optimized Fermi architecture of the GF108 chip and offers 96 cores, 16 TMUs and 4 ROPs. Each core is clocked twice as fast as the rest of the graphics chip. CUDA runtime version 5.5 has been used in the implementation.

Different experiments were performed to compare the parallel implementation execution time of the EN SR reconstruction method on GPU and CPU sequential implementations. Fig. 3 shows the speedup in execution time of the SR algorithm using different number of image patches, the nearest neighbors K is set to be 5. In this experiment, the LR patch size is set to be 3, the overlapping in LR image is 2 and the weight is set to be 4.

The speed of the GPU implementation compared to the serial one using CPU range from 3x for small images to 14x for large image patches using the naïve implementation and are ranged from 20x for small images to 30x for large image patches using the high performance implementation. Fig. 4 and 5 show the resulted HR images using bicubic, serial EN SR and parallel SN SR methods. Fig. 3 demonstrates the shared memory that was used in the EN parallel implementation outperforms the simple GPU and sequential implementation. The EN based super resolution reconstruction algorithm is considered a computationally economic choice especially in satellite image cases.

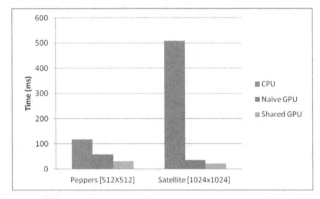

Fig. 3. Execution times (in seconds) of the EN SR algorithm on CPU, GPU for different image sizes

Table 1 illustrates the execution time for both sequential CPU implementation and GPU parallel implementation. CUDA simple implementation execution time is speeded up by approximately 14x for larger image sizes. In the shared memory implementation, the acceleration achieves about 24x faster than the sequential implementation. The EN method can be used in real-time applications with its GPU parallel implementation.

Fig. 4. Peppers HR image constructed using a) bicubic interpolation, b) serial implementation and c) GPU implementation respectively

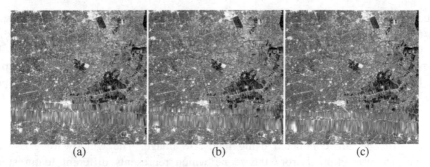

(a)   (b)   (c)

Fig. 5. Satellite window HR image constructed using a) bicubic interpolation, b) serial implementation and c) GPU implementation respectively

Table 1. Execution time of EN SR using CPU-sequential implementation, Simple-GPUand Shared-GPU implementations for different image sizes

Image	Time (ms)			Speed up	
	CPU	**Simple GPU**	**Shared GPU**	**Simple GPU**	**Shared GPU**
Peppers [512X512]	117.9991	58.292	31.02	2.0244	3.804
Satellite [1024x1024]	507.582	36.117	22.014	14.0534	23.0574

Peak signal-to-noise ratio [14] is commonly used to measure the peak error. The last experiment was adopted to compare the accuracy of the sequential implementation and the parallel implantation of the SR method with the classical method Bicubic using the PSNR. Table 2 shows the PSNR for the super-resolution images that are produced by the EN SR algorithm and the Bicubic classic method. The difference between PNSR values obtained from the parallel implementation and the serial Implementation is hardly noticeable.

Table 2. The accuracy of the EN SR Algorithm and bicubic method using PSNR for Peppers and satellite Image

Patch Size	PNSR		
	Bicubic	*CPU EN*	*GPU EN*
Peppers	33.217	34.87	33.05
Satellite	33.79	33.81	33.25

6   Conclusions

In this paper we presented a parallel implementation to EN based on NVIDIA's CUDA parallel computing architecture. The reconstruction method presented had successfully reconstructed a 3x high-resolution image from a low-resolution image. A comparative study has been done on serial, simple and shared memory implementation to the super-resolution reconstruction technique.

Different experiments are conducted using the synthetic image and both experimental results show that our shared memory GPU EN implementation speeds up by 4x over CPU serial implementation. However, the naïve implementation speeds up 2x. The accuracy of the HR image using serial EN SR outperforms the GPU parallel implementation, but the state-of-art method (bicubic interpolation) is considered better than the images obtained using GPU implementation.

Experiments are conducted using landsat-8 image. The window of 1024×1024 pixels size was clipped from the scene which represents different features; both simple and shared memory implementations were proposed. Experimental results show that our shared memory EN algorithm implementation speeds up by 24x over CPU serial implementation and the simple implementation speeds up 14x over the CPU serial implementation. The accuracy of the HR images reconstructed using EN SR parallel technique is better than the images obtained from the bicubic interpolation method.

References

1. Zhou, F., Yuan, T., Yang, W., Liao, Q.: Single-Image Super-Resolution Based on Compact KPCA Coding and Kernel Regression. IEEE Signal Processing Letters **22**(3), 336–340 (2015)
2. Chan, T.M., Zhang, J.P., Huang, H.: Neighbor embedding based super-resolution algorithm through edge detection and feature selection. Letters In Pattern Recognition, 494–502 (2009)
3. Zhang, K., Tao, D., Gao, X., Li, X., Xiong, Z.: Learning Multiple Linear Mappings for Efficient Single Image Super-Resolution. Transactions on IEEE Image Processing **24**(3), 846–861 (2015)
4. Zhou, F., Liao, Q.: Single-frame image super-resolution inspired by perceptual criteria. IET Image Processing **9**(1), 1–11 (2015)
5. Bhosale, G.G., Deshmukh, A.S., Medasani, S.S., Dhuli, R.: Image super-resolution using dictionaries and self-similarity. In: Proceeding of International Conference on Signal Processing and Communications (SPCOM), pp.1–6 (2014)
6. Yang, S., Wang, Z., Zhang, L., Wang, M.: Dual-Geometric Neighbor Embedding for Image Super Resolution with Sparse Tensor. IEEE Transactions on Image Processing **23**(7), 2793–2803 (2014)
7. Ramakanth, S.A., Babu, R.V.: Super resolution using a single image dictionary. In: IEEE International Conference on Electronics, Computing and Communication Technologies (IEEE CONECCT), pp. 1–6 (2014)
8. Chen, X., Qi, C.: Low-Rank Neighbor Embedding for Single Image Super-Resolution. IEEE Signal Processing Letters **21**(1), 79–82 (2014)
9. Juan, L., Jin, W., Shen, Y., Jin, L.: Dictionary learning for image super-resolution. In: 33rd Chinese Control Conference (CCC), pp. 7195–7199 (2014)
10. Chang, H., Yeung, D., Xiong, Y.: Super-resolution through neighbor embedding. In: Proceedings of the IEEE Computer Society Conference of Computer Vision and Pattern Recognition, vol. 1, pp. I (2004)

11. Bevilacqua, M., Roumy, A., Guillemot, C., Morel, M.-L.A.: Neighbor embedding based single image super-resolution using Semi-Nonnegative Matrix Factorization. In: Proceedings of the IEEE Acoustics, Speech and Signal Processing (ICASSP), pp.1289–1292 (2012)
12. Yu, W., Chen, S.: An Improved Neighbor Embedding Method to Super-resolution Reconstruction of a Single Image. Procedia Engineering **15**, 2418–2422 (2011)
13. Tsai, R.Y., Huang, T.S.: Multiframe image restoration and registration. In: Tsai, R.Y., Huang, T.S. (eds.) Advances in Computer Vision and Image Processing, vol. 1, pp. 317–339. JAI Press Inc. (1984)
14. Wang, Z., Bovik, A.C., Sheikh, H.R., Simoncelli, E.P.: Image quality assessment: From error visibility to structural similarity. IEEE Transactions on Image Processing **13**(4), 600–612 (2004)
15. USC-SIPI Image Database. http://sipi.usc.edu/database/
16. Vermote, E.F., Vogelmann, J., Wulder, M.A., Wynne, R.: Free access to Landsat imagery. Science **320**(5879), 10–11 (2008)

Performance Evaluation of Two Software for Analysis Through ROC Curves: Comp2ROC vs SPSS

Sara Coelho[1] and Ana C. Braga[2(✉)]

[1] MSc Student of Department of Informatics, University of Minho,
4710-057 Braga, Portugal
saraaccoelho@gmail.com
[2] ALGORITMI Centre, University of Minho,
4710-057 Braga, Portugal
acb@dps.uminho.pt

Abstract. Receiver Operating Characteristic (ROC) analysis is a powerful tool to evaluate, view and compare diagnostic tests by a discriminating way. Currently there several ROC analysis tools, but none is known, by containing all the features necessary for a full investigation.

In this work we determined the performance of Comp2ROC package (R package) and its functionality by performing comparative diagnostic systems based on empirical ROC curves for unpaired and paired samples. We compare this package with the functionality of IBM® SPSS® Statistics to analyse ROC curve, in order to determine whether it has better ability to execute both the level of performance as a result.

For illustrative purpose we use a random sample of clinical indexes used in neonatal intensive care to evaluate the risk of death for newborns with very low birth weight (VLBW) (< 1500g) and/or gestational age < 32 weeks.

Keywords: Comp2ROC · ROC curve · SPSS® (Statistical Package for Social Sciences)

1 Introduction

The ROC curve originally appeared derivative from difficulty to detect radar due to transmission quality, that is, the difficulty to distinguish a signal from noise [1]. After this appearance, the ROC curves were increasingly employed in various areas of knowledge, especially in experimental psychology [2], medicine, economy [3] and weather [4].

Usually, to evaluate performance of different diagnostic tests in medicine is used ROC analysis in particular through the index of area under curve (AUC) that provides the ability to compare two or more curves using a tangential approach to non-parametric statistics of Wilcoxon-Mann-Whitney [5].

The primary objective of this study is to check the performance of the package Comp2ROC with respect to its functionality, and compare this with of other

© Springer International Publishing Switzerland 2015
O. Gervasi et al. (Eds.): ICCSA 2015, Part II, LNCS 9156, pp. 144–156, 2015.
DOI: 10.1007/978-3-319-21407-8_11

available software that computes ROC analysis as IBM® SPSS® Statistics. The measures of performance evaluation were oriented in the statistical analysis, so as to have a way to compare the two software, verifying their results in terms of the ROC analysis, of interface user-friendly and of data structure.

1.1 ROC Curve and Performance Measures

Historically, the receiver operating characteristic (ROC) curve was introduced in World War II military radar operations as a means to characterize the ability of the operators to correctly identify friendly or hostile aircraft based on a radar signal. The loss incurred if a hostile aircraft is deemed friendly by mistake could be catastrophic, but at the same time military aircraft could not be sent to intercept an overwhelming number of benign vessels [9]. The ROC curve was devised as a graphical means to explore the trade-off's between these competing losses at various decision thresholds when a particular quantitative variable, X, is used to guide the decision (Fig. 1).

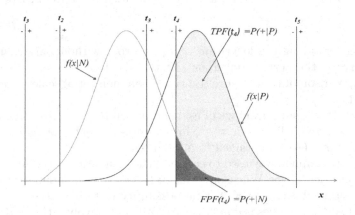

Fig. 1. Hypothetical example for evaluation of TPF and FPF at various decision thresholds (t_i) on X

In a single decision variable, changing the threshold necessarily results in monotonic changes in TPF and FPF. So, the classic empirical ROC curve is generated by plotting the TPF_i on the vertical axis, and FPF_i on the horizontal axis and connecting the data points, leading to a summary graph like that shown in Fig. 2.

The ROC graph has many advantages over single measurements of sensitivity (TPF) and specificity (1-FPF). The scales of the curve, TPF and FPF, are the basic measures of accuracy and are easily read from the plot of that displays all possible thresholds. Because sensitivity and specificity are independent of disease prevalence, so too is the ROC curve. The curve does not depend on the scale of the test results (ie, we can alter the test results by adding or subtracting

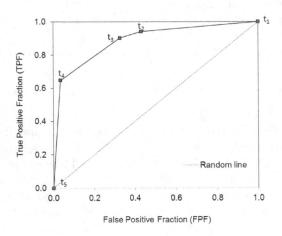

Fig. 2. Hypothetical example for generation of the ROC curve

a constant or taking the logarithm or square root without any change of the ROC curve) - the curve is invariant. Lastly, the ROC curve enables a direct visual comparison of two or more tests on a common set of scales at all possible thresholds.

One of the biggest advantage of using empirical ROC curves is the possibility to compare several diagnostic tests, for example, trough the value of the area under the curve (AUC) obtained from each test.

The most common way used to compute the estimative of the AUC is the statistical approach of the Wilcoxon-Mann-Whitney, which summarize each ROC curve in terms of area bellow it. One possibility to test weather the difference between two ROC curves is statistically significant, involves the AUC index.

Consider AUC_1 and AUC_2 the areas obtained from two ROC curves derived from two unpaired populations. The relevant hypothesis to test, H_0, is that the two data sets come from ROC curves with the same AUC:

$$H_0 : AUC_1 - AUC_2 = 0 \text{ vs } H_1 : AUC_1 - AUC_2 \neq 0$$

A method for testing the difference between two areas for unpaired samples is based on critical ratio Z [6]:

$$Z = \frac{\widehat{AUC_1} - \widehat{AUC_2}}{\sqrt{SE_1^2 + SE_2^2}} \sim N(0, 1) \tag{1}$$

If the samples are paired, H_0 can be tested using the follow ratio Z [6]:

$$Z = \frac{\widehat{AUC_1} - \widehat{AUC_2}}{\sqrt{SE_1^2 + SE_2^2 - 2rSE_1SE_2}} \sim N(0, 1) \tag{2}$$

were $\widehat{AUC_i}$ is the estimate of AUC for the diagnostic test i, and SE_i is the standard error for this. This Z ratio is valid if the ROC curves of each diagnostic test do not cross each other.

In previous work of Braga et al. [10] a methodology to compare two diagnostic tests based on empirical ROC curves when they cross each other was proposed. This methodology was implemented in a R package called Comp2ROC [11].

1.2 Comp2ROC Package

The Comp2ROC package allows to compare two diagnostic systems that cross each other based on AUC index and was developed in R language and it implementation use the methodology based on Z ratio and the methodology proposed by Braga et al. [10]. It works with paired or unpaired samples and produce graphics and text outputs according generated functions explained in [11]:

Table 1. Functions created on Comp2ROC

Function	Description
areatriangles	Triangle Areas
comp.roc.curves	Calculate distribution
comp.roc.delong	Calculate areas and stats
curvesegslope	Segment Slopes
curvesegsloperef	Segment Slopes to Reference Point
diffareatriangles	Difference Between Area Triangles
linedistance	Intersection Points
lineslope	Sampling Lines Slope
read.file	Read data from file
read.manually.introduced	Read data manually introduced
roc.curves.boot	Compare curves
roc.curves.plot	Plot ROC curves
rocboot.summary	Summary of Comparation
rocsampling	ROC Sampling
rocsampling.summary	Summary of ROC Sampling
save.file.summary	Save File

1.3 IBM® SPSS® Statistics Package

In IBM® SPSS® Statistics the algorithms for ROC analysis are implemented based on methodology of Hanley and MacNeil [5].

SPSS only considers quantitative diagnostic tests. The state variable can be of any type and indicates the true category to which a subject belongs. The value of the state variable indicates which category should be considered positive.

It only allow to compare paired samples with the same number of cases of the state variable. To do ROC curves analysis, it is necessary to create an indicator variable to define the groups to produce outputs individually.

It is assumed that increasing numbers on the rating scale represent the increasing belief that the subject belongs to one category, while decreasing numbers on the scale represent the increasing belief that the subject belongs to the other category. The user must choose which direction is positive. It is also assumed that the true category to which each subject belongs is known.

The IBM® SPSS® Statistics computes the estimative of AUC, standard errors and confidence intervals for each curve separately. It tests the significance relatively to the non-informative ROC curve (AUC=0.5).

2   Implementation

2.1   Software and Dataset

Initially, a lifting was made to perform the comparison of Comp2ROC package and IBM® SPSS® Statistics software and also was collected a set of available data in .csv format. The available dataset consists on evaluation of indexes of risk of death for newborns with very low birth-weight (VLBW), where the variables in study are the weight at birth and the CRIB (Clinical Risk Index for Babies). The weight vary inversely to the CRIB, i.e., a greater value in the scale indicates a negative result (alive), wile for CRIB a grater value in the scale indicates a positive value (deceased).

For this reason, in IBM® SPSS® Statistics software, the variable weight was transformed multiplying all values by −1. This transformation make possible to draw the two ROC curves (for CRIB and weight) in the same graph.

2.2   Comparison Rules

In order to have some basis for comparison between the two tools, we used some standards to evaluate their performances. The standards were chosen according to the usual model of using a certain software.

Resources to Execution. The data that are consumed by software for further processing was analyzed in terms of the type of supported files, structure of data files and how to use and handling them. The format of how the data is structured within these is also relevant as it has to be exactly in the required format, otherwise execution by software is impossible. Finally, we analyzed how these files and their data will be used and filled in applications analysis to indicate the level of simplicity and usability of the system, even for less experienced users.

Data Extraction. Each software was analyzed in terms of generated results. Their structure, organization of data, number of results, quality results information in terms of clarity and readability and interpretation. The type of output generated by the same is stored into a file, displayed only, or stored on database. If it is stored on file, we'll indicate the file type and location of writing (or save dialogue if applicable). If stored into a database, settings and environment variables required will be indicated. The organization of data and how they are arranged will also be examined, as a clear organization and information can save and lead to better conclusions after analyzing these.

Usability of Software. The way that the interface, usability and availability of information assists the user are very important factors of the softwares, since it is the first impact that leads to select a software or not. The second criterion involves the quality of the results, but if the application is not visually interesting or if you do not know how to use, no matter if you have a good algorithm and generates good results. The first impression directs the user to test the same, the provision of help buttons, informational messages and clear and objective notification leads to good handling of the application.

User Manual. The application's user manual should be one of the analysis in the assessment and review of software, as this is the most detailed documentation of all the capabilities, features, uses and application of standards. This concept is located between the application programmer and the end user, since the user is not privileged to contact with the developer, it becomes the only way to inform all the added value and the way to best use and take advantage of the software in question. As such, it is important to assess the level of detail of information, clarity and simplicity of texts but objectivity of writing used.

Study Results. This is one of the most important standards of assessment, due to integrity and accuracy.

It is also important due to the applied statistical methods, since several interpretations exist to calculate AUC as well as their main characteristics and limitations. It is essential to understand if it is possible to analyst paired data and unpaired data, and if it possible to carry out, parametric or non-parametric approaches.

3   Results

After an exhaustive study of algorithms for ROC curves analysis in IBM® SPSS® Statistics software and Comp2ROC, we can say that both obey the vast majority of the criteria described above, user help functions, tutorials, application demonstrations and hold both a solid user-friendly display, as seen in Table 2.

As regards the charging data, both provide the same ease of importing, charging can be done by hand or with a .csv file extension. To export data it could be done on the screen or in a data file.

In terms of results there are great differences. First, ROC curve analysis in IBM® SPSS® Statistics only make a comparison of each curve with the non-informative ROC (AUC=0.5). For illustrative purpose, using the data of newborns with VLBW, the data file type is in Fig. 3

Table 2. Comparison results for two packages analyzed

	SPSS	Comp2ROC
Resources for Execution	loads data file	loads data file
	load data manually	load data manually
Data Extraction	save to file	save to file
Usability of Software	help function	help function
	Tutorial	Tutorial
	Demonstration version	Demonstration version
User Manual	yes	yes

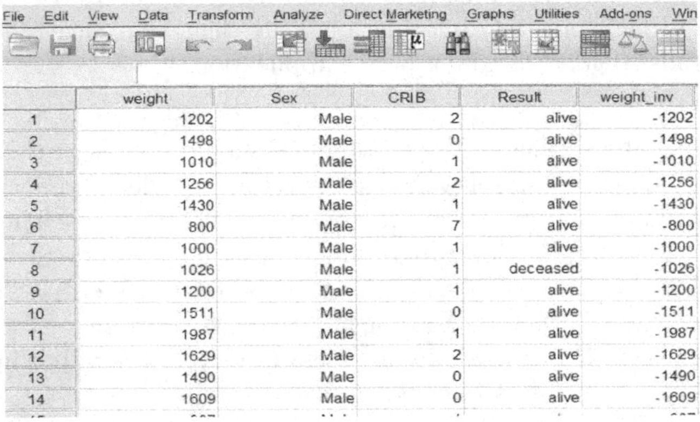

File Edit View Data Transform   Analyze Direct Marketing Graphs Utilities Add-ons Win

	weight	Sex	CRIB	Result	weight_inv
1	1202	Male	2	alive	-1202
2	1498	Male	0	alive	-1498
3	1010	Male	1	alive	-1010
4	1256	Male	2	alive	-1256
5	1430	Male	1	alive	-1430
6	800	Male	7	alive	-800
7	1000	Male	1	alive	-1000
8	1026	Male	1	deceased	-1026
9	1200	Male	1	alive	-1200
10	1511	Male	0	alive	-1511
11	1987	Male	1	alive	-1987
12	1629	Male	2	alive	-1629
13	1490	Male	0	alive	-1490
14	1609	Male	0	alive	-1609

Fig. 3. Example of data file in IBM® SPSS® Statistics

To run ROC analysis for unpaired samples, for example, to evaluate the performance of index CRIB for male and female VLBW infants we need to run the script with the following commands:

```
SORT CASES  BY Sex.
SPLIT FILE SEPARATE BY Sex.

ROC CRIB BY Result (1)
  /PLOT=CURVE(REFERENCE)
  /PRINT=SE
  /CRITERIA=CUTOFF(INCLUDE) TESTPOS(LARGE) DISTRIBUTION(FREE) CI(95)
  /MISSING=EXCLUDE.
```

The output produced, in terms of empirical ROC curves, is illustrated in Fig. 4 and results in Fig. 5:

As we can see in Fig. 5, the *sig* value is the p-value of the hypothesis test:

$$H_0 : AUC = 0.5 \text{ vs } H_1 : AUC \neq 0.5$$

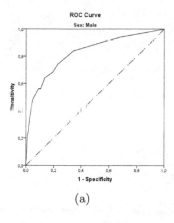

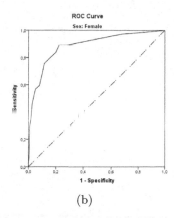

(a)   (b)

Fig. 4. Empirical ROC curves of index CRIB for male 4(a) and female 4(b) VLBW infants

Area Under the Curve[a]

Test Result Variable(s):   CRIB

			Asymptotic 95% Confidence Interval	
Area	Std. Error[b]	Asymptotic Sig.[c]	Lower Bound	Upper Bound
.888	.031	.000	.827	.948

The test result variable(s): CRIB has at least one tie between the positive actual state group and the negative actual state group. Statistics may be biased.[a]

a. Sex = Female

b. Under the nonparametric assumption

c. Null hypothesis: true area = 0.5

Fig. 5. Output results in IBM® SPSS® Statistics

It is impossible to compare directly the performance of CRIB for each sex of VLBW infants.

Taking into account the way to enter the unpaired data set in Comp2ROC, the results that are produced are illustrated in Fig. 6.

To run ROC analysis for paired samples, for example, to evaluate the performance of index CRIB and weight for VLBW infants we need to run the script with the following commands:

```
*Paired samples

ROC CRIB weight_inv BY Result (1)
  /PLOT=CURVE(REFERENCE)
  /PRINT=SE
  /CRITERIA=CUTOFF(INCLUDE) TESTPOS(LARGE) DISTRIBUTION(FREE) CI(95)
  /MISSING=EXCLUDE.
```

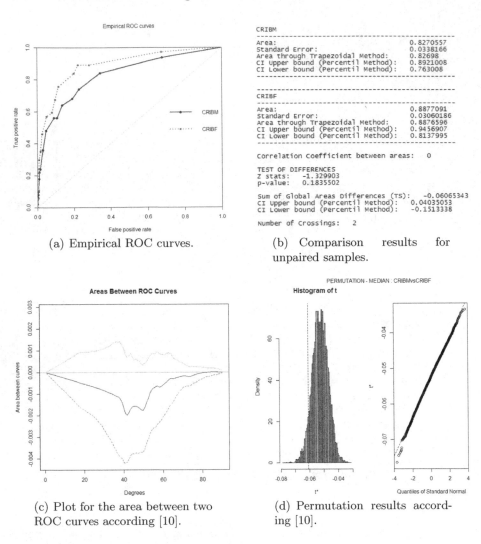

(a) Empirical ROC curves.

(b) Comparison results for unpaired samples.

(c) Plot for the area between two ROC curves according [10].

(d) Permutation results according [10].

Fig. 6. Comp2ROC output of index CRIB. Comparison between male and female VLBW infants.

The output produced, in terms of empirical ROC curves, is illustrated in Fig. 7 and results in Fig. 8:

For the same set of data and taking into account the format data for paired samples in Comp2ROC manual, the results obtained to compare the performance of indexes CRIB and weight are illustrated in Fig. 9.

In terms of methodology the Comp2ROC allows to compare two diagnostic systems based on AUC of the ROC curve when the two ROC curves do not cross and when they cross each other.

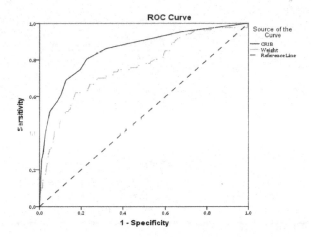

Fig. 7. Empirical ROC curves in IBM® SPSS® Statistics for paired samples

Area Under the Curve

Test Result Variable(s)	Area	Std. Error[a]	Asymptotic Sig.[b]	Asymptotic 95% Confidence Interval	
				Lower Bound	Upper Bound
CRIB	.854	.024	.000	.808	.900
Weight	.760	.029	.000	.703	.817

The test result variable(s): CRIB, Weight has at least one tie between the positive actual state group and the negative actual state group. Statistics may be biased.

a. Under the nonparametric assumption

b. Null hypothesis: true area = 0.5

Fig. 8. Output results in IBM® SPSS® Statistics for paired samples

The methodology implemented in Comp2ROC, with the graph of area produced in the output (Fig. 6(c) and Fig. 9(c)) enables to identify the regions of ROC space where one curve have better performance than the other. For example, based on the results for comparison of performance of CRIB and weight for VLBW infants the results (Fig. 9(b)) reveals that there are significant differences between the two AUCs in global (confidence interval by percentile method: [0.01849458, 0.1645742]) and the graph in Fig. 9(c) identify the region of ROC space where CRIB performs significantly better than weight (all tree lines above the central zero line).

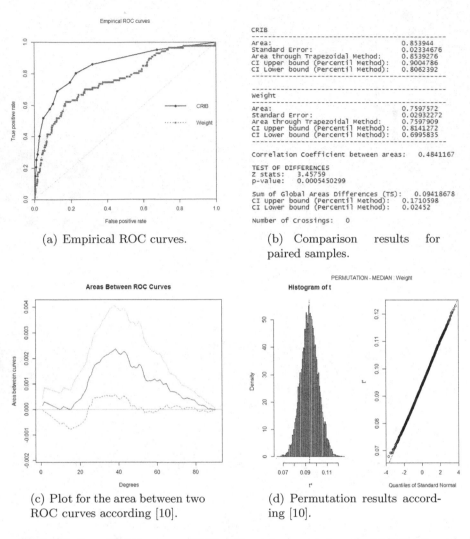

(a) Empirical ROC curves.

(b) Comparison results for paired samples.

(c) Plot for the area between two ROC curves according [10].

(d) Permutation results according [10].

Fig. 9. Comp2ROC output. CRIB and weight comparison for VLBW infants.

4   Conclusion

After the study we were able to mention the downside and benefits of each software and confirm that Comp2ROC provides a good optimization for ROC analysis for diagnostic performance study, and considering the analyzed criteria, Comp2ROC performs better.

The results show that both software provides a good support in terms of user help/info thought tutorials, usability, ease of data insertion and their changes, and a great advantage, that allow us to store the final results (output data) into a file or simply copy them to our clipboard. Although IBM® SPSS® Statistics it is not recommended its use for ROC analysis, because it doesn't allow to test two ROC curves. Also, it doesn't allow the calculation of the confidence intervals for the differences between two ROC curves and can only be produced more than one ROC curve, if and only if, the highest and lowest values of a register are associated in the same way, for example with an increased risk of disease. Finally IBM® SPSS® Statistics does not have a free license.

The Comp2ROC use the non-parametric method to compute AUC, SE and correlation coefficient for paired data, unlike IBM® SPSS® Statistics and the basic parameters can be calculated, also offers good flexibility in terms of graphics capability. One of the great advantage of this application, in terms of data sets, is the comparison between unpaired samples and finally and not the least the license is free. In terms of methodology implemented, Comp2ROC has the great advantage to enable the comparison between two ROC curves that cross each other. For all these reasons it is clear that the Comp2ROC meets the requirements, not only for the ROC analysis, but also in terms of licence and systems requirements.

Acknowledgments. The authors would like to thank the availability of the data by the Portuguese VLBW infants network.
This work has been supported by FCT - Fundação para a Ciência e Tecnologia within the Project Scope: UID/CEC/00319/2013.

References

1. Egan, J.P.: Signal detection theory and ROC analysis. Academic Press, New York (1975)
2. Metz, C.E.: Statistical analysis of ROC Data in evaluating diagnostic performance in multiple regression analysis: applications in the health sciences. In: Herbert, D.E., Myers, R.H. (eds.), vol. 13, pp. 365–384. American Institute of Physics (1986)
3. Gastwirth, J.L.: A general definition of the Lorenzcurve. Econometrica **39**(6), 1037–1039 (1971)
4. Mylne, K.R.: Decision-making from probability forecasts based on forecast value. Meteorological Applications **9**, 307–315 (2002)
5. Hanley, J.A., McNeil, B.J.: The Meaning and Use of the Area under a Receiver Operating Characteristic ROC Curve. Radiology **143**, 29–36 (1982)
6. Hanley, J.A., McNeil, B.J.: A method of comparing the areas under receiver operating characteristic curves derived from the same cases. Radiology **148**, 839–843 (1983)
7. Robin, X., Turk, N., Heinard, A., Tibertini, N., Lisacek, F., Sanchez, J.C., Müller, M.: pROC: an open-source package for R and S+ to analyze and compare ROC curves. BMC Bioinformatics **12**, 77 (2011)
8. Stephan, C., Wesseling, S., Schink, T., Jung, K.: Comparison of eight computer programs for receiver-operating characteristic analysis. Clinical Chemistry **49**(3), 4331–7439 (2003)
9. Brown, C.D., Davis, H.T.: Receiver operating characteristics curves and related decision measures: A tutorial. Chemometrics and Intelligent Laboratory Systems **80**, 241–738 (2006)

10. Braga, A., Costa, L., Oliveira, P.: An alternative method for global and partial comparasion of two diagnostic system based on ROC curves. Journal of Statistical Computation and Simulation **83**(2), 307–325 (2013)
11. Frade, H., Braga, A.C.: Comp2ROC: R package to compare two ROC curves. In: Mohamad, M.S., Nanni, L., Rocha, M.P., Fdez-Riverola, F. (eds.) 7th International Conference on Practical Applications of Computational Biology & Bioinformatics. AISC, vol. 222, pp. 127–135. Springer, Heidelberg (2013)

Adjusting Covariates in CRIB Score Index Using ROC Regression Analysis

Maria Filipa Mourão[1]([✉]), Ana C. Braga[2], Alexandra Almeida[3],
Gabriela Mimoso[4], and Pedro Nuno Oliveira[5]

[1] School of Technology and Management, Polytechnic Institute of Viana do Castelo,
4900-348 Viana do Castelo, Portugal
fmourao@estg.ipvc.pt
[2] ALGORITMI Centre, University of Minho, 4710-057 Braga, Portugal
acb@dps.uminho.pt
[3] Serviço de Neonatologia e Cuidados Intensivos Pediátricos,
Centro Materno Infantil do Norte, Centro Hospitalar do Porto,
4099-001 Porto, Portugal
maria.alexandra.almeida@gmail.com
[4] Serviço de Neonatologia B da Maternidade Bissaya Barreto,
Centro Hospitalar e Universitário de Coimbra, 3000-075 Coimbra, Portugal
gabriela.mimoso@gmail.com
[5] Biomedical Sciences Abel Salazar Institute, University of Porto,
4050-313 Porto, Portugal
pnoliveira@icbas.up.pt

Abstract. In medical studies, the receiver operating characteristic (ROC) curve is a tool of extensive use to analyze the discrimination capability of a diagnostic variable. In certain situations, the presence of related covariate, continuous or categorical, to the diagnostic variable can increase the discriminating power of the ROC curve [3].

The Clinical Risk Index for Babies (CRIB) scale, appeared in 1993 to predict the mortality of babies with very low birthweight (VLBW) and/or less than 32 weeks of gestation [2]. Braga and Oliveira [1] concluded that this index performs well in computing the risk of death for VLBW infants (< 1500 g).

In previous works, the authors studied the effect of the baby's sex [17] and the mother's age [18] on CRIB scale, using results of an intensive care unit of a Portuguese hospital.

In the present work, we propose to analyze the discriminative power of CRIB scale, using ROC regression analysis with GLM (Generalized Linear Models), in the classification of babies with and without the presence of covariates (newborn gender and mothers age).

This study is carried out using a random sample obtained from data collected during the period from 2010 − 2012. The data source was the "Portuguese VLBW infants network" that encompasses all newborns with less than 1500 g or 32 weeks of gestational age born in Portugal.

Keywords: Conditional ROC curve and CRIB · Nonparametric regression model · Covariates

© Springer International Publishing Switzerland 2015
O. Gervasi et al. (Eds.): ICCSA 2015, Part II, LNCS 9156, pp. 157–171, 2015.
DOI: 10.1007/978-3-319-21407-8_12

1   Introduction

1.1   CRIB Index

Assessment of newborn risk in neonatal intensive care units tends to rely on the risk of mortality adjusted for birthweight and gestational age. This is especially true when we refer to VLBW infants or newborns with a gestational age less than 32 weeks. In the 90s, neonatal scoring systems were developed taking into account other factors to predict newborn mortality. The use of scoring scales also allows us to standardize the patients severity, making hospital benchmarking easier.

Among the many existing scales to assess and classify the clinical status of very premature babies, the CRIB and SNAPPE II (Score for Neonatal Acute Physiology Perinatal Extension) are the most commonly used scales in Portugal.

The CRIB scale uses six variables (birth weight, gestational age, congenital malformations and three physiological measures), collected in the first 12 hours of life, to evaluate the clinical severity of newborns, which makes it less susceptible to the effect of treatments [2]. The result of this scale is based on a weighted sum of these six variables and their possible final value is between 0 and 23. The higher the results in the scale, the higher the probability of death.

1.2   ROC Curve

The ROC curve is a graph representation of true positive rates on the vertical axis and false positive rates on the horizontal axis for different values of a classification threshold. It can be interpreted as a curve that summarizes the information of the cumulative distribution functions of the scores of the two classes considered.

We take as a starting point the existence of two populations, a positive/abnormal population which we denote by P and a negative/normal population denoted by N, together with a classification rule for allocating the individuals by each of these populations. We assume this classification rule to be in some continuous function $t(x)$ of the random vector X of variables measured on each individual, conventionally arranged so that large values of the function are more indicative of population P and small ones more indicative of population N. Thus if x is the observed value of X for a particular individual and $t(x)$ is the function score for this individual, then he is allocated to population P or population N in case $t(x)$ exceeds or does not exceed some threshold c, respectively. Supposing that c is the value of the threshold in a particular classification rule, so as that an individual is allocated to population P if its classification score t exceeds c and to population N in case it doesn't exceed. In order to assess the efficacy of this classifier we need to calculate the probability of making an incorrect allocation. More specifically, we can define four probabilities and their associated rates for the classifier:

- the probability that an individual from P is correctly classified, i.e., the true positive fraction, $TPF = Prob(t > c|P)$;
- the probability that an individual from N is misclassified, i.e., the false positive fraction $FPF = Prob(t > c|N)$;

- the probability that an individual from N is correctly classified, i.e., the true negative fraction $TNF = Prob(t \leq c|N)$; and
- the probability that an individual from P is misclassified, i.e., the false negative fraction $FNF = Prob(t \leq c|P)$.

Given probability densities $Prob(t|P)$, $Prob(t|N)$, and the value c, numerical values lying between 0 and 1 can be obtained readily for these four rates and this gives a full description of the performance of the classifier.

Clearly, for a good performance, we require high true and low false fractions. However, this only appens for a particular choice of threshold c, and the best choice of this threshold is not generally known in advance but must be determined as part of the classifier construction. Varying c and evaluating the four probabilities mentioned above, will give full information on which this decision will be based and hence to assess the performance of the classifier. The ROC curve provides a much more easily accessible summary. It is the curve obtained on varying c, but when using just the true and false positive fractions and plotting (FPF, TPF) as points against orthogonal axes. Here FPF is the value on the horizontal axis (abscissa) and TPF is the value on the vertical axis (ordinate).

The purpose of the ROC curve is to provide an assessment of the classifier over the whole range of potential threshold values rather than at just a single one. Clearly, the value of a classifier can be judged by the extent to which the two distributions of its scores $Prob(t|P)$ and $Prob(t|N)$ differ. The more they differ, the lesser will there be any overlap between them and so the less likely will the incorrect allocations be made. Hence, the more successful will the classifier be in the making of correct decisions. Conversely, the more the two distributions arc alike, the more overlap there is between them and so the more likely for incorrect allocations to be made.

In practice, the ROC curve is a continuous curve which is between $(0, 0)$ and $(1, 1)$, situated at the upper triangle that defines the chart. The closer it is to the upper left corner of the graph, the closer we are to a situation of complete separation between the populations and therefore, the better the performance of the classifier.

When the score t is a continuous variable, it will possess density and distribution functions in each of the two populations. Let us denote its density and distribution functions in population N by f_N, F_N respectively, and in population P by f_P, F_P respectively. Thus,

$$Prob(t|N) = f_N(t) \qquad Prob(t|P) = f_P(t)$$
$$x_N(c) = 1 - F_N(c) \qquad x_P(c) = 1 - F_P(c)$$

Eliminating c from this last pair of equations, using standard mathematics, yields the form of the ROC curve

$$ROC(x) = 1 - F_P[F_N^{-1}(1 - x)] \quad \text{for } x \in (0, 1) \tag{1}$$

and the expression (1) will be the most convenient equation of the ROC curve to use with continuous scores.

We have seen above that the ROC is a convenient summary of the full set of information that would be needed for a comprehensive description of the performance of a classifier over all its possible threshold values. However, even such a summary may be too complicated in some circumstances, for instance, if a plot is difficult to produce or if very many different classifiers need to be compared, the interest therefore relies on obtaining deriving simpler summaries. Particular attention has been focused on single scalar values that might capture the essential features of a ROC curve, like the Area Under the ROC Curve (AUC). This is the most widely used summary index studied by Green and Swets [20], Bamber [21], Hanley and McNeil [12], [22] among others.

Simple geometry establishes the upper and lower bounds of AUC: for the case of perfect separation of P and N distributions, AUC is the area under the upper borders of the ROC (i.e., the area of a square of side 1). So the upper bound is 1.0, while for the case of random allocation AUC is the area under the chance diagonal (i.e., the area of a triangle whose base and height are both equal to 1) so the lower bound is 0.5. For all other cases, the formal definition is

$$AUC = \int_0^1 ROC(x)dx \tag{2}$$

One immediate interpretation of AUC follows from this definition, elementary calculus and probability theory, plus the fact that the total area of the ROC domain is 1.0: AUC is the average of true positive fractions, taken uniformly over all possible false positive fractions in the range $(0, 1)$. Another interpretation is as a linear transformation of the average misclassification rate, weighed by the mixture distribution of the true P and N classes [19]. We can assume from the above definition that, if A and B are two classifiers for witch the ROC curve for A is not inferior the ROC curve for B, then AUC for A must be greater than or equal to AUC for B. Unfortunately, the reverse implication is not true because of the possibility that the two curves can cross each other.

1.3   Estimating ROC Curve

For continuous scores t, as we have seen, the ROC curve can be expressed by equation (3).

The estimation problem is thus reduced to the estimation of this curve from the given data. To obtain the empirical estimator, we simply apply the relevant definitions above to the observed data. Thus, if n_P and n_N are the numbers of individuals in the samples from populations P and N respectively, and if $n_A(c)$ denotes the number of individuals in the sample from population A (where A is either N or P) whose classification scores are greater than c, then the empirical estimators of the true positive fraction $TPF = p(t > c|P)$ and false positive fraction $FPF = p(t > c|N)$ at the classifier threshold c are given by

$$\widehat{TPF} = \frac{n_P(c)}{n_P}$$

$$\widehat{FPF} = \frac{n_N(c)}{n_N}$$

Thus, plotting the set of values $(1 - \widehat{FPF})$ against c yields the empirical distribution function $\hat{F}_N(c)$ and doing the same for values $(1 - \widehat{TPF})$ yields the empirical distribution function $\hat{F}_P(c)$. The empirical ROC curve is then simply given by plotting the points $(\widehat{FPF}, \widehat{TPF})$ obtained on varying c with equation given by (3).

$$\widehat{ROC}(x) = 1 - \hat{F}_P[\hat{F}_N^{-1}(1-x)] \quad \text{for } x \in (0,1) \tag{3}$$

The expression (3) is called the empirical ROC curve.

2   ROC Curve and Covariates

Once a classifier t has been constructed from the vector X of principal variables and is in use for allocating individuals to one or the other of the populations N and P, confounding occurs in evaluating classification accuracy when there is a covariate which is associated with both the classifier and the binary outcome. For maximum benefit, such additional variable should be incorporated into any analysis involving the classifier. Let us denote a set of z covariates by the vector Z, recognizing that in many practical applications we may have just one covariate Z. In such cases, it is necessary to adjust the ROC curve and summaries derived therefrom before drawing any inferences. In particular, it will often be relevant to compute the ROC curve and allied summaries at particular values of the covariates Z, in order to relate these covariate-specific curves and summaries to sample members that have covariate values. Ignoring covariate values leads to the calculation of a single pooled ROC curve (the pooling being over all the possible covariate values). This traditional pooled ROC curve, which combines all case and control observations, regardless of the covariate value, is biased. Pepe [16] provides several important results linking the covariate-specific and the pooled ROC curves.

The ROC curve, in presence of covariates, can be considered for each value z of the covariate. Changes that occur in the curve, due to these values, might mean that the covariate has an effect on the discrimination power of the diagnostic test. The conditional ROC curve is defined as

$$ROC_z(x) = F_{PZ}[F_{NZ}^{-1}(x)] \quad \text{for } x \in (0,1) \tag{4}$$

where F_{PZ} and F_{NZ} are the conditional survival functions associated with subjects of populations P and N, respectively, and are estimated by

$$\hat{F}_{PZ}(c) = \frac{1}{n_P} \sum_{i=1}^{n_P} I(X_{Pi} \geq c) \tag{5}$$

$$\hat{F}_{NZ}(c) = \frac{1}{n_N} \sum_{i=1}^{n_N} I(X_{Ni} \geq c) \tag{6}$$

So, the covariate-adjusted ROC curve, is a measure of covariate-adjusted classification accuracy.

Adjusting ROC curves derived from continuous classification scores has been considered by several authors like Smith and Thompson [5], Pepe [6], [7], [8], Faraggi [9], Janes and Pepe [10], [11], Pepe and Cai [23] and Janes et al. [24]. Adjustment of summary values derived from the ROC curve has further been studied by Faraggi [9] and Dodd and Pepe [25], [26].

To obtain de covariate-adjusted ROC curve, two distinct approaches can be followed. The *induced adjustment*, proposed by Tolsteson and Begg [4], Zheng and Heagerty [13], Faraggi [9], in which the covariates have an effect on the diagnostic test is modeled in the two populations (P and N) separately and the ROC curve is then derived from the modified distributions; and *direct adjustment*, proposed by Pepe [8], Alonzo and Pepe [15], Cai [14], in which the effect of the covariates is modeled on the ROC curve itself. The authors, in [18], provide details about these two methodologies.

There are covariates that affect the classifier distribution among negatives. For example, center effects in multicenter studies may affect classifier observations. Other covariates may affect the inherent discriminatory accuracy of this (i.e., the ROC curve). As an example, disease severity can often affect the classifier accuracy, thus less severe positives can be more difficult to distinguish from negatives.

A separate ROC curve should be estimated for each covariate group that affects the discriminatory accuracy of the classifier. Covariate adjustment is often a necessary first step in estimating covariate-specific ROC curves in order to adjust the effects of the covariate on classifier observations among negatives.

In this work, we will use the STATA software to obtain the covariate-adjusted ROC curve.

3   ROC Regression with STATA Software

STATA software uses the ROC regression as methodology to model the classifier ROC curve as a function of covariates [8], [15]. Implementation proceeds in two steps:

1st: Modeling the distribution of the classifier among negatives, as a function of covariates, and calculating the case Percentile Values or specificity;

2nd: Modeling their cumulative distribution function (i.e., the ROC curve) as a function of covariates.

The result is an estimate of the ROC curve for the classifier as a function of covariates, or a covariate-specific ROC curve.

In STATA, the `rocreg` command is used to perform receiver operating characteristic (ROC) analyses with rating and discrete classification data under the presence of covariates.

This function can fit three models: a nonparametric model, a parametric probit model that uses the bootstrap for inference, and a parametric probit model fit using maximum likelihood.

The syntax to perform this analysis is given by:

`rocreg` *refvar classvar*[*classvars*] [*if*] [*in*] [, *np$_{options}$*]

The two variables *refvar* and *classvar* must be numeric. The reference variable indicates the true state of the observation – such as diseased (abnormal, P) and nondiseased (normal, N) – and must be coded as 1 and 0, respectively. The *refvar* coded as 0 can also be called the control population, while the *refvar* coded as 1 comprises the case population. The rating or outcome of the diagnostic test or test modality is recorded in *classvar*, which must be ordinal, with higher values indicating higher risk.

The covariate-adjusted ROC curve [24] at a given false-positive rate x is equivalent to the expected value of the covariate-specific ROC at x over all covariate combinations. When the covariates in question do not affect the case distribution of the classifier, the covariate-specific ROC will have the same value at each covariate combination. So here the covariate-adjusted ROC is equivalent to the covariate-specific ROC, regardless of covariate values.

4   Applications and Results

In this work, we intend to analyze and evaluate the discriminatory power of CRIB scale in the classification of risk of death for VLBW infants in Portugal. Available data refers to the period between 2010 and 2012, and we decided to select a random sample (approximately 50%) of original data collected by the RNMBP – Registo Nacional de Muito Baixo Peso (Portuguese VLBW infants network) for this period. We analyzed the performance of this scale without the inclusion of covariates and then considered their inclusion. We first perform ROC analysis for the classifier while adjusting for babies gender, then for maternal age and finally combining the information from both. This is done by specifying these variables in the *ctrlcov()* option. We adjust the covariates using a linear regression rule, by specifying *ctrlmodel(linear)*. This means that when a user of the diagnostic test chooses a threshold conditional on covariates, he assumes that the diagnostic test classifier has some linear dependence and equal variance, since their levels vary. Our cluster adjustment is made by specifying the *cluster()* option.

STATA software computes the percentile, or specificity, values empirically, and thus so do false-positive rates, $(1 - specificity)$. Also, the ROC curve values are empirically defined by the true-positive rates.

4.1   CRIB Index Without Covariates

In this section we verify the performance of CRIB scale in the classification of babies as "deceased" and "alive". To achieve its purpose, the study focused on the results presented in Table 1, which characterizes the grade given to newborns.

Table 1. Mortality rate

	Cases number	Percent
Deceased	166	10.45%
Alive	1423	89.55%
Total	1589	100.00%

From the results collected, 10.45% of the babies were classified as "deceased". The distribution of CRIB by result can be illustrated by the graph in Figure 1.

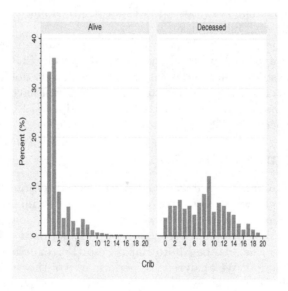

Fig. 1. CRIB classification by result

For these results, the empirical ROC curve was estimated, as is shown in Figure 2. Respective AUC, variability and confidence intervals were also obtained (Table 2).

From the estimated AUC value for the global empirical ROC curve (Figure 2) it is possible to observe that, without considering the effect of any the covariates, the CRIB index discriminates well between babies "alive" and "dead" in 87.9% of the cases.

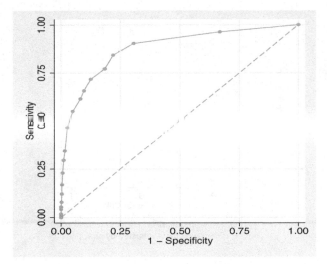

Fig. 2. Empirical ROC curve for CRIB index (output STATA)

Table 2. AUC's and Standard Errors results

	AUC	SE	95% LBCI	95% UBCI
CRIB	0.879	0.015	0.849	0.909

4.2   CRIB Index with Covariates

To verify if maternal age and the babie's sex contribute to the discrimination power of CRIB index, we propose to combine the CRIB information with additional information provided by these covariates. We use Nonparametric ROC estimation provided by STATA software to answer the research question:

"Does the discriminatory power of CRIB index increase with the inclusion of these covariates?"

We start by analyzing the contribution that each of the covariates provides in discriminatory power of the CRIB index, assuming that there is a linear relationship between the CRIB index and the covariate in question.

The conditioned ROC curve for the babie's sex is shown in Figure 3 and the associated results are in table 3. The AUC, SE and confidence intervals associated with ROC curve are shown in table 4.

When we include the sex of babies in the analysis, the AUC for conditioned ROC curve discriminate in 86.19% of the babies; less than the discriminates power without gender and higher standard error. Our covariate adjustment model shows that the newborn's sex has a negative effect on CRIB index.

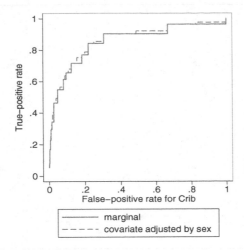

Fig. 3. Specific sex ROC curve (output STATA)

Table 3. Linear Regression results - CRIB conditioned by newborn sex

CRIB	Coefficient	SE	t-value	$P > |t|$	95% LBCI	95% UBCI
Sex	-0.0643	0.1306	-0.49	0.622	-0.3206	0.1919
Const	1.8934	0.2044	9.26	0.000	1.4923	2.2945

Table 4. AUC, SE and CI - CRIB conditioned by newborn sex

AUC	Bias	SE	95% LBCI	95% UBCI
0.8619	0.0022	0.1801	0.8266	0.8972
			0.8287	0.8977
			0.8234	0.8934

Let us now analyze the mother's age impact on CRIB index, assuming that there is a linear relationship between the CRIB index and this covariate. The specific CRIB ROC curve is shown in figure 4. The results for this linear regression are shown in table 5 and the correspondent AUC in table 6. When we include the mother's age in the analysis, the AUC for conditioned ROC curve discriminates in 87.98% of the babies, close to the AUC for CRIB without covariates. Despite the negative coefficient associated with this covariate in the regression model, it was no statistical significance ($p - value = 0.239 > 0.05$). It seems that the covariate has no impact in the discrimination power of the CRIB, in the towards of closeness of this AUC and the AUC of the CRIB ROC curve.

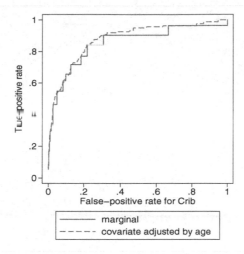

Fig. 4. Specific maternal age ROC curve (output STATA)

Table 5. Linear Regression results - CRIB conditioned by mother's age

CRIB	Coefficient	SE	$t-value$	$P > \|t\|$	95% LBCI	95% UBCI
Maternal age	-0.0100	0.0085	-1.18	0.239	-0.0269	0.0067
Const	2.1034	0.2689	7.82	0.000	1.5759	2.6309

Table 6. AUC, SE and CI - CRIB conditioned by mother's age

AUC	Bias	SE	95% LBCI	95% UBCI
0.8798	-0.0002	0.0152	0.8500	0.9097
			0.8487	0.9074
			0.8476	0.9062

Let us now combine the CRIB information with the information provided by these two covariates. We assume, once again, that there is a linear regression between the CRIB index and the two covariates. In figure 5, we show the specific CRIB ROC curve conditioned by the two covariates. Table 7 shows the results of the linear regression performed using STATA when we introduced the additional information of the two covariates. Table 8 shows the correspondent AUC, SE and confidence intervals.

Analyzing the results presented in Table 7, we can see that the coefficients for the two covariates are not different from zero ($p-value > 0.05$). So, we can say that there is no statistical influence of these covariates in the performance

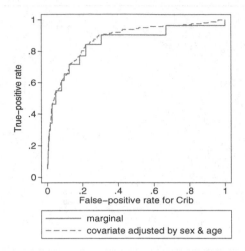

Fig. 5. Specific sex and maternal age ROC curve after ROC regression (output STATA)

Table 7. Linear Regression results - CRIB conditioned by sex and mother's age

CRIB	Coefficient	SE	$t-value$	$P > |t|$	95% LBCI	95% UBCI
Maternal age	-0.0101	0.0085	-1.18	0.236	-0.0269	0.0066
Sex	-0.0656	0.1306	-0.5000	0.6150	-0.3218	0.1905
Const	2.2018	0.3293	6.69	0.000	1.5558	2.8479

Table 8. AUC, SE and CI - CRIB conditioned by sex and mother's age

AUC	Bias	SE	95% LBCI	95% UBCI
0.8785	-0.0000471	0.0162	0.8468	0.9103
			0.8409	0.9084
			0.8385	0.9058

of the CRIB index. These could be confirmed by the values of AUC's for the adjusted and not adjusted ROC curve. Figure 6 illustrates this result.

5   Conclusions

From the estimated values of AUC for the global ROC curve it is possible to observe that, without considering the effect of covariates, the CRIB scale discriminates between "alive" and "dead" newborns in 87.9% of cases. This value, when considering the newborn sex as covariate, is 86.19% and with maternal age as covariate, it is 87.9%. We can also see that, standard-error (SE) associated

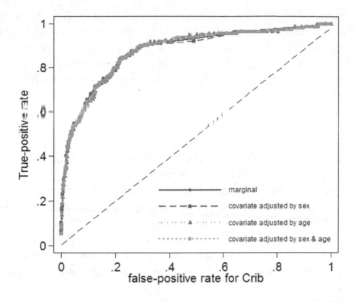

Fig. 6. All Specific ROC curves(output STATA)

to ROC curve, when considering the sex of babies as covariate, is greater when compared to the SE for the CRIB ROC curve and the SE for specific maternal age CRIB ROC curve.

From the results obtained, when we consider that there is a linear relationship between CRIB and the newborns gender, it is apparent that this covariate affects the discriminatory power of CRIB scale when used to classify babies. Our covariate adjustment model shows that the newborn sex has a negative effect on CRIB index, causing a decrease in the performance of CRIB scale. The standard error, in turn, is greater than the standard error of the ROC curve obtained without the covariate. Consequently, it also increases the dispersion of the pairs of values $(1 - specificity, sensitivity)$. However, when analyzing the overall quality of the regression model, we do not reject the hypothesis that the contribution of the baby's sex for the CRIB scale is zero ($t = -0.49$, $p - value = 0.622$).

When we analyze the mother's age impact on CRIB index and assume that there is a linear relationship between the CRIB scale and this covariate, the results for this linear regression, despite the negative coefficient associated with it in the regression model, it seems that this covariate has no impact in the discrimination power of the CRIB scale, in the towards vicinity of its AUC and the AUC of the CRIB ROC curve. The same occurs for the SE, which remains unchanged. In fact, when analyzing the overall quality of the regression model, we do not reject the hypothesis that the contribution of the mother's age for the CRIB scale is zero ($t = -1.18$, $p - value = 0.239$).

Finally, after analyzing the results, considering the two covariates, from the AUC associated with specific ROC curve, we can conclude that CRIB index discriminate in 87.85%. This discriminatory power is higher than that found when considering only sex as a covariate but it is lower when we consider only the mother's age as a covariate. Looking at dispersion present in AUC of conditioned ROC curve for these two covariates, we found that it is higher than the ROC curve of CRIB and than the specified ROC curve for maternal age. However, it is inferior to the dispersion present in the CRIB ROC curve conditioned by newborn sex. When Analyzing the overall quality of the regression model, we do not reject the hypothesis that the contribuition of the newborn's sex and the mother's age to the CRIB scale is zero ($t_{sex} = -0.50$, $p - value = 0.615$ and $t_{age} = -1.18$, $p - value = 0.236$). So, we may conclude that the discriminatory power of CRIB scale is not affected when we consider that baby's sex and mother's age may influence the classification of babies in "dead" and "alive".

To answer the research question, we can say that the performance of CRIB to predict the risk of death of newborns with very low birth-weight and/or gestational age < 32 weeks is not affected by the sex of the babies nor the age of their mothers.

Acknowledgments. The authors would like to thank the availability of the data by the Portuguese VLBW infants network.
This work has been supported by FCT - Fundação para a Ciência e Tecnologia within the Project Scope: UID/CEC/00319/2013.

References

1. Braga, A.C., Oliveira, P.: Diagnostic analysis based on ROC curves: theory and applications in medicine. International Journal of Health Care Quality Assurance, Emerald **16**(4), 191–198 (2003)
2. Dorling, J.S., Field, D.J., Manketelow, B.: Neonatal disease severity scoring systems. Arch. Dis. Child. Fetal Neonatal **90**, F11–F16 (2005)
3. López-de-Ullibarri, I., Cao, R., Cardaso-Suárez, C., Lado, M.J.: Nonparametric estimation of conditional ROC curves:application to discrimination tasks in computerized detection of early breast cancer. Computational Statistics & Data Analysis **52**(5), 2623–2631 (2008)
4. Tolsteson, A.N., Begg, C.B.: A general regression methodology for ROC curve estimation. Med. Decision Making **8**(3), 204–215 (1988)
5. Smith, D.J., Tompson, T.J.: Correcting for confounding in analising receiver operating characteristic curves. Biometrical Journal **38**, 357–863 (1996)
6. Pepe, M.S.: A regression modelling framework for receiver operating characteristic curve in medical diagnostic testing. Biometrika **84**, 595–608 (1997)
7. Pepe, M.S.: Three approaches to regression analysis of receiver operating characteristic curves for continuous tests results. Biometrics **54**, 124–135 (1998)
8. Pepe, M.S.: An interpretation for the ROC curve and inference using GLM procedures. Biometrics **56**, 352–359 (2000)
9. Faraggi, D.: Adjusting receiver operating characteristic curves and related indices for covariates. Journal of the Royal Statistical Society: Series D (The Statistician) **52**(2), 179–192 (2003)

10. Janes, H., Pepe, M.S.: Adjusting for covariates effects on classification accuracy using the covariate-adjusted ROC curve. UW Biostatistics Workin Paper Series. Working paper 283 (2006). http://biostats.bepress.com/uwbiostat/paper283
11. Janes, H., Pepe, M.S.: Adjusting for covariates in studies of diagnostic, screening, or prognostic markers: an old concept in a new setting. UW Biostatistics Working Paper Series. Working paper 310 (2007). http://biostats.bepress.com/uwbiostat/paper310
12. Hanley, J.A., McNeil, B.J.: A method of comparing the areas under receiver operating characteristic curves derived from the same cases. Radiology 148, 839–843 (1983)
13. Zheng, Y., Heagerty, P.J.: Semiparametric estimation of time-dependent ROC curves for longitudinal marker data. Biostatistics 4, 615–632 (2004)
14. Cai, T.: Semiparametric ROC regression analysis with placement values. Biostatistics 5, 45–60 (2004)
15. Alonzo, T.A., Pepe, M.S.: Distribution free ROC analysis using binary regression techniques. Biostatistics 3, 421–432 (2002)
16. Pepe, M.S.: The Statistical Evaluation of Medical Tests for Classification and Prediction. Oxford University Press, New York (2003)
17. Mourão, M.F., Braga, A.C., Oliveira, P.N.: CRIB conditional on gender: nonparametric ROC curve. International Journal in Health Care and Quality Assurance 27(8), 656–663 (2014)
18. Mourão, M.F., Braga, A.C., Oliveira, P.N.: Accommodating maternal age in CRIB scale: quantifying the effect on the classification. In: Murgante, B., Misra, S., Rocha, A.M.A.C., Torre, C., Rocha, J.G., Falcão, M.I., Taniar, D., Apduhan, B.O., Gervasi, O. (eds.) ICCSA 2014, Part III. LNCS, vol. 8581, pp. 566–579. Springer, Heidelberg (2014)
19. Krzanowski, W.J., Hand, D.J.: ROC curves for Continuous Data. Chapman and Hall (2009)
20. Green, D.M., Swets, J.A.: Signal Detection Theory and Psychophysics (rev.ed) Robert, E. Krieger Publishing Company (1974)
21. Bamber, D.: The area above the ordinal dominance graph and the area below the receiver operating characteristic graph. Journal of Mathematical Psychology. 12, 387–415 (1975)
22. Hanley, J.A., McNeill, B.J.: The meaning and use of the area under receiver operating characteristic (ROC) curve. Radiology 143(1), 29–36 (1982)
23. Pepe, M.S., Cai, T.: The analysis of placement values for evaluating discriminatory measures. Biometrics 60, 528–535 (2004)
24. Janes, H., Longton, G.M., Pepe, M.S.: Accommodating covariates in receiver operating characteristic analysis. Stata Journal 9, 17–39 (2009)
25. Dodd, L.E., Pepe, M.S.: Semiparametric regression for the area under the receiver operating characteristic curve. Journal of the American Statistical Association 98, 409–417 (2003)
26. Dodd, L.E., Pepe, M.S.: Partial AUC estimation and regression. Biometrics 59, 614–623 (2003)

Modeling Changes in Demands for Books with Elapsed Time from Publication

Kensuke Baba[1]([⊠]), Toshiro Minami[2], and Eisuke Ito[1]

[1] Kyushu University, Fukuoka, Japan
baba.kensuke.060@m.kyushu-u.ac.jp
[2] Kyushu Institute of Information Sciences, Dazaifu, Fukuoka, Japan

Abstract. Book selection in libraries should be conducted on the basis of analyses on circulation data. In addition to the number of loans, the change of the number over time can be a criterion for book selection. This paper proposes a model to represent the change of the number with the time from publication. The proposed model is applied to practical circulation data in a library and is evaluated in terms of the root mean square errors. As a result, the proposed model is suitable compared with the standard one. Additionally, this paper analyzes the differences of the changes between classified books which are obtained by the previous analysis. The information obtained by applying the proposed model to circulation data is expected to be utilized for book selection.

Keywords: Data mining · Library · Book selection · Circulation data · Obsolescence

1 Introduction

Book selection is one of the important operations for libraries. In usual university libraries in Japan, some books are selected according to static rules or by intuitions of a library staff. To meet the constantly changing demands of library patrons, the process of book selection should be evaluated with the circulation of the selected books, and improved on the basis of the results of the evaluation.

The straightforward way of using circulation data for book selection is considering the number of loans which can be thought to represent the demands for books. A simple idea of "good" book selection is that the selected books are used frequently as a result. To use circulation data on old books for selecting new books, the books should be classified in some senses so that new books are dealt with as the old books in the same category. Evans [1] investigated the number of loans for books classified by the subject of book selection in some academic libraries. Pritchard [6] also investigated the frequency of loans (that is, the turnover rate) for books classified by the way of the book selection. Koizumi [2] investigated the turnover rate in Keio University Library in Japan and clarified the difference between the ways of book selection.

For more detailed analyses, the change of the number of loans over time should be considered in addition to the number itself. For example, some books

© Springer International Publishing Switzerland 2015
O. Gervasi et al. (Eds.): ICCSA 2015, Part II, LNCS 9156, pp. 172–181, 2015.
DOI: 10.1007/978-3-319-21407-8_13

used for a long time should be selected even if the number of loans in a short term is small. The number of loans of a book usually decreases with the lapse of time. This decrease is called "obsolescence [4]". Matsui and Isono [5] investigated the turnover rate and confirmed its obsolescence in the library of Nara University in Japan. Analyzing the tendencies of obsolescence for closely classified books is expected to be useful for book selection. The tendencies should be quantified with a suitable obsolescence model for the analysis.

This paper proposes an obsolescence model for the number of loans with the time from the publication. The standard model of obsolescence is similar to the idea of the half-life of a radioisotope, that is, the target number is approximated to the t-th power of a constant (with an initial number) for the elapsed time t. Then, we can consider two kinds of elapsed time,

- The time from the accession of a book to the library and
- The time from the publication of a book.

The standard model cannot be applied straightforwardly to the number of loans when we consider the time from the publication as the elapsed time, because books are not loanable during the period from the publication to the accession of the books. Therefore, we multiplied the standard model by a function of t based on the probability of accession of books after the publication.

We applied the standard and the proposed models of obsolescence to a practical circulation data and verified the suitability. We used the loan records obtained in Kyushu University Library in Japan for the period from April 2013 to March 2014 (that is, the academic year of 2013 in Japan). We considered the two kinds of elapsed time. First, we applied the standard obsolescence model to the loan records with the time from the accession and confirmed the obsolescence. Next, we applied the standard and the proposed models with the time from the publication and investigated the root mean square errors for the two models. Additionally, we conducted the same analyses for closely classified books. We used Nippon Decimal Classification (NDC), the standard classification method in Japanese libraries.

Our study will suggest a viewpoint for improving the book selection process in university libraries. From the results of the analyses, we confirmed that our obsolescence model was more suitable than the standard one. Additionally, we obtained the characteristics of various subject fields in terms of the two kinds of obsolescence. We can estimate the demands for books more accurately by the obsolescence model.

The rest of this paper is organized as follows. Section 2 describes the collected loan records and introduces our obsolescence model. Section 3 reports the results of the analyses. Section 4 shows considerations about the results and future directions of our study.

2   Methods

The purpose of our study is to clarify the tendencies of obsolescence for several kinds of books. We considered two models for representing the changes of the number of loans quantitatively. Then, we applied the models to practical circulation data to evaluate the models.

2.1 Obsolescence Models

Obsolescence means the change (usually, decrease) of the number of use with the lapse of time. In this paper, we estimated the change of loans over the years from the distribution of the oldness of books in the records for one year. This kind of obsolescence is called synchronous [4]. We approximated the distribution of the oldness to a simple function (a model) and represented the obsolescence as a degree of decrease. We used two obsolescence models, the standard one and a novel one. The suitability of the models can be evaluated by considering errors in the approximation.

We considered two kinds of elapsed time, that is, the elapsed time from the accession to the library to the loan of a book (the *time from accession*) and the elapsed time from the publication to the loan (the *time from publication*).

By the standard model of obsolescence, the number of loans of books accessioned or published t units of time ago is represented by

$$f(t) = \frac{a}{b^t} \tag{1}$$

with two constants a and b. We call b the *decreasing rate*.

A naive use of the standard model causes a large error for the time from publication, because published books are not loanable until accession. The number of loans usually increases for a short term after the publication, although the function f decreases monotonically. A solution is considering the turnover rate, that is, the normalized number of loans by the total number of the accessioned books as Matsui and Isono [5]. In this solution, however, we need extra data about the total books in the library. We tried to use the unprocessed number of loans and added a kind of error function to the standard model. Fig. 1 illustrates the idea of our model. Let $e(t)$ be the probability that a book is accessioned until t units of time after the publication. Then, the number of loans of books published t units of time ago is

$$g(t) = \frac{a}{b^t} \cdot e(t) \tag{2}$$

with two constants a and b.

2.2 Collected Data

We used the loan records in Kyushu University Library for the period from April 2013 to March 2014. The total number of the records was 242,798. Each loan record includes the following attributes:

- Title of the book,
- Date of publication,

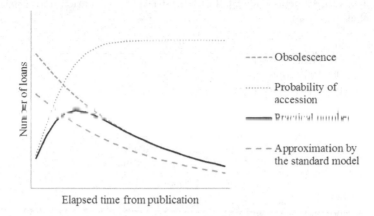

Fig. 1. The idea of the new obsolescence model for the number of loans with the time from publication

- Date of accession to the library, and
- Date of loan.

Additionally, some books have the attribute for classification. We focused on

- NDC number

which represents the subject field as a 3-digit number. Table 1 shows a part of the classification by NDC. In the collected data, 140,444 records (57.8% of the total records) had the NDC number.

Note that an NDC number can refer plural categories of different levels. In the rest of this paper, we call the 10 categories for the first-level classification (notated by the first digit) the *main categories*, and the categories for the second and the third levels (notated by the second and the third digits, respectively) the *subcategories* and the *subsubcategories*, respectively.

2.3 Analyses

We applied the two obsolescence models in Section 2.1 to the collected data in Section 2.2 and evaluated the suitability of the models.

Let t_p, t_a, and t_l be the dates of the publication, the accession, and the loan of a loan record in the collected data, respectively. Then, the times from accession and publication of the loan are $t_l - t_a$ and $t_l - t_p$, respectively. In the two models, t was quantized in units of years, specifically, we regarded t_d days as $\lceil t_d/365 \rceil$ years for the time t_d obtained from the dates in a record.

First, we applied the standard model to the collected data. We approximated the change of the number of loans with the time from accession by the function f in Eq. 1. We obtained discrete numbers $\ell(t)$ of loans for $t = 1, 2, \ldots, 34$ from

Table 1. The subject fields for NDC in the first-level classification and a part of the second-level classification

1st		2nd	
000	General works		
100	Philosophy		
200	General history		
300	Social sciences		
400	Natural sciences	400	Natural sciences (sub)
		410	Mathematics
		420	Physics
		430	Chemistry
		⋮	
500	Technology. Engineering		
600	Industry and commerce		
700	The arts. Fine arts		
800	Language		
900	Literature		

the data for the time from accession. Then, we optimized a and b in f by the generalized reduced gradient method [3] so that the root mean square errors between $\ell(t)$ and $f(t)$

$$\sqrt{\frac{\sum_{t=1}^{34} (\ell(t) - f(t))^2}{34}}$$

would be the smallest. We investigated the smallest root mean square errors and the decreasing rates b for

- The total records,
- The records with NDC number, and
- The records in the 10 main categories.

Next, we applied the standard and the modified models to the data. We approximated the number of loans with the time from publication by the function g in Eq. 2 in addition to f. We conducted the same analysis as the previous one. The length of the discrete values $\ell(t)$ for the time from publication was 117. The function e in Eq. 2 was estimated from the distribution of the times from publication to accession (that is, $t_a - t_p$) in the records. Let n be the number of the total records in the relevant category, and $h(t)$ the number of the books accessioned after t years from the publication. Then,

$$e(t) = \frac{1}{n} \int_0^t h(t).$$

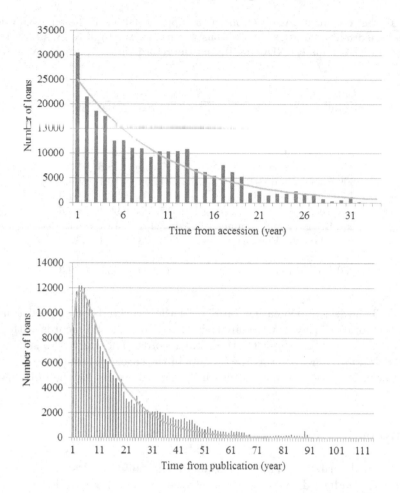

Fig. 2. The change of the number of loans with the time from accession (top) and publication (bottom)

3   Results

Fig. 2 shows the changes of the number of loans with the time from accession and publication for the total loan records. We can confirm the decrease of the number of loans over time, that is, obsolescence, in each graph. The change was approximated to the line by the least squares method with f in Eq. 1 and g in Eq. 2 for the top and the bottom graphs in Fig. 2, respectively.

Table 2 shows the root mean square errors in the approximation of the number of loans with the time from publication by f and g for the total records, the records with the NDC number, and the records in the 10 subcategories of NDC. The errors by g were smaller than those by f in the categories except for one.

Table 2. The root mean square errors in the approximation of the number of loans with the time from publication by functions f and g for the total records (Total), the records with NDC number (NDC total), and the records in the 10 main categories in NDC

Category	Num. loans	f	g
Total	242,798	529.9	383.6
NDC total	140,444	518.6	243.0
000 General works	6,386	23.1	22.1
100 Philosophy	9,140	28.7	21.2
200 General history	7,636	28.6	21.3
300 Social sciences	31,556	116.1	60.1
400 Natural sciences	38,459	190.9	94.0
500 Technology. Engineering	17,233	90.1	64.5
600 Industry and commerce	3,451	12.7	11.9
700 The arts. Fine arts	3,138	17.1	15.0
800 Language	11,746	104.7	72.4
900 Literature	11,699	35.6	39.3

Table 3 shows the decreasing rates with the two kinds of elapsed time for the 12 categories in the previous analysis. The rates for the records with NDC number were similar to those for the total records. Therefore, the subset was even in this sense. The rate was the highest in the category 800 (Language) and the lowest in 200 (General history) in the 10 subcategories for the both elapsed times.

4 Discussion

This section describes considerations about the main results in the previous section. We conducted extra analyses of obsolescence for detailed categories. Some problems still remain as future work.

4.1 Main Conclusion

We found that the proposed model represented the obsolescence, the change of the number of book loans over the time from publication, correctly in the sense of the root mean square errors compared with the standard model.

Additionally, we found the difference of the decreasing rates in the categorized books. We can expect that books in a category with a high (low) decreasing rate would be used for a short (resp. long) time after the accession or the publication. As a criterion for evaluating the decreasing rates, Fig. 3 simulates the change of the number of loans by Eq. 1 with two decreasing rates 1.05 and 1.15. For example, we can estimate that the number of loans of books in Language would be half 5 years after while the number be still about 80% in General history. The difference should be considered for selecting books in the library.

Table 3. The decreasing rates for the time from accession and publication for the total records (Total), the records with NDC number (NDC total), and the records in the 10 main categories in NDC

Category	Num. loans	Accession	Publication
Total	242,798	1.111	1.086
NDC total	140,444	1.102	1.088
000 General works	11,490	1.140	1.100
100 Philosophy	9,140	1.088	1.079
200 General history	7,636	**1.053**	**1.055**
300 Social sciences	31,556	1.133	1.110
400 Natural sciences	38,459	1.089	1.077
500 Technology. Engineering	17,233	1.100	1.086
600 Industry and commerce	3,451	1.099	1.090
700 The arts. Fine arts	3,138	1.093	1.085
800 Language	11,746	**1.152**	**1.111**
900 Literature	11,699	1.086	1.079

Table 4. The decreasing rates for the time from accession and publication for four subcategories and a subsubcategory in NDC

Category	Num. loans	Accession	Publication
320 Law	5,807	1.213	1.148
380 Customs, folklore and ethnology	5,286	1.063	1.061
400 Natural sciences (sub)	1,006	1.114	1.105
420 Physics	8,496	1.070	1.065
007 Information science	2,431	1.144	1.115

4.2   Key Findings

We also found the difference of the decreasing rates between detailed categories. We conducted the same analysis for the subcategories and some subsubcategories in NDC. In some categories, the number of the records was small and hence the error of the approximation was large. Table 4 shows some remarkable results of this extra analysis. In Table 3, the number of loans was large in 300 (Social sciences) and 400 (Natural sciences), and the two decreasing rates were high in 300 and low in 400. The rates for 320 (Law) were the highest and those for 380 (Customs, folklore and ethnology) were lowest in the 10 subcategories in 300 with more than 1,000 records, and the rates for 400 (the subcategory of Natural sciences) were the highest and those for 420 (Physics) were lowest in the main category 400 with more than 1,000 records. Additionally, we picked up 007 (Information science) which is categorized independently from the hierarchy in NDC.

The following situations are intimating that the hierarchy of NDC can stand improvement to meet the use of books. In Japan, we often argue about subject fields in a duality with social sciences and natural sciences. Actually, the numbers

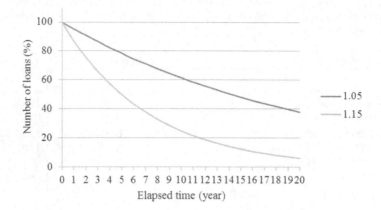

Fig. 3. The changes of the number of loans with the different decreasing rate

of the records in 300 (Social sciences) and 400 (Natural sciences) were large in the circulation data. The result that the decreasing rates in Social sciences were higher than Natural sciences was contrary to our prediction. According to the result of the extra analysis, the decreasing rate was high for 320 (Law), which corresponds with the intuitions of a staff in the library. Although the rates in Natural sciences were low, the category does not include the subject fields in Engineering or Computer science. The rates in 007 (Information science) were high compared with the subcategories in Natural sciences.

4.3   Future Directions

The loan records we used were for only one year. We are planning to evaluate the obsolescence models with loan records for a longer period. We are going to construct the models with old records and estimate the number of loans in new records, and then we can investigate a kind of suitability of the models as the accuracy of the estimation.

The function e in Eq. 2 was estimated from the practical data in the analysis. We can obtain a general model for obsolescence by replacing e as a general function such as the Gauss error function. However, a general model requires extra variables for approximation, and hence we need some techniques for optimization.

5   Conclusion

We proposed an obsolescence model for the number of book loans over the time from publication. We applied the proposed model and the standard model to practical circulation data in a library. The root mean square errors by the proposed model were smaller than the standard one. Therefore, we can conclude that the proposed model is suitable for representing obsolescence with the kind of elapsed time. Additionally, we obtained the differences of obsolescence between closely classified books. The information is expected to be useful for book selection in the library. The main idea of our analysis is clearly applicable to circulation data in other libraries.

Acknowledgments. This work was supported by Kyushu University Interdisciplinary Programs in Education and Projects in Research Development (P&P) from 2014 to 2015 and JSPS KAKENHI Grant Number 15K00310.

References

1. Evans, G.E.: Book selection and book collection usage in academic libraries. The Library Quarterly **40**(3), 297–308 (1970)
2. Koizumi, M.: Book selection by librarians and faculty through collection evaluation methods: A case study of keio university library in japan (in Japanese). Library and Information Science **63**, 41–59 (2010)
3. Lasdon, L.S., Fox, R.L., Ratner, M.W.: Nonlinear optimization using the generalized reduced gradient method. RAIRO - Operations Research - Recherche Operationnelle **8**(V3), 73–103 (1974)
4. Line, M.B.: Changes in the use of literature with time - obsolescence revisited. Library Trends **41**(4), 665–683 (1993)
5. Matsui, A., Isono, H.: An analytical survey of the loan transaction records using "turnover rates" and "circulation rates" - on some characteristics of the library material usage in nara university - (in Japanese). Memoirs of the Nara University **34**, 177–190 (2006)
6. Pritchard, S.J.: Purchase and use of monographs originally requested on interlibrary loan in a medical school library. Library Acquisitions: Practice & Theory **4**(2), 135–139 (1980)

Workshop on Computational Geometry and Security Applications (CGSA 2015)

Randomized Versus Deterministic Point Placement Algorithms: An Experimental Study

Asish Mukhopadhyay[(⊠)], Pijus Kumar Sarker,
and Kishore Kumar Varadharajan Kannan

School of Computer Science, University of Windsor, Windsor, ON N9B 3P4, Canada
{asishm,sarkerp,varadhak}@uwindsor.ca

Abstract. The point location problem is to determine the position of n distinct points on a line, up to translation and reflection by the fewest possible pairwise (adversarial) distance queries. In this paper we report on an experimental study of a number of deterministic point placement algorithms and an incremental randomized algorithm, with the goal of obtaining a greater insight into the behavior of these algorithms, particularly of the randomized one

Keywords: Computational geometry · Point-placement · Turnpike problem · Experimental algorithms

1 Introduction

The Point Placement Problem: Let $P = \{p_1, p_2, ..., p_n\}$ be a set of n distinct points on a line L. The point location problem is to determine the locations of the points uniquely (up to translation and reflection) by making the fewest possible pairwise distance queries of an adversary. The queries can be made in one or more rounds and are modeled as a graph whose nodes represent the points and there is an edge connecting two points if the distance between the corresponding points is being queried. The distances between the pairs of points returned by the adversary are exact.

A special version of this problem is when a query graph is presented with assigned edge lengths and all possible placements of its vertices are to be determined. In [1], this problem was solved for weakly triangulated graphs.

A classical version of this problem is the construction of the coordinates of a set of n points, given exact distances between all pairs of points (see [2], [3]). Algorithms exist that not only determine the coordinates but also the minimum dimension in which the points can be embedded (see [4]).

Motivation: The motivation for studying this problem stems from the fact that it arises in diverse areas of research, to wit computational biology, learning theory, computational geometry, etc.

A. Mukhopadhyay—Research supported by an NSERC Discovery Grant.

O. Gervasi et al. (Eds.): ICCSA 2015, Part II, LNCS 9156, pp. 185–196, 2015.
DOI: 10.1007/978-3-319-21407-8_14

In learning theory [5] this problem is one of learning a set of points on a line non-adaptively, when learning has to proceed based on a fixed set of given distances, or adaptively when learning proceeds in rounds, with the edges queried in one round depending on those queried in the previous rounds.

The version of this problem studied in Computational Geometry is known as the turnpike problem. The description is as follows. On an expressway stretching from town A to town B there are several gas exits; the distances between all pairs of exits are known. The problem is to determine the geometric locations of these exits. This problem was first studied by Skiena et $al.$ [6] who proposed a practical heuristic for the reconstruction. A polynomial time algorithm was given by Daurat et $al.$ [7].

In computational biology, it appears in the guise of the restriction site mapping problem. Biologists discovered that certain restriction enzymes cleave a DNA sequence at specific sites known as restriction sites. For example, it was discovered by Smith and Wilcox [8] that the restriction enzyme Hind II cleaves DNA sequences at the restriction sites GTGCAC or GTTAAC. In lab experiments, by means of fluorescent in situ hybridization (FISH experiments) biologists are able to measure the lengths of such cleaved DNA strings. Given the distances (measured by the number of intervening nucleotides) between all pairs of restriction sites, the task is to determine the exact locations of the restriction sites. The point location problem also has close ties with the probe location problem in computational biology (see [9])

The turnpike problem and the restriction mapping problem are identical, except for the unit of distance involved; in both of these we seek to fit a set of points to a given set of inter-point distances. As is well-known, the solution may not be unique and the running time is polynomial in the number of points. While the point placement problem, prima facie, bears a resemblance to these two problems it is different in its formulation - we are allowed to make pairwise distance queries among a distinct set of labeled points. It turns out that it is possible to determine a unique placement of the points up to translation and reflection in time that is linear in the number of points.

Overview of Contents: In the next section we briefly review some of the well-known deterministic algorithms and the only known incremental randomized algorithm. In the following section we report on the experimental results obtained by careful implementations of several deterministic algorithms and the incremental randomized algorithm. This is followed by a detailed discussion of the results and we conclude in the next section.

2   Overview of Some Current Point Placement Algorithms

Several algorithms are extant that work in one or more rounds. The current state of the art is summarized in Table 1.

Comment: The $9n/8$ lower bound on 2-round algorithms was proved in [10], improving the lower bound of $30n/29$ by Damaschke [5] and the subsequent

Table 1. *The current state of the art*

Algorithm	Rounds	Query Complexity		Time Complexity
		Upper Bound	Lower Bound	
3-cycle	1	$2n - 3$	$4n/3$	$O(n)$
4-cycle	2	$3n/2$	$9n/8$	$O(n)$
5-cycle	?	$4n/3 + O(\sqrt{n})$	$9n/8$	$O(n)$
5:5 jewel	2	$10n/7 + O(1)$	$9n/8$	$O(n)$
6:6 jewel	2	$4n/3 + O(1)$	$9n/8$	$O(n)$
3-path	2	$9n/7$	$9n/8$	$O(n)$
randomized	2	$n + O(n/\log n)$	?	$O(n^2/\log n)$

improvement to $17n/16$ by [11] and the further improvement to $12n/11$ by [12]. As for the lower bound on 1-round algorithms, the following result was proved in [5].

Theorem 1. *[5] The density of any line rigid graph is 4/3 with the exception of the jewel, $K_{2,3}$, K_3 and K_4^- (shown in Fig 1).*

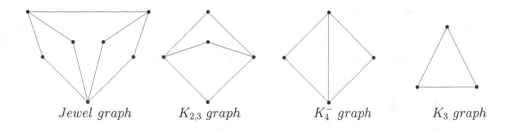

Jewel graph $K_{2,3}$ *graph* K_4^- *graph* K_3 *graph*

Fig. 1. Graphs quoted in Theorem1.

The density, multiplied by n, gives the lower bound of $4n/3$.

The simplest of all, the 3-cycle 1-round algorithm, has the query graph shown in Fig. 2:

The query complexity of this algorithm is $2n - 3$ self-evident as this is the number of edges in the graph. The 4-cycle 2-round algorithm is typical of the other 2-round algorithms listed in Table 1 and thus merits a brief description.

If $G = (V, E)$ is a query graph, an assignment l of lengths to the edges of G is said to be valid if there is a placement of the nodes V on a line such that the distances between adjacent nodes are consistent with l. We express this by the notation (G, l). By definition (G, l) is said to be line rigid if there is a unique placement up to translation and reflection, while G is said to be line rigid if (G, l) is line rigid for every valid l. A 3-cycle (or triangle) graph is line rigid,

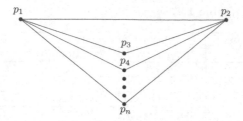

Fig. 2. Query graph using triangles.

which is why the 3-cycle algorithm needs only one round to fix the placement of all the points. A 4-cycle (or quadrilateral) is not line rigid, as there exists an assignment of lengths that makes it a parallelogram whose vertices have two different placements as in Fig. 3.

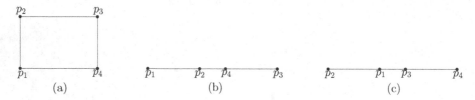

Fig. 3. Two different placements of a parallelogram $abcd$

4-cycle Algorithm

For this algorithm, the query graph presented to the adversary in the first round has the structure shown in Fig. 4.

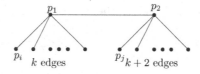

Fig. 4. Query graph for first round in a 2-round algorithm using quadrilaterals.

Making use of the following simple but useful observation,

Observation 1. *At most two points can be at the same distance from a given point p on a line L,*

in the second round we query edges connecting pairs of leaves, one from the group of size k and the other from the group of size $k + 2$, making quadrilaterals that are not parallelograms (the rigidity condition $|p_1 p_i| \neq |p_2 p_j|$ ensures that the quadrilateral $p_1 p_i p_j p_2$ is not a parallelogram).

5-cycle Algorithm

In the 5-cycle algorithm [11], the query graph submitted to the adversary in the first round is shown in Fig. 5.

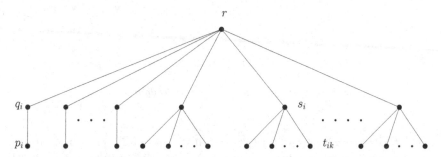

Fig. 5. *Query graph for the 5-cycle algorithm*

Each five cycle is completed by selecting edges to ensure that the following rigidity conditions are satisfied. For more details on this algorithm see [11].

1. $|p_i q_i| \neq |r s_j|$
2. $|p_i q_i| \neq |s_j t_{jk}|$
3. $|p_i q_i| \neq ||r s_j| \pm |s_j t_{jk}||$
4. $|s_j t_{jk}| \neq |q_i r|$
5. $|s_j t_{jk}| \neq ||p_i q_i| \pm |q_i r||$

3-path Algorithm

In the 3-path algorithm [13], the query graph submitted to the adversary in the first round is shown in Fig. 6.

In the second round, the algorithm select edges suitably to satisfy the following rigidity conditions.

1. $|p_1 p_2| \notin \{|r_1 s|, |r_2 s|, ||r_1 s| \pm |r_2 s||\}$,
2. $|p_2 p_3| \notin \{|r_2 s|, |r_3 s|, ||r_2 s| \pm |r_3 s||\}$,
3. $|p_3 p_1| \notin \{|r_3 s|, |r_1 s|, ||r_3 s| \pm |r_1 s||\}$,
4. $|p_1 q_1| \notin \{|r_1 s|, |r_2 s|, ||r_1 s| \pm |r_2 s||, ||p_1 p_2| \pm |r_1 s||, ||p_1 p_2| \pm |r_2 s||, ||p_1 p_3| \pm |r_1 s||, ||p_1 p_3| \pm |r_3 s||, ||p_1 p_2| \pm |r_1 s| \pm |r_2 s||, ||p_1 p_3| \pm |r_1 s| \pm |r_3 s||\}$,
5. $|p_2 q_2| \notin \{|r_1 s|, |r_2 s|, |p_1 q_1|, ||r_1 s| \pm |r_2 s||, ||p_1 p_2| \pm |r_1 s||, ||p_1 p_2| \pm |r_2 s||, ||p_2 p_3| \pm |r_2 s||, ||p_2 p_3| \pm |r_3 s||, ||p_1 q_1| \pm |r_1 s||, ||p_1 q_1| \pm |r_2 s||, ||p_1 p_2| \pm |r_1 s| \pm |r_2 s||, ||p_2 p_3| \pm |r_2 s| \pm |r_3 s||, ||p_1 q_1| \pm |r_1 s| \pm |r_2 s||, ||p_1 q_1| \pm |p_1 p_2| \pm |r_1 s||, ||p_1 q_1| \pm |p_1 p_2| \pm |r_2 s||, ||p_1 q_1| \pm |p_1 p_2| \pm |r_1 s| \pm |r_2 s||\}$,

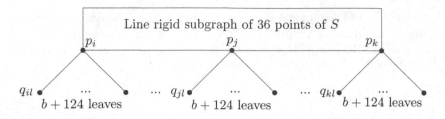

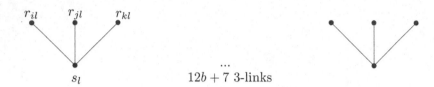

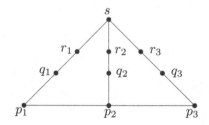

Fig. 6. *Query graph for the 3-path algorithm*

6. $|p_3q_3| \notin \{|r_1s|, |r_2s|, |r_3s|, |p_1q_1|, |p_2q_2|, ||r_2s| \pm |r_3s||, ||r_3s| \pm |r_1s||, ||p_1p_3| \pm |r_3s||, ||p_2p_3| \pm |r_3s||, ||p_1q_1| \pm |r_1s||, ||p_1q_1| \pm |r_3s||, ||p_2q_2| \pm |r_2s||, ||p_2q_2| \pm |r_3s||, ||p_1p_3| \pm |r_1s| \pm |r_3s||, ||p_2p_3| \pm |r_2s| \pm |r_3s||, ||p_1q_1| \pm |r_1s| \pm |r_3s||, ||p_2q_2| \pm |r_2s| \pm |r_3s||, ||p_1q_1| \pm |p_1p_3| \pm |r_3s||, ||p_2q_2| \pm |p_2p_3| \pm |r_3s||, ||p_1q_1| \pm |p_1p_3| \pm |r_1s| \pm |r_2s||, ||p_2q_2| \pm |p_2p_3| \pm |r_2s| \pm |r_3s||\}.$

on each 3-path component shown in Fig. 7. For more details on this algorithm see [13].

Fig. 7. *A 3-path component*

Randomized Algorithm

Damaschke [14] proposed an incremental randomized algorithm (for an introduction to randomized algorithms see [15]) that expands a set L of points whose positions have been fixed. The set L is initialized by picking an arbitrary point p_0 from S and setting it as the origin of the line on which the points lie. Relative to p_0 a random path $P = p_0p_1p_2...$ is incrementally constructed by choosing a point p_i at random from the set $S - L$, and measuring the distance $d(p_i, p_{i+1})$ for each $i = 0, 1, 2, ...$ Simultaneously, the algorithm maintains all possible signed sums $\pm d(p_0p_1) \pm d(p_1p_2) \pm \cdots \pm d(p_i, p_{i+1}) \cdots$, until for some p_{k+1} the signed sums are no longer all distinct.

If a signed sum that repeats is the actual distance of p_{k+1} from p_0, then the placement of p_k relative to p_{k+1} becomes ambiguous. We stop at this point, query the distance $d(p_0, p_k)$ and use the signed sum equal to this distance to fix the placements on L of all the points on the path from p_1 to p_k (in Damaschke's description the position of p_k is fixed relative to two points in L and the signed sum corresponding to this position is chosen to fix the placements of the other points on the path constructed thus far).

Resetting p_k as the new p_0 and p_{k+1} as the new p_1, the algorithm repeats until $L = S$.

Damaschke proved the following result.

Theorem 1. *[14] The above randomized algorithm for the point location problem has, for any instance, performance ratio $1 + O(1/\log n)$ with high probability.*

The term performance ratio is the number of distance queries divided by the number of points.

It is straightforward to turn this into a 2-round algorithm. Fix the placement of 2 points p_0 and p_1 and choose a random path $P = p_1p_2 \ldots p_n$ on all the remaining points to be placed and submit this query graph to the adversary. As before, we compute signed sums, stopping when two signed sums are equal when we have reached the point p_{k+1} on P. We resolve the ambiguity in the placement of p_{k+1} by adding edges from p_{k+1} to p_0 and p_1, whose lengths we will query in the second round. Continue as in the incremental algorithm from p_{k+1} on.

3 Experimental Results

We implemented all the four deterministic algorithms and the 2-round version of the incremental randomized algorithm, discussed in the previous section. The control parameters used for comparing their performances are: query complexity and time complexity. The results of the experiments for the deterministic algorithms are shown in the graphs below. In our experiments, we simulated an adversary by creating a linear layout and checking the placements of the points by the algorithms against this. This also solved the problem of ensuring a valid assignment of lengths to the queried edges. We will have more to say about this in the next section.

Predictably enough, the above chart shows that the behavior of the algorithms with respect to query complexity is consistent with the upper bounds for these algorithms shown in Table 1. Each of these algorithms were run on points sets of different sizes, up to 50000 points.

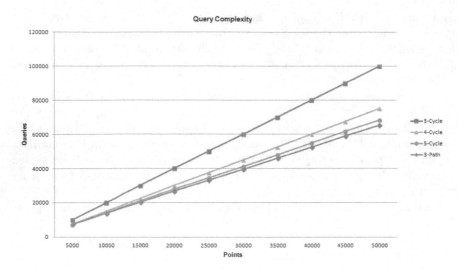

Fig. 8. *Query Complexity Graph*

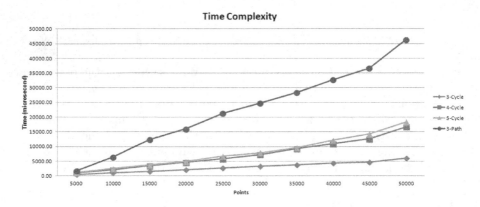

Fig. 9. *Time Complexity Graph*

Clearly, 3-cycle is consistently the fastest; but despite its complex structure the 3-path algorithm does well as compared to the 4-cycle and the 5-cycle algorithms. We have not included the performance of the randomized algorithm in the above graphs as it is incredibly slow and we ran it for point sets of size up to 16,000. Table 2 below shows its performance details.

4   Discussion

The behavior of the deterministic algorithms with respect to time complexity is opposite to their behavior with respect to query complexity. The growth-rate of

Table 2. Performance of 2-round randomized algorithm

Number of points	Number of Distance Queries	Running time (hrs:mins:secs)
2000	2382	0:10:41
4000	4712	0:57:32
6000	7048	2:25:38
8000	9348	5:27:53
10000	11000	8:38:34
12000	13999	13:25:24
14000	16282	18:34:58
16000	18625	23:19:40

the running time versus the size of the input point-set is also near-linear. Both results are as expected.

As reported, in none of the deterministic algorithms it was explicitly stated how to obtain an actual layout from the rigid graph constructed on the input point set. In our implementations we devised a signed-sum technique to generate a layout.

The assumption that an assignment of lengths is valid is a strong one and, as mentioned earlier, we circumvented this problem by creating a layout and reporting queried lengths based on this. The correctness of the placements of the points by an algorithm is verified by checking that it generates a layout identical to the one used to report queried lengths.

An algorithmic approach to the solution of this problem is based on constructing the Cayley-Menger matrix out of the squared distances of a query graph.

For a query graph with n vertices, the pre-distance matrix $D = [D_{ij}]$ is a symmetric matrix such that $D_{ij} = d_{ij}^2$, where d_{ij} is the distance between the vertices (points) i and j of the query graph. The Cayley-Menger matrix, $C = [C_{ij}]$ is a symmetric $(n + 1) \times (n + 1)$ matrix such $C_{0i} = C_{i0} = 1$ for $0 < i \leq n$, $C_{00} = 0$ and $C_{ij} = D_{ij}$ for $1 \leq i, j \leq n$ [16], [2].

The vertices of the query graph has a valid linear placement provided the rank of the matrix B is at most 3 (this is a special case of the result that there exists a d-dimensional embedding of the query graph if the rank of B is at most $d + 2$; our claim follows by setting $d = 1$) [2].

It's interesting to check this out for the query graph in Fig. 10 on 3 points. The Cayley-Menger matrix B for the above query graph is:

$$B = \begin{bmatrix} 0 & 1 & 1 & 1 \\ 1 & 0 & 1 & x^2 \\ 1 & 1 & 0 & 4 \\ 1 & x^2 & 4 & 0 \end{bmatrix},$$

where $x = d_{13}$, the unknown distance between the points p_1 and p_3.

By the above result, the 4×4 minor, $\det(\mathrm{B}) = 0$. This leads to the equation

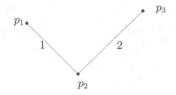

Fig. 10. *A query graph on 3 vertices*

$$x^4 - 10x^2 + 9 = 0$$

which has two solutions $x = 3$ and $x = 1$, corresponding to the two possible placements (embeddings) of the points p_1, p_2 and p_3. Assuming p_2 is placed to the right of p_1, in one of these placements p_3 is to the right of both p_1 and p_2; in the other, to the left of them both.

4.1   Deterministic versus Randomized

Table 2 lends credence to the claim by Damaschke [14] that the number of distance queries of the incremental randomized algorithm is bounded above by $O(n(1 + 1/\log n))$ in the worst case. Unfortunately, it is too slow to be run with very large inputs.

We suspect that the number of times signed sums become equal is intimately connected with the distribution of the points that we generate by pretending to be the adversary. To test this we generated the layout by picking a point at random in a fixed size interval, and picking the next random point in the same fixed-size interval whose left end point is the last point selected. In our experiments we varied this fixed interval from 5 units to 500000 units and reported the number of times we got equal signed sums for points sets of sizes varying from 20 to 1000. Interestingly enough, as can be seen from Table 3 below that the numbers decrease as the interval-size increases.

Table 3. Performance of incremental randomized algorithm for nearly uniform distributions

# of points	1-5	1-10	1-20	1-50	1-100	$1 - 10^3$	$1 - 10^4$	$1 - 5 * 10^4$	$1 - 10^5$	$1 - 5 * 10^5$
20	7	7	6	5	4	3	3	2	2	1
50	16	13	11	10	9	7	6	6	5	4
100	25	23	20	19	15	11	9	8	8	7
200	45	39	35	33	29	22	18	17	16	
400	78	70	61	56	49	41	39	34		
1000	167	149	140	123	111	94	82	76		

The incremental randomized algorithm is often held up as an example of simplicity in comparison to deterministic algorithms, like the 3-path one, for example. The above experiments paint a completely different picture. From a practical point of view, it is completely ineffective as it is essentially a brute-force algorithm. The 3-path algorithm, on the other hand, scores high on both parameters - low query complexity and low time complexity.

5 Conclusions

All algorithms have been implemented in C on a computer with the following configuration: Intel(R) Xeon(R) CPU, X7460 @ 2.66GHz OS: Ubuntu 12.04.5, Architecture: i686.

Further work can be done on several fronts. Particularly worthwhile is to conduct further experiments into the behavior of the randomized algorithm, specifically the influence of floating point arithmetic on keeping signed sums unequal. On the theoretical side, it might be interesting to come up with a completely different randomized algorithm - one that does not depend on maintaining an exponential number of signed sums.

References

1. Mukhopadhyay, A., Rao, S.V., Pardeshi, S., Gundlapalli, S.: Linear layouts of weakly triangulated graphs. In: Pal, S.P., Sadakane, K. (eds.) WALCOM 2014. LNCS, vol. 8344, pp. 322–336. Springer, Heidelberg (2014)
2. Young, G., Householder, A.S.: Discussion of a set of points in terms of their mutual distances. Psychometrika **3**(1), 19–22 (1938)
3. Blumenthal, L.M.: Theory and applications of distance geometry, 2nd edn. Chelsea, New York (1970)
4. Crippen, G., Havel, T.: Distance Geometry and Molecular Conformation. John Wiley and Sons (1988)
5. Damaschke, P.: Point placement on the line by distance data. Discrete Applied Mathematics **127**(1), 53–62 (2003)
6. Skiena, S.S., Smith, W.D., Lemke, P.: Reconstructing sets from interpoint distances (extended abstract). In: Proceedings of the Sixth Annual Symposium on Computational Geometry, SCG 1990, pp. 332–339. ACM, New York (1990)
7. Daurat, A., Gérard, Y., Nivat, M.: The chords' problem. Theor. Comput. Sci. **282**(2), 319–336 (2002)
8. Smith, H., Wilcox, K.: A restriction enzyme from hemophilus influenzae. i. purification and general properties. Journal of Molecular Biology **51**, 379–391 (1970)
9. Redstone, J., Ruzzo, W.L.: Algorithms for a simple point placement problem. In: Bongiovanni, G., Petreschi, R., Gambosi, G. (eds.) CIAC 2000. LNCS, vol. 1767, pp. 32–43. Springer, Heidelberg (2000)
10. Alam, M.S., Mukhopadhyay, A.: Three paths to point placement. In: Ganguly, S., Krishnamurti, R. (eds.) CALDAM 2015. LNCS, vol. 8959, pp. 33–44. Springer, Heidelberg (2015)

11. Chin, F.Y.L., Leung, H.C.M., Sung, W.K., Yiu, S.M.: The point placement problem on a line – improved bounds for pairwise distance queries. In: Giancarlo, R., Hannenhalli, S. (eds.) WABI 2007. LNCS (LNBI), vol. 4645, pp. 372–382. Springer, Heidelberg (2007)
12. Alam, M.S., Mukhopadhyay, A.: More on generalized jewels and the point placement problem. J. Graph Algorithms Appl. **18**(1), 133–173 (2014)
13. Alam, M.S., Mukhopadhyay, A.: Improved upper and lower bounds for the point placement problem. CoRR abs/1210.3833 (2012)
14. Damaschke, P.: Randomized vs. deterministic distance query strategies for point location on the line. Discrete Applied Mathematics **154**(3), 478–484 (2006)
15. Motwani, R., Raghavan, P.: Randomized Algorithms. Cambridge University Press, New York (1995)
16. Emiris, I.Z., Psarros, I.D.: Counting euclidean embeddings of rigid graphs. CoRR abs/1402.1484 (2014)

GPS Data Interpolation: Bezier Vs. Biarcs for Tracing Vehicle Trajectory

Rahul Vishen[1]($\boxtimes$), Marius C. Silaghi[1], and Joerg Denzinger[2]

[1] Florida Institute of Technology, Melbourne, FL, USA
rvishen2009@my.fit.edu, msilaghi@fit.edu
[2] University of Calgary, Calgary, AB, Canada
denzinge@cpsc.ucalgary.ca

Abstract. Our target is a driving simulator application that is designed to generate a simulation environment that is, in fact, a recreation of a prerecorded driving experience. The simulator does not just replicate the original scenery, but allows the users to maneuver within a recorded environment. The idea is to video-record the environment while driving a vehicle, recording the precise geolocation of each frame. During simulation, the precise geolocation of each frame is important for generating smooth transition from a viewpoint to another.

The assumption is that the recorded frames are tagged with their precise geolocations. However, in practice this is not true due to difference in sampling rates of the hardware systems involved. Cameras record images at a higher frequency in comparison to the GPS (Global Positioning System) data from a GPS unit. To tag frames with relatively accurate geolocations we present two interpolation techniques that trace the trajectory of a recoding vehicle using the GPS data. We then compare the effectiveness of the two techniques by drawing a comparison with respect to the ground truth based on error registered in position and orientation.

Keywords: GPS · Interpolation · Driving simulation · Bézier curves · Biarc · Trace trajectory

1 Introduction

Our target application is a driving simulator that is an attempt to move from a realistic experience towards a real one by creating a *driving circuit* using real world images that are recorded during an actual driving session [2,4,10,13]. Most simulators available today create a realistic environment through graphically designed virtual circuits. The designers attempt to represent the real environment to the best of their abilities. However, it is difficult to include the smallest of details that may still be of some importance. Visual cues are one of the factors that play an important role in the validation of driving simulations [12,15,17,22]. Studies suggest a direct correlation between the quality of the display environment and validation of simulation [16,22,26,30]. The driving

O. Gervasi et al. (Eds.): ICCSA 2015, Part II, LNCS 9156, pp. 197–208, 2015.
DOI: 10.1007/978-3-319-21407-8_15

circuits in our target application are previously recorded driving sequences where users can maneuver through as they would in the real life. A driving circuit is generated using a setup of several synchronized cameras mounted on a surface vehicle. During simulation, for the system to produce a smooth transition from a viewpoint to another knowing the precise geolocation of each frame is critical.

Common interpolation techniques focus on optimizing length, or on guaranteeing smoothness. However, the constraints of the interpolation problem we face greatly depend on the process of recording a driving circuit. Through interpolation of GPS data we attempt to trace the trajectory that may have been followed by the recording vehicle. In our studied simulator the recording vehicle is a four wheeled automobile which is subject to a limit on the maximum supported acceleration. For example, we know that a human being in such a vehicle can only sustain an acceleration of some $4.0g$ [32]. Therefore we have to be able to guarantee an upper bound on the trajectory curvature.

In Section 2 we give an overview of the work relevant to our problem. Section 3.1 and 3.2, present a technique that uses piecewise cubic Bézier curves based and biarc based interpolations techniques, respectively. In Section 4 we evaluate the techniques with respect to the ground truth.

2   Related Work

Previous attempts to build such simulators are presented in [2, 4, 10, 13]. A type of approach is used in the Aspen Movie Map [23], where four cameras are placed at a 90° angle interval on a circular disk. However, users experiencing simulation in the virtual environment do not have the ability to change their viewpoint. An improvement to this work is suggested in QuicTime VR [7]. The cameras capture an outward cylindrical projection [28] of the view around the vehicle, creating a cylindrical environment. We use a similar approach to recording a driving circuit. Circuit images are recorded while the vehicle is driven following a desired route. During simulation the recorded cylindrical projections allow the system to generate a virtual world where the users have the ability to look around. However, the users can only look around from a fixed point inside a cylindrical environment. To be able to navigate around in an environment, a solution using omni-directional cameras is suggested in [8, 14, 29, 33]. The common assumption is that the recorded frames are tagged with their precise geolocations [31]. Although the subproblem addressed in [33] does not require a GPS to allow transitions within a recorded viewpoint, for the solution to work with the full system, each recorded projection needs to be tagged with its precise geolocation. For example [14, 29], suggest using a GPS sensor to get precise geolocation for the outdoor image recordings. However, in practice this is not easy due to difference in sampling rates of the hardware systems involved. Cameras record images at a higher frequency in comparison to the data from a GPS unit. This makes an impediment for tagging every image projection with its precises geolocation. As a result, between any pair of consecutive GPS tagged cylindrical frames we have a set of projections that are not tagged with their precise geolocation.

Trajectory Compression: Essentially the problem we face here is of tracing the trajectory of a vehicle using only the GPS data, and represent the trace with a continuous curve. A relevant work is trajectory compression, a.k.a, trajectory encoding [20,21,36]. The problem of compression is to reduce the GPS data size by removing some recorded measurements. In order to reconstruct the original trajectory one uses some sort of interpolation techniques. In [36], a parametric cubic function is proposed that obtains a spline between any two spatiotemporal data points. Each spatiotemporal data point contains position and the time at which the data point is recorded. The data point also has the recorded velocity. A similar approach is used in [21] to encode a trajectory. However, the techniques compress the data maintaining trajectory within a given accuracy bound. Extension to this is presented in [20] where an attempt is made to improve accuracy using clothoids. Clothoids are curves where the curvature varies linearly with its length [24,35].

Path Planning: Another related work is that of path planning [1,9,11,34]. Curvature constrained path planning, in presence of obstacles, is studied in [1]. The algorithm determines an obstacle free path from a point A to point B, for a point robot with a bound on maximal curvature. In [9], navigation algorithms are suggested that guide a robot to visit a set of waypoints while adhering to corridor constraints. The algorithms use piecewise-Bézier-curve to represent a path between a pair of consecutive waypoints. The path segments are joined in a manner such that the result is a C^2 class continuous curvature curve.[1] Bézier curves are also used in [34] path smoothing. In addition, a bisection method based approach is suggested to subdivide Bézier curve segments in order to respect maximal curvature constraints. On the other hand a different approach is used in [11] exploring B-splines for real-time path smoothing.

Bézier Curves: Given a set of points, $(P_0, P_1, ..., P_N)$, parametric definition of a Bézier curve, of degree N, is given by [34]:

$$P(t) = \sum_{i=0}^{N} P_i \binom{N}{i} t^i (1-t)^{N-i}, t \in [0,1] \tag{1}$$

P_0 and P_N are the end points of the curve, the remaining intermediary points are the control points that do not necessarily fall on the curve. From interpolation perspective the major drawback of Bézier curves is that they only approximate the control points rather than pass through them, an important property for interpolation. This can be addressed by taking a piecewise approach as seen in [9,34]. Using the piecewise approach a curve is constructed by concatenating several Bézier segments. Although the resulting curve is a C^0 continuous curve, higher degree of continuity can be achieved by manipulation of control points as suggested in [9].

---

[1] In terms of parametric continuity, a C^n class curve is a curve whose first through n^{th} derivatives are continuous [3].

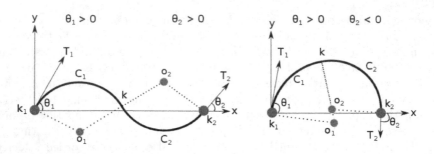

Fig. 1. S- and C-shaped biarcs and their tangents

Biarc Curves: Also know as piecewise-circular curves, Biarcs were introduced by Bolton in 1975 [6]. Biarcs have been extensively used for geometric modeling [5] [18] [25]. In certain application such as geometric modeling for computer aided manufacturing, machine tools are better suited to move along a circular path or a straight line [25]. A biarc is a curve consisting of two circular arcs that meet at a point where they share a common tangent vector. The shape of a biarc between two end points depends on the tangent vectors defined at these points and the vector connecting the end points. Biarcs have been classified into two categories, C-shaped, and S-shaped (See Figure 1). Figure 1 shows examples of the biarc curve fitting given two points along with their direction tangents. The biarc in this figure consists of circular arcs, C_1 and C_2, joining two end points, k_1 and k_2, which have associated tangent vectors, T_1 and T_2 respectively. The circular arcs C_1 and C_2 share a common tangent vector at point k. As shown in the figure, if θ_1 and θ_2 have the same sign, we have an S-shaped curve (Figure 1, left image). If the signs are opposite, we have a C-shaped curve (Figure 1, right image). Given the points k_1 and k_2 with their associated tangent vectors, T_1 and T_2 respectively, the problem of biarc curve fitting is to find the positions for the centers O_1 and O_2 of each circular arc C_1 and C_2 respectively. Versions of bi-arc algorithms are described in [19,27].

Unlike the trajectory compression problem, for our target driving simulator application we need higher positional and orientation accuracy. On the other we share two of the most common constraints found in the path planning problems - bounded curvature, and continuous curvature path. In this paper we propose two techniques that, given accurate GPS data, can be used to interpolate GPS coordinates for the previously untagged image projections. First, we present a piecewise cubic Bézier based interpolation technique where for each Bézier piece the placement of each of the two control points is a function of vehicle velocity as recorded by the GPS. Second, we present a piecewise Biarc based interpolation technique that, given recorded direction vectors for two GPS coordinates, guarantees a continuous curve with the minimal curvature by concatenating two arcs. We evaluate the results based on ground truth.

3 Interpolation Techniques

We are given a sequence of images, $(I_0, I_1, ..., I_Z)$, captured by a recording vehicle. Each image I_i is tagged with a unique GPS data record, $G_i = \langle P_i, V_i, t_i \rangle$. P_i · $\langle x_i, y_i \rangle$ is a GPS point (coordinate), where x_i is the longitude and y_i is the latitude, and V_i is the recorded velocity at time t_i. Due to movement and to difference between the data sampling frequencies of the camera and of the GPS sensor, some of the images are not tagged with GPS coordinates acquired within the last frame period. As a result, between a pair of images (I_l, I_n), $l < n$ that are tagged with their accurate GPS points P_l and P_n, we have a sequence of images, $(I_{l+1}, ..., I_m, ..., I_{n-1})$, with $P_m = P_l$, where P_m is the GPS point recorded for image I_m. The assumption here is that the GPS sensor has no positional error. The problem is to trace the trajectory of the recording vehicle in order to deduce the geolocations for the previously untagged images.

3.1 Bézier Based Trajectory Interpolation

We take the piecewise approach to tracing the trajectory. Each pair of consecutive GPS points, (P_i, P_{i+1}), for $i = 0, 1, ..., N - 1$, forms a single segment of the entire trajectory. Each segment, S_i, for $i = 1, 2, ..., N$, is a cubic Bézier curve that is defined by a sequence of four control points $(P_i, P_A^i, P_B^i, P_{i+1})$. Here P_i and P_{i+1} are a pair of consecutive GPS points, P_A^i and P_B^i are the intermediary control points. For each segment S_i, we compute the control points, P_A^i and P_B^i, such that when all the segments $(S_1, S_2, ..., S_N)$ are joined at points $(P_1, P_2, ..., P_{N-1})$ respectively, the trajectory is a C^1 class continuous curve. Each point P_i, except P_0 and P_N, has two associated control points P_B^{i-1} and P_A^i that are defined on a line parallel to $|P_{i-1}P_{i+1}|$ passing through P_i (See Figure 2a). This allows the concatenated trajectory to be a C^1 continuous curve. A similar approach is suggested in [9]. P_0 and P_N are the data end points and have only one associated control point, P_A^0 and P_B^{N-1} respectively. P_A^0 is placed on the line segment joining P_0 and P_1, and P_B^{N-1} is placed on the line segment joining P_{N-1} and P_N. The $||P_iP_B^{i-1}||$ and $||P_iP_A^i||$ is some fraction, f_i, of $||P_{i-1}P_i||$ and $||P_iP_{i+1}||$ respectively. The function is defined as:

$$f_i = F(s_i) = 0.5 \frac{s_i^2}{s_i^2 + x} \qquad (2)$$

The value of f_i is a parametrized function of speed, s_i recorded at P_i that can be tuned through the value of x, where $x >= 0$.

3.2 Biarc Based Trajectory Interpolation

Similar to the approach used in Bézier based technique, a trajectory is constructed by concatenating several segments. Each segment, $(S_i \mid i = 1, 2, ..., N)$, is a biarc curve that joins P_{i-1} and P_i. The algorithm we use for our experiments with biarc curves is described in [19]. Given a pair of consecutive GPS

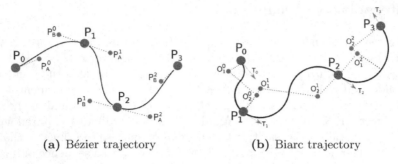

(a) Bézier trajectory   (b) Biarc trajectory

Fig. 2. Trajectory examples

points P_{i-1} and P_i along with their respective direction vectors T_{i-1} and T_i, the algorithm finds a biarc with minimal curvature difference between the two arcs. T_i is a shared direction vector between segments S_i and S_{i+1}, thus making the complete trajectory a C^1 continuous curve. The direction vector for each point P_i, except for P_0 and P_N, is set parallel to the direction vector $\overrightarrow{P_{i-1}P_{i+1}}$. For P_0 the direction vector is parallel to $\overrightarrow{P_0P_1}$, and for P_N the direction vector is parallel to $\overrightarrow{P_{N-1}P_N}$. Using the points and their corresponding direction vectors an intial piecewise biarc based trace of the trajectory is generated. We perform a hill-climbing search for a local optima while updating the direction vectors by $\pm\Delta$ radians ($\pm\Delta$ is reduced/halved on convergence up to a minimum value). An optima is reached when the global maximum curvature cannot be minimized. Each search iteration attempts to minimize the curvature difference between two consecutive biarc segments of a the trajectory.

4   Experiments

For evaluation of the proposed interpolation techniques we require highly accurate GPS data. High accuracy GPS units are expensive. Instead, we use an alternate approach to record geolocations that factors out GPS errors and the need to account for such errors. A recording vehicle marks the road surface while following a path at a constant speed of approximately 8 Kmph (5 mph). The marking represents the true trajectory of the vehicle on the Earth's surface, we call it the *ground truth* (See Figure 3). The ground truth is captured using a high altitude camera. Since we know the geographic location where the ground truth is recorded, using a GIS (Geographic Information System) tool, Google Earth, we are able to obtain a digitized version of the ground truth. Figure 4 shows the digitized version of the ground truth markings seen in Figure 3. The digitized version of the ground truth is a sequence of geolocations (latitude and longitude pair) with a sampling interval of 50 centimeters (See Figure 4). We want to be able to compare the two techniques with respect to their ability to reconstruct the trajectories in absence of the complete ground truth. To do so, a subset of the digitized ground truth, dark colored points in Figure 4, is used to simulate accurate input GPS points for our proposed interpolation techniques. This subset of GPS points can be sampled at some defined interval distance. Figure 4 gives two examples of the recorded ground truth and GPS points sampled at different intervals.

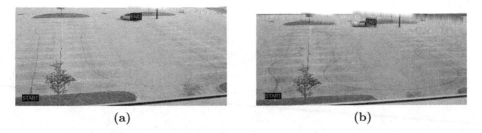

(a)   (b)

Fig. 3. Ground truth marked by a recording vehicle

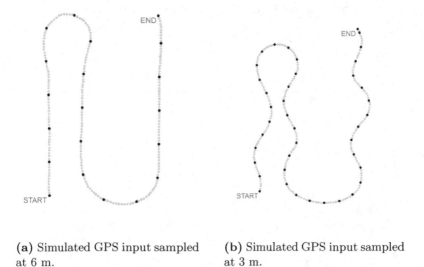

(a) Simulated GPS input sampled   (b) Simulated GPS input sampled
at 6 m.   at 3 m.

Fig. 4. Ground truth (light colored) and sample GPS input (dark colored)

The reconstructed trajectories are matched against the ground truth for accuracy in terms of position and orientation. Figure 5 shows an example of trajectory reconstruction using Bézier and biarc based techniques for a common input. The experiments are conducted using 296 different ground truths which are sampled at various interval distances to simulate GPS input sets for our techniques. We compare the root mean square error (positional and orientation) registered for

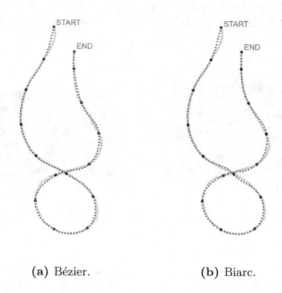

(a) Bézier.   (b) Biarc.

Fig. 5. Example of reconstructed trajectories (red colored)

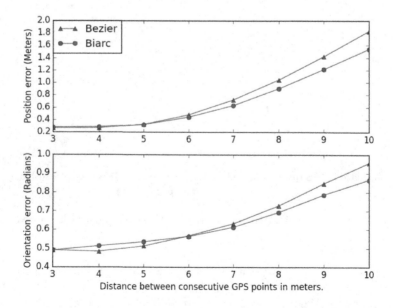

Fig. 6. Bézier and Biarc average RMS error

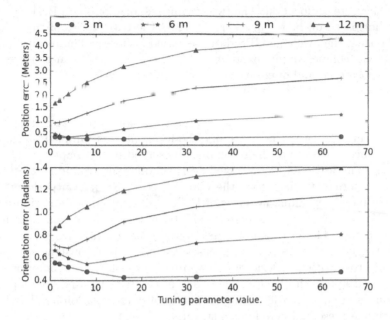

Fig. 7. Bézier tuning parameter

all the available ground truths when GPS points are sampled at different interval distances. Figure 6 shows the graph representing the average RMS (root mean square) errors registered in all the reconstructed trajectories with respect to position and orientation. To compute position and orientation error we compare the corresponding sample points from the ground truth and the reconstructed trajectory. The sample points are sampled at the distance of 0.5 meter. Position error for a single trajectory represents the average RMS displacement measured between the corresponding sample points from the ground truth and its reconstructed trajectory. Orientation error is the error in the bearing of the recording vehicle. Bearing of a point on the ground truth is the direction of movement registered by a recording vehicle at that point. In Section 3.1, we discuss the tuning parameter for Bézier technique (See Equation (2)). The parameter is used to fine tune reconstruction of trajectory when using the Bézier technique. Figure 7 shows the average errors registered at different values for the tuning parameter.

5   Conclusion

For our target application, the recording based driving simulator, we present in this paper two interpolation techniques addressing the need of tagging images with their accurate geolocation coordinates. The images and GPS data are recorded by a vehicle while moving along a defined path. Due to differences in sampling rate of the hardware systems involved, some images are not tagged with their accurate geolocation. The two techniques, one based on Bézier curves and the other based on biarc curves, reconstruct the trajectories of the recording vehicle using the available GPS data, assumed accurate. The reconstructed trajectories are compared with respect to the ground truth to measure errors in terms of positional and orientation accuracy. Ground truth of a trajectory is the digitized version of the actual trajectory marked by the recording vehicle on the earth's surface, the digitized version is extracted using a GIS tool. In absence of an accurate GPS device we choose to sample input GPS data from the ground truth. For a single trajectory the input data is sampled at various interval distances to evaluate the techniques presented in this paper.

The results in Section 4 are averaged over 296 different ground truth trajectories. Our experiments show that the biarc based technique outperforms the Bézier curve based technique in terms of both positional and orientation accuracy. The Bézier based technique performs better in terms of positional accuracy for interval distances lower than 6 meters. However, as the distance between two accurately tagged GPS data points increases, the errors in reconstruction of the original trajectory with the Bézier based approach grow faster in comparison to the biarc based approach. Another drawback of the Bézier based techniques is that it requires tuning in order to improve control point positioning. The biarc based technique presented here has no such requirement.

References

1. Agarwal, P.K., Biedl, T., Lazard, S., Robbins, S., Suri, S., Whitesides, S.: Curvature-constrained shortest paths in a convex polygon. SIAM Journal on Computing **31**(6), 1814–1851 (2002)
2. Allard, J.C., Deslypper, C., Saunier, C.: Method and device for training in the operation of moving vehicles, uS Patent 4,750,888, June 14, 1988
3. Barsky, B.A., DeRose, T.D.: Geometric continuity of parametric curves. Computer Science Division, University of California (1984)
4. Blanton, K.A., Finlay, W.M., Sinclair, M.J., Tumblin, J.E.: Method and apparatus for reproducing video images to simulate movement within a multi-dimensional space, uS Patent 4,752,836, June 21, 1988
5. Boissonnat, J.D., Cazals, F.: Smooth surface reconstruction via natural neighbour interpolation of distance functions. In: Proceedings of the Sixteenth Annual Symposium on Computational Geometry, pp. 223–232. ACM (2000)
6. Bolton, K.: Biarc curves. Computer-Aided Design **7**(2), 89–92 (1975)
7. Chen, S.E.: Quicktime vr: an image-based approach to virtual environment navigation. In: Proceedings of the 22nd Annual Conference on Computer Graphics and Interactive Techniques, pp. 29–38. ACM (1995)
8. Chen, S.E., Williams, L.: View interpolation for image synthesis. In: Proceedings of the 20th Annual Conference on Computer Graphics and Interactive Techniques, pp. 279–288. ACM (1993)
9. Choi, J.W., Curry, R., Elkaim, G.: Piecewise bezier curves path planning with continuous curvature constraint for autonomous driving. In: Machine Learning and Systems Engineering, pp. 31–45 (2010)

10. Deslypper, C.: Method for reading a recorded moving scene, in particular on a videodisk, and application of said method to driving simulators, uS Patent 4,939,587, July 3, 1990
11. Elbanhawi, M., Simic, M., Jazar, R.N.: Continuous path smoothing for car-like robots using b-spline curves. Journal of Intelligent & Robotic Systems, 1–34 (2015)
12. Dijgström, J., Johansson, E., Östlund, J.: Effects of visual and cognitive load in real and simulated motorway driving, Transportation Research Part F: Traffic Psychology and Behaviour 8(2), 97–120 (2005)
13. Foerst, R.: Driving simulator, uS Patent 4,383,827, May 17, 1983
14. Ikeuchi, K., Sakauchi, M., Kawasaki, H., Sato, I.: Constructing virtual cities by using panoramic images. International Journal of Computer Vision 58(3), 237–247 (2004)
15. Jamson, H.: Driving simulation validity: issues of field of view and resolution. In: Proceedings of the driving simulation conference, pp. 57–64 (2000)
16. Kaptein, N.A., Theeuwes, J., Van Der Horst, R.: Driving simulator validity: Some considerations. Transportation Research Record: Journal of the Transportation Research Board 1550(1), 30–36 (1996)
17. Kemeny, A., Panerai, F.: Evaluating perception in driving simulation experiments. Trends in cognitive sciences 7(1), 31–37 (2003)
18. Koc, B., Ma, Y., Lee, Y.: Smoothing stl files by max-fit biarc curves for rapid prototyping. Rapid Prototyping Journal 6(3), 186–205 (2000)
19. Koc, B., Lee, Y.S., Ma, Y.: Max-fit biarc fitting to stl models for rapid prototyping processes. In: Proceedings of the Sixth ACM Symposium on Solid Modeling and Applications (2000)
20. Koegel, M., Baselt, D., Mauve, M., Scheuermann, B.: A comparison of vehicular trajectory encoding techniques. In: Ad Hoc Networking Workshop (Med-Hoc-Net), 2011 The 10th IFIP Annual Mediterranean, pp. 87–94. IEEE (2011)
21. Koegel, M., Kiess, W., Kerper, M., Mauve, M.: Compact vehicular trajectory encoding. In: 2011 IEEE 73rd Vehicular Technology Conference (VTC Spring), pp. 1–5. IEEE (2011)
22. Levine, O.H., Mourant, R.R.: Effect of visual display parameters on driving performance in a virtual environments driving simulator. Proceedings of the Human Factors and Ergonomics Society Annual Meeting 40, 1136–1140 (1996). SAGE Publications
23. Lippman, A.: Movie-maps: an application of the optical videodisc to computer graphics. In: ACM SIGGRAPH Computer Graphics, vol. 14, pp. 32–42. ACM (1980)
24. Makino, H.: Clothoidal interpolation-a new tool for high-speed continuous path control. CIRP Annals-Manufacturing Technology 37(1), 25–28 (1988)
25. Moreton, D., Parkinson, D., Wu, W.: The application of a biarc technique in cnc machining. Computer-Aided Engineering Journal 8(2), 54–60 (1991)
26. Mullen, N., Charlton, J., Devlin, A., Bedard, M.: Simulator validity: Behaviors observed on the simulator and on the road (2011)
27. Rossignac, J.R., Requicha, A.A.G.: Piecewise-circular curves for geometric modeling. IBM Journal of Research Development, 31(3) (1987)
28. Snyder, J.P.: Map Projections - Working Manual, U.S. Geological Survey Professional Paper 1395, pp. 37–47. United States Government Printing Office, Washington, D.C. (1987)
29. Takahashi, T., Kawasaki, H., Ikeuchi, K., Sakauchi, M.: Arbitrary view position and direction rendering for large-scale scenes. In: Proceedings of IEEE Conference on Computer Vision and Pattern Recognition, vol. 2, pp. 296–303. IEEE (2000)

30. Thiffault, P., Bergeron, J.: Monotony of road environment and driver fatigue: a simulator study. Accident Analysis & Prevention **35**(3), 381–391 (2003)
31. Tomite, K., Yamazawa, K., Yokoya, N.: Arbitrary viewpoint rendering from multiple omnidirectional images for interactive walkthroughs. In: Proceedings of 16th International Conference on Pattern Recognition, vol. 3, pp. 987–990. IEEE (2002)
32. Voshell, M.: High acceleration and the human body (2004). http://csel.eng.ohio-state.edu/voshell/gforce.pdf
33. Wither, J., Tsai, Y.T., Azuma, R.: Indirect augmented reality. Computers & Graphics **35**(4), 810–822 (2011)
34. Yang, K., Jung, D., Sukkarieh, S.: Continuous curvature path-smoothing algorithm using cubic b zier spiral curves for non-holonomic robots. Advanced Robotics **27**(4), 247–258 (2013)
35. Yao, Z., Joneja, A.: Path generation for high speed machining using spiral curves. Computer-Aided Design & Applications **4**, 191–198 (2007)
36. Yu, B., Kim, S.H., Bailey, T., Gamboa, R.: Curve-based representation of moving object trajectory. In: Proceedings of the International Database Engineering and Application Symposium (2004)

Constrained k-Center Problem
on a Convex Polygon

Manjanna Basappa, Ramesh K. Jallu, and Gautam K. Das[(✉)]

Department of Mathematics, Indian Institute of Technology Guwahati,
Guwahati 781039, India
{manjanna,j.ramesh,gkd}@iitg.ernet.in

Abstract. In this paper, we consider a restricted covering problem, in which a convex polygon $\mathcal{P}$ with n vertices and an integer k are given, the objective is to cover the entire region of $\mathcal{P}$ using k congruent disks of minimum radius r_{opt}, centered on the boundary of $\mathcal{P}$. For $k \geq 7$ and any $\epsilon > 0$, we propose a $(1 + \frac{7}{k} + \frac{7\epsilon}{k} + \epsilon)$-factor approximation algorithm, which runs in $O(n(n + k)(|\log r_{opt}| + \log\lceil\frac{1}{\epsilon}\rceil))$ time. The previous best known approximation factor in the literature for the same problem is 1.8841 [H. Du and Y. Xu: An approximation algorithm for k-center problem on a convex polygon, J. Comb. Optim. (2014), 27(3), 504-518].

Keywords: Approximation algorithm · Convex Polygon Cover · Geometric disk cover

1 Introduction

Geometric covering is a well studied topic in computational geometry. Various types of covering problems include covering one type of geometric objects with minimum number of some other types of geometric objects. The wide range of geometric objects include points, lines, disks, squares, rectangles etc. One of the interesting covering problems is the k-center problem. In the k-center problem, a set of clients (e.g. mobile users, houses etc.) are distributed over a 2D region, the objective is to choose k locations for facilities (e.g. base stations for mobile networks, post office, warning sirens etc.) so that each client can get service from at least one facility with in minimum distance. Sometimes, we may need to restrict these facilities to be located only on the boundary of the given region, for example base station placement in a forbidden region (for a big lake) [5,12].

Let $\mathcal{P}$ be a given convex polygon. We use $\partial\mathcal{P}$ to denote the boundary of $\mathcal{P}$. Here, we consider the constrained k-center problem as follows:

> Constrained Convex Polygon Cover (CCPC): Given a convex polygon $\mathcal{P}$ and an integer k, the objective is to cover the entire region of $\mathcal{P}$ with k congruent disks of minimum radius r_{opt}, centered on $\partial\mathcal{P}$.

In Subsection 1.1, we discuss the related work. We present an algorithm for the decision version of the CCPC problem in Section 2. In Section. 3 we present an approximation algorithm for the CCPC problem. Finally, we conclude in Section 4.

© Springer International Publishing Switzerland 2015
O. Gervasi et al. (Eds.): ICCSA 2015, Part II, LNCS 9156, pp. 209–222, 2015.
DOI: 10.1007/978-3-319-21407-8_16

1.1 Related Work

Given a set $\mathcal{Q}$ of n points in the plane, the classical k-center problem is to cover $\mathcal{Q}$ with k congruent disks of radius as minimum as possible. The above problem is called discrete k-center problem when the centers of k disks are restricted to be chosen from a set of points, otherwise it is called the continuous k-center problem or simply the k-center problem. The various constrained versions of the k-center problem have been studied extensively in the literature. Hurtado et al. [8] considered Euclidean 1-center problem satisfying m linear constraints and proposed an $O(n + m)$ time algorithm for it. Bose and Toussaint [4] provided an $O((n + m) \log(n + m))$ time algorithm for the 1-center problem where the disk is centered on the boundary of a convex polygon and the objective is to cover the demand points that may lie either outside polygon or inside polygon. Brass et al. [3] studied the similar problem where the centers are constrained to lie on a straight line L. If the line L is given, their algorithm uses parametric search and runs in $O(n \log^2 n)$ time. Karmakar et al. [10] proposed three algorithms for this problem with time complexities $O(nk \log n)$, $O(nk + k^2 \log^3 n)$ and $O(n \log n + k \log^4 n)$ respectively. Using parametric search, Kim and Shin [11] solved 2-center problem for polygon in $O(n \log^2 n)$ time, where the two centers are restricted to be at some vertices of the polygon. Halperin et al. [7] also considered the version of 2-center problem where the two centers are restricted to lie outside the boundaries of disjoint simple polygons with a total of m edges. For this version of 2-center problem, Halperin et al. [7] gave an algorithm with expected time $O(m \log^2(mn) + mn \log^2 n \log(mn))$ and $(1 + \epsilon)$-approximation algorithm in time $O(\frac{1}{\epsilon} \log(\frac{1}{\epsilon})(m \log^2 m + n \log^2 n))$ or in randomized expected time $O(\frac{1}{\epsilon} \log(\frac{1}{\epsilon})((m + n \log n) \log(mn)))$, where $\epsilon > 0$ and n is the number of points to be covered. Suzuki and Drezner [13] investigated the p-center location problem for demand originating in an area and proposed heuristic procedures for this problem. Du and Xu [6] studied k-center problem on a convex polygon where the centers are restricted to lie on the boundary of the polygon and presented 1.8841-factor approximation algorithm, which runs in $O(nk)$ time, where n is the number of vertices in the polygon and k is the number of disks. Das et al. [5] provided $(1 + \epsilon)$-approximation algorithm for k-center problem on convex polygon where the centers are restricted to lie on a specified edge of the polygon. If the centers are restricted to be on the boundary of a convex polygon, Roy et al. [12] presented linear time algorithm and Das et al. [5] gave $O(n^2)$ time algorithm respectively for $k = 1, 2$. Das et al. [5] also discussed heuristic algorithm for the same problem, for $k \geq 3$.

Agarwal and Sharir [2] studied the Euclidean k-center problem. Here, a set $\mathcal{Q}$ of n points, a set $\mathcal{O}$ of m points, and an integer k are given. The objective is to find k disks centered at the points in $\mathcal{O}$ such that each point in $\mathcal{Q}$ is covered by at least one disk and the radius of the largest disk is minimum. This problem is known to be NP-hard [2]. For fixed k, Hwang et al. [9] presented a $m^{O(\sqrt{k})}$-time algorithm. Later, Agarwal and Procopiuc [1] gave $m^{O(k^{1-1/d})}$-time algorithm for the d-dimensional points.

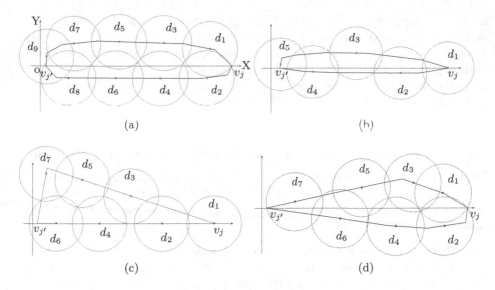

Fig. 1. Constrained placement of disks

1.2 Our Results

In this paper, we present an algorithm for covering a convex polygon $\mathcal{P}$ with k congruent disks of radius $r \leq (1 + \delta)r_{opt}$, centered on its boundary, where r_{opt} is the optimum radius, $\delta = \frac{7}{k} + \frac{7\epsilon}{k} + \epsilon$, $k \geq 7$ and any $\epsilon > 0$. The running time of our algorithm is $O(n(n + k)(|\log r_{opt}| + \log\lceil \frac{1}{\epsilon} \rceil))$, where n is the number of vertices of $\mathcal{P}$. The previous best known approximation factor for the same problem is 1.8841 [6]. The running time of 1.8841-factor approximation algorithm is $O(nk)$.

2 Decision Problem

In this section, we provide an algorithm to solve the decision version of CCPC problem approximately. The decision version of the problem is as follows:

k-COVER($\mathcal{P}$, k, r): Given a convex polygon $\mathcal{P}$, an integer k and a radius r, check whether $\mathcal{P}$ has a cover with k congruent disks of radius r centered on $\partial\mathcal{P}$.

Let $dist(p', p'')$ denote the Euclidean distance between two points p' and p''. For any two points p and q, $\overline{pq}$ denotes the line segment joining p and q. For any disk d_i, let ∂d_i be its boundary arc and the center of d_i be (x_i, y_i). Let the convex polygon $\mathcal{P}$ be placed lying to right of y-axis (see Fig. 1).

2.1 Preprocess

Let $\mathcal{P} = (v_1, v_2, \ldots, v_n)$ be a convex polygon. Here, we first perform alignment on $\mathcal{P}$ as follows: For each vertex v_j, $1 \leq j \leq n$, we identify a vertex $v_{j'}$ such that

$dist(v_j, v_{j'}) \geq dist(v_j, v_{j''})$ for all j'', $1 \leq j'' \leq n$ (in the case of more than one vertex, pick an arbitrary one). For each such pair of vertices $(v_j, v_{j'})$, we align $\mathcal{P}$ as follows: v_j and $v_{j'}$ are lying on x-axis and v_j is in right side of $v_{j'}$. For each such alignment, we apply the covering algorithm (Algorithm 1) to cover $\mathcal{P}$ by placing disks d_i ($i \geq 1$) on $\partial\mathcal{P}$ one by one from right to left, where the initial disk d_1 is either centered at v_j (see Fig. 1(b)) or centered on $\partial\mathcal{P}$ (in counter-clockwise direction of v_j) so that ∂d_1 passes through v_j (see Fig. 1(a)).

2.2 $(1 + \frac{7}{k})$-Factor Approximation Algorithm for Decision Problem

Definition 1. *Upper chain (resp. Lower chain) of $\mathcal{P}$ is the locus of points on $\partial\mathcal{P}$ starting from the right most vertex v_j of $\mathcal{P}$ to the left most vertex $v_{j'}$ in counter-clock (resp. clock) wise order. We use the notations U_c and L_c to denote upper chain and lower chain respectively.*

Definition 2. *Disk-constrained placement of disks is the placement of disks d_i ($i \geq 2$) from right to left, centered alternately on L_c and U_c of $\mathcal{P}$ such that the current disk d_i is centered at leftmost point on L_c (resp. U_c) satisfying (i) ∂d_i passes through the leftmost intersection point between the disks d_{i-1} and d_{i-2} for $i \geq 3$, and (ii) d_2 contains the vertex v_j or d_i contains the leftmost intersection point between d_{i-2} and L_c (resp. U_c) for $i \geq 3$ (see Fig. 1(a)).*

Definition 3. *Chain-constrained placement of disks is the placement of disks d_i ($i \geq 2$) from right to left, centered alternately on L_c and U_c of $\mathcal{P}$ such that the current disk d_i is centered at leftmost point on L_c (resp. U_c) satisfying the following condition: d_i contains the leftmost intersection point between (i) d_{i-1} and L_c, and (ii) d_{i-1} and U_c (see Fig. 1(b)).*

A high level description of our algorithm for the decision version of the problem is as follows: we first find the proper alignment of $\mathcal{P}$ as described before. Next we place congruent disks of radius r centered on $\partial\mathcal{P}$ by using the greedy strategies in definitions 2 and 3 (*constrained placement*) until $\mathcal{P}$ is covered by the union of these disks provided $\mathcal{P}$ is coverable by disks of radius r. Let ℓ be the number of congruent disks of radius r centered on $\partial\mathcal{P}$ such that the union of ℓ disks covers $\mathcal{P}$. Next, we increase the radius of first k disks, to $r' = r + \frac{(\ell-k)}{k}r$ and reposition their centers on $\partial\mathcal{P}$ such that every disk satisfies *constrained placement* requirement. Finally, we remove the disks placed after k-th disk. The detailed pseudo-code of our strategy is described in Algorithm 1, 2, 3 and 4. The outline of our algorithm for the decision problem is summarized by the following steps.

(a) Align the convex polygon $\mathcal{P}$ (as described in preprocess).
(b) Run Algorithm 1 which centers ℓ congruent disks of given radius r, on $\partial\mathcal{P}$ by *constrained placement* until $\mathcal{P}$ is covered by the union of these disks if at all $\mathcal{P}$ is coverable by disks of radius r.
(c) Run Algorithm 4 which resets the radius of first k disks to $r' = r + \frac{(\ell-k)}{k}r$, repositions their centers on $\partial\mathcal{P}$ such that every disk satisfies *constrained placement* requirement and finally removes the $(\ell - k)$ redundant disks.

Algorithm 1. ℓ-COVER$(\mathcal{P}, r)$

1: **Input:** Aligned convex polygon $\mathcal{P}$, radius r.
2: **Output:** Set $\mathcal{D} = \{d_1, d_2, \ldots, d_\ell\}$ of disks of radius r, centered on $\partial\mathcal{P}$ such that
   $\mathcal{P} \cap (\cup_{d \in \mathcal{D}} d) = \mathcal{P}$ if at all it is feasible to cover $\mathcal{P}$ with k disks of radius r.
3: Place the disk d_1 centered at v_j.
4: **if** d_2 can not be placed on lower chain by *chain-constrained placement* **then**
5:    Reposition d_1 such that ∂d_1 passes through v_j and d_1 is centered on upper chain.

6:    Place d_2 on lower chain such that ∂d_2 passes through v_j.
7:    Set $\mathcal{D} = \{d_1, d_2\}$
8: **else**
9:    Place d_2 on lower chain by *chain-constrained placement*
10:   Set $\mathcal{D} = \{d_1, d_2\}$
11: **end if**
12: $\mathcal{D} = constrained\_placement(\mathcal{P}, \mathcal{D}, r)$ //Call Algorithm 2
13: **if** $(\mathcal{P} \cap (\cup_{d \in \mathcal{D}} d) = \mathcal{P})$ **then**
14:   Set $\ell = |\mathcal{D}|$
15:   **Return** $(\mathcal{D}, \ell)$
16: **else**
17:   **Return** $(\emptyset, 0)$
18: **end if**

In Algorithm 2, if switching happens from *disk-constrained placement* to *chain-constrained placement*, then the current disk d_i covers the left intersection point between d_{i-1} and d_{i-2} in addition it covers left intersection points between U_c and the disk $(d_{i-1}$ or $d_{i-2})$ placed on U_c, and between L_c and the disk $(d_{i-1}$ or $d_{i-2})$ placed on L_c respectively (see Fig. 1(d)). On the other hand, if switching happens from *chain-constrained placement* to *disk-constrained placement*, then ∂d_i passes through the left intersection point (i) between U_c and the disk d_{i-1} if d_{i-1} is centered on L_c or (ii) between L_c and the disk d_{i-1} if d_{i-1} is centered on U_c (see Fig. 1(c)).

Whenever Algorithm 2 places a disk d_i, Algorithm 3 is invoked. If d_i is placed by *disk-constrained placement*, Algorithm 3 checks whether the remaining uncovered portion of $\mathcal{P}$ can be covered, by exhaustively placing at most 3 disks. If this portion of $\mathcal{P}$ is covered, Algorithm 3 returns these disks to Algorithm 2.

Notations: Let $\mathcal{D} = \{d_1, d_2, \ldots, d_\ell\}$ be the set of disks placed by Algorithm 1 and $\mathcal{D} \setminus \{d_{k+1}, d_{k+2}, \ldots, d_\ell\}$ be the set of disks retained to be centered on the boundary after increasing the radius and removing the redundant disks in Algorithm 4. Let $\mathcal{D}' = \{d_1', d_2', \ldots, d_k'\}$ be the set of disks in an optimal solution of k-COVER$(\mathcal{P}, k, r)$ and (x_i', y_i') be the center of the disk d_i' $(i = 1, 2, \ldots, k)$. Let $\alpha(i)$ denote the disk $d_s'(\in \mathcal{D}')$ such that $x_{i+2} \le x_s' \le x_i$ and centered on the same chain as d_i and d_{i+2} or $x_{i+3} \le x_s' \le x_{i+1}$ and centered on the same chain as d_{i+1} and d_{i+3}, where $1 \le s \le k$.

Lemma 1. *If the disks $d_1, d_2, \ldots, d_\ell$ are centered on $\partial\mathcal{P}$ by constrained placement (while-loop in line 4 of Algorithm 2), then $(\cup_{i=1}^{\ell} d_i) \cap \mathcal{P} = \mathcal{P}$.*

Algorithm 2. Constrained_placement($\mathcal{P}$, $\mathcal{D}$, r)

1: **Input:** Aligned convex polygon $\mathcal{P}$, set $\mathcal{D}$ of disks and radius r.
2: **Output:** Set $\mathcal{D} = \{d_1, d_2, \ldots\}$ of disks of radius r, centered on $\partial\mathcal{P}$ such that $\mathcal{P} \cap (\cup_{d \in \mathcal{D}} d) = \mathcal{P}$ if at all it is feasible to cover $\mathcal{P}$ with k disks of radius r.
3: Set $i = 3$
4: **while** $((\mathcal{P} \cap (\cup_{d \in \mathcal{D}} d) \neq \mathcal{P})$) **do**
5:     $\mathcal{D}^1 = non\_constrained\_placement(\mathcal{P}, \mathcal{D}, r)$ //Call Algorithm 3
6:     **if** $(\mathcal{P} \cap (\cup_{d \in \mathcal{D}^1} d) = \mathcal{P})$ **then**
7:         Set $\mathcal{D} = \mathcal{D}^1$
8:         **Return**($\mathcal{D}$)
9:     **end if**
10:    **if** d_{i-1} is centered on lower chain **then**
11:        **if** d_i can not be placed on upper chain by *chain-constrained placement* **then**
12:            **if** d_i can not be placed on upper chain by *disk-constrained placement* **then**
13:                **Return** ($\emptyset$)
14:            **end if**
15:            Place d_i on upper chain by *disk-constrained placement*
16:        **else**
17:            Place d_i on upper chain by *chain-constrained placement.*
18:        **end if**
19:    **else**
20:        **if** d_i can not be placed on lower chain by *chain-constrained placement* **then**
21:            **if** d_i can not be placed on lower chain by *disk-constrained placement* **then**
22:                **Return** ($\emptyset$)
23:            **end if**
24:            Place d_i on lower chain by *disk-constrained placement*
25:        **else**
26:            Place d_i on lower chain by *chain-constrained placement.*
27:        **end if**
28:    **end if**
29:    $\mathcal{D} = \mathcal{D} \cup \{d_i\}$, $i = i + 1$
30: **end while**
31: **Return**($\mathcal{D}$)

Proof. In every iteration of while-loop in line 4 of Algorithm 2, the disk d_i is placed on either U_c or L_c such that its ∂d_i passes through the left intersection point between d_{i-1} and d_{i-2} (see Fig. 1). Also d_i covers the nearest intersection point between $\partial\mathcal{P}$ and ∂d_{i-2} (if d_i, d_{i-1} and d_{i-2} are placed by *disk-constrained placement*) or d_i covers both the left intersection points between (i) previously placed disk and U_c, and (ii) previously placed disk and L_c (if d_i and d_{i-1} are placed by *chain-constrained placement*). Thus the lemma follows.   □

Lemma 2. *If the disks d_1, d_2, $\ldots$, d_ℓ are all placed only by chain-constrained placement in Algorithm 2, then $\ell \leq k + 1$.*

Proof. The proof follows from the fact that the center of at least one optimal disk must lie within disk d_i, for every $i \geq 2$, placed by *chain-constrained placement* in Algorithm 2 since x-coordinate of center of d_i is smallest (see Fig. 1(b)).   □

Algorithm 3. non_constrained_placement($\mathcal{P}, \mathcal{D}, r$)

1: **Input:** Aligned convex polygon $\mathcal{P}$, set $\mathcal{D}$ of disks and radius r.
2: **Output:** Set $\mathcal{D}^1 = \{d_1, d_2, \ldots\}$ of disks of radius r, centered on $\partial\mathcal{P}$ such that $(\mathcal{P} \cap (\cup_{d \in \mathcal{D}^1} d) = \mathcal{P})$ if uncovered region in $\mathcal{P}$ can be covered with at most 3 disks.
3: $i = |\mathcal{D}| + 1$, $\mathcal{D}^1 = \mathcal{D}$
4: **if** d_{i-1} is centered by *disk-constrained placement* **then**
5: $\quad$ Place a disk d of radius r centered at left intersection point between d_{i-1} and d_{i-2}.
6: $\quad$ Let $t_1, t_2, \ldots, t_m$ be the intersection points of $\partial\mathcal{P}$ with the disk d in the order on $\partial\mathcal{P}$ starting from center of d_{i-1} (t_1) to center of d_{i-2} (t_m).
7: $\quad$ **for** $(s = 2, 3, \ldots, m - 1)$ **do**
8: $\quad\quad$ Place the disk d_i centered at t_s and set $\mathcal{D}^1 = \mathcal{D}^1 \cup \{d_i\}$
9: $\quad\quad$ **if** $(\mathcal{P} \cap (\cup_{d \in \mathcal{D}^1} d) = \mathcal{P})$ **then**
10: $\quad\quad\quad$ **Return** $(\mathcal{D}^1)$
11: $\quad\quad$ **else**
12: $\quad\quad\quad$ **if** there are two uncovered components in $(\mathcal{P} - (\mathcal{P} \cap (\cup_{d \in \mathcal{D}^1} d)))$ **then**
13: $\quad\quad\quad\quad$ Let Δ_1 and Δ_2 be the uncovered components of $(\mathcal{P} - (\mathcal{P} \cap (\cup_{d \in \mathcal{D}^1} d)))$ lying below and above d_i respectively
14: $\quad\quad\quad\quad$ Place d_{i+1} and d_{i+2} centered on $\partial\mathcal{P}$ such that $(\Delta_1 \cap d_{i+1} = \Delta_1)$ and $(\Delta_2 \cap d_{i+2} = \Delta_2)$ (see Lemma 3)
15: $\quad\quad\quad\quad$ $\mathcal{D}^1 = \mathcal{D}^1 \cup \{d_{i+1}, d_{i+2}\}$
16: $\quad\quad\quad\quad$ **Return** $(\mathcal{D}^1)$
17: $\quad\quad\quad$ **end if**
18: $\quad\quad$ **end if**
19: $\quad$ **end for**
20: **end if**
21: **Return**($\mathcal{D}$)

Lemma 3. *At most three disks are required to be placed by non-constrained placement in Algorithm 3 to cover $\mathcal{P}$.*

Proof. In line 5 of Algorithm 2, Algorithm 3 is invoked to test whether the remaining uncovered portion of $\mathcal{P}$ can be covered with at most three disks (i.e., we have reached the end of $\mathcal{P}$). If the left intersection point of d_{i-1} and d_{i-2} needs to be covered by disk d_i using *non-constrained placement* (see Fig. 2), then the uncovered regions Δ_1 and Δ_2 (if at all existing) above and below d_i must be covered by at most one disk each, because of convexity of $\mathcal{P}$. $\qquad\square$

Algorithm 2 places as many as optimal number of disks plus one by *chain-constrained placement* only (see Lemma 2). Therefore, from now onwards we restrict our discussion to *disk-constrained placement*. For any consecutively placed disks d_i and d_{i+1} (without loss of generality, centered on L_c and U_c of $\mathcal{P}$ respectively) by *disk-constrained placement* in Algorithm 2, let p_1 and p_2 be the intersection points between ∂d_i and ∂d_{i+1} such that $x_{p_1} \leq x_{p_2}$ (for $i \geq 3$), where x_{p_1} and x_{p_2} are the x-coordinates of p_1 and p_2 respectively. Let d'_j and $d'_{j'}$ be two left-most disks centered on L_c and U_c of $\partial\mathcal{P}$ such that $x_i < x'_j$ and $x_{i+1} < x'_{j'}$ respectively. Let l_{p_1} be a vertical line passing through p_1. Let p'_1

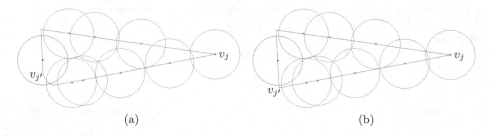

(a)   (b)

Fig. 2. Blue colored disk centered by non-constrained placement

and p'_2 be the left and right intersection points between $\partial d'_j$ and $\partial d'_{j'}$, such that $x_{p'_1} \leq x_{p'_2}$, where $x_{p'_1}$ and $x_{p'_2}$ are the x-coordinates of p'_1 and p'_2 respectively.

Observation 1. *For any consecutively placed disks d_i and d_{i+1} $(i \geq 3)$, p'_1 can not lie to the left of l_{p_1}.*

Proof. Assume that p'_1 is lying to the left of l_{p_1}. Then both d'_j and $d'_{j'}$ should intersect l_{p_1} and at least one of them should contain the point p_1. Without loss of generality let the disk containing p_1 be $d'_{j'}$ centered on U_c. Let q_1 and q_2 be the intersection points between ∂d_{i-1} and ∂d_{i-2} such that $x_{q_1} \leq x_{q_2}$. By *constrained placement* of disks, q_1 is lying to right of l_{p_1}. Let the centers of disks d_{i+1}, $d'_{j'}$ and d_{i-1} be labeled as x, y and z respectively. Let the horizontal line segments from x, y and z be incident at a, b and c respectively, on l_{p_1} (see Fig. 3). Let the length of line segments $dist(p_1, x) = r_1 = r$, $dist(p_1, y) = r_2$, $dist(p_1, z) = r_3$. By *constrained placement* of disks, $r_3 > r_1$ and by assumption $r_1 > r_2$. Therefore, $r_3 > r_1 > r_2$ and $\theta_1 < \theta_2 < \theta_3$, where $\theta_1 = \angle ap_1x$, $\theta_2 = \angle bp_1y$, $\theta_3 = \angle cp_1z$. Therefore there exists a reflex vertex between points x and z on U_c, contradicting that $\mathcal{P}$ is convex.   □

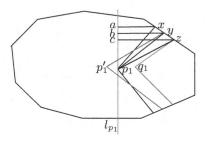

Fig. 3. Proof of Observation 1

In the lemmata 4, 5 and 6, we prove that there must be at least one optimal disk in every subsequence of consecutive four disks placed by *disk-constrained placement*, which leads to the result: at most $k+5$ disks are placed by Algorithm 1 to cover $\mathcal{P}$, where k is the optimal number of disks for a given radius r.

Lemma 4. *For any consecutively placed disks $d_i, d_{i+1}, d_{i+2}, d_{i+3}$, $(i \geq 3)$, there exist at least one disk d'_q centered at (x'_q, y'_q) such that $\alpha(i) = d'_q$, $1 \leq q \leq k$.*

Proof. Let $p_1 = (x_{p_1}, y_{p_1})$ and $p_2 = (x_{p_2}, y_{p_2})$ be the intersection points between ∂d_i and ∂d_{i+1} such that $x_{p_1} \leq x_{p_2}$. Let l_{p_1} be a vertical line passing through p_1 (see Fig. 4). On the contrary of the lemma, assume that no disk d'_q from optimal solution is centered on $\partial \mathcal{P}$ such that d'_q is centered on L_c and $x_{i+2} < x'_q < x_i$ or d'_q is centered on U_c and $x_{i+3} \leq x'_q \leq x_{i+1}$, $1 \leq q \leq k$ (see Fig. 4). Let d'_j and $d'_{j'}$ be two disks from optimal solution centered left most but to right of (x_i, y_i) and (x_{i+1}, y_{i+1}) respectively, that is, $x_i < x'_j$ and $x_{i+1} < x'_{j'}$. Note that (x_f, y_f) is the center of the disk d_f. Let p'_1 and p'_2 be the intersection points between $\partial d'_j$ and $\partial d'_{j'}$. By observation 1, p'_1 is lying to the right of vertical line l_{p_1}. If $dist(p'_1, (x_{i+2}, y_{i+2})) \geq r$, then there must be a disk d'_q centered on $\partial \mathcal{P}$ such that $x_{i+2} \leq x'_q$, otherwise the disk d'_q can not cover p'_1. If $dist(p'_1, (x_{i+2}, y_{i+2})) < r$, some disk d'_q has to be centered such that $x_{i+2} > x'_q$. This implies that the disk d'_q must not cover p_1, otherwise it would contradict that the disk d_{i+2} is centered at (x_{i+2}, y_{i+2}) as d_{i+2} could be moved with its new center is closer to y-axis than (x_{i+2}, y_{i+2}) while ∂d_{i+2} is still passing through p_1. Therefore, the distance between (x_{i+3}, y_{i+3}) and the left intersection point between $\partial d'_q$ and $\partial d'_{j'}$ is greater than r, implying that some other disk $d'_{q'}$ must be centered such that $x_{i+3} \leq x'_{q'}$. □

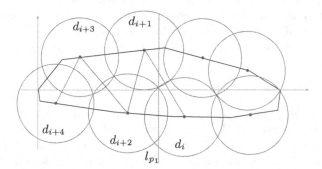

Fig. 4. Proof of Lemmata 4 and 5

Lemma 5. *If $\alpha(i) = d'_s$ and $\alpha(i+1) = d'_t$, then $s \neq t$, where $1 \leq s, t \leq k$.*

Proof. Let the disks d_i, d_{i+2}, d_{i+4} and d_{i+1}, d_{i+3} be centered on the L_c and the U_c of $\mathcal{P}$ respectively (see Fig. 4). By Lemma 4, there exist at least one disk d'_s centered on L_c between the centers of d_i and d_{i+2} or centered on U_c between the centers of d_{i+1} and d_{i+3}. Then, we consider the following two cases:

   (a) If the disk d'_s is placed on L_c between d_i and d_{i+2}, then by the same argument as in Lemma 4, for consecutively placed disks d_{i+1}, d_{i+2}, d_{i+3} and

d_{i+4}, there must be at least one disk d'_t such that $\alpha(i+1) = d'_t$, where $1 \le s, t \le k$ and $s \ne t$.

(b) If the disk d'_s is placed on U_c between d_{i+1} and d_{i+3}, then the center of the disk d'_t placed on L_c, covering the left intersection point between the disks d'_j and $d'_{j'}(= d'_s)$ (to the right of centers of d_i and d_{i+3} respectively) can not lie to the left of the center of d_{i+4} on the L_c since d_{i+4} do not intersect with d_i and d_{i+1} (otherwise there is no need of d_{i+2}). □

Lemma 6. *If k disks are sufficient to cover $\mathcal{P}$ entirely, then at most $k + 7$ disks are required to cover $\mathcal{P}$ by Algorithm 1.*

Proof. Case (i) if no switching occurs in Algorithm 2 and all ℓ disks are placed by *disk-constrained placement*: Let $\mathcal{D}_i$ be the set of disks in the optimal solution $\mathcal{D}'$ corresponding to disks d_i, d_{i+1}, d_{i+2} and d_{i+3} such that for every $d \in \mathcal{D}_i(\subseteq \mathcal{D}')$ is centered either between the centers of d_i and d_{i+2} and on the same chain as d_i and d_{i+2} or between the centers of d_{i+1} and d_{i+3} and on the same chain as d_{i+1} and d_{i+3}. By Lemma 5, $|\mathcal{D}_i \cap \mathcal{D}_{i+1}| < \max(|\mathcal{D}_i|, |\mathcal{D}_{i+1}|)$ and $|\mathcal{D}_i| \ge 1$ for $i \ge 3$. Then

$$k = |\mathcal{D}'| \ge \left| \bigcup_{i=3}^{\ell-3} \mathcal{D}_i \right| \ge \sum_{i=3}^{\ell-3} 1 = \ell - 5 \implies \ell \le k + 5.$$

Case (ii) if switching occurs in Algorithm 2 and a sequence of ℓ disks placed by Algorithm 1 are using *chain-constrained, disk-constrained* and *chain-constrained placement*: Let ℓ_1, ℓ_2 and ℓ_3 be the subsequences of consecutively placed disks by *chain-constrained, disk-constrained* and then *chain-constrained placement* again respectively. Then $\ell = \ell_1 + \ell_2 + \ell_3$. By lemma 2 and case (i), $\ell \le k + 7$. □

Lemma 7. *The running time of Algorithm 4 is $O(n(n + k))$.*

Proof. The for-loop in line 3 of Algorithm 4 runs (in the worst case) for every vertex of convex polygon $\mathcal{P}$. In every iteration of this for-loop, in line 4 we compute the farthest vertex $v_{j'}$ from a vertex v_j in $O(n)$ time. In line 6, we invoke Algorithm 1, which takes $O(n + \ell)$ time. Again lines 8-10 take $O(n + k)$ time. From Lemma 6 we know that $\ell \le k + 7$. Therefore, the running time of Algorithm 4 is $n(n + (n + k + 7) + (n + k)) = O(n(n + k))$ time. □

Theorem 1. *Algorithm 4 is $(1 + \frac{7}{k})$-factor approximation algorithm for $k \ge 7$.*

Proof. Let $\mathcal{D} = \{d_1, d_2, \ldots, d_\ell\}$ be the set of disks centered on $\partial\mathcal{P}$ to cover $\mathcal{P}$ by Algorithm 1. Now, the number of disks centered on L_c after the disk d_k is at most $\lceil \frac{(\ell-k)}{2} \rceil$ and the number of disks centered along L_c starting from the vertex v_j of $\mathcal{P}$ to the center of d_k is at most $\lceil \frac{k}{2} \rceil$. If the radius of these $\lceil \frac{k}{2} \rceil$ disks centered on L_c is increased by ρ such that the area covered by $\lceil \frac{(\ell-k)}{2} \rceil$ disks centered on L_c after d_k, is covered by these $\lceil \frac{k}{2} \rceil$ disks, then $\rho = \frac{(\lceil \frac{(\ell-k)}{2} \rceil)r}{\lceil \frac{k}{2} \rceil} = \lceil \frac{(\ell-k)}{k} \rceil r$. Let the radius of every disk $d_i \in \mathcal{D}$, for $1 \le i \le k$, be increased by an additive factor ρ,

---

Algorithm 4. k-COVER$(\mathcal{P}, k, r)$

1: **Input:** Convex polygon $\mathcal{P}$, a positive integer k and radius r.
2: **Output:** Set $\mathcal{D} = \{d_1, d_2, \ldots, d_k\}$ of disks of radius $r' \leq (1 + \frac{7}{k})r$, centered on $\partial\mathcal{P}$
   such that $\mathcal{P} \cap (\cup_{d \in \mathcal{D}} d) = \mathcal{P}$ if $\mathcal{P}$ is coverable with k disks of radius r.
3: **for** $(j = 1, 2, \ldots, n)$ **do**
4:     Compute the farthest vertex $v_{j'}$ from vertex v_j.
5:     Align $\mathcal{P}$ such that v_j and $v_{j'}$ are lying on x-axis, x-coordinate of v_j is greater
       than x-coordinate of $v_{j'}$ and the whole $\mathcal{P}$ lies to right of y-axis.
6:     $(\mathcal{D}, \ell)$= ℓ-COVER$(\mathcal{P}, r)$ //Run Algorithm 1
7:     **if** $((\mathcal{P} \cap (\cup_{d \in \mathcal{D}} d)) = \mathcal{P})$ **then**
8:         $\mathcal{D} = \mathcal{D} \backslash \{d_{k+1}, d_{k+2}, \ldots, d_\ell\}$
9:         Reset the radius of every $d \in \mathcal{D}$ to $r' = (1 + \frac{(\ell-k)}{k})r$
10:        Drag the centers of every disk $d \in \mathcal{D}$ from right to left towards y-axis such
           that every d satisfies *constrained placement* requirement.
11:        **Return** $(\mathcal{D}, r')$
12:    **end if**
13: **end for**
14: **Return** $(\emptyset, 0)$

---

where $\rho \leq \frac{7r}{k}$ because $(\ell - k) \leq 7$ by Lemma 6. The centers of the disks $d_1, d_2, d_3,$ $\ldots, d_k$ on $\partial\mathcal{P}$ are moved left such that the *constrained placement* requirement is satisfied by every disk (see Fig. 5). Then, the disks $d_{k+1}, d_{k+2}, \ldots, d_\ell$ will become redundant and can be removed. The radius $r' = r + \rho \leq (1 + \frac{7}{k})r$. $\square$

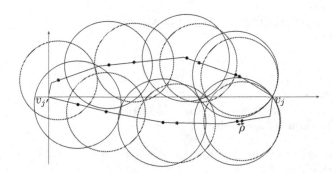

Fig. 5. Proof of Theorem 1

3   Constrained Convex Polygon Cover Problem

Here, we describe Algorithm 5 to solve CCPC problem. Algorithm 5 covers $\mathcal{P}$ with at most k disks of radius $r' \leq (1+\delta)r_{opt}$, centered on $\partial\mathcal{P}$, where $\delta = \frac{7}{k} + \frac{7\epsilon}{k} + \epsilon$ and r_{opt} is the optimum radius of k congruent disks centered on $\partial\mathcal{P}$, covering $\mathcal{P}$. To achieve this we first use the doubling technique as follows: if $r_{opt} > 1$, we invoke our algorithm (Algorithm 4) for decision problem with radius equal to 2^j for every $j = 1, 2, \ldots, j^*$ until cover of $\mathcal{P}$ is found for radius 2^{j^*}, where j^* is the smallest positive integer (lines 10-16 in Algorithm 5), otherwise we invoke Algorithm 4 with radius equal to 2^{-j} for every $j = 1, 2, \ldots, j^*$ till cover of $\mathcal{P}$ is found for radius 2^{-j^*}, where j^* is the largest positive integer (lines 4-9 in Algorithm 5). Therefore, r_{opt} belongs to either $[2^{j^*-1}, 2^{j^*}]$ or $[2^{-j^*-1}, 2^{-j^*}]$. Note that the size of this interval is at most r_{opt}. Let $[\mu, \nu]$ be the interval. Let $\gamma = \frac{\mu+\nu}{2}$. Now, we divide the interval $[\mu, \nu]$ into two intervals $[\mu, \gamma]$ and $[\gamma, \nu]$, and decide the interval that contains r_{opt}. Let this interval be $[\mu, \nu]$ and repeat the same process $\log\lceil\frac{1}{\epsilon}\rceil$ times.

Algorithm 5. COVER($\mathcal{P}, k$)

1: **Input:** Convex polygon $\mathcal{P}$, a positive integer k.
2: **Output:** Set $\mathcal{D} = \{ d_1, d_2, \ldots, d_k \}$ of k disks having equal radius
    $r' \leq (1 + \frac{7}{k} + \frac{7\epsilon}{k} + \epsilon)r_{opt}$, where $k \geq 7$ and an $\epsilon > 0$.
3: $(\mathcal{D}, r')$= k-COVER($\mathcal{P}, k, 1$) //Run Algorithm 4
4: **if** ($r' \neq 0$) **then**
5:    set $j = 1$
6:    **while** ($r' \neq 0$) **do**
7:        set $j = j - 1$
8:        $(\mathcal{D}, r')$= k-COVER($\mathcal{P}, k, 2^{j-1}$) //Run Algorithm 4
9:    **end while**
10: **else**
11:    set $j = 0$
12:    **while** ($r' = 0$) **do**
13:        set $j = j + 1$
14:        $(\mathcal{D}, r')$= k-COVER($\mathcal{P}, k, 2^j$) //Run Algorithm 4
15:    **end while**
16: **end if**
17: set $\mu = 2^{j-1}$, $\nu = 2^j$
18: **for** ($i = 1, 2, \ldots, \log\lceil\frac{1}{\epsilon}\rceil$) **do**
19:    set $\gamma = \frac{\mu+\nu}{2}$
20:    $(\mathcal{D}, r')$= k-COVER($\mathcal{P}, k, \gamma$) //Run Algorithm 4
21:    **if** ($r' \neq 0$) **then** set $\nu = \gamma$ **else** set $\mu = \gamma$
22: **end for**
23: $(\mathcal{D}, r')$= k-COVER($\mathcal{P}, k, \nu$) //Run Algorithm 4
24: **Return** $(\mathcal{D}, r')$

Lemma 8. $(\nu - \mu) \leq \epsilon r_{opt}$, where μ, ν are the values after for-loop in line 18 of Algorithm 5.

Proof. Initially $\mu = 2^{j^*-1}$ and $\nu = 2^{j^*}$, where $2^{j^*-1} \leq r_{opt} \leq 2^{j^*}$. After for-loop in line 18, the size of the interval is $(\nu - \mu) \leq \frac{(2^{j^*}-2^{j^*-1})}{2^{\log\lceil\frac{1}{\epsilon}\rceil}} \leq \frac{r_{opt}}{2^{\log\lceil\frac{1}{\epsilon}\rceil}} \leq \epsilon r_{opt}$. The same proof follows for initial values $\mu = 2^{-j^*-1}$ and $\nu = 2^{-j^*}$. $\square$

Theorem 2. *Algorithm 5 is $(1+\delta)$-approximation algorithm with running time $O(n(n+k)(|\log r_{opt}| + \log\lceil\frac{1}{\epsilon}\rceil))$, where $\delta = \frac{7}{k} + \frac{7\epsilon}{k} + \epsilon$, $k \geq 7$ and $\epsilon > 0$.*

Proof. The radius r_{opt} is initially made to lie in the interval $[2^{j^*-1}, 2^{j^*}]$ or $[2^{-j^*-1}, 2^{-j^*}]$ by while-loop at line 6 or line 12 of Algorithm 5. Then, after the for-loop in line 10 of Algorithm 5, we reduce this interval to $[\mu, \nu]$ such that $\mu \leq r_{opt} \leq \nu$ and $(\nu - \mu) \leq \epsilon r_{opt}$ (Lemma 8). Therefore, $\nu \leq \mu + \epsilon r_{opt} \leq r_{opt} + \epsilon r_{opt} \leq (1 + \epsilon)r_{opt}$. Line 23 in Algorithm 5 invokes Algorithm 4, which returns a set $\mathcal{D}$ of k disks of radius r', centered on $\partial\mathcal{P}$, where $r' \leq (1 + \frac{7}{k})\nu$ by Theorem 1. Hence, $r' \leq (1 + \frac{7}{k})\nu \leq (1 + \frac{7}{k})(1 + \epsilon)r_{opt} \leq (1 + \frac{7}{k} + \frac{7\epsilon}{k} + \epsilon)r_{opt}$. Algorithm 4 is invoked at most $|\log r_{opt}|$ times by while-loop and at most $\log\lceil\frac{1}{\epsilon}\rceil$ times by for-loop at lines 6 or 12 and 14 of Algorithm 5 respectively. The running time of Algorithm 4 is $O(n(n+k))$ (see Lemma 7). Hence, the running time is $n(n+k)|\log r_{opt}| + n(n+k)\log\lceil\frac{1}{\epsilon}\rceil = O(n(n+k)(|\log r_{opt}| + \log\lceil\frac{1}{\epsilon}\rceil))$. $\square$

4   Conclusion

In this paper, we describe an approximation algorithm for covering a convex polygon with k congruent disks centered on the boundary of the polygon. The approximation factor of the algorithm is $(1 + \frac{7}{k} + \frac{7\epsilon}{k} + \epsilon)$, where $k \geq 7$ and an $\epsilon > 0$. The approximation factor of the previous best known algorithm is 1.8841 [6]. Thus, for sufficiently large value of k, our algorithm is much better than the previous one. The complexity of this problem seems to be unknown as we did not find any prior work on the NP-hardness proof of this problem. As the future work, designing $(1 + \epsilon)$-approximation algorithm for every $\epsilon > 0$ (PTAS), can be considered and the hardness of the problem can be explored.

References

1. Agarwal, P.K., Procopiuc, C.: Exact and approximation algorithms for clustering. Algorithmica **33**, 201–226 (2002)
2. Agarwal, P.K., Sharir, M.: Efficient algorithm for geometric optimization. ACM Comp. Surv. **30**, 412–458 (1998)
3. Brass, P., Knauer, C., Na, H.S., Shin, C.S.: Computing k-centers on a line. CoRR abs/ 0902.3282 (2009)
4. Bose, P., Toussaint, G.: Computing the constrained euclidean, geodesic and link center of a simple polygon with applications. In: Proc. of Pacific Graphics International, pp. 102–112 (1996)
5. Das, G.K., Roy, S., Das, S., Nandy, S.C.: Variations of base station placement problem on the boundary of a convex region. Int. J. Found. Comput. Sci. **19**(2), 405–427 (2008)
6. Du, H., Xu, Y.: An approximation algorithm for k-center problem on a convex polygon. J. of Comb. Opt. **27**(3), 504–518 (2014)
7. Halperin, D., Sharir, M., Goldberg, K.: The 2-center problem with obstacles. J. Algorithms **42**, 109–134 (2002)
8. Hurtado, F., Sacriscan, V., Toussaint, G.: Facility location problems with constraints. Stud. Locat. Anal. **15**, 17–35 (2000)

9. Hwang, R., Lee, R., Chang, R.: The generalized searching over separators strategy to solve some NP-hard problems in sub-exponential time. Algorithmica **9**, 398–423 (1993)

10. Karmakar, A., Das, S., Nandy, S.C., Bhattacharya, B.K.: Some variations on constrained minimum enclosing circle problem. J. of Comb. Opt. **25**(2), 176–190 (2013)

11. Kim, S.K., Shin, C.-S.: Efficient algorithms for two-center problems for a convex polygon. In: Du, D.-Z., Eades, P., Sharma, A.K., Lin, X., Estivill-Castro, V. (eds.) COCOON 2000. LNCS, vol. 1858, pp. 299–309. Springer, Heidelberg (2000)

12. Roy, S., Bardhan, D., Das, S.: Base station placement on boundary of a convex polygon. J. Parallel Distrib. Comput. **68**, 265–273 (2008)

13. Suzuki, A., Drezner, Z.: The p-center location problem in area. Location Sci. **4**, 69–82 (1996)

Can Cancelable Biometrics Contribute to the Security Improvement of Biometric Authentication Systems?

Sanggyu Shin[(✉)] and Yoichi Seto

Advanced Institute of Industrial Technology, 1-10-40 Higashiooi,
Shinagawa-ku, Tokyo 140-0011, Japan
{shin,seto.yoichi}@aiit.ac.jp

Abstract. Cancelable biometrics techniques are considered from the viewpoint of data protection in authentication systems. Though various methods are proposed for this technology, the security criterion is indefinite. Moreover, there is still no work on the systematic study of the safety of biometric authentication systems. In this paper, cancelable biometric techniques are considered from the perspective of the safety of the system. Furthermore, the effect on the security precaution of the liveness detection techniques is verified by using Fault Tree Analysis, a risk evaluation method about data protection and spoofing prevention techniques.

Keywords: Cancelable biometrics · Biometric · Authentication system

1 Introduction

In biometric authentication systems, the biometric data that becomes the reference for comparison is called template data. When the template data leaks, problems such as the spoofing of other users and leakage of private data occur. These problems are usually addressed by the cryptography technology and the system technology of tamper-proof devices such as smart cards.

Various methods have been published for security techniques now known as cancelable biometrics [1]-[7]. However, there has been no systematic study on the criteria for security in biometric authentication systems.

The purpose of this paper is to discuss and evaluate the cancelable biometrics techniques from the viewpoint of ensuring safety of the systems. First, the location and the type where the threat is generated on the biometric authentication systems model are explored, and the meaning of cancelable biometrics as the countermeasure technology to those threats is described.

Next, we propose a scheme for evaluating the effectiveness of cancelable biometrics.

Finally, the effectiveness of the security precaution with the liveness detection techniques is verified by using FTA (Fault Tree Analysis), which is a quantitative risk analysis and is the evaluation approach about the data protection and the spoofing prevention technology. Furthermore, the superiority of the liveness detection technique is presented.

© Springer International Publishing Switzerland 2015
O. Gervasi et al. (Eds.): ICCSA 2015, Part II, LNCS 9156, pp. 223–232, 2015.
DOI: 10.1007/978-3-319-21407-8_17

2   Threats and Countermeasures Techniques of Biometric Authentication Systems

Fig. 1 shows the processing diagram of a biometric authentication system. At first, raw biometric data of an individual is captured from the user. For every input, biometric data changes due to the body and sensor device condition in the environment.

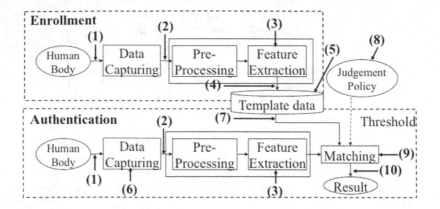

Fig. 1. Possible attack points in a biometric authentication system

In the enrollment process, the correction processing is included in the feature extraction from raw biometric data of an individual, and the extracted characteristics of individual is stored in the database as template data.

In the authentication process, the identification data that specifies the user is inputted from a sensor device and the corresponding template data is selected from the database. Two sets of features are matched, then the degree of similarity is obtained by a matching process. It is assumed that the authentication succeeds if the degree of similarity exceeds a threshold. As a result, the user can access the application.

The number in Fig. 1 shows the location under the threats of attack in a biometric authentication system. The threats at each location are explained as follows.

(1) Attack on sensor input of fake biometrics: The photograph of face, counterfeit fingerprint or signature are put on the sensor.
(2) Attack on the transfer data from sensor to feature extraction processing: Biometric data captured from the sensor is replaced with information attacking the network or the bus.
(3) Making replacement of extracted feature data: The feature extraction processing is attacked with the Trojan horse etc., and an arbitrary feature is set instead of an actual feature.

(4) Illegal conversion of body data: Body data is replaced with counterfeit data. It is very difficult to execute this attack because the feature extraction processing and the matching processing are often done in the same system. However, when an extracted feature data is transmitted to the matching processing by the Internet, this attack becomes possible by substituting the packet data.

(5) Tampering with stored template data: An illegal user makes the falsification of a template data stored in the database, such that an unfair user obtains an illegal attestation and gains access or a fair user obtains an illegal attestation and is denied access.

(6) Re-input of stored biometric data: Biometrics that remains on the sensor devices is automatically inputted again without user input.

(7) Attack on the transfer from template data storage to matching processing: When the template stored in the database is transferred to the matching processing through the communication channel, the template data is illegally changed.

(8) Replacement of threshold value: The threshold value is set to the given value by rewriting the judgment policy in order to get the intended result.

(9) Attack on matching process: The matching process is attacked, and the matching result is replaced with an arbitrary score.

(10) Substituting of final decision data: The judgment result of the authentication is substituted.

The countermeasures shown in Table 1 are effective ways to prevent the above-mentioned attacks. The threats in categories (4), (5) and (7) relate to the theft of template data and those in categories (1) and (6) relate to counterfeit use (spoofing) when biometric data is captured and attested.

Table 1. Threats and countermeasures

No	Threats	Countermeasures
(1)	Attack on sensor input of fake biometrics	Liveness detection
(2)	Attack to transfer data from sensor to feature extraction processing	Encryption
(3)	Making replacement of extracted feature data	Digital signature
(4)	Illegal conversion of body data	Encryption Cancelable biometrics
(5)	Tamper with stored template data	Physical security Cancelable biometrics
(6)	Re-input of stored biometric information	Liveness detection Challenge & Response
(7)	Attack to transfer from template data storage to matching processing	Encryption Cancelable biometrics
(8)	Tampering authentication parameters	
(9)	Replacement of threshold	Digital signature
(10)	Data	

For example, the development of countermeasure techniques for the following attacks will become important in the future.

- Study of fabricating a counterfeit fingerprint with cheap material such as the gelatin, which can be made for a short time.
- A method of the transformation of biometric data of protection and each application of biometric data by the encryption for the template protection and storing.
- A peculiar attack that counters dictionary attack to the biometric authentication.

Another problem includes the copying process when the data used for the biometric authentication loses reliability.

When reliability is lost, the authentic method such as using the key, the token, and the password, etc. can nullify these attestation devices as many times as you want. But, there is a limit in the number of times of nullification for biometrics.

In the security requirement for a biometric authentication system, the cancelable biometric techniques are not exclusive but are one of the measures technologies.

The problem of the template data leakage is divided into the problem of spoofing due to reuse and privacy concerns. Biometrics was originally exposed, and at 1:1 matching, the individual can be specified if there is a link to other information. Therefore, there is an opinion that biometrics is not privacy.

As for the problem of reuse, the measures technique when reusing such as the liveness detection technique is more effective than the encryption of data and nullification of data if biometrics can specify the individual by the link information.

It is necessary to examine effectiveness compared with the competing measures techniques, for instance, cryptography. T he analysis of the effectiveness of the measures technique is described in Chapter 4.

3   Only for Final Copy

3.1   Outline of the Cancelable Biometrics Techniques

Cancelable biometrics is a collective term of template protection techniques that nullifies it when an original biometric data is made invisible from the template not to be restorable in the biometric authentication systems.

There are two kinds templates. One is for image (signal) data input from the sensor and the other is for the feature data that used for processing. There are two kinds of nullification methods: based on encryption techniques and on image processing.

Fig. 2 shows typical processing flow of the cancelable biometrics.

The biometric data is distorted by the conversion processing using a one way transformation function in an enrollment or an authentication processing.

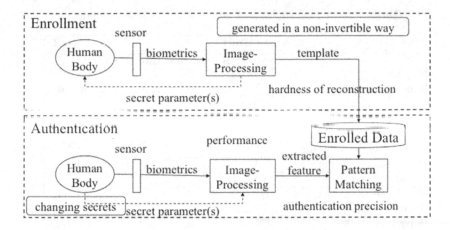

Fig. 2. The processing flow of cancelable biometrics

The template generated by a nonreciprocal method is stored in the smart card and the database.

The biometric data causes a different swerve at each registration. Therefore, when the reliability of biometric information is lost, the misinterpretation method used at that time is changed. Furthermore, it only has to register again based on a new conversion coefficient.

The technical details of a cancelable biometrics based on image processing is discussed thoroughly in [9].

The groupings by PET (Privacy Enhancing Technologies) [8]-[10] are classified techniques of privacy protection that can be applied to the systematization of cancelable biometrics. Techniques of cancelable biometrics and the dynamic key generation algorithm developed now can be classified into four categories (A to D) in reference to the classified techniques of PET, as shown in Table 2.

Category A is a method by noninvertible conversion. Conversion can be applied to both image (signal) and feature region. An example of conversion in the image area includes morphing and the block substitution. For example, the block structure is allocated in the block substitution according to a feature point in former image area, it arranges at every the block, and the scramble is done. However, not noninvertible conversion in this case but former image can often be computed reversibly.

Therefore, security strength is low, and the conversion in the image space is classified into category B.

Category C is the method to use information on fake data such as Fuzzy Vault.

Category D is the method of the data protection using the public key cryptography and zero knowledge proof technology. Furthermore, category D also proposes data protection techniques that use a general cryptographic protocol in ISO TC68 that is the technical committee intended for financial services.

Table 2. Grouping by PET of template protection techniques

Principle	Methods	Ref.
One-way function	A1. Non-reversible transformation of images	[6]
	A2. Non-reversible transformation of templates	
	B. Image morphing	
Common key cryptography	NA	
Public key cryptography	D1.	[5] [7]
Secret sharing function	NA	
Blind signature	NA	
Zero knowledge protocol	D2	[4]
Proxy network	NA	
Fake information	C1. Fuzzy vault	[1] [3]
	C2. Convolution random pattern	[2]
Privacy language	NA	

Currently, data protection technologies in biometric data are perceived to be immature. Therefore, the empty column in Table 2 represents template protection technologies that could be proposed in the future.

3.2 Evaluation of Validity of the Cancelable Biometrics

There are two technologies for the protection of template data: image processing based methods and cryptographic based methods.

Cryptographic template protection technology is open to the public, which allows evaluation of security strength by third parties. For example, FIPS140-2 is maintained as a safety standard at the implementation level.

Application Existence of an application where image processing-based template protection methods have the precedence over cryptographic-based.

On the other hand, from a technical point of view, image processing-based template protection methods have the following problems.

- Proper evaluation of the template protection technique is insufficient.
- The third party evaluation cannot be done because the algorithm and the interface are unpublished.
- The security strength evaluation scheme is not established, and an objective evaluation concerning a one way transformation, accuracy preservation and the processing performance is not done (Refer to Table 3).
- There is a possibility that it is technically immature as described in paragraph 3.1, and a more effective method will be developed in the future.
- There is no appropriately applied actual case.
- The individual data not in the database but in the smart card model is general from a viewpoint of a restriction of law and safety.

- Cancelable biometrics is a technology that assumes storage in a database, and a powerful application that can demonstrate superiority over template protection technology of cryptography based methods is not known.
- The priority is low in the viewpoint of the security risk. Details are presented in Chapter Chapter 4. Therefore, it is enough not to ensure that the template data is kept secret and discuss the safety of the system.

As a solution to these problems, security evaluation techniques will be established and an appropriate application will be developed.

Table 3. Evaluation axis of the cancelable biometrics

Items	Description
Hardness of reconstructing original data	Proof of being essentially unable to reconstruct original biometric data from converted data by a one-way function etc.
Preservation of authentication precision	Proof that authentication precision over converted data is no lower than that over pre-converted (original) data
Performance	Demonstration of the performance of a conversion algorithm for practical use
Application	Existence of an application where image processing-based template protection methods have the precedence over cryptographic-based.

4 Evaluation of Countermeasure Techniques for Securing the System

If biometrics have been exposed. the possibility that it is acquired by a malicious person without the consent of the user is high. Therefore, for the safety of a biometric authentication system, measures against reuse are more important than those against theft.

In reusing the template data, there are two kinds of abuse such as the capturing of the biometrics of the counterfeit and spoofing due to hacking to the system. The liveness detection function is necessary for the former and it is necessary for data transfer to be secure for the latter.

Applying the standardized cryptography technology for the data protection in the channel is an advantage in terms of cost and safety. Moreover, liveness detection measures that should be carried out when body information is acquired from the sensor, and image processing techniques become indispensable, as no alternative technology exists.

Therefore, from a practical viewpoint, it is thought that the development of liveness detection technology is a higher priority than the data protection technique of the image processing based cancelable biometrics.

In this chapter, the liveness detection technique is discussed as a spoofing prevention in the sensor aiming to prove the above-mentioned hypothesis quantitatively. Then, encryption and cancelable biometrics are taken up as data protection techniques in the channel and the database. The effectiveness of each technique is evaluated with respect to safety.

In this study, FTA (Fault Tree Analysis) is adopted as a quantitative risk evaluation method. FTA is a technique for making a logic diagram with a tree structure known as the Fault Tree that shows the causal relation of the generation process of the threat, the calculation of the probability of occurrence of the threat based on this logic diagram, and evaluation of the risk.

Applied to the security measures in the information system, FT is composed of making the threat a root[8], and uniting events derived according to the causal relation hierarchically by using the logic gate of the logical add and the logical product. The probability of occurrence of the root threat can be obtained by giving the probability of occurrence for each lower event after the FT is made.

Fig. 3 shows FT made for the spoofing attack. The probability of occurrence of a basic event of the liveness detection, the encryption, and the cancelable biometrics was given to FT under assumption that each technology was applied, and the probability of occurrence of the threat was calculated.

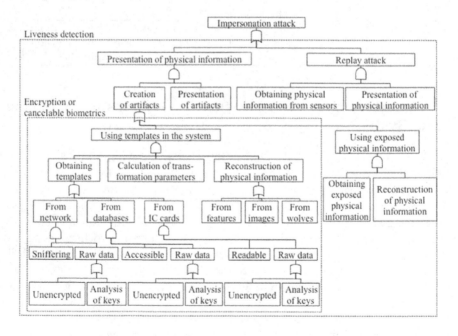

Fig. 3. Fault tree for an impersonation attack

Table 4 shows the probability of occurrence of the basic event used by this analysis. It was roughly distinguished that the probability of occurrence obtained an actual figure in three stages because it was difficult. The living body detection technology was assumed to be the one that it was possible to detect it surely when body information not to be alive was presented.

Table 4. Occurrence probabilities of basic events

Events	Prob.
Obtaining physical information from sensors	0.1
Obtaining exposed physical information outside the system	0.3
Reconstruction of physical information from exposed biological information	0.3
Derivation of transformation parameters of cancelable biometrics	0.1
Reconstruction of physical information from features	0.1
Reconstruction of physical information from images	0.3
Reconstruction of physical information from wolves	0.2
Analysis of encryption keys	0.01

Table 5 shows the numerical results. The risk type is usually defined by the product of the probability of occurrence and the size of the loss. Here it was assumed that the loss that occurred as a result of the spoofing was the same for each threat, and so the risk is evaluated only by the probability of occurrence.

Table 5. Probability of occurrence of threats

Applied technology	Prob. of occurrence of threats	Range of Reduction Probability
Nothing	0.33	-
Cryptography	0.18	0.15
Cancelable biometrics	0.20	0.13
Liveness detection	0	0.33

It is understood to have lowered the liveness detection in addition while the cryptography and the cancelable biometrics lower the probability of occurrence (0.15 and 0.13, respectively) under the situation in which measures are not done compared with the probability 0.33 that the threat is generated from the results in Table 5. The following conclusions can be derived from above.

- In these technologies, the liveness detection is the most effective.
- From viewpoint of prevention of reuse to effectiveness of this level cryptography and cancelable biometrics that is data protection technology.
- The cryptography technology is standardized, and it is an advantage if an objective evaluation approach for safety has been established.

5   Conclusions

The viewpoint of an effective methodology that secured the safety of biometric authentication systems was considered. The cancelable biometrics technique was systematized based on Privacy Enhancing Technology, and the possibility that a new technology will be developed in the future was shown in Chapter 4.

It proposed the evaluation items such as one way transformation, accuracy preservation and processing performance that clarified the effectiveness of the cancelable biometrics. The effectiveness of the technologies were quantitatively compared by using FTA method for the reuse problem in a biometric authentication system, when biometrics data had been leaked and exposed. According to the FTA analysis, when compared the cryptography and liveness detection, the effect of cancelable biometrics on the safety of biometric authentication systems was small.

Acknowledgments. This work was supported by JSPS KAKENHI Grant Number 25240017.

References

1. Clancy, T., Kiyavash, N., Lin, D.: Secure smartcard based fingerprint authentication. In: WBMA 2003 Proceedings of the 2003 ACM SIGMM Workshop on Biometrics Methods and Applications, pp. 45–52 (2003)
2. Hirata, S., Takahashi, K., Mimura, M.: Vulnerability analysis and improvement of cancelable biometrics for image matching. In: The 2007 Symposium on Cryptography and Information Security, 3C1-2, Japan (2007)
3. Juels, A., Sudan, M.: A fuzzy vault scheme. In: Proc. IEEE Int. Symp. Inf. Theory, pp. 408–413 (2002)
4. Kikuchi, H.: On security of asymmetric biometrics authentication. In: The 2006 Symposium on Cryptography and Information Security, 2D3-2, Japan (2006)
5. Kwon, T., Lee, J.: Practical digital signature generation using biometrics. In: Laganá, A., Gavrilova, M.L., Kumar, V., Mun, Y., Tan, C.J.K., Gervasi, O. (eds.) ICCSA 2004. LNCS, vol. 3043, pp. 728–737. Springer, Heidelberg (2004)
6. Ratha, N., Connell, J., Bolle, R.: Enhancing security and privacy in biometrics-based authentication systems. IBM systems J. **40**(3), 614–634 (2001)
7. Soutar, C., Roberg, D., Stoianov, A., Gilroy, R., Kumar, V.: Biometric encryption. http://www.bioscrypt.com/assets/Biometric_Encryption.pdf
8. Shimizu, S., Seto, Y.: A study on the effectiveness of cancelable biometric technology in biometric authentication systems. In: The 2008 Symposium on Cryptography and Information Security, 2B3-2, Japan (2008)
9. Shimizu, S., Seto, Y.: An evaluation of biometric template protection methods. In: The Asian Biometrics Consortium Conference and Exhibition 2007 (ABC2007), Invited talk, Singapore (2007)
10. Seto, Y.: Proposal to develop and operate for useful biometric application systems. The Journal of the Institute of Electronics, Information and Communication Engineers **90**(12), 1025–1030 (2007)

Gender Classification from Face Images Based on Gradient Directional Pattern (GDP)

Faisal Ahmed[1]([✉]), Padma Polash Paul[1], Patrick Wang[2],
and Marina Gavrilova[1]

[1] Department of Computer Science, University of Calgary, Calgary, AB, Canada
faahmed@ucalgary.ca
[2] College of Computer and Information Science,
Northeastern University, Boston, USA

Abstract. This paper presents an appearance-based facial feature
descriptor based on the gradient directional pattern (GDP) for gender classification from face images. The GDP operator encodes the
texture information of a local neighborhood by quantizing the gradient directions of the neighbors with respect to the center. The facial
feature descriptor is computed by first dividing the face image into a
number of sub-regions and then concatenating the individual GDP histograms computed from the corresponding sub-regions. Then, principal
component analysis (PCA) is applied on the obtained face descriptor
in order to reduce the feature dimensionality. We use a support vector
machine (SVM) for the classification task. Experimental analysis on a
large database comprising 1800 facial images shows promising results
for the proposed method, as compared to some well-known appearance-
based face descriptors.

Keywords: Gender classification · Gradient directional pattern
(GDP) · Principal component analysis (PCA) · Support vector machine
(SVM)

1 Introduction

In recent years, rapid progress in hardware and software technologies has triggered a growing demand for personalized interactions with consumer products
[11]. In this context of user-specific product and application design, automated
gender classification from facial image is particularly useful in order to offer
customized and user-friendly interactions [4,13]. Apart from that, gender classification has potential applications in surveillance and security systems, social
robotics, and human computer interaction [4]. It can also be used as an intermediate step in face recognition, which can potentially reduce the number of
candidates to be matched largely.

A typical gender classification system comprises two primary components: 1.
facial feature extraction and 2. classification. Extraction of discriminative and

© Springer International Publishing Switzerland 2015
O. Gervasi et al. (Eds.): ICCSA 2015, Part II, LNCS 9156, pp. 233–243, 2015.
DOI: 10.1007/978-3-319-21407-8_18

consistent facial features is an important task that determines the success of any gender classification system. Ideally, a good set of feature should minimize the intra-class variation and maximize the inter-class variation [2,18]. However, effective feature extraction from facial image is a challenging research problem due to the variability of facial images depending on different conditions. For example, facial appearance can largely change due to lighting variation, random noise, pose change, facial expression, low quality data, etc. [3,8,21]. Some early surveys on face recognition [24] and facial expression recognition [7] addressed these challenges. Later, different gender classification methods have been studied by Makinen and Raisamo, who presented comparative analysis of these methods in [6,15]. Existing gender classification methods found in the literature can broadly be categorized into two groups: 1. Geometric feature based methods, and 2. Appearance based methods [13]. While geometric feature based methods encode the geometric relationships among different facial components, appearance based methods rely on filter or filter bank to extract a holistic appearance representation of the facial image [2,10]. Early methods for gender classification were mostly based on geometric features. For example, Brunelli and Poggio [5] presented a gender classification method based on HyperBF network and a set of 16 geometric features. Later, another study by Abdi et al. [1] showed good recognition performance for both measurement based features (geometric feature) and pixel-based features (appearance features) using a radial basis function (RBF) kernel and a perceptron. However, effectiveness and consistency of geometric features critically depend on accurate detection of facial components. As a result, it is challenging to accommodate geometric feature based methods in any uncontrolled or changing environment [4]. In this context, appearance-based methods are more reliable since extraction of features does not require detection of facial components. Some well-known appearance-based methods include principal component analysis (PCA) [20], independent component analysis (ICA) [12], and Gabor wavelets [14].

In recent years, facial appearance descriptors based on the local binary pattern (LBP) [19] and its variants [3,8,21] have attained significant attention. Relatively low computational cost and robustness to illumination and pose variations are the major advantages of local texture patterns [9,22]. The LBP operator encodes the texture information of a small local neighborhood into a binary pattern, which acts as a template for detecting micro-level texture details. However, it is sensitive to large illumination variations and random noise [21]. Tan and Triggs [21] addressed this issue by introducing an extra level in their local ternary pattern encoding. Zhao et al. [23] proposed to use Sobel masks prior to applying LBP operator in order to facilitate more robust texture encoding. Local directional pattern (LDP) [8,10] introduced a different texture encoding approach that exploits eight-directional edge responses instead of gray levels. Although LDP yields better recognition performance than LBP, it tends to produce inconsistent texture response in smooth face regions [3,4].

In our previous work [2], we presented gradient directional pattern (GDP) for facial expression recognition, which achieved promising results compared to some existing works. In this paper, we propose the GDP operator and a facial descriptor based on the GDP codes and principal component analyis (PCA) for gender classification from face images. The proposed texture encoding scheme utilizes the gradient angle values that are more robust than gray scale values. As a result, the resultant GDP codes can more effectively represent the micro-level texture information, such as lines, corners, spots, edges, etc. In addition, the proposed method utilizes principal component analysis (PCA) to reduce the feature dimensionality. The effectiveness of the proposed method has been evaluated using a large database, comprising 900 male and 900 female facial images collected from the FERET face image database [17]. In our experiments, facial descriptor constructed using the GDP codes and PCA has achieved better recognition performance than some well-known appearance-based facial features.

2   Face Description Using Local Texture Patterns

Local binary pattern (LBP) proposed by Ojala et al. [16] is one of the pioneer works on local texture patterns. The idea behind the basic LBP operator is to represent the texture information of a small image region using a binary pattern. In order to achieve that, all the neighbors of the local region are compared against the center value. Any neighbor having a gray value greater or equal to the center is encoded as 1 and the rest are encoded as 0. The result is then concatenated to form an 8-bit binary pattern. This process is applied on the entire image using a window mechanism and thus a LBP encoded image representation is obtained. Formal definition of the LBP operator takes the following form:

$$LBP_{P,R}(x_c, y_c) = \sum_{p=0}^{P-1} s(i_p - i_c)2^p \qquad (1)$$

$$s(v) = \begin{cases} 1, v \geq 0 \\ 0, v < 0 \end{cases} \qquad (2)$$

Here, i_c is the gray value of the center pixel (x_c, y_c), i_p is the gray value of its neighbors, P is the number of neighbors and R is the radius of the neighborhood. The LBP histogram of the encoded image representation is then used as the feature vector. Figure 1 illustrates the basic LBP encoding process. However, using only the center gray value to threshold the local region makes the LBP operator unstable under the presence of non-monotonic illumination variations and random noise. Therefore, different variants of LBP encoding process address these issues. Here, we present a brief overview of some of the LBP variants.

Local ternary pattern (LTP) proposed by Tan and Triggs [21] extends the LBP to a 3-level encoding, which makes the encoded patterns more robust against random noise. In the LTP encoding process, instead of thresholding the local neighborhood against the center value, a threshold zone of width t about the center pixel C is created. Thus, neighbors having a gray value greater than $C + t$ are encoded as 1 and less than $C - t$ are encoded as -1. The rest of the neighbors which falls within the threshold region are encoded as 0. Thus, LTP

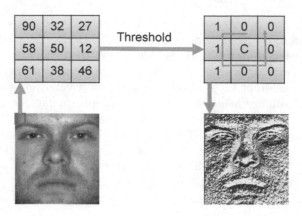

Fig. 1. Illustration of the basic LBP encoding. Here, the LBP binary code for pixel C is 01110000

forms a ternary pattern consisting 1, 0, and −1. This ternary pattern is then divided into corresponding positive and negative parts and treated as separate binary patterns.

Since gradients and edge response values are more robust than gray values, some texture encoding methods utilize gradients and edge response values for robust encoding of the local texture. Some examples are local directional pattern [10], directional ternary pattern [3], and Sobel-LBP [23]. In local directional pattern, eight directional edge responses are computed in a local neighborhood. Then, the K-most prominent edge response directions are encoded as 1 and the rest are encoded as 0. However, LDP also encodes smooth regions using the same process and thus, generates inconsistent texture responses in those regions. Ahmed and Kabir [3] addressed this issue in the directional ternary pattern by borrowing the concept of LTP and thus extending LDP to a ternary encoding scheme. Another method proposed by Zhao et al. [23] applies Sobel mask prior to applying LBP operator. This in turn, enhances the edge and texture information and facilitates a more robust description of local texture.

3   Proposed Method

3.1   Facial Texture Encoding Using Gradient Directional Pattern

Since image gradients are more robust than gray values, any texture operator that exploits the gradient values should be more robust in encoding the local texture information. In this paper, we propose to use gradient directional pattern (GDP) to encode the facial texture (face appearance). The main idea behind GDP is to utilize the gradient angle values of each pixel in a local neighborhood in order to isolate important micro-level texture information, such as edges, spots, corners, smooth regions, etc. The GDP operator follows the same encoding approach as LBP, except that the GDP operator thresholds the local region based on the gradient angle values. First, the direction of the gradient vector for each pixel is computed based on the following formula:

$$\alpha(i,j) = \tan^{-1}(G_i/G_j) \tag{3}$$

Here, $\alpha(i,j)$ is the gradient direction angle of the pixel (i,j), and G_i and G_j are the two elements of the gradient vector that can be obtained by applying the Sobel operator on the source image. The Sobel operator comprises a horizontal and a vertical mask which are convoluted with the image in order to obtain the values of G_i and G_j, respectively. The Sobel masks are shown in Fig. 2.

-1	-2	-1
0	0	0
1	2	1

Horizontal Mask

-1	0	1
-2	0	2
-1	0	1

Vertical Mask

Fig. 2. Sobel masks

After computing the gradient direction values, the GDP operator encodes a local 3×3 neighborhood by thresholding the neighbor gradient directions with respect to the center gradient α_c and a threshold value t. Any neighbor gradient angle α_i with a value $\alpha_c - t \le \alpha_i \le \alpha_c + t$ is encoded as 1 and the rest are encoded as 0. In practice, the GDP operator encodes the neighbors having a similar gradient direction with respect to the center as 1 and the rest of the neighbors as 0. Introducing the threshold t ensures consistent encoding in both high textured and smooth regions. The resultant bit values for each neighbor are then concatenated to form an 8-bit binary pattern and the corresponding decimal value is assigned to the center. This process is continued for every pixel in the facial image as a window mechanism and thus, a GDP encoded image representation is obtained. Formally, the GDP operator can be defined as:

$$GDP(x_c, y_c) = \sum_{p=0}^{P-1} s(GD_p, GD_c)2^p \tag{4}$$

$$s(GD_p, GD_c) = \begin{cases} 1, GD_c - t \le GD_p \le GD_c + t \\ 0, otherwise \end{cases} \tag{5}$$

Here, GD_c is the gradient direction angle of the center pixel (x_c, y_c), GD_p is the angles of its neighbors, and t is the threshold. Fig. 3 illustrates the basic GDP encoding method.

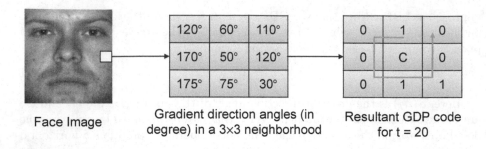

Face Image | Gradient direction angles (in degree) in a 3×3 neighborhood | Resultant GDP code for t = 20

Fig. 3. Illustration of the basic GDP encoding method. Here, the GDP code for the center C is 10001100

3.2 GDP Face Feature Descriptor

Applying the GDP-based texture encoding on a facial image results in an encoded image representation. A global histogram computed from this encoded image can be treated as a feature vector:

$$H_{GDP}(i) = \sum_{x=1}^{M} \sum_{y=1}^{N} f(GDP(x,y), i), where f(a,i) = \begin{cases} 1, a = i \\ 0, otherwise \end{cases} \quad (6)$$

Here, H_{GDP} is the GDP histogram of an $M \times N$ encoded image and i is the GDP code value. However, such a feature descriptor only contains the occurrence frequency of the GDP micro-patterns and fails to represent any locality information. Many researchers have argued that, inclusion of location information in the feature descriptor makes it more robust and informative, thus increasing the recognition performance. Hence, in order to incorporate some notion of location information with the GDP-based face feature descriptor, the whole face region is divided into some equal sub-regions and individual local GDP histograms computed for all the sub-regions are spatially concatenated to obtain a final feature descriptor. The process is illustrated in Fig. 4. In our experiments, the face images were divided into 7×7 sub-regions, as suggested in [4].

3.3 Feature Dimensionality Reduction Using PCA

For the GDP facial feature representation, the feature vector contains a total of $256 \times 7 \times 7 = 12544$ features for a face image partitioned into 7×7 sub-regions. Therefore, feature dimensionality reduction is an essential step in the proposed method. Ideally, a good feature representation should contain only the most relevant features that ensure high discrimination between classes. On the other hand, any redundant feature should be discarded. In addition, the trade-off between the number of selected features and the recognition performance is a concern. While selecting very low number of features might result in information loss and

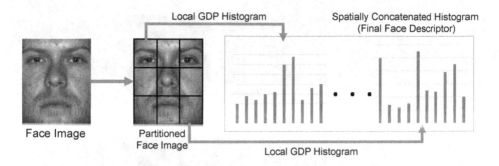

Fig. 4. Construction of the GDP face descriptor from a facial image

low recognition performance, having a large feature vector increases the computational complexity [10]. In this paper, we use principal component analysis (PCA) to reduce the feature dimensionality. In PCA, for a given feature matrix F, the eigenvectors are computed from the covariance of F, which are defined as the principal components (PCs). Each feature vector is then represented as an approximation, which is a linear combination of the top k eigenvectors. Thus, the feature matrix is transformed to a new representation, which is the reduced subspace. In practice, this matrix is a representation of the eigenspace that is defined by the eigenvectors. Here, every eigenvalue represents the corresponding axis of variance [10]. Thus, the original feature vector obtained from the GDP texture encoding is transformed into a lesser dimensional representation that minimizes redundancy and maximizes between-class discrimination.

4   Experiments and Results

To evaluate the effectiveness of the proposed method, a large dataset of 1800 face images were created. The images were selected from the facial recognition technology (FERET) face image database [17]. The FERET database is one of the largest available face recognition databases that comprises a total of 14,051 facial images collected from 1199 individuals. The face images have views ranging from frontal to left and right profile. In addition, different types of variations are introduced, such as age, facial expression, pose, and illumination variations. Our dataset comprises 900 male and 900 female frontal face images. The selected images were cropped from the original resolution of 512×768 pixels to 100×100 pixels. For the crop operation, the ground truth of the eyes was used. Figure 5 shows a few sample images from our dataset.

A support vector machine (SVM) was used for the classification task. The radial basis function (RBF) kernel was selected for the SVM. The performance of the proposed method can vary based on the threshold value t. Therefore, optimal parameter selection is an important step. In order to find the optimal t value, the performance of the proposed method has been evaluated for different values of t: 5, 10, 20, 30, and 40. A ten-fold cross validation was used to compute the

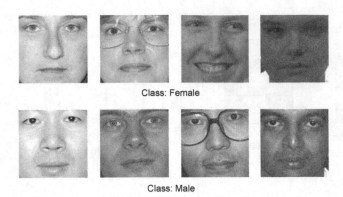

Class: Female

Class: Male

Fig. 5. Sample face images from the dataset

recognition rate. The ten-fold cross-validation is an iterative process where the dataset is divided into 10 equal subsets. At each iteration, the classifier is trained using 9 subsets and the remaining one subset is used for testing. This process is repeated for 10 times and each of the 10 subsets is used as a testing set once. The final recognition rate is obtained by averaging the recognition rates of all the iterations. Figure 6 shows the recognition performance of the proposed method for different threshold values. It can be observed that, the highest recognition rate is obtained for $t = 30$.

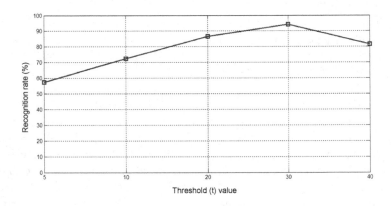

Fig. 6. Recognition performance of the proposed method for different threshold values

We also compare the performance of the proposed method with some well-known appearance-based face feature descriptors. The selected descriptors are: local binary pattern (LBP) [19], local ternary pattern (LTP) [21], and local directional pattern (LDP) [8]. Figure 7 and 8 shows the overall ten-fold cross validation rate (%) and individual recognition rates for male and female for

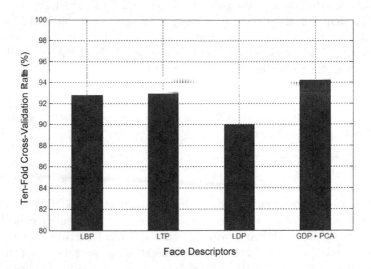

Fig. 7. Ten-fold cross-validation rates (%) for different face feature descriptors

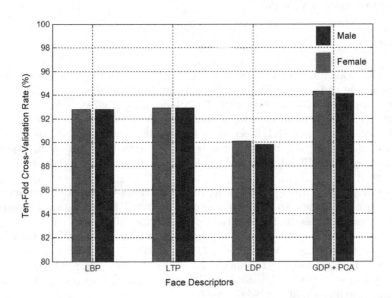

Fig. 8. Ten-fold cross-validation rates (%) for male and female recognition for different face feature descriptors

the selected face descriptors, respectively. It can be observed that, the proposed method achieves the highest recognition rate of 94.2% (Male 94.1% and Female 94.3%). LTP achieves the second best recognition rate of 92.89%. Based on the experimental results it can be said that, face description using gradient directional pattern (GDP) and principal component analysis (PCA) can effectively be applied in gender classification problem. The effectiveness of the proposed method is due to the utilization of the gradient direction values for encoding the local texture that facilitates a more robust description of the image microproperties. In addition, applying PCA results in a reduced feature space, which allows low computational cost.

5   Conclusion

In this paper, we propose to use gradient directional pattern (GDP) and principal component analysis (PCA) to represent facial texture for gender classification from face images. The GDP operator encodes the local texture of a neighborhood by quantizing the direction values of the gradient vector of each neighbor with respect to the center. Utilizing gradient direction values facilitates more informative and robust texture encoding, while use of PCA reduces the feature dimensionality. Experimental analysis with a large database reveals that, the proposed method can effectively represent and recognize gender based on facial texture. In the future, we plan to evaluate the proposed method for noisy and low resolution surveillance data.

References

1. Abdi, H., Valentin, D., Edelman, B., O'Toole, A.: More about the difference between men and women: Evidence from linear neural network and principal component approach. Neural Computing **7**(6), 1160–1164 (1995)
2. Ahmed, F.: Gradient directional pattern: a robust feature descriptor for facial expression recognition. IET Electronics Letters **48**(19), 1203–1204 (2012)
3. Ahmed, F., Kabir, M.H.: Directional ternary pattern (dtp) for facial expression recognition. In: IEEE International Conference on Consumer Electronics, pp. 265–266 (2012)
4. Ahmed, F., Kabir, M.H.: Facial feature representation with directional ternary pattern (dtp): application to gender classification. In: IEEE International Conference on Information Reuse and Integration, pp. 159–164 (2012)
5. Brunelli, R., Poggio, T.: Hyperbf networks for gender classification. In: DARPA Image Understanding Workship, pp. 311–314 (1992)
6. Makinen, E., Raisamo, R.: Evaluation of gender classification methods with automatically detected and aligned faces. IEEE Transactions on Pattern Analysis and Machine Intelligence **30**, 541–547 (2008)
7. Fasel, B., Luettin, J.: Automatic facial expression analysis: A survey. Pattern Recognition **36**(1), 259–275 (2003)
8. Jabid, T., Kabir, M.H., Chae, O.: Gender classification using local directional pattern (ldp). In: International Conference on Pattern Recognition, pp. 2162–2165 (2010)

9. Jabid, T., Kabir, M.H., Chae, O.: Local directional pattern for face recognition. In: IEEE International Conference on Consumer Electronics, pp. 329–330 (2010)
10. Jabid, T., Kabir, M.H., Chae, O.: Robust facial expression recognition based on local directional pattern. ETRI Journal 32(5), 784–794 (2010)
11. Jabid, T., Kabir, M.H., Chae, O.: Local directional pattern (ldp) for face recognition. International Journal of Innovative Computing, Information and Control 8(4), 2423–2437 (2012)
12. Jain, A., Huang, J.: Integrating independent components and linear discriminant analysis for gender classification. In: International Conference on Automatic Face and Gesture Recognition, pp. 159–163 (2004)
13. Kim, H.C., Kim, D., Ghahramani, Z., Bang, S.Y.: Appearance-based gender classification with gaussian processes. Pattern Recognition Letters 27(6), 618–626 (2006)
14. Lyons, M., Budynek, J., Plante, A., Akamatsu, S.: Classifying facial attributes using a 2-d gabor wavelet representation and discriminant analysis. In: International Conference on Automatic Face and Gesture Recognition, pp. 202–207 (2000)
15. Makinen, E., Raisamo, R.: An experimental comparison of gender classification methods. Pattern Recognition Letters 29, 1544–1556 (2008)
16. Ojala, T., Pietikainen, M., Maenpaa, T.: Multiresolution gray-scale and rotation invariant texture classification with local binary patterns. IEEE Transactions on Pattern Analysis and Machince Intelligence 24(7), 971–987 (2002)
17. Phillips, P., Wechler, H., Huang, J., Rauss, P.: The feret database and evaluation procedure for face recognition algorithms. Image and Vision Computing 16(10), 295–306 (1998)
18. Rivera, A.R., Castillo, J.R., Chae, O.: Local directional number pattern for face analysis: face and expression recognition. IEEE Transactions on Image Processing 22(5), 1740–1752 (2013)
19. Sun, N., Zheng, W., Sun, C., Zou, C., Zhao, L.: Gender classification based on boosting local binary pattern. In: Wang, J., Yi, Z., Zurada, J.M., Lu, B.-L., Yin, H. (eds.) ISNN 2006. LNCS, vol. 3972, pp. 194–201. Springer, Heidelberg (2006)
20. Sun, Z., Bebis, G., Yuan, X., Louis, S.J.: Genetic feature subset selection for gender classification: a comparison study. In: IEEE Workshop on Applications of Computer Vision, pp. 165–170 (2002)
21. Tan, X., Triggs, B.: Enhanced local texture feature sets for face recognition under difficult lighting conditions. In: Zhou, S.K., Zhao, W., Tang, X., Gong, S. (eds.) AMFG 2007. LNCS, vol. 4778, pp. 168–182. Springer, Heidelberg (2007)
22. Zhao, G., Pietikainen, M.: Boosted multi-resolution spatiotemporal descriptors for facial expression recognition. Pattern Recognition Letters 30(12), 1117–1127 (2009)
23. Zhao, S., Gao, Y., Zhang, B.: Sobel-lbp. In: IEEE International Conference on Image Processing, pp. 2144–2147 (2008)
24. Zhao, W., Chellappa, R., Phillips, P.J.: Face recognition: A literature survey. ACM Computing Survey 35(4), 399–458 (2003)

Smallest Maximum-Weight Circle for Weighted Points in the Plane

Sergey Bereg, Ovidiu Daescu$^{(\boxtimes)}$, Marko Zivanic, and Timothy Rozario

The University of Texas at Dallas, 800 W. Campbell Road, Richardson, TX, USA
{besp,daescu,mxz052000,tmr100020}@utdallas.edu

Abstract. Let P be a weighted set of points in the plane. In this paper we study the problem of computing a circle of smallest radius such that the total weight of the points covered by the circle is maximized. We present an algorithm with polynomial time depending on the number of points with positive and negative weight. We also consider a restricted version of the problem where the center of the circle should be on a given line and give an algorithm that runs in $O(n(m + n) \log(m + n))$ time using $O(m + n)$ space. The algorithm can report all k smallest maximal weight circles with an additional $O(k)$ space. Moreover, for this version, if all positively weighted points are required to be included within the circle then we prove a number of interesting properties and provide an algorithm that runs in $O((n + m) \log(n + m))$ time.

1 Introduction

Let P be a set of points in the plane. Let each point $p_i \in P$ have associated a real number w_i, called *weight*. For a circle C in the plane, let $\Phi(C)$ denote the total weight of points in $P \cap C$. Let Φ_{max} denote the maximum value $\Phi(C)$ taken over all circles in the plane. Let $\mathcal{S}$ denote the set of circles C such that $\Phi(C) = \Phi_{max}$. In this paper we address the three related problems. We start by discussing the problem of finding the smallest circle in $\mathcal{S}$. The corresponding circle is called the *maximum-weight circle* and denoted by C_{max}. See Fig. 1 for an illustration.

For the second problem we ask that the center of the circle is restricted to a given line. Thus, we are looking for the smallest circle that has its center on the line and maximizes the sum of the weights of the points of P that are within the circle. Without loss of generality, we assume the line is the x-axis.

For the third problem, the center is restricted to a given line (the x-axis), all points have unit absolute weight (i.e., each positively weighted point has weight one and each negatively weighted point has weight minus one), and we require that all positively weighted points are enclosed within the circle. We assume that not all points in P have positive weight since otherwise the problems would be trivial.

Daescu's research is partially supported by NSF award IIP1439718 and CPRIT award RP150164.

© Springer International Publishing Switzerland 2015
O. Gervasi et al. (Eds.): ICCSA 2015, Part II, LNCS 9156, pp. 244–253, 2015.
DOI: 10.1007/978-3-319-21407-8_19

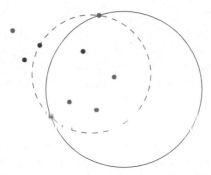

Fig. 1. The maximal weight circle for set P with 3 blue and 6 red points (dotted line)

All three problems can be modeled as problems on red-blue points. We call points with negative weight *blue* points and those with positive weight *red* points, and denote them by $\mathcal{B}$ and $\mathcal{R}$, respectively.

Let $\mathcal{R}$ be of size $|\mathcal{R}| = n$ and let $\mathcal{B}$ be of size $|\mathcal{B}| = m$. We assume that $\mathcal{R}$ and $\mathcal{B}$ cannot be separated by a circle, which can be decided in linear time and space [9]. If the points are separable with a circle then finding a smallest radius maximum-weight circle takes $O(m+n)$ time while finding a largest radius maximum weight circle takes $O((n+m)\log(n+m))$ time [9].

The problems arise naturally, for example for deploying wireless base stations in emergency and military applications, where points represent friends and foes, weights represent degrees of trust, and the circle radius relates to the broadcasting power of the base station. The line given in the last two problems could represent a portion of a street along which the base station could be placed.

2   Related Work

The problems we study are related to the red-blue minimum separating circle problem introduced by Daescu et al. [4], the weighted geometric set cover problem [7], the smallest k-enclosing circle problem [8], and the maximum weight planar boxes problem [1].

In the red-blue minimum (or maximum) separating circle problem a set $\mathcal{R}$ of size $|\mathcal{R}| = n$ and a set $\mathcal{B}$ of size $|\mathcal{B}| = m$ are given as part of the input and the goal is to find the smallest (or largest) radius circle that encloses all points in $\mathcal{R}$ while having the smallest number of points of $\mathcal{B}$ in its interior. Note that there is no weight associated with the points althought one could consider thered points having weight one and all blue points having weight minus one. This problem is thus a generalization of the third problem discussed in this paper.

In [4] they proved that the largest (maximum) separating circle can be found in $O(mn \log m + k(m + n)\log(m + n))$ time and $O(n + m)$ space, where $k = O(m)$ is the number of optimal solutions, by sweeping the edges of the farthest point voronoi diagram $FVD(\mathcal{R})$ of $\mathcal{R}$. They also provided two algorithms for finding the minimum separating circle, one with $O((m+n)\,logn + m^{1.5}\log^{O(1)} m)$

time and $O(m^{1.5} \log^{O(1)} m)$ space based on circular range queries, and one with $O(mn \log m + n \log n)$ time and $O(m + n)$ space based on sweeping $FVD(\mathcal{R})$. Kinetic solutions for this problem, where blue and red points could move with constant speed on linear trajectories, are presented by Cheung, Daescu, and Zivanic [5]. They present algorithms for three cases: when only one mobile blue point is allowed the locus of the center of the optimal circle has a complexity of $O(mn)$ and can be found in $O(mn \log(mn))$ time; when only one mobile red point is allowed the locus of the center of teh optimal circle has a complexity of $O(m^2 n)$ and can be found in $O(m^2 n \log m)$ time; and when multiple red and blue points are mobile the locus of the center of teh optimal circle has a complexity of $O(m^2 n^{2+\epsilon})$ and can be found in $O(m^2 n^{2+\epsilon} \log(mn) + mn^{3+\epsilon})$ time, where ϵ is an arbitrarily small positive constant.

Recently, Barbay et al.[1] considered a related problem, where circles are replaced by boxes. Given a set P of n points in d-dimensional space, where each point p in P is associated a real weight $w(p)$, the maximum-weight box problem is to find an axis-alligned box B maximizing $\sum_{p \in (P \cap B)} w(p)$. In the plane, they describe an algorithm with $O(n^2)$ worst case running time. They further show that the maximum-weight box problem can be solved in $O(n^d)$ time for any constant $d \geq 2$. If the points are partitioned in red and blue based on their weight (positive or negative), with n red points and m blue points, then they give an algorithm that runs in $O(n \min n, m)$ time.

In weighted geometric set cover problems, given a set of P elements (e.g., points) and B weighted sets (halfplanes, rectangles, circles) the goal is to find the minimum weight of sets of B that cover P. Weighted geometric set cover problems are in general NP-hard [7] and are related to the well known set cover problem. On the other hand, finding the smallest circle enclosing at least k points out of a set P of n points, with $k \leq n$, can be solved in polynomial time [8]. Some of these algorithms are based of the order-k Voronoi diagram or the $O(k)$ nearest neighbors for each point in P.

The red-blue set cover problem was introduced by Carr et al. [2] and its geometric version for unit squares was very recently considered by Chan and Hu [3]. In this version, given a red point set, a blue point set, and a set of objects (unit squares) the goal is to chose a subset of objects that cover all red points while minimizing teh number of blue points covered. They prove that this problem is NP-hard even when the objects are unit squares in the plane and give a polynomial-time approximation scheme.

We are not aware of any previous work that directly addresses the smallest maximum-weight circle problem. For the unrestricted version, where the center of the circle can be anywhere in the plane, we present a solution that takes $O(m^2(m + n) \log(m + n))$ time and $O(m + n)$ space. We notice however that there is work on computing certain discrepancies of points that can be adapted to solve this problem. In [6], they study algorithms for finding the maximal discrepancy of certain shapes on N points relative to the area of the shape, including the case of half spaces. By performing a standard lifting, the first problem discussed in this paper becomes the halfspace problem in R^3, for which they have an $O(N^3)$ time, $O(N)$ space algorithm (Theorem 5). We point out that our algorithm is different, since it works in the plane. It is also easy to understand and to implement and we have implemented an applet to illustrate the algorithm. Moreover, our algorithm can be considerably faster than that in [6] when $m << N$ or $n << N$, where $N = m + n$ is the total number of points. This situation may appear in spatial scan statistics where typically $m << n$.

For the second problem, where the center of teh circle is restricted to a given line, we give an algorithm that runs in $O(n(m + n) \log(m + n))$ time using $O(m + n)$ space. The algorithm can report all k smallest maximal weight circles with an additional $O(k)$ space.

For the third problem, where all positively weighted points have to be included in the circle, we prove a number of interesting properties and provide an algorithm that runs in $O((n + m) \log(n + m))$ time.

3   General Problem

In this section we present our solution for the general problem of computing smallest minimum-weight circle. We start by proving some properties of the optimal solution.

Lemma 1. *The smallest maximal weight circle must have at least two points of $\mathcal{R}$ on its boundary. The largest maximal weight circle must have at least two points of $\mathcal{B}$ on its boundary.*

Proof. Let C_{max} be a smallest maximal weight circle. If there are no red points on the boundary of C_{max} then we can keep the same center while making the radius smaller, until the circle passes through a red point. If there is one red point on the boundary then we can move teh center of the circle towards the red point, along the line segment joining the center and red point, until the circle passes through a second red point. Thus, at least two red points must be on teh boundary of C_{max}. A similar argument holds for proving that the largest maximal weight circle must have at least two points of $\mathcal{B}$ on its boundary.   □

Lemma 1 leads to the following solution for finding the smallest maximal weight circle. Consider a bisector line l_{ij} defined by two points p_i and p_j of $\mathcal{R}$, $i, j \in \{1, 2, \ldots, n\}$, $i \neq j$. An event point ocurs when a red or blue point enters or exits the circle centered on l_{ij} and passing through p_i and p_j when the center of circle moves along l_{ij}. There can be $O(m + n)$ event points on l_{ij}, which can be found and sorted in $O((m + n) \log(m + n))$ time. Once the red and blue points within one such circle are known their weighted sum can be maintained in constant time while sweeping the event points.

Note that the radius of the sweeping circle is unimodal. This leads to the following Lemma.

Lemma 2. *There can be at most two smallest maximal weight circles defined by p_i and p_j.*

Proof. Once a maximal weight circle is found by sweeping the bisector line l_{ij} defined by points p_i and p_j of $\mathcal{R}$, while increasing the circle radius, any other maximal weight circle will correspond to a circle of larger radius.   □

A smalest maximal weight circle can then be reported by scanning the $O(n^2)$ bisectors of the points in $\mathcal{R}$ and keeping track of the weight of the circles defined by various event points. This procedure actually finds and reports **all** smallest maximal weight circles.

Lemma 3. *A smallest maximal weight circle can be found in $O(n^2(m + n)\log(m + n))$ time using $O(m + n)$ space. Additional $O(k)$ space is needed for reporting all smallest maximal weight circles, where k is the number of such circles.*

Note that a similar procedure can be used to find the largest maximal weight circle by reversing the roles of $\mathcal{R}$ and $\mathcal{B}$.

4   Restricted Bichromatic Problem

One can expect to improve the running time if the center of the maximum-weight circle is restricted to a subset of the plane. We consider the problem under the restriction that the circle center must be on a given line l. The smallest maximum-weight circle in this case has two points from $\mathcal{R} \cup \mathcal{B}$ on its boundary and at least one of them is red.

The algorithm is similar to the algorithm for the general problem. We consider the given line instead of the bisector of two red points. For a red point p_i, we cnsider all circles centered on l and having p_i on the boundary. Again the radius of the sweeping circle is unimodal and the sweep can be done in $O((m + n)\log(m + n))$ time. This leads to the following lemma.

Lemma 4. *A smallest maximum-weight circle with center on a given line can be found in $O(n(m+n)\log(m+n))$ time using $O(m+n)$ space. Additional $O(k)$ space is needed for reporting all smallest maximal weight circles, where k is the number of such circles.*

5   Minimum Separating Circle

In this section we consider the restricted version of the minimum separating circle problem [4]. We assume that the center of the circle is restricted to a line. This problem can also be viewed as a special case of the smallest minimum-weight circle problem where the weights of red points are significantly large and the weights of blue points are equal. More precisely, we assume that, for any red point p_i, $w_i > -mw$ where $w < 0$ is the weight of a blue point. The minimum-weight circle in this case contains all red points and minimum number of blue points. Notice that the smallest such circle now may contain only one red point on its boundary, see Fig. 2. In general, there are three posibilities for the optimal circle descibed in the following lemma.

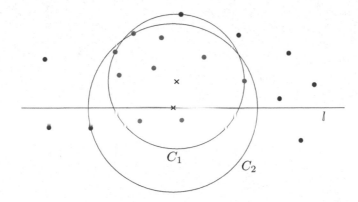

Fig. 2. Circle C_1 is the optimal separating circle for 9 red points and 9 blue points. Circle C_2 is the optimal separating circle if the center is restricted to line l.

Lemma 5. *Let C be the smallest radius circle centered at l such that C encloses all the points of $\mathcal{R}$ and C encloses the smallest possible number of points of $\mathcal{B}$. Then one of the following conditions holds*

1. *There are two red points on the boundary of C.*
2. *There is a red point and a blue point on the boundary of C.*
3. *There is only one red point p and no blue points on the the boundary of C, and segment pq is perpendicular to line l, where q is the center of C.*

Proof. Since C is the smallest radius circle centered at a point on l such that C encloses all red points and the minimum number of blue points, it must have a red point on the boundary of C. Let p be a red pont on C and let q be the center of C. If we can move the center of C slightly toward the projection of p onto l while keeping p on the circle, then either a red point will be no longer within the circle or a new blue point enters the circle. These are the first two cases of the lemma. If the center of C cannot move, then it is the third case of the lemma.   □

5.1   Algorithm

Without loss of generality we can assume that line l is horizontal and is actually the x-axis. Our algorithm is kinetic in the sense that we move the circle center q from $-\infty$ to $+\infty$. We maintain circle C of smallest radius and enclosing all red points. When the motion "begins" the circle is the vertical line with equation $x = x_{max}$ where x_{max} is the maximum x-coordinate of a red point, see Fig. 3. In the end of the motion the circle corresponds to the vertical line with equation $x = x_{min}$ where x_{min} is the smallest x-coordinate of a red point. For example, a blue point with x-coordinate less than x_{min} will be counted at the beginning and will be not counted at the end of the motion. This point is actually of *type* I in our terminology.

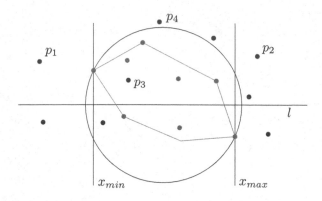

Fig. 3. Four types of blue points

Lemma 6. *There are four types of blue points with respect to the motion of circle C (the center moves along x-axis). Let p be a blue point and let $S(p)$ be the set of centers of C such that $p \in C$. Then*

I) $S(p) = (-\infty, x_0]$ for some x_0, or
II) $S(p) = [x_0, +\infty)$ for some x_0, or
III) $S(p) = (-\infty, +\infty)$, or
IV) $S(p) = (-\infty, x_0] \cup [x_1, +\infty)$ for some $x_0 < x_1$.

We call the endpoints of the intervals for blue points, *counting events*, since the number of blue points changes when center of C passes any of these points. There are other events affecting C where C contains more than one red point on the boundary. We call them *circle events*. They can be computed as intersection points of l and the farthest-point Voronoi diagram for $\mathcal{R}$.

Proof of Lemma 6. Let q be the center of C. Suppose that $x(p) > x_{max}$. Then $p \notin C$ at the beginning of the motion (when $x(q)$ is near $-\infty$). We show that p is of type II, i.e. p enters C one time (when $x(q) = x_0$) and C contains p afterwards (when $x(q) \geq x_0$). Consider red point p' on the boundary of C when p enters C. Draw the vertical line l' through p'. Point p lies on the the arc of C (dashed circle in Fig. 4) on the right side of this vertical line. Therefore p will be covered until the red point defining C changes.

Suppose that p'' is the next red point (after p') on the boundary of C under the motion of C. Then p'' must be on the left side of l'. This is because the red points must be in the intersection of all circles C determined by p', see Fig. 4. Therefore, we have $x(p) > x(p'')$ and the same argument can be applied to p''. The proof for type II is completed.

We call the event where blue point enters C as *enter event*. Similarly, one can show that, for any blue point p with x-coordinate less than x_{min}, there is only one *exit event* when p leaves C. Therefore p is of type I. The argument for blue points of type III and IV is similar. $\square$

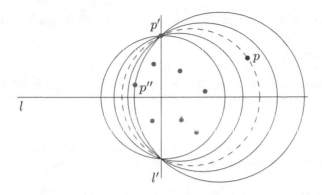

Fig. 4. The motion of C when it is determined by p'. The dashed circle is the first circle containing p.

Lemma 7. *Set $\mathcal{R}$ of red points can be preprocessed in $O(n \log n)$ time such that, for any blue point (it can be viewed a query) its counting events can be computed in $O(\log n)$ time.*

Proof. We preprocess set $\mathcal{R}$ as follows. First, we compute the farthest-point Voronoi diagram $FVD(\mathcal{R})$. Then we compute the intersection points of $FVD(\mathcal{R})$ with line l. These points determine the circle events. Let L be the sorted list of the circle events.

Consider a blue point $p \in \mathcal{B}$. If p is of type I (i.e. $x(p) < x_{min}$) then a simple binary search on L can be applied to find two consecutive circle points c_1 and c_2 such that the circle centered at c_1 contains p and the circle centered at c_2 does not contain p. Suppose that red points p_1 and p_2 determine event at c_1 and red points p_2 and p_3 determine event at c_2. Then the enter event for p is determined by red point p_2 and can be computed as the intersection of l and the bisector of p_2 and p.

Similarly we can compute exit events for blue points of type II (points p with $x(p) > x_{max}$). For blue points p with $x_{min} < x(p) < x_{max}$, we need to detect their types (III or IV). For blue points of type IV, we also need to compute x_0 and x_1 values. We detect the type of a blue point p using binary search on L. Let c_1 and c_2 be two two consecutive circle points in L. Suppose that c_1 is determined by red points p_1 and p_2 and c_2 is determined by red points p_2 and p_3. We test whether circle C centered at points between c_1 and c_2 always contain p. If it does not always contain p, then p is of type IV and the corresponding entry and exit events can be computed as before using binary search on L. Suppose that p is always covered by C centered between c_1 and c_2. If $x(p_2) > x(p)$ then continue binary search to the left of c_1, otherwise continue binary search to the right of c_2.

The preprocessing time is $O(n \log n)$ since we compute the Voronoi diagram and sort the circle events. The query time is $O(\log n)$ since we use binary search. The proof of correctness is omitted. $\square$

The algorithm. Accoring to Lemma 5 an optimal circle can apear at

(1) a circle event, or
(2) a counting event, or
(3) its center is the projection of a red point to p the line l.

There are two possiblities in case 3: point p is either the highest point in $\mathcal{R}$ or it is the lowest point in $\mathcal{R}$. If it is the highest point, then it is the only point with the maximum y-coordinate. It can be computed in $O(n)$ time and its weight can be computed in $O(n+m)$ time by testing every red and blue point whether it is within the circle or not. Similarly the lowest point can be checked in $O(n+m)$ time. It remains to compute the circles for the circle events and the counting events.

After the preprocessing of $\mathcal{R}$ and $\mathcal{B}$, the algorithm use all the events (circle events and count events) to compute the smallest circle C that contains all the red points and the smallest (possible) number of blue points. The kinetic algorithm "moves" the center of circle C along l from $-\infty$ to ∞ and process the events. First, the events are sorted by x-coordinate. The algorithm maintain circle C and the number of blue points covered by C.

For each interval between two consecutive event points we compute the smallest radius circle centered in this interval and determined by the current red point. Then the smallest radius circle corresponding to the interval with minimum number of blue points is computed.

We implemented an algorithm for tracking a circle containing all the red points and computing a circle with the minimum number of blue points[1].

Theorem 1. *A smallest separating circle can be found in $O((m+n)\log n)$ time using $O(m+n)$ space.*

6   Conclusion

We considered the restricted bichromatic smallest separating circle problem and the general and restricted smallest maximum-weight circle problem, where the restriction is a line containing the center of the desired circle. It is natural to ask about different restrictions, for example, when the center of C belongs to a curve or a polygonal line.

Lemma 2 implies an $O(n^2)$ upper bound on the number of optimal solutions. On the other hand, an $\Omega(n)$ lower bound can be easily derived from the lower bound on the minimum separating circle problem [4]. We leave bridging the gap between the two bounds as an open problem.

---

[1] The Java applet can be run at
http://www.utdallas.edu/~sxb027100/applet/redblue-circle

References

1. Barbay, J., Chan, T.M., Navarro, G., Pérez-Lantero, P.: Maximum-weight planar boxes in $O(n^2)$ time (and better). Inf. Process. Lett. **114**(8), 437–445 (2014)
2. Carr, R.D., Doddi, S., Konjevod, G., Marathe, M.: On the red-blue set cover problem. In: Proceedings of ACM-SIAM Symposium on Discrete Algorithms (SODA), pp. 345–353 (2000)
3. Chan, T.M., Hu, N.: Geometric red blue set cover for unit squares and related problems. Comput. Geom. **48**(5), 380–385 (2015)
4. Bitner, S., Cheung, Y.K., Daescu, O.: Minimum separating circle for bichromatic points in the plane. In: Proceedings of the 7th International Symposium on Voronoi Diagrams in Science and Engineering, pp. 50–55 (2010)
5. Cheung, Y.K., Daescu, O., Zivanic, M.: Kinetic red-blue minimum separating circle. In: Wang, W., Zhu, X., Du, D.-Z. (eds.) COCOA 2011. LNCS, vol. 6831, pp. 448–463. Springer, Heidelberg (2011)
6. Dobkin, D.P., Eppstein, D., Mitchell, D.P.: Computing the discrepancy with applications to supersampling patterns. ACM Transactions on Graphics **15**, 354–376 (1996)
7. Har-Peled, S., Lee, M.: Weighted geometric set cover problems revisited. Journal of Computational Geometry **3**(1), 65–85 (2012)
8. Har-Peled, S., Mazumdar, S.: Fast algorithms for computing the smallest k-enclosing circle. Algorithmica **41**(3), 147–157 (2005)
9. O'Rourke, J., Kosaraju, S., Megiddo, N.: Computing circular separability. Discrete. Computational Geometry **1**, 105–113 (1986)

Workshop on Computational Algorithms for Sustainability Assessment (CLASS 2015)

Modeling Socially Synergistic Behavior in Autonomous Agents

Shagun Akarsh[1], Rajdeep Niyogi[1]([✉]), and Alfredo Milani[2,3]

[1] Department of Computer Science and Engineering, Indian Institute of Technology Roorkee, Roorkee 247667, India
shagun.akarsh@gmail.com, rajdpfec@iitr.ac.in
[2] University of Perugia, Perugia, Italy
[3] Department of Computer Science, Hong Kong Baptist University, Kowloon, Hong Kong
milani@unipg.it

Abstract. The "Tragedy of the Commons" (TOC) is a problem in which the sustainability of the society (group of agents) reduces due to self-interested individual agents. Many areas of interest to society like climate change, fisheries management, preservation of rainforests exhibit this phenomenon. We have focused on understanding what is the degree of sacrifice that an agent can make so that the sustainability of the society can be extended. For this we first make a mathematical modeling of the TOC dilemma. Next we propose three types of algorithms corresponding to the different behaviors of the agents. First we assume that the agents are interested in their individual gains only. In this case the society survives for the least amount of time. In the second approach we assume that the agents make decisions based on the resource availability, individual gains or a combination of both. Here an agent's behavior takes into account the welfare of the society to some extent. Thus now the society survives for a longer period of time compared to that in the previous case. In the third approach we define a measure of social awareness of the agents. This measure is indicative of the degree of sacrifice the agent is willing to make. Now the society performs considerably better than the second case. We have experimentally validated these results. Our study shows that if the agents are willing to sacrifice for some period of time, the sustainability of the society increases considerably.

1 Introduction

The "Tragedy of the Commons" (TOC)[3,6,7] is a problem in which the sustainability of the society (group of agents) reduces due to self-interested individual agents using a shared resource (a commons). This problem first appeared in the seminal paper of Hardin in 1968 [3]. Many areas of interest to society like climate change, fisheries management, preservation of rainforests exhibit this phenomenon [4].

Researchers in the area of Distributed Artificial Intelligence (DAI) and Multi-agent systems [1,2,4,8–10] have also addressed the TOC problem. In [4], how

© Springer International Publishing Switzerland 2015
O. Gervasi et al. (Eds.): ICCSA 2015, Part II, LNCS 9156, pp. 257–272, 2015.
DOI: 10.1007/978-3-319-21407-8_20

the TOC is applicable to a DAI setting is studied. In [2] an algorithm is suggested to give an optimal resource utilization by the individuals (agents) of the society, where the agents have only local information. In [1] the performance of the society is studied when aspiration levels are associated with an individual. An aspiration level corresponds to the satisficing return for an individual. Such an aspiration level is adjusted based on past experience.

Any attempt to avoid the tragedy of the commons should incorporate in to the decision making process of an agent the following: the individual gains as well as the social welfare. However these two aspects often conflict. This issue has been addressed in [5] in the context of designing socially intelligent agents, although the TOC problem is not studied in the paper. In [5] a framework is proposed for making socially acceptable decisions.

Consider a society where a public good is available for free (or very little cost) to the members of the society. If there is no law associated for the utilization of the public good, an individual of the society would like to act in a manner that maximizes its utility of the public good. From an individual perspective this is the best decision. However if all individuals act in the like manner, the public good would soon get depleted due to the synergistic behavior and so the society collapses. Thus laws are necessary for the proper functioning of a society. When there is a law in effect it entails a member to abide by it. TOC is concerned with the situation when there is no such formal law or rule. This is where the behavior of an individual comes in to effect that should consider (i) its utility from the public good and (ii) the depletion rate of the public good. If only (i) is considered we are faced with what is called the tragedy of the commons. When (ii) is taken into account the decisions are *to some extent* based on the welfare of the society.

If we view the public good as a resource, the survival of the society depends directly on the rate of depletion of the resource. The slower the rate of depletion, the longer the time of survival of the society. Although the algorithms developed in [1,2] can use the shared resource optimally, the issue of survival time of the society is not considered. In this paper we consider socially motivated agents. The agents make decisions that consider the welfare of the society. This helps the society to survive for a longer period of time compared to the situation when the agents would have acted for their individual gains only.

This paper is organized as follows. In Section 2 we give a model of the TOC problem. We present an algorithm that corresponds to the behavior of self interested agents and give some experimental results. In Section 3 we present an algorithm that corresponds to the behavior of an agent that is socially motivated and give some experimental results. In Section 4 we define a parameter to quantify an agent's willingness to sacrifice for the society. We present an algorithm that corresponds to socially motivated agents and give some experimental results. In Section 5 we analyze the validity of the results. We conclude in Section 6.

2   Modeling the Tragedy of the Commons

The tragedy of the commons as developed in Hardin [3] is as follows.

*"Picture a pasture open to all. It is to be expected that each herds-
man will try to keep as many cattle as possible on the commons.
Such an arrangement may work reasonably satisfactorily for centuries because tribal wars, poaching, and disease keep the numbers
of both man and beast well below the carrying capacity of the land.
Finally, however, comes the day of reckoning, that is, the day when
the long desired goal of social stability becomes a reality. At this
point, the inherent logic of the commons remorselessly generates
tragedy [3]."*

We assume that there are a finite number of agents (herdsmen), that we refer
to as the society. We denote the agents by the numbers $1, 2, \ldots, n$. Each agent
i has g_i^t number of cattle at time t. Total number of cattle grazing in the field
at time t is denoted by G^t. Each cattle consumes Q unit(s) of grass (commons).
The field, shared by the cattle of the agents, is available for free to the agents.
We refer to the field as the shared resource (initially it is R units).

Resource Available (RA) at time t is denoted as RA^t, ($RA^0 = R$). RA^t
denotes the shared resource available to the cattle for consumption and is calculated using equation (3). All agents increase their cattle from time to time
to increase their profit. $incRate_i$ denotes the rate at which an agent i incre-
ments its number of cattle. Profit of agent i at time t is denoted by $profit_i^t$ and
is calculated using equation (2). Profit of an agent is directly proportional to
the number of cattle it uses. Total profit of the society at time t is denoted as
$totalProfit^t$ and is calculated using equation (5). This is the sum of the profits
of the agents till the time t. Resource consumed (RC) in time interval $(t-1, t]$
is denoted as RC_{t-1}^t. Resource consumed during any time interval is calculated
using equation (4).

Initial Conditions: An agent i at time $t = 0$ has g_i^0 number of cattle, $1 \leq i \leq n$.
Initial resource available: $RA^0 = R$. We make the following assumptions. The
shared resource is non-renewable and quantifiable. Agents do not communicate
with each other. No cost is incurred to an agent when it increases the number
cattle.

$$G^t = \sum_{i=1}^{n} g_i^t \tag{1}$$

$$profit_i^t = profit_i^{t-1} + Q \times g_i^t \tag{2}$$

$$RA^t = RA^{t-1} - RC_{t-1}^t \tag{3}$$

$$RC_{t-1}^t = \sum_{i=1}^{n} g_i^t \times Q \tag{4}$$

$$totalProfit^t = \sum_{i=1}^{n} \sum_{t=1}^{t} profit_i^t \tag{5}$$

2.1   Goodness Index: Motivation

We wish to define a goodness-index of the society. The following factors influence such a index.

1. The survival time of the society (denoted by K).
2. The total profit received by the society during the survival time.
3. Total number of cattle used during the survival time. It indicates how efficiently the society was able to utilize the resource available.
4. Initial resource available to the society.

We want the society to survive for a longer period of time without compromising much on the total profit. The intuition behind the definition of the index is to obtain a measure of the success of the society that is achieved by utilizing the initial resource available and the number of cattle used. Synergy refers to the interaction of two or more agents so that their combined effect is greater than the sum of their individual effects. It is the combined effect of the society that is of utmost importance even though the agents of the society may not directly interact. Thus we define θ as:

$$\theta = \frac{totalProfit^K \times K}{RA^0 \times G^K} \tag{6}$$

We shall use this index to compare the behavior of the agents in different settings. In the following we propose three types of algorithms corresponding to the different behaviors of the agents. In the next subsection we give the first algorithm where we assume that the agents are interested in their individual gains only.

2.2   Algorithm 1: Behavior of Self Interested Agents

This algorithm corresponds to the situation where the agents are interested in their individual gains only. The algorithm terminates when the resource gets exhausted. When the algorithm terminates we obtain the total profit and the survival time of the society.

```
1      t := 0
2      while(RAᵗ > 0)
3          for each agent i, profitᵢᵗ := profitᵢᵗ⁻¹ + Q × gᵢᵗ
4          totalProfitᵗ := totalProfitᵗ⁻¹ + ∑ⁿᵢ₌₁ profitᵢᵗ
5          RCᵗₜ₋₁ := ∑ⁿᵢ₌₁ (gᵢᵗ × Q)
6          RAᵗ := RAᵗ⁻¹ − RCᵗₜ₋₁
7          t := t + 1
8          for each agent i, gᵢᵗ := gᵢᵗ⁻¹ + rand₀,₁ × incRateᵢ
```

2.3   Experimental Results

For the experiments we took $n = 100$, $Q = 1$, $RA^0 = 10000$. The initial number of cattle is obtained as: $g_i^t = rand_{0,1} \times 5$, where $0 < rand_{0,1} < 1$ is a randomly generated value. Increment rate: $0 < incRate_i \leq 5$.

All the algorithms are implemented using C++11 with gcc version 4.7.2 by including the *random* header file. We used the pseudo-random number Engine Mersenne Twister 19937 generator (class)mt19937, and uniform_real_distribution for generating the random numbers. For the details we refer to [17].

The experimental results for all the algorithms are obtained as follows. For each run, a distinct seed value is assigned to each agent. This corresponds to the different behaviors of the agents. Table 1 shows the seed values for different runs for the agent numbered 10. Consider row 1 of Table 1. The values corresponding to the different attributes are obtained by taking $seed = 0.827114$ for $i = 10$ and other seed values for all the other agents. Only 5 seed values are shown in Table 1. However we have conducted the experiment for several seed values. The average obtained from these runs is taken to plot the graphs as in Figure 1. Figure 1 shows the results for $incRate_i = 5$ for all agents. The values obtained are as follows: $totalProfit = 10869$, survival time is 9 units, total number of cattle used is 2363, and $\theta = 4.13971 \times 10^{-3}$.

Table 1. Results for Algorithm 1 for different 'seed' values for an agent

$seed_{i=10}$	$g_{i=10}^{t=0}$	$profit_{i=10}^K$	K	G^K	$totalProfit^K$	θ
0.827114	4	137	9	2471	11488	0.00418422
0.525986	3	145	9	2533	11640	0.00413581
0.230025	1	111	9	2510	11619	0.00416618
0.850985	4	112	9	2417	11227	0.00418051
0.728975	3	173	9	2516	11469	0.00410258

From the experimental results we find that the total profit of the society increases until the resource gets exhausted. The survival time of the society decreases on increasing the increment rate.

3   Socially Motivated Agents

In the previous section we considered self-interested agents. Now we consider agents that are socially motivated. That is the agents make decisions that consider the welfare of the society. Unlike the case where an agent was always increasing the cattle, now the agent considers the resource available and the degree to which its profit has been achieved. The motivation being that the decision should be such that helps the society to survive for a longer period of time.

In the previous case we assumed that incrementing cattle does not incur any cost. Now we associate a cost for purchasing new cattle (X units per cattle).

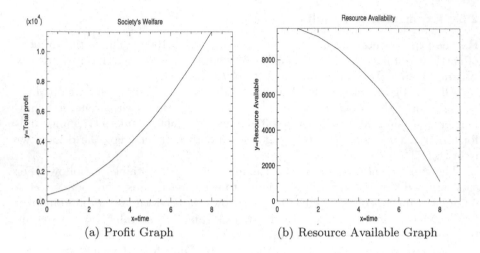

Fig. 1. Results for Algorithm 1 where $incRate = 5$ for all agents

An agent may increase the number of cattle based on (a) the total resource available (case 1), (b) its individual profit (case 2), or (c) combination of both (a) and (b) (case 3).

We assume that shared resource is quantifiable and information about shared resource is available to all agents.

3.1 Algorithm 2: Behavior of Socially Motivated Agents

```
1      t := 0
2      while(RA^t > 0)
3          for each agent i, profit_i^t := profit_i^{t-1} + Q × g_i^t − (X × g_i^t)
4          totalProfit^t := totalProfit^{t-1} + Σ_{i=1}^n profit_i^t
5          RC_{t-1}^t := Σ_{i=1}^n (g_i^t × Q)
6          RA^t := RA^{t-1} − RC_{t-1}^t
7          t := t + 1
8          for each agent i, g_i^t := g_i^{t-1} + increment_i()
```

where function $increment_i()$ is defined as:

Case 1: The decision is taken based on the total resource available.

```
1      if(RA^t > th_{2,i}) then
2          if (incRate_i < INC) then
3              incRate_i := incRate_i + 1
4          else if (RA^t ≥ th_{1,i} and RA^t < th_{2,i}) then
5              incRate_i := INC/2
6          else if(RA^t ≥ th_{2,i}) then
```

7          $incRate_i := incRate_i \div 2$
8   return $rand_{0,1} \times incRate_i$

Case 2: The decision is taken based only on the individual profit obtained.

1   if($profit_i^t < th_{1,i}$) then
2          $incRate_i := incRate_i + 1$
3   else if ($profit_i^t \geq th_{1,i}$ and $profit_i^t < th_{2,i}$) then
4          $incRate_i := incRate_i - rand_{0,1} * \alpha$
5   else if($profit_i^t \geq th_{2,i}$) then
6          $incRate_i := incRate_i \div 2$
7   return $rand_{0,1} \times incRate_i$

Case 3:
The decision is taken based on both the resource availability and agent's individual profit.

1   if ($RA^t \geq th_{3,i}$) then
2   if ($profit_i^t < th_{1,i}$) then
3               $incRate_i := incRate_i + 1$
4   else if ($profit_i^t \geq th_{1,i}$ and $profit_i^t < th_{2,i}$) then
5               $incRate_i := incRate_i - \alpha \times rand_{0,1}$
6   else $incRate_i := incRate_i \div 4$
7   else if ($RA^t \geq th_{4,i}$ and $RA^t < th_{3,i}$) then
8   if ($profit_i^t < th_{1,i}$) then
9               $incRate_i := incRate_i - 1$
10   else if ($profit_i^t \geq th_{1,i}$ and $profit_i^t < th_{2,i}$) then
11              $incRate_i := incRate_i \div 2$
13   else $incRate_i := incRate \div 4$
14   else if ($RA^t < th_{4,i}$) then
15   if ($profit_i^t < th_{1,i}$) then
16              $incRate := incRate_i \div 4$
19   return $rand_{0,1} \times incRate_i$

Table 2. Results for Algorithm 2 for different 'seed' values for an agent

$seed_{i=10}$	$g_{i=10}^{t=0}$	$profit_{i=10}^K$	K	G^K	$totalProfit^K$	θ
0.702449	3	83	15	689	67930	0.147888
0.957374	4	94	16	679	77612.3	0.182886
0.48069	1	83	23	459	112311	0.562777
0.648213	3	147	22	476	108222	0.500186
0.276817	1	121	22	466	108325	0.511406

3.2   Experimental Results

Following results were observed when using the algorithm given in section 3.1. For the experiments we took $n = 100$, $Q = 1$, $RA^0 = 10000$. The initial number of cattle is obtained as: $g_i^t = rand_{0,1} \times 5$, where $0 < rand_{0,1} < 1$ is a randomly generated value. The experimental results are obtained as outlined in section 2.3. Increment rate: $incRate_i$ is initialized to 5 for all the experiments. α is the rate at which cattle is decreased; taken as 2.

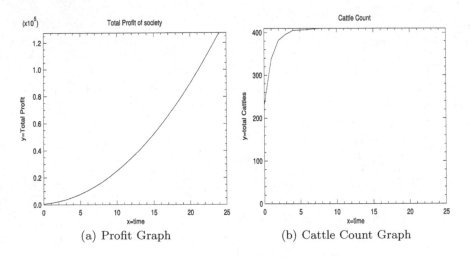

(a) Profit Graph   (b) Cattle Count Graph

Fig. 2. Results of Algorithm 2: Case 1

Table 2 shows the seed values for different runs for the agent numbered 10. The average obtained from these runs is taken to plot the graphs as in Figures 2, 3, 4.

1. For Case 1, $INC = 5$ is used as the initial maximum increment rate.
   $$th_{1,i} \in (0.5RA^0, 0.65RA^0), \quad th_{2,i} \in (0.85RA^0, 1.0RA^0).$$
   These are threshold values on the resource available. Figure 2 shows the results for Case 1 of the algorithm. $totalProfit = 68096.2$, total cattle used $= 700$ $time = 15$, $\theta = 0.14592$.
2. For Case 2,
   $$th_{1,i} = 0.1 \times \frac{RA^0}{n}, \quad th_{2,i} = 0.5 \times \frac{RA^0}{n}$$
   Figure 3 shows the results for Case 2 of the algorithm. $totalProfit = 127434$, total cattle used $= 410$ $time = 24$, $\theta = 0.777034$.
3. For Case 3,
   $$th_{1,i} = 0.1 \times \frac{RA^0}{n}, \quad th_{2,i} = 0.5 \times \frac{RA^0}{n}, \quad th_{3,i} \in (0.5R^0, 0.65RA^0), \quad th_{4,i} \in$$
   $(0.85R^0, 1.0RA^0)$
   Figure 4 shows the results for Case 3 of the algorithm. $totalProfit = 84012.6$, total cattle used $= 564$, $time = 18$, $\theta = 0.268125$.

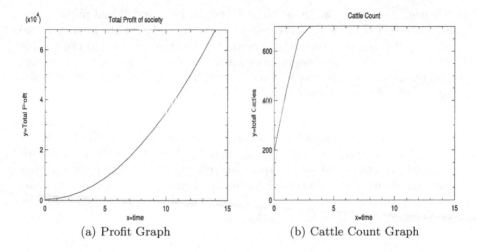

(a) Profit Graph   (b) Cattle Count Graph

Fig. 3. Results of Algorithm 2: Case 2

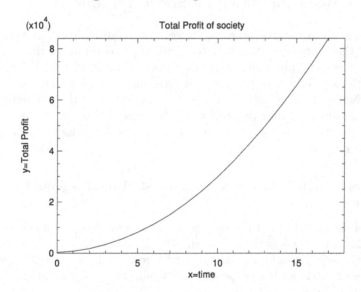

Fig. 4. Results of Algorithm 2: Case 3

We have used some threshold values on the resource available. These are denoted by $th_{1,i}, th_{2,i}, th_{3,i}, th_{4,i}$. The values have been obtained experimentally.

3.3 Inference

1. The theta values have improved considerably in case 1 as compared to that obtained for the algorithm in section 2.2.

2. In case 2 the agents make decisions based on their individual profits. However, they set their own requirements thresholds and alter the increment rate reasonably, thus increasing the survival time of the society and the θ value.

3. In case 3, as both the parameters are used for incrementing, the θ-value slightly improves over that for case 1 as the resources quickly get exhausted before any of the agent reaches its second satisfaction level of profits.

From the behavior of the agents given in sections 2.2 and 3.1, we get the idea that if the behavior of all the agents can be controlled in a manner such that they consider the welfare of the society, then at the cost of their personal loss they can contribute to the society that can sustain for a longer period of time. Information about shared resource is hard to calculate. In general, the information is not available to all the agents. Thus agents should be capable of taking decisions based only on their local information.

4   Agent Behavior Based on Social Parameter

We consider an agent to have two levels of requirements. The first level (we denote this by $need_i$) reflects the bare minimum requirement that is required for carrying out the minimal operations. The second level (we denote this by $greed_i$) is the aspired greed level. We now introduce a social parameter s_i, (a real number whose value lies between 0 and 1) that indicates the willingness of an agent to contribute toward the welfare of the society. Thus an agent may even decrease its cattle at any point of time with a view for enhancing the survival time of the society.

4.1   Algorithm 3: Behavior of Socially Motivated Agents Based on a Social Parameter

1. If $need_i$ is not reached then the agent will increase the number of cattle by a random number less than some threshold value.
2. Once the $need_i$ is achieved but less than $greed_i$, the agent will increase its number of cattle with a probability $1 - s_i$ and decrease the number of cattle by probability s_i.
3. Once $greed_i$ is achieved the agent first decreases its greed level and then decreases its number of cattle with probability 1.

The pseudo-code for the algorithm is given below:

```
1 t := 0
2 while(RAᵗ > 0)
3      for each agent i, profitᵗᵢ := profitᵗ⁻¹ᵢ + Q × gᵗᵢ
4      totalProfitᵗ := totalProfitᵗ⁻¹ + Σⁿᵢ₌₁ profitᵗᵢ
5      RCᵗ := Σⁿᵢ₌₁ (gᵗᵢ × Q)
6      RAᵗ := RAᵗ⁻¹ − RCᵗₜ₋₁
```

7      $t := t + 1$
8   for each agent i, $g_i^t := updateCattle(i)$

where function $updateCattle(i)$ is defined as:
1   if $(g_i^t < need_i)$ then
2          $g_i := g_i + 1$
3   else if $(g_i^t \geq need_i$ and $g_i^t \leq greed_i)$ then
4          $rv := rand_{0,1}$
5   if$(rv < s_i)$ then
6              $g_i^t := g_i^t \div (1 + s_i)$
7   else $g_i := g_i + 1$
8   else if $(g_i^t \geq greed_i)$ then
9              $greed_i := (greed_i + need_i) \div 2$
10             $g_i := g_i \div (1.5 + s_i)$
11   return g_i

4.2   Experimental Results

For the experiments we took $n = 100$, $Q = 1$, $RA^0 = 10000$. The initial number of cattle is obtained as: $g_i^t = rand_{0,1} \times 5$, where $0 < rand_{0,1} < 1$ is a randomly generated value. The experimental results are obtained as outlined in section 2.3.

The initial value of g_i is: $0 \leq g_i \leq 5$. The other values taken for the experiments are:

$s_i \in (0.0, 1.0)$
$need_i \in (0.0, \frac{0.01 \times RA^0}{n})$
$greed_i \in (\frac{0.70 \times RA^0}{n}, \frac{0.85 \times RA^0}{n})$

Table 3 shows the seed values for different runs for the agent numbered 10. The average obtained from these runs is taken to plot the graphs as in Figures 5, 6, 7.

Table 3. Results for Algorithm 3 for different 'seed' values for an agent

$seed_{i=10}$	$g_{i=10}^{t=0}$	$profit_{i=10}^K$	K	G^K	$totalProfit^K$	θ
0.935899	3	209	243	10016	1.43E+06	3.46935
0.616323	2	97	262	10038	1.47E+06	3.83682
0.557808	1	100	291	9993	1.58E+06	4.60121
0.454304	3	110	307	9985	1.66E+06	5.10386
0.0535557	4	61	303	10031	1.64E+06	4.95384

Figure 5 shows the results of the algorithm in section 4.1 when $s \in (0.1, 1.0)$ $totalProfit = 1.34E + 006$, total cattle used= 10038, $time = 237$, $\theta = 3.1611$.

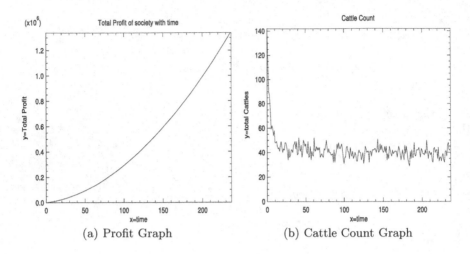

Fig. 5. Results of Algorithm 3 when decision is based on social parameter $s \in (0.1, 1.0)$

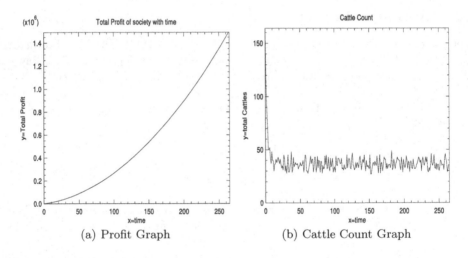

Fig. 6. Results of Algorithm 3 where decision is based on social parameter $s \in (0.3, 1.0)$

Figure 6 and Figure 7 shows the results of the algorithm when $s \in (0.3, 1.0)$ $totalProfit = 1.49E + 006$, total cattle used is 10015, $time = 265$, $\theta = 3.9465$.

When $s \in (0.6, 1.0)$, $totalProfit = 1.67E + 006$, total cattle used is 10002, $time = 305$, and $\theta = 5.09486$; when $s \in (0.9, 1.0)$ $totalProfit = 1.63E + 006$, total cattle used is 10028, $time = 300$, and $\theta = 4.87298$.

4.3   Inference

1. The total profit of the society increases quickly up to a threshold value and then varies slightly around it.

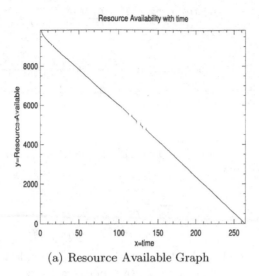

(a) Resource Available Graph

Fig. 7. Resource Available using Algorithm 3 where decision is based on social parameter $s \in (0.3, 1.0)$

2. When the s parameter is increased from $(0.1,1.0)$ to $(0.3,1.0)$, the survival time of the society increases from 237 units to 265 units. This shows that if all the agents are ready to sacrifice for at least 30% of the time then the overall survival time of society increases by 11.81% as compared to the situation when all the agents sacrifice for at least 10% of time.

3. On increasing value of s the survival time of the society increases. The value of θ also increases, implying that the society's welfare also increases. But value of θ saturates when all the agents are sacrificing for most of the time.

5   Discussion

We give an analytical argument of the behavior of the agents (for different cases) based on the parameter θ.

Let us first compare the θ values corresponding to Algorithm 1 and Algorithm 2.

$$\theta_1 = \frac{totalProfit^{K_1} \times K_1}{RA^0 \times G^{K_1}} \qquad (7)$$

$$\theta_2 = \frac{totalProfit^{K_2} \times K_2}{RA^0 \times G^{K_2}} \qquad (8)$$

Now lets see the value of $\frac{\theta_2}{\theta_1}$ using equations (7) and (8). So it can be written as:

$$\frac{\theta_2}{\theta_1} = \frac{totalProfit^{K_2}}{totalProfit^{K_1}} \times \frac{K_2}{K_1} \times \frac{G^{K_1}}{G^{K_2}} \qquad (9)$$

$$= \left(\frac{totalProfit^{K_2}}{G^{K_2}} \div \frac{totalProfit^{K_1}}{G^{K_1}}\right) \times \frac{K_2}{K_1}$$
$$= \frac{K_2}{K_1}$$

Since total profit is proportional to $\frac{Q \times G^K}{G^K}$, so $\frac{totalProfit^{K_2}}{G^{K_2}}$ gives Q. Hence the ratio of $\frac{totalProfit^{K_2}}{G^{K_2}}$ and $\frac{totalProfit^{K_1}}{G^{K_1}}$ equals to 1. The ratio $\frac{K_2}{K_1} > 1$ because in the second algorithm the number of cattle is a non-decreasing function while in the first algorithm it is monotonically increasing. Thus $K_2 > K_1$. So $\frac{\theta_2}{\theta_1} > 1$ and thus $\theta_2 > \theta_1$.

Similarly we can compare θ_1 and θ_3 and obtain $\theta_1 < \theta_3$. Now we can compare θ_2 and θ_3 using the above results.

$$\frac{\theta_3}{\theta_2} = \frac{\theta_3}{\theta_1} \div \frac{\theta_2}{\theta_1} \tag{10}$$

$$= \left(\frac{totalProfit^{K_3}}{G^{K_3}} \div \frac{totalProfit^{K_2}}{G^{K_2}}\right) \times \frac{K_3}{K_2}$$

In algorithm 3 the number of cattle can also decrease while in algorithm 2 the cattle count is non-decreasing and thus the society will survive for longer period of time in case of algorithm 3, i.e., $K_3 > K_2$ and thus $\theta_3 > \theta_2$. Combining the above results, we get $\theta_3 > \theta_2 > \theta_1$. Thus when the actions of the agents consider the welfare of the society, the goodness index of the society increases. This implies that the survival period of the society increases. The summary of the experimental results for the three algorithms is given in Table 4. For algorithm 3 the θ value is maximum, followed by that for algorithm 2. The θ value for algorithm 1 is the least. Thus the values obtained from the experiments validate the above relation $\theta_3 > \theta_2 > \theta_1$.

Table 4. Comparison of θ values for the Algorithms

	K	θ
Algorithm 1	9	0.004
Algorithm 2	15 to 23	0.14 to 0.56
Algorithm 3	243 to 307	3.5 to 5.1

We have experimented and evaluated the proposed solution using an agent framework which has been implemented in C++. For efficiency reasons a special purpose implementation has been preferred with respect to more popular general purpose agent simulation frameworks like Netlogo [14], CORMAS [15], or MaDKIT [16].

We considered agents that are socially motivated. Reputation of agents [11][1] is an important parameter that helps solve the tragedy of the commons. It would be interesting to compare our approach with that in [11]. In our model we did not consider interaction of agents. This is motivated by the fact that in many real world scenarios (eg, too many cars limiting the flow of traffic on a

---

[1] This work [11] was pointed out by an anonymous reviewer.

highway [12] or several automobile designers, each making increasing demand on the car battery [13]) the agents simply cannot interact. However if we allow the agents to interact by sharing information, it may be helpful for the society.

6   Conclusion

In this paper we made an experimental analysis of the tragedy of commons dilemma. We gave a mathematical modeling of the tragedy of commons problem. In order to understand this problem for different behaviors of the agents we designed three algorithms. First we assumed that the agents are interested in their individual gains only. Then we assumed that the agents make decisions based on the resource availability, individual gains or combination of both. Here an agent's behavior takes into account the welfare of the society. Thus now the society survives for a longer time compared to that in previous case. For the third type of behavior, we defined a measure of greediness of the agents. Now the agents make decision based on this parameter. Now the society performs much better than the second case. From our study we conclude that if the agents are willing to sacrifice (reduce the number of cattle) for some period of time, the sustainability of the society increases considerably. As part of our future work we would like to study group decision making when both utility as well as social parameter are considered. We would like to see the scope of interaction of the agents and study how it helps the society.

Acknowledgments. The authors thank the anonymous reviewers of ICCSA for their valuable comments and suggestions for improving the paper.

References

1. Sen, O., Sen, S.: Averting the tragedy of commons by adaptive aspiration levels. In: PRIMA, pp 355–370 (2010)
2. Saha, S., Sen, S.: Local decisions procedures for avoiding the tragedy of commons. In: Das, S.R., Das, S.K. (eds.) IWDC 2003. LNCS, vol. 2918, pp. 311–320. Springer, Heidelberg (2003)
3. Hardin, G.: The Tragedy of the Commons. Science **162**(3859), 1243–1248 (1968)
4. Turner, R.M.: The tragedy of the commons and distributed AI systems. In: 12th International Workshop on Distributed AI, pp. 379–390 (1993)
5. Hogg, L.M.J., Jennings, N.R.: Socially Intelligent Reasoning for Autonomous Agents. IEEE Trans. SMC(A) **31**(5), 381–393 (2001)
6. Lopez, L., Almansa, G.D.R., Paquelet, S., Fernandez, A.: A mathematical model for the TCP Tragedy of the Commons. Theoretical Computer Science **343**(1–2) (2005)
7. Diekert, F.K.: The tragedy of the commons from a game-theoretic perspective. Sustainability **4**, 1776–1786 (2012)
8. Killingback, T., Doebeli, M., Hauert, C.: Diversity of cooperation in the Tragedy of Commons. Biological Theory **5**, 3–6 (2010)

9. Doebeli, M., Hauert, C.: Models of cooperation based on the Prisoner's dilemma and the snowdrift game. Ecology Letters **8**, 748–766 (2005)
10. Castelfranchi, C.: Modeling social action of AI agents. Artificial Intelligence **103**(1–2) (1998)
11. Milinsky, M., Semmann, D., Krambeck, H.-J.: Reputation helps solve the 'tragedy of the commons'. Letters to Nature, Nature **415**, 424–426 (2002)
12. Senge, P., Kleiner, A., Roberts, C., Ross, R., Smith, B.: The Fifth Discipline Fieldbook. Doubleday, New York (1994)
13. Braun, W.: The System Archetypes (2002)
14. Wilensky, U.: NetLogo. Center for connected learning and computer-based modelling (CCL). Northwestern University, USA. http://ccl.northwestern.edu/netlogo/
15. CORMAS. http://cormas.cirad.fr/indexeng.htm
16. MadKIT. http://www.madkit.org/
17. Pseudo-random numbers. http://www.cplusplus.com/reference/random/

Geostatistics Applied to the Geoprospective

GP-SET. Krige Method Theory and Application

Stéphane Bourrelly[✉]

UMR 7300 ESPACE (CNRS), Nice University,
98 Boulevard Edouard Herriot, BP 3209 06204, Nice, France
s.bourrelly@hotmail.fr, stephane.bourrelly@univ-amu.fr

Abstract. Big Data provides the ability to describe many living environment dimensions, from administrative data. Territorial scales allow decision-making but are not sufficiently accurate to establish effective policies tailored to the needs of sustainable development.

Geographic Information Systems (GIS) can integrate any kinds of data. Unfortunately GIS are still used without exactly knowing the methods implemented, or having geographic knowledge of the phenomena studied. Very different results can be obtained with a same dataset. The inappropriate use of GIS is damaging for prospective studies derived from spatial analysis.

The goal of geoprospective is to develop robust methods to address these challenges. GP-SET.krige makes two spatiotemporal indicators that accurately model the spatial spread of human phenomena and their uncertainty. It is based on univariable geostatistics. Applied to census data, GP-SET.krige precisely models the potential to have a demographic growth in the next ten years.

Keywords: Geoprospective · Kriging · Spatial modelling · Demographic growth · Decision-making

1 Introduction

Current Big Data provides access to many geographical indicators describing all aspects of living environments (Vitolo C., Elkhatib Y. et al., 2015).

The technology of Geographic Information Systems (GIS) enables the integration of any kinds of data and calculates indicators, in order to model the spatiotemporal spread of many phenomena (Raper J., 2000).

The ergonomic design of GIS might be a problem. In fact, they are often used without understanding the methods implemented, or having expert knowledge of geography on the phenomena studied. But "button pushed" is the opposite of robustness. Therefore, GIS can be damaging especially for prospective studies derived from spatial analysis (Donnay J-P., 2005).

One of the objectives of geoprospective is to realise spatial analysis, enabling the accurate targeting areas where future challenges were occurred. The methods designed must provide significant maps for stakeholders to help with decision-making. The spatial uncertainty of prospective modelling must be taken into account in order to consider the effectiveness of current and future policies (Voiron C., 2012).

© Springer International Publishing Switzerland 2015
O. Gervasi et al. (Eds.): ICCSA 2015, Part II, LNCS 9156, pp. 273–287, 2015.
DOI: 10.1007/978-3-319-21407-8_21

This paper proposes a significant method for modelling geoprospective phenomena at micro-scales: GP-SET.krige. It is based on the combination of univariable geostatistics. Among all the spatial interpolation methods, only geostatistics incorporate both autocorrelation and anisotropy of phenomena, as well as the assessment of spatial uncertainty (Gaetan C., Guyon X., 2010).

For illustrating GP-SET.krige and ensuring its reproducibility; it is applied to the spatial modelling of the land potential to undergo a demographic growth. Particular attention is paid to the adequacy of geographic concepts with theoretical backgrounds and hypotheses of geostatic models. In geoprospective, the probability of demographic increase can be clearly forecasted, at the scale of communes, from the GP-SET indicator (Bourrelly S., Voiron C. 2012)

Demographic growth is unavoidable in developed countries. It is a socio-economic leverage when it is accompanied by the creation of sanitation networks, transport infrastructures, universities, hospitals... But these facilities are expensive. Therefore it is necessary to accurately target the places where they should be located. Thus commune scale is too crude to be effective (Turner B., Lambin E. et al., 2007).

The future potential of demographic growth is modelled at the scale of communes by GP-SET(2016) indicator. Geostatistics are used to estimate this potential at micro-scales; and to assess the modelling uncertainty. Results are analysed and the significance is discussed. The strengths and limits of the method are identified. Then the expected added values are declined in the decision-making process.

2   Material and Hypotheses

2.1   GP-SET(2016) Indicator Source

GP-SET indicator is defined as a geographical random variable X_{c_i}. It represents the probability for the commune c_i to undergo a demographic increase in 2016 (Bourrelly S., Voiron C., 2012).

The GP-SET method is reproducible in any countries where administrative data are available. Here, it has been applied to data from the last six French censuses, i.e. 1968, 1975, 1982, 1990, 1999, 2006 (INSEE, 2012).

The GP-SET(2016) was computed for all communes, located within the 6 departments of PACA region, in south-eastern France. It consists of two steps. i. population density adjustment from the laws of probability. Models used are the log-normal distributions. The adjustment is verified by performing fit testing, i.e. Khi² and Kolmogorov-Smirnov & Massart (DasGupta A., 2008). ii. average demographic growth rates are calculated; and they have been geometrically applied within the geoprospective trend scenario. Then parameters of log-normal distributions in 2016 are assessed through regression models from those estimated previously (Bourrelly S., Voiron C., 2012). Modelling results are presented in Fig.1.

Population growth is a complex phenomenon. It is affected by socio-economic and environmental factors that are very different according to the territories. The spatial average is a biased estimation of mathematical expectation, because spatial distribution of

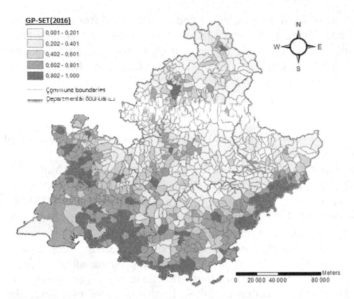

Fig. 1. Geoprospective mapping of GP-SET(2016)

demographic increases contains extreme values at the scale of communes. Furthermore the phenomenon spread is different according to directions; this finding means that spatial autocorrelations exist at different scales (Bourrelly S., Voiron C., 2012).

2.2   State of Art on Spatial Interpolation for Improving Source Accuracy

Spatial interpolation gives the ability to estimate at micro-scales all unknown values in space, from a sample of geo-localised variables. Those interpolation methods are either deterministic or stochastic.

Deterministic methods interpolate unknown values from those available into neighbourhood, in using moving averages, Kernel approximations, or inverse distance weighted. But model parameters as well as the shape and size of the neighbourhood window are subjectively defined; namely without actually taking into account the data. In fact, these techniques are not advised for the benefit of barycentric ones, as Thiessen, Voronoï or Dirichlet. Although they are specified from spatial observations, the results are typically imprecise and contained within the convex hull of known sites. Finally splines, as the thin plate spline, are the only deterministic methods that are really significant. However they do not allow a direct estimation of the interpolation error, despite the fact that uncertainty analysis is a seminal stake in the spatial modelling field (Wang X-B., Liu L. et al., 2004 ; Gaetan C., Guyon X., 2010).

The more common stochastic methods are geographically weighted regressions. They interpolate unknown values by taking into account the spatial distribution of sample data, and they assess the associated error. Unfortunately neighbourhood characteristics are specified by statistical criteria without taking into account the data location. Conversely geostatistics, and especially kriging methods, combine the main advantages of all deterministic and stochastic methods set out. They are BLUP: the best linear unbiased predictors. These methods are ideal for geo-localised data. Above all, the shape and size of the neighbourhood are estimated from the spatial distribution of sample data, through the variograms. Those instruments also allow one the characterisation of the stationary and anisotropy of the spatial phenomena (Matheron G., 1989 ; Wang X-B., Liu L. et al., 2004; Gaetan C., Guyon X., 2010).

2.3   Underlying Hypotheses and Concepts Required for Using Kriging

GP-SET.Krige is a geostatistic extension of the GP-SET method. His goal is to interpolate the GP-SET (2016) values, in order to model prospective demographic growth at infra-municipal scale.

However, kriging only runs with regionalised variables z_{s_i} (Matheron G., 1989). Consequently, GP-SET indicators (2016) must be "wisely" geo-located.

In order to convert x_{c_i} into z_{s_i} the first were geo-localised at the urban centres.

This assumption implies that the spread of demographic growth is contiguous in space and slightly dependent on administrative boundaries.

Central place theory explains that urbanization takes place from centre of the commune to its periphery (Christaller W., 1933). However the phenomenon is not isotropic and its intensity is different according to directions. This finding has been already observed in the study area (Dauphiné A., Voiron C., 1988). Furthermore, the maximum likelihood of a demographic increase is reached in spaces located at an intermediate distance between the urban centre and its periphery. In fact, this spatial autocorrelation process depends on socio-demographic contexts in the vicinity as well as the location of former urban structures; it is the path dependence (Hamilton I., Dimitrovska A. et al., 2005)

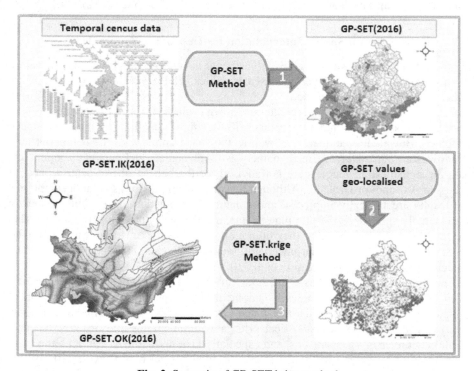

Fig. 2. Synoptic of GP-SET.krige method

Kriging concepts incorporate all these theoretical bases. First, it is an exact estimator, therefore at the historical centre, the value is bounded. The variograms model the autocorrelation and spatial anisotropy of phenomena. In fact, demographic growth also increases to preferred directions, according to neighbouring socio-economic contexts. Thus interpolated values at intermediate locations may be stronger than those geo-localised at the urban centre.

GP-SET.Krige method uses Ordinary Kriging (OK) to interpolate GP SET(2016) indicators, from GP-SET.KO (2016). The Indicator Kriging (IK) then estimates the certainty linked to the high values of GP-SET.OK (2016), from GP-SET.IK (2016).

3   Method

Potential demographic increases spread over geographic space with a contiguous and anisotropic pattern. Variograms model these features at different spatial scales (Matheron G., 1989).

3.1   Variogram

The variogram $\gamma(h)$ - instead the semi-variogram - is the suitable instrument for kriging. This mathematical function summarises spatial information contained in z_{s_i}. It is estimated through:

$$\hat{\gamma}(h) = \frac{1}{2 \cdot N(h)} \cdot \sum_{i=1}^{n} \sum_{j \neq i} \left(z_{s_i} - z_{s_j} \right)^2 \cdot \mathbb{1}_{(h_{ij} \leq h_{crit})} \tag{1}$$

Variogram is the variance of the difference between z_{s_i} and z_{s_j} localised at an Euclidean distance h_{ij} shorter than a critical distance h_{crit}. This bound is limited to half the maximum distance between all available geo-localised data. In order to be easily interpreted squared differences of field values are aggregated by lag distances h. N(h) is the point quantity involved into each lag distance (Matheron G., 1989).

When a phenomenon is stationary, $\hat{\gamma}(h)$ is fitted to a theoretical model; and kriging can carry out with z_{s_i}. On chart, the stationarity presents $\hat{\gamma}(h)$ as an increasing function of h - (Fig.3).

Theoretical sill σ_a^2 is the value when the variance is reached; beyond there is no longer autocorrelation. Theoretical origin of variogram is a nugget effect σ_0^2. It reflects random perturbations at micro-scales. This original variability is due to non-uniform spatial distribution and outliers of the sampled values. The range: a is the distance where σ_a^2 is reached. There are two kinds of variograms: those with a fixed sill or those with an asymptotic sill – Fig.3. Gaussian model describes a great spatial continuity of shape; its sill is never actually reached (Matheron G., 1989).

Directional empirical variograms $\hat{\gamma}(h; \theta_k)$ involve values localised in a specific direction $\theta_k \in [0°;180°]$, and for an angle of tolerance whose range value is usually $\varphi \in [30°;45°]$. The $\hat{\gamma}(h; \theta_k)$ are used to detect anisotropies. Typically two types of anisotropy are distinguished (Gaetan C., Guyon X., 2010).

Geometric anisotropy is characterised by directional variograms having the same values of nuggets and sills in all directions. Nevertheless, sills are reached at different ranges a_k. Geometric anisotropy is corrected with an isovalue matrix specified from an ellipse, whose semimajor axis is theoretically equal to the largest directional range a_g, oriented in this direction θ_g, among all a_k. Semiminor axis is the greater distance in the orthogonal direction a_p. The anisotropy factor is defined such as $F_a = a_g/a_p$. In practice geometric anisotropy is corrected if $F_a \geq 1.5$ and the size sample overcomes 50 values (Gaetan C., Guyon X., 2010).

Zonal anisotropy is characterised by $\hat{\gamma}(h; \theta_k)$ reaching their sills σ_k^2 at various ranges a_k according to directions θ_k. It is a drift. In this case $\hat{\gamma}(h)$ is an increased function of h. Drift is removed by fitting a layer trend to z_{s_i}. The directional variograms-derived $\hat{\gamma}'(h; \theta_k)$ are computed from residuals with the k-order surface trend; until the variograms-derived are stationary (Gaetan C., Guyon X., 2010).

In this way, empirical variograms allow fitting to theoretical models $\tilde{\gamma}(h; \vartheta_k)$. Among the variety of documented functions; the most used are identified – Fig.3.

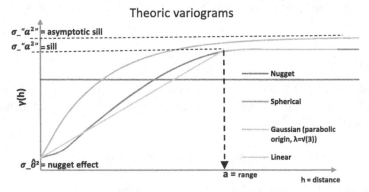

Fig. 3. Common theoretical variograms

When the phenomenon is stationary; directional models $\tilde{\gamma}(h; \theta_k, \vartheta_k)$ are combined into an omnidirectional variogram $\tilde{\gamma}(h; \vartheta)$. Model parameters ϑ are usually calibrated with the weighted least squares (WLS) criterion, according to Cressie's advice. Thus estimator parameters are obtained $\{\hat{\vartheta} = \hat{a}, \hat{\sigma}_a^2, \hat{\sigma}_0^2\}$. The specification of $\tilde{\gamma}(h; \hat{\vartheta})$ solves the kriging systems (Gaetan C., Guyon X., 2010).

3.2   Ordinary Kriging

Ordinary Kriging (OK) is a kind of improved geographically weighted regression. Like the deterministic component, a part of the stochastic component becomes estimable from the variogram. The rest of the stochastic component is assumed as an error function. Kriging estimator is carried out with available data z_{s_i}. It is linear, unbiased and has a minimum variance.

$$z_{s_0} = \sum_{\forall i \in V(s_0)} \lambda_i \cdot z_{s_i} \qquad (2)$$

OK is also exact. Therefore the known values correspond exactly with those inter-polated. Mean and variance are assumed invariants into the kriging neighbourhood; $\forall\, i \in V(s_o) \stackrel{def}{=} \{[i] = [1], \dots, [n_o]\}$. Weights λ_i increase with the distance between the known sites s_i and the one of interpolating s_o. In this way interpolation involves only $z_{s_i} \subseteq V(s_n)$; it is the screening effect. In practice a significant interpolation requires using at least 8 z_{s_i} included into the neighbourhood, but do not exceed 20. Estimation is more efficient when spatial distribution of z_{s_i} is uniform into $V(s_n)$. Unbiased assessment induces that the $\sum_{i \in V(s_o)} \lambda_i = 1$. The variance minimisation enables the specification of λ vector from $\hat{\lambda}$. In fact, it is a convex function that be minimised $-\lambda^t \Gamma_{ij} \lambda + 2\lambda \gamma_{io}$. Typically a Lagrangian μ is introduced in order to solve the system; which become $L(\lambda; \mu) = -\lambda^t \Gamma_{ij} \lambda + 2\lambda^t \gamma_{io} + 2\mu(\lambda \mathbb{I}_{n_o} - 1)$ - with $\mathbb{I}_{n_o} = (1, \dots, 1)^t$. When all partial differential equations are equal zero $L(\lambda; \mu)/\partial \lambda = 0$; weights are computed $\hat{\lambda}_\mu = \Gamma_{ij,1}^{-1} \gamma_{io,1}$ - (Gaetan C., Guyon X., 2010)

Ordinary kriging value estimator is:

$$\hat{z}_{s_o} = \sum_{\forall i \in V(s_o)} \hat{\lambda}_i \cdot z_{s_i} \tag{3}$$

Ordinary kriging variance estimator is:

$$\hat{\sigma}_{z_{s_o}}^2 = \sum_{\forall i \in V(s_o)} \hat{\lambda}_i \cdot \tilde{\gamma}(h_{io}) + \hat{\mu} \tag{4}$$

3.3   Indicator Kriging

Indicator Kriging (IK) assesses the likelihood that z_{s_o} is equal or greater than a threshold τ. IK estimator is also punctual, exact, unbiased and has a minimum vari-ance. It is carried out with z_{s_i} transformed through an indicator function, such as:

$$i_{(s_i|\tau)} = \mathbb{1}_{\{z_{s_i} \geq \tau\}} = \begin{cases} 1 & \text{if} \{z_{s_i} \geq \tau\} \\ 0 & \text{else} \end{cases} ; \; \forall\, i = \{1, \dots, n\} \tag{5}$$

Unknown values into the field $i_{(s_o|\tau)}$ are estimated from $\hat{i}_{(s_o|\tau)}$, which is fitted to available data. When assuming that $i_{(s_i|\tau)}$ are stationary in space, so estimation process is the same as OK: computation of empirical variograms, selection and calibration of a theoretical model, then specification of the kriging neighbourhood $V(s_o)$.

The values estimated from IK can be interpreted as certainty measures such as:

$$\mathbb{E}\left[\hat{i}_{(s_o|\tau)}\right] \approx 1 - \mathbb{P}\left(\hat{z}_{s_o} \geq \tau \middle| z_{s_1}, \dots, z_{s_{n_o}}\right) \tag{6}$$

However, few spatial predictions may be greater than one, or to be negatives, because the weights of IK are real numbers.

The transformation performed from $\mathfrak{i}_{(s_i|\tau)}$ involves an information loss. As consideration IK estimator is not affected by outliers, so it gives the ability to define areas of certainty from isolines. This certainty is linked to the high values interpolated from OK, by only taking into account the spatial distribution of data (Gaetan C., Guyon X., 2010).

3.4   Cross Validation

Cross-validation (cv) process evaluates the suitability for the kriging models specified and the accuracy of interpolations.

Since kriging interpolations are exacts, they cannot be compared with sample data. Various validation techniques exist.

In order to accurately evaluate the model quality, the most suited process compares the spatial interpolations with them of a test set. However it is often impossible. Indeed, here the spatial density of the GP-SET(2016) indicators geo-localised is too low for dividing the original dataset in a training set and test set. Nevertheless cv is the most relevant strategy when the dataset could not be split.

The cv scheme considers each observed value z_{s_i} as unknown and forecasts it, from all other sample values. Thereby cv-interpolated values $z_{s_i}^*$ and their cv-standard deviation $\tilde{\sigma}_{z_i}^*$ are computed. Cv-indicators for assessing the quality of kriging models are performed to cv-residuals $e_i^* = z_{s_i} - z_{s_i}^*$ and standardised cv-residuals $\varepsilon_i^* = e_i^*/\tilde{\sigma}_{z_i}^*$. For models correctly specified, the quality indicators tend toward the following values (Gaetan C., Guyon X., 2010).

The best model is the one whose Standard deviation* and Average Kriging sd* values are lowest. Consequently, the choice of the final model needs put in competition several kriging models (Gaetan C., Guyon X., 2010).

Table 1. Cv-indicators and expected values for kriging models correctly specified

Cross-Validation	Residuals	Residuals standardised
Mean*	$\frac{1}{n} \cdot \sum_{i=1}^{n} e_i^* \approx 0$	$\frac{1}{n}\sum_{i=1}^{n} \varepsilon_i^* \approx 0$
Standard deviation*	$\sqrt{\frac{1}{n}\sum_{i=1}^{n}(e_i^*)^2} = \text{"smal"}$	$\sqrt{\frac{1}{n}\sum_{i=1}^{n}(\varepsilon_i^*)^2} \approx 1$
Average Kriging sd*	$\frac{1}{n}\sum_{io=1}^{n} \tilde{\sigma}_{z_i}^* \approx \{\text{standard deviation}^*\}$	
Correlation coefficient*	$\left|\hat{\rho}(z_{s_i}; z^*_{(s_i)})\right| \approx 1$	

4   Results

Potential demographic growth is impacted by many auxiliary cofactors. Thus random processes are occurred at different spatial scales; inducing anisotropic spreads of the phenomenon. For this reason a great attention is paid to the variogram modelling.

4.1   Geostatistic Applications for Geoprospective Modelling

Non-uniform spatial distribution of z_{s_i} and extreme values characterizing few larger metropolitan cities (Nice, Marseille) or little villages (Annot, Cipières) – Fig.6, introduces noise into the empirical variograms.

Maximum distance between sample values geo-localised is $h_{max} = 253\,600$ meters. For OK critical distance $h_{crit.OK}$ to estimate variogram, is reduced at 1/3 of h_{max}. Beyond h_{crit}, spatial interactions are computed between z_{s_i} and z_{s_j} separated at least by one department. For IK the $h_{crit.IK}$ is even limited at 1/5 of h_{max}, since apparent sills, of directional variograms $\hat{\gamma}(h, \theta_k)$, are systematically reached before this bound – Fig.5.

The $\hat{\gamma}(h, \theta_k)$ detect anisotropies and characterise their type. They are represented from bins, the red points on chart – Fig.4 and 5. Binning process groups pairs of z_{s_i}, z_{s_j} located into a lag distance h in a specific direction θ_k, and for an angle of tolerance set φ. Directional variograms computed on z_{s_i} highlight a zonal anisotropy; their sill and their range are reached for different values – Fig.4 left. So as to remove the drift, z_{s_i} were fitted to a polynomial trend surface, and then differentiate one time in order to get a stationary phenomenon. The omnidirectional variogram $\hat{\gamma}'(h)$ is carried out with residuals – Fig. 4 right. Conversely with $\hat{\gamma}(h)$, immediately carries out with the $i_{(s_i|\tau \geq 0.75)}$ – Fig 5 left.

Variogram modelling from indicator function $i_{(s_i|\tau \geq 0.75)}$ is usually easier than this using the original regionalised variable z_{s_i}. Above all, human phenomena are known for having erratic spatial forms.

IK provides the ability to define areas and assign into them a probability at the high values forecasted from OK. Here the threshold of indicator function was set at $\tau \geq 0.75$, that is the empirical third quartile of z_{s_i}.

The tow omnidirectional variograms are modelled from the sum of a Gaussian function and a nugget, such as:

$$\tilde{\gamma}(h; \theta_k, \vartheta_k) = \sigma_0^2 + \sigma_k^2 \cdot \left(1 - e^{-\left(\frac{h}{\lambda \cdot a_k}\right)^2} \right) \tag{7}$$

On the one hand as sigmoidal shape of $\tilde{\gamma}(h; \theta_k, \vartheta_k)$ characterizes this phenomena which has an important spatial continuity. On the other hand, although sills σ_k^2 are never been actually reached, their growth rates perform from $\hat{\gamma}(h; \theta_k)$ are lower than those compute on h^2 (Matheron G., 1989).

Desert rose aspect of the superposition of directional variograms $\tilde{\gamma}(h; \theta_k)$ on chart; highlights a geometric anisotropy. Nugget and sill values are equivalent in all directions. But, the same $\hat{\sigma}_a^2$ is reached at different ranges $\hat{a}_k$.

Thus neighbourhood windows have to ellipsoidal shapes, whose parameters of orientation $\hat{\theta}_a$ and semimajor axe $\hat{a}$ are defined from $\tilde{\gamma}(h; \hat{\vartheta})$ – Fig.4 and 5. Geometric anisotropies are always corrected as $F_a > 1.5$, (Wang X-B., Liu L. et al., 2004).

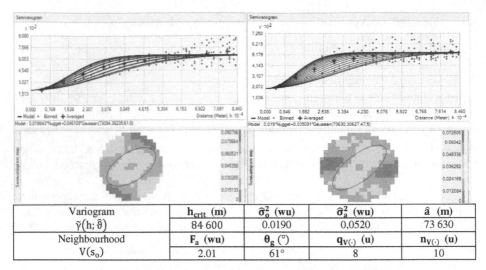

Variogram	h_{crit} (m)	$\hat{\sigma}_o^2$ (wu)	$\hat{\sigma}_a^2$ (wu)	$\hat{a}$ (m)
$\tilde{\gamma}(h; \hat{\vartheta})$	84 600	0.0190	0,0520	73 630
Neighbourhood	F_a (wu)	θ_g (°)	$q_{V(\cdot)}$ (u)	$n_{V(\cdot)}$ (u)
$V(s_o)$	2.01	61°	8	10

Fig. 4. Omnivariogram modelling and neighbourhood for OK

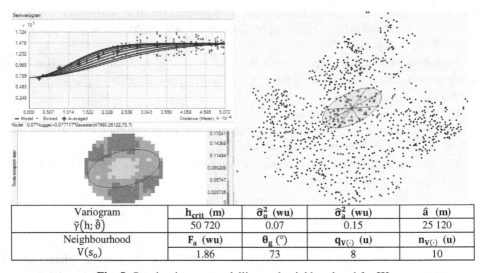

Variogram	h_{crit} (m)	$\hat{\sigma}_o^2$ (wu)	$\hat{\sigma}_a^2$ (wu)	$\hat{a}$ (m)
$\tilde{\gamma}(h; \hat{\vartheta})$	50 720	0.07	0.15	25 120
Neighbourhood	F_a (wu)	θ_g (°)	$q_{V(\cdot)}$ (u)	$n_{V(\cdot)}$ (u)
$V(s_o)$	1.86	73	8	10

Fig. 5. Omnivariogram modelling and neighbourhood for IK

Thereby, during the specification of omnidirectional variograms $\tilde{\gamma}(h; \hat{\vartheta})$ care should be taken to ensure that, there are at least 5 averages for the aggregating of bins per lag distance, before reaching the asymptotic sill – Fig.4 and 5.

Nuggets $\hat{\sigma}_o^2$ of $\tilde{\gamma}(h; \hat{\vartheta})$ were reduced, compared to those calibrated with the WLS criterion. They were greater than apparent nuggets - which are usually overestimated when geo-localised data are non-uniformly distributed (Gaetan C., Guyon X., 2010).

Parameters of the neighbourhood window $V(s_o)$ for OK are the same as for IK – Fig.4 and 5. They are specified according to omnivariograms. However axe dimensions of ellipsoidal windows were divided by 2, due to the spatial distribution density of z_{s_i}. The $V(s_o)$ are decomposed into $s_{V(\cdot)}$ sectors for increasing the uniformity selection of neighbourhood z_{s_i}. In order to guarantee significant interpolations, the neighbourhood sizes - namely z_{s_i} involved - are set from $n_{V(\cdot)}$.

The cv process allowed choosing kriging models correctly specified. Cv-indicators presented are the best values get among all the models tested – Table.2

Table 2. Results of cv process according to kriging models used

Cross-Validation indicators	KO	KI
Mean*	0,0003	0,0015
Mean standardized*	-0,0008	0,0036
Standard Deviation*	0,1435	0,213
Standard Deviation*(standardize)	0,992	1,013
Average Kriging sd*	0,144	0,225
Correlation coefficient*	0.82	0,80

4.2   Cartographic Results of Geoprospective Modelling

GP-SET.KO(2016) models at micro-scales the spatial potential of areas to undergo a demographic growth in 2016 – Fig.6.

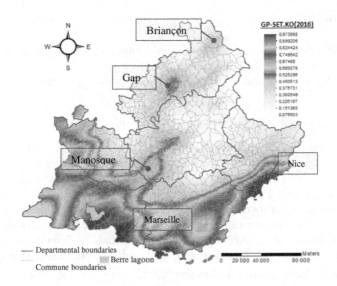

Fig. 6. Geoprospective mapping of GP-SET.KO(2016)

Geographic zones wherein high demographic growths are forecasted are accurately highlighted from cool colours –Fig.6.

So as to assess the certainty of the geoprospective modelling from OK, the probability modelling obtained from IK is draped over the top. Areas defined from isolines, derived from GP-SET.IK(2016) indicators, assign likelihood for the GP-SET.KO(2016) values greater than 0.75 – Fig.7.

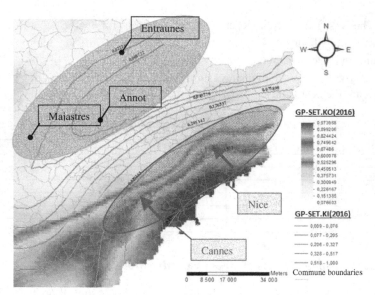

Fig. 7. Zoom in the geoprospective mapping of GP-SET.OK(2016) and GP-SET.IK(2016)

5   Discussion

The cv-validation ensures the statistical significance of geoprospective modelling. Good results indirectly verify original assumptions for the anisotropy and autocorrelation patterns of the spatial spreads of potential demographic growth.

5.1   Spatial Analysis and Geoprospective Considerations

More the communes are located in the hinterland, i.e. in the North, more GP-SET.KO(2016) take low values – Fig.6. The likelihood of undergoing a high demographic growth in the (green) area bounded by Entraunes, Annot and Majastre is almost null - as shown GP-SET.KI(2016) – Fig.7.

This rural zone tends towards depopulation phenomenon for the benefit of communes located in the Mediterranean coastline, where the likelihood to obtain important population increases reaches 85%. High further potentials of demographic growth are spread in space with a preferential direction, i.e. Southwest-Northeast. This spatial trend was previously denoted from variograms carried out with z_{s_i} for OK, and with $i_{(s_i|\tau \geq 0.75)}$ for IK. Now this specific spatial pattern is displayed on maps, especially between the communes of Nice and Cannes and those surrounding – Fig.6.

The slight difference in orientation between omnivariograms $\hat{\theta}_{g.OK} \approx 61°$ and $\hat{\theta}_{g.IK} \approx 73$, means the singularity of potential demographic growth modelled around the communes of Gap, Manosque and Briançon. In fact, those communes are located in the hinterland, but also along the Durance River. They benefit from important urban structures and favourable economic contexts, induced by great industrial or mercantile activities inherited from the past (Hamilton I., Dimitrovska A. et al., 2005).

Geoprospective modelling of potential demographic growth in the northwest of Marseille, Cannes and Nice, reflects the orthogonal directions of omnivariograms, computed for OK and IK – Fig.3, 4 and 6. The spatial certainty of high potentials is modelled with the isolines derived from GP-SET.IK(2016) values. In the red zone the likelihood, linked to GP-SET.K0(2016) ≥ 0.75, are between 60% and 100% – Fig.7.

The reticular attraction pattern of these zones encompasses areas, where the road time to access the metropolis is less than one hour, and which have advantageous cultural or socio-economic positions. However the spatial phenomenon spread is contained by the weakness of urban structures and social disadvantages of territories beyond this temporal boundary (Fusco G., Caglioni M., 2011).

Areas with a high potential to undergo demographic growth in the future are represented by dark colours. However it is neither in the black areas where the population increases will be highest; nor where the major stakes will occur. In fact these spaces are already heavily urbanised; they cannot undergo further urban densifications. Here the population increase will be low. Nevertheless it will be more important in peripheries or out of urban areas, into adjacent zones (Christaller W., 1933). Those geographic zones are represented in light grey or in dark blue – Fig.6. They are mainly those included in the isolines GP-SET.IK(2016) having high values – Fig.7. Unfortunately these attractive areas do not supply the urban structures suitable for supporting population growth and ensuring their development. Thus, substantial investments would be necessary for undertaking large public projects, such as transport infrastructures, sanitation and water networks, universities, medical facilities and commercial zones (Turner B., Lambin E. et al., 2007).

5.2   Strengths

GP-SET.KO(2016) and GP-SET.KI(2016) indicators are suitable for geoprospective analysis of further demographic growth. Nugget effects indicate stochastic processes at micro-scales, which alone encompass zones within a radius of 2.5km. But low nugget values mean low uncertainties associated with the modelling process. Furthermore, spatial anisotropic correlations are also taken into account within a maximum scope of 30,5km – Fig.4 and 5.

The idea in using IK is to give a better spatial certainty of geoprospective forecasts, in a way that makes the modelling process more robust. Indeed, OK standard deviations $\hat{\sigma}_{z_{s_0}}$ are not significant since their values rise where interpolated values $\hat{z}_{s_0}$ increase and they are necessarily low where lower values are predicted (Gaetan C., Guyon X., 2010). Thus GP-SET.KI(2016) is the only way of accurately estimating the spatial certainty.

Here OK and IK are used in a univariate analysis. Their application in this geoprospective issue, for forecasting the future spatial potential of demographic growth, confirms geographers' findings, obtained in the past. Geostatistic concepts converge with expert knowledge.

GP-SET.krige provides indicators at infra-municipal scales in order to increase the decision-making and accuracy of urban planning. This method can be reproduced anywhere and applied to all phenomena described from administrative data, if they might be wisely geo-localised according to documented hypotheses.

5.3   Limits

GP-SET.Krige significance is validated from a cv. Although this strategy is admitted, ideally to confirm results, it would be better using a test set (DasGupta, 2008).

Furthermore, indicators provided from GP-SEP.krige are not sufficient to satisfy the expectations of suitable development. Indeed, the spread of potential demographic growth is not necessarily continuous at infra-municipal scale. For promoting the development of suitable urban infrastructures, in order to support demographic growth; this geoprospective issue must consider spatial auxiliary cofactors. In fact, indicators proposed must be crossed into multivariate spatial analyses, by superposing GIS layers which model current constraints as topographic reliefs, hydrographic networks, climatic changes, as well as both the current land use plans, local economies, social activities and their future evolutions (Turner B., Lambin E. et al., 2007).

In order to further improve the effectiveness of the geoprospective modelling process for decision-making, it should laudable to merge spatial indicators representing influent cofactors, to create a more synthetic indicator. For instance by using the fuzzy set theory applied to multivariate spatial analysis (Dubois D., Prade H., 2004). The combination of these kinds of spatial indicators, with other composite indicators merging the certainty linked to univariable geoprospective forecasts - as GP-SET.KI(2016) – through the decision theory applied using Monte-Carlo methods. These should provide the ability to assess the risk-benefit of envisaged policies (Kwaka Y-H., Ingallb L., 2007).

6   Conclusion

The combination of univariable geostatistics provides significant modelling for several phenomena studied in geoprospective. Thus GP-SET.krige is suitable for aiding in decision-making. To guarantee the statistical significance of this method; ongoing work aims to verify the geoprospective modelling presented, by using a test set. The indicators GP-SET.OK(2016) and GP-SET.IK(2016) will be recomputed from census outcomes in 2016. They will then be compared with those forecasted.

A heuristic evolution of the method is also in progress. It carries out with cokriging (COK) that is the more robust multivariable geostatistic method (Gaetan C., Guyon X., 2010). The goal is to perform a more accurate and more consistent geoprospective model of the potential demographic growth. GP-SET.COK(2016) indicators simultaneously integrate spatial key cofactors, namely economic rates and land prices.

Acknowledgement. The author extends his gratitude to the Pr. Serra for his complete review, the Pr. Auquier for his advice and the financing, Pr. Voiron for his support, as well as my friend the professional proofreader Lepinte.

References

1. Bourrelly, S., Voiron, C. The GP-SET method. spatial and temporal probabilistic model for geoprospective. In: AGILE 2012. Lecture Notes in Geoinformation and Cartography, vol. 7418, pp. 287–303. Spinger, Avignon, April 2012
2. Christaller, W.: Die zentralen Örte in Süddeutschland. Prentice Hall, Englewood Cliffs (1933), Trad. G. W. in 1966. Inam Finaher
3. DasGupta, A.. Asymptotic Theory of Statistics and Probability. Springer Texts in Statistics, vol. XXVIII. Springer, Heidelberg (2008)
4. Dauphiné, A., Voiron, C.: Variogrammes et structures spatiales. Reclus, Montpellier (1988)
5. Donnay, J.-P.: Les Systèmes d'Information Géographique (SIG). Préliminaires à un usage dans l'enseignement. Bulletin de la Société géographique de Liège 45, 45–52 (2005)
6. Dubois, D., Prade, H.: On the use of aggregation operations in information fusion processes. Fuzzy Sets and Systems 142, 143–161 (2004)
7. Gaetan, C., Guyon, X.: Spatial Statistics and Modeling. Springuer, Berlin (2010)
8. Fusco, G., Caglioni, M.: Hierarchical clustering through spatial interaction data. the case of commuting flows in south-eastern france. In: Proceedings Computational Science and Its Applications, Part I: ICCSA 2011. Springer, Santander, June 2011
9. Hamilton, I., Dimitrovska, A., et al.: Transformation of Cities in Central and Eastern Europe, Towards Globalization. United Nations University Press, New York (2005)
10. INSEE. National statistics: Population (2015). (see in 2012) http://www.insee.fr
11. Kwaka, Y.-H., Ingallb, L.: Exploring Monte Carlo Simulation Applications for Project Management. Risk Management 9, 44–57 (2007)
12. Matheron, G.: Estimating and choosing. Springer, Berlin (1989)
13. Raper, J.: Multidimensional Geographic Information Science, 2nd edn. Taylor & Francis, London (2000). 2005
14. Turner, B., Lambin, E., et al.: The emergence of land change science for global environmental change and sustainability. Proceedings of the National Academy of Science of the USA 104(52), 20666–20671 (2007)
15. Vitolo, C., Elkhatib, Y., et al.: Web technologies for environmental Big Data. Environmental Modelling & Software 63, 185–198 (2015)
16. Voiron, C.: L'anticipation du changement en prospective et des changements spatiaux en géoprospective. L'Espace Géographique 41, 99–110 (2012)
17. Wang, X.-B., Liu, L., et al.: Application of Spatial Interpolation in GIS. Journal of Chongqing Jianzhu University 26, 35–39 (2004)

Workshop on Chemistry and Materials Sciences and Technologies (CMST 2015)

Angular Distributions of Fragment Ions Produced by Coulomb Explosion of Simple Molecular Dications of Astrochemical Interest

Stefano Falcinelli[1,2(✉)], Marzio Rosi[1], Pietro Candori[1], Franco Vecchiocattivi[1],
James M. Farrar[2], Konstantinos S. Kalogerakis[3], Fernando Pirani[4], Nadia Balucani[4],
Michele Alagia[5], Robert Richter[6], and Stefano Stranges[5,7]

[1] Department of Civil and Environmental Engineering, University of Perugia,
Via G. Duranti 93, 06125 Perugia, Italy
{stefano,vecchio}@dyn.unipg.it, marzio@unipg.it
[2] Department of Chemistry, University of Rochester, Rochester, NY 14627, USA
stefano@dyn.unipg.it, james.farrar@rochester.edu
[3] Aeronomy Program, SRI International, Menlo Park, CA 94025, USA
ksk@sri.com
[4] Department of Chemistry, Biology and Biotechnologies, University of Perugia,
Via Elce di Sotto, 8, 06123 Perugia, Italy
pirani@dyn.unipg.it, nadia.balucani@unipg.it
[5] IOM CNR Laboratorio TASC, 34012, Trieste, Italy
alagiam@elettra.trieste.it, stefano.stranges@uniroma1.it
[6] Sincrotrone Trieste, Area Science Park, 34149, Basovizza, Trieste, Italy
robert.richter@elettra.trieste.it
[7] Department of Chemistry and Drug Technology, University of Rome
''La Sapienza'', 00185, Rome, Italy
stefano.stranges@uniroma1.it

Abstract. The double photoionization of N_2O and C_2H_2 molecules by linearly polarized light in the 30-50 eV energy range has been studied by coupling ion imaging and electron-ion-ion coincidence techniques. In the case of N_2O, for the two possible dissociative processes leading to $N^+ + NO^+$ and $O^+ + N_2^+$, anisotropic angular distributions of ionic fragments have been measured, indicating that N_2O ionizes when its axis is parallel to the light polarization vector and the fragments are separating in a time shorter than the dication rotational period. In the case of C_2H_2, the two-body dissociation reactions producing $C_2H^+ + H^+$ and $CH_2^+ + C^+$ have shown almost isotropic angular distributions, indicating the double photoionization occurs when acetylene is mainly oriented perpendicularly to the light polarization vector. The analysis of results based on a Monte Carlo trajectory simulation provides: i) the metastable N_2O^{2+} and $C_2H_2^{2+}$ dication lifetimes, ii) the kinetic energy release (KER) distribution for the final ions resulting from the Coulomb explosion, and iii) the anisotropy β parameter values as a function of the investigated photon energy extracted from the measured ionic angular distributions.

Keywords: Double photoionization · Molecular dications · Synchrotron radiation · Monte Carlo simulation · Atmospheric escape · Astrochemistry

© Springer International Publishing Switzerland 2015
O. Gervasi et al. (Eds.): ICCSA 2015, Part II, LNCS 9156, pp. 291–307, 2015.
DOI: 10.1007/978-3-319-21407-8_22

1   Introduction

The existence of gaseous free ions in the upper terrestrial atmosphere was recognized following the experiments of Marconi, who succeeded in transmitting radio waves from the United Kingdom to Canada. Later, the ion CH^+ was detected in space practically at the same time of the detection of the first neutral radical species CH and CN [1]. At that time, it was a common opinion that complex molecules could not survive the harsh environments of the interstellar objects, which were impinged on by cosmic rays and energetic radiation from nearby stars. At present, more than 150 molecules have been detected in interstellar and circumstellar objects (for a continuous update see http://www.astrochymist.org/astrochymist_ism.html), among which there are 20 cations and 7 anions. Following the detection of molecular ions in the interstellar medium (ISM), it was recognized that ion-molecule reactions play a pivotal role in the chemical evolution of interstellar clouds, where the low number density and low-temperature conditions mostly prevent neutral chemistry. This is because ion-molecule reactions are typically barrierless, while neutral-neutral reactions are normally characterized by a certain activation energy that cannot be surmounted at the very low temperatures typical of diffuse and dense clouds (10-100 K). The first chemical models of interstellar clouds [1] relied completely on ion-molecule reactions to explain the detection of most molecular species. Dissociative recombination with free electrons is believed to be responsible for the final conversion of the ions into the observed neutral species. The chemistry of molecular clouds is particularly interesting because solar systems originate from them and, therefore, the molecules synthesized in ion-molecule and neutral-neutral reactions, as well as on dust icy grains, can be inherited by the planets or comets of the newly formed solar system. In turn, these molecules may constitute the basic building blocks of increasingly more complex molecules thought to have preceded the emergence of life on our planet [2,3].

Ions are also extremely important in the upper atmosphere of planets, where they govern the chemistry of ionospheres [4,5]. In particular, the ionosphere chemistry of Titan has recently been revealed to be extremely active by the instruments on board Cassini [6,7]. Finally, molecular ions have also been detected in comet tails [1].

In space, ions are formed in various ways, the importance of which depends on the specific conditions of the extraterrestrial environment considered [8,9,10]. The interaction of neutral molecules with cosmic rays, UV photons, X-rays and other phenomena such as shock waves are all important processes for their production. In particular, the absorption of UV photons with an energy content higher than the ionization potential of the absorbing species can induce ionization (for most species, extreme-UV or far-UV photons are necessary). Cosmic rays are also significant since they are ubiquitous and carry a large energy content (up to 100 GeV). They consist of protons, alpha particles, electrons, γ-rays and (to a small extent) also heavier nuclei (such as C^{6+}). Cosmic rays are very penetrating and can induce ionization in objects that are completely opaque to UV photons. X-rays are relatively abundant in several regions, such as active galactic nuclei, young stellar objects, and planetary nebulae with hot central stars. Absorption of X-rays induces the ejection of a core electron followed by the Auger emission of another electron, producing doubly charged species, which have been suggested to play a role in the envelope of young stellar objects [11] and upper planetary atmospheres [12,13,14,15].

2   Experimental

The present experiments were performed at the ELETTRA Synchrotron Light Laboratory (Trieste, Italy) using the ARPES (Angle-Resolved Photoemission Spectroscopy) end station of the Gas Phase Beamline. Details about the beamline and the end station have been already reported elsewhere [16] and the apparatus used for the experiment discussed here has also been described previously [17,18]. Therefore, only some features relevant for the present investigation are outlined here.

The monochromatic energy from the selected synchrotron light beam crosses an effusive molecular beam of N_2O and C_2H_2 neutral precursors, and the product ions are then detected in coincidence with photoelectrons. The molecular beam of N_2O and C_2H_2 molecules and the VUV light beam cross at right angles, and the light polarization vector is parallel to the synchrotron ring plane and perpendicular to the time-of-flight direction of detected ions. Incident photon fluxes and the gas pressure are monitored, and the ion yield has been corrected for flux changes of pressure and photon, when the photon energy was scanned. The gas inlet effusive source is supplied with a mixture of the molecule under study (N_2O and C_2H_2) with helium, allowing the normalization of all ion signals (measured at each photon energy) to the total ion yield of helium at that energy. In order to record photoions in coincidence with photoelectrons, we used the electron-ion-ion coincidence technique. Our extraction and detection system was discussed in detail in previous papers [17,18] and was assembled following the design described by Lavollée [19]. All the experimental components were controlled by a computer used to record the experimental data. The incident photon flux and the gas pressure have been monitored and stored in separate acquisition channels [19,20].

Nitrous oxide and acetylene, from a commercial cylinder at room temperature, were supplied to a needle effusive beam source. The used N_2O and C_2H_2 gases had a 99.99% and ~99.0% nominal purity, respectively. Nitrous oxide was used without any further treatment, whereas acetylene was used after a ~193 K cold trap purging to remove acetone impurities. The performance of the cold trap was verified by recording mass spectra. An adjustable leak valve along the input gas pipe line was used in order to control the gas flow, which was monitored by checking the pressure in the main vacuum chamber.

3   Computational Analysis by Monte Carlo Trajectory Simulation and Angular Distributions of Product Ions

In Fig. 1 are shown coincidence spectra recorded at a 39.0 eV photon energy in the double photoionization studies for both systems here presented. On the left side of the figure, in the upper panel, are the reported mass spectrum and the associated ion–ion time of flight correlation of ions produced by single and double photoionization of N_2O. In this type of plot, which is typical of double photoionization experiments, the two time-of-flight values are shown for a pair of ions produced in the same photoionization event and define a point (see for instance Ref. [21]). The diagonal weak traces are false and spurious coincidences that have been neglected in the present analysis. All product ions are evident together with some background peaks. An enlargement of the most relevant part of such a diagram is shown in the left lower panel of the same figure, where it is possible to distinguish the ionic products of the double photoioniza-tion detected in coincidence, which are $NO^+ + N^+$ and $N_2^+ + O^+$. The time correlation diagram also shows some diagonal weak traces of false and spurious coincidences, which have been simply neglected in the present analysis. A typical trace due to the formation of a metastable N_2O^{2+} dication is also evident. The right side of Fig. 1 shows the coincidence plot for the double photoionization of acetylene at the same photon energy of 39.0 eV. The points in the figure represent the coincidence events as a function of the arrival time of the first ion, t_1, and of the second ion, t_2. In the lower panel, the typical spot for H^+/C_2H^+ coincidences is seen together with a tail that charac-terizes the presence of a $C_2H_2^{2+}$ dication metastable state dissociating in H^+ and C_2H^+. Also evident in the figure are C^+/CH_2^+ coincidences and the region con-taining all overlapping CH^+/CH^+ coincidences, $C_2H_2^{2+}$ dications, and CH^+ ions pro-duced by single ionization of acetylene—all contribute to the same mass-to-charge ratio. Analysis of coincidence distribution maps like those shown in Fig. 1 allows the kinetic energy of the two products released into the two ionic fragments to be obtained by a simple analysis of the ion intensity maps based on the method sug-gested by Lundqvist et al. [22]. In particular, this target can be reached by examining the dimensions and shapes of the peak for each ion pair in the coincidence spectra measured at all the investigated photon energies. An important experimental result is the distribution of the dot density along the area of the "tail" and the "V-shaped" traces clearly visible in the coincidence plots of Fig. 1 for both investi-gated systems. A careful analysis of such kind of distributions as a function of the arrival time differences (t_2-t_1) of the fragment ions to the ion-position-sensitive MCP detector generated by Coulomb explosion of the molecular N_2O^{2+} or $C_2H_2^{2+}$ dications, can be performed by using the method developed by Field and Eland to calculate the lifetime of such metastable species [23]. Alternately, in order to obtain the lifetime of the metastable N_2O^{2+} and $C_2H_2^{2+}$ dications, these authors have applied also a Monte Carlo trajectory simulation [23]. In this work we have analyzed our recorded data using both methods. We have developed a proper computer routine in order to eva-luate the t_1 and t_2 distribution once the kinetic energy released (KER) by the two ionic fragments and the dissociation lifetime, τ, of the metastable dication have been fixed. For a detailed description of such a routine we refer to a previous published paper [7]. In our computational procedure a simulation of the experimental distribution of the coincidences dot density $I(t_2-t_1)$ is performed, adjusting KER and τ and evaluating the standard deviation as a reliability level of the simulation. A typical $I(t_2-t_1)$, related to the double photoionization of acetylene at a 39.0 eV photon energy, is shown in Fig. 2 (upper panel). In this figure the dots are the experimental results and the full line is obtained with a simulation with 10^6 trajectories. We analyzed the plot in order to ob-tain the lifetime of the metastable dication during the $H^+ + C_2H^+$ dissociation. To do that, we plotted the number of coincidence points along the track of the tail as a func-tion of the (t_2-t_1) difference, and we fitted the data by a Monte Carlo simulation of the ion trajectories in the mass spectrometer [24,25]. In the lower panel of the same figure, the mean square values, χ^2, of the difference between the experimental data and the simulation, are plotted as a function of the dication lifetime; the minimum deviation is obtained for a lifetime of 108 ± 22 ns. The dissociation of the $C_2H_2^{+2}$ dications, leading to the formation of $C^+ + CH_2^+$ product ions, does not show a tail in the coincidence plot (see Fig. 1, right), indicating that the reaction occurs in a time shorter than the channel coincide time of our apparatus, that is ~50 ns for the conditions of the present experiment [17].

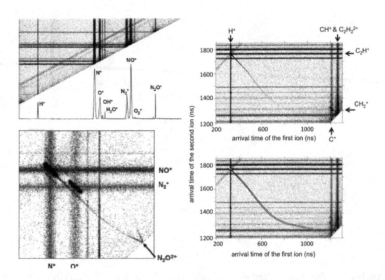

Fig. 1. Left side: (upper panel) the mass spectrum and ion–ion time-of-flight correlation of ions produced by single and double photoionization of N_2O at 39 eV. In the lower panel is the extended plot of the portion of the ion–ion time of flight correlation diagram in which it is possible to distinguish the N^+/NO^+ and N_2^+/O^+ coincidence spots and the trace for the metastable N_2O^{2+} dication. Right side: the photoion-photoion coincidence plot for the double photoionization of C_2H_2 at 39 eV. In the upper part the rough data are reported, while in the lower part the same data are shown with a contour line around the coincidence spots for the three possible two-body dissociation processes (see reactions (5)-(7) in Section 4.2) and around the typical tail, indicating a metastable $(C_2H_2^{2+})^*$ precursor.

It is important to note that the analysis here discussed, and based on the recorded ion imaging coincidence plots, allows us to evaluate also the kinetic energy distribution of the final product ions as discussed in recent papers [7,26]. Finally, the ion image analysis also provides the angular distribution of product ions with respect to the polarization vector direction. It is well known [27,28] that such angular distributions can provide valuable information about the dissociation dynamics and are usually represented by $I(\theta)\sin\theta$

$$I(\vartheta)\sin(\vartheta) = \frac{\sigma_{tot}}{4\pi}\left[1 + \frac{\beta}{2}\left(3\cos^2\vartheta - 1\right)\right]$$

(1)

where $I(\theta)$ and σ_{tot} are the differential and the total cross section of the process, respectively, while θ is the angle between the velocity vector of the fragment ion and the light polarization vector. The β parameter, also called anisotropy parameter, ranges from -1, for the emission of product ions along a direction perpendicular to the polarization vector, up to a value of 2 for a parallel direction. Isotropic distribution of fragment ions is characterized by a value $\beta=0$. The recorded angular distributions obtained in the dissociative double photoionization of N_2O and C_2H_2 molecules will be presented and discussed in the next section.

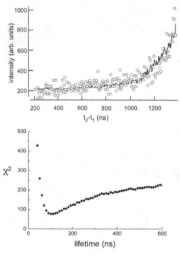

Fig. 2. Analysis of the "tail" in the coincidence plot of Fig. 1 (right side) in order to derive the lifetime of the dissociation reaction leading to $H^+ + C_2H^+$. The upper panel shows the intensity of coincidences along the tail track as a function of the t_2-t_1 difference (open circles) with the fit (continuous line) by a Monte Carlo trajectory calculation. The lower panel shows the mean square deviation, χ^2, between the simulation and the experimental results as a function of the lifetime used in the simulation (see the text).

4   Discussion

Nitrous oxide and acetylene are simple molecules of interest for interstellar medium (ISM) and planetary atmospheres, not only for the Earth but also for other planets of the solar system like Mars, Venus, and Titan, the largest satellite of Saturn. The presence of N_2O in the ISM has been demonstrated by microwave spectroscopy, whereas C_2H_2 was detected by IR spectroscopic measurements [29]. On the other hand, the importance of nitrous oxide is well known in the Earth's atmosphere, where it is an important greenhouse gas and participates in ozone depletion [30,31]. Moreover, acetylene has been detected as a minor component in Titan's atmosphere (about 10 ppb) [32]. Finally, looking at the importance of acetylene, we note: i) this species is found as a minor component in the atmospheres of gas giants like the planet Jupiter and in comets; ii) photochemical experiments have demonstrated that acetylene is a likely precursor of C_2 species, a widely observed component of comets; and iii) it has been proposed that polymerization of acetylene in cometary impact on planetary atmospheres may be responsible for the formation of polycyclic aromatic hydrocarbons (PAHs), which may in turn determine the characteristic colors of the atmospheres of Jupiter and Titan [33-36]. The presence of VUV photons in these environments makes highly probable double photoionization of these molecular species and their subsequent dissociation leads to ionic fragments with a high kinetic energy content of several eV. This translational energy is sufficient for some of these fragments to escape from the upper atmosphere of Mars and Titan into space, a phenomenon we discuss in the following subsections.

4.1   Double Photoionization of Nitrous Oxide

Recently, we have reported the photon energy dependence of the two dissociative processes leading to $N^+ + NO^+$ and $O^+ + N_2^+$ products in the 30–40 eV energy range, finding that the former exhibits a probability about four times higher than the latter [18,37]. In the whole investigated energy range, we observed the two dissociation channels reported below with a threshold energy of about 32.2 eV:

$$N_2O + h\nu \rightarrow (N_2O^{2+})^* + 2e^- \rightarrow N_2^+ + O^+ + 2e^- \tag{2}$$

$$N_2O + h\nu \rightarrow (N_2O^{2+})^* + 2e^- \rightarrow N^+ + NO^+ + 2e^- \tag{3}$$

Where $(N_2O^{2+})^*$ indicates an intermediate "short-lived" molecular dication. No stable N_2O^{2+} molecular dications were observed in our experiment. In the present paper, we discuss the double photoionization experiments by reviewing our recorded 2D images of N^+/NO^+ and N_2^+/O^+ coincident ions in the 30–50 eV photon energy range, and discussing these results in a comparative way with those recently obtained in the case of the double photoionization of acetylene molecules (see Section 4.2). Maps are shown in the left panel of Fig. 3.

The x axis corresponds to the direction of the light polarization vector in coincidence intensity maps measured at a photon energy of 39 eV. This evident anisotropy indicates that N_2O molecules ionize when their axis is parallel to the light polarization vector, and the fragment ions are separating in a time shorter than the dication rotational period. When the energy decreases below such a threshold, the anisotropy also decreases and β becomes zero at 32 eV, indicating an isotropic angular distribution.

There are two effects that can smooth the angular dependences like that in Fig. 3 (right panel). One can be ascribed to the finite scattering volume given by the crossing between molecular and light beams. The second is due to the formation of possible metastable dications dissociating with a lifetime comparable to the characteristic ion flight time of the experimental setup. These ions, as discussed in a previous paper [37] do not produce a well defined time-of-flight peak, but increase background pulses. Therefore, the obtained values of β are to be considered as lower limits.

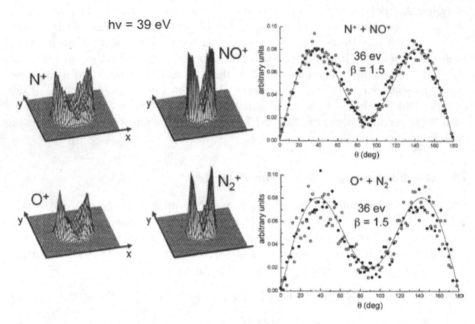

Fig. 3. Left panel: The intensity maps, recorded at a photon energy of 39 eV, for N^+ ions corre-lated with NO^+, and for O^+ ions correlated with N_2^+ (the light polarization vector is parallel to the x axis). Right panel: The angular intensity distributions of products of the two dissociative channels leading to N^++NO^+ and $O^++N_2^+$ on the plane parallel to the light beam direction, as measured at a photon energy of 36 eV. In both plots, the intensities of the two ionic products are reported as open circles for the atomic ions (N^+ or O^+) and as full circles for the diatomic ions (NO^+ or N_2^+). The curve is a fitting obtained by setting $\beta = 1.5$.

The anisotropy of the angular distribution of ionic fragments in the dissociative double photoionization of N_2O can be regarded as the combined effect of two main conditions: (i) the adsorption of the VUV photon occurs when the molecule has the most favorable alignment of the orbitals involved in the double photoionization; and (ii) the whole double photoionization process, followed by the separation of two ionic fragments, must occur in a time shorter than the rotational period of the dication. Therefore, at those energies at which the product ions exhibit an evident spatial aniso-tropy, the two above conditions have to be completely fulfilled. On the other hand, at lower energies, around 32 eV, the two dissociative channels are still open, but disso-ciative products are isotropically distributed and therefore at least one of the two con-ditions is not fulfilled. In the present system, above the vertical threshold, the two singly charged product ions are separating because of the strong Coulomb repulsion, and therefore one expects a rapid separation. In the low energy range, below the ver-tical threshold, the ionization occurs indirectly. One can argue, for instance, that a singly charged ion could be formed, which then evolves and ejects a second electron, followed by dissociation—the whole process should occur so slowly that the triatomic complex has enough time to rotate [21]. Practically, the final ion products ejected are randomly distributed because the system loses the memory of the initial alignment of the precursor state. However, the dissociative process, leading to N^++NO^+ (above 33 eV) and to $O^++N_2^+$ (above 34 eV), should occur faster than the typical rotation period of the N_2O^{2+} dication (expected to be in the order of 10^{-11} s).

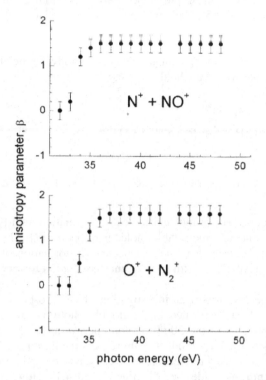

Fig. 4. The anisotropy β parameter values as a function of the photon energy obtained from the analysis of the angular distribution of ion fragments in the two dissociative channels leading to N^++NO^+ and $O^++N_2^+$ (see the text)

Therefore, the present observation of a very short lifetime of the produced dication is not very surprising when formed above the vertical threshold, where Coulombic repulsion controls the dissociation dynamics. However, the present experiment shows that such a characteristic still occurs below the threshold as well, where the ionization is through an indirect mechanism.

As mentioned in Section 3, an additional interesting information obtained by the analysis of distribution maps is the kinetic energy of the two products released into the two ion fragments. In a recent paper [7], we have shown that the product ions coming from reactions (2) and (3) are characterized by a total KER of about 5.0-6.0 eV. This translational energy is compatible with their possible escape from the upper atmosphere of Mars and Titan [7].

4.2   Double Photoionization of Acetylene

Besides the planetary atmospheric chemistry mentioned above [32-36], the ionization of acetylene molecules is an important process in plasma devices [38], flames [39], and semiconductor manufacturing [40]. The double photoionization using VUV synchrotron radiation in the 32-65 eV photon energy range produces ethyne dication with subsequent dissociation reactions. Our measurements, in agreement with previous experimental determinations [41], have shown the formation of a stable molecular dication in the investigated energy range. Three two-body dissociation reactions are prominent with the following threshold energies:

$$C_2H_2 + h\nu \rightarrow C_2H_2^{2+} + 2e^- \qquad\qquad\qquad h\nu \geq 31.7 \text{ eV} \qquad (4)$$

$$C_2H_2 + h\nu \rightarrow (C_2H_2^{2+})^* + 2e^- \rightarrow H^+ + C_2H^+ \qquad h\nu \geq 33.8 \text{ eV} \qquad (5)$$

$$C_2H_2 + h\nu \rightarrow (C_2H_2^{2+})^* + 2e^- \rightarrow CH^+ + CH^+ \qquad h\nu \geq 34.0 \text{ eV} \qquad (6)$$

$$C_2H_2 + h\nu \rightarrow (C_2H_2^{2+})^* + 2e^- \rightarrow (H_2CC^{2+})^* \rightarrow C^+ + CH_2^+ \quad h\nu \geq 34.0 \text{ eV} \quad (7)$$

where $C_2H_2^{2+}$ indicates a stable molecular dication formation, and $(C_2H_2^{2+})^*$ refers to doubly charged intermediate metastable molecular species [7]. The reaction (7) is a proton-transfer-rearrangement reaction that occurs via formation of the intermediate vinylidene dication, $(H_2CC^{2+})^*$, with subsequent dissociation toward the C^+ and C_2H^+ ion products [7,41].

In the present paper, we report an investigation of the two-body dissociative reactions of the $C_2H_2^{+2}$ dication formed in the double photoionization of acetylene by linearly polarized photons in the 30-50 photon energy range, detecting angular and energy distributions of final ion products with respect the light polarization vector. To collect the data, we have used the technique already applied and described for double photoionization of nitrous oxide (see Sections 3 and 4.1). However, in the present case, the double ionization cross section appears to be rather small [41] and the ion signal is weak; therefore, in a previous experiment we have studied the reactions (4)–(7) at 39.0 eV only, an energy at which all the above three channels are open [20]. After that, we have extended our study for different photon energy values in the 30-50 eV range, recording the angular distributions of the final product ions (with about 14 hours of accumulation for each investigated energy) for 15 different photon energy values during a 12 days beam time at ELETTRA Synchrotron Radiation Facility in Trieste (Italy) last October 2014.

In the first experiment, we have analyzed the coincidence plot obtained for 39 eV photons (Fig. 1, right) by using the computational procedure illustrated in Section 3. In this way we were able to obtain the lifetime of the metastable dication $(C_2H_2^{2+})^*$ during the $H^+ + C_2H^+$ dissociation as shown in Fig. 2, in which the mean square deviation, χ^2, of the difference between the experimental data and the Monte Carlo simulation as a function of the used lifetime reached its minimum for a lifetime of 108±22 ns—a value in reasonable agreement with the value of 80 ns previously reported by Thissen et al. [41]. As we discussed in Section 3, the ion image analysis also provides the angular distribution of product ions for the three two-body fragmentation reactions

(5) and (7) with respect the polarization vector direction. The center of mass distributions for the H^+/C_2H^+ and C^+/CH_2^+ ion products are reported in Fig. 5 in the left and right panels, respectively, where they are analyzed in terms of the β parameter as well. For H^+/C_2H^+, the best fit is provided by β =− 0.19, while for the C^+/CH_2^+ ion products it is β =− 0.47.

In order to better understand the information content of the angular distribution of product ions we observed for reactions (5) and (7), we stress this angular distribution derives from the combination of three main factors: (i) the stereo specificity of the photoionization; (ii) the lifetime of the dissociation; and (iii) the rotational state of the molecule. As described in Section 2, in our experimental apparatus the molecules cross the light beam with their velocity vector almost parallel to the polarization vector and, when they interact with the light, they rotate with their rotational axis (or the rotational angular momentum) randomly oriented. Therefore, the usual formula that correlates the β parameter with the $\sigma_{\parallel}/\sigma_{\perp}$ ratio,

$$\frac{\sigma_{\parallel}}{\sigma_{\perp}} = \frac{1+\beta}{2-\beta} \qquad (8)$$

where $\sigma_{\parallel}$ and $\sigma_{\perp}$ are respectively the parallel and the perpendicular cross section for the process [17], has to be considered with some caution because the observed β parameter is the result of a convolution of several contributions. In other words, what is measured in this kind of experiment is the outgoing direction of product ions and this is obviously related to the $\sigma_{\parallel}/\sigma_{\perp}$. However, the rotational period of the neutral molecule, its dication, and the lifetime of the dication before dissociation have the tendency to redistribute the products in a wider cone. In the limiting case of a sufficiently long lifetime, the angular distribution of products becomes isotropic, even though the ionization event is anisotropic.

In the distribution in the left panel of Fig. 5, which refers to reaction (5), we note that the experimental data are quite close to the behavior expected for an isotropic (β = 0) distribution, although they are all systematically lower than the β = 0 distribution. This indicates that the lifetime of the $(C_2H_2^{2+})^*$ dication's metastable state in this reaction is long enough in comparison with the rotational period to give an isotropic distribution, but there must be an additional faster component producing a negative value of the effective β. The effect is larger in the case of the distribution reported in the right panel of Fig. 5 and related to reaction (7). It is well known this reaction is very fast [42-44] and occurs more quickly than the usual rotation period of the precursor dication, which is on the order of picoseconds.

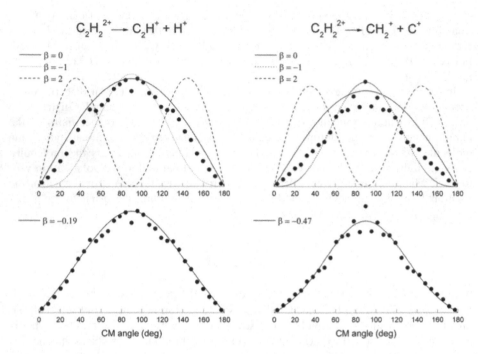

Fig. 5. Left panel: Angular distribution, I(θ), of $C_2H^+ + H^+$ product ions in the center of mass frame. In the upper panel, the experimental distribution is compared with limiting situations (β =−1, 0, 2), while, in the lower panel, it is compared with the best fit β =−0.19 value. Right panel: Angular distribution, I(θ), of $CH_2^+ + C^+$ product ions in the center of mass frame. In the upper panel, the experimental distribution is compared with limiting situations (β =−1, 0, 2), while, in the lower panel, it is compared with the best fit β =−0.47 value.

The results indicate rotation and lifetime have the tendency to smooth the angular distribution, and therefore the negative values of β given here have to be considered an upper limit to the intrinsic β value. However, it seems clear photoionization has the propensity to occur with the molecular axis oriented perpendicular to the light polarization vector. In the case of acetylene molecules, the molecular axis is also the cylindrical symmetry axis of the electron density distribution. Since the molecules are rotating, we can assert that the ionization mainly occurs when their rotational angular momentum vector is parallel to the light polarization vector; however, when the rotational angular momentum vector has a different orientation, the ionization probability is lower. The other two systems we have studied so far with this technique, CO_2 [17,45] and N_2O (presented in Section 4.1) are also linear molecules, and the ionization occurs with the molecular axis oriented almost parallel to the light polarization vector. More recently, we have measured the angular distributions for the three possible two-body fragmentation channels (reactions 5-7) coming out from $C_2H_2^{2+}$ dication explosion at different photon energies, obtaining the anisotropy β parameter values reported in Fig. 6.

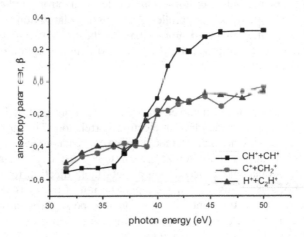

Fig. 6. The anisotropy β parameter values as a function of the photon energy obtained from the analysis of the angular distribution of ion fragments in the three dissociative channels leading to $CH^+ + CH^+$, $C^+ + CH_2^+$, and $H^+ + C_2H^+$ (see the text)

In the Fig. 6 it is possible to appreciate a quite evident change in the β parameter related to the $CH^+ + CH^+$ fragmentation reaction, passing from a value of about -0.55 at a photon energy of 32.0 eV up to 0.32 for a photon energy higher than 45.0 eV. This should be an indication that for reaction (6) can be operative two mechanisms involving several electronic states of the intermediate $(C_2H_2^{2+})^*$ molecular dication and giving rise to a different microscopic dynamical behavior during the fragmentation process. The analysis of these recent data is still in progress in our laboratory and needs a detailed theoretical investigation of the potential energy surface of the $C_2H_2^{2+}$ system, besides the energetics and structure of its fragmentation products as already done in previous investigated systems [24,25].

Finally, as already noted in the case of the double photoionization of nitrous oxide (see Section 4.1), the computational method described in Section 3 allow us to obtain the kinetic energy release of the ion fragment products of reactions (5)-(7). Recently, we have found that the product ions coming from reactions (5)-(7) are characterized by a total KER ranging between 1.5 and 4.0 eV that is sufficient to allow this species participating in the atmospheric escape from Mars and Titan [7].

5   Conclusions

This paper describes the angular distribution of fragment ions produced by double photoionization of nitrous oxide and acetylene, two simple molecules of interest in astrochemistry, using tunable and polarized synchrotron radiation.

In the case of N_2O, it is possible to summarize the conclusions from the analysis of the measured angular distribution of the ionic products in the following point: in a wide range of energy, the angular distributions are rather anisotropic, indicating anisotropy in VUV absorption and a lifetime comparable or shorter than 10^{-11} s for dissociation into two singly charged ions. This evident anisotropy indicates that N_2O molecules ionize when their axis is parallel to the light polarization vector, and the fragment ions are separating in a time shorter than the dication rotational period.

In electronic photoexcitation processes, the polarized light absorption probability varies with the symmetry of the molecular orbitals involved in the transition [46]. In particular, in single photoionization, the light absorption probability depends on the symmetry of the bound- and free-electron wave functions. In double photoionization, the situation is more complicated because of the involvement of two electrons and the relevant wave functions, which must include the electron-electron correlation. The strong anisotropy reported in our work sheds light on the stereodynamics of such a process.

In the dissociative double photoionization of C_2H_2, the deprotonation reaction forming $C_2H^++H^+$ product ions proceeds via a metastable dication formation with a measured lifetime of 108 ± 22 ns. The angular distribution of products with respect to the polarization direction of the light has been found to be slightly narrower than the one expected for an isotropic distribution, indicating the presence of a fast component together with the metastable state dissociation. The ionization has a higher probability when the neutral molecule is oriented perpendicularly to the polarization direction. For reaction (7), the one occurring through the formation of a vinylidene dication, H_2CC^{2+}, a lifetime shorter than the typical rotational period of the acetylene molecule (of the order of about 10^{-12} s) has been observed. Moreover, the $H_2C^++C^+$ dissociation products are collected in a direction mainly perpendicular to the light polarization direction. Reaction (6), producing CH^++CH^+ fragment ions, has also been observed, but it was not possible to analyze the data in detail because of the limitation of the experimental method, as discussed in Section 3. A systematic investigation in a broader range of energies appears to be very productive to obtain detailed information on the dynamics of the dissociative double photoionization processes in acetylene, just above the threshold. As already mentioned in Section 4.2, we performed an experiment for different photon energy values in the 30-50 eV range in our last beam time at the Gas Phase Beamline of the ELETTRA Synchrotron Facility (Trieste, Italy), and our analysis of these recent data is still in progress and will be the subject of publication in a future paper.

In conclusion, the present work indicates that the coupling of ion imaging technique with photoelectron-photoion-photoion coincidence can be a very powerful tool also for investigating details of the dynamics of dissociative double photoionization of molecules, since it provides: (i) the angular distribution of product ions; (ii) the evaluation of the lifetime of the intermediate metastable dication (N_2O^{2+} and $C_2H_2^{2+}$) before its Coulomb explosion towards the fragment final ions production; and (iii) the values of the kinetic energy released to products.

These types of experiments allow understanding of dissociative double photoionization processes induced by VUV and EUV photons leading to the production of fragment ions with a high kinetic energy content, and they may provide important insights into ion species escape from the atmospheres of some planets like Venus, Mars, and Titan [15]. These processes occur via formation of intermediate molecular dications that can dissociate by Coulomb explosion toward the formation of two fragment ions with a kinetic energy release of several eV, which is much larger than the limiting thermal escape velocity. As already obtained in pioneering studies concerning the production and characterization of metastable molecular dications [47,48], in the case of the double photoionization of N_2O and C_2H_2 molecules, the fragment product ions O^+, N^+, N_2^+, NO^+, CH_2^+, CH^+, C^+, H^+ have a translational energy content ranging between 1.0 and 6.0 eV [7], that is large enough to allow their escape process from the upper atmospheres of Mars and/or Titan.

Acknowledgments Financial contributions from the "Fondazione Cassa di Risparmio di Perugia" is gratefully acknowledged. This material is based, in part, upon work supported by the U.S. National Science Foundation under Award number AST-1410297.

References

1. Larsson, M., Geppert, W.D., Nyman, G.: Rep. Prog. Phys. **75**, 066901 (2012)
2. Balucani, N.: Int. J. Mol. Sci. **10**, 2304–2335 (2009)
3. Rosi, M., Falcinelli, S., Balucani, N., Casavecchia, P., Leonori, F., Skouteris, D.: Theoretical study of reactions relevant for atmospheric models of titan: interaction of excited nitrogen atoms with small hydrocarbons. In: Murgante, B., Gervasi, O., Misra, S., Nedjah, N., Rocha, A.M.A., Taniar, D., Apduhan, B.O. (eds.) ICCSA 2012, Part I. LNCS, vol. 7333, pp. 331–344. Springer, Heidelberg (2012)
4. Falcinelli, S., Pirani, F., Vecchiocattivi, F.: Atmosphere **6**(3), 299–317 (2015)
5. Alagia, M., Balucani, N., Candori, P., Falcinelli, S., Richter, R., Rosi, M., Pirani, F., Stranges, S., Vecchiocattivi, F.: Rendiconti Lincei Scienze Fisiche e Naturali **24**, 53–65 (2013)
6. Vuitton, V., Dutuit, O., Smith, M.A., Balucani, N.: Chemistry of Titan's atmosphere. In: Mueller-Wodarg et al. (eds.) Titan: Surface, Atmosphere and Magnetosphere. Cambridge University Press (2013). Chapter 7
7. Falcinelli, S., Rosi, M., Candori, P., Vecchiocattivi, F., Farrar, J.M., Pirani, F., Balucani, N., Alagia, M., Richter, R., Stranges, S.: The escape probability of some ions from mars and titan ionospheres. In: Murgante, B., Misra, S., Rocha, A.M.A., Torre, C., Rocha, J.G., Falcão, M.I., Taniar, D., Apduhan, B.O., Gervasi, O. (eds.) ICCSA 2014, Part I. LNCS, vol. 8579, pp. 554–570. Springer, Heidelberg (2014)
8. Falcinelli, S.: Penning ionization of simple molecules and their possible role in planetary atmospheres. In: Batzias et al. (Eds.): Recent Advances in Energy, Environment and Financial Planning – Mathematics and Computers in Science and Engineering Series Vol. 35, pp. 84–92. © WSEAS press 2014 (2014). ISSN: 2227-4588; ISBN 978-960-474-400-8
9. Pei, L., Carrascosa, E., Yang, N., Falcinelli, S., Farrar, J.M.: The Journal of Physical Chemistry Letters **6**, 1684–1689 (2015)
10. Candori, P., Falcinelli, S., Pirani, F., Tarantelli, F., Vecchiocattivi, F.: Chemical Physics Letters **436**, 322–326 (2007)
11. Stauber, P., Doty, S.D., van Dishoeck, E.F., Benz, A.O.: Astron. Astrophys. **440**, 949–966 (2005)
12. Thissen, R., Witasse, O., Dutuit, O., Wedlund, C.S., Gronoff, G., Lilensten, J.: Phys. Chem. Chem. Phys. **13**, 18264–18287 (2011)
13. Schio, L., Li, C., Monti, S., Salén, P., Yatsyna, V., Feifel, R., Alagia, M., Richter, R., Falcinelli, S., Stranges, S., Zhaunerchyk, V., Carravetta, V.: Physical Chemistry Chemical Physics **17**(14), 9040–9048 (2015)
14. Alagia, M., Bodo, E., Decleva, P., Falcinelli, S., Ponzi, A., Richter, R., Stranges, S.: Physical Chemistry Chemical Physics **15**(4), 1310–1318 (2013)

15. Falcinelli, S., Rosi, M., Candori, P., Farrar, J.M., Vecchiocattivi, F., Pirani, F., Balucani, N., Alagia, M., Richter, R., Stranges, S.: Planetary and Space Science **99**, 149–157 (2014)
16. Blyth, R.R., Delaunay, R., Zitnik, M., et al.: J. Electron Spectrosc. Relat. Phenom., 101–103, 959 (1999)
17. Alagia, M., Candori, P., Falcinelli, S., Lavollèe, M., Pirani, F., Richter, R., Stranges, S., Vecchiocattivi, F.: Phys. Chem. Chem. Phys. **12**, 5389–5395 (2010)
18. Alagia, M., Candori, P., Falcinelli, S., Lavollée, M., Pirani, F., Richter, R., Stranges, S., Vecchiocattivi, F.: J. Chem. Phys. **126**, 201101 (2007)
19. Lavollée, M.: Rev. Sci. Instrum. **70**, 2968 (1990)
20. Alagia, M., Callegari, C., Candori, P., Falcinelli, S., Pirani, F., Richter, R., Stranges, S., Vecchiocattivi, F.: J. Chem. Phys. **136**, 204302 (2012)
21. Taylor, S., Eland, J.H.D., Hochlaf, M.: J. Chem. Phys. **124**, 204319 (2006)
22. Lundqvist, M., Baltzer, P., Edvardsson, D., Karlsson, L., Wannberg, B.: Phys. Rev. Lett. **75**, 1058 (1995)
23. Field, T.A., Eland, J.H.D.: Chem. Phys. Lett. **211**, 436 (1993)
24. Alagia, M., Candori, P., Falcinelli, S., Mundim, M.S.P., Pirani, F., Richter, R., Rosi, M., Stranges, S., Vecchiocattivi, F.: J. Chem. Phys. **135**, 144304 (2011)
25. Alagia, M., Candori, P., Falcinelli, S., Pirani, F., Pedrosa Mundim, M.S., Richter, R., Rosi, M., Stranges, S., Vecchiocattivi, F.: Phys. Chem. Chem. Phys. **13**, 8245 (2011)
26. Falcinelli, S., Candori, P., Bettoni, M., Pirani, F., Vecchiocattivi, F.: Journal of Physical Chemistry A, Isolated Molecules, Clusters, Radicals, and Ions; Environmental Chemistry, Geochemistry, and Astrochemistry; Theory **118**(33), 6501–6506 (2014)
27. Zare, R.N.: Mol. Photochem. **4**, 1 (1972)
28. Dehmer, J.L., Dill, D.: Phys. Rev. A **18**, 164 (1978)
29. Kaiser, R.I.: Chem. Rev. **102**, 1309 (2002)
30. Biondini, F., Brunetti, B.G., Candori, P., De Angelis, F., Falcinelli, S., Tarantelli, F., Teixidor, M.M., Pirani, F., Vecchiocattivi, F.: J. Chem. Phys. **122**, 164307 (2005)
31. Biondini, F., Brunetti, B.G., Candori, P., De Angelis, F., Falcinelli, S., Tarantelli, F., Pirani, F., Vecchiocattivi, F.: J. Chem. Phys. **122**, 164308 (2005)
32. Cravens, T.E., Robertson, I.P., Waite Jr., J.H., Yelle, R.V., Kasprzak, W.T., Keller, C.N., Ledvina, S.A., Niemann, H.B., Luhmann, J.G., McNutt, R.L., Ip, W.-H., De La Haye, V., Mueller-Wodarg, I., Wahlund, J.-E., Anicich, V.G., Vuitton, V.: Geophys. Res. Lett. **33**, L07105 (2006)
33. Brooke, T.Y., Tokunaga, A.T., Weaver, H.A., Crovisier, J., Bockelee-Morvan, D., Crisp, D.: Nature **383**, 606 (1996)
34. Cernicharo, J., Heras, A.M., Pardo, J.R., Tielens, A.G.G.M., Guélin, M., Dartois, E., Neri, R., Waters, L.B.F.M.: Astrophys. J. **546**, L123 (2001)
35. Woods, P.M., Millar, T.J., Zijlstra, A.A., Herbst, E.: Astrophys. J. **574**, L167 (2002)
36. Momoh, P.O., Abrash, S.A., Mabrouki, R., El-Shall, M.S.: J. Am. Chem Soc. **128**, 12408–12409 (2006)
37. Alagia, M., Candori, P., Falcinelli, S., Lavollée, M., Pirani, F., Richter, R., Stranges, S., Vecchiocattivi, F.: Chem. Phys. Lett. **432**, 398 (2006)
38. Janev, R.K.: Atomic and Molecular Processes in Fusion Edge Plasmas. Plenum, New York (1995)
39. Larionova, I.A., Fialkov, B.S., Kalinich, K.Y., Fialkov, A.B., Ospanov, B.S.: Combust. Explos. Shock Waves **29**, 341 (1993)
40. Gordillo-Vazquez, F.J., Albella, J.M.: Plasma Sources Sci. Technol. **11**, 498 (2002)
41. Thissen, R., Delwiche, J., Robbe, J.M., Duflot, D., Flament, J.P., Eland, J.H.D.: J. Chem. Phys. **99**, 6590 (1993)

42. Hishikawa, A., Matsuda, A., Fushitani, M., Takahashi, E.J.: Phys. Rev. Lett. **99**, 258302 (2007)
43. Flammini, R., Fainelli, E., Maracci, F., Avaldi, L.: Phys. Rev. A **77**, 044701 (2008)
44. Laksman, J., Céolin, D., Gisselbrecht, M., Canton, S.E., Sorensen, S.L.: J. Chem. Phys. **131**, 244305 (2009)
45. Alagia, M., Candori, P., Falcinelli, S., Lavollée, M., Pirani, F., Richter, R., Stranges, S., Vecchiocattivi, F.: J. Phys. Chem. A **113**, 14755–14759 (2009)
46. Ashfold, M.N.R., Nahler, N.H., Orr-Ewing, A.J., Vieuxmaire, O.P.J., Toomes R.L., Kitsopoulos, T.N., Garcia, I.A., Chestakov, D.A., Wu, S.M., Parker, D.H.: Phys. Chem. Chem. Phys. **8**, 26–53 (2006)
47. Falcinelli, S., Fernandez-Alonso, F., Kalogerakis, K., Zare, R.N.: Molecular Physics **88**, 663–672 (1996)
48. Tosi, P., Correale, R., Lu, W., Falcinelli, S., Bassi, D.: Physical Review Letters **82**, 450–452 (1999)

Chemical Characterization of "Coco de Mer" (Lodoicea Maldivica) Fruit: Phytosterols and Fatty Acids Composition

Stefano Falcinelli[1(✉),2], Marta Bettoni[1,2], Federico Giorgini[2],
Martino Giorgini[2], and Bartolomeo Sebastiani[3]

[1] Department of Civil and Environmental Engineering, University of Perugia,
Via G. Duranti 93, 06125 Perugia, Italy
stefano.falcinelli@unipg.it
[2] Vis Medicatrix Naturae S.r.l., 50034 Marradi, FI, Italy
stefano.falcinelli@unipg.it
[3] Department of Chemistry Biology and Biotechnologies, University of Perugia,
Via Elce di Sotto, 8, 06123 Perugia, Italy
bartolomeo.sebastiani@unipg.it

Abstract. This work reports the first attempt for detection and identification of chemical compounds in fruit kernel of Lodoicea Maldivica coco nucifera palm. The analysis was performed by GC-MS technique to determine phytosterol and fatty acid composition profiles in internal and external pulp. Total phytosterol content was almost constant in both kernel coco-nut (24.6 µg/g for the external and 22.5 µg/g for the internal portion). The fatty acid pattern has been determined. The composition was characterized by seven saturated acids ranged from C14:0 (myristic) to C20:0 (arachidic) and two monounsaturated acids, the palmitoleic (C16:1, ω7) and the oleic (C18:1, ω9). Palmitic acid (C16:0) was the predominant one with contribution of about 49% followed by pentadecanoic, stearic (C18:0) and myristic acids (C14:0) in all two examined kernel parts. Despite its remarkable and widely known alimentary use and interesting peculiarities, Lodoicea Maldivica and its coconut (commonly named as "Coco de mer") has not received sufficient scientific research attention. The analytical study here presented intends to fill this gap, with particular attention to highlight the health and food safety use of the fruit, and the possible presence of chemical compounds interesting from a nutritional and pharmacological point of view.

Keywords: GC-MS analysis · Gas chromatography · Mass spectrometry · Fruit kernel · Phytosterols · Fatty acids · "Coco de mer" · Lodoicea maldivica

1 Introduction

The Lodoicea Maldivica, commonly named "Coco de mer" or "Double Coco-Nut", is a native palm of the Seychelles islands in the Indian Ocean. The relatively small native flora of these islands contains many endemic plants including six monospecific genera palm species, one of which is the famous "Coco de mer". This kind of palm,

O. Gervasi et al. (Eds.): ICCSA 2015, Part II, LNCS 9156, pp. 308–323, 2015.
DOI: 10.1007/978-3-319-21407-8_23

which is enormous reaching about 40 meters of height, grows exclusively in the Vallée de Mai, in the Praslin and Curieuse islands, a "World Heritage Site" designed by UNESCO in 1983 because of its unique forest vegetation, where is the largest population of these plants [1]. Unlike other Seychelles palms, the male and female flowers of the "Coco de mer" are borne on separate trees. While the female trees bear the nuts, which grow for about seven years before they fall, the male trees grow enormous catkins (up to 1 m long, making them the longest in the world), giant phallus shaped tubes studded with hundreds of delicate yellow flowers that give off a musky odour (see Fig. 1). The nut from the female palm, commonly named "Coco de mer" shows an impressive resemblance with the female pubes, because it has generally two lobes suggesting a double coconut (see Fig. 1). Because of the strong similarity of the male flowers and seed (coconut) female to genital organs of human beings, a Seychelles legend says that during a full moon the male and female "Coco de mer" trees are walking around the forest in order to mate. The male tree of these palm lives longer than the female one. Savage and Ashton in their studies on the structure of palm populations [2] tentatively concluded that age of the tree rarely exceeds 300 years. The about 4000 palms of the "Coco de mer" are protected and despite the importance of the site, the local government allows to sold with high price all hundreds of coco-nuts picked every year, e.g. 701 nuts in 2003 [1]. These palms are considered as the last witness of the old continent "Gondwana", formed by Africa, Madagascar and India, which 65 millions of years ago split into a number of lands leaving Seychelles alone. As we have mentioned above, the flowers of the Lodoicea Maldivica are borne in enormous fleshy spadices (spikes), the male and female on distinct plants. "Coco de mer" fruits, which are among the largest seeds known (about 50 cm long and up to 30 kg of weight), take about seven years to mature. They have a fleshy and fibrous envelope surrounding a hard nut-like portion which is generally two-lobed, suggesting a large double coco-nut (see Fig. 1). The contents of the nut are edible as in the coco-nut. The empty fruits (after germination of the seed) are found floating in the Indian Ocean, and were known long before the palm was discovered, giving rise to various mythical stories as to their origin. In the ancient times, people attribute to "Coco de mer" both supernatural and earthly powers, since was used as aphrodisiac in the Middle Ages. Various traditional herbalists of Seychelles hadhelped to create many uses for this particular double coconut, such as energizing, stimulating and aphrodisiac beverages originally used by the Seychellois people to improve the performance of their everyday lives. The dried "Coco de mer" kernel is exported in various parts of the world, especially the Middle East and China, where it has been used for centuries in traditional medicine and in "ayurvedics", both as an aphrodisiac and for rejuvenating cosmetic creams and to treat other pathologies such as coughs. Despite the characteristics mentioned above, it has never been studied from a chemical point of view. However, although this palm is well known and has been strictly protected for over 25 years, scientific investigations of the stand structure, species diversity and regeneration have been published only in 2005 by Fleischmann and co-workers [1]. More recently, papers have been published concerning characterization [3] and sustainable harvesting of "Coco de mer" [4]. The authors

Fig. 1. Left side: Lodoicea Maldivica (Coco de mer) palms. These plants can reach about 40 meters of height. In general, the male tree lives longer than the female one, rarely exceeding an age of 300 years. Right side: in the upper panel is reported an image of the nut from the female palm, commonly named "Coco de mer". Its peculiarity is the impressive resemblance with the female pubes, because it has generally two lobes suggesting a double coconut. For this reason in the ancient times people attribute to it supernatural and heartly powers, since was used as aphrodisiac in the Middle Ages. It is among the largest known seed (about 50 cm long and up to 30 kg of weight) and take about 7 years to ripen. In the lower panel is shown the male "Coco de mer" inflorescence: this kind of catkins can reach up to one meter in length, making them the longest in the world.

emphasize that Lodoicea Maldivica has not received sufficient scientific research attention although its remarkable and widely known alimentary use and interesting peculiarities. The analytical study here presented intends to fill this gap, with particular attention to highlight the healthy and food safety use of the fruit, and the possible presence of chemical compounds interesting from a nutritional and pharmacological point of view.

2   Experimental

The analytical characterization of the lipidic portion of Lodoicea Maldivica fruit pulp has been investigated by mass spectrometry (MS) detection of the analytes after their gas chromatographic (GC) separation (GC-MS technique), to define the main fruit quality characteristics from a chemical point of view.

The employment of MS in chromatography contributes to greatly expand its applications and uses because MS is a powerful and universal diagnostic technique able in principle to detect and recognize any chemical compound. Actually, MS is one of the most widely used analytical chemistry technique, being applied both in research and analytical studies, as for example in fundamental [5-8], environmental [9,10], combustion [11-13] and atmospheric chemistry [14-18], in bio-medical applications [19], and in astrochemistry [20-23].

2.1   Reagents and Standard Solutions

All the used reagents were of analytical grade purity. The extraction solvent was a mixture of hexane-acetone (1:1, v/v); both reagents were purchased from Sigma-Aldrich (Deisenhofen, Germany) respectively ≥98% and ≥99% pure. For the derivatisation procedure of the lipidic fraction a BSTFA (bis-trimethylsilyl-trifluoroacet amide)-TMCS (trimethylchlorosilane) (99:1, v/v) reagent solution supplied by Supelco (Bellefonte, Pennsylvania, USA) and pyridine ACS reagent purchased from Fluka (Buchs, Switzerland) ≥99.8% pure were used. The solution used for SPE purification of the derivatised samples were a mixture of hexane-methyl tert-butyl ether (99:1, v/v) with reagents obtained from Sigma-Aldrich and J.T. Baker respectively ≥98% and ≥99% ultra resi-analysed pure. The following standard solutions in chloroform were purchased from Supelco (Milano, Italy): β-sitosterol (100µg/ml) and stigmasterol (10 mg/ml); the campesterol in white crystals from soybean were supplied by Sigma-Aldrich approximately 65% pure, whereas fatty acids solid standard ranged from C12 to C22 were supplied by Matreya (Inc. Pleasant Gap, PA). To determine linear retention index a n-alkane calibration standard solution C6-C44, (ASTM method D2887), obtained from Superchrom (Milano, Italy) was used.

2.2   Apparatus

High-performance dispersing instrument Ultra Turrax T25 was used to ensure perfect complete homogenization coconut samples. To perform normal-phase SPE extraction method, plastic syringe barrels (4 ml) were handmade packed with 1 g silica gel (Bondesil-SI 40µm, Varian), previously conditioning overnight at 130°C and capped with polymeric frits. Before to be used, the SPE cartridges were pre-conditionated with the same solvent mixture used for derivatized lipidic fractions. Sample analyses were performed using a Chrompack CP 3800 gas chromatograph equipped with 1079 split-splitless injector and coupled to an ion trap detector Saturn 2200 (Walnut Creek, CA, USA) operating in positive electron and liquid chemical ionization mode. The data acquisition and the quantitative analysis were achieved with Varian Saturn GC/MS Workstation system software, version 5.41 (Walnut Creek, CA, USA).

2.3   Plant Material

The fruit of Lodoicea Maldivica was obtained by plants grown in the Vallée de Mai National Park, in the Praslin islands of Seychelles Archipelago in the Indian Ocean and harvested at advanced maturation degree (seven years). The fruit was frozen and stored   up at -10°C from Seychelles islands to Italy and up to the analysis.

2.4   Sample Preparation

The "Coco de mer", after defrosting, was husked and split using a knife. The analyzed fruit, a coco-nut of 5.4 kg of weight and about 32 cm long (polar and equatorial diameter 7 and 24 cm respectively), was characterized by the presence of two different portions of the internal pulp so-called endosperm tissue: a more hard and external kernel of 25 mm thickness of grey-white colour (named in the text as sample A), covering one soft and whitish inner part of about 15 mm thickness (sample B). The coco-nut was completely lacking in liquid inside (watery endosperm), indicating a probable advanced degree of maturation for the fruit. The samples, consisting of the two distinct portions of the coco-nut kernel corresponding to different pulp consistency, were treated following standard procedures published by ISTISAN 97/23 report [4]. Both external layer harder and softer inside part of the fruit were scooped by a stainless steel spoon and care was taken to not scrape off the peel; after they were ground with "Ultra Turrax" T25. The fresh fruit pulp was finely grounded, thoroughly homogenized, stored in clean glass jar and frozen at -18°C in the dark until being analyzed the aroma flavor, sterol and fatty acid compositions. Before lipidic extractions, the two pulp samples (A and B) were dried at 40°C for 2 h, with the purpose to eliminate the watery part, in order to maximize the extractive yield of the organic fraction. The representative share of both samples, submitted to analysis, has been gotten by a quantity of around 100 g proceeding according to the technique of "quartatura" followin ISTISAN 97/23 method [24]. This work was carried out within a 90 day period after harvest.

2.5   Extraction and Derivatisation

For the analysis of the lipidic portion of the fruit, the extraction was carried out on aliquot of 13.0 g, sample A, and of 13.3 g, sample B, for 24 h under reflux in a Soxhlet extractor with 200 ml hexane-acetone (1:1, v/v) solution [25]. Both crude lipid extracts were concentred under reduced pressure in a rotary evaporator (Büchi, Postfach, Switzerland) to the volume of 2 ml and transferred into conical flask, reduced to few μl under N_2 gentle stream at 30°C and, without saponification, directly derivatisated at 70°C for 15 min with 50 μl Pirydine and 40 μl BSTFA-TMCS (99:1, v/v) reagent solution to give the trimethylsilyl (TMS) derivatives. Finally the extracts were loaded, cleaned up onto silica gel SPE column, eluted with 4.5 ml hexane-methyl tert-buthyl ether (99:1, v/v) solution [25] and 1.0 ml of both samples A and B were analyzed.

2.6   GC-MS Conditions

The GC-MS analysis was achieved using the following operative conditions: i) a low-bleed/MS CP-Sil8 CB capillary column (30 m x 0.25 mm, 0.25 μm film thickness - Chrompack, Middelburg, The Netherlands); 1ml/min. He carrier gas constant flow; ii) a 90°÷290°C thermal program with a temperature rise of 12°C/min; iii) a temperature of 260°C for the injector operating in the splitless mode; iv) ITD temperature of 150 °C; electron impact ionization mode at 70 eV and EM voltage 2000 V, full scan acquisition in the 100÷600 uma range at 1 sec/scan [26].

The separated compounds were identified by their mass spectra generated by electron ionisation. Component identification was confirmed by comparison of experimental collected mass spectra with those in the NIST98 and Wiley5 mass spectral library databases. Selected ions were used for the sterol and fatty acid quantitative analysis. Quantification of analytes was performed by external standard calibration The components not qualitatively confirmed by comparison with database mass spectra were considered tentatively identified (NI).

2.7   Analyte Recovery and Repeatability

Both recovery and repeatability of the analytical procedure were tested and verified to minimize variation in accurate and precise quantification of sterol and fatty acid fractions. Three different samples were taken in triplicate and three injections were conducted using the same sample. Recoveries as high as 90% were calculated and reproducibility of replicate measurements agreed to within ± 10%.

3   The Computational Procedure for Gc-Ms Chemical Compound Identification

In the computational method that we have used for extracting individual component spectra from GC/MS data files and then using these spectra to identify target compounds by matching spectra in a reference library, a commercial software was employed. The employed software is the "Automated Mass Spectrometry Deconvolution and Identification System" (AMDIS) – version 2.65. The AMDIS software extracts pure component spectra from complex GC/MS or LC/MS data files and searches against specialized libraries or the commonly used NIST library. This module was developed at the "National Institute of Standards and Technology" (NIST) USA. AMDIS can operate as a "black box" chemical identifier, displaying all identifications that meet a user-selectable degree of confidence. The chemical compounds identification can be aided by internal standards and retention times. In our case we can compare the obtained mass spectra of the analyzed chemical compounds with the Electron Ionization (EI) mass spectral library that is interfaced with our GC/MS-MS apparatus. It consists of NIST 11 MS/MS Library containing more than 250,000 spectra of about 230,000 unique compounds (see http://www.sisweb.com/software/ms/nistsearch.htm). In this commercial library, besides spectra, typical data include name, formula, molecular structure, molecular weight, CAS number, list of peaks, synonyms, and estimated and/or measured retention index. In our computational procedure, used to analyze and to identify the chemical compounds, we are able also to built directly our own mass spectra library. In this way we can use the AMDIS software as a processing tool for our collected GC/MS data files. In such a way we are able to collect and store mass spectra of any chemical compound and in specific experimental conditions. Concerning the gas phase retention data for compounds common to the EI and NIST

Retention Data collection, they are available with links to the EI library. This involves more than 300,000 Kovats retention indices and corresponding GC methods, column conditions and literature citations for about 75,000 compounds. We can also implement our library data determining retention indices for chemical compounds of our interest in any specific experimental condition used for the analysis. For such a purpose we have built a specific algorithm base on the Van den Dool and Kratz equation [27], reported below (see Equation (1)). For the present work, the linear retention index (I^T) values for analytes were calculated by using the retention time data obtained by analyzing a series of normal alkanes ranging from C5 to C44 under identical chromatographic conditions used for the analysis of the volatile aroma compound fraction that will be the subject of a future paper. The I^T values are non-isothermal Kovats retention indices because temperature-programmed gas chromatography was used. As already mentioned, the (I^T) values were calculated using direct retention time numbers ($t_{Ri}{}^T$) instead of their logarithm, according to Van den Dool and Kratz equation [27]:

$$I^T = 100 \left[\frac{t_{Ri}{}^T - t_{Rz}{}^T}{t_{R(z+1)}{}^T - t_{Rz}{}^T} + z \right]$$

(1)

In general, a relatively easy computational approach to extract reliable spectra in GC/MS and GC/MS/MS techniques is the so called "backfolding" procedure, consisting in subtracting adjacent scans [28]. An advantage of this approach is that it does not explicitly require maximization. However, it does not account for ion counting noise or peak shape, so is unlikely to adequately identify weak components. For such a reason the used procedure to evaluate the noise in the recorded GC/MS data is of crucial importance. Therefore it is essential to perform a proper and rigorous noise analysis, whose the main step is to extract the so called "noise factor" from the GC/MS data file. In this effort we have followed the suggestions by Stein and Scott [29], considering that our particles detector is an electron multiplier which generates signals that fluctuate by an average amount proportional to the square root of the signal intensity [30]. The knowledge of this proportionality factor allows the simple estimation of the magnitude of this type of noise for any signal strength, and in our case we can define the "noise factor", N_f, as follows:

$$N_f = \frac{R_D}{\sqrt{I}}$$

(2)

where R_D is the average random deviation and I is the recorded signal by the detector. In principle, N_f may be obtained from measured levels of random signal fluctuation during instrument tuning. However, this information is not generally available from instrument data systems. Therefore, N_f is derived for each data file from ion-chromatographic regions of relatively constant signal intensity. In the estimation of the noise factor, each ion chromatogram, as well as the total ion chromatogram (TIC), is divided into segments of thirteen scans. If any abundance in a segment is zero, the segment is rejected. For each accepted segment, a mean abundance is computed and the number of times that this mean value is "crossed" within the segment is counted (crossings occur for adjacent mass spectral scans where one abundance is above the mean and other abundance is below the mean). If the number of crossings is less than one-half the number scans in the segment (6 or less), the segment is rejected. For each accepted judgment, the "median" deviation from the mean abundance for that segment is found [29]. This deviation is divided by the square root of the mean abundance for that segment to obtain a sample N_f value, which is then saved. After processing the entire data file, the "median" of these sample N_f values is taken as the characteristic N_f value for the entire GC/MS data file. The use of medians in place of means (simple averages) and the crossing criterion serve to reject high N_f values arising from real chromatographic components. In this paper the square root of a signal multiplied by N_f is the magnitude of this signal in "noise units". One noise unit represents the typical scan-to-scan variation arising from ion-counting noise at a given abundance level. Testing with data files from properly tuned instruments, Slein and Scott have demonstrated that N_f was independent of both signal intensity level and m/z value and that run-to-run consistency for data files acquired on a single instrument was good (N_f variations of less than 10%) [29]. Over a wide range of well-tuned commercial mass spectrometers, including quadupole and ion trap instruments, that is the case of our used GC/MS device, N_f fell in the range 0.5 to 10. However, some dependence on signal strength was noticed at low signal levels in the presence of large amounts of spurious signal. Proper signal threshold setting eliminated this problem, and no adverse effects attributable to the averaging of multiplier signals ("centroiding") were noted [29].

4   Results and Discussion

As mentioned above, there are no previously published data on Lodoicea Maldivica fruit regarding its chemical characterization. Here we report an analytical study concerning the lipidic fraction characterization of this kind of coconut, in order to present a comprehensive overview of free sterol and fatty acid profiles, useful to demonstrate the healthy and food safety use of the fruit.

4.1   Free Sterols

The crude lipid extract amounts of external (sample A) and internal (sample B) coconut pulps were estimate to be respectively 28.6 mg (2.2 mg/g dry weight) and 37.5 mg (2.8 mg/g dry weight). The chromatograms reported in Fig. 2 are the TIC (total ion current) comparative profiles of both analyzed samples and show three different families of organic compounds: the peaks of sugars, fatty acids and free phytosterols are evident going from low to high retention time values, t_r, respectively ranging from 7.1 to 12.5 min, from 12.5 to 22.5 and from 22.5 to 30.0 minutes. The sugar fraction will not be object for discussion in this article. The predominant sterol class in vegetable oils is the 4-desmethyl sterols. Plant sterols and cholesterol are products of the isoprenoid biosynthetic pathway, derived from squalene compound. The dedicated pathway to sterol synthesis in photosynthetic plants occurs at the squalene stage through the activity of squalene synthetase [31]. In fact in all two analysed samples, the squalene molecule at t_r 20.5 (corresponding molecular ion m/z 410) at concentration trace level were identified. Plant sterols have a structure similar to cholesterol and they can be present as free or esterified forms being their quantitative proportions variable depending on the vegetable matrix [32]. Concerning the free form phytosterol family, chromatographic analysis of the unsaponifiable fraction of coconut showed three identifiable phytosterol peaks (see Fig. 2).

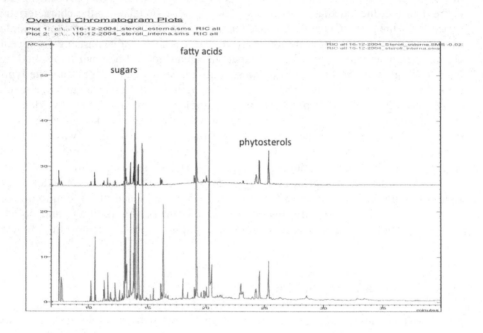

Fig. 2. Reconstructed total ion current comparative GC-MS profiles of the TMS derivative compounds in the two kernel parts (A = external, B = internal) of "Coco de mer" fruit showing three different family of organic compounds

These chemical compounds, identified by GC-MS technique, were confirmed by comparing their absolute retention time values and MS fragments with those of TMS derivative pure standards: campesterol at t_r 24.3 (molecular ion corresponding m/z 472), stigmasterol (t_r 24.6, m/z 484) and finally β-sitosterol which is the highest peak on the right side of the chromatogram (t_r 25.3, m/z 486). Cholesterol was also identified at t_r 23.2 (molecular ion corresponding m/z 458) in both samples and his identity was confirmed by comparison to database mass spectra. A keto-steroidal compound with molecular ion of 382 m/z, not TMS derivative, was revealed at t_r 24.0, only in the internal pulp matter (sample B). The identity of this keto-compound, confirmed by comparison with library data, was the cholesten-3,5-dien-7-one. This identification was corroborated by many studies on phytosterol products and particularly the identified compound traced back to cholesterol oxidation product [33]. The analysis of the sterol oxidation fraction is mostly focused on cholesterol, since the toxicity of some of these products has been asserted. Although the products of cholesterol oxidation are well established, much less is known about the plant sterol products. The free sterols are the most abundant components present in the sterol fractions of vegetable oils. These organic substances are of particular interest because the phytosterols are chemical compounds with well known pharmacological activity and nutritional properties. Plant sterols have a structure similar to cholesterol and therefore reduce cholesterol absorption. In particular they are able to inhibit the uptake of cholesterol from the diet and therefore to lower the high concentration levels of this sterol in the plasma low-density lipoproteins (LDL), resulting in significant reduction in the risk of heart disease [34]. This effect have been known since the 1950s [35]. Although many studies have been performed, the mechanism of cholesterol-lowering effect of phytosterols is not fully understood. Today there is an increasing interest about the possible commercial sources of plant sterols and they are commonly added as medicinal "nutraceuticals" to margarines and other foods in the USA as well as in some European countries, for example in Finland (since 1995) as well as the plant stanol ester margarine Benecol [31,36]. Furthermore, there are evidences that β-sitosterol may have a significantly estrogenic and anti-estrogenic activity [37]. However, in addition to the blood cholesterol-lowering effect, phytosterols are also considered to have antioxidative, anti-inflammatory, anti-bacterial, anti-ulcerative, anti-atherosclerotic and anti-tumor effects in the animals and humans [38]. For these reasons we performed a quantitative determination of the phytosterols identified in both analyzed samples of double coconut. Synoptic Table 1 exhibits the summary of individual quantitative results obtained, both in external kernel and internal pulp with the following concentration units: μg/g of dry sample and of lipidic extract. As expected the predominant phytosterols in all two parts of coconut are the β-sitosterol and the stigmasterol, with comparable quantitative values, as observed in most of vegetable oils [39]. In the sample A for β-sitosterol and stigmasterol compounds we obtained values of concentration of 10.5 μg/g and 9.0 μg/g (dry weight) respectively, whereas in the case of campesterol, the observed concentration value was lower: 2.8 μg/g (dry weight). Comparable sterols amounts were also obtained for the internal pulp (sample B): 11.6 μg/g for β-sitosterol, 10.0 μg/g for stigmasterol, and 3.1 μg/g (dry weight) for campesterol. Semi-quantitative considerations can be found relatively to the cholesterol and cholesten-3,5-dien-7-one compounds. The first molecule contribute for less than 1% and the second molecule for less than 5% at the total identified sterols fraction. The low percent trace content of cholesterol in both double coconut parts confirm the data reported by Phillips et al. [39], in edible oils. No substantial differences were detected in free sterols composition between the two analyzed samples (A and B), both referred to grams of pulp and to milligrams of lipid extracts. All three individual free sterols concentration values were showed higher in external kernel than in internal pulp of 10 percent when referred to lipidic extract. Probably this occurred because the external kernel was more hard and dried respect to the internal part, and consequently relatively more concentrated. These analytical data were in agreement with those from literature, where β-sitosterol and campesterol have percent values ranging respectively from 40.2 to 93.2% and from 2.6 to 38.6%, as pointed out in the sterol fractions of commodity vegetable oils like coconut, cocoa butter and other kind of palms [40]. However, the percentage of stigmasterol results in the literature data ranging from

0.0 to 31.0% [40]. This situation is probably occurred because the literature reference values are derived from published results achieved considering variations in different total sterol contents and in the specificity of the used analytical method, that may place our stigmasterol experimental values outside the literature ranges.

Table 1. Sterols composition of external and internal kernel of "Coco de mer" and individual percentage contribution

Sterol	external kernel (sample A)			internal kernel (sample B)		
	µg/g d.w.[a]	µg/g fat	% of total	µg/g d.w.[a]	µg/g fat	% of total
Campesterol	2.8	1.3	12.6	3.1	1.1	12.5
Stigmasterol	9.0	4.2	40.3	10.0	3.5	40.5
β-sitosterol	10.5	4.9	47.1	11.6	4.0	47.0
[a] *d.w. = dry weight*						

Furthermore it's surprising that the GC-MS sample analysis excludes the presence of other plant sterols like Δ^5-avensterol, that is absent in both pulp parts, whereas it should be present in coconut fruits in percent values ranged from 1.5 to 29.0% [40]. In fact in their study Ham and co-workers [40] found high percentage of Δ^5-avensterol only in the unique case of coconut oil among most edible seed oils. The measurable amounts of this sterol was estimate in the rest of vegetable oils rather lower (1.5-18.8%) [25,40]. One possible explanation on this disagreement about the sterol quantitative pattern would be probably due to the different and unusual structural characteristics of this singular palm fruit, compared to more widespread coconut therefore to genetic factors [31]. Besides further variation in coconut pulp composition would seem also to be depending on several other factors such as fruit maturity degree, different processing conditions of fruits and nature of soil on which the coconut is grown [31].

4.2   Fatty Acids

Concerning fatty acid composition from both external and internal pulp samples the individual compounds were identified and quantified as listed in synoptic Table 2. The average values were expressed both as ng/g of dry net matter and of lipidic extract. The fatty acid qualitative composition was characterized by seven saturated acids (SFA) ranging from C14:0 (myristic acid) to C20:0 (arachidic acid) and two monounsaturated acids (MUFA) the palmitoleic (C16:1, ω7) and the oleic acid (C18:1, ω9). In the external kernel the nonadecanoic (C19:0) and the arachidic acid (C20:0) were absent, probably because their concentration were outside the limit of the instrumental detection. It should be noted that in both examined samples (A and B) the polyunsaturated fatty acid fraction (PUFA) was totally absent, while monounsaturated content resulted to amount to about 7.8 and 1.0 % of the total fatty acid concentration for sample A and B respectively. With a similar absolute amount of MUFA, the percent contribution of this fraction was higher in the external endosperm part than internal. Therefore the saturated/unsaturated calculated ratio was high like as

11.8 for external part and very high, 96.3, for the internal. That confirm the general characteristics of coconut nature. In fact this fruit is rich in medium and long chain saturated fatty acids, but deficient in essential fatty acids, ω3 and ω6 polyunsaturated fatty acids. Consequently, it has been confirmed that the saturated fatty acids were predominant in both pulp parts of "Coco de mer". The abundance of saturated fatty acids confirm the literature data for coconut oil and its classification as a hypercholesterolemic fat compared to PUFA rich oils [41]. It is interesting to note that our study did not point out the presence of medium chain fatty acids (MCFA) ranged from C8:0 to C 12:0, in spite of literature data [42,43]. Neither it is confirmed the presence of linoleic (C18:2, ω6) and linolenic (C18:3, ω3) acids as found and reported in the study performed by Dong-Sun and coworkers [43] concerning fatty acids in various commercial vegetable oils. Moreover the lauric acid (C12:0), which shows a great anti-viral activity, normally present in endosperm meat of coconut as predominant fatty acid at level of roughly 40% of all SFA [42], was totally absent in the examined "Coco de mer" fruit. On the contrary the myristic acid was present at lower level respect to literature data (~20%), about 12.6% and 7.8% for the sample A and B respectively. In general lauric (C12:0) together myristic (C14:0) acids, mainly present in coconut and palm kernel oil, appear to be the principal saturated fatty acids responsible for raise plasma cholesterol [44]. In comparison, MUFA decrease the "bad cholesterol" LDL-cholesterol (LDL-C). Also PUFA decrease LDL-C, whereas intakes of ω-6 PUFA may have adverse effects on "good cholesterol" high density lipoprotein-cholesterol (HDL-C) [45]. From a quantitative point of view the internal part pulp of "Coco de mer" presents higher fatty acid contribution than external pulp and the concentration of total fatty acids amount to 1392.0 and 143.9 ng/g dry net respectively. The SFA experimental profiles present a mono-modal distribution of concentrations with a maximum corresponding to palmitic acid: 70.8 and 689.3 ng/g dry net in sample A and B respectively. This fatty acid is the main dominant of the total amount with a contribution of about 49.3% in external and 49.6% in internal pulp. As reported in Table 2, pentadecanoic acid results to be the second most abundant fatty acid component: 17.5% in the sample A and 15.2% in the sample B (25.2 and 211.6 ng/g dry net respectively). At lower concentration levels, there are other fraction components like stearic acid (9.2% in the sample A and 13.0% in the sample B), myristic acid (12.6% and 7.8%), margaric acid (3.6% and 7.3%), nonadecanoic acid (3.9% only in the sample B), arachidic acid (2.2%, only sample B), oleic acid (4.8% and 0.7%) and palmitoleic acid (3.0% and 0.3%). The percentage contribution of margaric, stearic, nonadecanoic and arachidic acid increase from external to internal pulp while for all remaining other acids we observe a decrease, as obtained in the case of MUFA. The great disagreement revealed in the fatty acid profile of "Coco de mer" and in the individual percentage contribution respect to the other varieties of coconut can be explained with the same considerations made for free sterols concerning the genetic characteristics of studied fruit, its age and the particular site where it was grown. Therefore it is also important to notice that not only the absolute fatty acid amount but also their relative composition may be highly different in distinct parts of the endosperm kernel.

Table 2. Fatty acid composition of external and internal kernel of "Coco de mer" and individual percentage contribution

Fatty acid	external kernel (sample A)			internal kernel (sampleB)		
	ng/g d.w.[a]	ng/g fat	% of total	ng/g d.w.[a]	ng/g fat	% of total
Saturated						
C14:0 *myristic*	18.2	8.5	12.6	108.9	37.7	7.8
C15:0	25.2	11.7	17.5	211.6	73.4	15.2
C16:0 *palmitic*	70.8	32.9	49.3	689.3	238.9	49.6
C17:0 *margaric*	5.2	2.4	3.6	101.4	35.2	7.3
C18:0 *stearic*	13.3	6.2	9.2	181.2	62.8	13.0
C19:0	-	-	-	54.3	18.8	3.9
C20:0 *arachidic*	-	-	-	31.0	10.7	2.2
Monounsaturated						
C16:1 ω7 *palmitoleic*	4.3	2.2	3.0	4.7	1.5	0.3
C18:1 ω9 *oleic*	6.9	3.2	4.8	9.6	3.3	0.7
[a] *d.w. = dry weight*						

5   Conclusions

The results of this study can be taken as a preliminary data base for further investigations on the "Coco de mer" fruits. This work reports for the first time the phytosterol and fatty acid GC-MS profiles, although the analytical data were referred to only one coconut fruit. Because this limitation data are not available on quantitative changes in phytosterol and fatty acid contents during the different ripe stages (the fruit take about ten years to ripen) and before/after harvest. The semi-quantitative approach adopted in these preliminary studies allows to establish the relative concentration of different biological volatile organic compounds for each fruit parts under investigation. The examined fruit showed a very low amount of unsaponifiable sterols. These steroidal compounds are not endogenously synthesized in the human body, but derived exclusively from dietary sources. The estimated daily dietary intakes of plant sterols required to obtain a 10-15% lowering in blood low-density lipoprotein (LDL) cholesterol is about 1.5-3.0 g/day, consequently the "Coco de mer" is not indicate for such purpose. The fatty acid composition of lipidic fraction of analyzed fruit shows remarkable differences from common coconut meat, with palmitic acid (C16:0) being the predominant fatty acid (~50%) in both pulp parts. The dominant presence of saturated fatty acids and the absence of polyunsaturated fatty acid fraction (PUFA) in the kernel fruit confirm its classification as a hypercholesterolemic. In conclusion, no substantial and significant qualitative and quantitative compositional differences were observed in the two analyzed kernel parts and the results obtained in this study would seem to indicate that the "Coco de mer" has a nutritional and pharmacological value that needs some additional investigations. Therefore our analytical study confirm that "Coco de mer" is an edible coconut that can be used for alimentary purposes without any adverse indication for its healthy and food safety use. Further studies could concern the chemical characterization of the fruit to different stadiums of maturation to better clarify its possible alimentary and pharmacological potentialities.

Acknowledgments. Financial support by the "VIS MEDICATRIX NATURAE s.r.l." – Marradi (FI – Italy) is gratefully acknowledged by the authors.

References

1. Fleischmann, K., Edward, P.J., Ramseier, D., Kollmann, J.: African Journal of Ecology **43**, 291–301 (2005)
2. Savage, A.J.P., Ashton, P.S.: Biotropica **15**, 15–25 (1983)
3. Falcinelli, S., Giorgini, M., Sebastiani, B.: Applied Engineering Sciences, Ed. By Wei Deng - Taylor & Francis Group, London, Chapter 19, pp. 99–104 (2015). ISBN: 978-1-138-02649-0
4. Rist, L., Kaiser-Bunbury, C.N., Fleischer-Dogley, F., Edwards, P., Bunbury, N., Ghazoul, J.: Forest Ecology and Management **260**, 2224–2231 (2010)
5. Cappelletti, D., Bartocci, A., Grandinetti, F., Falcinelli, S., Belpassi, L., Tarantelli, F., Pirani, F.: Chemistry A European Journal **21**, 6234–6240 (2015)
6. Cappelletti, D., Candori, P., Falcinelli, S., Albertì, M., Pirani, F.: Chemical Physics Letters **545**, 14–20 (2012)
7. Brunetti, B., Candori, P., Cappelletti, D., Falcinelli, S., Pirani, F., Stranges, D., Vecchiocattvivi, F.: Chem. Phys. Lett., 539–540, 19 (2012)
8. Balucani, N., Bartocci, A., Brunetti, B., Candori, P., Falcinelli, S., Pirani, F., Palazzetti, F., Vecchiocattivi, F.: Chemical Physics Letters **54**, 634–39 (2012)
9. Falcinelli, S., Fernandez-Alonso, F., Kalogerakis, K., Zare, R.N.: Molecular Physics **88**, 663–672 (1996)
10. Tosi, P., Correale, R., Lu, W., Falcinelli, S., Bassi, D.: Physical Review Letters **82**, 450–452 (1999)
11. Balucani, N., Leonori, F., Nevrly, V., Falcinelli, S., Bergeat, A., Stranges, D., Casavecchia, P.: Chemical Physics Letters **602**, 58–62 (2014)
12. Cavallotti, C., Leonori, F., Balucani, N., Nevrly, V., Bergeat, A., Falcinelli, S., Vanuzzo, G., Piergiorgio Casavecchia, P.: The Journal of Physical Chemistry Letters **5**(23), 4213–4218 (2014)
13. Leonori, F., Balucani, N., Nevrly, V., Bergeat, A., Falcinelli, S., Vanuzzo, G., Casavecchia, P., Cavallotti, C.: The Journal of Physiscal Chemistry C Nanomaterials and Interfaces (in press) (2015). doi:10.1021/jp512670y
14. Falcinelli, S., Pirani, F., Vecchiocattivi, F.: Atmosphere **6**(3), 299–317 (2015)
15. Falcinelli, S., Rosi, M., Candori, P., Vecchiocattivi, F., Farrar, J.M., Pirani, F., Balucani, N., Alagia, M., Richter, R., Stranges, S.: The escape probability of some ions from mars and titan ionospheres. In: Murgante, B., Misra, S., Rocha, A.M.A., Torre, C., Rocha, J.G., Falcão, M.I., Taniar, D., Apduhan, B.O., Gervasi, O. (eds.) ICCSA 2014, Part I. LNCS, vol. 8579, pp. 554–570. Springer, Heidelberg (2014)
16. Falcinelli, S., Bartocci, A., Candori, P., Pirani, F., Vecchiocattivi, F.: Chemical Physics Letters **614**, 171–175 (2014)

17. Falcinelli, S., Candori, P., Bettoni, M., Pirani, F., Vecchiocattivi, F.: Journal of Physical Chemistry A, Isolated Molecules, Clusters, Radicals, and Ions; Environmental Chemistry, Geochemistry, and Astrochemistry; Theory 118(33), 6501–6506 (2014)

18. Brunetti, B.G., Candori, P., Falcinelli, S., Pirani, F., Vecchiocattivi, F.: Journal of Chemical Physics 139(16), 164305 (2013)

19. Eberlin, L.S., Tibshirani, R.J., Zhang, J., Longacre, T.A., Berry, G.J., Bingham, D.B., Norton, J.A., Zare, R.N., Poultsides, G.A.: Proceedings of the National Academy of Sciences 111(7), 2436–2441 (2014)

20. Alagia, M., Bodo, E., Decleva, P., Falcinelli, S., Ponzi, A., Richter, R., Stranges, S.: Physical Chemistry Chemical Physics 15(4), 1310–1318 (2013)

21. Falcinelli, S., Rosi, M., Candori, P., Farrar, J.M., Vecchiocattivi, F., Pirani, F., Balucani, N., Alagia, M., Richter, R., Stranges, S.: Planetary and Space Science 99, 149–157 (2014)

22. Schio, L., Li, C., Monti, S., Salén, P., Yatsyna, V., Feifel, R., Alagia, M., Richter, R., Falcinelli, S., Stranges, S., Zhaunerchyk, V., Carravetta, V.: Physical Chemistry Chemical Physics 17(14), 9040–9048 (2015)

23. Pei, L., Carrascosa, E., Yang, N., Falcinelli, S., Farrar, J.M.: The Journal of Physical Chemistry Letters 6, 1684–1689 (2015)

24. Rapporti ISTISAN 97/23: Multiresidue analytical procedures for pesticides residues in vegetable products. Rome: Italian National Institute for Health (1997). ISSN 1123-2117

25. Abidi, S.L.: Journal of Chromatography A 935, 173–201 (2001)

26. Falcinelli, S., Giorgini, M., Sebastiani, B.: Abstract of Papers of The American Chemical Society, vol. 248, Meeting Abstract: 244-AGFD - 248th National Meeting of the American-Chemical-Society (ACS), San Francisco, CA (USA), August 13, 2014. ISSN: 0065-7727

27. Van Den Dool, H., Kratz, P.D.: J. Chromatography A 11, 463–471 (1963)

28. Pool, W.G., Leeuw, J.W., van de Graaf, B J.: Mass Spectrom. 32, 438–443 (1997)

29. Stein, S.E., Scott, D.R.: J. Am. Soc. Mass Spectrom. 5, 859–866 (1994)

30. Peterson, D.W., Hayes, J.M.: In: Contemporary Topics in Analytical and Clinical Chemistry, vol. 3, pp. 217–251. Plenum Publishing (1978)

31. Piironen, V., Lindsay, D.G., Miettinen, T.A., Toivo, J., Lampi, A.M.: Journal of the Science of Food and Agriculture 80, 939–966 (2000)

32. McKevith, B.: Nutrition Bulletin 30, 13–26 (2005)

33. Rodríguez-Estrada, M.T., Frega, N., Lercker, G.: Grasas y Aceites 53(1), 76–83 (2002)

34. Cleeman, J.I., Grundy, S.M., Becker, D., Clark, L.T., Cooper, R.S., Denke, M.A.: Journal of American Medical Association 285, 2486–2497 (2001)

35. Peterson, D.W.: Proceedings of the Society for Experimental Biology and Medicine 78, 143–147 (1951)

36. Mellentin, J.: The future of functional foods (pagg. 54,56,58,60,62), Nutraceutical World, November 2005

37. Di Silverio, F., D'Eramo, G., Lubrano, C., Flammia, G.P., Sciarra, A., Palma, E., Caponera, M., Sciarpa, F.: European Urology 21, 309–314 (1992)

38. Beveridge, T.H.J., Li, T.S.C., Drover, J.C.G.: Journal of Agricultural and Food Chemistry 50(4), 744–750 (2002)

39. Phillips, K.M., Ruggio, D.M., Toivo, J.I., Swank, M.A., Simpkins, A.H.: Journal of Food Composition and Analysis 15, 123–142 (2002)

40. Ham, B., Butleer, B., Thionville, P.: LC-GC Magazine 18(11), 1174–1181 (2000)

41. Garcia-Fuentes, E., Gil-Villarino, A., Zafra, M.F., Garcia-Peregrin, E.: Comparative Biochemistry and Physiology 133(B), 269–275 (2002)

42. Rao, R., Lokesh, B.R.: Molecular and Cellular Biochemistry **248**, 25–33 (2003)
43. Dong-Sun, L., Bong-Soo, N., Sun-Young, B., Kun, K.: Analytica Chimica Acta **358**, 163–175 (1998)
44. Gil-Villarino, A., Torres, M.I., Garcia-Peregrin, E., Zafra, M.F.: Comparative Biochemistry and Physiology **117C**(3), 243–250 (1997)
45. Clarke, R., Frost, C., Collins, R., Appleby, P., Peto, R.: British Medical Journal **314**, 112–117 (1997)

A Trial User, Resources and Services Quality Evaluation for Grid Communities Sustainability

Carlo Manuali[1,2(✉)] and Antonio Laganà[1,2]

[1] Consorzio Interuniversitario Nazionale per la Scienza
E Tecnologia Dei Materiali (INSTM), Firenze, Italy
{carlo,lag}@unipg.it
[2] Department of Chemistry, Biology and Biotechnology,
University of Perugia, Perugia, Italy

Abstract. We discuss here the possibility of utilizing the innovative features of GriF (a Grid Framework allowing the guided management, monitoring and results recollection of computational simulations under the form of advanced Grid Services) to the end of empowering the Chemistry, Molecular and Materials Sciences and Technologies communities with a set of quality evaluation functions. For this purpose, the enhanced options offered by GriF in terms of Quality Evaluation have been tested on the trial Virtual Research Environment of the COMPCHEM Virtual Organization for ranking resources, users and services. The results obtained are analysed with the purpose of addressing the Sustainability problems of Virtual Research Communities.

1 Introduction

The progress towards Virtual Research Environments (VRE)s is at present taking significant steps forward thanks to the financial support of EU to several distributed computing projects. As a matter of fact, in moving from distributed computing individual users to Virtual Organizations (VO)s[1], and from VOs to Virtual Research Communities (VRC)s[2], it has become increasingly more

---

[1] VOs are groups of researchers bearing similar scientific interests and requirements being able to work collaboratively with other members. VO members share resources (e.g. data, software, expertise, CPU and storage space) regardless of their geographical location and join a VO to the end of using the Grid computing resources provided by the resource provider. According to the VO's requirements and goals, EGI [1] provides support, services and tools allowing them to make the most of their resources [2].

[2] VRCs are self-organised research communities which give individuals within their community a clear mandate to represent the interests of their research field within the EGI ecosystem. They can include one or more VOs and act as the main communication channel between the researchers they represent and EGI [3].

© Springer International Publishing Switzerland 2015
O. Gervasi et al. (Eds.): ICCSA 2015, Part II, LNCS 9156, pp. 324–338, 2015.
DOI: 10.1007/978-3-319-21407-8_24

important to properly define and test the features of the associate VREs[3] as well as the related implementation strategy.

In the specific case of the Chemistry, Molecular and Materials Sciences and Technologies (CMMST) communities this is more difficult to achieve than in the case of well structured aggregates of academic research laboratories having common scientific goals and being grafted on a common distributed ICT platform (or a common experimental or operational frame). Nonetheless, as further commented below, CMMST communities are well suited to benefit from the highly collaborative environment of a VRE thanks to the complementarity of many aspects of the related researches and applications. Therefore, even if it is impossible to establish at present a unitary CMMST scientific organization, it is possible to make partners and stakeholders of the related communities operate within a single VRE in which users and resource providers are encouraged to compete through collaboration (collaborative competition) and to define a roadmap for its implementation (as it has been proposed in the project 676589 (internal reference number: SEP-210240103) submitted to the EINFRA-9-2015-1 call [5]).

More in detail the goals of the project are:

1. Orchestrate and support the activities of the CMMST communities in using the existing e-infrastructure resources and tools (including High Performance (HPC), High Throughput (HTC) and Cloud Computing (CC) technologies) in order to boost mutual interactions;

2. Apply and adapt e-Infrastructure instruments (like gateways, workflows and Quality of Service (QoS) functionalities) enabling an intra- and trans-community "a la carte" selection and composition of distributed hardware and software of the CMMST disciplines.

3. Exploit the existing e-Infrastructure instruments to the end of developing more advanced theoretical and computational chemistry methods, producing further technological innovation and enhancing related knowledge management procedures allowing the partner research laboratories to implement higher level of complexity collaborative multi-scale interdisciplinary applications interlacing different areas of molecular and materials science and technology expertise;

4. Structure Quality Evaluation (QE) mechanisms (and in particular the QoS and Quality of User (QoU) ones) to the end of enhancing the already mentioned collaborative competition among the CMMST users in building services developed by the community members, establishing quality based credit

---

[3] VRE are e-infrastructures used for e-science operations that maximise coordination and identify commonalities across the european research infrastructures as well as common solutions to problems, which can then be implemented by the involved communities thanks to synergies across member states, while minimising overlap of effort. VREs can automate aspects of data recording, including metadata, using customisable workflows [4].

mechanisms[4] for proactive users [7] and designing a systematic sustainable market oriented business model.

Instrumental to the achievement of the above mentioned goals is the effort reported in the present paper and concerned with a trial implementation of a VRE for the COMPCHEM VO based on the Grid Framework GriF [8,9]. Accordingly, the paper is articulated as follows: in section 2 enhanced options offered by the GriF to quantify synergies are described; in section 3 Quality Evaluation indicators are formulated and discussed; in section 4 a Resource Ranking indicator is formulated and discussed; in sections 5 and 6 quantitative measurements of the introduced indicators are reported and discussed. Some conclusions are outlined in section 7.

2   GriF: A Booster for Collaboration

As a first step towards the introduction of QE parameters for the collaboration implemented within a synergistic e-infrastructure, COMPCHEM membership has been articulated in four levels subdivided, when appropriate, in *Active* (AU) and *Passive* (PU) subclasses (see Table 1). This was aimed at classifying the level of participation of the members of a community to its activities [10] and at quantifying the support given by the Virtual Organization members to the other members. Accordingly, at the entry level 1 of Table 1, the COMPCHEM members can either use the programs made available on the e-infrastructure by other users (see level *1. Passive user* of Table 1) or can implement and run for free their own codes (see level *1. Active user* of Table 1). Level 1, however, is in a way peculiar because it implies a prevalent nature of pre-membership status and is meant to be limited in time. During that period it is evaluated if the membership is either worth to be continued and promoted to higher levels for free operations on the e-infrastructure or to be converted into a different non-free regime. Already at this level, in fact, the (minimum) amount of competences needed to run (and also update in the case of *1. Active users*) the codes allows an evaluation of the usefulness of the continuation of the membership. The COMPCHEM membership for which is important to quantify the quality is that starting from levels *2. Software provider* and *3. Hardware provider* of Table 1. Such memberships can be achieved only after making available to the other users either at least one of the codes implemented at level *1. Active user* or a subset of machines. In the case of *2. Software provider* a stable version of the code has to be validated and maintained together with all the necessary GUIs for its user-friendly usage by other researchers. For the purpose of the present work

---

[4] The accurate determination of QoU and QoS parameters out of the outcomes of extended statistical analyses allows the quantification of the Terms Of EXchange (TOEX) [6] between the activities performed and the credits awarded in return (as well as between the credits redeemed and the services or the financial resources provided in return). The awarding/redeeming fluxes of credits is intended to aid the creation of a community economy useful to enhance research and innovation.

we further identify within the AUs the Software Developers (SD) which create
new software tools and programs and within the Active *2. Software providers*
the Service providers (SP) as discussed in ref. [9]. In the case of *3. Hardware
provider* a cluster of machines has to be permanently grafted to the network
and maintained for use by other researchers. Both levels of provision need to
rely also on an adequate level of user support.

Table 1. Levels of Membership in COMPCHEM

Membership Level	Short Description
1. User	*Passive*: Runs programs implemented by other VO members.
	Active: Implements at least one program for personal usage.
2. Software Provider	*Passive*: Implements at least one program for use by other members.
	Active: Manages at least one implemented program for collaborative usage.
3. Infrastructure Provider	*Passive*: Confers to the infrastructure at least a small cluster of processors.
	Active: Contributes to deploy and manage the infrastructure.
4. Manager (Stakeholder)	Takes part to the development and the management of the VO.

As a matter of fact, at both *2. Software provider* and *3. Hardware provider*
levels, it is useful to quantify the effort spent by the members of the community
in providing services to the other members. Typically, the involvement of the
community members in using and developing Scientific Gateways to the end of
determining common requirements for high-level computing tasks of data processing (e.g. user-driven workflows and data-driven pipelines), for multi-scale and
multi-competence simulation workflows, for any other stand-alone computing
activities and related components, tools and interfaces built on top of the existing technological solutions (frameworks and gateways) will increasingly involve
collaborative competition (synergism) that can be spotted by appropriate tools.
In other words, such tools are the ground on which the already mentioned "synergistic" approach model inspiring the mentioned CMMST VRE project operates
and supports a transition to a common implementation of tools, programs and
libraries enabling a concerted use of both the existing users' data and applications. These tools are also the technology evolution driver that increasingly
enables the users to get proper access to the required resources of the underlying e-infrastructures, the implementation of multiplatform workflows for the use
cases, the facilitation of the interoperability among different workflow systems
and the evolution towards PaaS (Platform as a Service) and SaaS (Software as
a Service) solutions [11].

In this, as already mentioned, GriF and its functionalities play, indeed, a key role. For this reason they have been used by us both to test an embryo synergistic environment within the COMPCHEM VO as a state-of-the-art procedure supporting a fine grain optimization of the involvement of the users in producing services and of the use of the e-infrastructure resources. GriF [12] is structured, in fact, as a Service Oriented Architecture (SOA) [13] Grid Framework aimed at running both on the EGI Grid and various HPC platforms multi-purpose applications. The SOA-based organisation of GriF consists of two JAVA servers (YR (Yet a Registry) and YP (Yet a Provider)) and a JAVA client (YC (Yet a Consumer)). The first server YR is based on the standard UDDI (Universal Description, Definition, and Integration) protocol. Users inspect YR to the end of finding the appropriate YPs. The second server YP makes use of the Simple Object Access Protocol (SOAP) which is the XML-based messaging format established as transmission framework for inter-service communication via HTTP or HTTPS. YP holds the Grid Services of the VO. Both YR and YP make use of WSDL (Web Services Description Language) to describe the services provided and to reference self-describing interfaces which are available in platform-independent XML documents. The JAVA multi-platform client YC needs not the issuing of Grid Certificates (security is granted by the access procedure in which username and password have to be specified on YC). The selected YP takes also care of running the jobs on the associated User Interface (UI), of managing their status and of notifying the users upon completion, leveraging on a Robot Certificate strategy. YC is weakly coupled with respect to the Grid Middleware and implements all the extensions and the protocols mentioned above in order to correctly interface the Grid Services offered. It supports the management of large result files (even Gigabytes) empowering Quality-based Single, Parameter Study and Workflow job approaches.

Accordingly, GriF is extremely useful in keeping track of the usage of memory, of the engagement of the cpu, of the consumption of wall time and of the distribution of computational tasks over networked resources when assembling applications of higher complexity and exploiting the collaboration between the members of the community. GriF does this by capturing, out of the data produced by the Grid monitoring sensors, the information relevant to properly manage the computational resources and applications so as to articulate sequential, concurrent or alternative quality paths thanks to its robust SOA and Web Service organisation. This allows GriF to support collaboration among researchers bearing complementary expertise and wishing to articulate their computational applications as a set of multiple tasks. This means also that GriF can be of invaluable help in facilitating the optimal usage of memory, a reduced engagement of the cpu, a minimal consumption of wall time and an optimized distribution of distributed computational tasks when assembling complex applications and exploiting for example the collaboration between the experimentalists and the experts in various (complementary) theoretical areas. Moreover, GriF provides a user-friendly environment allowing the exploitation of the innovative features of networked computing with no need for the user to master related technicalities.

In addition, in order to simplify the management of related operations, GriF has been structured so as not to require the use of Grid Certificates (by adopting Robot Certificates [14]), the handling of binary programs and the explicit indication of the specific Grid resources to use. Further features of GriF are those allowing also the adoption of a natural-like language when carrying out some key operations. The advantage of this approach becomes particularly apparent when the related scientific problems have an intrinsic complexity that does not allow a straightforward concurrent distribution of the computational tasks as is the case, in fact, of most of the CMMST computational procedures. In most CMMST use cases, in fact, alternative articulations of the codes and runtime Framework-side interventions (like those related to the computing resources match-making for meeting the specific request of services) need to be performed.

3   Towards a Quantitative Estimate of Quality

The above mentioned features of GriF make it a suitable candidate for developing the mandatory high-level data-management functionalities required by QE. In this, the already mentioned complex nature of the calculations targeted by COMPCHEM, offered an ideal ground for testing the effectiveness of the GriF QE mechanisms and its effectiveness in supporting the synergistic model for:

1. Enabling users to ask for Grid Services by specifying as keywords high-level capabilities rather than low level ones as, for example, memory size, cpu/wall time and storage capacity;
2. Allowing managers to activate a (semi) automatic selection of the most appropriate low-level capabilities and the adoption of different running policies;
3. Offering resource selection functionalities rather than pure discovery ones [15];

It is important to emphasize here that the version of GriF implemented for our investigation bears already, over the pure Service Discovery (in which VO users searching for Grid Services are provided with an unranked list of matchings), the advantage of offering the possibility of carrying out Service Selection (in which VO users are provided with a list of matching resources ranked on a QoS basis). Based on this ground, GriF already provides the basic Service Orchestration features requested by the CMMST-VRE project which foster the establishing of collaborative/competitive operational modalities and enhance interaction between users and providers to the end of collaborating in carrying out higher level of complexity computational activities (such as those related to use of the Generalized Electrons Nuclei Intra-inter-molecular Universal Simulator (GENIUS) for simulations of innovative CMMST technologies). Yet, the richness of a proper quantification of QE parameters (as will be detailed later) and its experimentation through the use of appropriate sets of non-functional relationships [7] is the core subject of this paper. It is worth mentioning here, also, that we have considered as a Grid Service any set of collaborative Web Services of GriF running on the Grid by sharing a common distributed goal. We want to further comment that such development stimulates a significant extent of collaborative competition (that is at the core of the synergistic model) and that the possibility of quantifying QE parameters as indicators for the success of Grid Service providers and users is crucial for estimating the usability and the usefulness of a VRE. However, the implementation of appropriate QE procedures for Grid platforms is a real challenge because of the high heterogeneity of the competences and resources of the related communities. This is even more apparent when considering the variety of standards and protocols involved and adopted (like SOAP, REST [16], UDDI [17] and WSDL [18]) for the wide range of Web Services being currently deployed for various purposes by all major Web Service providers.

The QE parameters adopted by us for ranking the ability of Service Providers to meet the needs of Grid Services are inspired by the work of Mani and Nagarajan that quantifies QoS in terms of Accessibility (S_{acc}), Integrity (S_{int}), Reliability (S_{rel}), Security (S_{sec}) and Performance (S_{per}) of the Web Services [19]. Accordingly, the overall QoS is formulated in COMPCHEM VO as in ref. [7]:

$$QoS = w_0 S_{acc} + w_1 S_{int} + w_2 S_{rel} + w_3 S_{sec} + w_4 S_{per} \tag{1}$$

where $w_{(i=0..4)}$ are weights adopted by the Quality Board (QB) of the VO with the S parameters being in turn defined in terms of Time-To-Repair, number of Grid Service calls, number of errors occurred in the time interval considered, number of Grid jobs, absence of rollback, encryption, authentication and authorization support, Average Latency (AL), Throughput (TR) and time t as defined in ref [7].

Similarly, the QE parameters adopted by us for ranking the COMPCHEM users' QoU are the ratio between the number of compilations and the number of executions (U_{cx}), the fraction of custom running modality (U_{cu}), the amount of resources consumed per retrieved job results (U_{cm}), the grid efficiency (U_{ge}), the produced feedbacks (U_{fb}). Accordingly the overall QoU is formulated as:

$$QoU = z_0 U_{cx} + z_1 U_{cu} + z_2 U_{cm} + z_3 U_{ge} + z_4 U_{fb} \tag{2}$$

where $z_{(i=0..4)}$ are weights adopted by the QB of the VO with the U parameters being defined in terms of number of compilations, number of executions, number of custom runs, number of retrieved results, number of messages (feedbacks) generated by the user, average cpu time, and average memory consumed per retrieved job as defined in ref [7].

It is worth noticing here that the first three parameters (U_{cx}, U_{cu} and U_{cm}) are profile dependent while the second ones (U_{ge} and U_{fb}) can be taken as profile independent. In order to explicit that we have scaled the coefficients of the first three terms (z_0, z_1 and z_2) by multiplying them for the profile factors (p_0, p_1 and p_2, respectively) depending on whether the user is PU, AU, SP or SD.

4   Ranking of the Resources

In addition to QoS and QoU taken from ref. [7] other parameters are important for a complete definition of Communities Sustainability (ComS). Among them, the selection of the machines of the Grid platform with respect to their availability and reliability with respect to the different sections of the considered scientific calculations that usually goes under the name of Resource Ranking (RR). RR implementation is the real leap forward made through the investigation presented here. RR ability is based on the fact that for each subjob belonging to the same collection a different Computing Element (CE) queue q is used according to the so called 'Ranking' feature of GriF [20]. In particular, we define as 'Ranking' the ability of GriF of evaluating, by making use of in house designed adaptive algorithms, the Quality of each q (depending on various variables) accessible by the considered VO. When a user chooses to utilize the GriF Ranking feature, GriF automatically selects (querying the database) the q better suited to be chosen for each subjob.

Therefore, the global Ranking of a q is given by the following general formula:

$$R(q) = R_1 + R_2 \qquad ; \qquad R_1 > 0 \qquad (3)$$

where R_1 is a positive integer number representing the initial positional place of the related q resulting from the Grid itself. R_1 is obtained by fully relying on the already available Grid middleware facilities[5]. In particular, GriF runs a fake Grid job using some proper GLUE 2.0 Schema [21] attributes (in particular the Requirements and Rank ones) carrying out different attempts (creating at runtime a proper Job Description Language (JDL) file specifying each time different requirements and related order, from highest to lowest values) trying to estimate:

1. The qs (ordered from lowest to highest cpu time) having the (maximum − current) number of running jobs > 0, the number of free cpus > 0, the number of waiting jobs equal to 0 and ensuring a wall time of at least 2 days;
2. Or (if no q matches the previous criterion) the qs (ordered as in the previous case) having the number of free cpus > 0, the number of waiting jobs equal to 0 and ensuring a wall time of at least 2 days;
3. Or (if no q matches the previous criterion) the qs (ordered as in the previous case) having the number of waiting jobs equal to 0 and ensuring a wall time of at least 2 days;
4. Or (if no q matches the previous criterion) the qs (ordered from the lowest to the highest number of waiting jobs) having the number of free cpus > 0 and ensuring a wall time of at least 2 days;
5. Or (if no q matches the previous criterion) the qs (ordered as in the previous case) ensuring a wall time of at least 2 days;
6. Otherwise: the last queues ranked list is adopted.

---

[5] In this respect, a request (ref. EGI #3404) has been made to *EGI.eu* also about the possibility of making related information more reliable especially with respect to free cpus, concurrent jobs and memory available for each CE queue, that will be also useful for a wiser calculation of QoU.

This means that, in order to be considered by GriF, a q has to be first returned by the Grid itself (in fact, in Eq. 3, $R_1 > 0$ is the necessary and sufficient condition for a q to own a $R(q)$ value and then to be considered active). In other words, by using specific Grid middleware facilities one can select which qs can be considered. Then GriF determines the qs position in a quality-ordered rank list according to the following formula:

$$R_2 = K_P * \frac{n_f}{n_t} - \left(B[Q] + \frac{1}{K_{AL} * \frac{\sum_{i=1}^{n_d}(wt_i - ct_i)}{n_d}} \right) \tag{4}$$

where K_P is a constant weighing the ratio between the failed (n_f) and the total (n_t) number of jobs run by a q (that we call here the *Performance* (P) of a q) and K_{AL} is a constant weighing the AL time. Typically, $K_P = \frac{1}{K_{AL}}$ (in our implementation we set 100 and 0.01, respectively). AL is defined as the sum over the n_d "Done" jobs of the differences between wall time (wt_i) and cpu time (ct_i) of each i "Done" job divided by n_d. B is a boxed constant depending on AL that can assume the values of a vector Q(q1, q2, q3, q4) depending on the boundary values given in the vector L(l1, l2, l3) with both Q and L values (arbitrarily though reasonably) chosen by QB according to the resources management policy[6].

In particular, when running Single jobs the best q in the list will be invoked; when running Parameter Study or Workflow jobs different qs will be used according to a Round Robin strategy depending by the number of subjobs that have been automatically generated by the GriF engine; when re-sending past jobs fresh ranked qs will be used. Moreover, in the used extended version of GriF a new Ranking option that offers the possibility of running a Parameter Study job by making use of a single q, has been also implemented. A a matter of fact, the present version of GriF in this case chooses, among the queues ensuring at least 1-day long executions, the q having the maximum number of cpus available and minimizing the risk of failure. Accordingly, if the same q is able to run parallel applications exporting this information [21], it can be easily used by GriF for running local parallel programs (making use, for example, of MPI and OpenMP libraries) due to the fact that all the parallel communications between the subjobs will be of the intra-cluster type.

5   QoS and QoU of the COMPCHEM Community

In order to carry out a quantitative study of the above discussed parameters we have performed a one year (2014) long monitoring experiment of the performances of the COMPCHEM members activities for the three different instances

---

[6] In the running version of GriF, we set $Q = (10, 3, 0, -10)$ and $L = (5, 60, 1440)$. Accordingly, $B = 10$ when AL $\leq$ 5 minutes (*fast* CE queue class), $B = 3$ when $5 <$ AL $\leq$ 60 minutes (*normal* CE queue class), $B = 0$ when $1 <$ AL $\leq$ 24 hours (*slow* CE queue class) and $B = -10$ otherwise (*not available* CE queue class).

of the General Purpose (GP) Grid Service offered by GriF. GP is, in general, aimed at running various kinds of computational applications on the Grid. Typically its most useful outcomes are those for Single (first instance, for running a single program using a single input), Parameter Study (second instance, for concurrently running the same workflow program using different initial inputs) and complex Workflow-based (third instance, for concurrently running the same workflow starting by different initial inputs in a way that can be considered as a sequential execution of multiple programs in which the output of one run is the input of the next), and these are the ones we did monitor.

The resulting yearly averaged QoS parameters S_{acc}, S_{int}, S_{rel}, S_{sec} and S_{per} for the three mentioned different implementations of GP in production state are given in Table 2.

Table 2. Yearly averaged QoS parameters for the year 2014 of the GP Grid Service

	Single	*Parameter Study*	*Workflow*
S_{acc}	0.99	0.95	0.99
S_{int}	0.50	0.50	0.50
S_{rel}	0.99	0.97	0.99
S_{sec}	0.60	0.60	0.60
S_{per}	8.41	17.48	160.94
QoS	**11.50**	**20.50**	**164.03**

As expected, the total QoS value (bottom row) is much higher for the Workflow type GP case (one Grid Service call may correspond to several Grid jobs but also subjobs), given in the rightmost column, than for the other ones while those of both the Single and Parameter Study GP cases (with the latter being higher than the former) are not that different. It should be noticed here that, even if Parameter Study runs the jobs concurrently, the related QoS is not dramatically larger than that of a Single one, as it should be expected. This is a result of the ranking action of GriF that, as we shall discuss later, allocates to Single jobs more efficient Grid resources due to its preference to guarantee higher value resources to Single runs in terms of shortest possible AL and reliability.

The same measurements were performed also for the QoU of the COMPCHEM members. The values of the QoU parameters measured in the period ranging from June 1, 2013 to May 31, 2014 for some representative members are shown in Table 3. As apparent from the Table (where the results are shown at increasing values of total QoU (bottom row) in going from the left to the right hand side column) the highest ranking user is *Andrea* totalling a QoU evaluation of 774 for running customized versions of classical trajectory codes (*Andrea* is a moderate AU that has implemented on the Grid a classical trajectory code for a computationally intensive computing campaign for energy interconversion in gas phase molecular collisions). Other COMPCHEM members rank lower (see the other columns of the Table) due to the their lower demand of custom running. Two additional observations need to be made at this point. The first one is

Table 3. The yearly QoU value for a subset of COMPCHEM users

	Andrea	Domenico	Dimitris	Carlo	Student
U_{cx}	0.	0.	0.	0.	0.
U_{cu}	7747.	1000.	300.	89.	95.
U_{cm}	0.	0.	0.	0.	0.
U_{ge}	0.	0.	0.	0.	0.
U_{fb}	0.	0.	0.	2.	0.
QoU	**774.**	**100.**	**30.**	**9.**	**9.**

that *Carlo* (one of the two lowest custom running users) shows a non negligible value for the feedback provided U_{fb} which can be considered typical of a SD (who, obviously, does not issue significant requests of memory and/or cpu time although interacting with the system). The second one is that the student makes really little use of the Grid services made available by GriF.

Another note is in order for commenting the large number of zero values for the QoU parameters. Unfortunately, this is due to the fact that the tools for providing the value of some QoU basic information are not yet implemented on the Grid[7].

Table 4. Coefficients used in the QoU formulation for the various COMPCHEM user profiles

User Profiles	p_0	p_1	p_2
Passive User (PU)	10	0.1	0.1
Active User (AU)	1	0.1	0.1
Service Provider (SP)	0.1	1	1
Software Developer (SD)	0.1	10	10

For completeness one has to mention here also that some of the introduced parameters are still tentative and are not properly calibrated. Accordingly (see Table 4) the value of the p_i parameters to be used as scaling factors for the profile-dependent coefficients of equation 2 have been taken to be $p_0 = 10$ (indeed, low values of U_{cx} are more likely for PUs because they compile less than other users), $p_1 = 0.1$ and $p_2 = 0.1$ (indeed, high values of both U_{cu} and U_{cm} are more likely for PUs (and Active Users, AUs) because they run on the Grid more than the others).

---

[7] In particular, U_{cx} is 0 due to the fact the Grid does not allow source code compilations to be performed on its nodes directly, U_{cm} and U_{ge} are 0 due to the fact the Grid does not save information as the virtual memory consumed per job and the cpu time elapsed per job, respectively. In this respect, the already mentioned request (ref. EGI #3404) has also been made to *EGI.eu* about the possibility of providing and making available related information.

6   Costs, Credits and Productivity

By having tested the QoS and QoU parameters on the available monitoring
information of the Grid, we undertook the further step of testing the RR for
the used e-infrastructure compute resources again by relying on the information
available on the Grid. For our measurements we relied on the EGI computing
resources made available to COMPCHEM by the Italian, French, Iberian, Polish
and Greek NGIs (National Grid Infrastructure)s. The resulting networked platform consists of about 30000 cpu cores (having typically 2GB ram each) and of
about 3 PB of disk space in total. Moreover, depending on the job types provided by the users (which, as already mentioned, can be typically run in Single,
Parameter Study or Workflow modality) a different set of CEs was selected each
time by GriF according to the RR feature described above. The results of the
ranking procedure are shown on the screenshot of Fig. 1.

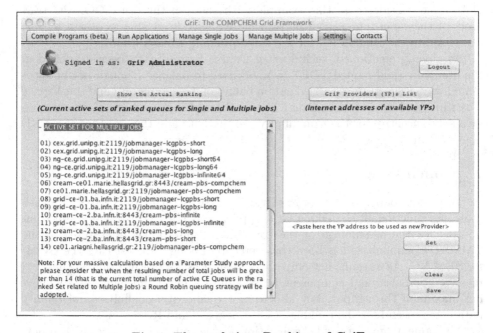

Fig. 1. The real-time Ranking of GriF

As apparent form the Figure for the test considered (Multiple Jobs) a set of
14 machine queues belonging to *unipg*, *hellasgrid*, and *infn* have been selected.
Accordingly, Single jobs will be always run by using the first available queue
(RR is recalculated at regular intervals) while Parameter Study and Workflow
jobs will use a number of queues depending on the number of real jobs to be
send on the Grid (when the number of jobs is larger than 14, that is the current number of active CE queues in the ranked set related to multiple jobs, a
Round Robin queuing strategy will be adopted). The possibility of ranking the

machines to use has the clear advantage of both allowing the running the application on the machines best matching the user needs and estimating of the best price/performance "virtual cost" of the use of the e-infrastructure as a whole. This allows to assign to the estimated total QoS the most appropriate estimate of *Virtual Costs*. At the same time in a *Virtual Balance* such *Virtual Costs* can be matched by the *Virtual Credits* that the user may have earned by offering services on the e-infrastructure.

It has to be also noticed here, however, that one will be able to associate a truly reliable cost, for the different implementations of the Grid Service GP considered in this example, only after comparing the resulting QoS value also with those of other Grid Services of the same type (although a limiting value can be given for it as suggested by the maximum value for S_{acc}, S_{int}, S_{rel} and S_{sec}) provided by different GriF sites (in this respect, S_{int} and S_{sec} return the same values for all the three instances due to the fact that they correspond to Grid site-dependent parameters). As a matter of fact, after comparing the obtained QoU value considered in this example with those of the other COMPCHEM users of the same type shown in Table 3, we have been able to assign a *High* class level to the *Andrea* PU for the corresponding *Credits* award.

As already mentioned, such *Virtual Credits* have already been described in the literature (and named TOEX [6]) and used to balance (though not in automated way) the activities carried out on behalf of the community and the *Virtual Costs* incurred by a user. Such "credit award & redemption system" promises to be a solid incentive for the Community productivity and the real core of a VRE "business model".

7   Conclusions

In the present paper the use of the Grid Framework GriF to enrich VREs with tools and functionalities making Grid applications truly user-friendly and composable for the assemblage of the complex computational procedures of the CMMST Communities has been considered by taking as a test case the COMPCHEM VO. As a result, not only it has been possible to carry out on the Grid massive computational campaigns by spending a minimum effort and achieving maximum throughput, but also to test quantitative definitions of quality parameters useful for the ranking of Grid Services and Resources.

The same parameters and classifications turned out to be of particular importance also for profiling the applications and taking the resulting ranking as a basis for rewarding the work carried out by the Community members on its behalf.

This turned out also to be useful for the formulation of the proposal submitted last January to EINFRA and aimed at establishing a CMMST VRE. In particular it was useful for designing the testing of new quality evaluators: namely the Quality of Computing consisting of an evaluation of computing-related objects of the Grid Middleware (and of different mixed e-infrastructures, e.g. the CC mentioned before, if any) belonging to a given Community and the Quality of Provider consisting of an average evaluation of related Hardware characteristics offered.

Acknowledgments. The authors acknowledge financial support from the project EGI Inspire for building the CMMST collaborative distributed network. Thanks are also due for funding to MIUR and MASTER-UP srl.

Computer time allocation has been obtained through the COMPCHEM VO of EGI. To this end, the VRE will leverage on both the ICT expertise of EGI (and associated NGIs) partners, INFN, SZTAKI, ENEA, University of Westminster, University of Perugia and the external support of other technology providers like EUDAT, ELIXIR and CINECA in order to exploit the innovative features of the presently available e-infrastructures and related innovative tools (which have already shown to be highly beneficial for research and innovation progress in other domains).

References

1. The European Grid Initiative (EGI), cited 04 May 2015. http://www.egi.eu/infrastructure
2. EGI Virtual Organisations, cited 04 May 2015. http://www.egi.eu/community/vos
3. EGI Virtual Research Communities, cited 04 May 2015. http://www.egi.eu/community/vos/vrcs
4. Voss, A., Procter, R.: Virtual Research Environments in Scholarly Work and Communications. Library Hi Tech **27**(2), 174–190 (2009). doi:10.1108/07378830910968146, 04 May 2015
5. Research and Innovation actions. Chemistry, Molecular & Materials Sciences and Technologies Virtual Research Environment (CMMST-VRE), cited 04 May 2015. http://www.hpc.unipg.it/ojs/index.php/virtlcomm/article/view/1
6. Laganà, A., Crocchianti, S., Faginas Lago, N., Riganelli, A., Manuali, C., Schanze, S.: From Computer Assisted to Grid Empowered Teaching and Learning Activities in Higher Chemistry Education in Innovative Methods in Teaching and Learning Chemistry in Higher Education. In: Eilks, I., Byers, B. (eds) RSC Publishing, pp. 153–190 (2009). ISBN 978-1-84755-958-6
7. Manuali, C., Laganà, A.: A grid credit system empowering virtual research communities sustainability. In: Murgante, B., Gervasi, O., Iglesias, A., Taniar, D., Apduhan, B.O. (eds.) ICCSA 2011, Part III. LNCS, vol. 6784, pp. 397–411. Springer, Heidelberg (2011)
8. Manuali, C., Laganà, A., Rampino, S.: GriF: A Grid Framework for a Web Service Approach to Reactive Scattering. Computer Physics Communications **181**, 1179–1185 (2010)
9. Manuali, C., Laganà, A.: GriF: A New Collaborative Framework for a Web Service Approach to Grid Empowered Calculations. Future Generation Computer Systems **27**(3), 315–318 (2011)
10. Laganá, A., Riganelli, A., Gervasi, O.: On the structuring of the computational chemistry virtual organization COMPCHEM. In: Gavrilova, M.L., Gervasi, O., Kumar, V., Tan, C.J.K., Taniar, D., Laganá, A., Mun, Y., Choo, H. (eds.) ICCSA 2006. LNCS, vol. 3980, pp. 665–674. Springer, Heidelberg (2006)
11. Turner, M., Budgen, D., Brereton, P.: Turning software into a service. Computer **36**(10), 38–44 (2003)
12. GriF: The Grid Framework, cited 04 May 2015. http://www.hpc.unipg.it/grif
13. The Web Services Architecture, W3C Working Group, (2004), cited 04 May 2015. http://www.w3.org/TR/ws-arch

14. INFN Certification Authority, (2010), cited 04 May 2015. http://security.fi.infn. it/CA/docs
15. Karta, K.: An Investigation on Personalized Collaborative Filtering for Web Service, Honours Programme of the School of Computer Science and Software Engineering. University of Western Australia (2005)
16. Rest Specification, cited 04 May 2015. http://www.restdoc.org/spec.html
17. Universal Description, Discovery and Integration (UDDI) 3.0.2 (2005), cited 04 May 2015. https://www.oasis-open.org/standards#uddiv3.0.2
18. Web Service Description Language (WSDL) 1.1 (2001), cited 04 May 2015. http:// www.w3.org/TR/wsdl
19. Mani, A., Nagarajan, A.: Understanding Quality of Service for Web Services. Improving the performance of your Web services (2002), cited 04 May 2015. http:// www.ibm.com/developerworks/webservices/library/ws-quality.html
20. Manuali, C.: A Grid Knowledge Management System aimed at Virtual Research Communities Sustainability based on Quality Evaluation, Ph.D. Thesis, Department of Mathematics and Informatics, University of Perugia (IT), 14 Feb 2011, cited 04 May 2015. http://www.unipg.it/carlo/PhD_Thesis.pdf
21. EGI Profile for the use of the GLUE 2.0 Information Schema, cited 04 May 2015. https://documents.egi.eu/document/1324

Exploiting Structural Properties During Carbon Nanotube Simulation

Michael Burger[1]([⊠]), Christian Bischof[2], Christian Schröppel[3],
and Jens Wackerfuß[3]

[1] Graduate School of Computational Engineering, TU Darmstadt,
Dolivostr. 15, 64293 Darmstadt, Germany
`burger@gsc.tu-darmstadt.de`
[2] Institute for Scientific Computing, TU Darmstadt, Mornewegstr. 30,
64293 Darmstadt, Germany
`christian.bischof@tu-darmstadt.de`
[3] Institute for Structural Analysis, University of Kassel, Mönchebergstr. 7,
34109 Kassel, Germany
`{schroeppel,wackerfuss}@uni-kassel.de`

Abstract. In this paper, we present a novel matrix-free algorithm for
the simulation of the mechanical behavior of carbon nanotubes (CNTs).
For small deformations, this algorithm is capable of exploiting the inherent symmetry within CNT structures. The symmetry information is
encoded with a graph algebra (GA) construction process and preserved
within a tuple based atom-indexing. The exploitation of symmetry leads
to a reduction of the needed calculations by a factor of more than 100
in the case of larger CNTs. Combining the usage of symmetry information with a new potential caching mechanism, our software is able to
store even large tubes in a compressed way with only a few megabytes
of data. Altogether, our implementation allows a matrix-free, resource-
aware simulation of CNTs. For larger cases it is only about the factor
1.45 - 1.6 slower than the reference solution with a fully assembled stiffness matrix, but consumes twelve times less memory. Also first results
of the parallelization of our new algorithm are presented.

Keywords: Carbon nanotubes · Simulation · Symmetry-exploiting ·
Parallelization · Matrix-free solver · Graph algebra

1 Introduction and Related Work

One can imagine carbon nanotubes (CNTs) as a rolled up sheet formed by
hexagonal carbon rings. Figure 1 shows a CNT that is rolled up around the red
x-axis. The center of the coordinate system lies in the middle of the CNT. The
synthesis of Y-shaped junctions allows the composition of CNTs to structures of
higher order called super carbon nanotubes (SCNTs), see [1]. These structures
are highly hierarchical and symmetric as long they are not deformed. Such a
SCNT is depicted in figure 2. For the simulation of small deformations based on

© Springer International Publishing Switzerland 2015
O. Gervasi et al. (Eds.): ICCSA 2015, Part II, LNCS 9156, pp. 339–354, 2015.
DOI: 10.1007/978-3-319-21407-8_25

the linear theory of elasticity, the symmetry of undeformed carbon nanotubes can be exploited by the methods presented in this paper. All principles that are described are demonstrated on CNTs, but are directly applicable to SCNTs.

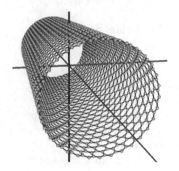

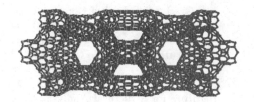

Fig. 1. Carbon nanotube   **Fig. 2.** Super carbon nanotube

CNTs are an active field of research because they possess a lot of properties which make them an interesting alternative to existing materials and techniques in different research fields. Their mechanical characteristics make them a promising material for high strength, lightweight material applications [2]. Yu et al. simulate in [2] the behavior of CNT ropes. They also present a list of literature describing other important properties of CNTs.

There already exist approaches that try to reduce the amount of data during calculations with matrices by exploiting some properties of these matrices like symmetry or repeating blocks/lines. For example, Kangwai et al [3] propose a strategy for coping with symmetric structures based on mathematical group representation theory. They present a technique for block-diagonalizing the stiffness matrix which takes the geometry into account. After defining coordinate systems, the symmetry operations can be represented by matrices and used for calculations. The goal is to split the whole problem into a number of smaller problems, and avoid the creation of the full stiffness matrix. So it is possible to reduce the complexity of the model (and the data to be kept available) as well as to reduce the computational effort that is required. This approach was applied by Koohestani [4] in a modified form to different examples, e.g. the buckling of structures. In contrast to those approaches that work on the matrix as primary abstraction level, our algorithm directly determines and exploits the symmetry information of the CNT structures within a graph algebra (GA) developed by Schröppel and Wackerfuß [5]. Their work was motivated by the need to simulate very large and high-order CNTs. In this case, the assembly of the stiffness matrix becomes infeasible and a matrix-free approach must be developed.

The paper is structured as follows: Section 2 introduces the algorithm that is underlying our calculation. Afterward, section 3 summarizes the graph algebra employed and the construction process of our CNTs. The following two sections, 4 and 5, provide theoretical considerations on the coding of symmetry information within the GA tuple-system and efficient ways of speeding up the calculations by a cache for intermediate results. Sections 6 and 7 describe in detail the implementation of our new algorithm and the performance results achieved. Section 8 summarizes results and outlines possible future developments.

2   Algorithm for Simulating Mechanical Behavior of Carbon Nanotubes

Our software simulates the mechanical behavior of undeformed CNTs, based on the linear theory of elasticity. During the simulation, the forces that are acting between the atoms are modeled by the so called Dreiding potential [6]. Only the atoms in the immediate neighborhood of each atom are taken into account by this potential. These are all atoms that are at most three edges away from the center atom of which the neighborhood is constructed. This neighborhood consists of at most 19 different atoms within a hexagonal grid, including the center atom. In the linearization around the initial configuration, the first and second derivatives of these Dreiding potentials are described by the residual vector r and the stiffness matrix $\mathbf{K}$. Together with the displacement-vector Δx, this leads to the equation system $\mathbf{K} * \Delta x - r = 0$ that must be solved. Boundary conditions are applied to selected atoms. These will modify the stiffness matrix $\mathbf{K}$ and the residual vector r. Details for the algorithm can be found in [7]. Previous work employed a direct solver, but to avoid the need for the construction of $\mathbf{K}$, a conjugate gradient method with the diagonal matrix as preconditioner (PCG) is employed.

During each iteration of the PCG algorithm, the matrix-vector multiplication (MVM) $\mathbf{K} * \Delta x$ has to be performed. The *reference solver* in this paper follows the traditional approach of explicitly assembling and saving the stiffness matrix $\mathbf{K}$ and doing the MVM of $\mathbf{K} * \Delta x$ within the compressed-row-storage (CRS) data-format. Because of the need to store $\mathbf{K}$, this approach is not feasible for very large tubes, but allows us to gauge the tradeoffs for our on-the-fly computation of the MVM.

In earlier work [8], walking the graph, and assembling the matrix-vector product on-the-fly, reduced memory consumption by a factor of 12, but the need to recompute potentials and their derivatives every iteration increased serial runtime by up to three orders of magnitude. This approach is called the *matrix-free solver* in the following.

In this paper, we present algorithmic improvements that drastically reduce the runtime overhead by exploiting inherent structural properties of the CNT structures, especially the symmetry, while preserving the savings in memory.

3   Carbon Nanotube Description with Graph Algebra

3.1   Graph Algebra Definition

To create a model of a CNT, we employ the graph algebra approach that is presented in [9] and discussed in detail in [5]. Here, we briefly summarize the basics.

Our CNTs are modeled as a graph $G = (V, E)$. This graph consists of a set of vertices V, representing the carbon atoms, and a set of edges E that models the chemical bonds between atoms. The edges are directed and thus an edge can be written as an ordered pair of two vertices (v_s, v_t) where v_s is the start and v_t is the terminal of the edge. We also define two functions on the edges. Function σ returns the start vertex of an edge, function τ the terminal vertex of the corresponding edge, respectively. Thus an edge e can also be denoted as $(\sigma(e), \tau(e))$. Additionally, σ and τ can be applied to a whole edge-set E as well. In that case, σ returns the set of all the vertices that form the start of at least one edge in G. This is analogous for τ and the end vertices.

The main difference to existing graph algebra concepts is the naming of the nodes in the graphs. Instead of letters or integers, Wackerfuß and Schröppel employ tuples of integers to identify the nodes. These tuples are ordered sets of integers of length n: $(x_n, x_{n-1}, ..., x_2, x_1)$. This length is called the *dimension* of a tuple and also the dimension of the node in a CNT represented by the tuple. Since we assume that all graphs only contain vertices with equal dimension, we define the dimension of the graph equal to the dimension of its vertices. We call x_n the leading entry of the tuple. Starting to count at the lowest entry, $t[m]$ denotes the value at position m of tuple t.

These tuples emerge during the construction process that is outlined in section 3.2. These tuples are a powerful abstraction because they encode and preserve information about geometry and symmetry of the vertices. We will exploit this information in order to reduce the computations when simulating the tubes.

In addition, we need several operations that are defined on these graphs to construct the CNT. The first one is the graph union, denoted by $\cup$. The union of two graphs $G_1 = (V_1, E_1)$ and $G_2 = (V_2, E_2)$ is defined as:

$$G_1 \cup G_2 = (V_1 \cup V_2, E_1 \cup E_2)$$

So every vertex and every edge that is contained either in G_1 or G_2 will be contained in the result graph. In the same way the intersection $\cap$ of graphs G_1 and G_2 is calculated as:

$$G_1 \cap G_2 = (V_1 \cap V_2, E_1 \cap E_2)$$

This operation keeps those vertices and edges in the result graph that are simultaneously member of G_1 and G_2.

The only operation that raises the tuple level is the *categorical product*, denoted as $\otimes$. Let us assume that we have two graphs G_1 and G_2, with tuple dimension n_1 and n_2, respectively. The categorical product of G_1 and G_2 creates a result graph G_r whose node-set V_r is the cross-product of the two input node-sets. So its dimension n_r is determined by $n_1 + n_2$. The edges of the result graph are constructed by the following rule: Let $e_1 \in E_1$ and $e_2 \in E_2$, then the resulting edge $e_r \in E_r$ is determined by:

$$e_r = ((\sigma(e_1) \times \sigma(e_2)), (\tau(e_1) \times \tau(e_2)))$$

In that case $\times$ does a concatenation of all possible pairs of the involved tuples $\sigma(e_1)$ and $\sigma(e_2)$ as well as for $\tau(e_1)$ and $\tau(e_2)$, respectively. A very simple example of a categorical product is illustrated in figure 3. We observe:

- G_1 and G_2 are both of dimension 1, so G_r must be of dimension 2.
- G_1 has one edge, G_2 has two edges, so G_r will have $1 * 2 = 2$ edges.
- Nodes of $V_1 \times V_2$ will have incident edges and thus will be connected in G_r, when they fulfill

$$v \in (\sigma(E_1) \times \sigma(E_2)) \cup (\tau(E_1) \times \tau(E_2)).$$

Only those incident nodes will be part of the resulting CNT. All others can be ignored and removed.

These facts can be generalized in order to deduce some characteristics of the result graph in advance. This knowledge is exploited during the implementation of the construction process and especially for the development of the data-structure.

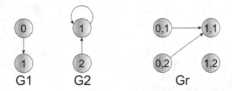

Fig. 3. Categorical product of G_1 and G_2 resulting in G_r

We also define two unary operators on the graphs. The first one is the opposite of a graph and denoted by a $*$ after the graph identifier, e.g. G_o^*. This operator creates the complementary graph. This means that all edges within a graph will be flipped in the opposite graph. The operator v transforms a directed graph into an undirected one and is defined by: $v(G) \simeq (G \cup G^*)$.

3.2   CNT Construction Process

Three base graphs form the fundamental building blocks of our construction process:

- Path Graph [P-graph]: A chain of vertices with length l is created, where each node is connected to its successor.
- Cyclic Graph [Cy-graph]: Create a P-graph of length l and connect additionally the endpoint of the chain to the starting point.
- Self-connected Graph [D-graph]: Create l vertices and connect each vertex only to itself.

The construction of a tube can be imagined as the formation of a flat graphene sheet that is then rolled up to a tube. It is important to note that our construction algorithm decouples the graph algebra operations from the geometry.

The creation of a tube consists of four different intermediate graphs numbered S_1 until S_4. S_1 only consists of two non-connected vertices 0 and 1. S_2 takes S_1 as its input. It is depicted in figure 4 and defined as:

$$S_2 = D_2 \otimes (S_1 \cup v(P_2)) \tag{1}$$

The right side of the categorical product connects the two vertices of S_1 with each other in both directions. This is done by taking a path of length 2 and creating the undirected version of it by using operator v. The D-graph on the left side of the categorical product creates two non-connected duplicates of the right side. If one would apply a D_3 instead of D_2 this would result in three non-connected duplicates.

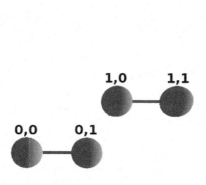

Fig. 4. Graph S_2

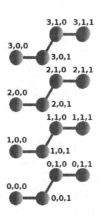

Fig. 5. Graph S_3

S_3, shown in figure 5, and defined as

$$S_3 = D_d \otimes (S_2 \cup v(P_2 \otimes P_2^*)) \ , \ d = 4 \tag{2}$$

has the same structure. On the right side, the duplicates that have been created in the last step get connected, while the categorical product with D_d builds d duplicates of the resulting right side. In our example, we set $d = 4$.

The graph algebra description for S_4 is shown in equation 3

$$\begin{aligned} S_4 = \ &(D_l \otimes S_3) \cup \\ &v((P_l * (D_d \cup Cy_d) \otimes P_2^* \otimes P_2^*) \cup \\ &(D_l \otimes Cy_d \otimes P_2^* \otimes P_2)) \ , \ d = 4 \ , \ l = 5 \end{aligned} \tag{3}$$

and the corresponding graph is depicted in figure 6. Equation 3 looks more complicated than those for S_2 and S_3, but it is similar to them. The first line creates l copies of the right side, with $l = 5$ in this case. The second and third line of equation 3 connect the copies of S_3 to hexagons and create the flat sheet. The original S_3 is contained on the left side of figure 6 with a leading zero at each tuple. S_2 is also visible at the bottom left of the figure. The tuples of these nodes now have two additional leading zeros. Really new in equation 3 is the appearance of cyclic graphs. Here additional edges are constructed that connect the top of the sheet with the bottom and vice versa. They are needed because after rolling up the tube the bottom vertices will be direct neighbors of the top ones. These lines can be seen in figure 6 and connect for example $(0, 3, 1, 0)$ and $(0, 0, 0, 1)$ as well as $(2, 3, 1, 1)$ and $(3, 0, 0, 0)$. Afterward, the CNT is cut to the appropriate size. This operation can be formulated in graph algebraic terms, but we omit the formula due to size constraints. The cutting is visualized in figure 6 by the red vertical lines. This cutting process ensures that there are only complete hexagons in the resulting structure. This means, for example, that node $(0, 0, 0, 0)$ will be removed, while $(0, 0, 0, 1)$ is kept.

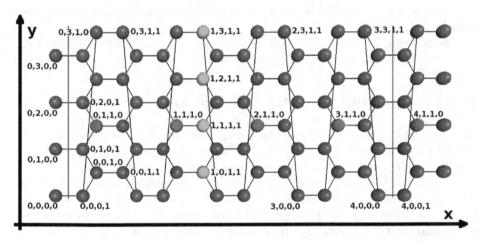

Fig. 6. Planar projection of the graph S_4 associated with the small $(4, 8)$ tube and described by equation 3. The red vertical lines indicate the cutting lines. One group of translational symmetric atoms is marked in magenta, one group of rotational symmetric atoms in cyan.

We see that d defines the diameter of the tube, while l sets the length. For technical reasons, tubes are characterized by two parameters (d_0, l_0), where $d = d_0$ and $l_0 = (l - 1) * 2$. For the tube constructed within the figures 4, 5 and 6 that means that we are dealing with a $(4, 8)$ tube there. This notation will be kept throughout the rest of the paper in order to identify CNTs.

3.3 Applying Geometry to the Graph Structure

The last step is the application of the geometry to the structure. To that end, the tuple t of each node is processed from the lowest to the leading tuple entry and a specified geometrical transformation is applied to the position of the vertex. For example, for the first three tuple entries in figure 6, these transformations apply:

1. Move along the x-axis: The length is defined by the multiplication of the value of the tuple at this position, denoted as $t[0]$, and a predefined base-length l_{base}. The result will be doubled, thus: $2 * t[0] * l_{base}$. For example $(0,0,0,1)$ results from (0,0,0,0) by moving $2 * 1 * l_{base}$ along the x-axis.
2. Do two moves: One along the x-axis with $x_{move} = 3*t[1]*l_{base}$ and one along the y-axis with $\sqrt{3} * t[1] * l_{base}$. This transformation is for example applied between the nodes $(0,0,0,0)$ and $(0,0,1,0)$ with $t[1] = 1$.
3. Move along y-axis with length $\sqrt{3}*2*t[2]*l_{base}$: Examples for these transformations are $(0,1,0,0)$ and $(0,2,0,0)$ which are derived from $(0,0,0,0)$ with $t[2] = 1$ and $t[2] = 2$, respectively.

The positions of the atoms in figures 4 and 5 also visualize these transformations.

The last step folds the sheet around the x-axis with the help of trigonometric functions. The result is shown in figures 7 and 8.

4 Deduction of Symmetry Information

From the tuple system we are able to deduce translational and rotational symmetry information between the nodes. This information can be directly exploited to minimize the amount of calculations during the solving process.

Translational symmetry can be identified by looking at the leading entry of the tuples. This is indicated in figure 6. The magenta atoms only differ in their leading tuple entry. So in principle the same forces are acting between these atoms and their neighbors. But figure 6 also shows a second prerequisite that is needed for symmetry. Atom $(0,1,1,0)$ does not have a complete neighborhood with 19 atoms. Its direct neighbors $(0,2,0,1)$ and $(0,1,0,1)$ only have two neighbors and not three. In contrast to this, $(1,1,1,0)$ and $(2,1,1,0)$ have full occupied neighborhoods. Thus the potentials differ between $(0,1,1,0)$ and the other two, but are identical for $(1,1,1,0)$ and $(2,1,1,0)$. $(3,1,1,0)$ has a non-complete neighborhood again and differs in the potential from the others. $(4,1,1,0)$ will be removed by the cutting process. So it is sufficient to calculate the potentials for $(0,1,1,0)$, $(3,1,1,0)$ and $(1,1,1,0)$, while the values for the latter one can be directly reused for atom $(2,1,1,0)$. The atom with the lowest leading tuple entry that is symmetric to other ones will be called the *symmetry base node* in the following. Nodes that are symmetric to a symmetry base node are called *symmetric nodes*. All other nodes are *non-symmetric nodes*.

We observe a critical point: The longer the tube, the more possibilities it contains to exploit translational symmetry. To estimate the impact of translational symmetry, we assess the number of symmetry base, symmetric and non-symmetric nodes, denoted by c_{base}, c_{sym}, and c_{nosym}, respectively. c_{atoms} denotes the number of atoms and can be calculated by $l_0 * d_0 * 2$:

- *Symmetry base nodes.* $c_{base} = 1 * d_0$, since d_0 determines how many symmetric copies of graph S_2 are created and S_2 consists of 4 nodes.
- *Symmetric nodes:* We know that $l = l_0/2 + 1$ copies of S_3 are created, which contain the symmetric copies of S_2 each. Through cutting and leaving some nodes that have no complete neighborhoods at the new border, we lose about three copies of S_3 for symmetric reuse. The first copy with complete neighborhoods are the symmetry base nodes. Thus $l - 4$ copies are symmetric nodes, and $c_{sym} = (l - 4) * c_{base}$.
- *Non-symmetric nodes:* $c_{nosym} = c_{atoms} - (c_{sym} + c_{base})$.

All symmetry base nodes have complete neighborhoods. Thus $19 * c_{base}$ potential calculations have to be done. The neighborhood of the non-symmetric nodes is occupied by 10 different atoms on average. So we can estimate the factor, by which computations can be reduced through exploitation of symmetry by:

$$\frac{19 * c_{atoms}}{19 * c_{base} + 10 * c_{nosym}}$$

with the numerator assessing the amount of calculations without taking symmetries into account and the denominator the amount of calculations with exploiting symmetry. Plugging in the two parameters d_0 and l_0 that determine the tube we arrive at $\frac{19}{86} * l_0$.

Figure 7 shows a $(16, 16)$ tube. Here, the yellow atom that lies nearest to the view-point is the symmetry base node for which the potentials are calculated. These values can be applied to all other yellow atoms. The same reasoning applies for the red atoms. For this $(16, 16)$ tube the actually needed potential calculations per MVM reduce by the factor of 3.4 from 4896 to 1451. The actual calculation demand for the larger $(256, 512)$ tube is reduced by a factor of 112. Table 1 summarizes the reduction of the calculation demand for different CNTs, conforming our estimates of in the order of $\frac{19}{86} * l_0$.

Table 1. Summary of the calculation reduction that is achievable by exploiting translational symmetry

CNT configuration	(64,64)	(128,128)	(128,256)	(256,256)	(256,512)	(512,512)
atoms	$8.2 * 10^3$	$3.3 * 10^4$	$6.6 * 10^4$	$1.3 * 10^5$	$2.6 * 10^5$	$5.2 * 10^5$
potentials (no symmetry)	$8.1 * 10^4$	$3.2 * 10^5$	$6.4 * 10^5$	$1.3 * 10^6$	$2.6 * 10^6$	$5.2 * 10^6$
potentials (symmetry)	$5.8 * 10^3$	$1.2 * 10^4$	$1.2 * 10^4$	$2.3 * 10^4$	$2.3 * 10^4$	$4.7 * 10^4$
factor (observed)	13	27	56	56	112	112
factor (estimated)	14	28	56	56	113	113

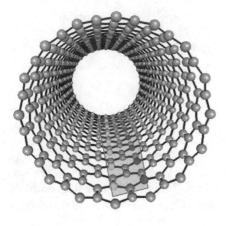

Fig. 7. Two symmetry base nodes and their translational-symmetric counterparts within a $(16, 16)$ tube

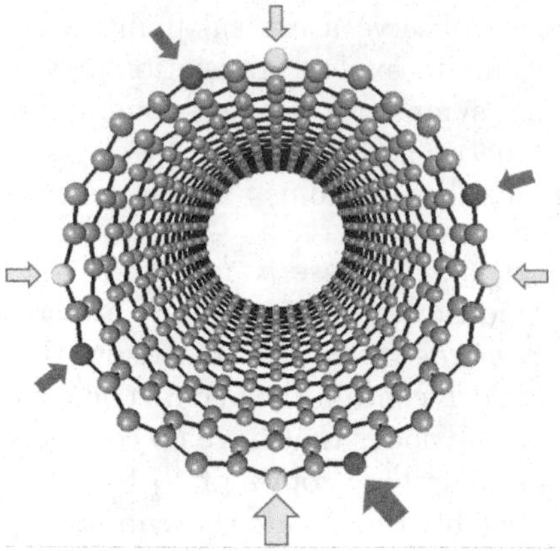

Fig. 8. Two symmetry base nodes and their rotational-symmetric counterparts within a $(16, 16)$ tube

Another type of symmetry is rotational symmetry. During the construction the tube is wrapped around the x-axis. Thus the atoms possess rotational symmetry in the direction of the y-axis. The corresponding tuple position for this type of symmetry is the third entry. In figure 6, the cyan colored atoms $(1, 0, 1, 1)$, $(1, 1, 1, 1)$, $(1, 2, 1, 1)$ and $(1, 3, 1, 1)$ are related by rotations around the x-axis. Choosing $(1, 0, 1, 1)$ as base-symmetry node means that $(1, 1, 1, 1)$ is rotational symmetric by $90°$, $(1, 2, 1, 1)$ by $180°$ and $(1, 3, 1, 1)$ by $270°$ in the rolled up tube. These rotations have to be applied during the potential calculations. In contrast to the translational symmetry, this implies additional computational work. For the three mentioned cases the application of the rotations results only in little additional effort. They are visualized for the $(16, 16)$ tube in figure 8. The yellow atoms in figure 8 are symmetric by rotation. The symmetry base node is marked by a larger yellow arrow (center, bottom). The same applies for the red nodes. To apply the transformation, some modifications to the matrices that represent the second derivative of the potentials, called $d2$, need to be applied. If we assume for the original potential $d2$ the matrix:

$$d2 = \begin{pmatrix} xx & xy & xz \\ yx & yy & yz \\ zx & zy & zz \end{pmatrix}$$

then the following matrices correspond to the symmetric nodes:

$$d2_{180°} = \begin{pmatrix} xx & -xy & -xz \\ -yx & yy & yz \\ -zx & zy & zz \end{pmatrix}$$

and

$$d2_{90°} = \begin{pmatrix} xx & -xz & xy \\ -zx & zz & -zy \\ yx & -yz & yy \end{pmatrix} \qquad d2_{270°} = \begin{pmatrix} xx & xz & -xy \\ zx & zz & -zy \\ -yx & -yz & yy \end{pmatrix}$$

That is, for these three rotations it is sufficient to swap and invert some entries of the original matrix. Since transformations based on other angles result in additional computational load and will introduce rounding errors, we limit ourselves to the three mentioned angles. Hence, in contrast to the translational symmetry the potential gain by exploiting rotational symmetry does not increase with the size of the tube.

5   Compressed Data Caching

Our *matrix-free solver* presented in [8] is able to reduce the memory consumption during the routine by the factor of 12 in comparison to the *reference solver* that assembles the matrix $\mathbf{K}$. As a result, some memory may be available for us to use. In order to design a resource-optimized application, we developed an additional caching routine: A predefined portion of RAM is used to store contributions to $\mathbf{K}$ temporarily, avoiding the need for recomputation in the next MVM. To maximize the efficiency of this procedure, values that belong to symmetric nodes are saved with higher priority. If the cache is exhausted, the following contributions are calculated anew during each subsequent iteration. Although the cache is actually storing immediate 3×3 contributions to the stiffness matrix, we call it the *potential cache* for the sake of brevity.

Together with the usage of symmetry, this becomes a very powerful instrument. Because of the fact that the translational symmetry is able to reduce the number of atoms for which properties need to be calculated, the information required to describe a tube is drastically decreased: For the $(256, 512)$ tube, the demand for saving the whole stiffness matrix $\mathbf{K}$ is over 200 MB plus the vectors r and Δx. In contrast, when exploiting symmetry, less than 2 MB need to be stored for the contributions to $\mathbf{K}$, plus again the vectors r and Δx as well as the diagonal of $\mathbf{K}$ as preconditioner. All vectors are of length $3 * n$, where n is the number of atoms.

6   Implementation

For the exploitation of symmetry, an efficient way to identify symmetric nodes is needed. Our newly developed data-structure is based on an implicit tree that is constructed on the base of the tuples, and the knowledge of their maximum extent. Extent means the maximum value that appears per tuple position. This tree allows a fast calculation of a serial index that is used to directly access the nodes within a one-dimensional vector. Thus, after calculating this index only one memory access is needed in order to fetch the node. This can be viewed as a custom tree-based hashing mechanism.

6.1   Translational Symmetry

The new MVM algorithm consists of a preparation part and then runs through all the nodes, distinguishing different node types. The preparation part is done once per solving process and distinguishes the nodes based on two different types of categories. The first categorization distinguishes between the nodes which have to be directly visited and whose potentials have to be calculated, and those which are indirectly processed by reusing symmetric potentials. The second categorization distinguishes the nodes which have symmetric counterparts within the model, and those which have not. Two Boolean arrays are instantiated that keep this information during the whole PCG algorithm. This avoids checking these properties for each iteration, but comes with a memory overhead of $\frac{2*n}{8}$ bytes. The achieved speedup and simplified algorithmic structure justifies this additional memory consumption.

For each MVM there are two different types of nodes. The first case handles all symmetry base nodes. For each of these symmetry base nodes, all nodes in their neighborhood are visited and their potential is calculated, to compute the contribution of the element to the result vector. Then, the corresponding symmetric nodes and their respective neighborhood are identified. For these, the values are reused to determine their contribution to the vector r. For each symmetric node a check of the boundary conditions (see section 2) is necessary. If the symmetric node is constrained by a boundary condition, its contribution must be modified before it is being used for the calculation. Hence, during the treatment of symmetric nodes, a copy of the 3×3 contributions must be saved in order to restore those 3×3 matrices that were modified by boundary conditions. In a second case, all remaining nodes are covered by the standard algorithm from the *matrix-free solver* presented in section 2 and the MVM $\mathbf{K} * \Delta x = r$ is calculated.

6.2   Rotational Symmetry

The exploitation of rotational symmetry is implemented within a separate routine and isolated from the translational symmetry at this time. Like in the translational symmetry case, a preparing step marks those atoms that have symmetrical counterparts and those that must be visited during the MVM. In contrast to the translational symmetry, this step involves more caution. This results from the fact that an atom can be symmetric to more than one other atom with different angles. For example, in figure 6 the node $(1, 2, 1, 1)$ has $(1, 0, 1, 1)$ as symmetry base atom for a rotation of $180°$ and at the same time $(1, 1, 1, 1)$ for a rotation of $90°$. So it must be defined in advance which types of symmetry should be taken into account and avoid visiting atoms multiple times. This information is again stored in Boolean arrays of size $\frac{2*n}{8}$ bytes. Then each atom that needs to be visited is processed and for those that are symmetry base nodes the symmetric counterparts are also managed.

6.3   Parallelization

Since the calculation of all 3×3 contributions is independent of each other, we parallelize the corresponding part of code with OpenMP. In order to have a reasonable amount of work per loop iteration per thread, we apply the line-based approach presented in [7]. Therefore, each thread calculates all contributions that belong to the center atom n_{center} it is assigned to, i.e. all three lines of the stiffness matrix $\mathbf{K}$ that belong to n_{center}.

7   Results

7.1   Test System

The measurements were taken on single, exclusively used nodes of the Lichtenberg Cluster at Technische Universität Darmstadt[1]. The nodes are dual socket machines equipped with two Intel Xeon E5-2670 CPUs with eight cores each. The board has 32 gigabyte of shared DDR3 ECC RAM. The Intel C++ compiler in version 15 SP2 with optimization level Ox (i.e. full optimization) is employed. The type of vectorization is set to host-dependent to support our vectorized calculation of the stiffness contributions (see [8]). The native OpenMP implementation provided by Intel is used.

7.2   Performance Results

We ran different performance tests, combining the symmetry-exploiting approach with the capability of caching and a variable size of cache memory. Since the impact of translational symmetry increases with the model size, we focus on this type of symmetry. We also measured the scaling behavior of our OpenMP parallelization.

Figure 9 summarizes the performance results for three different solvers running with one thread: First, the *reference solver* which assembles the stiffness matrix $\mathbf{K}$, our *matrix-free solver* that does not take symmetry into account, and our new symmetry-exploiting solver that also avoids the assembly of $\mathbf{K}$. At this point, no potential cache is employed.

Figure 9 reveals that the symmetric versions deliver a remarkable speedup. While the runtime increase between *reference* and *matrix-free solver* is always a factor over 100, the new symmetry-exploiting version is only slower by a factor about 2.5 for larger tubes like $(256, 512)$ or $(512, 512)$. The symmetry-exploiting solver speeds up the calculation by a factor up to 40 compared to the *matrix-free solver*. From table 1 one can see that theoretically a factor of more than 110 is possible. But it is clear that identifying and accessing the information for symmetric nodes causes additional overhead during the calculation. Measurements performed with the code analysis tool Intel VTune Amplifier 2015 confirm that

---

[1] http://www.hhlr.tu-darmstadt.de/

Performance results (Symmetry)

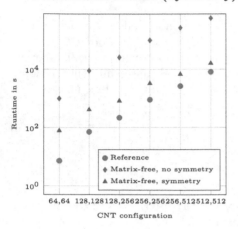

Performance results (Caching)

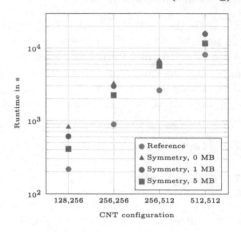

Fig. 9. Runtime comparison of reference, matrix-free and matrix-free symmetry-exploiting version running with 1 thread

Fig. 10. Runtime comparison of the reference solver and the symmetry-exploiting version with different amount of potential cache running with 1 thread

most of the overhead time is caused by the searching for nodes within the neighborhood of symmetric nodes in order to find the corresponding neighbor-atom and assign the symmetric potential. Tests with the rotational symmetry reveal a similar picture. For example, when using only the symmetry related to 90° rotations, it is possible to reduce the calculation count by the factor of 2, but the performance increase is a factor of 1.66.

The effect of our caching mechanism for larger configurations is shown in figure 10. The measurements validate our theoretical estimation from section 4 that it is possible to store all potentials needed for the calculation with the symmetry-exploiting solver within a few megabytes. For a user-defined potential cache size of 5 MB, even the information needed for a $(512, 512)$ tube with its $5.2 * 10^5$ atoms can be stored completely. The caching mechanism reduces the runtime further. So the execution time difference between the *reference solver* and the symmetry-exploiting version with caching is a factor of 1.45 to 1.6 now, while retaining the memory saving factor of 12 from the *matrix-free solver*. The speedup of our new symmetry-exploiting version compared to the matrix-free solver from [8] is shown in figure 11. The computation is always more than 50 times faster.

The scaling behavior of our parallelized algorithm is shown in figure 12. It summarizes the total runtime of the symmetry-exploiting version with different potential cache sizes and varying thread number. The scaling behavior for 2 threads is nearly ideal in all the cases. From 2 to 4 threads the runtime decreases by a factor of more than 1.5 which is still acceptable. But thereupon, the improvement factor decreases even further and from 8 to 16 threads, we do not obtain a performance improvement at all. This behavior is the result of the

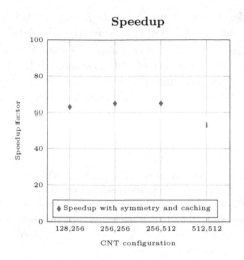

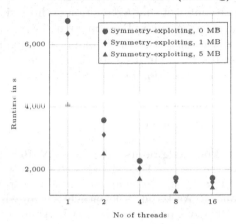

Fig. 11. Speedup of symmetry-exploiting version with 5 MB potential cache in comparison to the original matrix-free solver

Fig. 12. Runtime comparison of the matrix-free symmetry-exploiting version with varying cache size and thread number for the $(256, 512)$ tube

data management and potential cache accessing overhead, combined with the fact that the different cases during the treatment of all nodes cause different workloads for each loop iteration.

8   Conclusion and Outlook

We presented a new algorithm for the simulation of CNTs. The goal is to avoid the construction of the complete stiffness matrix. This was done in earlier work by walking the graph and assembling the MVM on-the-fly. It reduced memory consumption by a factor of 12, but the need to recompute the potentials every time increased the serial runtime by up to three orders of magnitude. To reduce this overhead, we now exploit the symmetry within the CNT structure by taking advantage of the tuple system that emerges during the graph algebra construction process. Thus, our new algorithm is able to reduce the amount of calculations by a factor of 112 for larger CNTs and the runtime by a factor of 40. Combining this fact with a caching mechanism for immediate results enables the storage of complete tubes with only a few megabytes. This in turn allows us to simulate CNTs in less than twice the runtime of the *reference solver*, in $\frac{1}{12}$ the memory. This allows simulating even very large tubes on commodity desktop computers. The performance of the *reference solver* which assembles the matrix just once cannot be reached at the moment because of the aforementioned management overhead for exploiting symmetry and because of the access time for the cached values. These access patterns are much more irregular than those within a compressed row storage (CRS) MVM of the *reference solver*.

We are planning to address these performance issues in future research. We are confident that a reduction of this serial overhead will also improve the parallel scaling behavior. We note that, in principle, we can reduce memory consumption even further by avoiding the explicit construction of the diagonal of $\mathbf{K}$, instead recomputing it when needed. Overall, we thus are well positioned for the computation of large SCNTs, where symmetry will play an even larger role. We are also confident that this approach is inherently well suited to upcoming hardware platforms, in which memory becomes more expensive, while FLOPs become cheaper.

Acknowledgments. The work of M. Burger is supported by the 'Excellence Initiative' of the German Federal and State Governments and the Graduate School of Computational Engineering at Technische Universität Darmstadt. The work of C. Schröppel and J. Wackerfuß is financially supported by the Deutsche Forschungsgemeinschaft (DFG) via the Emmy Noether Program under Grant No. Wa 2514/3-1. This support is gratefully acknowledged.

References

1. Coluci, V.R., Galvão, D.S., Jorio, A.: Geometric and electronic structure of carbon nanotube networks: 'super'-carbon nanotubes. Nanotechnology **17**(3), 617 (2006)
2. Yu, M.-F., Files, B.S., Arepalli, S., Ruoff, R.S.: Tensile loading of ropes of single wall carbon nanotubes and their mechanical properties. Phys. Rev. Lett. **84**, 5552–5555 (2000)
3. Kangwai, R.D., Guest, S.D., Pellegrino, S.: An introduction to the analysis of symmetric structures. Computers & Structures **71**(6), 671–688 (1999)
4. Koohestani, K.: Exploitation of symmetry in graphs with applications to finite and boundary elements analysis. International Journal for Numerical Methods in Engineering **90**(2), 152–176 (2012)
5. Schröppel, C., Wackerfuß, J.: Constructing meshes based on hierarchically symmetric graphs. Submitted for publication (2015)
6. Mayo, S.L., Olafson, B.D., Goddard, W.A.: DREIDING: A generic force field for molecular simulations. Journal of Physical Chemistry **94**, 8897–8909 (1990)
7. Wackerfuß, J.: Molecular mechanics in the context of the finite element method. International Journal for Numerical Methods in Engineering **77**(7), 969–997 (2009)
8. Burger, M., Bischof, C., Schröppel, C., Wackerfuß, J.: A unified and memory efficient framework for simulating mechanical behavior of carbon nanotubes. In: Proceedings of the International Conference on Computational Science (2015). In print
9. Schröppel, C., Wackerfuß, J.: Algebraic graph theory and its applications for mesh generation. PAMM **12**(1), 663–664 (2012)

Ion-Water Cluster Molecular Dynamics Using a Semiempirical Intermolecular Potential

Noelia Faginas-Lago[1](✉), Margarita Albertí[2],
Antonio Laganà[1], and Andrea Lombardi[1]

[1] Dipartimento di Chimica, Biologia e Biotecnologie,
Università di Perugia, Perugia, Italy
piovro@gmail.com
[2] IQTCUB, Departament de Química Física,
Universitat de Barcelona, Barcelona, Spain

Abstract. Classical Molecular Dynamics (MD) simulations have been performed to describe structural and dynamical properties of the water clusters forming around the Na^+ and K^+. The dynamics of K^+ and Na^+ was investigated for small water clusters $[K(H_2O)_n]^+$ and $[Na(H_2O)_n]^+$ (n = 3 - 8), isolated in gas phase following the structure transformation through isomerizations between the accessible energy minima. The extent to which a classical molecular simulation accurately predicts properties depends on the quality of the force field used to model the interactions in the fluid. This has been explored by exploiting the flexibility of the Improved Lennard-Jones (ILJ) function in describing the long-range interaction of ionic water systems.

Keywords: Molecular Dynamics · Empirical potential energy surface · Ion-water clusters · DL_POLY

1 Introduction

Aqueous solvation of metal ions plays a central role in the chemical mechanisms of biological signaling, in the bioavailability of toxic elements within the environment, in the design of new catalysts and in extraction processes, as well as in many phenomena relevant to technology [1–6]. For example, a problem of current interest focuses on the selectivity of biological ion channels; it seems clear that the selective transport of K^+ relative to Na^+ ions in potassium channels [7–9] depends on details of the ion hydration structure that might differ for K^+ relative to Na^+ ions.

A key feature of solvation is the structural organization of water about the ion, which includes the positions of the oxygen atoms and the influence of the ion upon the water's H-bonding network [10].

A fingerprint of the structure of these systems is the coordination number of the ion in liquid water, a number that in terms of the organization of water around the ion as spherical shells, determines how many water molecules are contained in each of them [11].

© Springer International Publishing Switzerland 2015
O. Gervasi et al. (Eds.): ICCSA 2015, Part II, LNCS 9156, pp. 355–370, 2015.
DOI: 10.1007/978-3-319-21407-8_26

The measurements of this property of ions in water is to date still controversial, due to uncertainties in related experimental measurements and in theoretical simulations. Experimental efforts to define the hydration structure of K^+, by means of neutron diffraction [12,13] and extended X-ray absorption fine structure spectroscopy (EXAFS) experiments [14], gave coordination number values varying between 4 and 8 [15]. Also, for Na^+ ion, neutron diffraction [16,17] and spectroscopic [18] methods produce a hydration number ranging between 4 and 6 [19]. To the end of understing such data one has to account accurately for ion-water and water-water interactions. For example, the complex interplay of ion-water and water-water interactions is the mechanism ensuring the functioning of sodium-potassium pump of living cells [20]. In this respect, fundamental is the understanding of the intermolecular interactions, and the modeling of competing non-covalent forces.

The main goal of this work goes, indeed, in that direction by investigating the geometry optimization and binding energy of clusters $[Na(H_2O)_n]^+$ and $[K(H_2O)_n]^+$ (where $n=$ 3 - 8) using classical Molecular Dynamics assisted by ab initio electronic structure calculations [21]. These simulations were analyzed with respect to structural parameters such as radial distribution functions and coordination number distributions.

The original potential model used here is based on a formulation of the non electrostatic approach to the intermolecular interaction that exploits the decomposition of the molecular polarisability [22] into effective components associated with atoms, bonds or groups of atoms of the involved molecules. This type of contribution to the intermolecular energy was already applied in the past to the investigation of several neutral [23–31] and ionic [32–34] systems, often involving weak interactions [23,35], difficult to calculate. The adequacy of such potential energy functions to describe several intermolecular systems was proved by comparing energy and geometry predictions at several configurations with ab initio calculations. In particular, the study of the alkali ion systems [36,37] was useful to quantify the role, that chemical contributions play in such aggregates.

The paper is organized as follows: in Sec. 2, we outline the construction of the potential energy function. We give in Sec. 3 the details of the Molecular Dynamics simulations. Results are presented in Sec. 4 and concluding remarks are given in Sec. 5.

2   Potential Energy Surface

To obtain a potential energy surface suitable for running MD simulations, the M^+-$(H_2O)_n$ ($M^+=K^+$, Na^+) intermolecular interaction energy, V, is decomposed in terms of ion-molecule and molecule-molecule pair contributions, as follows:

$$V = \sum_{k=1}^{n} V_{M^+-(H_2O)_k} \tag{1}$$
$$+ \sum_{k=1}^{n-1} \sum_{j>k}^{n} V_{(H_2O)_k-(H_2O)_j}$$

where n is the total number of water molecules and all intermolecular terms in Eq. 1 include both electrostatic and non electrostatic contributions. The non electrostatic contributions to the sum of Eq. 1 are evaluated by assigning a value of polarizability (denoted hereinafter as α) to the two interacting centers of each term, so as to account for both the strength of the induced dipoles (the attraction) and the average atomic and molecular sizes (exchange - repulsion). A detailed account of the use of polarizabilities for this purpose is given in Refs. [24,38] and references therein. As it has been mentioned above, in this case the H_2O molecule has a low value of polarizability and the quantities α_{M^+} (M^+ = K^+, Na^+) and α_{H_2O} have neither been decomposed as a sum of contributions nor displaced with respect to the ion and molecule positions, meaning that, besides K^+ and Na^+, also H_2O is considered in our model as a single interaction centre placed coincident with the O atom and bearing the total value of α_{H_2O}. This finds its rationale in the fact, that a model considering the presence of more interaction centres (i.e. bond and effective atom polarisabilities) leads to about the same structural and energetic properties of the ion-water interaction [34,39].

The model, however, could be improved by taking into account, if needed, the anisotropy in water-water interaction by explicitly considering the water molecule bonds as interacting centres. The pairwise interaction contributions between centres placed on different molecules, are described by means of the Improved Lennard Jones (ILJ) function V_{ILJ} [40,41],

$$V_{ILJ} = \varepsilon \left[\frac{m}{n(r) - m} \left(\frac{r_0}{r} \right)^{n(r)} - \frac{n(r)}{n(r) - m} \left(\frac{r_0}{r} \right)^{m} \right] \tag{2}$$

with

$$n(r) = \beta + 4.0 \left(\frac{r}{r_0} \right)^2 \tag{3}$$

used to describe other systems interacting with water [42].

The reliability of the ILJ function given above has been validated by reproducing the highly accurate scattering data obtained from experiments performed under high angular and energy resolution conditions [41]. Further reliability tests were performed by comparing calculated vibrational spacings with experimental values and calculated interaction energies at short-range with those obtained from the inversion of gaseous transport properties. The analysis, extended also to systems involving ions, suggests that the ILJ potential model can be used to estimate the behaviour of a variety of systems and can help to assess the different role of the leading interaction components [43].

In Eq. 2, r is the distance between the interaction centres and ε and r_0 represent the interaction well depth and equilibrium distance, respectively. When effective atoms are considered, ε and r_0 are directly obtained from basic physical properties of the interaction centres. In our case we have the K^+-H_2O, Na^+-H_2O and H_2O-H_2O pairs. The interaction centres are placed on the cation and on the oxygen atom of the water molecule and ε and r_0 are calculated from the polarisability and charge of the potassium and sodium ions and the average polarisability of water [44].

A key feature of the ILJ function (Eq. 2) is the additional β paramete parameterr, that corrects most of the inadequacies of the well known Buckingham (exp,6) and Lennard Jones (LJ) models [40], for which alternative forms have yet been proposed [45]. The proposed alternative models, in fact, although satisfactorily reproducing the mid-range features of the potential well, fail in describing accurately both the short-range repulsion and the long-range attraction. Moreover, the introduction of the parameter β, of the ILJ model by allowing the portable use of the same values of ε and r_0 for the same interactions centres in different environments, adds the necessary flexibility to the V_{ILJ} function in comparison with the V_{LJ} one [46]. A wise use of β (the only adjustable parameter in Eq. 2), permits to incorporate some additional effects, for instance charge transfer, in an effective way [47]. The parameter m is set equal to 4 or 6 to describe ion-neutral and neutral-neutral interactions, respectively. The parameters ε, r_0 and β adopted for K^+-H_2O, Na^+-H_2O and H_2O-H_2O interaction pairs are reported in Table 1.

Table 1. Values of well depth (ε), equilibrium distance (r_0) and parameter β defining the K^+-H_2O, Na^+-H_2O and H_2O-H_2O non electrostatic energy contributions

	ε / meV	r_0 / Å	β
K^+-(H_2O)	102.10	3.161	7.0
Na^+-(H_2O)	151.89	2.732	6.0
(H_2O)-(H_2O)	9.060	3.730	7.5

The electrostatic interaction contributions contained in each term of Eq. 1 are calculated by placing on any single water molecule a set of point charge whose distribution reproduces the H_2O dipole and quadrupole moments and by applying to them the Coulomb law.

It is important to mention here that the dipole moment of the H_2O monomer has been again considered only as an effective model parameter, related to the true dipole moment of H_2O in the ionic aggregate but not necessarily coincident with it. The parameterization of the H_2O-H_2O non electrostatic interaction contribution was performed using scaling laws exploiting the overall molecular polarizability. The electrostatic charge distribution for the dimer was derived from its dipole moment, equal to 2.1 D [48]. The potential energy function allows to obtain second virial coefficients in excellent agreement with experimental results [49]. By using the same parameters of the potential and only increasing slightly the dipole moment of the monomer, radial distribution functions were calculated for both rigid and flexible ensembles.

As a matter of fact, in our simulations the values 2.1 and 2.07 D (for Na^+ and K^+, respectively) have been used in order to best reproduce the information available from the literature [50]. To reproduce the dipole moment of the water molecules we placed charges of -0.74726 a.u. and -0.7366 a.u. on the O atom and of 0.37363 and 0.3683 a.u. on the H atoms for Na^+ and K^+ in solution, respectively. We have considered [51] the OH bond distances as equal to 1.00 Å and the HOH angle as equal to 109.47 degrees

The water molecules in simulations have been considered flexible using an harmonic potential function to describes the dependence of the internal molecular potential energy on the atom displacements from the equilibrium positions, for each mode of vibration. In this case, the intramolecular potential functions for the flexible water are given for the bond and angle interactions by a harmonic model potential.

3 Molecular Dynamics: Calculations

3.1 The Simulation Protocol

Classical molecular dynamics simulations were performed using the DL_POLY [52] molecular dynamics simulation package. We performed classical MD simulations of M^+-$(H_2O)_n$, $n=$ 3 - 8, using structures optimized by means of ab initio calculations, (DFT/B3LYP method) taken from literature [21] for both ions. For each optimized structure, we performed simulations with increasing temperature. A microcanonical ensemble (NVE) of particles, where the number of particles, N, volume, V, and total energy, E, are conserved, has been considered. The total energy, E, is expressed as a sum of potential and kinetic energies. The first one is decomposed in non electrostatic and electrostatic contributions and its mean value at the end of the trajectory is represented by the average configuration energy E_{cfg} ($E_{cfg} = E_{nel} + E_{el}$). The kinetic energy at each step, E_{kin}, allows to determine the instantaneous temperature, T, whose mean values can be calculated at the end of the simulation. The total time interval for each simulation trajectory was set equal to 2 ns after equilibration at each temperature. We implemented the Improved Lennard-Jones potential function and used itand used it to treat all the intermolecular interactions in M^+-$(H_2O)_n$, $n=$ 3 - 8. For each ion, statistical data were collected for 2 ns production run. The hydration system consists of flexible water molecules (bond and angle vibrations). This simulation time is sufficiently long to allow observation of isomerisation processes and fragmentation of both ions-water clusters. Most of the calculations have been performed at values of total energy that correspond to the 10 K-100 K temperature range. A specific value of T can be achieved by allowing a temperature equilibration of the system, with the corresponding results excluded in the statistical analysis performed at the end of the trajectory. The behavior of the aggregates has been investigated at low and high temperatures. At low temperatures, if the system is not far from the equilibrium configuration, the energy value obtained from an extrapolation at T = 0 K should be close to that at equilibrium and the corresponding configurations can be considered as equilibrium-like structures. By increasing T, other configurations far away from equilibrium can be reached and the system can surmount isomerisation barriers.

The dynamics simulations reveal interesting features in the isomerization of the ion-water cluster structures. We report here results on the thermal stability and dynamical behavior of the K^+ and Na^+ water clusters classical molecular dynamics.

3.2 The Potassium-Water Clusters K^+-$(H_2O)_n$ $(n=3$ - $8)$

The K^+-H_2O interaction was previously tested by performing MD simulations of ionic water solutions, for which the same ILJ parameters but different charge distribution (derived from the dipole moment of bulk water) was used [37].

From the structures optimized by ab initio calculation (see Fig. 1), (DFT/B3LYP method) we observed the isomerization of the $[K(H_2O)_3]^+$ cluster from a structure with symmetry (C_{2v}) to one with symmetry (D_3), displacing one of the water molecules from a longer distance of 4.5 Å (from the second hydration shell) a shorter one by placing K^+ at the bond distance of 2.75 Å (first hydration shell), after 1.27 ns, (see Fig. 2 reporting the evolution of K^+-O bond

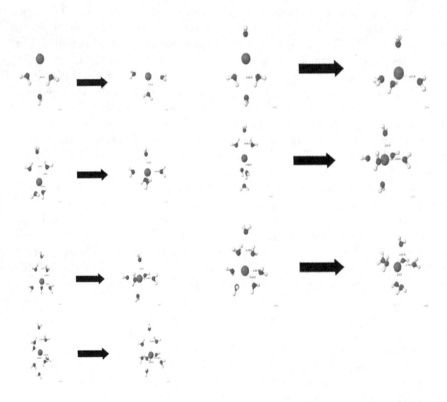

Fig. 1. $[K$-$(H_2O)]_n^+$ $(n=3$ - $8)$ initial and final isomerization structutes. (See Sec. 3.2)

K[(H₂0)ₙ]⁺

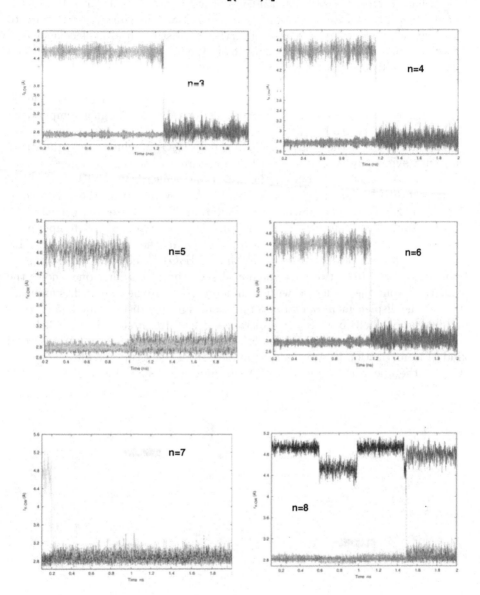

Fig. 2. Time evolution of the distance from K^+ to the O atom of H_2O for the $[K\text{-}(H_2O)]_n^+$ ($n=3$ - 8) clusters at different values of the temperature T (different molecules are represented by different colours).

distance during the simulation). MD simulations at 50 K of the $[K(H_2O)_n]^+$ cluster have been run using as initial configuration a C_2 one obtained from DFT calculations [21]. An isomerization from the initial C_2 symmetry structure to a S_4 symmetry structure has been observed, consisting in a displacement of a water molecule from a distance of 4.6 Å from the K^+ atom (typical of the second hydration shell) to a distance of 2.8 Å . (characteristic of the first hydration shell (see [11])). The isomerization occurs after 1.16 ns. The time evolution of the K^+-O distance that reveals the isomerization is shown in Fig. 2. Due to the lower configurational energy of the S_4 structure, the cluster has a higher temperature after isomerization (see Fig. 5).

Fig. 1, too, shows the isomerization of the $[K(H_2O)_5]^+$ cluster from the structure with symmetry (C_2) to one with a structure of the same symmetry. This follows a displacement of one of the water molecules from a distance of 4.6 Å from the K^+ atom to a distance of 2.8 Å, by reporting the time evolution of the K^+-O distance during the simulation of the cluster performed at 60 K. The isomerization occurs after 1 ns and is accompanied by a temperature increase to higher values due to the occupation of another minimum on the PES with lower configurational energy. For the $[K(H_2O)_6]^+$ cluster, two isomerizations were found from two different initial structures. The first one, shows the transition from the structure with symmetry (C_2) to the one with symmetry (C_4) by the displacement of two water molecules from a distance of 4.45 Å (second hydration shell) from K^+ to a distance of 2.85 Å (first hydration shell), as shown by the time evolution of the K^+-O bond distance in Fig. 2. The second kind of isomerization is given by a transition from the structure with symmetry (C_1) to the one with symmetry (C_4) through the displacement of one of the

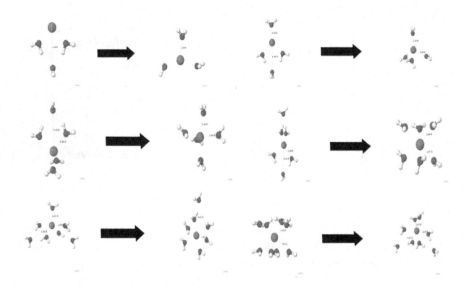

Fig. 3. $[Na-(H_2O)]_n^+$ ($n=3$ - 8) initial and final isomerization structures. (See Sec. 3.3)

Na[(H₂0)ₙ]⁺

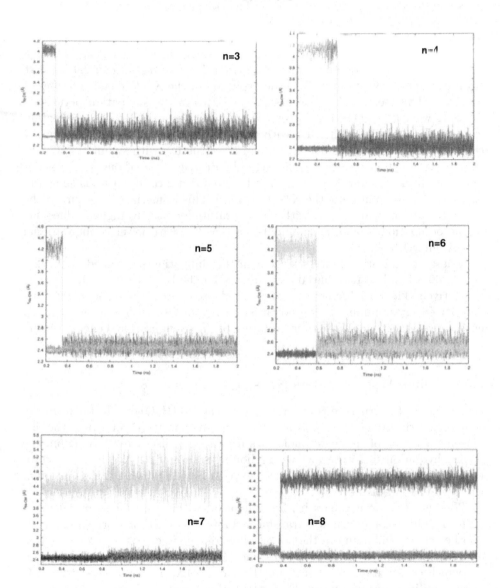

Fig. 4. Evolution of the distance from Na^+ to the O atom of H_2O for the $[Na\text{-}(H_2O)]_n^+$ (n=3 - 8) at different values of T (different molecules are represented by different colours).

water molecules from a longer distance of 4.8 Å (second hydration shell) from K^+ to a distance of 2.85 Å (first hydration shell). These isomerizations occur in the simulations performed at 25 and 80 K, respectively, and correspond to a temperature increase to higher values due to the occupation of minima on the PES with lower configurational energy. The transitions occur after 1.36 ns and 0.55 ns (see Figs. 1 and 2) respectively. The $[K(H_2O)_7]^+$ cluster undergoes isomerization from the structure with symmetry (C_2) to another structure with the same symmetry, following the displacement of one of the water molecules from a distance of 4.8 Å (second hydration shell) from K^+ to a distance of 2.9 Å (first hydration shell), as shown in the time evolution of the K^+-O distance reported in Fig. 2. This isomerization occurs after 0.02 ns in the simulation performed at 75 K with a temperature increase to higher values due to the occupation of another minimum on the PES with lower configurational energy. Finally, for the cluster $[K(H_2O)_8]^+$, a transition from the structure with symmetry (C_2) to the one with symmetry of (C_1) through the displacement of one of the water molecules from a distance of 5 Å from K^+ to a distance of 2.9 Å is shown in Fig. 2 the time evolution of the K^+-O distance. This isomerization occurs in the simulation performed at 15 K with a temperature increase to higher values due to the occupation of another minimum on the PES with lower configurational energy, after 1.5 ns.

These isomerization processes occur in the simulation performed at T= 25, 35, 50, 60, 75 and 80 K as initial temperature for the K^+-$(H_2O)_n$ with n =3 - 8, respectively. The initial structure transformed into a more stable one by reaching another energy minimum of the potential energy surface with a lower potential energy (see Fig. 2, where the time evolution of the configurational energy during the simulation is shown).

3.3   Sodium-Water Clusters Na^+-$(H_2O)_n$ $(n=3 - 8)$

Fig. 3 shows the structure transformation in the $[Na(H_2O)_3]^+$ cluster from the structure with symmetry (C_{2v}) to the one with symmetry (D_3) through the displacement of one of the water molecules from longer distance 4.1 Å (from the second hydration shell) the Na^+ with a bond distance of 2.4 Å (the first hydration shell) after 0.31 ns simulation (as shown in Fig. 4 following the evolution of the Na^+-O bond distance).

This isomerization occurs in the simulation with an initial temperature of 15 K. The initial structure was transformed into a more stable corresponding to another energy minimum in the potential energy surface with a lower potential energy while the total energy is constant (due to the use of the NVE ensemble which conserve the total energy), the same behaviour as discussed for K^+. When, adding one more molecule of water, the $[Na(H_2O)_4]^+$ cluster transform from a structure with symmetry (C_2) to the one with symmetry (S_4) through the displacement of one of the water molecules from longer distance 4.1 Å (second hydration shell) Na^+ with a bond distance of 2.4 Å (first hydration shell) after 0.61 ns (as shown in Fig. 4 by the time evolution of the Na^+-O bond distance).

This isomerization occurs in the simulation performed at 25 K with a temperature jump to higher values due to the occupation of another minimum on the PES with lower configurational energy at 0.61 ns. For the structure transformation in the $[Na(H_2O)_5]^+$ cluster from the structure with symmetry (C_2) to the other with symmetry (S_3) through the displacement of two water molecules from longer distance 4.1 Å (second hydration shell) Na^+ with bond distance 2.4 Å (first hydration shell). In this case the isomerization occurs at 40 K and 0.58 ns. Unlike to the K^+-water clusters in the case of the $[Na(H_2O)_6]^+$ one we found only one isomerization. The aggregate structure passes from a structure with symmetry (C_2) to the one with symmetry (S_3) including the displacement of one of the water molecules from the longer distance of 4.4 Å (second hydration shell) the Na^+ with bond distance 2.4 Å (first hydration shell) at 40 K well before the reach 1/2 ns of the simulation. The structure transformation in the $[Na(H_2O)_7]^+$ cluster from the structure with symmetry (C_2) to the other with symmetry (C_1) taking account of the displacement of one water molecules from longer distance 4.4 Å (second hydration shell) the Na^+ with a bond distance of 2.4 Å (first hydration shell) at 70 K and before 1 ns. Meanwhile, we observed the displacement of two of the water molecules from shorter distance 2.6 Å (first hydration shell) away from the Na^+ with a bond distance of 4.4 Å (second hydration shell) while the other six bonds move to slightly shorter distance contracting the radius of the first shell to 2.5 Å for the $[Na(H_2O)_8]^+$ cluster. That is indicating the tendency of the first shell to dissociate reducing to only six water molecules in agreement with the coordination numbers obtained from the classical MD calculation of Na^+ in liquid water. This isomerization occurs in the simulation performed at 35 K and very fast, at 0.38 ns of the simulation. All, the behaviours can be observed in the Figs. 3 and 4.

4   Conclusions

Our MD calculations have focused on the investigation on whether the Improved Lennard-Jones potential is suited to accurately simulate the K^+ and Na^+ water cluster dynamics and structures by matching the stable structures predicted by DFT calculations as a results of the evolution over PES through the isomerization. We have found, indeed, that, under the considered conditions, the $[Na(H_2O)_8]^+$ cluster tends to dissociate from a one shell structure to a two shell structure with coordination number 6 because of the smaller size of the Na^+ ion with respect to that of K^+. In addition, we have been able to estimate the temperature effect of the isomerization between these two different structures by progressively adding a water molecule to the initial $[K(H_2O)_n]^+$ and $[Na(H_2O)_n]^+$ clusters as nicely confirmed by Fig. 5 that shows the change of the isomerization temperature in $[K(H_2O)_n]^+$ and $[Na(H_2O)_n]^+$ clusters by increasing of the number of water molecules (n). As a matter of fact, the figure, shows that for K^+, the isomerization temperature increases until $n=6$ indicating more configurational energy released and converted to kinetic energy heating the system, and, after that, the temperature starts decreasing meaning less released

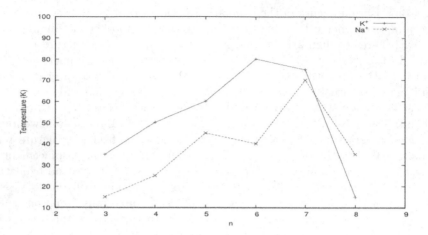

Fig. 5. Isomerization temperature as a function of the number of water molecules (n) in $[K(H_2O)_n]^+$ and $[Na(H_2O)_n]^+$ clusters

energy, and less energy gained. The higher energy jump occurs at $n=6$, around the calculated coordination number in the K^+ water liquid system, while in the case of the Na^+ the isomerization temperature increases until $n=5$ indicating more configurational energy released and moving to more stable structures. The temperature for Na^+ at $n=6$ is almost the same at $n=5$ indicating that the most stable structure is one of these two structures. After that, for higher n values we has seen the tendency of the $[Na(H_2O)_8]^+$ complex to shrink the first hydration shell to 6 water molecules and release the other two water molecules to the second hydration shell.

Acknowledgments. M. Albertí acknowledges financial support from the Ministerio de Educación y Ciencia (Spain, Projects CTQ2013-41307-P) and to the Comissionat per a Universitats i Recerca del DIUE (Generalitat de Catalunya, Project 201456R25). The Centre de Serveis Científics i Acadèmics de Catalunya CESCA and Fundació Catalana per a la Recerca are also acknowledged for the allocated supercomputing time. Noelia Faginas Lago acknowledges financial support from MIUR PRIN 2008 (contract 2008KJX4SN 003), Phys4entry FP72007-2013 (contract 242311). The computing for this project was performed at the OU Supercomputing Center for Education & Research (OSCER) at the University of Oklahoma (OU). A. Lombardi also acknowledges financial support to MIUR-PRIN 2010-2011 (contract 2010ERFKXL 002). Thanks are also due to INSTM, IGI and the COMPCHEM virtual organization for the allocation of computing time.

References

1. Dunand, F., Helm, L., Merbach, A.: Solvent exchange on metal ions. Advances in Inorganic Chemistry **54**, 1–69 (2003)
2. Helm, L., Merbach, A.E.: Applications of advanced experimental techniques: high pressure nmr and computer simulations. J. Chem. Soc., Dalton Trans. **5**, 633–641 (2002)

3. Helm, L., Merbach, A.E.: Inorganic and bioinorganic solvent exchange mechanisms. Chemical Reviews **105**, 1923–1960 (2005). PMID: 15941206
4. Helm, L., Nicolle, G.M., Merbach, A.E.: Water and proton exchanges processes on metal ions. Adv. Inorg. Chem. **57**, 327–379 (2005)
5. Lincoln, S.F., Merbach, A.E.: Substitution reactions of solvated metal ions. **42**, 1–88 (1995)
6. Marcus, Y.: Effect of ions on the structure of water: Structure making and breaking. Chemical Reviews **109**, 1346–1370 (2009). PMID: 19432440
7. Doyle, D.A., Cabral, J.M., Pfuetzner, R.A., Kuo, A., Gulbis, J.M., Cohen, S.T., Chait, B.T., MacKinnon, R.: The structure of the potassium channel: Molecular basis of k+ conduction and selectivity. Science **280**, 69–77 (1998)
8. Guidoni, L., Torre, V., Carloni, P.: Potassium and sodium binding to the outer mouth of the k+ channel. Biochemistry **38**, 8599–8604 (1999). PMID: 10393534
9. Laio, A., Torre, V.: Physical origin of selectivity in ionic channels of biological membranes. Biophysical Journal **76**, 129–148 (1999)
10. Mooney, B.L., Corrales, L.R., Clark, A.E.: Novel analysis of cation solvation using a graph theoretic approach. The Journal of Physical Chemistry B **116**, 4263–4275 (2012). PMID: 22417120
11. Faginas-Lago, N., Lombardi, A., Albertí, M., Grossi, G.: Accurate analytic inter-molecular potential for the simulation of na$^+$ and k$^+$ ion hydration in liquid water. Journal of Molecular Liquids **204**, 192–197 (2015)
12. Neilson, G.W., Mason, P.E., Ramos, S., Sullivan, D.: Neutron and x–ray scattering studies of hydration in aqueous solutions. Philosophical Transactions of the Royal Society of London A: Mathematical, Physical and Engineering Sciences **359**, 1575–1591 (2001)
13. Soper, A.K., Weckström, K.: Ion solvation and water structure in potassium halide aqueous solutions. Biophysical Chemistry **124**(3), 180–191 (2006). http://www.sciencedirect.com/science/article/pii/S0301462206001207
14. Glezakou, V.A., Chen, Y., Fulton, J., Schenter, G., Dang, L.: Electronic structure, statistical mechanical simulations, and exafs spectroscopy of aqueous potassium. Theoretical Chemistry Accounts **115**, 86–99 (2006)
15. Varma, S., Rempe, S.B.: Coordination numbers of alkali metal ions in aqueous solutions. Biophysical Chemistry **124**, 192–199 (2006). Ion Hydration Special Issue
16. Caminiti, R., Licheri, G., Paschina, G., Piccaluga, G., Pinna, G.: Interactions and structure in aqueous nano3 solutions. The Journal of Chemical Physics **72**, 4522–4528 (1980)
17. Skipper, N.T., Neilson, G.W.: X-ray and neutron diffraction studies on concentrated aqueous solutions of sodium nitrate and silver nitrate. J. Phys. Condens. Matter **1**, 4141–4154 (1989)
18. Michaelian, K.H., Moskovits, M.: Tetrahedral hydration of ions in solution. Nature **273**, 135–136 (1978)
19. Ohtaki, H., Radnai, T.: Structure and dynamics of hydrated ions. Chemical Reviews **93**, 1157–1204 (1993)
20. Köpfer, D.A., Song, C., Gruene, T., Sheldrick, G.M., Zachariae, U., de Groot, B.L.: Ion permeation in k+ channels occurs by direct coulomb knock-on. Science **346**, 352–355 (2014)
21. Shafei, R.A.M.: Theoretical Study of Ion Water Interaction Ab Initio and Classical Molecular Dynamics, September 2014
22. Pirani, F., Cappelletti, D., Liuti, G.: Range, strength and anisotropy of intermolecular forces in atom-molecule systems: an atom-bond pairwise additivity approach. Chem. Phys. Lett. **350**, 286–296 (2001)

23. Albertí, M., Castro, A., Laganà, A., Pirani, F., Porrini, M., Cappelletti, D.: Properties of an atom-bond additive representation of the interaction for benzene-argon clusters. Chemical Physics Letters **392**, 514–520 (2004)

24. Bartolomei, M., Pirani, F., Lagan, A., Lombardi, A.: A full dimensional grid empowered simulation of the $co_2 + co_2$ processes. Journal of Computational Chemistry **33**, 1806–1819 (2012)

25. Lombardi, A., Faginas-Lago, N., Pacifici, L., Costantini, A.: Modeling of energy transfer from vibrationally excited co_2 molecules: Cross sections and probabilities for kinetic modeling of atmospheres, flows, and plasmas. J. Phys. Chem. A **117**, 11430–11440 (2013)

26. Lago, N.F., Albertí, M., Laganà, A., Lombardi, A., Pacifici, L., Costantini, A.: The molecular stirrer catalytic effect in methane ice formation. In: Murgante, B., Misra, S., Rocha, A.M.A.C., Torre, C., Rocha, J.G., Falcão, M.I., Taniar, D., Apduhan, B.O., Gervasi, O. (eds.) ICCSA 2014, Part I. LNCS, vol. 8579, pp. 585–600. Springer, Heidelberg (2014)

27. Lombardi, A., Laganà, A., Pirani, F., Palazzetti, F., Lago, N.F.: Carbon oxides in gas flows and earth and planetary atmospheres: state-to-state simulations of energy transfer and dissociation reactions. In: Murgante, B., Misra, S., Carlini, M., Torre, C.M., Nguyen, H.-Q., Taniar, D., Apduhan, B.O., Gervasi, O. (eds.) ICCSA 2013, Part II. LNCS, vol. 7972, pp. 17–31. Springer, Heidelberg (2013)

28. Falcinelli, S., Rosi, M., Candori, P., Vecchiocattivi, F., Bartocci, A., Lombardi, A., Lago, N.F., Pirani, F.: Modeling the intermolecular interactions and characterization of the dynamics of collisional autoionization processes. In: Murgante, B., Misra, S., Carlini, M., Torre, C.M., Nguyen, H.-Q., Taniar, D., Apduhan, B.O., Gervasi, O. (eds.) ICCSA 2013, Part I. LNCS, vol. 7971, pp. 69–83. Springer, Heidelberg (2013)

29. Lombardi, A., Lago, N.F., Laganà, A., Pirani, F., Falcinelli, S.: A bond-bond portable approach to intermolecular interactions: simulations for N-methylacetamide and carbon dioxide dimers. In: Murgante, B., Gervasi, O., Misra, S., Nedjah, N., Rocha, A.M.A.C., Taniar, D., Apduhan, B.O. (eds.) ICCSA 2012, Part I. LNCS, vol. 7333, pp. 387–400. Springer, Heidelberg (2012)

30. Laganà, A., Lombardi, A., Pirani, F., Gamallo, P., Sayos, R., Armenise, I., Cacciatore, M., Esposito, F., Rutigliano, M.: Molecular physics of elementary processes relevant to hypersonics: atom-molecule, molecule-molecule and atom-surface processes. The Open Plasma Physics Journal **7**, 48 (2014)

31. Faginas-Lago, N., Albertí, M., Costantini, A., Laganà, A., Lombardi, A., Pacifici, L.: An innovative synergistic grid approach to the computational study of protein aggregation mechanisms. Journal of Molecular Modeling **20**, 2226 (2014)

32. Albert, M., Castro, A., Lagan, A., Moix, M., Pirani, F., Cappelletti, D., Liuti, G.: A molecular dynamics investigation of rare-gas solvated cation-benzene clusters using a new model potential. The Journal of Physical Chemistry A **109**, 2906–2911 (2005). PMID: 16833608

33. Albertí, M., Aguilar, A., Lucas, J., Pirani, F.: Static and dynamic properties of anionic intermolecular aggregates: the i-benzene-ar_n case. Theoretical Chemistry Accounts **123**, 21–27 (2009)

34. Faginas-Lago, N., Albertí, M., Laganà, A., Lombardi, A.: Water $(H_2O)_m$ or benzene $(C_6H_6)_n$ Aggregates to Solvate the K^+? In: Murgante, B., Misra, S., Carlini, M., Torre, C.M., Nguyen, H.-Q., Taniar, D., Apduhan, B.O., Gervasi, O. (eds.) ICCSA 2013, Part I. LNCS, vol. 7971, pp. 1–15. Springer, Heidelberg (2013)

35. Albertí, M.: Rare gas-benzene-rare gas interactions: Structural properties and dynamic behavior. The Journal of Physical Chemistry A **114**, 2266–2274 (2010). PMID: 20104928

36. Albertí, M., Aguilar, A., Lucas, J.M., Pirani, F., Cappelletti, D., Coletti, C., Re, N.: Atom-bond pairwise additive representation for cation-benzene potential energy surfaces: An ab initio validation study. The Journal of Physical Chemistry A **110**, 9002 9010 (2006) PMID: 16830404

37. Coletti, C., Re, N.: Theoretical study of alkali cation benzene complexes: Poten tial energy surfaces and binding energies with improved results for rubidium and cesium. J. Phys. Chem. A **110**, 6563 (2006)

38. Liuti, G., Pirani, F.: Regularities in van der waals forces: correlation between the potential parameters and polarizability. Chem. Phys. Lett. **122**, 245 (1985)

39. Albertí, M., Lago, N.F.: Competitive solvation of K^+ by C_6H_6 and H_2O in the K^+-(C_6H_6) n-(H_2O)m (n = 1–4; m = 1–6) aggregates. European Physical Journal D **67**(4), art. no. 73 (2013). http://www.scopus.com/inward/record.url?eid=2-s2.0-84879071541&partnerID=40&md5=d471977fe0c15ea412f5330514ec038c

40. Pirani, F., Albertí, M., Castro, A., Moix, M., Cappelletti, D.: Atom-bond pairwise additive representation for intermolecular potential energy surfaces. Chem. Phys. Lett. **394**, 37 (2004)

41. Pirani, P., Brizi, S., Roncaratti, L., Casavecchia, P., Cappelletti, D., Vecchiocattivi, F.: Beyond the lennard-jones model: a simple and accurate potential function probed by high resolution scattering data useful for molecular dynamics simulations. Phys. Chem. Chem. Phys. **10**, 5489 (2008)

42. Albertí, M., Faginas-Lago, N., Laganà, A., Pirani, F.: A portable intermolecular potential for molecular dynamics studies of nma-nma and nma-h2o aggregates. Phys. Chem. Chem. Phys. **13**, 8422–8432 (2011)

43. Albertí, M., Aguilar, A., Lucas, J.M., Pirani, F., Coletti, C., Re, N.: Atom-bond pairwise additive representation for halide-benzene potential energy surfaces: an ab initio validation study. J. Phys. Chem. A **113**, 14606 (2009)

44. Albertí, M., Aguilar, A., Cappelletti, D., Laganà, A., Pirani, F.: On the development of an effective model potential to describe ater interaction in neutral and ionic clusters. Int. J. Mass Spec. **280**, 50–56 (2009)

45. Halgren, T.A.: The representation of van der waals (vdw) interactions in molecular mechanics force fields: potential form, combination rules, and vdw parameters. J. Am. Chem. Soc. **114**, 7827 (1992)

46. Albertí, M., Aguilar, A., Lucas, J.M., Pirani, F.: A generalized formulation of ion-electron interactions: Role of the nonelectrostatic component and probe of the potential parameter transferability. J. Phys. Chem. A **114**, 11964–11970 (2010)

47. Albertí, M., Aguilar, A., Lucas, J.M., Pirani, F.: Competitive role of ch4-chxi44 and ch-π interactions in c_6h_6-$(ch_4)_n$ aggregates: The transition from dimer to cluster features. The Journal of Physical Chemistry A **116**, 5480–5490 (2012). PMID: 22591040

48. Gregory, J.K., Clary, D.C., Liu, K., Brown, M.G., Saykally, R.J.: **275**, 814 (1997)

49. Albertí, M., Aguilar, A., Bartolomei, M., Cappelletti, D., Laganà, A., Lucas, J., Pirani, F.: A study to improve the van der waals component of the interaction in water clusters. Phys. Script. **78**, 058108 (2008)

50. Ikeda, T., Boero, M., Terakura, K.: J. Chem. Phys. **126**, 034801 (2007)
51. Manion, J.A., Huie, R.E., Levin Jr., R.D., D.R.B., Orkin, V.L., Tsang, W., McGivern, W.S., Hudgens, J.W., Knyazev, V.D., Atkinson, D.B., Chai, E., Tereza, A.M., Lin, C.Y., Allison, T.C., Mallard, W.G., Westley, F., Herron, J.T., Hampson, R.F., Frizzell, D.H.: Nist chemical kinetics database, nist standard reference database 17 (2013)
52. Smith, W., Yong, C., Rodger, P.: Dl_poly: Application to molecular simulation. Molecular Simulation **28**, 385–471 (2002)

Exchange of Learning Objects Between a Learning Management System and a Federation of Science Distributed Repositories

Simonetta Pallottelli[1(✉)], Sergio Tasso[1(✉)], Marina Rui[2], Antonio Laganà[3], and Ioannis Kozaris[4]

[1] Department of Mathematics and Computer Science, University of Perugia, via Vanvitelli, 1, 06123 Perugia, Italy
{simonetta.pallottelli,sergio.tasso}@unipg.it
[2] Department of Chemistry and Industrial Chemistry, University of Genoa, via Dodecaneso, 31, 16146 Genova, Italy
marina@chimica.unige.it
[3] Department of Chemistry, University of Perugia, via Elce di Sotto, 8, 06123 Perugia, Italy
lagana05@gmail.com
[4] Department of Chemistry, Aristotle University of Thessaloniki, University Campus, 54012 Thessaloniki, Greece
ikozaris@chem.auth.gr

Abstract. The paper illustrates G-Lorep, a tool devoted both to dynamic sharing of learning materials among the members of a Virtual Education Community and to the implementing of a collaborative mechanism of increasing and improving learning objects. The purpose of integrating a federation of distributed repositories with an Learning Management System (LMS), allowing the sharing of information among different platforms, is to provide the capability of uploading/ downloading files on/from a common server where the LMS is installed and made available to a federated repository. As a use case the activities related to a set of chemistry learning objects generated on the LMS platform of a federated Content Management System (CMS) by the University of Genoa and used by other European Universities are considered.

Keywords: Repository · Learning objects · G-Lorep · E-studium · Aulaweb

1 Introduction

The new technologies, with advanced tools and interoperable environments, allow to overcome, both in training and information, the problem of the, any time and any place, presence to deliver and benefit from services of e-learning and repository.

In these environments, repository platforms and e-learning systems can share content of various nature. In particular, repository platforms being provided with a powerful and flexible structure, allow the complete management of Learning Object Metadata (LOM), while e-learning systems being targeted to the management of

© Springer International Publishing Switzerland 2015
O. Gervasi et al. (Eds.): ICCSA 2015, Part II, LNCS 9156, pp. 371–383, 2015.
DOI: 10.1007/978-3-319-21407-8_27

online courses, are able to handle a large variety of information and activities such as blogs, message boards, quizzes, usage statistics, etc.

Both environments can be interconnected and interoperable, not only to switch between them by means of a single sign-on, but also, and above all, to exchange LOM [1,2]. This is because traceability and reusability of contents are fundamental and essential features of the repository design.

The present scenario of e-learning content planning involves the application of a development model characterized by serializing the materials with the main objective of getting self-consistent learning units (typically learning objects) in SCORM format.

In prevailing practice, however, although compliant with reusability, accessibility, interoperability and durability characteristics, laid down by the standard, these materials show, often, critical issues that are connected to the production process.

The G-Lorep system [3], in order to facilitate the production stages of the learning objects (LO)s, proposes to adopt the solution of sharing and distributing the learning content for reuse, after appropriate cataloging and indexing [4,5].

2   The Currently Adopted Environment

The project G-Lorep (Grid Learning Object Repository) [6,7,8,9,10] centered on the Mathematics and Computer Science Department of the University of Perugia and on the Department of Chemistry, Biology and Biotechnology of the same University gathers together also other Chemistry related Departments of Italian and foreign Universities, as the ones of Genoa, Thessaloniki and Vienna. The project is aimed at implementing a federation of repositories, in which the information is processed by LOs and is supported by a cluster of SMEs (Small and Medium Enterprises) coordinated by ECTN.

The objective is to integrate various software tools so as to allow G-Lorep to become more powerful and efficient and become as well the reference product of the Virtual Research environment proposed for funding at the recent EINFRA-9-2015 Horizon 2020 call [11].

The federation of distributed and shared repository G-LOREP was designed (see Fig.1) with the purpose:

- of making available educational and scientific content to the teaching and learning community at large;
- of providing the possibility of dynamic improvement of the available materials by properly storing and cataloguing revised versions.

As shown in Fig. 1 the federation consists of repositories bearing equivalent dignity, responsibilities, duties and functions, which provide each other with complete information about available LOs. As for the federation structure, a hybrid solution was chosen, exploiting the advantages of both centralized and distributed architectures.

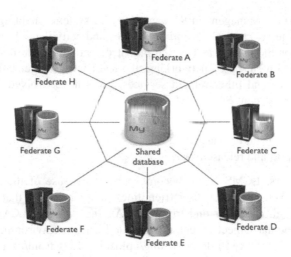

Fig. 1. The G_Lorep network architecture

The tools available for the management of content platforms and services are manifold and various. For our purposes, let's consider in detail the following two categories.

1) CMSs (Content Management System)s [12] are server-side systems based on a persistent database for managing content focused websites. CMS tools consist in general of:

- a back-end section, to organize and oversee the production of content;
- a front-end section, to enable content and services fruition.

CMS has become an essential tool for managing of website contents.

G-lorep makes use of the modular CMS Drupal [13,14] for collection, management and traceability of information on distributed and collaborative networks. Drupal is an open source platform that is regarded by the Web community as a truly powerful, flexible and customizable tool. It is a Free Software released under the GNU GPL and maintained and developed by a community of thousands of users and developers. It is modular and highly flexible. There are more than 300 modules to extend its functionalities. It is entirely developed in PHP and is multiplatform.

It should be emphasized that the modularity of Drupal, consists of a collection of modules performing only primary functions. However, Drupal developers provide APIs (Application Programming Interface)s to create additional modules for extending its functionalities.

Compared to other CMSs, Drupal is characterized by the suitability of its content management to insert a variety of content types with the particular property that their common features are nodes. All the content types inherit the characteristics of their node and have a set of common attributes, such as author, creation date and title.

2) LMSs (Learning Management System)s [15] are systems enabling the delivery of e-learning courses, providing online services and activities. They integrate several tools for the design and development of activities, the delivery of online courses and content-based networking, the feedback management and traceability of activities and other related services. LMS have evolved into Learning Content Management Systems (LCMS)s that bear the functionalities needed for a direct management of contents. Namely:

- creating and managing LOs;
- import and release LOs.

In recent years, LCMSs have become popular because of their reliability and security, as well as of their user friendliness and simplicity, that ensure benefits to both the administrator and the users. Among different LCMSs we adopted Moodle (Modular Object-Oriented Dynamic Learning Environment) [16,17] because of its flexibility in downloading/uploading LOs from/to a LMS to/from a CMS.

Moodle is a complete environment for modular, dynamic learning, it is object-oriented and fully open source. Moodle is a widespread e-learning environment that allows to organize and manage online courses, providing students with social tools like a forum, a blog and a chat, a wiki, a glossary and quizzes.

Since version 2.0, distributed towards the end of the year 2010, Moodle, through a substantial rewriting of the code, has introduced several new features, including:

- access to external repositories, such as Dropbox, Amazon S3, Google Drive;
- support to Web Services thanks to its ability to communicate with other web services using standard communication protocols.

As to resources and materials sharing, any digital content as Word, Powerpoint, Flash, PDF, Video etc. can be managed.

3   G-Lorep and the Perugia E-studium System

To the end of delivering e-learning services to several different study courses the University of Perugia has designed and implemented the "E-studium" system [18] on the Moodle platform.

E-studium provides students and teachers, in a user friendly manner, with a variety of online activities and services that, alongside the traditional support to lecturing, increases the possibility of delivering and profiting from organized educational materials, of keeping track of the activities whose outcomes are then capitalized, reviewed, revised and reused with no time or place limits.

"E-studium", whose structure is shown in Fig. 2, can be used in various ways, starting from simple board notice holder and ending with a highly sophisticated self-education [19,20] modality. As shown by the figure the teaching (left hand side panel) and learning (right hand side panel) are highly interleaved by the system functionalities. Lately, a configuration of this type has been implemented at the Genoa University on an educational portal based on Moodle 2 and named AulaWeb as will be illustrated in detail in Section 6.

Both Moodle (i.e. E-studium and AulaWeb) and Drupal can manage learning and scientific contents (including the sharing of LOM) through the integration of both the LMS and the CMS for transferring contents.

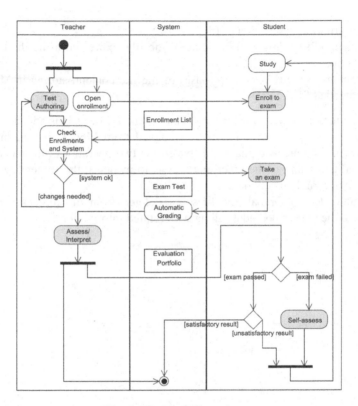

Fig. 2. The E-studium schema

4   Moodle and Drupal Communication

In G-Lorep the implementation of the above mentioned functionalities are delegated to the module Moodledata whose structure is given in Fig. 3. Moodledata includes two main functions:

- Moodledata Download: to download the files from Moodle to Drupal;
- Moodledata Upload: to upload on Drupal the files previously downloaded from Moodle.

On the Drupal side, settings are changed into the database of G-Lorep through the introduction of a new table (also named Moodledata). The provision of the table is crucial for the module as it brings the following information about the files downloaded from Moodle:

- id: unique numeric identifier of the file, automatically generated;
- course: name of the course (Moodle) in which the file was created;
- name: name of file;
- author: author of file;
- description: description of file;
- timemodified: last modified date of the file, expressed using the Unix timestamp;
- contenthash: SHA1 coding (hash) of the file content, storing in Moodle the uploaded files.

In the filesystem a folder, named moodledata_cache (see Fig.3) is added, in order to provide a temporary storage for files downloaded from Moodle in pending upload. This creates an intermediate phase in between the two operations (downloading from one part and uploading to the other one), so as to evidence the content being elaborated for the user and prevent data loss.

When uploaded in Drupal, the files are deleted from the directory moodledata_cache and the table moodledata along with related information.

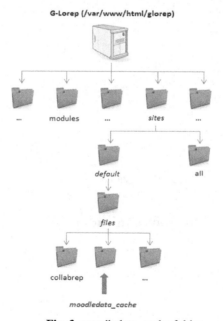

Fig. 3. moodledata_cache folder

Before searching and downloading files, the Moodle Database Connection and SSH Connection must be configured using the «Moodle Data Settings» page in the Drupal Administrator Menu.

The first part of the configuration of Moodle Data Settings deals with the database configuration and refers to the Moodle database used to find the files for download. On the contrary, the second part of Moodle Data Settings deals with the configuration of SSH and FTP connections needed to download the files.

It should be stressed, here, that the page Moodledata Downloads, if not properly configured in Moodle Data Settings, would fail to be loaded when trying to access it. The search for the files to download is accomplished using information retrieval techniques, i.e. searching for a file by typing a string that represents either the exact name of the file or part of its name. It is possible as well, to rely on an advanced search method that allows to specify additional parameters such as:

- the course name;
- the LO author;
- last modified date (in the format YYYY-MM-DD).

The files matching the entered search criteria are shown in the screenshot of fig. 4. It is important to specify that, to get all the files uploaded to Moodle, in the folder Moodledata, one should simply start the search without entering any other information.

The file thus obtained can be selected/deselected and then downloaded.

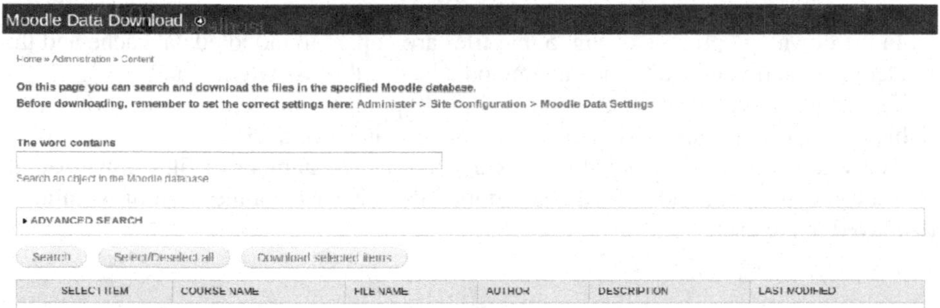

Fig. 4. Moodle Data Download: search

5   Moodle Resource Management

The database tables that are queried by the module Moodledata (specified by the prefix "mdl_") are [21,22]:

- mdl_resource: location of the files individually loaded in Moodle;
- mdl_folder: location of the stored folders in Moodle;
- mdl_files: location for all uploaded files in Moodle.

Additional tables are:

- mdl_modules: in which are stored all kinds of modules, including resource, folder, etc.;
- mdl_context: as "context", Moodle means context, environment, such as the entire system, a course, a particular activity. The type is given by the field contextlevel [23].

The context level that covers the modules (including resources, files and folders) is "70" and contains:

- mdl_course: in which all the information about the courses run by Moodle are stored;
- mdl_course_modules: in which the modules referring to a course are stored.

A sketch of the Moodledata pattern is given in Fig. 5. According to such pattern the selected files are downloaded and stored in the storage folder of Moodle created during installation (that has been called moodledata in the filedir subfolder).

The files are stored according to the coding SHA1 hash of their content. This means that any files with particular content is stored only once, regardless of the number of times that will be loaded with different names.

The SHA1 hash of each file can be recognized from the field in the table files contenthash of the Moodle database.

Assuming that a file has SHA1 hash
081371cb102fa559e81993fddc230c79205232ce, its ties will be stored in:
moodledata/filedir/08/13/081371cb102fa559e81993fddc230c79205232ce [16].

In the download process of Fig. 5, the files are copied to moodledata_cache and the related information added to the table moodledata of the G-LOREP database.

The download takes place by opening an FTP point to point channel, allowing a reliable and efficient transmission even for a large amount of data.

The whole operation ends with a message stating the number of files downloaded. When trying to download files already in moodledata_cache, a message of warning is displayed to the user.

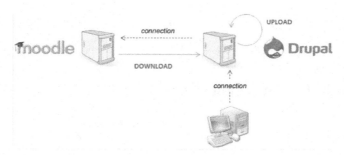

Fig. 5. Moodledata pattern

The second action of the moodledata pattern deals with the upload module in Drupal of files downloaded from Moodle. During the upload one can create a new node of the Linkable Object type with the selected files attached.

This requires the following information to be entered:

- title: title of the new node that is being created. The default value is "Moodledata Upload";
- body: description of the new node;
- keywords: (optional) useful keywords to describe the new Linkable Object.

To complete the process, there are menus allowing to enter information associated with Linkable Object compliant with the IEEE LOM. Also the list of files which can be attached to Linkable Object is shown. They are the files previously downloaded from Moodle for which information are taken from the database in moodledata (name, author, description, last modified).

The upload of selected files is started through the appropriate buttons "select / deselect". This creates a new node of type Linkable Object, to which, all the information entered along with attached files are assigned. These files are copied into the table collabrep (sites/default/files/collabrep) and subsequently deleted from moodledata_cache. At the same time related records are removed from the table moodledata of the GLOREP database and it is possible to display the new Learning Objects created in the Administration-> Content.

6   A Use Case: G-Lorep and the Genoa AulaWeb System

Here, it is reported a practical example of how to operate for transferring files from the Moodle environment (named AulaWeb [24] at the University of Genoa Educational System) to Glorep. As already stated, first, the connection between the Moodle database (DB) and the server hosting Glorep was established. Then local SSH connection settings as well as DB settings are filled in and saved.

At present, for security reasons, the system allows only the Glorep administrator to move the files. However, there are plans for extending this option to other users.

Once the selected Moodle DB and Glorep are connected, the Moodle Data Download page lists all the files contained in this specific Moodle database. To improve performances, the recommended structure of Moodle has a DB for each of its sites. Fig. 6 shows part of the list produced by the Molecular Spectroscopy course along with the description of all files as it is given on the Moodle site. By the side of Glorep it is possible to select the required file and to upload it by applying the Glorep taxonomy, as can be seen in fig. 7 where the information that the file "Symmetry and group theory" has correctly moved from Moodle to Glorep via the Collabrep module is shown. The file appears labeled as "new" on top of the Dashboard page reporting the recent content; the file is now moved to Glorep and can be managed according to its capabilities, as shown in fig. 8.

Fig. 9 shows the Glorep-unige (the one implemented at University of Genoa) displaying the file recently moved with the information provided under the Glorep taxonomy.

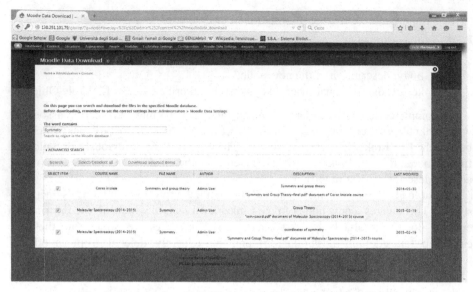

Fig. 6. An example of Moodledata download: a module of Glorep

Fig. 7. The example of Moodledata upload on Glorep

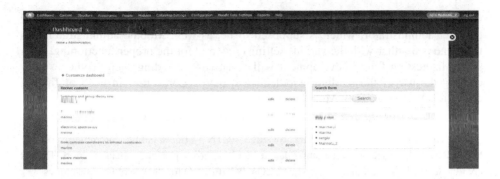

Fig. 8. Recent content

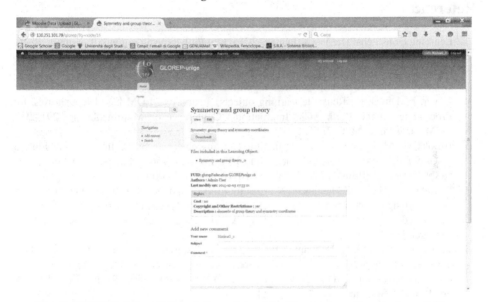

Fig. 9. GLOREP-unige with the information of the "Symmetry and group theory" file

7   Conclusion and Future Work

The module realized and integrated on Drupal within the project discussed in the present paper offers the possibility of downloading files resident in Moodle and moving them within the Federation of G-Lorep implemented for a Europe wide Federation of Science Distributed Repositories. The net result is an increased sharing of resources and dissemination of produced LOs.

Using the new moodledata module, and other Drupal modules implemented ad hoc for the project (Collabrep, Linkable Object, searchLO and tax_assistant) G-Lorep achieved its full functionality. To this end the paper discusses the main features of a use case taken care by the University of Genoa (Italy) and have been utilized and updated by other European Universities.

As for the module moodledata, currently each module functionality is reserved for the user administrator, which is supposed to be, as well, Moodle administrator or a user who is familiar with the various settings required for the proper functioning.

An interesting future development will consist in extending such privileges concerning the use of the module moodledata to any G-Lorep user in need of finding educational materials from Moodle.

Acknowledgements. The authors acknowledge ECTN (VEC standing committee) and the EC2E2N 2 LLP project for stimulating debates and providing partial financial support. Thanks are due also to EGI and IGI and the related COMPCHEM VO for the use of Grid resources.

References

1. Learning Object characteristics, January 2015. http://it.wikipedia.org/wiki/Learning_object
2. Learning Object Metadata, January 2015. http://it.wikipedia.org/wiki/Learning_Object_Metadata
3. G-Lorep, January 2015. http://glorep.unipg.it
4. Never, F., Duval, E.: Reusable learning objects: a survey of LOM-based repositories. In: Proceedings of the Tenth ACM International Conference on Multimedia, pp. 291–294. ACM, New York © (2002)
5. Brooks, C., McCalla, G.: Towards flexible learning object metadata. International Journal of Continuing Engineering Education and Life Long Learning, Inderscience Publishers, ISSN 1560-4624 (Print), 1741–5055, vol. 16, Number 1-2/2006
6. Pallottelli, S., Tasso, S., Pannacci, N., Costantini, A., Lago, N.F.: Distributed and collaborative learning objects repositories on grid networks. In: Taniar, D., Gervasi, O., Murgante, B., Pardede, E., Apduhan, B.O. (eds.) ICCSA 2010, Part IV. LNCS, vol. 6019, pp. 29–40. Springer, Heidelberg (2010)
7. Tasso, S., Pallottelli, S., Bastianini, R., Lagana, A.: Federation of distributed and collaborative repositories and its application on science learning objects. In: Murgante, B., Gervasi, O., Iglesias, A., Taniar, D., Apduhan, B.O. (eds.) ICCSA 2011, Part III. LNCS, vol. 6784, pp. 466–478. Springer, Heidelberg (2011)
8. Tasso, S., Pallottelli, S., Ferroni, M., Bastianini, R., Laganà, A.: Taxonomy management in a federation of distributed repositories: a chemistry use case. In: Murgante, B., Gervasi, O., Misra, S., Nedjah, N., Rocha, A.M.A., Taniar, D., Apduhan, B.O. (eds.) ICCSA 2012, Part I. LNCS, vol. 7333, pp. 358–370. Springer, Heidelberg (2012)
9. Tasso, S., Pallottelli, S., Ciavi, G., Bastianini, R., Laganà, A.: An efficient taxonomy assistant for a federation of science distributed repositories: a chemistry use case. In: Murgante, B., Misra, S., Carlini, M., Torre, C.M., Nguyen, H.-Q., Taniar, D., Apduhan, B.O., Gervasi, O. (eds.) ICCSA 2013, Part I. LNCS, vol. 7971, pp. 96–109. Springer, Heidelberg (2013)
10. Tasso, S., Pallottelli, S., Rui, M., Laganá, A.: Learning objects efficient handling in a federation of science distributed repositories. In: Murgante, B., Misra, S., Rocha, A.M.A., Torre, C., Rocha, J.G., Falcão, M.I., Taniar, D., Apduhan, B.O., Gervasi, O. (eds.) ICCSA 2014, Part I. LNCS, vol. 8579, pp. 615–626. Springer, Heidelberg (2014)
11. Laganà, A.: Horizon 2020 proposal for Research and Innovation actions Chemistry, Molecular & Materials Sciences and Technologies Virtual Research Environment (CMMST-VRE) Call EINFRA-9-2015

12. Content Management System, January 2015. http://it.wikipedia.org/wiki/Content_Management_System
13. Drupal, January 2015. http://drupal.org
14. System requirements, January 2015. https://drupal.org/requirements
15. LMS, January 2015. http://it.wikipedia.org/wiki/Learning_Management_System
16. Moodle, January 2015. http://docs.moodle.org/26/en/Features
17. Moodle 2.5, January 2015. https://docs.moodle.org/dev/Moodle_2.5_release_notes
18. E-STUDIUM Project, Department of Mathematics and Computer Science, University of Perugia, Italy, March 2015. https://estudium.unipg.it
19. Falcinelli, E., Gori, C., Jassó, J., Milani, A., Pallottelli, S.: E-studium: an Italian experience of blended e-learning, DEXA 2007. In: 18th Int. Conference on Database and Expert Systems Applications, pp.658–662, September 3–7, 2007
20. Jasso, J., Milani, A., Pallottelli, S.: Blended e-Learning: survey of on-line student assessment, DEXA 2008. In: 19th International Workshop on Database and Expert Systems Application, pp. 626–630, Sepember 1–5, 2008
21. File API migration - Database tables, January 2015. http://docs.moodle.org/dev/Resource_module_le_API_migration
22. Files table - File storage, January 2015. http://docs.moodle.org/dev/File_API_internals
23. Context, January 2015. http://docs.moodle.org/dev/Database_schema_introduction
24. The educational system of the University of Genoa based on Moodle. http://www.aulaweb.unige.it/

A Theoretical Investigation of 1-Butanol Unimolecular Decomposition

Leonardo Pacifici[1]([✉]), Noelia Faginas-Lago[1], Andrea Lombardi[1],
Nadia Balucani[1], Domenico Stranges[2], Stefano Falcinelli[3], and M. Rosi[3]

[1] Department of Chemistry, Biology and Biotechnologies,
University of Perugia, via Elce di Sotto, 8, 06123 Perugia, Italy
`xleopac@gmail.com`
[2] Department of Chemistry, University of Rome La Sapienza,
Piazzale Aldo Moro, 5, 00185 Roma, Italy
[3] Department of Civil and Environmental Engineering,
University of Perugia, via G. Duranti 93, 06125 Perugia, Italy

Abstract. Electronic structure calculations of the stationary points of
the potential energy surface associated to the unimolecular decomposition of 1-butanol have been performed with the aim to characterize the
pyrolysis mechanism under combustion conditions. The new results compare well with those of previous work concerning the C-C bond breaking
channels and the H_2 or H_2O elimination channels. The channels leading to H emission have been characterized for the first time. This study
will be of support to a new experimental characterization of 1-butanol
pyrolysis by means of the flash pyrolysis technique coupled to mass spectrometric detection.

1 Introduction

Among bio-alcohols, bio-butanol is considered a very promising biofuel candidate
because of its relatively high energy content, low water absorption, high misci-
bility with conventional fuels, and the possibility of being used in conventional
engines [1,2]. Because of that, numerous experimental and theoretical studies
on butanol combustion have been performed with the aim of characterizing the
global combustion properties, such as heat release and CO_2 emission (see, for
instance, [1–3] and references therein). Indeed, it has become clear from the
study of combustion systems that the fuel molecule structure strongly affects
the fuel performance and the production of pollutants in trace amounts [1,4].
In particular, in the case of most biofuels the production of specific oxygenated
compounds, including undesired combustion emissions (such as highly toxic alde-
hydes and ketones), cannot be understood without an in-depth characterization
of the elementary reaction sequence which accounts for the global transformation [1]. Among the various molecular processes, unimolecular high temperature
decomposition of butanol is important under the high temperature conditions of
combustion environments [2]. This has motivated numerous experimental and
theoretical studies on the pyrolysis of butanol (including all four structural

© Springer International Publishing Switzerland 2015
O. Gervasi et al. (Eds.): ICCSA 2015, Part II, LNCS 9156, pp. 384–393, 2015.
DOI: 10.1007/978-3-319-21407-8_28

isomers 1-butanol, 2-butanol, iso-butanol, and tert-butanol (see [2] and references therein)).

In a very recent study, 1-butanol pyrolysis was investigated in a flow reactor in a pressure range from 5 to 760 Torr and synchrotron VUV photoionization mass spectrometry was used to identify pyrolysis products [3]. In the same study, the rate constants of unimolecular reactions were calculated with the variable reaction coordinate-transition-state theory (VRC-TST) and the Rice-Ramsperger-Kassel-Marcus (RRKM) theory coupled with the master equation method [3]. In spite of the use of such a sophisticate experimental technique, it is still unclear whether some observed species, like butanal, are primary products of the unimolecular decomposition of 1-butanol. In our laboratory, we are now performing a new series of experiments based on the flash pyrolysis technique [5,6] and mass spectrometric detection with the aim to address the nature of the products formed by 1-butanol unimolecular decomposition. With this technique, the very limited residence time inside a SiC tube of 2.5 cm, which can be resistively heated at temperatures as high as 1500 K, will allow us to assess the yield of the primary pyrolysis products [7–9]. The interpretation of the experimental results requires the assistance of RRKM calculations performed under the conditions of our experiments, as previously done for other systems [10–14]. For this reason we have performed new electronic structure calculations at the B3LYP and CCSD(T) levels of theory of the stationary points necessary to describe the unimolecular decomposition of 1-butanol. In this contribution these new results are reported and compared with those previously obtained by Cai et al [3].

2   Computational Details

The potential energy surface of the species of interest was investigated by locating the lowest stationary points at the B3LYP [15,16] level of theory in conjunction with the correlation consistent valence polarized set aug-cc-pVTZ [17–19]. At the same level of theory we have computed the harmonic vibrational frequencies in order to check the nature of the stationary points, i.e. minimum, if all the frequencies are real, saddle point if there is one, and only one, imaginary frequency. The assignment of the saddle points was performed using intrinsic reaction coordinate (IRC) calculations [20,21]. The energy of the main stationary points was computed also at the higher level of calculation CCSD(T) [22–24] using the same basis set (aug-cc-pVTZ). It is worth pointing out that in the literature we can found that B3LYP and CCSD(T) levels of theory can be applied for larger systems (i.e gas storage over carbonaceous nanomaterials) in a good agreement with other computational techniques [25]. Both the B3LYP and the CCSD(T) energies were corrected to 298.15 K by adding the zero point energy (ZPE) and the thermal corrections computed using the scaled harmonic vibrational frequencies evaluated at B3LYP/aug-cc-pVTZ level. Corrections to other temperatures were performed by following the same procedure. All calculations were done using Gaussian 09 [26] while the analysis of the vibrational frequencies was performed using Molekel [27,28], an interactive, three-dimensional molecular graphics package allowing to animate and display vibrational modes.

Some of the calculations have been carried out on the machines of EGI (European Grid Infrastructure) [29] and on the 64-bit based Linux Cluster of the Herla (cgcw.herla.unipg.it) INSTM research unit of Perugia (IT) [30,31].

3   Results and Discussion

The optimized structure of the most stable isomer of 1-butanol is shown in Figure 1 and denoted as isomer (a). We have found two degenerate isomers, (a) and (a′), which differ only in the HOCC dihedral angle.

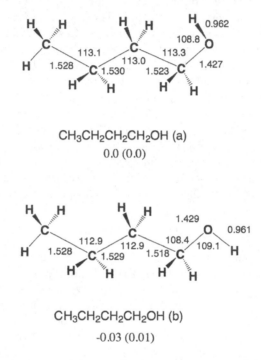

CH₃CH₂CH₂CH₂OH (a)

0.0 (0.0)

CH₃CH₂CH₂CH₂OH (b)

-0.03 (0.01)

Fig. 1. B3LYP optimized geometries (Å and °) and relative energies (kcal/mol) at 298.15 K of minima of 1-butanol; CCSD(T) relative energies are reported in parentheses

These two minima are connected by a transition state (TS) which is only 0.8 kcal/mol higher than the minima at 298.15 K. We computed the same value both at B3LYP and CCSD(T) level of calculation. A second minimum (b), characterized by a dihedral angle HOCC of 180.0 degrees, is almost degenerate with minimum (a). Minimum (b) is slightly more stable than minimum (a) at B3LYP level by 0.03 kcal/mol; at CCSD(T) level of calculation, however, minimum (a) is more stable than minimum (b) by 0.01 kcal/mol. The isomerization of (a) into (b) is very easy since the barrier is only 0.6 kcal/mol at 298.15 K, both at B3LYP and CCSD(T) levels of theory. Considering these energy values we can say that the rotation of the OH group with respect to the CO bond is almost free

Table 1. Enthalpy changes and barrier heights (BH) (kcal/mol, 298.15 K) computed at the B3LYP/aug-cc-pVTZ and CCSD(T)/aug-cc-pVTZ levels of theory for selected dissociation and isomerization processes for the system $CH_3CH_2CH_2CH_2OH$

Process	$\Delta^\circ_{298}H$ b3lyp	$\Delta^\circ_{298}H$ ccsd(t)	BH b3lyp	BH ccsd(t)
$CH_3CH_2CH_2CH_2OH$ (a) $\rightarrow$ $CH_3CH_2CH_2CH_2OH$ (b)	-0.03	0.01	0.6	0.6
$CH_3CH_2CH_2CH_2OH$ (a) $\rightarrow$ $CH_3CH_2CH_2CH_2OH$ (a')	0.0	0.0	0.8	0.8
$CH_3CH_2CH_2CH_2OH$ $\rightarrow$ $CH_2CH_2CH_3+CH_2OH$	82.0	88.0	90.0	100.0
$CH_3CH_2CH_2CH_2OH$ $\rightarrow$ $CH_3CH_2CH_3+CH_2O$	9.5	14.7	83.9	87.1
$CH_3CH_2CH_2CH_2OH$ $\rightarrow$ $CH_3CH_2CHCH_2+H_2O$	5.6	9.1	78.5	81.2
$CH_3CH_2CH_2CH_2OH$ $\rightarrow$ $CH_3CH_2CH_2CHO+H_2$	14.2	17.4	82.3	85.1
$CH_3CH_2CH_2CH_2OH$ $\rightarrow$ $CH_2CH_2C_2OH+CH_3$	81.3	88.2		
$CH_3CH_2CH_2CH_2OH$ $\rightarrow$ $CH_3CH_2CH_2+CH_2OH$	76.2	85.4		
$CH_3CH_2CH_2CH_2OH$ $\rightarrow$ $CH_3CH_2+CH_2CH_2OH$	79.0	88.4		
$CH_3CH_2CH_2CH_2OH$ $\rightarrow$ $CH_3CH_2CH_2CH_2+OH$	87.5	92.4		
$CH_3CH_2CH_2CH_2OH$ $\rightarrow$ $CH_2CH_2CH_2CH_2OH+H$	98.7	100.6		
$CH_3CH_2CH_2CH_2OH$ $\rightarrow$ $CH_3CHCH_2CH_2OH+H$	94.8	98.0		
$CH_3CH_2CH_2CH_2OH$ $\rightarrow$ $CH_3CH_2CHCH_2OH+H$	95.9	99.3		
$CH_3CH_2CH_2CH_2OH$ $\rightarrow$ $CH_3CH_2CH_2CHOH+H$	91.2	94.4		
$CH_3CH_2CH_2CH_2OH$ $\rightarrow$ $CH_3CH_2CH_2CH_2O+H$	99.2	103.3		

and, therefore, we can refer to only one minimum for 1-butanol. The enthalpy changes and barrier heights at 298.15 K for the main dissociation processes of 1-butanol are reported in Table 1.

Some of these processes imply the presence of a transition state and the related geometries of the saddle points are shown in Figure 2, while in Figure 3 we show the geometries of the fragments produced in the hydrogen atom loss processes. The influence of the temperature on the values reported in Table 1 is not very pronounced.

In fact, the barrier height for the dissociation of 1-butanol into $CH_3CH_2CH_2CH_2$ and water is 80.8 kcal/mol at 0 K, 81.2 kcal/mol at 298.15 K and 81.7 kcal/mol at 1000 K, at the CCSD(T) level of calculation. For these reasons, we can refer to 298.15 K as in Figure 4 and Figure 5, where we have reported a schematic representation of the main dissociation channels of 1-butanol. For sake of clarity, only the CCSD(T) energies are shown in the figures. In particular, in Figure 4, we have reported the dissociation processes involving a transition state, while in Figure 5 we have reported the dissociation processes which do not show a transition state, since the geometrical rearrangement before the bond breaking is not pronounced.

The comparison of our results with those of Cai et al. [3] points out significant differences. The first one is concerned with the theoretical approaches adopted: whereas we used a B3LYP/aug-cc-pVTZ followed by a CCSD(T) level of theory they made use of a HF/6-311G(d,p) for the geometry optimization and evaluation of frequencies followed by a QCISD/6-311G(d,p) calculation.

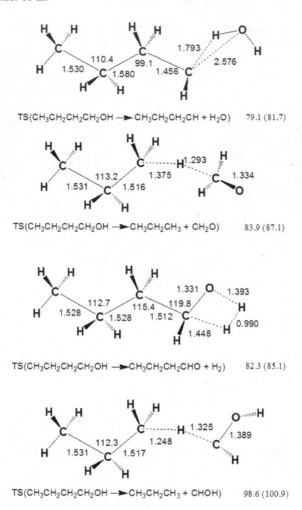

Fig. 2. B3LYP optimized geometries (Å and °) and relative energies (kcal/mol) at 298.15 K of saddle points, relevant for the dissociation of 1-butanol; CCSD(T) relative energies are reported in parentheses

The outcomes of the calculations do not differ dramatically in terms of energy values as well as in the presence and locations of some stationary points. Our calculations provide the same four fragmentation sets of products ($CH_3CH_2CHCH_2 + H_2O$, $CH_3CH_2CH_3 + CH_2O$, $CH_3CH_2CH_3 + CHOH$ and $CH_3CH_2CH_2CHO + H_2$) of Cai et al., as depicted in Fig. 4. The energy of fragmentation channels found in this work differs only by a few kcal/mol (the most significant difference is the one associated to the $CH_3CH_2CH_3 + CHOH$ channel, ca. 5 kcal/mol). Moreover, the energy of the related transition state of 109.9 kcal/mol (our results) does not agree with that of Cai et al. of 84.34 kcal/mol. Another significant difference is present in the channel leading to $CH_3CH_2CHCH_2 + H_2O$: our results indicate a single transition state of 81.1

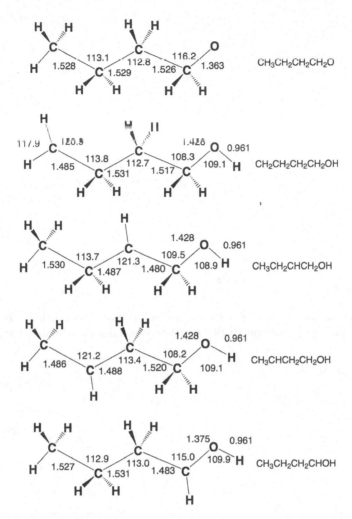

Fig. 3. B3LYP optimized geometries (Å and °) of the fragments produced by H atom emission channels

kcal/mol while they found a first TS of 66.73 kcal/mol (directly leading to $CH_3CH_2CHCH_2+H_2O$) and a second one (of 80.58 kcal/mol) followed by a $CH_3CH_2CH_2CH+H_2O$ minimum (not found in our calculations) located at 84.89 kcal/mol. This intermediate is followed in the PES of Cai et al. by a transition state (lower in energy with respect to the already mentioned intermediate) located at 82.51 kcal/mol and leading to the $CH_3CH_2CHCH_2+H_2O$ products. In conclusion, in our calculations we do not find a minimum between

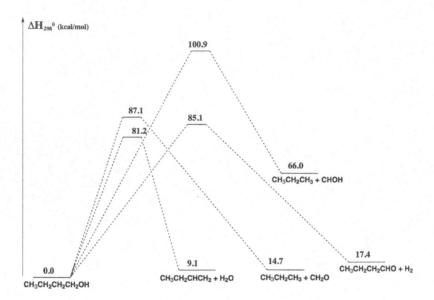

Fig. 4. Schematic representation of the 1-butanol dissociation channels showing an exit barrier. For simplicity, only the CCSD(T) relative energies (kcal/mol) are reported.

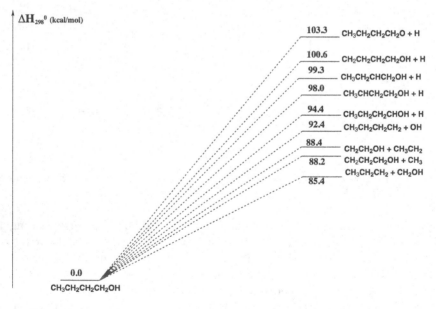

Fig. 5. Schematic representation of the 1-butanol dissociation channels which do not show an exit barrier. For simplicity, only the CCSD(T) relative energies (kcal/mol) are reported.

two transition states for this channel as Cai et al [3]. With the noted exceptions, however, overall there is an excellent agreement with the work of Cai et al. on the four channels describing the direct unimolecular dissociation of 1-butanol, even if the theoretical methods adopted are quite different. In fact, as can be seen in Tab. 1, the channels leading to $CH_3CH_2CH_2+CH_2OH$ (85.4 kcal/mol), $CH_2CH_2C_2OH+CH_3$ (88.2 kcal/mol), $CH_3CH_2+CH_2CH_2OH$ (88.4 kcal/mol) and $CH_2CH_2CH_2CH_2+OH$ (92.4 kcal/mol) show substantially the same energy levels found by Cai et al. (85.40 kcal/mol, 88.25 kcal/mol, 88.6 kcal/mol and 93.06 kcal/mol, respectively).

Our calculations have also been extended to the H-elimination channels, which were not considered by Cai et al. We have identified 5 channels for the direct elimination of an H atom depending of the chemical type of H considered in the parent molecule. The channel lowest in energy is the one associated to the H emission from the CH_2 group located at the beginning of the C skeleton; the highest in energy is the one associated to the H emission from the O-H group. Even though the H-emission channels are higher in energy compared to the other channels, we cannot exclude that they give a contribute at the high temperature conditions of flames. RRKM and experimental results will help to address this issue.

4 Conclusions

A new set of electronic structure calculations on the pathways associated to unimolecular decomposition of 1-butanol has been presented. The agreement with a previous, recent study is excellent, with the exception of the pathway leading to the most important set of products, that is 1-butene + H_2O. An entirely new set of data have been calculated for the H emission channels which were not considered in previous studies. The present data will be used in RRKM calculations to estimate the yield of the various fragmentation channels and to assist the interpretation of new experimental results obtained by means of the flash pyrolysis technique with mass spectrometric detection. These experiments are currently under way. The final purpose is to obtain a convincing description of the micromechanism of 1-butanol unimolecular dissociation. Such a detailed description is necessary to build realistic combustion models where also the minor processes leading to pollutant production in trace amounts need to be understood and, eventually, controlled and mitigated.

Acknowledgments. This work was financially supported by "Fondazione Cassa di Risparmio di Perugia (Codice Progetto: 2014.0253.021 Ricerca Scientifica e Tecnologica)". Noelia Faginas-Lago acknowledges financial support from Phys4entry FP72007-2013 (contract 242311) and EGI Inspire. A. Lombardi also acknowledges financial support to MIUR-PRIN 2010-2011 (contract 2010ERFKXL 002). Thanks are also due to INSTM, IGI and the COMPCHEM virtual organization for the allocation of computing time.

References

1. Kohse-Höinghaus, K., Oßwald, P., Cool, T., Kasper, T., Hansen, N., Qi, F., Westbrook, C.K., Westmoreland, P.R.: Angewandte Chemie International Edition **54**, 3572 (2010)
2. Harper, M.R., Van Geeem, K.M., Pyl, S.P., Marin, G.B., Green, W.H.: Combust. Flame **158**, 16 (2011)
3. Cai, J., Zhang, L., Zhang, F., Wang, Z., Cheng, Z., Yuan, W., Qi, F.: Energy&fuels **26**, 5550 (2012)
4. Balucani, N., Leonori, F., Casavecchia, P.: Energy **43**, 5550 (2012)
5. Vasiliou, A., Nimlos, M.R., Daily, J.W., Ellison, G.B.: J. Phys. Chem. A **113**, 8540 (2009)
6. Urness, K.N., Guan, Q., Golan, A., Daily, J.W., Nimlos, M.R., Stanton, J.F., Ahmed, M., Ellison, G.B.: J. Chem. Phys. **139**, 124305 (2013)
7. O'Keefe, P., Scotti, G., Stranges, D., Rodrigues, P., Barros, M.T., Costa, M.L.: J. Phys. Chem. A, **112**, 3086 (2008)
8. Morton, T.H., Weber, K.H., Zhang, J.: Int. J. Mass Spectrom. **306**, 210 (2011)
9. Balucani, N., Leonori, F., Bergeat, A., Petrucci, R., Casavecchia, P.: Phys. Chem. Chem. Phys. **13**, 8322 (2011)
10. Leonori, F., Petrucci, R., Balucani, N., Casavecchia, P., Rosi, M., Skouteris, D., Berteloite, C., Le Picard, S., Canosa, A., Sims, I.R.: J. Phys. Chem. A **113**, 15328 (2009)
11. Balucani, N., Skouteris, D., Leonori, F., Petrucci, R., Hamberg, M., Geppert, W.D., Casavecchia, P., Rosi, M.: J. Phys. Chem. A **116**, 10467 (2012)
12. Balucani, N., Leonori, F., Petrucci, R., Stazi, M., Skouteris, D., Rosi, M., Casavecchia, P.: Faraday Disc. **147**, 189 (2010)
13. Leonori, F., Skouteris, D., Petrucci, R., Casavecchia, P., Rosi, M., Balucani, N.: J. Chem. Phys. **138**, 024311 (2013)
14. Balucani, N., Bergeat, A., Cartechini, L., Volpi, G.G., Casavecchia, P., Skouteris, D., Rosi, M.: J. Phys. Chem. A **113**, 11138 (2009)
15. Becke, A.D.: J. Chem. Phys. **98**, 5648 (1993)
16. Stephens, P.J., Devlin, F.J., Chabalowski, C.F., Frisch, M.J.: J. Chem. Phys. **98**, 11623 (1994)
17. Dunning Jr., T.H.: J. Chem. Phys. **90**, 1007 (1989)
18. Woon, D.E., Dunning Jr., T.H.: J. Chem. Phys. **98**, 1358 (1993)
19. Kendall, R.A., Dunning Jr., T.H., Harrison, R.J.: J. Chem. Phys. **96**, 6796 (1992)
20. Gonzales, C., Schlegel, H.B.: J. Chem. Phys. **90**, 2154 (1989)
21. Gonzales, C., Schlegel, H.B.: J. Chem. Phys. **94**, 5523 (1990)
22. Bartlett, R.J.: Annu. Rev. Phys. Chem. **32**, 359 (1981)
23. Raghavachari, K., Trucks, G.W., Pople, J.A., Head-Gordon, M.: Chem. Phys. Lett. **157**, 479 (1989)
24. Olsen, J., Jorgensen, P., Koch, H., Balkova, A., Bartlett, R.J.: J. Chem. Phys. **104**, 8007 (1996)
25. Yeamin, M., Faginas-Lago, N., Albertì, M., Cuesta, I., Sánchez-Marìn, J., De Merás, A.S.: RSC Advances **4**(97), 54447 (2014)

26. Frisch, M.J., Trucks, G.W., Schlegel, H.B., Scuseria, G.E., Robb, M.A., Cheeseman, J.R., Scalmani, G., Barone, V., Mennucci, B., Petersson, G.A., Nakatsuji, H., Caricato, M., Li, X., Hratchian, H.P., Izmaylov, A.F., Bloino, J., Zheng, G., Sonnenberg, J.L., Hada, M., Ehara, M., Toyota, K., Fukuda, R., Hasegawa, J., Ishida, M., Nakajima, T., Honda, Y., Kitao, O., Nakai, H., Vreven, T., Montgomery Jr., J.A., Peralta, J.E., Ogliaro, F., Bearpark, M., Heyd, J.J., Brothers, E., Kudin, K.N., Staroverov, V.N., Kobayashi, R., Normand, J., Raghavachari, K., Rendell, A., Burant, J.C., Iyengar, S.S., Tomasi, J., Cossi, M., Rega, N., Millam, J.M., Klene, M., Knox, J.E., Cross, J.B., Bakken, V., Adamo, C., Jaramillo, J., Gomperts, R., Stratmann, R.E., Yazyev, O., Austin, A.J., Cammi, R., Pomelli, C., Ochterski, J.W., Martin, R.L., Morokuma, K., Zakrzewski, V.G., Voth, G.A., Salvador, P., Dannenberg, J.J., Dapprich, S., Daniels, A.D., Farkas, O., Foresman, J.B., Ortiz, J.V., Cioslowski, J., Fox, D.J.: Gaussian?09 Revision D.01. Gaussian Inc., Wallingford CT (2009)
27. Flukiger, P., Lüthi, H.P., Portmann, S., Weber, J.: Swiss Center for Scientific Computing, Manno (2000)
28. Portmann, S., Lüthi, H.P.: Chimia **54**, 54 (2000)
29. www.egi.eu (last accessed: February 25, 2015)
30. http://mccw.hpc.thch.unipg.it/ganglia (last accessd: March 20, 2015)
31. Pacifici, L., Manuali, C., Costantini, A., Vitillaro, G. and Laganà, A. (5.2014.13). https://www3.compchem.unipg.it/ojs/index.php/ojs/article/view/103   (last accessd: January 10, 2015)

Workshop on Computational Optimization and Applications (COA 2015)

SDP in Inventory Control: Non-stationary Demand and Service Level Constraints

Karin G.J. Pauls-Worm[1] and Eligius M.T. Hendrix[2]($\boxtimes$)

[1] Operations Research and Logistics, Wageningen University
Wageningen, The Netherlands
karin.pauls@wur.nl
[2] Computer Architecture, Universidad de Málaga, Málaga, Spain
eligius@uma.es

Abstract. Inventory control implies dynamic decision making. Therefore, dynamic programming seems an appropriate approach to look for order policies. For finite horizon planning, the implementation of service level constraints provides a big challenge. This paper illustrates with small instances the implementation of stochastic dynamic programming (SDP) to derive order policies in a straightforward way for systems with non-stationary demand and service level constraints. The small instances allow to perform a full enumeration of possible policies and show that the SDP derived policies are not necessarily optimal.

Keywords: Stochastic dynamic programming · Inventory control · Non-stationary demand · Service level constraint

1 Introduction

This study is motivated by practical cases of inventory management. In the practical case, a retailer or a producer faces non-stationary demand for a product and has to determine when and how much to order or to produce to meet a certain service level. The decision maker, hereafter manager, uses a periodic review, meaning that on fixed moments in time, e.g. every day or every week, the manager decides on the order quantity. Many products in retail face a non-stationary demand [10]. The decision on the order quantity is inherently a multi-stage problem, and therefore Stochastic Dynamic Programming (SDP) seems an appropriate approach to attempt to solve the problem [1]. An earlier application of SDP to a problem with non-stationary demand and a service level constraint [8] has shown that it does not necessarily generate an optimal solution for this case. This motivates a further study on this phenomenon.

SDP has been used for perishable products with stationary demand by [13] to derive an optimal order-up-to policy. His work was extended by [7], who also used SDP. [11] studied a system for non-perishable products with combined variable deterministic and non-stationary stochastic demand using SDP to find an optimal policy. The above models typically use penalty costs when demand

© Springer International Publishing Switzerland 2015
O. Gervasi et al. (Eds.): ICCSA 2015, Part II, LNCS 9156, pp. 397–412, 2015.
DOI: 10.1007/978-3-319-21407-8_29

exceeds the stock level. In practice however, often a service level constraint is imposed. [4] show that in general it is not possible to transform a service level model into a cost model, above a certain "critical" service level.

In this paper, we consider two types of service levels used in practice, the α-service level, also called cycle service level, and the β-service level, also called fill rate. [5] define cycle service level as the probability of not having a stock-out in a replenishment cycle. A replenishment cycle is the time between two orders. The fill rate indicates that a predefined percentage of the demand per replenishment cycle has to be fulfilled from stock. [4] define the difference between mean service level constraints and minimal service level constraints. Mean service level constraints measure the service level over the time horizon, whereas minimal service level constraints measure the service level in every period. [9] show for inventory systems with stationary demand the conditions for a one-to-one relation between cost and service models, where they consider a mean service level. [2] formulated a lost-sales inventory model with an average fill rate requirement as a constrained dynamic programming problem for stationary demand. They solved the problem with Lagrange relaxation by a value-iteration algorithm to find optimal replenishment policies. In this paper we focus on minimal service level constraints for the non-stationary demand case and we investigate whether SDP is a suitable method if a service level constraint applies. One stylized example is elaborated for different situations. It shows how in general SDP handles service level constraints and does not necessarily provide the optimal solution.

Section 2 introduces a stylized example of variable demand. It has a finite time horizon of 6 periods representing the possibility of a manager in retail to order at most six times a week. This simple instance shows the application of Dynamic Programming (DP) to solve the deterministic variable discrete demand case. In Section 3 the example is extended to a stochastic uniform discrete demand case and solved with SDP to generate $Q_t(I)$ order policies, where the order quantity Q depends on the inventory level I on hand. Section 4 adds an α-service level constraint and Section 5 a fill rate constraint. The derived SDP solution for both cases is compared to an optimal policy fulfilling the service level constraints found by full enumeration. Section 6 illustrates the SDP approach for a continuous distribution of demand by a Gamma distribution. Section 7 concludes the paper.

2   Dynamic Programming

We consider a small inventory problem with a deterministic variable demand. The problem has a finite time horizon of $T = 6$ and demand is $d_t = 3, 1, 2, 4, 3, 2$. The set-up cost is $k = 5$ and holding cost $h = 1$. Holding cost is paid over the inventory that is carried over to the next period. The unit variable cost is $c = 0$ and the starting inventory is $I_0 = 0$. An order is placed and delivered at the beginning of the period, demand occurs during the period, and the inventory level is calculated at the end of the period. The question is when to order and how much. This problem can be formulated as follows:

$$\min TC = \sum_{t=1}^{T} g(Q_t) + hI_t \tag{1}$$

where procurement cost is given by the function

$$g(x) = k + cx, \quad \text{if } x > 0, \text{ and } g(0) = 0. \tag{2}$$

Subject to

$$I_t = I_{t-1} + Q_t - d_t, \; t = 1, \ldots, T \tag{3}$$

$$I_0 = 0 \tag{4}$$

$$I_t, Q_t \geq 0, \; t = 1, \ldots, T \tag{5}$$

The objective function (1) represents the total relevant costs, consisting of ordering and holding costs. In (2) the ordering costs are specified by a fixed set-up cost when an order is placed and variable procurement cost. Equations (3) are balance equations stating that the inventory level at the end of period t equals the inventory level at the end of the period before, increased with the order quantity minus the demand in period t where (4) gives the initial inventory level. Equations (5) are non-negativity constraints.

[12] derived properties of the optimal solution to this problem and developed an efficient algorithm to solve the problem. The algorithm they developed is a modification of Dynamic Programming. This paper applies DP in order to develop towards SDP with service level constraints.

In the example, the first order Q_1 fulfils the accumulated total demand for a number of future periods from one to six, i.e. $Q_1 = 3, 4, 6, 10, 13$ or 15. After demand $d_1 = 3$ is realized, the possible inventory level is $I_1 = 0, 1, 3, 7, 10$ or 12. In period $t = 2$, demand $d_2 = 1$ will be fulfilled from the available inventory, or, in case $I_1 = 0$, there will be an order to fulfil demand of period 2 up to 6, or to fulfil demand of period 2 up to 5, 4 or 3 or just period 2. The same reasoning holds for periods 3 to 5, so either $Q_t = 0$, or $Q_t = \sum_{j=t}^{N} d_j$ for some $N, t \leq N \leq T$.

The beginning of a period, when the decision about the order quantity has to be made, is called a *stage* in DP. A stage starts with the inventory level at the end of the previous period which is called the *state*. The set of possible values is called the state space. DP has the property that the optimal decision on time t (stage t) with state I_{t-1}, does not depend on the decisions made leading to state I_{t-1}. This means that the problem can be decomposed into sub-problems. The stages and states of this small instance are depicted in Fig. 1. A path from the start (stage 1) to the end of period 6 models the timing and quantity of ordering and the inventory levels at the end and beginning of each period (stage). Given the possibility to order for all future periods, we have $T! = 720$ possible paths for this inventory problem. DP generally starts with a backward procedure, followed by a forward procedure. The backward procedure assigns a valuation to each state, called a value function $V_t(I)$ to each state in the system. Stage 6 ends with $I_6 = 0$ at the end of the time horizon. $V_t(I)$ equals the costs that have to be made from stage t at state I to the end of the final stage 6 with state $I = 0$. The so-called recursive Bellman equation for the value function $V_t(I)$ can be written as

$$V_t(I) = \min_Q \{g(Q) + h \cdot (I + Q - d_t) + V_{t+1}(I + Q - d_t)\} \text{ for } t = T, \ldots, 1. \quad (6)$$

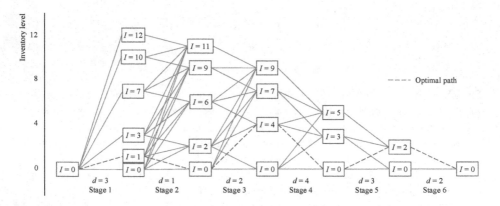

Fig. 1. All possible states of inventory levels and all possible paths

We illustrate the backward procedure for this small instance. Consider $V_6(2) = 0$. Holding cost is paid at the end of the previous period, so no cost has to be paid in this state; no order takes place. $V_6(0) = k + cQ_6 = 5 + 0 = 5$. An order of $Q_6 = d_6 = 2$ is placed at a cost of $k = 5$.

$V_5(5) = h \cdot (5 + 0 - 3) + V_6(5 + 0 - 3) = 2 + 0 = 2$. No order takes place, but at the end of stage 5, there is an inventory level of 2 that incurs holding cost. $V_5(3) = \min_{Q \in \{0,2\}} h \cdot (3 + Q - 3) + V_6(3 + Q - 3) = \min\{(0 + 5), (7 + 0)\} = 5$. For $I_4 = 3$, the choice is not to order, or to order for demand of period 6. The minimum cost corresponds to not ordering.

$V_5(0) = \min_{Q \in \{3,5\}} \{k + h \cdot (0 + Q - 3) + V_6(0 + Q - 3)\} = \min\{5 + 0 + 5, 5 + 2 + 0\} = 7$. Here the minimum costs are obtained for $Q_5 = d_5 + d_6$.

Continuing this process provides finally the minimum total cost of $V_1(0) = 22$. The path leading to the minimum cost can be determined in a forward procedure. The minimum cost at $V_1(0)$ is obtained via $I_1 = 1, I_2 = 0, I_3 = 4, I_4 = 0, I_5 = 2$ and $I_6 = 0$, so the optimal ordering path is $Q_1 = 4, Q_2 = 0, Q_3 = 6, Q_4 = 0, Q_5 = 5$ and $Q_6 = 0$ represented by the dashed red path in Fig. 1.

3   Stochastic Dynamic Programming

We continue the example from Section 2, but now demand is stochastic and non-stationary, with expected demand $\mu_t = 3, 1, 2, 4, 3, 2$. Demand has a discrete

Uniform distribution with $d_t \sim U\{0, 2\mu_t\}$. All demand has to be met. Notice that one can set this target as the support of the distribution is finite. Demand in period $t = 1$ can take the values $d_{1j} \in \{0, 1, 2, 3, 4, 5, 6\}$, where the index is $j = 1, \ldots, N_t$ with $N_t = 2\mu_t + 1$. Every outcome d_{tj} has probability of occurrence $p_t = \frac{1}{N_t}$, so $p_1 = \frac{1}{7}, p_2 = \frac{1}{3}$, etc. The maximum possible inventory level equals $I_{\max} = \sum_{t=1}^{T} 2\mu_t = 30$. In this model, this number bounds the state space.

In the optimal solution of the deterministic example, ordering only takes place when the inventory level equals zero, and the inventory level is always the sum of demand in upcoming periods. In this stochastic example demand is uncertain, so ordering might be necessary even if the inventory is not zero. That results in many more possible inventory paths. The value function for this problem is defined by

$$V_t(I) = \min_{Q \in F_t(I)} \left[g(Q) + p_t \left(h \sum_{j=1}^{N_t} (I + Q - d_{tj}) \right)^+ + p_t \sum_{j=1}^{N_t} V_{t+1}(I + Q - d_{tj})^+ \right], \quad (7)$$

where the feasible area of Q is given by

$$F_t(I) = [0, (d_{tN_t} + \ldots + d_{TN_T} - I)^+] \tag{8}$$

and $x^+ = \max\{x, 0\}$. Optimal order quantity as function of inventory level I is

$$Q_t(I) = \arg \min_{Q \in F_t(I)} \left[g(Q) + h \cdot p_t \sum_{j=1}^{N_t} (I + Q - d_{tj})^+ + p_t \sum_{j=1}^{N_t} V_{t+1}(I + Q - d_{tj})^+ \right]. \tag{9}$$

The SDP approach has been implemented in MATLAB. Table 1 shows the generated optimal order quantity $Q_t(I)$ for each stage and inventory level. One can observe that the optimal order quantities follow in fact from order-up-to levels,

Table 1. SDP order quantities given the inventory level

I_{t-1}	Q_1	Q_2	Q_3	Q_4	Q_5	Q_6
0	6	2	4	8	6	4
1	0	1	3	7	5	3
2	0	0	2	6	4	2
3	0	0	1	5	3	1
4	0	0	0	4	2	0
5	0	0	0	3	1	0
6	0	0	0	2	0	0
7	0	0	0	1	0	0
8	0	0	0	0	0	0

i.e. there are values S_t, such that $Q_t(I) = (S_t - I)^+$. As all demand has to be fulfilled, the order-up-to levels have the size $S_t = 2\mu_t$. The expected total costs are $V_1(0) = 38.49$. The introduction of uncertainty gives a 75% cost increase when compared to the corresponding deterministic variant of the problem. The objective function value of expected costs of the optimal order quantities $Q_t(I)$ follows from the generation of all possible demand paths as illustrated in Fig. 2. There are $N_1 \times N_2 \times N_3 \times N_4 \times N_5 \times N_6 = 33075$ possible paths. The evaluation computes exactly the same expected total costs of $E(TC) = 38.49$ as the value of $V_1(0)$ in the SDP approach.

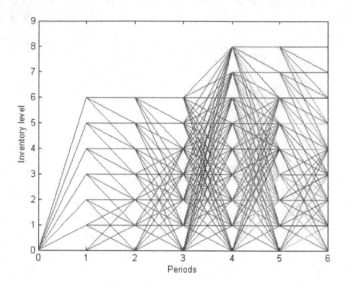

Fig. 2. Possible inventory development given demand paths

4   SDP with an α-Service Level Constraint

In Section 3, 100% of the demand has to be fulfilled. Now we add an α-service level constraint to the problem. We define the α-service level as a minimum probability α of not being out of stock at the end of a period, so

$$P(I_t \geq 0) \geq \alpha, \ t = 1, \ldots, T. \tag{10}$$

This means that at the start of a period holds

$$P(I_{t-1} + Q_t - d_t \geq 0) \geq \alpha \implies P(d_t \leq I_{t-1} + Q_t) \geq \alpha, \ t = 1, \ldots, T. \tag{11}$$

The total quantity $I_{t-1} + Q_t$ can be seen as a basic order quantity, which we will call $\hat{Q}_t$. The cumulative distribution function of $d \sim U\{a, b\}$ is

$$F(k) := P(d \leq k) = \frac{\lfloor k \rfloor - a + 1}{b - a + 1}$$

for $k \in \{a, a+1, \ldots, b-1, b\}$. For $d_t \sim U\{a, b\}$ we have $a = 0, b = 2\mu_t$ and $b - a + 1 = N_t = 2\mu_t + 1$. Elaboration of (11) for this case gives

$$P(d_t \le \hat{Q}_t) = \frac{\lfloor \hat{Q}_t \rfloor + 1}{N_t} \ge \alpha, \ t = 1, \ldots, T. \tag{12}$$

This results in

$$\hat{Q}_t = \lceil \alpha N_t \rceil, \ t = 1, \ldots, T. \tag{13}$$

The basic order quantity is a lower bound for the order quantity, or the starting inventory in a period, necessary to meet the α-service level requirement.

Consider the case of Section 3 with $k = 5$ or $k = 50$, $\alpha = 0.8$ and expected demand $\mu_t = 3, 1, 2, 4, 3, 2$. Recursion 7 is still valid, but the feasible area is now

$$F_t(I) = [\hat{Q}_t, (d_{dtN_t} + \ldots + d_{TN_T} - I)^+]. \tag{14}$$

Demand that is not met will be lost. An SDP approach has been implemented in MATLAB. The results of the SDP approach and the evaluated service levels are listed in Table 2.

Table 2. SDP order policy and its reached service levels (required $\alpha = 0.8$)

		$k = 5; E(TC) = 37.95$							$k = 50; E(TC) = 129.01$				
$\hat{Q}_t$	5	2	4	7	5	4		5	2	4	7	5	4
I_{t-1}	Q_1	Q_2	Q_3	Q_4	Q_5	Q_6	I_{t-1}	Q_1	Q_2	Q_3	Q_4	Q_5	Q_6
0	5	2	4	7	5	4	0	18	16	16	15	10	4
1	0	1	3	6	4	3	1	0	15	15	14	9	3
2	0	0	2	5	3	2	2	0	0	14	13	8	2
3	0	0	1	4	2	1	3	0	0	13	12	7	1
4	0	0	0	3	1	0	4	0	0	0	11	6	0
5	0	0	0	2	0	0	5	0	0	0	10	0	0
6	0	0	0	1	0	0	6	0	0	0	9	0	0
7	0	0	0	0	0	0	7	0	0	0	0	0	0
8	0	0	0	0	0	0	:	0	0	0	0	0	0
9	0	0	0	0	0	0	18	0	0	0	0	0	0
service	0.86	1.00	1.00	0.89	0.89	1.00		1.00	1.00	1.00	0.999	0.989	1.00

Table 2 shows that in case of $k = 5$ one orders on average (almost) every period where Q_t at inventory level $I_{t-1} = 0$ equals the basic order quantity $\hat{Q}_t$. Basic order quantity $\hat{Q}_t$ and the order quantities Q_t behave as order-up-to levels. In case $k = 50$, one can observe in the different periods different reorder points. When the inventory level is at the reorder point or lower, one orders up to the level $Q_t(0)$, otherwise there is no order. E.g. in period 4, the reorder point equals 3. If $I = 4$ or more, this is enough to meet demand and service level in period 4. If $I = 3$ or less, there is an order to meet demand and service level in period 4 and the upcoming periods. In three of the six periods, the service levels are slightly higher than required. The difference is due to the discrete and small

SDP alpha service level

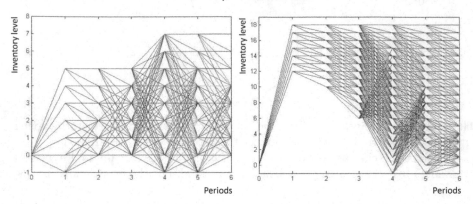

Fig. 3. Inventory development for all demand paths of the SDP solution with α-service level constraint for $k = 5$ (left) and $k = 50$ (right). Inventory levels below zero represent lost sales.

numbers of demand. In the other periods, as well as in all periods of the case of $k = 50$, the evaluated service levels are much higher.

Fig. 3 shows inventory development for all possible demand paths. Also inventory levels below zero (lost sales) are depicted. These are virtual inventory levels to show out-of-stocks, but lost sales is assumed, so the inventory level never drops below zero. Notice that the different inventory levels have different chances of occurrence. From all possible starting inventory levels, no matter how small the chance of occurrence, SDP wants to meet the α-service level requirement. This results in an overall over-achievement of the service level. The SDP approach due to the definition of the feasible area tightens the α-service level constraint in (10) towards the conditional variant

$$P(I_t \geq 0 | I_{t-1}) \geq \alpha, \ t = 1, \ldots, T. \tag{15}$$

The SDP policies taught us that an optimal policy may consist of order-up-to levels for every period t, i.e. there is an order-up-to value S_t for each period such that the policy chooses as order quantity

$$Q_t(I) = (S_t - I)^+, \ t = 1, \ldots, T. \tag{16}$$

Although such a policy does not necessarily lead to the lowest possible expected costs, we can look for the best values such that the α-service level constraint (10) is met. The optimal order-up-to level policy can then be compared to the SDP policy. The small instance allows us to perform a full enumeration to find the best feasible values for $(S_1, \ldots, S_T)$. The enumeration procedure uses bounds for the order-up-to levels S_t, being

$$S_1 \in \left[\hat{Q}_1, \sum_{t=1}^{T} 2\mu_t\right] \text{ and } S_t \in \left[0, \sum_{j=t}^{T} 2\mu_j\right], \ t = 2, \ldots, T. \tag{17}$$

Algorithm 1. Order-up-to($\mu_1, \ldots, \mu_T, \alpha$): Determines S_t for $d_t \sim U\{0, 2\mu_t\}$ fulfilling (10) via (16)

1: Determine all possible demand path for $d_1, \ldots, d_T$
2: Determine upper bounds on S_t
3: Evaluate all combinations $S_1, \ldots, S_T$ within the bounds and keep the lowest cost solution

Systematically S_t is lowered and checked for feasibility of the service level constraint. The minimum expected total costs order-up-to levels are determined in Algorithm 1. The number of possible combinations of the order-up-to levels within this range is large, about $\Pi_{t=1}^{T} \sum_{j=t}^{T} N_j$. For the case this are around 10^7 combinations. To compare, SDP evaluates in each stage $t = 1, \ldots, T$, for each value of I less than $\sum_{j=t}^{T} N_j$ possibilities for the value of Q. The generated order

Table 3. Optimal order-up-to policy and its reached service levels (required $\alpha = 0.8$)

	\multicolumn{6}{c}{$k = 5; E(TC) = 32.79$}		\multicolumn{6}{c}{$k = 50; E(TC) = 108.37$}
$\hat{Q}_t$	5	2	4	7	5	4	5	2	4	7	5	4
I_{t-1}	Q_1	Q_2	Q_3	Q_4	Q_5	Q_6	I_{t-1} Q_1	Q_2	Q_3	Q_4	Q_5	Q_6
0	6	0	3	8	4	3	0 18	0	0	7	0	0
1	0	0	2	7	3	2	1 0	0	0	6	0	0
2	0	0	1	6	2	1	2 0	0	0	5	0	0
3	0	0	0	5	1	0	3 0	0	0	4	0	0
4	0	0	0	4	0	0	4 0	0	0	3	0	0
5	0	0	0	3	0	0	5 0	0	0	2	0	0
6	0	0	0	2	0	0	6 0	0	0	1	0	0
7	0	0	0	1	0	0	7 0	0	0	0	0	0
:	0	0	0	0	0	0	: 0	0	0	0	0	0
10	0	0	0	0	0	0	18 0	0	0	0	0	0
service	1.00	0.86	0.86	1.00	0.83	0.86	1.00 1.00	1.00	0.996	0.90	0.80	

policy is listed in Table 3. The expected total costs E(TC) of the SDP policy are respectively 12.7% and 19.0% higher than those of the optimal order-up-to policy. Fig. 4 shows the inventory development for all possible demand paths. Compared to the SDP solution, more out-of-stocks are allowed represented by a negative inventory level.

5   Fill Rate Constraint

Instead of an α-service level constraint we now consider a β-service level or fill rate constraint in the problem. The fill rate indicates that a predefined percentage of the demand per period has to be fulfilled from stock. Demand that cannot be fulfilled from stock is lost. Lost sales is defined as

$$X_t = (d_t - I_{t-1} - Q_t)^+, \ t = 1, \ldots, T. \tag{18}$$

Opt alpha service level

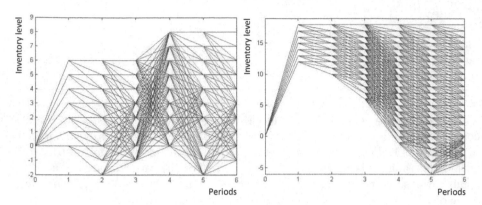

Fig. 4. Inventory development for all demand paths of the order-up-to policy with α-service level constraint for $k = 5$ (left) and $k = 50$ (right)

The fill rate constraint we consider is defined for each period

$$E(X_t) \leq (1 - \beta)\mu_t, \ t = 1, \dots, T. \tag{19}$$

The fill rate constraint requires that the expected shortage is smaller or equal to a fraction $(1 - \beta)$ of the expected demand per period. For the SDP approach, now the feasible set $F_t(I)$ should take (19) into account.

Table 4. SDP order policy and its reached fill rates (required $\beta = 0.8$)

	$k = 5; E(TC) = 32.305$							$k = 50; E(TC) = 122.92$					
I_{t-1}	Q_1	Q_2	Q_3	Q_4	Q_5	Q_6	I_{t-1}	Q_1	Q_2	Q_3	Q_4	Q_5	Q_6
0	4	5	3	5	4	3	0	17	15	15	14	9	3
1	0	4	2	4	3	2	1	0	14	14	13	8	2
2	0	0	1	3	2	1	2	0	0	13	12	7	1
3	0	0	0	2	1	0	3	0	0	0	11	6	0
4	0	0	0	1	0	0	4	0	0	0	10	0	0
5	0	0	0	0	0	0	5	0	0	0	0	0	0
:	0	0	0	0	0	0	:	0	0	0	0	0	0
7	0	0	0	0	0	0	17	0	0	0	0	0	0
fill rate	0.86	1.00	0.94	0.83	0.87	0.92		1.00	1.00	1.00	0.99	0.98	0.97

This problem has been solved by an SDP approach implemented and evaluated in MATLAB. The results of the SDP model and the evaluated fill rates are listed in Table 4. Fig. 5 shows all the demand paths in cases $k = 5$ and $k = 50$. Negative inventory levels represent out-of-stocks related to lost sales. As can be observed, also with a fill rate constraint the order quantity $Q_t(I)$ follows an

SDP beta service level

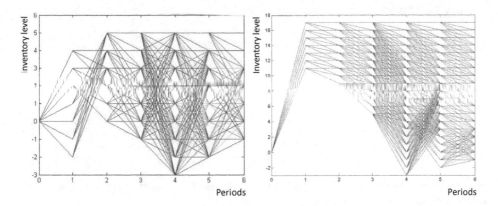

Fig. 5. Inventory development for all demand paths of the SDP policy with fill rate constraint for $k = 5$ (left) and $k = 50$ (right)

order-up-to level policy and when k = 50, for every period there is a reorder point, i.e. an inventory level above which no order takes place. The evaluated fill rates are in most periods well above the required fill rate of $\beta = 0.80$. As described in Section 4, from all possible starting inventory levels, SDP wants to meet the fill rate requirement. This results in an overall over-achievement of the service level. The fill rate constraint we want to meet is described by (19). Instead, the SDP approach meets the conditional constraint

$$E(X_t|I_{t-1}) \le (1 - \beta)\mu_t, \ t = 1, \dots, T. \tag{20}$$

Essentially the same algorithm as described in Section 4 can be used to find the optimal order-up-to value solution satisfying (19). The results are listed in

Table 5. Optimal order-up-to policy and reached fill rates (required $\beta = 0.8$)

	$k = 5; E(TC) = 30.03$						$k = 50; E(TC) = 111.81$					
I_{t-1}	Q_1	Q_2	Q_3	Q_4	Q_5	Q_6	I_{t-1}	Q_1	Q_2	Q_3	Q_4	Q_5	Q_6
0	6	0	2	7	4	2	0	18	0	0	7	3	0
1	0	0	1	6	3	1	1	0	0	0	6	2	0
2	0	0	0	5	2	0	2	0	0	0	5	1	0
3	0	0	0	4	1	0	3	0	0	0	4	0	0
4	0	0	0	3	0	0	4	0	0	0	3	0	0
5	0	0	0	2	0	0	5	0	0	0	2	0	0
6	0	0	0	1	0	0	6	0	0	0	1	0	0
7	0	0	0	0	0	0	7	0	0	0	0	0	0
8	0	0	0	0	0	0	:	0	0	0	0	0	0
9	0	0	0	0	0	0	18	0	0	0	0	0	0
fill rate	1.00	0.81	0.81	0.97	0.90	0.80		1.00	1.00	1.00	0.999	0.95	0.80

Opt beta service level

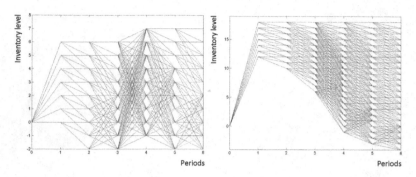

Fig. 6. Inventory development for all demand paths of the order-op-to policy with fill rate constraint for $k = 5$ (left) and $k = 50$ (right)

Table 5 and shown in Fig. 6. The expected total costs $E(TC)$ of the SDP policy are respectively 7.6% and 9.9% higher than of the optimal order-up-to policy. In this example, with a fill rate constraint the SDP solution is closer to the optimal order-up-to policy than with an α-service level constraint.

6   Continuous Demand

The use of the discrete Uniform distribution to model the demand enabled us to calculate order policies with SDP and to evaluate the policies with all possible demand paths. However, the discrete Uniform distribution is not very realistic to model demand. In cases of small demand numbers, the Gamma distribution can be suitable [3]. We consider the Gamma distribution $d \sim \Gamma(K, \theta)$, where K is the shape parameter and θ is the scale parameter. The expected value of

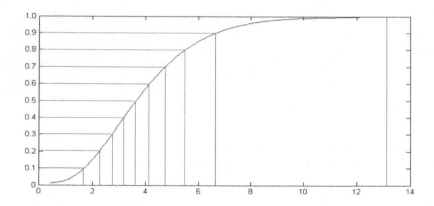

Fig. 7. Discretization of Gamma distribution, $\mu = 4$ and $p = 0.1$

the Gamma distribution is $E[d] = \theta K$. We assume $\theta = 1$, then $K = \mu$. As before, the expected demand $\mu_t = 3, 1, 2, 4, 3, 2$. Let now $d_t \sim \Gamma(\mu_t, 1)$ with G_t as corresponding cumulative distribution function (cdf).

To apply SDP, demand has to be discretized. A common method is to discretize the space of possible outcomes of the stochastic demand δ_{tj} by using quantiles of the Gamma distribution. An equidistant grid is taken over the range $[0, 1]$ in N steps, each with probability $p = \frac{1}{N}$. The discrete outcomes are

$$\delta_{tj} = G_t^{-1}(j \times p), j = 1, \ldots, N$$

[6]. Each outcome has the same probability of occurrence and the outcome space is truncated. Because of the skewness of the Gamma distribution, there are more outcomes with demand smaller than or around μ_t, than greater than μ_t. In Fig. 8 this principle is illustrated for $N = 10$ steps with $p = 0.1$.

After discretization, value function (7) can be reformulated into

$$V_t(I) = \min_{Q \in F_t(I)} \left[g(Q) + h \cdot p \sum_{j=1}^{N_t} (I + Q - \delta_{tj})^+ + p \sum_{j=1}^{N_t} V_{t+1}(I + Q - \delta_{tj})^+ \right] \tag{21}$$

with feasible area

$$F_t(I) = [0, (\delta_{tN_t} + \ldots + \delta_{TN_T} - I)^+]. \tag{22}$$

The feasible set $F_t(I)$ should take the fill rate constraint (19) into account, so at least in period 1, the actual feasible set is smaller. However, there is no easy analytical expression for the lower bound in case of a Gamma-distributed fill rate. The number of possible demand paths depends on the number of steps in the grid. In this case we use a grid of $N = 100$ steps in every period, so there are $100^T = 10^{12}$ possible paths. SDP still generates a solution within a second, because each stage, only 100 outcomes have to be considered, but the solution

Table 6. SDP order quantities given inventory level, Gamma-distributed demand, simulated costs and reached fill rates (required $\beta = 0.8$)

I_{t-1}	Q_1	Q_2	Q_3	Q_4	Q_5	Q_6	I_{t-1}	Q_1	Q_2	Q_3	Q_4	Q_5	Q_6
	$k = 5; V_1(0) = 31.82$							$k = 50; V_1(0) = 120.59$					
	SimAvg(TC) = 33.88							SimAvg(TC) = 123.99					
0	5.12	3.72	2.50	4.05	5.01	2.50	0	16.78	14.66	14.34	13.00	8.80	2.50
1	0	2.72	1.50	3.05	4.01	1.50	1	0	13.66	13.34	12.00	7.80	1.50
2	0	0	0.50	2.05	3.01	0.50	2	0	0	12.34	11.00	6.80	0.50
3	0	0	0	1.05	2.01	0	3	0	0	0	10.00	5.80	0
4	0	0	0	0.05	0	0	4	0	0	0	9	0	0
5	0	0	0	0	0	0	5	0	0	0	0	0	0
:	0	0	0	0	0	0	:	0	0	0	0	0	0
7	0	0	0	0	0	0	17	0	0	0	0	0	0
fill rate	0.95	0.95	0.86	0.81	0.94	0.85		1.00	1.00	1.00	0.99	0.98	0.94

cannot be evaluated any more by generating all paths, it has to be simulated. Notice that the demand outcomes lead to inventory levels that do not have an integer (grid) value. This means that the value of $V_{t+1}(I + Q - \delta_{tj})^+$ in (21) has to be found by interpolation for each demand outcome δ_{tj} in order the get the expected value for an order choice Q.

SDP Gamma fill rate

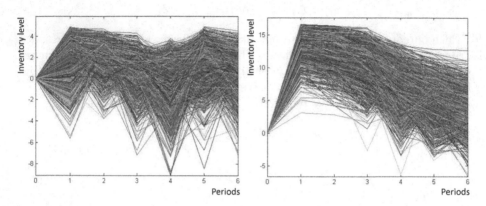

Fig. 8. Inventory development for sampled demand from Gamma distribution of the SDP policy with fill rate constraint, $k = 5$ (left) and $k = 50$ (right).

For the final evaluation of the policy, 20000 random Gamma distributed demand paths are simulated. In Table 6 the SDP order policies are listed, together with the results of the simulation, for $k = 5$ and $k = 50$. Due to the continuous distribution of the demand, the optimal order quantities do no longer have an integer value and are found by a one-dimensional nonlinear optimization over the interval $F_t(I)$ for a set of integer valued grid points of the inventory. The order quantities $Q_t(I)$ follow an order-up-to level policy, with a reorder point in some periods. In case of $k = 5$, only in period 4 the fill rate is close to the requirement. In the other periods, and in case of $k = 50$, an over-achievement of the fill rate can be observed, as in the discrete model in Section 5.

Algorithm 1 is based on enumerating integer values for order-up-to levels S_t, whereas in the continuous case they can take a continuous value. Moreover, the number of demand paths is not finite, so we rely on sample paths. The optimization of S_t was based on nonlinear optimization using the FMINCON solver of MATLAB and using 50000 sample paths to evaluate the costs and to estimate the fill rate. Table 7 gives the resulting solution of the order-up-to levels.

The simulated total costs SimAvg(TC) of the SDP policy are respectively 11.6% and 17.5% higher than of the order-up-to level policy. The reached fill rates of the order-up-to level policy are close to the requirements. This specific example proves that the over-achievement of the fill rate constraint by SDP due to the conditional treatment, also occurs for continuously distributed demand.

Table 7. Optimal order-up-to level quantities for inventory level, Gamma-distributed demand, simulated costs and reached fill rates (required $\beta = 0.8$)

I_{t-1}	Q_1	Q_2	Q_3	Q_4	Q_5	Q_6	I_{t-1}	Q_1	Q_2	Q_3	Q_4	Q_5	Q_6
	\multicolumn{6}{c}{$k = 5$; SimAvg(TC) $= 30.35$}		\multicolumn{6}{c}{$k = 50$; SimAvg(TC) $= 105.54$}
0	5.53	0	2.03	3.92	6.13	0.23	0	17.67	0.66	3.06	0	0	0
1	0	0	1.03	2.92	5.13	0	1	0	0	2.06	0	0	0
2	0	0	0.03	1.92	4.13	0	2	0	0	1.06	0	0	0
3	0	0	0	0.92	3.13	0	3	0	0	0.06	0	0	0
4	0	0	0	0	2.13	0	4	0	0	0	0	0	0
5	0	0	0	0	1.13	0	5	0	0	0	0	0	0
6	0	0	0	0	0.13	0	:	0	0	0	0	0	0
7	0	0	0	0	0	0	17	0	0	0	0	0	0
fill rate	0.96	0.80	0.80	0.80	0.98	0.81		1.00	1.00	1.00	1.00	0.93	0.80

7 Conclusions

By means of a stylized example of an inventory management problem that is relevant for practice we study whether SDP is a suitable method if a minimal service level constraint applies and demand is non-stationary. The example has a fixed time horizon is six periods. We started to solve a deterministic problem, where the variable demand equals the expected demand in the stochastic cases. Introducing uncertainty through a Uniform distribution and requiring that all demand has to be fulfilled gives a 75% cost increase. Subsequently we introduced an α-service level constraint and a fill rate constraint per period, being minimal service level constraints.

Evaluation of the SDP order policies shows that SDP generates order policies that overachieve the service level requirements. This is due to the conditional behavior of the service level constraints. From all possible starting inventory levels, no matter how small the chance of occurrence, SDP has to meet the service level requirement per period. We could compare the SDP policy with an order-up-to level policy, which in principle has less choice for the order quantity. The corresponding order-up-to levels were generated by full enumeration meeting the service level requirement, resulting in a policy with lower costs and service levels that are at the end of the replenishment cycle close to the required service level.

Finally, continuously distributed demand is modeled with a more realistic Gamma distribution and a fill rate constraint. Compared to the uniform distribution, this results in lower order quantities and lower costs, but also in an over-achievement of the service level requirement.

Future research may aim at investigating the effect of the over-achievement of SDP of service levels for other demand distributions. Due to the conditional fulfilment of the service levels, we observe this over-achievement. However this effect may be less for other type of distributions and therefore SDP might generate acceptable solutions.

To conclude: SDP generates order policies that meet a conditional service level constraint. Given a minimal expected service level requirement, the resulting policy is not necessarily optimal.

Acknowledgments. This work has been funded by grants from the Spanish Ministry (TIN2012-37483-C03-01) and Junta de Andalucía (P11-TIC-7176), in part financed by the European Regional Development Fund (ERDF).

References

1. Bellman, R., Stanley, L.E.: Functional equations in dynamic programming. Aequationes Mathematicae **17**(1), 1–18 (1978)
2. Bijvank, M., Vis, I.F.A.: Lost-sales inventory systems with a service level criterion. European Journal of Operational Research **220**(3), 610–618 (2012)
3. Burgin, T.A.: The gamma distribution and inventory control. Operational Research Quarterly **26**(3), 507–525 (2012)
4. Chen, F.Y., Krass, D.: Inventory models with minimal service level constraints. European Journal of Operational Research **134**(1), 120–140 (2001)
5. Chopra, S., Meindl, P.: Supply Chain Management: Strategy, Planning, and Operation. Pearson, New Yersey (2010)
6. van Dijk, D., Hendrix, E.M.T., Haijema, R., Groeneveld, R.A., van Ierland, E.C.: On solving a bi-level stochastic dynamic programming model for analyzing fisheries policies: Fishermen behavior and optimal fish quota. Ecological Modelling **272**, 68–75 (2014)
7. Fries, B.: Optimal ordering policy for a perishable commodity with a fixed lifetime. Operations Research **23**(1), 46–61 (1975)
8. Hendrix, E.M.T., Haijema, R., Rossi, R., Pauls-Worm, K.G.J.: On solving a stochastic programming model for perishable inventory control. In: Murgante, B., Gervasi, O., Misra, S., Nedjah, N., Rocha, A.M.A.C., Taniar, D., Apduhan, B.O. (eds.) ICCSA 2012, Part III. LNCS, vol. 7335, pp. 45–56. Springer, Heidelberg (2012)
9. van Houtum, G.J., Zijm, W.H.M.: On the relationship between cost and service models for general inventory systems. Statistica Neerlandica **54**(2), 127–147 (2000)
10. Pauls-Worm, K.G.J., Hendrix, E.M.T., Haijema, R., van der Vorst, J.G.A.J.: An MILP approximation for ordering perishable products with non-stationary demand and service level constraints. International Journal of Production Economics **157**, 133–146 (2014)
11. Sobel, M.J., Zhang, R.Q.: Inventory policies for systems with stochastic and deterministic demand. Operations Research **49**(1), 157–162 (2001)
12. Wagner, H.M., Whitin, T.M.: Dynamic version of the economic lot size model. Management Science **5**(1), 89–96 (1958)
13. van Zyl, G.: Inventory control for perishable commodities. Ph.D. thesis, University of North Carolina, Chapel Hill (1964)

Fast Heuristics for Integrated Planning and Scheduling

Jürgen Rietz, Cláudio Alves[(✉)], and José Valério de Carvalho

Centro Algoritmi, Escola de Engenharia,
Universidade Do Minho, 4710-057 Braga, Portugal
Juergen_Rietz@gmx.de, {claudio,vc}@dps.uminho.pt

Abstract. In this paper, we address the integrated planning and scheduling problem on parallel machines in which a set of jobs with release and due-dates have to be assigned first to consecutive time periods within the planning horizon, and then scheduled on the available machines. We explore in particular different alternative low complexity heuristics. The importance of job sequencing in the performance of these heuristics is analyzed, and a new property characterizing the optimal solutions of the problem is described. We also present a heuristic that yields optimal solutions for specific instances of the problem, and local exchange procedures that proved to be effective. To the best of our knowledge, these are the first contributions concerning the heuristic solution of this integrated planning and scheduling problem through low complexity procedures. To evaluate performance of these heuristics, we report on extensive computational experiments on benchmark instances of the literature.

Keywords: Planning · Scheduling · Heuristics · Low complexity

1 Introduction

The development of computational models and methods for integrated and real-time optimization is one of the key challenges in the field of supply chain management. To reduce their costs and improve their operational performance, the companies have to keep low inventory levels. As a consequence, it is fundamental that planning and scheduling is made in an integrated and coordinated way from procurement to delivery planning [5]. From a complementary standpoint, real-time optimization is used to address the inherent variability that characterizes any real system. It involves recomputing the plans and schedules with a high frequency to account for unpredicted events and new data that becomes available during execution. The goal is to reduce the planning cycles, and respond effectively to new requests without compromising customer service. For this purpose, the availability of efficient computational methods is essential.

The potential benefit of integrated and real-time optimization is widely recognized. Despite this clear practical relevance, research on integrated and real-time optimization at the operational level is very recent [1,2,4,10]. In this paper, we

© Springer International Publishing Switzerland 2015
O. Gervasi et al. (Eds.): ICCSA 2015, Part II, LNCS 9156, pp. 413–428, 2015.
DOI: 10.1007/978-3-319-21407-8_30

address the integrated planning and scheduling problem discussed in [7] in a setting that is characterized by a set of parallel identical machines whose activity has to be planned along a given time horizon. The planning part of the problem consists in determining first the set of jobs that will be done in each one of the time periods that are considered. At a second stage, the goal is to assign the jobs to a set of machines such that all the jobs are processed before the end of the corresponding time period. As in [7], we will assume that the processing time of the jobs are always shorter than the length of the time periods, and that each job must be completed in a single time period. Each job has a release-date and a due-date, and the objective is to determine the plan and schedule that minimizes the total costs related to the penalties incurred whenever a job is completed before of after its due-date.

In [7], the authors describe two exact solution approaches for this problem. The first one is based on the solution of a compact integer programming model that assigns jobs to machines and time periods. The second one consists in a decomposition based approach relying on an integer programming model where the planning part of the problem is considered. The feasibility of the schedules is enforced at a second stage using cutting planes, while integer solutions are sought through a branch-and-cut method.

In this paper, we explore different heuristic approaches of low complexity for the integrated planning and scheduling problem. To the best of our knowledge, this is the first contribution towards the heuristic and efficient resolution of this problem reported in the literature. We describe different constructive approaches, together with an heuristic that leads to optimal solutions for specific instances, and a local exchange procedure to improve the solutions generated by the previous methods. The importance of job sequencing is analyzed, and we prove also an important property that characterizes the optimal solutions of this problem. Additionally, we report on an extensive set of computational experiments on benchmark instances from the literature.

The paper is organized as follows. In Section 2, we introduce formally the definition and notation related to the integrated planning and scheduling problem addressed in this paper. In Sections 3 and 4, we describe and analyze a set of alternative constructive heuristics for the problem. A local search heuristic is presented in Section 5. Computational results are presented and discussed in Section 6, and some final conclusions are drawn in Section 7.

2   The Integrated Planning and Scheduling Problem

The integrated planning and scheduling problem is defined formally as follows. We are given a set $T = \{1, \ldots, \tau\}$ of time periods, such that the length of each time period is equal to $P \in \mathbb{N}$, and a set M of identical machines that are available on each of these τ time periods. Additionally, there is a set N of jobs with processing times $p_j \in \mathbb{N} \setminus \{0\}$, release-dates $r_j \in T$, due-dates $d_j \geq r_j$ with $d_j \in T$, and penalty factors $e_j, \ell_j \geq 0$ for fulfilling the job $j \in N$ earlier or later than the due-date, respectively. The set of feasible jobs for a time period $t \in T$

is $N_t := \{j \in N : r_j \leq t\}$. If a job $j \in N_t$ is finished in time period $t \in T$, then the penalty equals

$$w_t^j := e_j * \max\{0, d_j - t\} + \ell_j * \max\{0, t - d_j\}. \tag{1}$$

The problem consists in planning in which time period $t \in T$ each job will be done, and in assigning the chosen jobs to the machines, such that the sum of the processing times of the jobs to be done on any machine does not exceed the length P of the time periods, with $P \geq \max\{p_j : j \in N\}$. The objective is the minimization of the sum of all penalties.

Scheduling the assignment of machines and jobs for a given time period $t \in T$ is equivalent to solving a one-dimensional cutting stock problem, and checking if its objective function value does not exceed $|M|$, the number of available machines. Therefore, the problem is $\mathcal{NP}$-hard in the strong sense [3]. Hence, solving the problem up to optimality may be out of reach for medium and large scale instances. Moreover, even the search for initial feasible solutions may be highly time-consuming.

In this paper, we describe and analyze different fast heuristics for the integrated planning and scheduling problem, and we provide several insights that may guide the practitioner towards efficiently solving large scale instances. Note that the approaches described in this paper allow for a complete resolution of the integrated planning and scheduling problem, and they are in this sense different from the heuristics used by Kis and Kovács in [7] in their hierarchical approaches. In the latter, heuristics are used with the only purpose of testing if all jobs chosen for a certain time period can be assigned to the machines in a feasible way.

3   Fast Constructive Heuristics

In this section, we describe different polynomial heuristics based on constructive approaches. We analyze the impact of some natural priority rules on the performance of these heuristics, and we prove an important property of the optimal solutions of the problem that distinguishes it from standard scheduling problems based on the minimization of the *makespan*, *i.e.* the total time required to process all the jobs.

3.1   Definition

Assume that the jobs have been ordered according to a given priority rule. After sorting, the jobs must be assigned to machines and time periods in a feasible way. The sorting procedure has the complexity $\mathcal{O}(|N| * \ln |N|)$. After that, there are basically four variants H0–H3 of assigning the jobs one by one to free machines. Suppose that the job $j \in N$ is the next job to be assigned.

- H0: Assign the job j to the first machine with enough free capacity in the first possible time period $t \geq d_j$. As a consequence, the jobs are always postponed, whenever necessary.

- H1: Look in the first possible time period $t \geq d_j$ for the best fitting machine, *i.e.* for a machine with minimal remaining capacity not below p_j, and assign the job j to this machine in time period t;

- H2: Assign the job j first according to the rule H0. If this decision yields a penalty, then try to find the first fitting machine when job j is scheduled earlier than demanded. Formally, let c be the cost for assigning job j according to the rule H0, and set $t := d_j$. As long as the reassignment $t := t - 1$ yields $t \geq r_j$, and penalties $(d_j - t) * e_j < c$, a usable machine in the time period t is looked for, and if such a machine is found then c is replaced by the lower penalty, such that the loop is left.

- H3: Similarly, assign the job j first according to the rule H1. If scheduling the job j earlier than demanded decreases the objective function value, then look for feasible time periods like in the variant H2, and choose the best fitting machine in the corresponding time period.

The complexity of these four heuristics is linear in $|N|$ for a fixed number $|M|$ of machines and number τ of time periods. A very coarse estimation of the complexity would be $\mathcal{O}(|N| * |M| * \tau)$, but jobs need to be moved only as long as other jobs already occupied the given resources. Therefore, the complexity can be bounded by $\mathcal{O}(|N| * \min\{|M| * \tau, |N|\})$.

3.2   Evaluating the Impact of Job Sequencing

The chosen priority order may influence significantly the quality of the results obtained using H0–H3. In [7], Kis and Kovács sorted the jobs by decreasing processing times. Their objective was to check if all jobs, which were scheduled on a certain time period, could be assigned in a feasible way to the machines. This objective is totally different from ours. This "longest processing time first" rule yields very bad results in most of the benchmark test instances used in our computational experiments. In almost neglectable few cases, this rule led to improvements in comparison with a priority rule based on the relative values of the penalties. The rationale behind this latter ordering rule is the following. Usually, there are too many jobs to schedule them all in the time period when they are due. Therefore, some jobs must be postponed, while only a few jobs may be scheduled earlier than demanded. The aim is the minimization of penalties due to moving jobs to other time periods. These penalties vary absolutely, and also relatively compared to the required processing times. A job with a very long processing time and a high penalty may occupy so much resources that other jobs with short processing times and moderate penalties cost in the sum more than the long job, if they must be postponed. Therefore, a better priority rule for the heuristics described above consists in ordering the jobs by their relative penalties ℓ_j / p_j. In the case of a tie, the job with the longer processing time gets the higher priority. In this section, we further analyze the impact of job ordering on the quality of the results obtained.

Let $U \in \mathbb{N}$, with $U \gg 1$, be a general value such that the penalty factor ℓ_j for postponing a job j can become at most U. Furthermore, recall that we have $p_j, \ell_j \in \mathbb{N}$, $p_j > 0$ for all $j \in N$. Let z_H be the objective value obtained by one of the heuristics H0–H3, and let z_O be the optimal objective function value. Since z_O may be equal to 0, it makes little sense to discuss the ratio z_H/z_O. To overcome this difficulty, one may use $(z_H + 1)/(z_O + 1)$ or $\max\{1, z_H\}/\max\{1, z_O\}$. Since the optimal objective function value cannot be worse than the value obtained by a heuristic, this ratio must be at least 1. Hence, one may use an appropriate non-decreasing function $f : \mathbb{R} \to \mathbb{R}$ with $f(0) > 0$, and calculate $f(z_H)/f(z_O)$ instead of z_H/z_O.

As mentioned above, the heuristics H0–H3 depend strongly on the sequence of the jobs to be assigned. On the contrary, the distinction between first and best fit has less influence as Johnson *et al.* [6] noticed already in the case of the one-dimensional bin packing problem. For the sake of clearness, we will assume that $r_j = d_j = 1$ for all jobs. Then the heuristics H2 and H3 are equivalent to H0 and H1, respectively, and all jobs get the chance to be scheduled in the first time period, but not earlier. In the sequel, we analyze through examples the performance of the heuristics when some standard job sequences are chosen.

Consider the "longest processing time first" priority rule in which the postponing penalties are totally disregarded. The following example shows that this ordering of the jobs may lead to very poor results. This example can be compared to a special extreme case instance of the problem.

Example 1. Let $P := 100$, $|M| := 25$, $\tau \geq 2$, $N := \{1, \ldots, 50\}$ and $p_j = 50 + j$. Moreover, let $\ell_j := U$ for $1 \leq j \leq 25$, and $\ell_j := 1$ for $26 \leq j \leq 50$. The heuristics choose the long jobs with processing times 100, 99, $\ldots$, 76 first, and assign them to the first time period, while the expensive jobs regarding the postponing penalty are scheduled on the second time period. That leads to $z_H = 25 * U$. The optimal choice would have been opposite, putting the long but cheap jobs into the second time period, and assigning the short but expensive jobs to the first time period, yielding $z_O = 25$. The optimality of this choice is clear, because the jobs are pairwise incompatible on a machine in the same time period due to their lengths. One gets $z_H/z_O = U$, and this ratio can exceed in principle every fixed bound.

Another common priority rule would consist in scheduling first the jobs j with the absolute maximal penalty factor ℓ_j. This approach may also lead to bad results as illustrated in the following example.

Example 2. Let $|M| \in \mathbb{N} \setminus \{0\}$ machines and two time periods be given. Assume that there are $|M|$ jobs with processing times P and postponing penalties $\ell_j = U$, and $|M| * P$ other jobs with processing times 1 and penalty factors $\ell_j = U - 1$ to be scheduled. The heuristics assign the long jobs to the first time period and the short jobs to the second time period, resulting in $z_H = |M| * P * (U - 1)$. In this case, scheduling the long jobs in the second time period and the short jobs in the first one would yield $z_O = |M| * U$. Hence, we have $z_H/z_O = P * (U - 1)/U = P * (1 - \frac{1}{U}) \approx P$, which is also a very bad ratio.

3.3   Other Properties

In this section, we show in what extent the integrated planning and scheduling problem may differ from standard scheduling problems whose objective is the minimization of the *makespan*. For this purpose, we resort to the heuristics introduced above. We show in particular through appropriate examples that minimizing the *makespan*, *i.e.* the number of time periods used, is not necessarily compatible with the minimization of the total sum of all penalties.

From this point forward, we will assume that the jobs are ordered by relative costs for postponing jobs, *i.e.* jobs with the highest ratio ℓ_j/p_j will be scheduled first. In the case of a tie, the job with the longer processing time is the first to be scheduled. As claimed above, this ordering of the jobs yields results that outperform in the majority of the cases those obtained with the standard rules discussed in the Section 3.2. The following example shows that even under the constraint $2 \le P/p_j < 4$ for all jobs $j \in N$, optimizing the *makespan* is different from minimizing the total penalties, and that the difference in the number of used time periods may be arbitrary large.

Example 3. Let $n \in \mathbb{N} \setminus \{0\}$ be a given constant. Let $P := 7$, $|M| := 1$, $|N| := 18n$, $\tau := 7n$ and $r_j := d_j := 1$ for all $j \in N$. For $j = 1, \ldots, 6n$ there are long jobs with $p_j := 3$ and $\ell_j := 3n$. The remaining jobs are short with $p_j := 2$, $\ell_j := 1$ for $6n < j \le 18n$. Minimizing the *makespan* requires that $6n$ time periods are used, each with one long and two short jobs. The total penalties for this schedule are

$$\sum_{j=0}^{6n-1} j * (\ell_{j+1} + \ell_{j+1+6n} + \ell_{j+1+12n}) = \sum_{j=0}^{6n-1} j * (3n + 2) = 54n^3 + 27n^2 - 6n.$$ Less

penalties are caused for example by scheduling the long jobs as early as possible, namely jobs j and $j + 3n$ on the jth time period ($j = 1, \ldots, 3n$), and the short jobs after them on the time periods $3n + 1, \ldots, 7n$, such that in all the used $7n$ periods the single machine is spare for one time unit. The sum of the penalties

is $\displaystyle\sum_{j=0}^{3n-1} j * 6n + \sum_{j=3n}^{7n-1} j * 3 = 27n^3 + 51n^2 - 6n$. This schedule causes $3n^2 * (9n - 8)$

units less penalties. Nevertheless, this schedule is generally not optimal.

Example 3 is a simple example, where minimizing the *makespan* leads to a schedule with always three jobs on the machine. In the following proposition, we generalize this result by showing that, for almost any $k \in \mathbb{N}$, similar situations as in Example 3 may occur, where any k jobs fit on one machine, but not any $k + 2$ jobs. Moreover, both minimizing the *makespan* and using one of the heuristics H0–H3 would schedule always $k+1$ jobs on the machine in the same time period, while optimal schedules need more time periods.

Proposition 1. *For every $k \in \mathbb{N} \setminus \{0, 1\}$, there are instances of the integrated planning and scheduling problem with $|M| = 1$, $r_j = d_j = 1$ and $P/k \ge p_j > P/(k + 2)$, for all jobs $j \in N$, such that the heuristics H0–H3 find a schedule where in all used time periods the machine is totally busy with exactly $k+1$ jobs, while optimal schedules require at least one time period more.*

Proof. The examples of this proof are built from the instances described in [8]. For the sake of shortness, proving examples for $k < 8$ are omitted, and hence, we will assume $k \geq 8$. Let $N := \{1, \ldots, 4k + 4\}$ and $\tau := 5$. Since $r_j = d_j = 1$ for all $j \in N$, the penalty factors e_j have no influence. Let $P := 10k^2 + 6k$, $p_1 := p_{k+1} := 10k + 6$, $p_2 := p_{2k+1} := 10k + 4$, $p_j := 10k - 4$, for $j \in \{3, \ldots, 4k\} \setminus \{k, k+1, 2k, 2k+1, 3k, 3k+1\}$, $p_k := p_{4k+1} := 10k - 11$, $p_{2k} := p_{4k+2} := 10k - 7$, $p_{3k} := p_{4k+3} := 10k - 6$, $p_{3k+1} := p_{4k+4} := 10k - 2$, and $\ell_j := 10p_j - \lfloor \frac{j-1}{k} \rfloor$, for $1 \leq j \leq 4k$ and $\ell_j := 1$ for $4k < j \leq 4k + 4$.

One gets for all $i, j \in N$, with $i < j$, that $\ell_i/p_i \geq \ell_j/p_j$ and if $\ell_i/p_i = \ell_j/p_j$, hence $p_i \geq p_j$. Moreover, $P/k \geq p_j > P/(k + 2)$ for all $j \in N$ can easily be verified. The heuristics H0–H3 schedule the jobs $1, \ldots, k, 4k + 1$ on the first time period, the jobs $k+1, \ldots, 2k, 4k+2$ on the second, the jobs $2k+1, \ldots, 3k, 4k+3$ on the third, and the remaining jobs $3k + 1, \ldots, 4k, 4k + 4$ on the fourth time period, such that the machine is completely busy. The heuristics yield the penalty $z_H = 0 * (\ell_1 + \cdots + \ell_k + \ell_{4k+1}) + 1 * (\ell_{k+1} + \cdots + \ell_{2k} + \ell_{4k+2}) + 2 * (\ell_{2k+1} + \cdots + \ell_{3k} + \ell_{4k+3}) + 3 * (\ell_{3k+1} + \cdots + \ell_{4k} + \ell_{4k+4}) = 60P - 614k + 256$.

We show the existence of a better plan, and also that minimizing the *makespan* does not allow any other plan than the heuristic one. Another feasible solution is to schedule the following sets of jobs on the first, second, $\ldots$, fifth time period in this sequence: $\{1, 3, \ldots, k, k+2, k+3\}$, $\{2, k+4, \ldots, 2k, 2k+2, 2k+3, 3k\}$, $\{2k+4, \ldots, 3k-1, 3k+1, \ldots, 3k+5\}$, $\{k+1, 2k+1, 3k+6, \ldots, 4k+3\}$, and $\{4k + 4\}$ with 1, 1, 2, $p_3 - 2$ and $P - p_3 - 2$ unused time units, respectively. The given schedule yields a solution value $z = 0 * (\ell_1 + \ell_3 + \cdots + \ell_k + \ell_{k+2} + \ell_{k+3}) + 1 * (\ell_2 + \ell_{k+4} + \cdots + \ell_{2k} + \ell_{2k+2} + \ell_{2k+3} + \ell_{3k}) + 2 * (\ell_{2k+4} + \cdots + \ell_{3k-1} + \ell_{3k+1} + \cdots + \ell_{3k+5}) + 3 * (\ell_{k+1} + \ell_{2k+1} + \ell_{3k+6} + \cdots + \ell_{4k+3}) + 4 * \ell_{4k+4} = 60P - 1214k + 882$, and hence the heuristic solution is not optimal.

The processing times p_j, $j \in N$, and the length P do not allow any other combination of jobs on the single machine as in the heuristic, such that no idle time remains. To prove this, suppose that exactly $k + 1$ jobs $j \in N$ are scheduled on one time period. The differences $p_j - p_3$ are used instead of the processing times p_j. That leads to the following problem: given items of sizes $+10, +8, +2, 0, -2, -3, -7$, each of them given twice (except the size 0), exactly four bins of capacity 4 should be filled. If the items of size 0 are neglected, then $4 = 10 + 8 - 2 * 7 = 10 - 2 * 3 = 8 - 2 * 2 = 2 * 2$ represents the unique solution, because the odd numbers -3 and -7 can together be used only in even number per bin, such that only the -2 remains as negative number. Hence, one bin must be used with $4 = 8 - 2 * 2$, and another with $4 = 2 * 2$. Therefore, every other combination of jobs requires at least five time periods. $\square$

Even under the conditions of Proposition 1, the integrated planning and scheduling problem remains more difficult than the one-dimensional bin packing problem. If there are more machines or the release and due-dates of the jobs are more general, the deviation between minimizing the total sum of the penalties and minimizing the *makespan* will expectably increase.

4   A Heuristic Based on the Transportation Problem

In this section, we describe a new heuristic for the integrated planning and scheduling problem that yields the optimal solution of the problem for specific instances. The first step of this heuristic consists in determining the longest jobs, such that for a fixed number $k \in \mathbb{N}$ exactly k jobs fit on a machine in one time period, but no combination of $k + 1$ of these jobs fits. The second step will be to schedule these longest jobs in an optimal way, if the other jobs are neglected. At the end, the shorter jobs are added by using one of the heuristics H0–H3. Let

$$k := \min_{j \in N} \lfloor P/p_j \rfloor, \tag{2}$$

and initially $L := \{j \in N : p_j > P/(k+1)\}$. Clearly, exactly k jobs with processing times $p_j \in \left(\frac{P}{k+1}, \frac{P}{k} \right]$ may be scheduled on the same machine in the same time period. As long as $L \neq N$, and the job $j \in N \setminus L$ with the longest processing time p_j cannot be combined with any set of k jobs of L on the same machine in any time period due to the processing times, one may set $L := L \cup \{j\}$. Hence, any $k+1$ jobs $j \in L$ still do not fit on one machine. The following example illustrates how L is determined.

Example 4. Given $P := 75$, the processing times of jobs in descending order are $35, 31, 29, 25, 23, 22, \ldots$ Then, we have $k = 2$. Since $75 < 31 + 29 + 25$ and also $75 < 29 + 25 + 23$, the set L contains the first five jobs, *i.e.* any three jobs of L do not fit on one machine in spite of $23 < P/3$. The job with processing time 22 cannot be included in L, because $75 \geq 25 + 23 + 22$.

The objective consists in scheduling all jobs $j \in L$ such that the total penalties for these jobs are minimized, *i.e.* each job $j \in L$ shall be assigned in an optimal way to one of the $|M| * \tau$ possible pairs of machine and time period. Since at most k jobs of L fit on one machine in a time period, this can be modelled as a transportation problem as follows. Suppose that there are $|L|$ producers who offer their single unit of a good to $|M| * \tau$ customers. The transportation cost for the good equals its penalty (1) if it is feasible. Infeasible assignments, where a job is scheduled before its release date, get infinite penalty. Each customer demands exactly k units of that good. If necessary, fictitious jobs without penalties are added to L in the required number. If one has $|L| > k * |M| * \tau$, then the instance is infeasible. Indeed, it is impossible to assign the jobs to the machines and time periods, because there are too many jobs with long processing times and only k jobs per machine and time period can be fulfilled. After assigning these $|L|$ largest jobs, the remaining ones may be assigned by applying one of the heuristics H0–H3. Algorithm 1 describes formally this method.

The following proposition shows that Algorithm 1 yields the optimal solution of the integrated planning and scheduling problem for specific instances.

Proposition 2. *Consider an instance of the integrated planning and scheduling problem in which $\lfloor P/p_j \rfloor$ is the same for all jobs $j \in N$. Then, Algorithm 1 yields the optimal solution of this instance with small complexity for fixed τ and $|M|$.*

Algorithm 1. Heuristic Based on the Transportation Problem

Input:
 Set M of machines; length P and number τ of time periods;
 Set N of jobs j with release dates r_j, due dates d_j, processing times p_j, and
 penalty factors e_j, ℓ_j;
Output:
 A feasible schedule for all jobs;
Auxiliary variables:
 h, k, $n \in \mathbb{N}$;
Sort the jobs in non decreasing order of p_j.
Let $k := \lfloor P/p_{|N|} \rfloor$;
if $k \geq |N|$ **then** let $k := |N|$ and $n := 0$;
else
 | Let $n := |N| - k$;
 | **while** $(k+1) * p_n > P$ **do**
 | | $n := n - 1$;
 | | **if** $n = 0$ **then goto** Label;
 | **end**
 | $h := P - \sum\limits_{j=n+1}^{n+k} p_j$;
 | **while** $n > 0 \land h < p_n$ **do**
 | | $h := h + p_{n+k} - p_n$;
 | | $n := n - 1$;
 | **end**
end
Label: Solve the transportation problem instance with producers $n + 1, \ldots, |N|$ and $|M| * \tau$
customers. Schedule the jobs $n + 1, \ldots, |N|$ according to the corresponding optimal
transportation plan;
while $n > 0$ **do**
 | Schedule job n according to one of the heuristics H0–H3;
 | $n := n - 1$;
end

The same holds, if N does not contain $k + 1$ jobs with a sum of their processing times less than or equal to P, with k defined as in (2).

Proof. If N does not contain $k + 1$ jobs with a sum of their processing times less than or equal to P, we have that either $k \geq |N|$ or $k < |N|$, but the $k + 1$ shortest jobs do not fit on one machine. In any case, at most k jobs can be put on one machine in a time period. If $k \geq |N|$, then $L = N$. Otherwise, the algorithm checks if $\lfloor P/p_n \rfloor = k$. If one has to jump to the label due to $n = 0$, then $\lfloor P/p_j \rfloor$ is the same for all jobs $j \in N$. Otherwise, h is set to the remaining time, when the k shortest jobs of L are put onto one machine. As long as $h < p_n$, including job n to the set L still does not allow any combination of $k + 1$ jobs on a machine. The loop with "while $n > 0 \land h < p_n$ do" recalculates h accordingly. The prerequisites of the proposition are fulfilled if and only if the label is reached with $n = 0$. The complexity of the entire statement "if $k \geq |N|$ then ..." is at most $\mathcal{O}(|N|)$. The most complex part is the solution of the transportation problem, which may be reformulated as an assignment problem. Its cost matrix has $|M| * \tau * k$ rows and columns. Therefore, the complexity of this optimization is at most $\mathcal{O}((|M| * \tau * k)^3)$. If $k \geq |N|$, then all jobs fit on a single machine, such that setting $k := |N|$ does not change anything except that it bounds k, and therefore the complexity to at most $\mathcal{O}((|M| * \tau * |N|)^3)$. Hence, if the prerequisites are fulfilled, then all jobs are assigned according to the optimal transportation plan, and the complexity remains polynomial.   □

5   A Local Exchange Heuristic

Given a feasible solution, one may improve the quality of the plan and schedule by applying the following local exchange procedure. Let $i, j \in N$ be jobs to be scheduled in different time periods $t_i, t_j \in T$. If $t_i \geq r_j \wedge t_j \geq r_i$, and if the capacities of the used machines allow it, then an exchange of the jobs i, j is feasible, *i.e.* job i is moved to the machine and time period of j and job j is moved similarly to the place of i. These exchanges may be performed if they improve the objective function value, until no more pairs of jobs allow for an improvement by exchanging them. In our computational study, we considered in particular the following two strategies:

- L0: Use the first pair of jobs for the exchange;
- L1: Use the best pair of jobs for the exchange.

If the exchanged jobs have different processing times, then the capacity of the machine, from which the job with the shorter processing time was removed, is used better, while on the other machine, from which the longer job was removed, more capacity becomes available. Hence, another job can move to that machine, possibly making enough space free for further movements of jobs.

This local exchange heuristic is described formally in Algorithm 2. Each job j is assigned a number $m_j \in \{0, \ldots, |M| * \tau - 1\}$, which identifies the time period $t_j = 1 + \lfloor m_j/|M| \rfloor$ that is used, and the corresponding machine $1 + m_j - |M| * \lfloor m_j/|M| \rfloor \in \{1, \ldots, |M|\}$. The expression $m_j - |M| * \lfloor m_j/|M| \rfloor$ is the remainder when m_j is divided by $|M|$. Since the machines are numbered as $1, \ldots, |M|$, we had to add 1. The search for the first or best pair of jobs to be exchanged is done between Label_1 and Label_2. In the repeat-loop, a single job is moved from one machine and time period to the other, where more capacity became available, if that movement is feasible, and if it improves the objective function value. The complexity of the loop between Label_1 and Label_2 is quadratic in $|N|$, and each repetition in the later repeat-loop has a linear complexity in $|N|$.

6   Computational Experiments

To evaluate and compare the performance of the heuristics described in this paper, we conducted an extensive set of computational experiments on 675 benchmark instances of the literature [7]. The instances are divided into three classes A, B and C. Within each class, there are five different instances for a fixed number $|N|$ of jobs, $|M|$ of machines and τ_0 of time periods. For class A, we have $|N| \in \{100, 150, 200, 250, 300\}$, while in class B and C, the number of jobs varied respectively as follows: $|N| \in \{50, 100, 150, 200, 250, 300\}$ and $|N| \in \{40, 60, 80, 100\}$. In each class, the number of machines and time periods were equal to $|M| \in \{2, 6, 10\}$ and $\tau_0 \in \{2, 6, 10\}$, respectively. For all the

Algorithm 2. Local Exchange Heuristic

Input:

    Set M of machines; length P and number τ of time periods;

    Set N of jobs j with release dates r_j, due dates d_j, processing times p_j, and penalty factors e_j, ℓ_j;

    Feasible schedule for all jobs with a vector $m \in \{0, \ldots, |M| * \tau - 1\}^{|N|}$;

    $w \in \{0, 1\}$, with $w = 1$ if the best pair of jobs for exchanges shall be chosen;

Output:

    A feasible and possibly improved schedule for all jobs;

Auxiliary variables:

    $i, j, s, t_1, t_2, v_0, v_1, v_2 \in \mathbb{N}$ and vector $\mathbf{c} \in \mathbb{N}^{|M|*\tau}$ for free machine capacities;

for $i := 1, \ldots, |M| * \tau$ **do** $c_i := P$;

foreach $i \in N$ **do** $c_{1+m_i} := c_{1+m_i} - p_i$;

$v_0 := 0$;

Label_1:

foreach $i \in N$ **do**

    $t_1 := 1 + \lfloor m_i/|M| \rfloor$;

    for $j := i - 1$ **downto** 1 **do**

        $t_2 := 1 + \lfloor m_j/|M| \rfloor$;

        if $p_i - p_j \leq c_{1+m_j} \wedge p_j - p_i \leq c_{1+m_i} \wedge r_i \leq t_2 \wedge r_j \leq t_1$ **then**

            $s := w_{t_1}^i + w_{t_2}^j - w_{t_2}^i - w_{t_1}^j$ according to (1);

            if $s > v_0$ **then**

                $v_0 := s$, $v_1 := i$, $v_2 := j$;

                if $w = 0$ **then goto** Label_2;

            end

        end

    end

end

if $v_0 = 0$ **then return**;

Label_2: /* Exchange jobs v_1 and v_2.   */

$i := v_1$, $j := v_2$, $s := m_i$, $c_{1+s} := c_{1+s} + p_i - p_j$, $c_{1+m_j} := c_{1+m_j} + p_j - p_i$, $m_i := m_j$, $m_j := s$;

if $p_i < p_j$ **then** $s := m_i$;

repeat

    $t_1 := 1 + \lfloor s/|M| \rfloor$;   /* A job may be moved to this time period. */

    $v_0 := 0$;

    foreach $i \in N$ **do**

        if $p_i \leq c_{1+s} \wedge r_i \leq t_1$ **then**

            $t_2 := w_{1+\lfloor m_i/|M| \rfloor}^i - w_{t_1}^i$ according to (1);

            if $t_2 > v_0$ **then** $v_0 := t_2$, $j := i$;

        end

    end

    if $v_0 = 0$ **then goto** Label_1;

    $c_{1+m_j} := c_{1+m_j} + p_j$, $t_2 := m_j$, $m_j := s$, $c_{1+s} := c_{1+s} - p_j$, $s := t_2$;

until *false*;

---

instances, the length P of each time period was set to 100. Class A is characterized by short jobs with $p_j < P/3$ only. Class C consisted of only long jobs with $p_j > P/3$, while in class B half of the jobs are long and half are short. To guarantee that all the instances are feasible, Kis and Kovács [7] considered additional time periods where the jobs could be finished late, such that $\tau = \tau_0 + \lceil \frac{2}{P*|M|} * \sum_{j \in N} p_j \rceil$. Note that in [7], the authors did not report any result for the largest instances of class A, B and C, while here, we considered the whole set of instances provided by these authors. The tests were conducted on a PC with 3.7 GB of RAM, and processors Intel Core i5 running with 2.4 GHz.

In Tables 1-9, we report on the average computational results obtained with the different heuristics on sets of 15 instances with a common number $|N|$ of jobs and $|M|$ of machines from the classes A, B and C. The quality of the solutions is compared to the solutions provided by an exact solution approach for the problem described in [9], given that lower and upper bounds for these instances were not provided in [7]. This exact method ran under a time limit of 1200 seconds per instance. Note that the computing times required to run the heuristics are typically very low (a few milliseconds), and hence, they are not reported in the tables. The meaning of the columns in the tables is the following:

- *best lb*: average best lower bound from the exact method described in [9];
- *best ub*: average best upper bound from the exact method described in [9];
- *gap*: average optimality gap (%) between the previous bounds, *i.e.* the average among all the considered instances i of $(ub_i - lb_i)/ub_i$ with ub_i and lb_i denoting respectively an upper and lower bound for the instance i;
- H0–H3: average solution value obtained with heuristics H0–H3, respectively;
- L0–L1: average solution value obtained with the local exchange procedures L0–L1, with the initial solution being the best among those from H0–H3;
- L0+–L1+: average solution value obtained with the local exchange procedures L0–L1, with the initial solution computed using the heuristic based on the transportation problem;
- v_{best}: average best solution value given by the methods referred to in the corresponding table;
- *rg*: relative gap (%) between v_{best} and *best ub*, *i.e.* $(v_{best} - best\ ub)/v_{best}$;
- *best*: number of times the corresponding method provided the best solution among the heuristics considered in the corresponding table;
- *largest relative post. cost first* and *largest absolute post. cost first*: indicates that the jobs are ordered respectively according to the "largest relative post-poning cost first" and the "largest absolute postponing cost first" rule in the heuristics that are listed in the associated columns.

In Tables 1-3, we report on the results for the instances of class A. Table 1 shows that the average best solution given by H0–H3 is in general very near to best solution provided by the exact solution method, which has in turn a small optimality gap for most of the instances of this class. Among the four heuristics, H3 is the one that provides more consistently the best results. From Tables 2 and 3, we observe that the local exchange procedures lead to solutions that are better in many cases to the best upper bound given by the exact method. Furthermore, the results confirm the observation we made above that the "largest relative postponing cost first" sequencing rule yields typically the best solutions.

The results for the instances of class B are reported in Tables 4-6. In this case, the heuristics H0–H3 found solutions with gaps compared to the best upper bounds provided by the exact method that are usually larger than in class A. However, in many different subsets, these heuristics generated solutions whose quality is equivalent to these best upper bounds. Moreover, the local

Table 1. Computational results for the heuristics H0–H3: instances of class A

$|N|$	$|M|$	best lb	best ub	gap	H0	H1	H2	H3	v_{best}	rg	H0 best	H1 best	H2 best	H3 best
100	2	449,3	458,0	3,1	562,0	560,5	529,8	530,1	521,5	13,9	6	7	7	7
100	6	35,9	35,9	0,1	56,5	56,3	46,2	45,7	45,7	27,1	10	10	13	15
100	10	7,0	7,0	0,0	12,3	12,2	11,9	11,7	11,7	67,6	12	12	14	15
150	2	1432,4	1617,7	26,7	1746,6	1746,3	1655,0	1651,3	1640,5	1,4	2	3	8	9
150	6	128,2	128,8	0,3	152,3	151,9	160,9	158,1	148,0	14,9	10	11	8	9
150	10	38,1	38,1	0,0	55,7	55,7	49,9	49,8	49,5	29,7	11	11	13	14
200	2	2975,7	3319,3	20,2	3440,3	3439,5	3337,7	3341,7	3010,0	0,0	5	6	5	5
200	6	345,2	365,9	11,0	000,0	302,7	415,4	414,7	300,1	0,1	12	13	5	6
200	10	96,5	97,8	0,5	124,9	124,3	118,3	117,9	117,5	20,1	10	10	13	14
250	2	4438,6	5605,6	7,2	5857,9	5856,7	5649,3	5639,0	5605,6	0,0	3	5	5	8
250	6	678,7	758,3	21,7	788,7	785,3	809,0	805,5	771,1	1,7	3	9	3	5
250	10	207,2	218,3	1,8	241,9	241,5	257,8	257,5	239,1	9,5	11	12	8	9
300	2	0,0	9079,3	-	9383,3	9376,1	9128,6	9102,7	9079,3	0,0	4	5	4	8
300	6	1185,0	1360,2	48,1	1422,5	1420,2	1402,0	1396,9	1366,7	0,5	5	7	4	5
300	10	415,6	448,3	17,3	470,7	469,9	500,2	499,3	467,5	4,3	9	11	7	7

Table 2. Computational results for the heuristics L0–L1: instances of class A (Part I)

$|N|$	$|M|$	best lb	best ub	gap	largest relative post. cost first L0+	L1+	L0	L1	largest absolute post. cost first L0+	L1+	L0	L1
100	2	449,3	458,0	3,1	616,1	641,9	500,2	498,8	791,9	792,1	598,7	603,0
100	6	35,9	35,9	0,1	50,5	48,5	42,0	42,1	68,7	70,9	50,1	50,1
100	10	7,0	7,0	0,0	11,9	12,0	11,0	11,0	15,0	15,3	13,5	13,5
150	2	1432,4	1617,7	26,7	1871,8	1968,4	1545,4	1541,1	2572,4	2571,8	1872,3	1869,6
150	6	128,2	128,8	0,3	175,2	174,3	143,3	143,3	240,5	243,9	171,1	170,3
150	10	38,1	38,1	0,0	61,4	58,1	45,5	45,4	75,3	77,5	51,0	51,2
200	2	2975,7	3319,3	20,2	3836,7	3968,5	3153,1	3146,6	5421,8	5425,5	3962,7	3965,8
200	6	345,2	365,9	11,0	482,5	481,1	381,3	381,5	725,5	723,5	470,2	471,9
200	10	96,5	97,8	0,5	133,5	136,2	105,9	106,6	201,6	197,6	132,7	133,1
250	2	4438,6	5605,6	7,2	6613,3	6957,1	5401,9	5404,6	9607,9	9652,3	6962,1	6949,3
250	6	678,7	758,3	21,7	930,1	978,5	735,1	734,9	1323,7	1327,5	948,2	946,3
250	10	207,2	218,3	1,8	289,8	295,6	230,0	230,3	389,1	396,8	292,7	293,3
300	2	0,0	9079,3	-	10632,9	11122,4	8729,5	8713,7	15155,1	15161,3	11333,5	11324,7
300	6	1185,0	1360,2	48,1	1616,9	1666,8	1302,0	1301,5	2382,7	2380,8	1666,1	1660,3
300	10	415,6	448,3	17,3	584,5	574,6	455,7	456,4	857,6	862,9	574,7	574,3

Table 3. Computational results for the heuristics L0–L1: instances of class A (Part II)

$|N|$	$|M|$	v_{best}	rg	larg. rel. post. cost first L0+ best	L1+ best	L0 best	L1 best	larg. abs. post. cost first L0+ best	L1+ best	L0 best	L1 best
100	2	496,1	8,3	2	1	4	10	0	0	2	2
100	6	41,5	15,4	9	9	12	12	10	10	8	8
100	10	10,2	45,7	12	13	12	12	11	11	11	11
150	2	1534,7	-5,1	0	0	5	11	0	0	0	0
150	6	142,0	10,2	4	4	12	12	5	5	5	6
150	10	45,3	18,7	9	9	12	14	10	10	10	10
200	2	3140,5	-5,4	0	0	6	9	0	0	0	0
200	6	375,4	2,6	3	2	7	6	1	1	2	2
200	10	105,7	8,0	10	10	14	11	10	10	10	10
250	2	5391,8	-3,8	0	0	11	5	0	0	0	0
250	6	729,8	-3,8	0	0	7	6	1	0	2	3
250	10	226,4	3,7	6	6	11	9	5	5	6	6
300	2	8709,4	-4,1	0	0	6	9	0	0	0	0
300	6	1294,2	-4,9	1	0	8	7	0	0	1	1
300	10	454,4	1,4	4	4	7	9	3	3	3	4

exchange procedures could even improve these average upper bounds in 5 subsets of instances. In general, the comparative results given in Tables 4-6 confirm most of the observations made for the instances of class A.

Table 4. Computational results for the heuristics H0–H3: instances of class B

$|N|$	$|M|$	best lb	best ub	gap	H0	H1	H2	H3	v_{best}	rg	H0 best	H1 best	H2 best	H3 best
50	2	346,1	346,1	0,0	515,5	515,7	477,7	470,4	463,1	33,8	4	3	7	11
50	6	31,3	31,3	0,0	46,5	46,1	49,1	48,7	44,9	43,4	9	11	11	12
50	10	8,9	8,9	0,0	16,3	16,3	14,1	14,1	14,1	57,5	10	10	15	15
100	2	1956,9	1974,9	1,2	2505,5	2504,3	2333,9	2339,7	2326,3	17,8	2	2	11	9
100	6	285,1	285,4	0,2	425,2	420,5	404,7	395,8	391,5	37,2	0	4	7	11
100	10	89,6	89,6	0,0	130,4	129,5	135,4	134,7	125,6	40,2	8	9	8	8
150	2	5498,9	6234,7	14,8	6573,9	6541,9	6291,9	6266,8	6258,7	0,4	3	3	3	9
150	6	875,3	879,0	0,7	1196,5	1184,7	1140,9	1128,8	1119,0	27,3	1	4	3	9
150	10	329,8	330,2	0,2	458,5	452,9	458,0	448,0	433,9	31,4	4	7	2	8
200	2	10808,8	12011,1	12,1	12321,5	12289,7	12068,1	12060,9	12011,1	0,0	1	4	5	7
200	6	1997,6	2015,8	1,5	2616,5	2599,8	2512,7	2481,3	2464,6	22,3	1	4	2	8
200	10	711,1	713,0	0,7	975,1	963,7	946,7	931,1	918,4	28,8	0	5	2	10
250	2	2755,7	19765,7	5,7	20201,9	20198,6	19844,5	19776,1	19765,7	0,0	3	1	2	11
250	6	3570,7	3912,9	13,7	4482,1	4441,0	4329,1	4307,5	4265,8	9,0	1	4	3	8
250	10	1401,1	1413,8	1,6	1938,3	1906,1	1849,0	1824,7	1798,8	27,2	1	4	2	9
300	2	0,0	28786,7	-	29249,3	29217,3	28909,7	28825,6	28786,7	0,0	0	3	4	9
300	6	5761,0	6577,0	18,6	7008,9	6967,9	6745,7	6682,5	6649,9	1,1	0	6	1	8
300	10	2446,1	2668,5	10,8	3297,0	3261,9	3094,5	3068,6	3047,7	14,2	0	4	3	8

Table 5. Computational results for the heuristics L0–L1: instances of class B (Part I)

$|N|$	$|M|$	best lb	best ub	gap	largest relative post. cost first L0+	L1+	L0	L1	largest absolute post. cost first L0+	L1+	L0	L1
50	2	346,1	346,1	0,0	517,4	519,1	427,9	428,7	535,2	533,1	438,0	437,9
50	6	31,3	31,3	0,0	53,0	52,6	42,7	42,4	53,3	55,5	43,1	43,1
50	10	8,9	8,9	0,0	14,5	14,5	13,1	13,1	15,5	15,5	14,8	14,8
100	2	1956,9	1974,9	1,2	2969,5	3032,1	2252,7	2250,5	3088,6	3091,8	2501,3	2498,6
100	6	285,1	285,4	0,2	465,7	466,7	363,3	364,5	489,4	490,9	402,3	403,5
100	10	89,6	89,6	0,0	147,5	151,6	120,8	121,1	158,4	158,3	115,5	115,5
150	2	5498,9	6234,7	14,8	8037,5	8152,9	6037,6	6036,9	8589,1	8589,4	7185,7	7186,3
150	6	875,3	879,0	0,7	1407,5	1411,3	1057,5	1051,3	1494,3	1499,9	1175,1	1175,9
150	10	329,8	330,2	0,2	521,4	535,8	412,0	414,4	591,1	590,6	452,0	452,2
200	2	10808,8	12011,1	12,1	15295,5	15384,7	11646,7	11649,3	16226,9	16222,0	14066,3	14059,4
200	6	1997,6	2015,8	1,5	3066,1	3064,6	2346,9	2339,1	3368,8	3371,9	2712,9	2709,1
200	10	711,1	713,0	0,7	1144,7	1135,0	873,9	874,9	1266,5	1267,0	999,4	999,1
250	2	2755,7	19765,7	-	25036,7	25269,7	19109,7	19178,2	27246,2	27236,9	23918,7	23922,5
250	6	3570,7	3912,9	13,7	5390,9	5462,1	4052,0	4069,5	5894,5	5890,5	4803,6	4796,1
250	10	1401,1	1413,8	1,6	2137,4	2167,5	1674,9	1673,7	2368,1	2370,8	1931,5	1927,8
300	2	0,0	28786,7	-	36383,7	36642,0	28044,8	28114,0	39763,5	39777,9	35352,7	35382,7
300	6	5761,0	6577,0	18,6	8234,3	8391,1	6399,5	6406,2	9159,7	9155,6	7834,9	7832,9
300	10	2446,1	2668,5	10,8	3694,2	3718,3	2853,8	2862,1	4096,2	4100,3	3414,1	3408,9

Table 6. Computational results for the heuristics L0–L1: instances of class B (Part II)

$|N|$	$|M|$	v_{best}	rg	larg. rel. post. cost first L0+ best	L1+ best	L0 best	L1 best	larg. abs. post. cost first L0+ best	L1+ best	L0 best	L1 best
50	2	412,7	19,2	0	0	6	4	0	1	7	7
50	6	40,2	28,3	7	7	8	8	6	6	9	9
50	10	12,2	36,6	10	12	12	12	10	10	12	12
100	2	2235,2	13,2	0	0	6	8	0	0	3	3
100	6	349,7	22,5	0	1	7	7	0	0	6	5
100	10	112,8	25,9	6	5	4	4	5	5	12	12
150	2	6020,7	-3,4	0	0	8	7	0	0	0	0
150	6	1040,9	18,4	0	0	5	10	0	0	3	3
150	10	400,1	21,2	0	1	6	5	0	0	6	5
200	2	11620,1	-3,3	0	0	7	8	0	0	0	0
200	6	2327,9	15,5	0	0	2	10	0	0	2	2
200	10	853,3	19,7	0	0	6	6	0	0	4	5
250	2	19081,6	-3,5	0	0	8	7	0	0	0	0
250	6	4046,8	3,4	0	0	11	5	0	0	0	0
250	10	1664,3	17,7	0	0	5	9	0	0	0	3
300	2	28026,3	-2,6	0	0	11	4	0	0	0	0
300	6	6387,3	-2,9	0	0	10	6	0	0	0	0
300	10	2840,9	6,5	0	1	8	3	0	0	1	2

Table 7. Computational results for the heuristics H0–H3: instances of class C

											H0	H1	H2	H3
$|N|$	$|M|$	best lb	best ub	gap	H0	H1	H2	H3	v_{best}	rg	best	best	best	best
40	2	741,1	741,1	0,0	945,9	944,1	875,2	872,2	871,1	17,5	4	4	10	13
40	6	86,1	86,1	0,0	108,3	108,3	105,1	104,5	102,5	19,1	7	7	11	11
40	10	20,7	20,7	0,0	29,5	29,5	27,0	27,0	26,9	29,9	10	10	14	14
60	2	1868,4	1868,4	0,0	2244,7	2241,9	2132,9	2125,5	2115,0	13,2	4	5	8	9
60	6	291,4	291,4	0,0	382,7	379,7	365,7	361,5	356,5	22,4	4	6	7	8
60	10	91,7	91,7	0,0	123,0	122,1	117,3	116,3	114,1	24,5	6	7	10	11
80	2	3770,3	3770,3	0,0	4280,1	4290,1	1000,0	1011,5	1176,3	10,5	5	5	10	9
80	6	653,3	653,0	0,0	856,5	853,1	810,2	804,6	794,4	21,0	4	6	6	4
80	10	218,8	218,8	0,0	290,2	287,7	283,9	282,2	273,1	24,8	2	3	10	10
100	2	6473,7	6473,7	0,0	7385,7	7385,6	7101,1	7096,1	7095,5	9,6	2	2	11	14
100	6	1140,8	1140,8	0,0	1484,1	1475,5	1418,8	1410,3	1399,6	22,7	2	4	6	10
100	10	420,8	420,8	0,0	555,3	550,5	553,1	546,5	533,5	26,8	5	7	5	7

Table 8. Computational results for heuristics L0–L1: instances of class C (Part I)

					largest relative post. cost first				largest absolute post. cost first			
$|N|$	$|M|$	best lb	best ub	gap	L0+	L1+	L0	L1	L0+	L1+	L0	L1
50	2	741,1	741,1	0,0	913,5	913,7	809,7	810,7	911,0	912,5	830,3	830,9
50	6	86,1	86,1	0,0	112,2	113,0	95,3	95,9	113,3	113,5	96,5	96,5
50	10	20,7	20,7	0,0	28,3	28,2	25,7	25,7	28,3	28,2	25,8	25,8
100	2	1868,4	1868,4	0,0	2218,7	2237,5	1979,7	1977,2	2211,9	2225,9	2066,1	2060,9
100	6	291,4	291,4	0,0	359,9	358,7	332,5	332,5	352,5	351,5	338,9	338,6
100	10	91,7	91,7	0,0	116,1	116,1	107,0	107,2	112,7	112,5	103,1	103,0
150	2	3779,3	3779,3	0,0	4493,5	4493,9	3978,5	3981,9	4443,3	4454,9	4186,8	4184,7
150	6	653,3	653,0	0,0	798,0	799,9	729,7	732,1	796,8	796,9	766,5	765,5
150	10	218,8	218,8	0,0	256,3	256,9	252,8	252,2	260,9	261,2	260,1	258,1
200	2	6473,7	6473,7	0,0	7785,1	7783,5	6707,4	6734,3	7725,5	7727,3	7210,4	7207,5
200	6	1140,8	1140,8	0,0	1415,2	1417,2	1272,5	1273,4	1406,5	1405,6	1334,4	1338,2
200	10	420,8	420,8	0,0	539,3	538,6	493,8	493,5	538,3	539,5	496,6	496,0

Table 9. Computational results for the heuristics L0–L1: instances of class C (Part II)

				larg. rel. post. cost first				larg. abs. post. cost first			
				L0+	L1+	L0	L1	L0+	L1+	L0	L1
$|N|$	$|M|$	v_{best}	rg	best	best	best	best	best	best	best	best
50	2	798,4	7,7	1	0	8	4	1	0	4	5
50	6	91,7	6,6	5	5	7	5	6	5	9	9
50	10	23,1	11,6	11	11	10	10	11	11	6	6
100	2	1968,7	5,4	0	0	8	8	0	0	0	2
100	6	318,5	9,3	2	5	8	6	2	5	2	3
100	10	99,5	8,5	6	6	5	5	5	5	8	8
150	2	3965,5	4,9	0	0	10	7	1	1	0	0
150	6	712,2	9,0	2	1	6	5	2	2	1	3
150	10	243,4	11,2	4	4	6	7	4	4	4	5
200	2	6689,3	3,3	0	0	10	6	0	0	0	0
200	6	1249,3	9,5	2	3	5	4	2	2	3	2
200	10	485,0	15,3	2	2	5	5	3	3	6	5

In Tables 7-9, we provide the results obtained for the instances of class C. Note that, for these instances, the exact solution method always finds the optimal solution within the computing time limit. The average gap between the best solutions provided by H0–H3 is equal to 20,2%, while it goes down to 8,5% when the local exchange procedure is considered. Again, the "largest relative postponing cost first" sequencing rule yields in general the best results, while H3 provides typically better results than the H0, H1 and H2.

7 Conclusions

We described and analyzed the first low complexity heuristics for the integrated planning and scheduling problem. Different methods were proposed including constructive heuristics, an heuristic based on the transportation problem that yields optimal solutions for specific instances, and a local exchange procedure that proved to be effective in improving the quality of the solutions. The results of extensive computational experiments attest the capacity of these approaches in generating good quality solutions in a very small amount of time.

Acknowledgments. This work was supported by FEDER funding through the Programa Operacional Factores de Competitividade - COMPETE and by national funding through the Portuguese Science and Technology Foundation (FCT) in the scope of the project PTDC/EGE-GES/116676/2010 (FCOMP-01-0124-FEDER-020430), and by FCT within the project scope: UID/CEC/00319/2013.

References

1. Armstrong, R., Gao, S., Lei, L.: A 0-inventory production and distribution problem with a fixed customer sequence. Annals of Operations Research **159**, 395–414 (2008)
2. Chang, Y., Lee, C.: Machine scheduling with job delivery coordination. European Journal of Operational Research **158**, 470–487 (2004)
3. Garey, M., Johnson, D.: Strong NP-completeness results: motivation, examples, and implications. Journal of the ACM **25**, 499–508 (1978)
4. Geismar, H., Laporte, G., Lei, L., Sriskandarajah, C.: The integrated production and transportation scheduling problem for a product with a short life span and non-instantaneous transp. time. INFORMS Journal on Computing **20**, 21–33 (2008)
5. Grossmann, I., Furman, K.: Challenges in enterprise-wide optimization for the process industries. In: Chaovalitwongse, W., Furman, K., Pardalos, P. (eds.) Optimization and Logistics Challenges in the Enterprise, pp. 3–60. Springer (2009)
6. Johnson, D., Demers, A., Ullman, J., Garey, M., Graham, R.: Worst case performance bounds for simple one-dimensional packing algorithms. SIAM Journal on Computing **3**, 299–325 (1974)
7. Kis, T., Kovács, A.: A cutting plane approach for integrated planning and scheduling. Computers and Operations Research **39**, 320–327 (2012)
8. Rietz, J.: Untersuchungen zu MIRUP für Vektorpackprobleme. Ph.D. thesis, Technical University for Mining and Technology of Freiberg (2003)
9. Rietz, J., Alves, C., Carvalho, J.: A new exact approach for the integrated planning and scheduling problem. Working paper (2015)
10. Wang, X., Cheng, T.: Machine scheduling with an availability constraint and job delivery coordination. Naval Research Logistics **54**, 11–20 (2007)

On Computing Order Quantities for Perishable Inventory Control with Non-stationary Demand

Alejandro G. Alcoba[1]([✉]), Eligius M.T. Hendrix[1], Inmaculada García[1],
Gloria Ortega[2], and Karin G.J. Pauls-Worm[3], and Rene Haijema[3]

[1] Computer Architecture, Universidad de Málaga, Malaga, Spain
{agutierrez,eligius,igarciaf}@uma.es
[2] Informatics, University of Almería, Agrifood Campus
of International Excellence, ceiA3, Almería, Spain
gloriaortega@ual.es
[3] Operations Research and Logistics, Wageningen University,
Wageningen, The Netherlands
{karin.pauls,rene.haijema}@wur.nl

Abstract. The determination of order quantities in an inventory control
problem of perishable products with non-stationary demand can be formulated as a Mixed Integer Nonlinear Programming problem (MINLP).
One challenge is to deal with the β-service level constraint in terms of the
loss function. This paper studies the properties of the optimal solution
and derives specific algorithms to determine optimal quantities.

Keywords: Inventory control · Perishable products · MINLP · Loss
function · Monte Carlo

1 Introduction

The basis of our study is a fill rate variant of a Stochastic Programming (SP)
model presented in [4] for a practical production planning problem over a finite
horizon of T periods of a perishable product with a fixed shelf life of J periods.
Items of age J cannot be used in the next period and are considered waste.
Demand is stochastic and non-stationary, as it changes over time. To keep waste
due to out-dating low, one issues the oldest product first, i.e. FIFO (first in, first
out) issuance. Literature provides many ways to deal with perishable products,
order policies and backlogging, e.g. [1,3,7]. The model we investigate aims to
guarantee a β-service level constraint; the supplier guarantees expected shortage
not to exceed $(1 - \beta)\%$ of the expected demand for every period. Not fulfilled
demand is lost.

   The solution for such a model is a so-called order policy. We consider a policy
with a list of order periods Y with order quantities Q_t. The question is to derive
a set of order quantities that leads to minimum cost and fulfils the service level
constraint.

   Section 2 introduces the model under study. Properties of the order quantities
are studied in Section 3. Section 4 then describes several algorithms to determine

© Springer International Publishing Switzerland 2015
O. Gervasi et al. (Eds.): ICCSA 2015, Part II, LNCS 9156, pp. 429–444, 2015.
DOI: 10.1007/978-3-319-21407-8_31

solutions, including an optimal one. Sections 5 and 6 analyse the effect of the value of the parameters of the objective function on the optimal delivery orders. Finally, Section 7 concludes and summarizes the findings.

2   Stochastic Programming Model

The model describes the inventory development over T periods of products with J periods of shelf life, where t is the time index and j is the age of the product in inventory. The stochastic demand implies that the model has random inventory variables $\boldsymbol{I}_{jt}$ apart from the initial fixed levels I_{j0}. The model has to keep track of the lost sales $\boldsymbol{X}_t$ in periods where demand exceeds the available amount of product. Moreover, typical is the amount of product that perishes and becomes waste, $\boldsymbol{I}_{Jt}$. In the notation, $E(.)$ is the expected value operator. Moreover, we use $x^+ = max\{x, 0\}$. A formal description of the SP model from [4] is given.

Indices

t period index, $t = 1, \ldots, T$, with T the time horizon
j age index, $j = 1, \ldots, J$, with J the fixed shelf life

Data

d_t Normally distributed demand with mean value $\mu_t > 0$ and variance $(cv \times \mu_t)^2$ where cv is a given coefficient of variation.
k fixed ordering cost, $k > 0$
c unit procurement cost, $c > 0$
h unit inventory cost, $h > 0$
w unit disposal cost, is negative when having a salvage value, $c > -w$
β service level, $0 < \beta < 1$

Variables

$Q_t \geq 0$ ordered and delivered quantity at the beginning of period t
$Y_t \in \{0, 1\}$ setup of order
$\boldsymbol{X}_t$ lost sales in period t
$\boldsymbol{I}_{jt}$ Inventory of age j at end of period t, initial inventory fixed $I_{j0} = 0$,
            $\boldsymbol{I}_{jt} \geq 0$ for $j = 1, \ldots, J$.
The total expected costs over the finite horizon is to be minimized.

$$f(Q) = \sum_{t=1}^{T} \left(C(Q_t) + E \left(h \sum_{j=1}^{J-1} \boldsymbol{I}_{jt} + w \boldsymbol{I}_{Jt} \right) \right), \tag{1}$$

where procurement cost is given by the function

$$C(x) = k + cx, \quad \text{if } x > 0, \text{ and } C(0) = 0. \tag{2}$$

The FIFO dynamics of inventory of items of age j starts by defining waste $(j = J)$

$$I_{Jt} = (I_{J-1,t-1} - d_t)^+, \ t = 1, \ldots, T \tag{3}$$

followed by the inventory of items with age $1 < j < J$ that still can be used in the next period:

$$I_{jt} = \left(I_{j-1,t-1} - (d_t - \sum_{i=j}^{J} I_{i,t-1})^+ \right)^+ , \ t - 1, \ldots, T, j - 2, \ldots, J \quad 1. \tag{4}$$

and finally the incoming and freshest products, $j = 1$:

$$I_{1t} = \left(Q_t - (d_t - \sum_{j=1}^{J-1} I_{j,t-1})^+ \right)^+ , \ t = 1, \ldots, T. \tag{5}$$

Lost sales for period t is defined by

$$X_t = \left(d_t - \sum_{j=1}^{J-1} I_{j,t-1} - Q_t \right)^+ . \tag{6}$$

The service level constraint for every period is

$$E(X_t) \le (1 - \beta)\mu_t, \ t = 1, \ldots, T. \tag{7}$$

Notice that given the stochastic variables of demand, the expectation of lost sales only depends on the ordered quantities $Q : (Q_1, \ldots, Q_T)$. For a MINLP approach, we can also express these constraints as follows

$$g_t(Q) = E(X_t) - (1 - \beta)\mu_t \le 0, \ t = 1, \ldots, T. \tag{8}$$

We consider a simple order policy, where the decision maker should provide an integer vector $Y = (Y_1, \ldots, Y_T) \subset \{0, 1\}^T$ of order periods and order quantities Q_t with $t = 1, \ldots, T$ where $Y_t = 0$ implies $Q_t = 0$. Finding the best values of the (continuous) order quantities Q_t and the corresponding optimal (integer) order timing Y can be considered a MINLP problem. In summary, we have a $MINLP(Q)$ problem: $\min_Q f(Q)$ subject to the inventory development (3), (4), (5), (6) and $g_t(Q) \le 0, \ t = 1, \ldots, T$.

3   Replenishment Cycles and Basic Order Quantities

We study several theoretical properties of the order quantities Q and the list of order periods Y. We first focus on the concept of replenishment cycles in Section 3.1 and determine in which cases a so-called basic order quantity defines the optimal order quantity in Section 3.2. Section 3.3 then derives properties of the optimal quantities given a timing vector Y, which can be used in the design of specific algorithms in Section 4.

3.1 Feasible Replenishment Cycles

Literature on inventory control (e.g. [7]) applies the concept of a replenishment cycle, i.e. the length of the period R for which the order of size Q is meant. For stationary demand, the replenishment cycle is fixed, but for non-stationary demand the optimal replenishment cycle may depend on the period.

Definition 1. *Given a list of order periods $Y \in \{0,1\}^T$ and $M = \sum_{t=1}^T Y_t$. The order timing vector $A \in \mathbb{N}^M$ for Y gives the periods $A_i < A_{i+1}$ where $Y_{A_i} = 1$.*

Definition 2. *Given a list of order periods $Y \in \{0,1\}^T$ with total number of orders $M = \sum_{t=1}^T Y_t$. Replenishment cycle $R_i = A_{i+1} - A_i$, $i = 1, \ldots, M-1$ and $R_M = T - A_M + 1$.*

Notice that for the perishable case with a shelf life J, to fulfil the service level constraint, practically the replenishment cycle cannot be larger than the shelf life J; so $1 \leq R_i \leq J$.

Lemma 1. *Let Y be an order timing vector of the SP model, i.e. $Y_t = 0 \Rightarrow Q_t = 0$. Y provides an infeasible solution of the SP model, if it contains more than $J - 1$ consecutive zeros.*

This means that a feasible order timing vector Y does not contain a consecutive series with more than $J - 1$ zeros.

Definition 3. *Let F_T be the set of feasible order timing vectors Y of length T.*

For the given model the number of elements $|F_T|$, for $T \leq J$ is equal to 2^{T-1}. Given the assumption of zero starting inventory, an element $Y \in F_T$ must start with $Y(1) = 1$, otherwise demand will not be satisfied for the first period. From here, any combination is valid, since there are no more than $J - 1$ elements to consider. The case $T = J + 1$ is slightly different: for the first period again $Y(1) = 1$ is needed, and for the rest of elements there are 2^J possible cases, and only the one with J zeros is not feasible, so $|F_{J+1}| = 2^J - 1$.

For $T > J + 1$ the number of feasible series can be derived recursively according to the following proposition.

Proposition 1. *The number of elements $|F_T|$ of the set F_T of feasible order timings of T periods and a shelf life J with $J + 1 < T$ follows the recursive rule*

$$|F_{T+1}| = 2|F_T| - |F_{T-J}| \tag{9}$$

with the initial terms $|F_t| = 2^{t-1}$, $t < J + 1$ and $|F_{J+1}| = 2^J - 1$

Proof. Consider $p = (p_1, \ldots, p_T) \in F_T$, then obviously $(p_1, \ldots, p_T, 1) \in F_{T+1}$. We now focus on the case of $(p, 0) = (p_1, \ldots, p_T, 0)$. Its feasibility is determined by the values of $p_{T-(J-2)}, \ldots, p_{T-1}, p_T$. Let $\mathbf{0}$ be now an all zeros vector with $J - 1$ elements. If $(p_{T-(J-2)}, \ldots, p_{T-1}, p_T) = \mathbf{0}$, then $(p, 0)$ is not feasible. But how many feasible $p \in F_T$ exist with this feature?

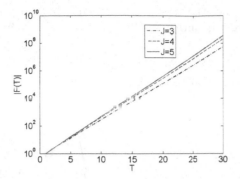

Fig. 1. $|F(T)|$ as a function of T and several values of the shelf life J

Given $(p_{T-(J-2)}, \ldots, p_{T-1}, p_T) = \mathbf{0}$ and $p \in F_T$, we must have $p_{T-J+1} = 1$. There are $|F_{T-J}|$ feasible timing vectors $(p_1, \ldots, p_{T-J})$ with $(p_1, \ldots, p_{T-J}, 1, \mathbf{0})$ feasible and $(p_1, \ldots, p_{T-J}, 1, \mathbf{0}, \mathbf{0})$ infeasible. From this follows the final result, $|F_{T+1}| = 2|F_T| - |F_{T-J}|$. □

The proof of Proposition 1 gives a constructive method to determine all feasible policies recursively. Moreover, (9) shows that the space of feasible policies grows exponentially with T as illustrated in Figure 1.

3.2 Basic Order Quantities

We focus now on the minimum order quantity at period $t \in \{1, \ldots, T\}$ to cover demand for the next r periods $t, t+1, \ldots, t+r-1$ that is just sufficient according to the service level constraint where at period t no older items are in stock.

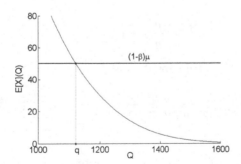

Fig. 2. One period loss function $E(X|Q)$ for $d \sim N(1950, 0.25 \cdot 1950)$ and corresponding basic order quantity q: $E(X|q) = (1 - \beta)\mu$

Definition 4. *Basic order quantity* $\overline{Q}_{r,t}$ *is the amount for period t that fulfils constraint (7) for the next r periods: $t, t+1, \ldots, t+r-1$.*

We first consider $\overline{Q}_{1t}$. Let the replenishment cycle be one period $R = 1$, zero inventory and q the order quantity. In this case, lost sales X from expression (6) can be simplified to:

$$X = (d - q)^+. \tag{10}$$

Let φ be the density function (pdf) of d and Φ be the corresponding cumulative distribution function (cdf). Then the so-called loss function expressing the expected lost sales as a function of q is

$$L(q) = E(X) = E\left((d - q)^+\right) = \int_q^\infty (x - q)\varphi(x)dx, \tag{11}$$

where the symbol x is used here as the argument in the integral. The cost function (1) is monotonously increasing in the order quantity q and $L(q)$ decreases, so constraint (7) is binding for the optimal value of q such that $L(q) = (1 - \beta)\mu$ as illustrated in Figure 2. Since demand is normally distributed, there is no closed-form expression for the first order loss function. Some approximations for the loss function can be found in [3],[5],[6],[8]. From a root-finding perspective, there are several ways to proceed (see [2]). For instance, one can use the derivative of loss function $L'(q) = \int_{-\infty}^q \varphi(x)dx - 1 = \Phi(q) - 1$ to approximate q using *Newton-Raphson* method. The following theoretical result shows that for the described model, the determination of q has to be done only once.

Lemma 2. *Let $d \sim N(\mu, cv \times \mu)$ and φ be the pdf and Φ the cdf of the standard normal distribution. The solution of $L(q) = (1 - \beta)\mu$ fulfils $q = \mu(1 + cv \times \hat{x})$ where $\hat{x}$ solves $\varphi(\hat{x}) - (1 - \Phi(\hat{x}))\hat{x} = \frac{1-\beta}{cv}$.*

Proof. Using the results in [5] for $d \sim N(\mu, cv \times \mu)$, the loss function can be expressed as

$$L(q) = cv \times \mu \left(\varphi\left(\frac{q - \mu}{cv \cdot \mu}\right) - \left(1 - \Phi\left(\frac{q - \mu}{cv \cdot \mu}\right)\right) \frac{q - \mu}{cv \cdot \mu} \right). \tag{12}$$

The Equation $L(q) = (1 - \beta)\mu$ substituting $q = \mu(1 + cv \times \hat{x})$ implies

$$\varphi\left(\frac{q - \mu}{cv \cdot \mu}\right) - \left(1 - \Phi\left(\frac{q - \mu}{cv \cdot \mu}\right)\right) \frac{q - \mu}{cv \cdot \mu} = \varphi(\hat{x}) - (1 - \Phi(\hat{x}))\hat{x} = \frac{1 - \beta}{cv}. \tag{13}$$

$\square$

The basic order quantity $\overline{Q}_{1t} = \mu_t(1 + cv \times \hat{x})$ provides an upper bound on the order quantity Q_t if $R_t = 1$, because inventory may be available. Figure 3 shows the relation of $\hat{x}$ with parameters cv and β. Its value indicates how much bigger (smaller if negative) the basic quantity is compared to μ.

The basic order quantities for longer replenishment cycles look more complicated. For instance, $R_t = 2$ implies

$$E\left(d_{t+1} - (\overline{Q}_{2t} - d_t)^+\right)^+ = (1 - \beta)\mu_{t+1}.$$

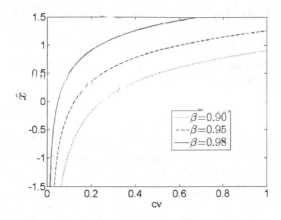

Fig. 3. Solution $\hat{x}$ of (13) as function of cv for several values of $\beta = 0.90$, 0.95, 0.98

These expressions become more cumbersome with the size of the replenishment period. To solve them we need a different approach. Consider a replenishment cycle of r periods, from t_1 to t_2 with $t_2 = t_1 + r - 1$ and $\boldsymbol{d_{t_1,t_2}} = \boldsymbol{d_{t_1}} + \ldots + \boldsymbol{d_{t_2}}$. Let φ_{t_1,t_2} be the density function (pdf) of $\boldsymbol{d_{t_1,t_2}}$ and Φ_{t_1,t_2} be the corresponding cumulative distribution function (cdf). As we are considering that demand is normally distributed, the distribution of $\boldsymbol{d_{t_1,t_2}}$ is normal as well, with expected value $\mu = \mu_{t_1} + \ldots + \mu_{t_2}$ and $\sigma = cv\sqrt{\mu_{t_1}^2 + \ldots + \mu_{t_2}^2}$. Starting with q units at period t_1, the expected value of the sum of lost sales from periods t_1 to t_2, $SL_{t_1,t_2}(q)$ is determined by

$$SL_{t_1,t_2}(q) = E\left(\sum_{i=t_1}^{i=t_2} X_i\right) = E\left((\boldsymbol{d_{t_1,t_2}} - q)^+\right) = \int_q^\infty (x - q)\varphi_{t_1,t_2}(x)dx. \quad (14)$$

Following the considerations in [5], $SL_{t_1,t_2}(q)$ can be expressed as

$$SL_{t_1,t_2}(q) = \mu - q + \sigma\varphi\left(\frac{q-\mu}{\sigma}\right) + \Phi\left(\frac{q-\mu}{\sigma}\right)(q - \mu). \quad (15)$$

Now consider the expected lost sales $L_{t_1,t_2}(q)$ in period t_2 when q fresh units are available at the beginning of period t_1. The quantity follows from subtracting from the expected total lost sales up to period t_2 in (14) the expected value of total lost sales from periods t_1 to $t_2 - 1$

$$L_{t_1,t_2}(q) = SL_{t_1,t_2}(q) - SL_{t_1,t_2-1}(q) = \quad (16)$$

$$= \int_q^\infty (x - q)\left(\varphi_{t_1,t_2}(x) - \varphi_{t_1,t_2-1}(x)\right)dx.$$

To find the value q_{rt} for which the service level constraint is just fulfilled in period $t_2 = t + r - 1$ when having q_{rt} fresh units at the beginning of period t

follows from solving $L(q_{1t}) = (1 - \beta)\mu_t$ according to Lemma 2 and for $r > 1$ one can solve

$$SL_{t,t+r-1}(q_{rt}) - SL_{t,t+r-2}(q_{rt}) = (1 - \beta)\mu_{t+r-1} \tag{17}$$

for q_{rt}. The basic order quantity $\overline{Q}_{rt}$ now follows from taking the maximum of these quantities $\overline{Q}_{rt} = \max_{j=1,\dots,r} q_{jt}$, such that the service level constraint is fulfilled in all periods of the replenishment cycle. The values of the basic order quantities in Table 1 follow from the parameter values in Example 1.

Example 1. Consider an instance of the problem with $T = 12$ periods and shelf life $J = 3$. Random variables of demand for every period are expressed in terms of their mean values μ_t given in last row of Table 1 and the coefficient of variation is given by $cv = 0.25$. The required service level is $\beta = 95\%$, such that the value of $\hat{x}$ in Lemma 2 is $\hat{x} = 0.493$. This value can be used, by Lemma 2, to determine the values for $\overline{Q}_{1t}$ in the first row of Table 1. Solving (17) and taking $\overline{Q}_{rt} = \max_{j=1,\dots,r} q_{jt}$ provides values for $\overline{Q}_{rt}$ for $r > 1$ in Table 1. The values are rounded up.

Table 1. Basic order quantities

r\t	1	2	3	4	5	6	7	8	9	10	11	12
1	**898**	1067	224	1010	898	**168**	730	898	**1010**	336	168	673
2	1952	**1450**	1211	**1925**	1225	885	**1618**	1898	1458	529	**825**	0
3	2372	2277	2131	2287	1812	1769	2596	2386	1666	1145	0	0
μ_t	800	950	200	900	800	150	650	800	900	300	150	600

3.3 Optimal Quantities for a Given Y

So far, we have described a method to find a feasible Q vector for a given Y, using the so called Basic order quantities. Now we can define a method to optimize the objective function (1) subject to a given timing vector Y.

We consider now the properties of the MINLP problem described in Section 2. We first focus on the monotonicity of the objective function in the order quantities when $w > -c$ applies.

Lemma 3. *Given an order timing vector Y with corresponding values A and M according to Definition 1, if $Q^*(Y)$ is an optimal solution of (1),(2),(3), (4), (5), (6) and (8) with $Q_t = 0$ if $Y_t = 0$ then $g_T(Q^*) = 0$.*

Proof. Assume Q^* is optimal and $g_T(Q^*) < 0$. Due to the inventory dynamics and g_T being a continuous function in its arguments, $\exists \epsilon > 0$ such that $Q = Q^* - \epsilon e_{A_M}$ is a feasible solution with $g_T(Q) = 0$. Due to the FIFO dynamics $I_{jt}(Q) \leq I_{jt}(Q^*), j = 1,\dots,J$ and specifically $I_{Jt}(Q^*) - I_{Jt}(Q) \leq \epsilon$. As also $\sum Q_t - \sum Q_t^* = \epsilon$ we have that $f(Q) \leq f(Q^*) - (c - w)\epsilon < f(Q^*)$ which contradicts Q^* can be optimal with $g_T(Q^*) < 0$. $\square$

Proposition 2. *Given an order timing vector Y with corresponding values A according to Definition 1 and $M = \sum Y_t$. If $Q^*(Y)$ is an optimal solution of (1),(2),(3), (4), (5), (6) and (8) with $Q_t = 0$ if $Y_t = 0$ then (8) is binding for $t = A_i - 1, i = 2, \ldots, M$ and $t = T$:*

$$g_{A_2-1}(Q^*) = \ldots = g_{A_M-1}(Q^*) = g_T(Q^*) = 0. \tag{18}$$

Proof. Follows from applying the proof of Lemma 3 in a sequential way. Lemma 3 has shown that $g_T(Q^*) = 0$. Assume Q^* is optimal with $g_{A_M-1}(Q^*) < 0$. As g_{A_M-1} is a continuous function in Q, $\exists \epsilon > 0$ such that for $Q = Q^* - \epsilon e_{A_M-1}$ we still have $g_{A_M-1}(Q) < 0$. If $R_{M-1} = J$, we have that less items perish at the end of the cycle for Q than for Q^* and therefore $f(Q) < f(Q^*)$ leading to a contradiction.

However, if $R_{M-1} < J$ less inventory $\mathcal{I}(Q) = \sum_{j=R_{M-1}}^{J-1} I_j(Q)$ is available at the end of the cycle and Q_M^* has to be increased to meet the constraint for the last period. [5] show that the so called complementary loss function $E(\mathcal{I}(Q))$ is strictly convex with partial derivatives less than 1. This implies that $\delta = E(\mathcal{I}(Q^*)) - E(\mathcal{I}(Q)) < \epsilon$. Consequently there exists a feasible solution $Q^* - \epsilon e_{A_M-1} + \delta e_{A_M}$ that has lower cost than Q^*. This contradicts the assumption. By following this reasoning backward, it is shown that for all periods at the end of a replenishment cycle constraint (8) has to be binding.   □

Finding (18) implicitly gives a way to determine $Q^*(Y)$. Minimise the quantity of the first cycle up to constraint (7) is fulfilled and proceeding with the next cycle until the last one. Notice that given Y, the optimal quantities do not depend on the parameter values of the objective function (1).

4   Algorithms for Generating Order Quantities

The derived basic order quantities can now be used to obtain feasible solutions and to improve them towards optimal solutions.

4.1   Generating a Feasible Solution for a Given Y

Given a list of order periods Y one can determine a feasible solution Q based on $\overline{Q}_{r,t}$. After determination of A and R from Definitions 1 and 2, the appropriate values can be assigned. Example 2 shows a case for a vector Y.

Example 2. Consider the data and variables of Example 1. One can determine a feasible solution Q_t for a given timing vector Y using the basic order quantities $\overline{Q}_{r,t}$. For timing vector $Y = (1, 1, 0, 1, 0, 1, 1, 0, 1, 1, 1, 0)$, vectors A and R are defined by: $A = (1, 2, 4, 6, 7, 9, 10, 11)$ and $R = (1, 2, 2, 1, 2, 1, 1, 2)$. Selecting the appropriate basic order quantities results into $Q = (898, 1450, 0, 1925, 0, 168, 1618, 0, 1010, 336, 825, 0)$ as illustrated in Table 1.

Algorithm 1. MinQ(Y): Optimal order quantity Q for Y

Require: Orders timing Y, fixed shelf life J
Ensure: Optimal order quantity Q for Y

1: Determine replenishment periods $A(Y)$, Definition 1
2: Determine corresponding replenishment cycles R, Definition 2
3: **for** $i = 1$ to $\sum_t Y_t$ **do**
4:     **if** $(A_i = 1 \text{ or } A_i - A_{i-1} = J)$ **then**
5:        $Q_i = Q_{R_i, A_i}$;   #minimum value for constraint (7)
6:        $(z, I_{A_i + R_i - 1}) =$ **floss**$(Q_i, A_i, A_i + R_i - 1, 0)$ # Alg. 2 updates inventory I
7:     **else**
8:        $Q_i =$ **Ordervalue**$(A_i, R_i, I_{A_i - 1})$ # Algorithm 3
9:     **end if**
10: **end for**

4.2 Generating the Optimal Solution for a Given Y

The process of optimizing quantities for a timing vector Y is sketched at Algorithm 1. Proposition 2 shows with (18) that the quantities for the order periods of Y can be calculated sequentially from $A_1 = 1$ to A_M, $M = \sum_{t=1}^{T} Y_t$. If a

---

Algorithm 2. floss(q, t_1, t_2, I): Monte Carlo sim, estimates $E(\boldsymbol{X})$, updates $\boldsymbol{I_{t_2}}$

Require: Order quantity (q), time window $[t_1, t_2]$, number of sample paths N, sample
     starting inventory $I_{jn}, j = 1, \ldots, J - 1, n = 1, \ldots, N$
Ensure: Estimate Z of $E(X_{t_2})$, a sample of I_{t_2}

1: **for** $n = 1$ to N **do**
2:     $I_{1,t,n} = \left(q - (d_{t_1,n} - \sum_{j=1}^{J-1} I_{jn})^+ \right)^+$;
3:     **for** $t = t_1$ to t_2 **do**
4:        **for** $j = 2$ to $j = J - 1$ **do**
5:           $I_{j,t,n} = \left(I_{j-1,t-1,n} - (d_{t,n} - \sum_{k=j}^{J-1} I_{k,t-1,n})^+ \right)^+$;   #Update inventory ($I$)
6:        **end for**
7:     **end for**
8:     $Z = \frac{1}{N} \sum_{n=1}^{N} \left(d_{t_2,n} - \sum_{j=1}^{J-1} I_{j,t_2-1,n} - q \right)^+$;
9: **end for**

---

replenishment period A_i is preceded by $J - 1$ periods without replenishment, the inventory perishes and starting inventory at A_i is equal to zero. For those cases basic order quantity Q_{R_i, A_i} with R_i the length of the replenishment cycle i, is optimal. When period A_i is preceded by less than $J - 1$ no-replenishment periods, the optimal order quantity is lower than the basic order quantity, due to random inventory $\boldsymbol{I}_{1, A_i - 1} \cdots, \boldsymbol{I}_{J-1, A_i - 1}$. The minimum order quantity that just fulfils (7) can be determined by simulation.

Algorithm 2 estimates expected lost sales and determines the end inventory based on Monte Carlo simulation. Function **floss**(q, t_1, t_2, I) consists of simulating inventory for N demand paths and determines lost sales of period t_2 given order quantity q at period t_1. In Algorithm 1 (line 6), when starting inventory is zero and the order quantity is a basic order quantity (line 5), Algorithm 2 is also run to determine the inventory I at the end of the cycle, i.e. period t_2. The value of N must be large enough to provide accurate estimates for the expected lost sales.

To determine the order quantity such that (8) is binding when the starting inventory is nonzero, an approach can be used as sketched in Algorithm 3 based on the secant method (lines 5-8). It takes replenishment cycle $[t_1, t_2 = t_1 + r - 1]$ as input, where r is the length of the cycle and a starting inventory I. Iteratively, function **floss**(q, t_1, t_2, I) is evaluated to determine expected loss at the end of the cycle.

Algorithm 3. Ordervalue(t_1, t_2, I): Determines an order quantity fulfilling (7)

Require: Time window $[t_1, t_2]$, sample starting inventory $I_{jn}, j = 1, \ldots, J - 1, n = 1, \ldots, N$, required fill rate β, accuracy ϵ

Ensure: Order quantity q

1: $K = (1 - \beta)\mu_{t_2}$;   #Target level
2: Choose two initial order quantities q_1 and q_2;   #Secant method
3: $(z_1, I_{t_2}) = $ **floss**(q_1, t_1, t_2, I) and $(z_2, I_{t_2}) = $ **floss**(q_2, t_1, t_2, I);   # Algorithm 2
4: **while** $|z_2 - K| > \epsilon$ **do**
5:      $q = q_1 + \frac{(K - z_1)(q_2 - q_1)}{z_2 - z_1}$;
6:      $z_1 = z_2$;
7:      $z_2 = $ **floss**(q, t_1, t_2, I);   # Algorithm 2
8:      $q_1 = q_2; q_2 = q$;
9: **end while**

Example 3. Consider timing vector Y of Example 2 and corresponding feasible order quantities $Q = (898, 1450, 0, 1925, 0, 168, 1618, 0, 1010, 336, 825, 0)$, Algorithm 3 reduces these values to $Q = (898, 1379, 0, 1670, 0, 120, 1536, 0, 879, 273, 726, 0)$.

4.3   Algorithm for the Global Optimum Solution

Given Algorithm 1 to optimize quantities Q given timing vector Y, the question is how to find the optimal timing Y^* of the MINLP problem of Section 2. This is done by a systematic enumeration of feasible orders timing vectors Y and evaluating the objective function (1). By Lemma 1, infeasible orders timing vectors can be discarded. Furthermore, a lower bound on the cost function can be used to leave out more orders timing vectors for which the lower bound is higher than the cost of the best solution found thus far.

Determination of a Lower Bound on the Cost of Function for Each Y.
Let Y be a feasible timing vector. The ordering cost for Y is known to be $k \sum_{t=1}^{T} Y_t$.
Suppose t is an order period for a cycle $R > 1$. To fulfil the β service level, the minimum remaining stock to be stored at periods $t, t+1, \ldots, t+R-1$ is given by $\overline{Q}_{R,t+1}, \ldots, \overline{Q}_{1,t+R}$ respectively. Notice that the basic order quantity is the minimum amount needed to fulfil demand for a sequence of periods. Based on these considerations, a lower bound on the cost function for Y can be calculated.

Algorithm 4. AllY(): Evaluating all feasible timing vectors Y

Require: All feasible Y's
Ensure: The optimal timing vector (Y^*) and order quantities (Q^*)

  1: mincost=∞
  2: **for all** Y **do**
  3:    **if** lower bound (Y) < mincost **then**
  4:      $Q_Y = $ **MinQ**(Y);   # Algorithm 1
  5:      $C(Q_Y)$;   #Determine the cost of Q_Y
  6:      **if** $C(Q_Y)$ <mincost **then**
  7:        mincost=$C(Q_Y)$;
  8:        $Y^* = Y$;
  9:      **end if**
10:    **end if**
11: **end for**

Determination of the Optimal Timing Vector Y^* and Order Quantities $Q(Y^*)$. Algorithm 4 determines the optimal timing vector Y and the optimal production quantities by enumerating and testing feasible timing. Using the lower bound described some or many of the feasible policies can be discarded, depending on the cost parameters. For each feasible Y, such that *lower bound*(Y) < *mincost*, Algorithm 1 provides the optimal production quantities $Q^* = Q(Y^*)$. Alternatively step 4 can also be done by a standard NLP algorithm.

    The computational burden of Algorithm 4 is related to the number of feasible policies Y, which is exponential in T, i.e. $O(2^T)$. Moreover, the hardness for each timing Y is related to the number simulated periods by Algorithm 3, which is also bounded by T. Evaluation of **floss** requires simulating N demand paths (Algorithm 2). Summarizing, the total computational burden for finding the best timing vector Y^* and corresponding orders $Q(Y^*)$ is in the order $O(N \cdot T \cdot 2^T)$.

5   Parameter Sensitivity of the Optimal Solution

This section analyses the effect of the value of the parameters of the objective function on the optimal orders quantities. This analysis will help to better understand the cost associated to the service level quality (value of β) as well as the uncertainty of demand (measured by cv).

5.1   Effects of Parameters β and cv

First, we focus on the parameters that determine the basic order quantities to fulfil the β-service level constraint. For a replenishment cycle of length R, the basic order quantity depends on parameters β and cv. Section 3.2 shows that the loss function for the last period of the cycle approaches zero, i.e. $\lim_{q \to \infty} L(q) = 0$. The basic order quantity q, for which $L(q) = (1 - \beta)\mu$, typically goes up with the requirement $\beta \in [0, 1)$ and $q \to \infty$ when $\beta \to 1$ as illustrated in Figure 4.

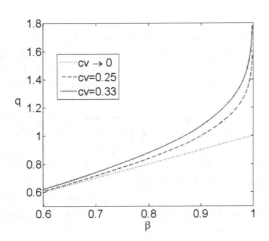

Fig. 4. Basic order quantity q as function of β for $cv \to 0$ (blue), $cv = 0.25$ (green) and $cv = 0.33$ (red)

Similarly basic order q tends to be $(1 - \beta)\mu$ when $cv \to 0$, as the normal distribution approximates a degenerate distribution at μ. Figure 4 also shows this effect for demand centred at $\mu = 1$ and different values of cv: the blue line illustrates the degenerate case and the green and red curves represent q as a function of β, for $cv = 0.25$ and $cv = 0.33$ respectively. As cv grows, the distribution spreads and the order quantity has to be larger to meet the fill rate.

Obviously, when dealing with more than one period and inventory variables from previous periods the analytical expression of the loss function is different, but results show that the trend is the same.

5.2   Cost Parameter Sensitivity

A surprising result from Proposition 2 is that given a timing vector Y and the assumptions on the cost parameters (positive and $c > -w$), the optimal quantities $Q^*(Y)$ do not depend on the values of the cost parameters of (1). However, like in most inventory models (see [7]) there is a trade off between the order cost k and the holding cost which causes the optimal timing vector Y^* to depend on specific cost parameter values.

When order cost $k = 0$ or relatively low compared to other cost parameters, the optimal timing vector tends to correspond to ordering at every period. Something similar happens when k is relatively high compared with the rest of parameters. In this case the optimal timing vector is to spread the orders as much as possible, every J periods. This is also valid, if the distribution of demand for one or some periods has a much higher mean value than the rest. The challenging situation occurs when k has a value in between the two extremes.

6   Experiments

We now evaluate the effectiveness and efficiency of Algorithm 4 using two different implementations of step 4 to obtain the optimal order quantity Q for a given Y. The first one is the derived Algorithm 1 $MinQ(Y)$ using the presented properties of the problem. The second is the general purpose nonlinear optimization solver *fmincon* of Matlab.

Example 1 has been used as base case (see Subsection 3.2) considering varying values of the parameters: fill rate ($\beta = 0.90, 0.95, 0.98$), the coefficient of variation ($cv = 0.10, 0.25, 0.33$) and the ordering cost ($k = 500, 2000$). The values of the other cost parameters have been set to $c = 2$, $h = 0.5$ and $w = 0$ for all the runs. For the evaluation of the costs, $N = 5000$ demand paths (Monte Carlo simulations) are used. Notice that the same pseudo random numbers for the demand have been used, allowing comparison of the obtained order policy for both solvers. For the same reason, *fmincon* and Algorithm 3 use the same termination tolerance ($\epsilon = 10^{-6}$) for constraint (8).

Using these settings, for each feasible timing vector Y, the order quantities computed by both methods match up to at least the first five digits. According to Proposition 2, the procedure Algorithm 3 also should lead to the optimal solution. The fact that a practical general purpose solver reaches the same optimum shows that the underlying problem does not exhibit numerical instabilities.

For every possible combination of the considered values of β, cv and k, the optimal value of the objective function (1) and the number of orders of the optimal Y^* have been calculated. Figure 5 shows the optimal cost versus the number of orders ($\sum Y^*$) for each considered problem. This figure illustrates the following aspects:

1. Naturally, a lower value of k provides lower optimal total costs.
2. The value of the optimal cost increases with the requirement β as well as with the coefficient of variation cv. This is a consequence of the effect of β and cv on the basic order quantities, as shown in Figure 4.
3. The number of ordering periods for Y^* increases as k decreases.
4. An increase of β results in a reduction of the number of ordering periods for Y^*. This is more evident for the cases where $k = 500$, because there is a greater chance to reduce the aforementioned number of ordering periods to its limit of 4 orders for the case $T = 12$ and $J = 3$.

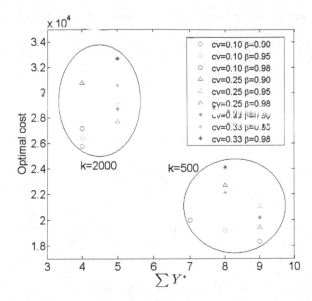

Fig. 5. Optimal cost and number of ordering periods for every instance. Circles correspond to $cv = 0.10$, triangles to $cv = 0.25$ and asterisks to $cv = 0.33$. The values of $\beta = 0.90$, 0.95 and 0.98 is represented by colors red, green and blue, respectively. k identifies the ordering cost. Outcomes have been grouped according to the value of k.

With respect to efficiency, using Algorithm 1 (MinQ(Y)) instead of standard nonlinear optimization solver *fmincon* reduces the computational time by a factor of 20. We specifically measured this for Algorithm 4 without considering the existence of a lower bound function (line 3), optimizing all feasible policies. The main reason of the reduction is that MinQ(Y) uses all derived specific properties of the MINLP problem.

7   Conclusions

A MINLP model has been presented to determine order quantities for a perishable product inventory control problem. Basic order quantities can be determined to provide feasible production quantities for different delivery policies of the model. Theoretical properties of feasible order timings and of the NLP problem when the timing Y is fixed have been derived. Based on that, a method has been developed to finding optimal quantities for the objective function (1) when the timing is given. For real applications, when T is not very large, this method is able to find the optimal solution for the problem by exhaustive search of feasible policies, using a lower bound function on cost to reduce enumeration. Moreover, obtained results have shown that the dedicated method can reduce the runtime in a factor of 20 with respect to the general purpose Matlab function *fmincon* for solving NLP problems, while keeping the high quality of the optimal solutions.

As the NLP optimization given a timing vector Y can be solved independently for each timing vector, a challenge for future investigation is to address the parallelization of the method. Such parallelization facilitates obtaining more accurate results due to a possible increase of the number of Monte Carlo simulations to solve a particular problem and simultaneously reduces the runtime.

Acknowledgments. This paper has been supported by The Spanish Ministry (TIN2012-37483) and Junta de Andalucía (P11-TIC-7176), in part financed by the European Regional Development Fund (ERDF). The study is co-funded by the TIFN (project RE002).

References

1. Hedjar, R., Bounkhel, M., Tadj, L.: Predictive control of periodic-review production inventory systems with deteriorating items. TOP **12**(1), 193–208 (2004)
2. Hendrix, E.M.T., Toth, B.G.: Introduction to Nonlinear and Global Optimization. Springer, New York (2010)
3. Kurawarwala, A.A., Matsuo, H.: Forecasting and inventory management of short life-cycle products. Operations Research **44**, 131–150 (1996)
4. Pauls-Worm, K.G.J., Hendrix, E.M.T., Haijema, R., van der Vorst, J.G.A.J.: An MILP approximation for ordering perishable products with non-stationary demand and service level constraints. International Journal of Production Economics **157**, 133–146 (2014)
5. Rossi, R., Tarim, S.A., Prestwich, S., Hnich, B.: Piecewise linear lower and upper bounds for the standard normal first order loss function. Applied Mathematics and Computation **231**, 489–502 (2014)
6. Schrijver, S.K.D., Aghezzaf, E.H., Vanmaele, H.: Double precision rational approximation algorithm for the inverse standard normal first order loss function. Applied Mathematics and Computation **219**(3), 1375–1382 (2012)
7. Silver, E.A., Pyke, D.F., Peterson, R.: Inventory Management and Production Planning and Scheduling. Wiley (1998)
8. Waissi, G.R., Rossin, D.F.: A sigmoid approximation of the standard normal integral. Applied Mathematics and Computation **77**(1), 91–95 (1996)

Heuristics for Longest Edge Selection
in Simplicial Branch and Bound

Juan F.R. Herrera[1][(✉)], Leocadio G. Casado[1], Eligius M.T. Hendrix[2],
and Inmaculada García[2]

[1] University of Almeria (ceiA3), Almeria, Spain
{juanfrh,leo}@ual.es
[2] Universidad de Málaga, Málaga, Spain
{eligius,igarciaf}@uma.es

Abstract. Simplicial partitions are suitable to divide a bounded area in
branch and bound. In the iterative refinement process, a popular strategy is to divide simplices by their longest edge, thus avoiding needle-
shaped simplices. A range of possibilities arises in higher dimensions
where the number of longest edges in a simplex is greater than one. The
behaviour of the search and the resulting binary search tree depend on
the selected longest edge. In this work, we investigate different rules to
select a longest edge and study the resulting efficiency of the branch and
bound algorithm.

Keywords: Bisection · Branching rule · Branch and bound · Global
Optimization · Simplices

1 Introduction

Global Optimization (GO) searches for global optima of a nonlinear function on
a non-empty domain that may have local nonglobal minima. Several methods
can be used to find the solution. Within deterministic methods, the branch
and bound method (B&B) guarantees to find a global minimum point up to
a guaranteed accuracy δ. This method iteratively divides the search space into
subsets and discards those that are proven not to contain a global solution. Five
rules define the method:

Branching Rule. It determines how to divide a problem into subproblems.
Bounding Rule. It defines how to obtain upper and/or lower bounds of the
objective function on subproblems.
Selection Rule. It chooses a subproblem among all subproblems stored in a
working set.
Rejection Rule. It discards subproblems which are proven not to contain a
global solution.
Termination Rule. It defines when the given accuracy has been reached. Once
a subproblem meets this criterion, it is not further divided. Otherwise, it is
stored in the working set.

© Springer International Publishing Switzerland 2015
O. Gervasi et al. (Eds.): ICCSA 2015, Part II, LNCS 9156, pp. 445–456, 2015.
DOI: 10.1007/978-3-319-21407-8_32

Every B&B rule plays an important role on the efficiency of the algorithm. Careless decisions in one of the rules may lead to inefficient algorithms. This work focuses on the efficiency of the branching rule using longest edge bisection within simplicial B&B optimization methods.

For some problems like mixture design, the search space is a regular simplex [5]. Here, we focus on box-constrained problems, where the search space is an n-dimensional hyper-rectangle that can be partitioned into a set of non-overlapping n-simplices. An n-simplex is a convex hull of $n+1$ affinely independent vertices. A simplex is a polyhedron in a multidimensional space, which has the minimal number of vertices. Therefore simplicial partitions are preferable in GO when the values of the objective function at all vertices of partitions are used to evaluate subregions.

A recent study shows how the number of generated sub-simplices varies when different heuristics are applied in the iterative bisection of a regular n-simplex [1] when dimensions are higher than 2. In that study, the complete binary tree is built by bisecting the heuristically-selected longest edge of a sub-simplex until the width, determined by the length of their longest edge, is smaller or equal to a given accuracy ϵ. A large reduction in the number of generated sub-simplices and therefore the size of the binary tree is achieved when one deviates from a heuristic that simply bisects the first longest edge found in terms of vertex indexation.

In this context, our initial question was about the effect when one applies different heuristics to simplicial B&B on a box-constrained area, where the initial search region is not a regular simplex and the termination criterion is based on the bounding rule. In a previous study [7], we showed that the number of evaluated simplices can be reduced by not selecting the first longest edge, but that which has the smallest sum of angles with the other edges. That study generated the upper part of a binary tree running a Lipschitz B&B with a rough accuracy. We focus now on new heuristics and investigate their efficiency for a B&B algorithm that reaches at least 5% of the function range as accuracy. The question is which of the rules are most effective when the dimension of the problem is going up.

Section 2 briefly explains the main features of the used simplicial B&B algorithm. Section 3 describes the studied edge selection heuristics. The resulting search tree is compared numerically in Section 4 and Section 5 summarizes the findings.

2 Simplicial B&B Method for Multidimensional GO

We focus on the multidimensional box-constrained global optimization problem. The goal is to find at least one global minimum point x^* of

$$f^* = f(x^*) = \min_{x \in X} f(x), \tag{1}$$

where the feasible area $X \subset \mathbb{R}^n$ is a nonempty box-constrained area, i.e. it has simple upper and lower bounds for each variable. The function f is not required

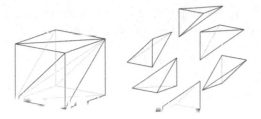

Fig. 1. Division of a hypercube into six irregular simplices

to be differentiable nor (Lipschitz) continuous. We will see in this section how one can subdivide the search space and derive simple bounds on the simplicial subsets. All ingredients are then collected into an algorithm.

Initial Space

Most B&B methods use hyper-rectangular partitions. However, other types of partitions may be more suitable for some optimization problems. Compared to the use of rectangular partitions, simplicial partitions are convenient when the feasible region is a polytope [11]. Optimization problems with linear constraints are examples where feasible regions are polytopes which can be vertex triangulated.

For the use of simplicial partitions, the feasible region is partitioned into simplices. There are two methods: over-covering and face-to-face vertex triangulation. The first strategy covers the hyper-rectangle by one simplex, that can be a regular one. The disadvantage of this method is that the search space is bigger than the feasible area and some regions can be out of the function definition. The most preferable initial covering is face-to-face vertex triangulation. It involves partitioning the feasible region into a finite number of n-dimensional simplices with vertices that are also the vertices of the feasible region. A standard method [12] is triangulation into $n!$ simplices. All simplices share the diagonal of the feasible region and have the same hyper-volume. Figure 1 depicts a hypercube of dimension three partitioned into six irregular simplices.

Bounding and Rejection Rules

Consider the objective function f with a global minimum f^* on box-constrained area X. Given a global minimum point x^*, let scalar K be such that

$$K \geq \max_{x \in X} \frac{|f(x) - f^*|}{\|x - x^*\|}, \tag{2}$$

where $\| \cdot \|$ denotes the Euclidean norm. Although this is not essential, we will work with Euclidean distance. The function $f^* + K\|x - x^*\|$ is an upper fitting according to [2] for an arbitrary $x \in X$. Consider a set of evaluated points $x_i \in X$ with function values $f_i = f(x_i)$, then the area below

$$\varphi(x) = \max_i \{f_i - K\|x - x_i\|\} \tag{3}$$

cannot contain the global minimum (x^*, f^*). Let $f^U = \min_i f_i$ be the best function value of all evaluated points, i.e., an upper bound of f^*. Then the area $\{x \in X : \varphi(x) > f^U\}$ cannot contain the global minimum point x^*.

Now consider a simplex S with evaluated vertices $v_0, v_1, \ldots, v_n$, where $f_i = f(v_i)$. To determine the existence of optimal solution x^* in S, each evaluated vertex (v_i, f_i) provides a cutting cone:

$$\varphi_i(x) := f_i - K\|x - v_i\|. \tag{4}$$

Let Φ be defined by

$$\Phi(S) = \min_{x \in S} \max_i \varphi_i(x). \tag{5}$$

If $f^U < \Phi(S)$, then simplex S cannot contain the global minimum point x^*, and therefore S can be rejected. Notice that $\Phi(S)$ is a lower bound of f^* if S contains the minimum point x^*.

Equation (5) is not easy to determine as shown by Mladineo [10]. Therefore, alternative lower bounds of (5) can be generated in a faster way. We use two of them and take the best (highest) value.

An easy-to-evaluate case is to consider the best value of $\min_{x \in S} \varphi_i(x)$ over the vertices i. This results in a lower bound

$$\underline{\Phi}_1(S) = \max_i \{f_i - K \max_j \|v_j - v_i\|\}. \tag{6}$$

The second lower bound is based on the more elaborate analysis of infeasibility spheres in [3] and developed to non-optimality spheres in [6]. It says that S cannot contain an optimal point if it is covered completely by so-called non-optimality spheres. According to [3], if there exists a point $c \in S$ such that

$$f_i - K\|c - v_i\| > f^U \quad i = 0, \ldots, n, \tag{7}$$

then S is completely covered and cannot contain x^*. This means that any interior point c of S provides a lower bound $\min_i \{f_i - K \max_j \|c - v_i\|\}$. Instead of trying to optimize the lower bound over c, we generate an easy-to-produce weighted average based on the radii of the spheres. Consider that $f_i > f^U$, otherwise S can contain an optimum point. Let

$$\lambda_i = \frac{K}{f_i - f^U} \tag{8}$$

and take

$$c = \frac{1}{\sum_j \lambda_j} \sum_i \lambda_i v_i. \tag{9}$$

Algorithm 1. Simplicial B&B algorithm, bisection

Require: X, f, K, δ
1: Partition X into simplices S_k, $k = 1, \ldots, n!$
2: Start the working list as $\Lambda := \{S_k : k = 1, \ldots, n!\}$
3: The set of evaluated vertices $V := \{v_i \in S_k \in \Lambda\}$
4: Set $f^U := \min_{v \in V} f(v)$ and $x^U := \arg\min_{v \in V} f(v)$
5: Determine lower bounds $f_k^L = f^L(S_k)$ based on K
6: **while** $\Lambda \neq \emptyset$ **do**
7:     Extract a simplex $S = S_k$ from Λ
8:     Bisect S into $S1$ and $S2$ generating x
9:     **if** $x \notin V$ **then**
10:        Add x to V
11:        **if** $f(x) < f^U$ **then**
12:            Set $f^U := f(x)$ and $x^U := x$
13:            Remove all S_k from Λ with $f_k^L > f^U - \delta$
14:        **end if**
15:    **end if**
16:    Determine lower bounds $LB(S1)$ and $LB(S2)$
17:    Store $S1$ in Λ if $f^L(S1) \leq f^U - \delta$
18:    Store $S2$ in Λ if $f^L(S2) \leq f^U - \delta$
19: **end while**
20: **return** x^U, f^U

A second lower bound based on (7) is

$$\underline{\Phi}_2(S) = \min_i \{f_i - K\|c - v_i\|\}. \tag{10}$$

The final lower bound we consider in this paper for the B&B is the best value $f^L(S) = \max\{\underline{\Phi}_1(S), \underline{\Phi}_2(S)\}$.

Selection and Termination Rules

The algorithm performs a depth-first search by selecting the sub-simplex with the smallest $f^L(S)$ value among those generated in the last division, until the final accuracy is reached or both new sub-simplices are rejected. In general, depth-first search minimizes the memory requirement of the algorithm. A simplex S is discarded when $f^L(S) + \delta > f^U$ for an accuracy $\delta > 0$. The steps of the B&B algorithm are described in Algorithm 1.

3   Longest Edge Bisection

Literature discusses many methods to subdivide a simplex [8]. One of them is the Longest Edge Bisection (LEB), which is a popular way of iterative division in the finite element method, since it is very simple and can easily be applied in higher dimensions [4]. This method consists of splitting a simplex using the hyperplane that connects the middle point of the longest edge of a simplex with the opposite vertices. This is illustrated in Figure 2 that also shows that in higher dimensions there can be several longest edges. For our study, we should notice that due to the initial partition as sketched in Figure 1, for $n = 3$ the longest edge is unique in all generated subsets. This means that to observe what happens with a choice of the longest edge to the search tree, we should focus on dimensions higher than 3. We formulate several rules to select the longest edge.

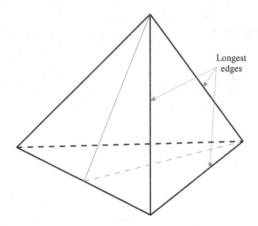

Fig. 2. Longest Edge Bisection generating a sub-simplex with three longest edges

The most common edge selection rule in LEB is the following:

LEB₁. Natural coding implicitly selects a longest edge being the first one found. The sequence depends on the coding and storing of the vertices and edges, i.e. the index number assigned to each vertex of the simplex. When a simplex is split into two new sub-simplices, the new vertex of each sub-simplex has the same index as the one it substitutes.

Our preliminary studies show the existence of many sub-simplices having more than one longest edge when LEB is used as iterative partition rule in a simplicial B&B algorithm.

In order to reduce the search tree size, other heuristics for selecting the longest edge in the division of a regular n-simplex are investigated to be used to simplicial B&B algorithms. They are summarized below:

LEB$_\alpha$. For each vertex in a longest edge, the sum of the angles between edges ending at that vertex is determined and the longest edge corresponding to the smallest sum is selected.

LEB$_C$. Bisects the longest edge with the largest distance from its middle point to the centroid of the simplex.

LEB$_M$. Determines the distance from a longest edge midpoint to the other vertices. It then selects that longest edge that has the maximum sum of distances to the other vertices.

LEB$_W$. Selects an edge that has not been involved in many bisections yet via a weight system. The initial set of evaluated vertices (line 3 of Algorithm 1) are set to $w_i := 0$. A new vertex v_i (generated by the branching rule, line 8 of Algorithm 1) is initiated with weight $w_i := 1$. Each time vertex v_i belongs to a divided edge, its weight is updated to $w_i := w_i + 1 \mod n$. For each longest edge defined by vertices (v_i, v_j), the two weights (w_i, w_j) are summed and the one with smallest sum is selected.

The research goal is to determine a LEB rule that minimizes the search tree generated by a simplicial B&B algorithm measured as the total number of generated nodes (subsimplices).

4   Comparison of the Selection Rules

A set of several test functions has been built to measure the tree generated by the set of LEB strategies discussed in the previous section. A complete suite of test functions can be found in [9]. From this set, we select a subset of functions that allow varying the dimension of the problem. We remind the reader that the often-used low dimensional instances are not appropriate to measure the difference of the generated tree as for dimensions $n \leq 3$ there is no choice on the selected longest edge to be bisected.

For each test function, at least one global minimum point is known and we determined the sharpest value of parameter K in (2) that we could find using a multistart approach. For instances like *MaxMod* or *Zakharov*, a value for K can be determined analytically. The data of the corresponding test-bed is given in Table 1. A description of the test instances with the considered minimum point and the function range $[f^*, \overline{f}]$ on the given domain is provided in the appendix, where $\overline{f}$ is the maximum function value on the domain. The depth of the generated B&B tree is mainly determined by the accuracy δ. To obtain reasonable size trees, in the experiments the value of the accuracy δ is set on $\delta = 0.05\,(\overline{f} - f^*)$.

Table 2 shows the numerical results for a search domain defined in a four-dimensional space. The computational effort is captured in terms of the number of generated and evaluated simplices in the corresponding B&B tree, see lines 5 and 16 of Algorithm 1. Column LEB$_1$ denotes the total number of evaluated simplices and the other columns provide the reduction with respect to LEB$_1$ generated by the remaining rules, expressed as a percentage. Rules LEB$_\alpha$ and LEB$_M$ provide higher reductions than LEB$_C$ and LEB$_W$, which in some cases perform more simplex evaluations than LEB$_1$. The interesting aspect is that selection rule LEB$_M$ is easier to generate than selection rule LEB$_\alpha$ in terms of computational operations.

A side result not related to our original question is the effectiveness of the bounding rule. We measured that the more sophisticated bound $\underline{\Phi}_2$ is lower for 75% of the evaluated sub-simplices than the simpler lower bound $\underline{\Phi}_1$.

Table 3 contains the results for dimension $n = 5$. In this case, LEB$_M$ provides a higher reduction than the rest of rules. LEB$_\alpha$ shows reductions similar to LEB$_C$

Table 1. Test instances for dimension $n = 4, 5, 6$, and the corresponding K_n values

No.	Test problem	Domain	K_4	K_5	K_6
1	Ackley	$[-30, 30]^n$	5.4	4.9	4.4
2	Dixon & Price	$[-10, 10]^n$	21,646.5	29,086.1	37,248.8
3	Holzman	$[-10, 10]^n$	5,196.2	7,000.0	9,000.0
4	MaxMod	$[-10, 10]^n$	1.0	1.0	1.0
5	Perm	$[-n, n]^n$	183,998.1	31,159,684.8	7,746,536,437.2
6	Pinter	$[-10, 10]^n$	60.0	75.3	88.7
7	Quintic	$[-10, 10]^n$	29,712.0	33,219.0	36,389.7
8	Rastrigin	$[-5.12, 5.12]^n$	91.8	102.7	112.5
9	Rosenbrock	$[-5, 10]^n$	168,005.3	190,517.7	210,669.4
10	Schwefel 1.2	$[-10, 10]^n$	176.8	176.8	444.6
11	Zakharov	$[-5, 10]^n$	312,645.0	1,415,285.7	4,962,758.1

Table 2. Experimental results for $n = 4$, number of evaluated simplices by LEB_1 and the reduction by the other rules

No.	Test problem	LEB_1	LEB_α	LEB_C	LEB_M	LEB_W
1	Ackley	8,467,608	35%	−3%	35%	0%
2	Dixon & Price	2,312,132	28%	8%	28%	5%
3	Holzman	3,419,030	25%	4%	25%	2%
4	MaxMod	5,742,672	54%	9%	54%	5%
5	Perm	39,987,438	3%	2%	31%	0%
6	Pinter	4,896,640	26%	−2%	26%	−1%
7	Quintic	2,527,376	29%	8%	29%	6%
8	Rastrigin	117,620,808	12%	0%	13%	1%
9	Rosenbrock	2,806,400	27%	0%	27%	−1%
10	Schwefel 1.2	1,857,322	30%	6%	30%	3%
11	Zakharov	5,299,018	30%	−6%	30%	−3%

and LEB_W, in contrast with dimension $n = 4$ reported in Table 2. Interesting is also the contrast with earlier findings where the rules were applied for refining a unit simplex in [1]. There the LEB_α rule clearly dominates other rules.

Table 4 provides numerical results for a subset of the test instances that could be run within a reasonable computation time for dimension $n = 6$. One can observe that LEB_M provides a larger reduction than the other rules.

In all of the experiments we found that the longest edge is not unique in about 70% of the sub-simplices providing the opportunity for a selection rule. Moreover, heuristics LEB_α and LEB_W sometimes provide not unique criterion values to the multiple longest edges. This calls for a second criterion to be evaluated. Interesting enough, it appeared that additional criteria lead to worse reductions, such that one best takes the first edge with the smallest sum of angles or smallest sum of weights for heuristics LEB_α and LEB_W respectively.

Table 3. Experimental results for $n = 5$, number of evaluated simplices by LEB_1 and the reduction by the other rules

No.	Test problem	LEB_1	LEB_α	LEB_C	LEB_M	LEB_W
1	Ackley	1,010,945,400	1%	6%	26%	7%
2	Dixon & Price	123,575,850	9%	14%	25%	13%
3	Holzman	911,1006,634	10%	12%	24%	11%
4	MaxMod	1,877,094,680	12%	7%	20%	3%
5	Perm	166,831,502	13%	5%	19%	3%
6	Pinter	989,052,844	10%	5%	20%	4%
7	Quintic	261,009,818	9%	13%	25%	14%
8	Rastrigin	23,085,565,464	4%	7%	18%	8%
9	Rosenbrock	175,613,436	0%	7%	22%	7%
10	Schwefel 1.2	87,628,502	12%	13%	25%	10%
11	Zakharov	603,678,276	0%	3%	18%	5%

Table 4. Experimental results for $n = 6$, number of evaluated simplices by LEB_1 and the reduction by the other rules

No.	Test problem	LEB_1	LEB_α	LEB_C	LEB_M	LEB_W
2	Dixon & Price	8,203,852,060	9%	14%	24%	13%
3	Holzman	16,628,924,978	17%	13%	26%	6%
7	Quintic	24,424,636,700	16%	12%	26%	4%
9	Rosenbrock	11,689,814,082	10%	9%	24%	6%
10	Schwefel 1.2	5,154,080,906	20%	16%	37%	20%

5   Conclusion

The question in this paper is how different selection rules for selecting the appropriate longest edge in simplicial B&B algorithms may influence the size of the generated search tree. To investigate this question, a simplicial B&B algorithm is used where non-optimal area is cut away via the concept of an upper fitting. Evaluating five LEB heuristics on a set of test instances in dimensions $n = 4, 5, 6$, shows that a rule called LEB_M gives the best performance. The search tree can be reduced up to about 25% compared to the rule that selects the first longest edge as stored in an implementation of the simplicial B&B.

Acknowledgments. This work has been funded by grants from the Spanish Ministry (TIN2012-37483) and Junta de Andalucía (P11-TIC-7176 and P12-TIC-301), in part financed by the European Regional Development Fund (ERDF). J.F.R. Herrera is a fellow of the Spanish FPU program.

Appendix: Function Definitions

The test function description is given with the considered minimum point x^*, minimum value f^* and maximum value $\overline{f}$ over the domain in Table 1 for dimensions $n = 4, 5, 6$.

Ackley

$$f(x) = -20\exp\left(-0.2\sqrt{\frac{1}{n}\sum_{i=1}^{n} x_i^2} - \exp\left(\frac{1}{n}\sum_{i=1}^{n}\cos(2\pi x_i)\right)\right) + 20 + e$$

$x^* = 0,\ f^* = 0,\ \overline{f}_4 = 22.2,\ \overline{f}_5 = 22.2,\ \overline{f}_6 = 22.2$

Dixon and Price

$$f(x) = (x_1 - 1)^2 + \sum_{i=2}^{D} i(2x_i^2 - x_{i-1})^2$$

$x_i^* = 2^{\frac{2-2^i}{2^i}},\ f^* = 0,\ \overline{f}_4 = 397{,}021,\ \overline{f}_5 = 617{,}521,\ \overline{f}_6 = 88{,}212$

Holzman

$$f(x) = \sum_{i=1}^{n} i x_i^4$$

$x^* = 0,\ f^* = 0,\ \overline{f}_4 = 100{,}000,\ \overline{f}_5 = 150{,}000,\ \overline{f}_6 = 210{,}000$

MaxMod

$$f(x) = \max(|x_i|)$$

$x^* = 0,\ f^* = 0,\ \overline{f}_4 = 10,\ \overline{f}_5 = 10,\ \overline{f}_6 = 10$

Perm

$$f(x) = \sum_{i=1}^{n}\left(\sum_{j=1}^{n}(j^i + \beta)\left(\left(\frac{x_j}{j}\right)^i - 1\right)\right)^2$$

$x_i^* = i,\ f^* = 0,\ \overline{f}_4 = 809{,}249,\ \overline{f}_5 = 476{,}712{,}082,\ \overline{f}_6 = 59{,}926{,}724{,}566$

Pinter

$$f(x) = \sum_{i=1}^{n} ix_i^2 + \sum_{i=1}^{n} 20i \sin^2 A + \sum_{i=1}^{n} i \log 10(1 + iB^2)$$

where

$$\begin{cases} A = (x_{i-1} \sin i_i + \sin w_{i+1}) \\ B = (x_{i-1}^2 - 2x_i + 3x_{i+1} - \cos x_i + 1) \end{cases}$$

where $x_0 = x_n$ and $x_{n+1} = x_0$.
$x^* = 0$, $f^* = 0$, $\overline{f}_4 = 500$, $\overline{f}_5 = 625$, $\overline{f}_6 = 751$

Quintic

$$f(x) = \sum_{i=1}^{n} |x_i^5 - 3x_i^4 + 4x_i^3 + 2x_i^2 - 10x_i - 4|$$

$x_i^* = -1$, $f^* = 0$, $\overline{f}_4 = 532{,}816$, $\overline{f}_5 = 668{,}520$, $\overline{f}_6 = 802{,}224$

Rastrigin

$$f(x) = 10n + \sum_{i=1}^{n} (x_i^2 - 10 \cos(2\pi x_i))$$

$x^* = 0$, $f^* = 0$, $\overline{f}_4 = 153$, $\overline{f}_5 = 190$, $\overline{f}_6 = 231$

Rosenbrock

$$f(x) = \sum_{i=1}^{n-1} [100 \left(x_{i+1} - x_i^2\right)^2 + (x_i - 1)^2]$$

$x_i^* = 1$, $f^* = 0$, $\overline{f}_4 = 2{,}722{,}743$, $\overline{f}_5 = 3{,}532{,}824$, $\overline{f}_6 = 4{,}342{,}905$

Schwefel 1.2

$$f(x) = \sum_{i=1}^{n} \left(\sum_{j=1}^{i} x_i\right)^2$$

$x^* = 0$, $f^* = 0$, $\overline{f}_4 = 3{,}000$, $\overline{f}_5 = 5{,}500$, $\overline{f}_6 = 9{,}100$

Zakharov

$$f(x) = \sum_{i=1}^{n} x_i^2 + \left(\frac{1}{2} \sum_{i=1}^{n} ix_i \right)^2 + \left(\frac{1}{2} \sum_{i=1}^{n} ix_i \right)^4$$

$x^* = 0$, $f^* = 0$, $\overline{f}_4 = 6{,}252{,}900$, $\overline{f}_5 = 31{,}646{,}750$, $\overline{f}_6 = 121{,}562{,}250$

References

1. Aparicio, G., Casado, L.G., Hendrix, E.M.T., García, I., Toth, B.G.: On computational aspects of a regular n-simplex bisection. In: 2013 Eighth International Conference on P2P, Parallel, Grid, Cloud and Internet Computing (3PGCIC), pp. 513–518 (2013)
2. Baritompa, W.: Customizing methods for global optimization, a geometric viewpoint. Journal of Global Optimization **3**(2), 193–212 (1993)
3. Casado, L.G., Hendrix, E.M.T., García, I.: Infeasibility spheres for finding robust solutions of blending problems with quadratic constraints. Journal of Global Optimization **39**(4), 577–593 (2007)
4. Hannukainen, A., Korotov, S., Křížek, M.: On numerical regularity of the face-to-face longest-edge bisection algorithm for tetrahedral partitions. Science of Computer Programming **90**, 34–41 (2014)
5. Hendrix, E.M.T., Casado, L.G., García, I.: The semi-continuous quadratic mixture design problem: Description and branch-and-bound approach. Eur. J. Oper. Res. **191**(3), 803–815 (2008)
6. Hendrix, E.M.T., Casado, L.G., Amaral, P.: Global optimization simplex bisection revisited based on considerations by Reiner Horst. In: Murgante, B., Gervasi, O., Misra, S., Nedjah, N., Rocha, A.M.A.C., Taniar, D., Apduhan, B.O. (eds.) ICCSA 2012, Part III. LNCS, vol. 7335, pp. 159–173. Springer, Heidelberg (2012)
7. Herrera, J.F.R., Casado, L.G., Hendrix, E.M.T., García, I.: On simplicial longest edge bisection in Lipschitz global optimization. In: Murgante, B., et al. (eds.) ICCSA 2014, Part II. LNCS, vol. 8580, pp. 104–114. Springer, Heidelberg (2014)
8. Horst, R., Tuy, H.: Global Optimization (Deterministic Approaches). Springer, Berlin (1990)
9. Jamil, M., Yang, X.: A literature survey of benchmark functions for global optimization problems. Int. Journal of Mathematical Modelling and Numerical Optimisation **4**(2), 150–194 (2013)
10. Mladineo, R.H.: An algorithm for finding the global maximum of a multimodal multivariate function. Math. Program. **34**, 188–200 (1986)
11. Paulavičius, R., Žilinskas, J.: Simplicial Global Optimization. SpringerBriefs in Optimization. Springer New York (2014)
12. Todd, M.J.: The computation of fixed points and applications. Lecture Notes in Economics and Mathematical Systems, vol. 24. Springer-Verlag (1976)

An Improvement of the Greedy Algorithm for the $(n^2 - 1)$-Puzzle

Kaede Utsunomiya and Yuichi Asahiro[✉]

Department of Information Science, Kyushu Sangyo University,
2-3-1 Matsukadai, Fukuoka 813-8503, Japan
asahiro@is.kyusan-u.ac.jp

Abstract. For $n \geq 4$, the $(n^2 - 1)$-PUZZLE is a generalization of the well known 15-Puzzle. Ratner et al. showed that finding a sequence of moves of minimum length for the $(n^2 - 1)$-PUZZLE is NP-hard, and many researches have been devoted to it. For the $(n^2 - 1)$-PUZZLE, a real-time algorithm is proposed by Parberry, which completes the puzzle in at most $5n^3 - 9n^2/2 + 19n/2 - 89$ moves and needs $O(1)$ computation time per move, although there is no guarantee that the number of moves is minimum. In this paper, we follow the direction of the research by Parberry, and present an algorithm, which is obtained by modifying Parberry's algorithm and giving a tight analysis. The number of moves by the new algorithm is smaller; it needs at most $5n^3 - 21n^2/2 + 35n/2 - 141$ moves.

1 Introduction

The 15-PUZZLE is one of the well known puzzles, defined as follows. We are given a board of size 4×4 and 15 tiles having a distinct number from $\{1, 2, \ldots, 15\}$ on the board as an initial configuration. The objective of the 15-PUZZLE is to achieve the final configuration in the right of Fig. 1, by repeatedly moving tiles to the blank space. Classical result in [3] shows how to verify existence of a solution, i.e., a sequence of moves from the given initial configuration to the final one.

The $(n^2 - 1)$-PUZZLE is a generalization of the 15-PUZZLE for $n \geq 4$, in which the size of the board is $n \times n$ and $n^2 - 1$ tiles on the board are numbered from $\{1, 2, \ldots, n^2 - 1\}$. Because of their popularity, the 15-PUZZLE and the $(n^2 - 1)$-PUZZLE are sometimes chosen as target problems, to which researchers apply their proposed methods in order to express their importance, good performance, and so on. We refer only a few of the previous researches here: Reinefeld applied a scheme in IDA* to the 8-PUZZLE[7]. A genetic algorithm for $(n^2 - 1)$-PUZZLE is proposed by Bhasin et al[1]. Brüngger et al. developed a library for parallel search, and applied it to the 15-PUZZLE[2]. Levitin et al. discuss on using puzzles including the $(n^2 - 1)$-PUZZLE as topics to learn how to design algorithms[4].

From the theoretical point of view, finding a shortest solution, i.e., a sequence of moves of minimum length, for the $(n^2 - 1)$-PUZZLE is NP-hard, shown by

This work was supported by JSPS KAKENHI Grant Number 25330018.

O. Gervasi et al. (Eds.): ICCSA 2015, Part II, LNCS 9156, pp. 457–473, 2015.
DOI: 10.1007/978-3-319-21407-8_33

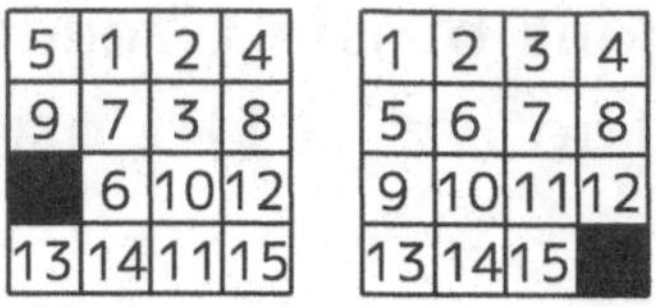

Fig. 1. 15-PUZZLE: the left is an example initial configuration and the right is the final configuration, where the blank is in black

Ratner et al[6]. They also proposed a (fairly large) constant factor approximation algorithm. In general, an approximation algorithm with good approximation ratio is rather involved, and running time might not be small, even if it is polynomial. Compared to it, the algorithm proposed by Parberry[5] combines two simple strategies, greedy and divide-and-conquer methods, and has good properties: It is moderately simple to implement, its running time is quite fast, i.e., constant time per move, and the maximum number of moves to achieve a given final configuration is bounded above by $5n^3 - 9n^2/2 + 19n/2 - 89$. Note that this number is polynomial in n, however it does not mean that this number is optimal nor this is guaranteed to be sufficiently close (e.g., constant factor) to the optimal one. Nevertheless, since there is a pair of two configurations, which requires $\Omega(n^3)$ moves from one to the other, it works well for such hard instances.

In this paper, we improve the algorithm in [5] by following their framework, modifying some procedures, and giving tight analyses. The proposed algorithm needs at most $5n^3 - 21n^2/2 + 35n/2 - 141$ moves, which is smaller than that of the algorithm in [5] by $6n^2 - 8n + 52$. In addition, the running time is $O(1)$ per move, as well. Here we would like to insist the importance of reducing the number of moves maintaining the algorithm's simplicity because the number is not optimal while we want to complete the puzzle as soon as possible.

The paper is organized as follows. In Sect. 2, we first give notations used in this paper. Section 3 including 6 subsections is the main part of the paper and presents the proposed algorithm. Finally we conclude in Sect. 4.

2   Notations

Consider a board of size $n \times n$ having n rows and n columns, and $(n^2 - 1)$ tiles on it. The position in the ith row and the jth column on the board is referred as (i, j). Each tile occupies one position (i, j) for some i and j on a board and has a distinct number from $\{1, 2, \ldots, n^2 - 1\}$. One position on the board is not occupied by any tile, which is called the blank. For simplicity, the blank is denoted by β.

Let $r(t)$ and $c(t)$ denote the row and column positions of a tile t, respectively. Also, $r(\beta)$ and $c(\beta)$ are of β. We can *move* a tile t which is next to β, to the position $(r(\beta), c(\beta))$ of β, by which we *swap* the positions of t and β. Sometimes we say that we *move* β to the position $(r(t), c(t))$ of a tile t which is next to β, although what we really move is the tile t. A *configuration* is a pair of the board and a sequence of tiles and β in a fixed order of the positions on the board, e.g., $(1,1),(1,2),\ldots,(n,n)$. The objective of the $(n^2 - 1)$-PUZZLE is, given an initial configuration and a final one for a board of size $n \times n$, to find a solution of minimum length from the initial configuration to the final one. Here, a solution is a sequence of moves of tiles. A direction of a move and a relative position are represented by the corresponding letters N(North), E(East), S(South), and W(West). For example, moving a tile t to $(row(t) - 1, col(t) + 1)$ is called *NE-move*, and $(r(t) + 1, c(t))$ is $S(South)$ of t. The tile at (i, j) is denoted by $tile(i, j)$. Then, $target(i, j)$ represents the tile whose position in the final configuration is (i, j), while it can be located at any position until the final configuration is achieved. In addition, (i, j) is called $target(i, j)$'s *home* or *final position*. The final position of β is assumed to be in the range of $(n - 2, n - 2)$ through (n, n)[1]. The *remaining part* of a board or a configuration is a rectangular range including the blank β and the tiles that have not been placed to their final positions.

The *(Manhatten) distance* $d(i_1, j_1, i_2, j_2)$ between (i_1, j_1) and (i_2, j_2) is defined as $|i_2 - i_1| + |j_2 - j_1|$. One can easily see that the number of moves to move β (or, a tile t) to (i, j) is (at least) $d(row(\beta), col(\beta), i, j)$ (or, $d(row(t), col(t), i, j)$). This is an important property of the moves. For a procedure **proc** moving β and/or a tile, #**proc** is the number of moves needed to complete the procedure. Let $p(n)$ denotes $(n \bmod 2)/2$, i.e., $p(n) = 1/2$ if n is odd and $p(n) = 0$ otherwise.

3   The Improved Greedy Algorithm

In this section, we present the improved greedy algorithm and its analysis with making remarks on the modifications to the algorithm in [5] and its analysis.

3.1   Overview

This section provides an overview of the proposed algorithm pu and the main result. The description of the algorithm is given in Fig. 2. Its main framework

1: **if** $n = s$ **then**
2:     Apply a known algorithm for the pair of configurations I and F.
3: **else**
4:     Apply 1st_row() to the pair (I, F) of configuration I and F.
5:     Apply row_corner() to (I, F)
6:     Apply 1st_column() to (I, F)
7:     Apply column_corner() to (I, F)
8:     $I' \leftarrow$ the remaining part of the current configuration, i.e., $(2, 2)$ through (n, n).
9:     $F' \leftarrow$ the part of F corresponding to I', i.e., $(2, 2)$ through (n, n) in F.
10:     Call pu$(I', F', n - 1, s)$
11: **end if**

Fig. 2. Algorithm pu(I, F, n, s) for an initial configuration I of size $n \times n$, a final configuration F of size $n \times n$, and a constant s

---

[1] [5] also assumes it.

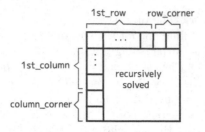

Fig. 3. Overview of the algorithm pu using four procedures: the procedure 1st_row arranges $(1,1)$ through $(1, n-2)$, row_corner arranges $(1, n-1)$ and $(1, n)$, 1st_column arranges $(2,1)$ through $(n-2, 1)$, and column_corner arranges $(n-1, 1)$ and $(n, 1)$

follows that of the algorithm in [5]: Arrange the first row and the first column, then apply the algorithm recursively to the remaining part of the board. See Fig. 3 for which part each of the procedures 1st_row, row_corner, 1st_column, and column_corner arranges. We modify the procedures for the first row and the first column in [5], for which detailed expositions will be given in the following subsections. Briefly noting here, all of the four procedures 1st_row(), row_corner(), 1st_column(), and column_corner() are different from the procedures of the same objectives in [5], which is stated in the following remark.

Remark 1. *All of the four procedures 1st_row, row_corner, 1st_column, and column_corner are different from those of the same objectives in [5].*

We obtain the following main theorem.

Theorem 1. $\#pu(I, F, n, 3) \leq 5n^3 - 21n^2/2 + 35n/2 - 59.$

Before giving its proof, we would like to make a remark on what are different here from [5] in terms of analysis to obtain this theorem, although the detailed descriptions of the procedures are deferred to the following subsections.

Remark 2. *We take care of the parity of n, although [5] does not.*

Then the proof of Theorem 1 is given as follows.

Proof (of Theorem 1). As will be shown in Propositions 10, 16, 20, and Corollary 21, it hold $\#\texttt{1st\_row}() \leq 15n^2/2 - 29n + 36 + p(n)$, $\#\texttt{row\_corner}() \leq 16n - 26$, $\#\texttt{1st\_column}() \leq 15n^2/2 - 39n + 52 + 3p(n)$, and $\#\texttt{column\_corner}() \leq 16n - 30$. Let $T_e(n)$ and $T_o(n)$ denote $\#\texttt{pu}$ for even n and odd n, respectively. Based on these, we have:

$$T_e(n) = \#\texttt{1st\_row}() + \#\texttt{row\_corner}() + \#\texttt{1st\_column}() + \#\texttt{column\_corner}()$$
$$+T_o(n-1)$$
$$\leq 15n^2 - 36n + 32 + T_o(n-1), \text{ and also} \tag{1}$$
$$T_o(n) \leq 15n^2 - 36n + 34 + T_e(n-1) \tag{2}$$

Combining these two, the upper bounds become as follows.

$$T_e(n) < (15n^2 - 36n + 32) + (15(n-1)^2 - 36(n-1) + 34) + T_e(n-2)$$
$$\leq 30n^2 - 102n + 117 + T_e(n-2), \text{ and also}$$
$$T_o(n) \leq (15n^2 - 36n + 34) + (15(n-1)^2 - 36(n-1) + 32) + T_o(n-2)$$
$$\leq 30n^2 - 102n + 117 + T_o(n-2)$$

By solving these recurrence formulas, we have.

$$T_e(n) \leq 5n^3 - \frac{21}{2}n^2 + \frac{35}{2}n - 222 + T_e(4) \quad \text{and} \tag{3}$$

$$T_o(n) \leq 5n^3 - \frac{21}{2}n^2 + \frac{35}{2}n - 93 + T_o(3). \tag{4}$$

From the result in [8], $T_o(3) \leq 34$, which implies that $T_e(4) \leq (15 \cdot 4^2 - 36 \cdot 4 + 32) + T_o(3) \leq 125$ by (1). Therefore the upper bound is given by $T_o(n)$, that is, $\#\text{pu}(I, F, n, 3) \leq 5n^3 - 21n^2/2 + 35n/2 - 59.$ □

In the above, the result in [8] is used, which is also used in [5]. For comparison, relying on the same result[8] has a reasonable meaning. However, we would like to plug new results into the analysis in order to reduce the number of moves:

Remark 3. *The results in [7] and [2] are put into the proof, while [5] uses [8].*

Based on [7], the upper bound on the number of moves is slightly improved:

Corollary 2. $\#pu(I, F, n, 3) \leq 5n^3 - 21n^2/2 + 35n/2 - 62.$

Proof. It holds $T_o(3) \leq 31$[7]. Replacing 34 with 31 in the last paragraph of the proof of Theorem 1, we obtain $\#\text{pu}(I, F, n, 3) \leq 5n^3 - 21n^2/2 + 35n/2 - 62.$ □

According to [2], it holds $T_e(4) \leq 80$. Since it is obtained by parallel computation, it may be questionable that we can easily utilize such a method as a sub-routine. However, if we can use it, we can improve the bound further:

Corollary 3. $\#pu(I, F, n, 4) \leq 5n^3 - 21n^2/2 + 35n/2 - 141.$

Proof. Plugging the bound $T_e(4) \leq 80$ into (3) derives $T_e(n) \leq 5n^3 - 21n^2/2 + 35n/2 - 142$. For odd n, (4) is interpreted as $T_o(n) \leq 5n^3 - 21n^2/2 + 35n/2 - 450 + T_o(5)$. Since $T_o(5) \leq 15 \cdot 5^2 - 36 \cdot 5 + 34 + T_e(4) \leq 309$ by (2), $T_o(n) \leq 5n^3 - 21n^2/2 + 35n/2 - 141$ holds, which is worse than the case for even n.

3.2 Moving the Blank

For completeness of the paper, this section first presents two very simple procedures blank_V (Fig. 4, Left) and blank_H (Fig. 4, Right). The procedure blank_V (or blank_H) moves β vertically (or horizontally) to $(i, c(\beta))$ (or $(r(\beta), j)$). The argument given to each of these procedures is only the row- or column-position. We observe the following proposition based on the notion of distance:

```
1: if i ≥ r(β) then
2:     Δ ← +1
3: else
4:     Δ ← −1
5: end if
6: for j = 1 to |i − r(β)| do
7:     move β once to (r(β) + Δ, c(β))
8: end for
```

```
1: if j ≥ c(β) then
2:     Δ ← +1
3: else
4:     Δ ← −1
5: end if
6: for k = 1 to |j − c(β)| do
7:     move β once to (r(β), c(β) + Δ)
8: end for
```

Fig. 4. (Left) blank_V(i) moving β vertically to the ith row (Right) blank_H(j) moving β horizontally to the jth column

```
1: blank_H(j)
2: blank_V(i)
```

Fig. 5. blank_HV(i, j) moving β horizontally to $(r(\beta), j)$ and then vertically to (i, j)

```
1: blank_V(i)
2: blank_H(j)
```

Fig. 6. blank_VH(i, j) moving β vertically to $(i, c(\beta))$ and then horizontally to (i, j)

Proposition 4. #*blank_V*(i) = $|i − r(\beta)|$ *and* #*blank_H*(j) = $|j − c(\beta)|$.

The procedures blank_HV and blank_VH in Figs. 5 and 6 move β to (i, j), where the difference is the order of vertical and horizontal moves. The proposition below shows the number of moves by these procedures based on Proposition 4.

Proposition 5. #*blank_HV*(i, j) = #*blank_VH*(i, j) = $|i − r(\beta)| + |j − c(\beta)|$.

The next procedure blank_S_tile(t) in Fig. 7 moves β to South of a tile t. This is one of the modifications we made to the algorithm in [5] as in Remark 4 below. We note that by this procedure the tile t may be N-moved once in Step 4 or 9. Because of this modification, we cannot simply borrow the analyses in [5], so that we need to develop procedures and show several propositions.

Remark 4. *The algorithm in [5] moves β to East of a tile to be moved, while we basically place β at South of it[2].*

The reason that this modification reduces the number of moves is intuitively explained as follows. As we will see in the next subsection, a diagonal move of a tile is more effective than a horizontal or a vertical one in terms of reducing the distance between the tile and its final position. With a diagonal move constituted by six moves, we reduce the distance by two, i.e., we spend three moves to reduce the distance by one. On the other hand, a horizontal and a vertical move spends five moves to reduce the distance by one. Thus a sequence of moves including as

---

[2] We move β to East of some tiles in Sect. 3.5.

1: **if** $r(t) < r(\beta)$ **then**
2:     blank_HV$(r(t) + 1, c(t))$
3: **else if** $r(t) > r(\beta)$ **then**
4:     blank_HV$(r(t), c(t))$
5: **else if** $r(t) \leq n - 1$ **then**
6:     blank_VH$(r(\beta) + 1, c(t))$
7: **else**
8:     blank_V$(r(\beta) - 1)$
9:     blank_HV$(r(t), c(t))$
10: **end if**

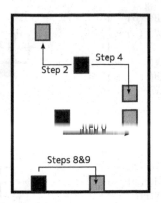

Fig. 7. blank_S_tile(t), which moves β to a South of a tile t (Left: pseudocode, Right: picture on how to move β in each case, where β is in black and the tile t is shaded)

many diagonal moves as possible is a good choice to design a better algorithm, and placing β to South of a tile does it. From Propositions 4 and 5, the number of moves by blank_Stile(t) is estimated as follows.

Proposition 6. $\#blank\_Stile(t) = |c(t) - c(\beta)| + f(t)$, *where*

$$f(t) = \begin{cases} |r(t) - r(\beta) + 1| & \text{if } r(t) < r(\beta), \\ |r(t) - r(\beta)| & \text{if } r(t) > r(\beta), \\ 1 & \text{if } r(t) = r(\beta) \leq n - 1, \text{ and} \\ 2 & \text{otherwise, i.e., if } r(\beta) = r(t) = n. \end{cases}$$

3.3 Basic Moves

We define six procedures to move a tile diagonally, horizontally, and vertically, such that the position of the blank β is South of the tile initially, and also at the end of the sequence of moves. Figures 8 and 9 show NW-move and NE-move of a tile, respectively (each of them is given as a pair of pseudocode and a figure depicting its move). W-move and E-move of a tile are defined in Fig.10, in which reading the pseudocode from the top to the bottom or looking the figures from left to right gives W-move, and the opposite order defines E-move. Similarly, Fig. 11 defines N-move and S-move of a tile. As in Remark 5 below, they are different from the ones moving a tile in [5]. Note that every one of those moves only changes the positions of the tiles in a small range of the board, that is, 2×3 or 2×2 around the target tile.

Remark 5. *Since the algorithm in [5] places the blank at East of the tile to be moved, [5] provides a sequence of moves to achieve diagonal, vertical and horizontal moves of a tile in which the locations of the blank at the beginning and also at the end are East of the tile. We need to define diagonal, horizontal, and vertical moves according to our change placing the blank at South of the tile.*

1: blank_H($c(\beta) - 1$)
2: blank_V($r(\beta) - 1$)
3: blank_H($c(\beta) + 1$)
4: blank_V($r(\beta) - 1$)
5: blank_H($c(\beta) - 1$)
6: blank_V($r(\beta) + 1$)

Fig. 8. `tile_NW_blankS`(t), moving the shaded tile t, where β is in black

1: blank_H($c(\beta) + 1$)
2: blank_V($r(\beta) - 1$)
3: blank_H($c(\beta) - 1$)
4: blank_V($r(\beta) - 1$)
5: blank_H($c(\beta) + 1$)
6: blank_V($r(\beta) + 1$)

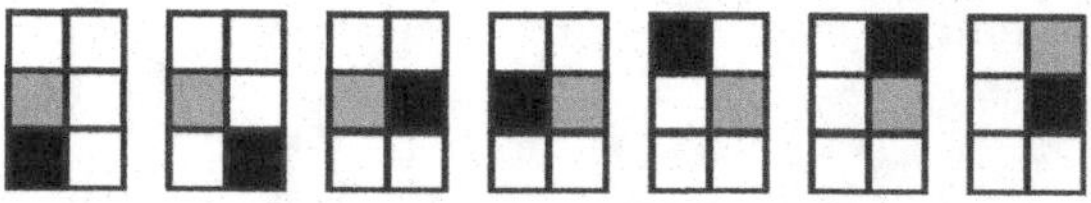

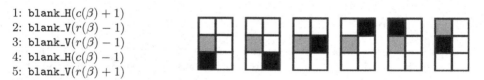

Fig. 9. `tile_NE_blankS`(t), moving the shaded tile t, where β is in black

1: blank_H($c(\beta) - 1$)
2: blank_V($r(\beta) - 1$)
3: blank_H($c(\beta) + 1$)
4: blank_V($r(\beta) + 1$)
5: blank_H($c(\beta) - 1$)

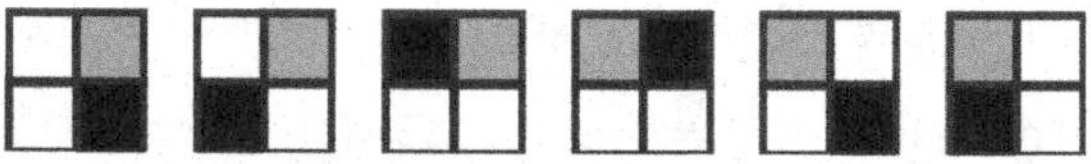

Fig. 10. `tile_W_blankS`(t) (from top to bottom in the pseudocode, or left to right in the figure) and `tile_E_blankS`(t) (from bottom to top in the pseudocode, or right to left in the figure), moving the shaded tile t, where β is in black

1: blank_H($c(\beta) + 1$)
2: blank_V($r(\beta) - 1$)
3: blank_V($r(\beta) - 1$)
4: blank_H($c(\beta) - 1$)
5: blank_V($r(\beta) + 1$)

Fig. 11. `tile_N_blankS`(t) (from top to bottom in the pseudocode, or left to right in the figure) and `tile_S_blankS`(t) (from bottom to top in the pseudocode, or right to left in the figure), moving the shaded tile t, where β is in black

3.4 Arranging the First Row

The procedure `move_tile_Sblank` in Fig. 12 moves $t = target(1, k)$ to its home $(1, k)$ assuming that $target(1, c)$ are home for $1 \le c \le k - 1$. This procedure first moves β to South of t, and then moves t by diagonal moves from Steps 3 to 10. Finally t is moved to its home by horizontal and/or vertical moves. This procedure mostly performs only one of vertical and horizontal moves repeatedly after the diagonal moves, e.g., a tile is moved by NW-moves and then followed by N-moves. One exception is that, for a case that $c(t) < k$ (Steps 8-15

1: $t \leftarrow target(1, k)$
2: blank_S_tile(t)
3: if $c(t) \geq k$ then
4:    repeat
5:       tile_NW_blankS(t)
6:      until $r(t) = 1$ or $c(t) = k$ holds
7: else
8:    repeat
9:       tile_NE_blankS(t)
10:     until $r(t) = 2$ or $c(t) = k$ holds
11:    if $r(t) = 2$ then
12:      repeat
13:        tile_E_blankS(t)
14:       until $c(t) = k$ holds
15:    end if
16: end if
17: if $r(t) = 1$ then
18:    repeat
19:       tile_W_blankS(t)
20:    until $col(t) = k$ holds
21: else
22:    repeat
23:       tile_N_blankS(t)
24:    until $row(t) = 1$ holds
25: end if

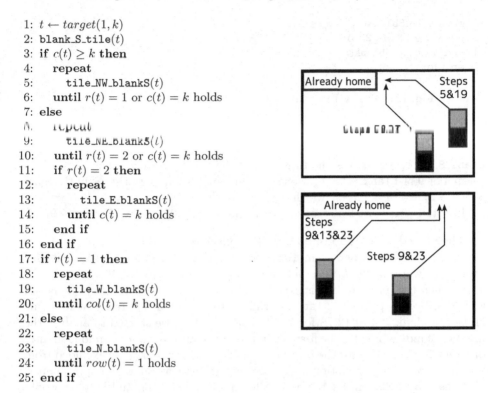

Fig. 12. move_tile_Sblank(k) moving $target(1, k)$ to its home $(1, k)$, where $target(1, k)$ is shaded and β is in black

will be done), this procedure may move the tile by three types of moves, NE-moves, E-moves and then N-moves. This is because that we now assume that $target(1, c)$ are home for $1 \leq c \leq k - 1$ and do not move them. An upper bound of #move_tile_Sblank(1) is estimated in the next proposition.

Proposition 7. #*move_tile_Sblank*$(1) \leq 8n - 9$.

Proof. Let $t = target(1, 1)$, $i_0 = r(target(1, 1))$ and $j_0 = c(target(1, 1))$ when the procedure starts. The number of moves #blank_S_tile$(t) = |c(t) - c(\beta)| + f(t)$ is given by Proposition 6. Since t and β can be located anywhere on board at the beginning, $f(t)$ in the estimation of #blank_S_tile(t) takes the maximum if $r(t) = n$ and $r(\beta) = 1$. Thus when t is located at (n, n) and β is at $(1, 1)$, blank_S_tile spends the maximum number $2n - 2$ of moves. Note that this situation also gives the worst case in the remainder part of the procedure below.

Since blank_S_tile may N-move the target tile t, let $i_1 = r(target(1, 1))$ and $j_1 = c(target(1, 1))$ right after Step 2. Notice that the condition $c(t) \geq k$ is true in this case because $k = 1$ and $j_1 \geq 1$, i.e., Steps 4 to 6 are executed. From Steps 4 to 6, the number of moves is based on the number of diagonal NW-moves, that is, $6 \min\{i_1 - 1, j_1 - 1\}$. The number of horizontal (W-) or vertical (N-) moves in Steps 17 to 25 is then $\max\{i_1 - 1, j_1 - 1\} - \min\{i_1 - 1, j_1 - 1\} = |i_1 - j_1|$, so

```
1: move_tile_Sblank(1)
2: for k = 2 to n − 2 do
3:    move_tile_Sblank(k)
4: end for
```

Fig. 13. 1st_row() by which $target(1, k)$ is home for $1 \leq k \leq n - 2$

that $5|i_1 - j_1|$ moves are spent. The total number of moves in Steps 3 to 25 is thus $6 \min\{i_1 - 1, j_1 - 1\} + 5|i_1 - j_1|$. Here $\min\{i_1 - 1, j_1 - 1\} \leq n - 2$ holds, since blank_S_tile(t) N-moves the target tile t if $i_0 = n$ in Step 2, i.e., $i_1 \leq n - 1$ (Step 4 or 9 of blank_S_tile). Hence the maximum number of steps in this part is $6(n - 2) + 5 \cdot 1 = 6n - 7$ which is given by the case $i_0 = j_0 = n$.

In total, #move_tile_Sblank(1) is at most $(2n - 2) + (6n - 7) = 8n - 9$.   □

The procedure move_tile_Sblank is applied to $target(1, 1), target(1, 2), \ldots,$ $target(1, n - 2)$ in order as in 1st_row (See Fig. 13). One can see that this procedure working for k does not move any tile at $(1, c)$ for $1 \leq c \leq k - 1$. Thus when this procedure finished, $target(1, k)$ for $1 \leq k \leq n - 2$ are home. In addition, by this procedure 1st_row, right after $target(1, k)$ arrives at its home $(1, k)$ the blank β is at $(2, k)$ for $1 \leq k \leq n - 2$. Since we started to analyze the number of moves only for the first tile in the above and its upper bound is slightly different from the one for the other tiles, move_tile_Sblank(1) is excluded from the for-loop in the description of the procedure. Next we show the upper bound of #move_tile_Sblank for the other tiles than $target(1, 1)$, and here we would like to note on one of the modifications to the analysis in [5]:

Remark 6. *Although the estimation in [5] has been done without consideration on the parity of n , we take care of it[3].*

The upper bounds on the number #move_tile_Sblank(k) in 1st_row() are shown for $2 \leq k \leq n - 2$ in the following two propositions, the former is for the left half of the first row, and the latter is for the right half of it.

Proposition 8. *For $2 \leq k \leq \lfloor (n + 1)/2 \rfloor$, it holds that #move_tile_Sblank(k) $\leq 8n - 2k - 11$ in 1st_row().*

Proof. The discussion is similar to the one in the proof of Proposition 7. One difference is that β is located at $(2, k - 1)$, when move_tile_Sblank(k) starts.

When t is at (n, n), #blank_S_tile(t) takes the maximum, i.e., $|c(t) - c(\beta)| + |r(t) - r(\beta)| = (n - (k - 1)) + (n - 2) = 2n - k - 1$. Then the diagonal moves of t are done $\min\{i_1 - 1, j_1 - k\}$ times, where (i_1, j_1) is the position of t after blank_S_tile(t) in Step 2 of move_tile_Sblank. Since $j_1 - k = n - k \leq n - 2 = i_1 - 1$, $n - k$ diagonal (NW- or NE-) moves and $k - 2$ horizontal (W-) or vertical (N-) moves are done at the worst case. Here, one relatively complicated case is that the tile will be moved by NE-moves in Step 9, E-moves in Step 13, and then N-moves in Step 23. This case happens if $3 \leq r(t) + c(t) \leq k + 1$ when

---

[3] This is a restatement of Remark 2.

the procedure starts, implying that at most $k-2$ NE-moves will be done and the total number of moves is smaller than the above case. As a result, at most $6(n-k)+5(k-2)=6n-k-10$ moves are spent.

Note that $\lfloor (n+1)/2 \rfloor$ in the statement of this proposition takes care of the parity of n. If n is odd, $(1, \lfloor (n+1)/2 \rfloor)$ is precisely the middle of the first row, for which the other worst case happens when the target tile t is at $(n,1)$; the number of moves needed for this case is exactly the same as the above. In total, $\#$move_tile_Sblank$(k) \leq (2n-k-1)+(6n-k-10)=8n-2k-11.$ □

Proposition 9. *For* $\lceil (n+1)/2 \rceil \leq k \leq n-2$, *it holds* $\#$move_tile_Sblank$(k) \leq 6n+2k-15$ *in* 1st_row().

Proof. Similar to the proof of Proposition 8, the worst case occurs when t is at $(n,1)$ (as noted in the last part of the proof of Proposition 8). According to the location of t, $\#$blank_S_tile$(t) = |c(t)-c(\beta)|+|r(t)-r(\beta)| = |1-(k-1)|+|n-2| = n+k-4$. Then the diagonal (NW- or NE-) moves of the target tile t is done $\min\{i_1-1, k-j_1\}$ times, where (i_1, j_1) is the position of t after blank_S_tile(t) in Step 2 of move_tile_Sblank. Since $k-j_1 = k-1 < n-2 = i_1-1$, $k-1$ diagonal (NW- or NE-) moves and $n-k-2$ horizontal (W-) or vertical (N-) moves are done at the worst case. The discussion for the case that Step 13 is done is similar to the part in the proof of the previous proposition. As a result, $6(k-1)+5(n-k-2)=5n+k-11$ moves are spent in this part. In total, $\#$move_tile_Sblank$(k) \leq (n+k-4)+(5n+k-11) = 6n+2k-15$.   □

As a summary, $\#$1st_row() is estimated as follows.

Proposition 10. $\#$1st_row$() \leq 15n^2/2 - 29n + 36 + p(n)$.

Proof. The upper bound is obtained by simply summing up the upper bounds in Propositions 7, 8 and 9, while we divide it into two cases based on the parity of n: For even n, $\#$1st_row$() \leq (8n-9)+\sum_{k=2}^{n/2}(8n-2k-11)+\sum_{k=n/2+1}^{n-2}(6n+2k-15) = 15n^2/2 - 29n + 36$. For odd n, $\#$1st_row$() \leq (8n-9)+\sum_{k=2}^{(n+1)/2}(8n-2k-11)+\sum_{k=(n+1)/2+1}^{n-2}(6n+2k-15) = 15n^2/2 - 29n + 73/2$.   □

3.5   Handling the Corner

Direct use of the procedure move_tile_Sblank is not appropriate for the tiles $target(1, n-1)$ and $target(1, n)$; if we put $target(1, n-1)$ into $(1, n-1)$ and do not to move it anymore, we are not able to move the tile $target(1, n)$ to $(1, n)$. Hence we need a special treatment for these two tiles.

To handle the tiles $target(1, n-1)$ and $target(1, n)$, we first design a procedure rescue_corner in Fig. 14. This procedure moves $target(1, n)$ to $(1, n-1)$ if it is at $(1, n)$, assuming that β is at $(2, n-2)$, and then moves β back to $(2, n-2)$. Without this procedure, in the next part of the algorithm, the tile $target(1, n)$ at $(1, n)$ will reside at $(1, n-1)$ and $target(1, n-1)$ will be put to $(1, n)$, and then it requires more steps than the proposed algorithm, in order to move them to their own homes. Here we make a remark on this part:

```
1: blank_HV(1, n − 1)
2: blank_H(n)
3: blank_VH(2, n − 2)
```
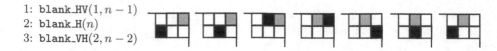

Fig. 14. rescue_corner() moving the tile $target(1, n)$ to $(1, n − 1)$ assuming that β is located at $(2, n − 2)$, where $target(1, n)$ is shaded and β is in black

```
1: if c(t) ≤ c(β) then
2:     blank_HV(r(t), c(t) + 1)
3: else if r(t) ≥ 2 then
4:     blank_VH(r(t), c(t)).
5: else if c(t) = n − 1 then
6:     blank_HV(1, n)
7: else
8:     blank_HV(1, n − 1)
9:     blank_H(n)
10: end if
```

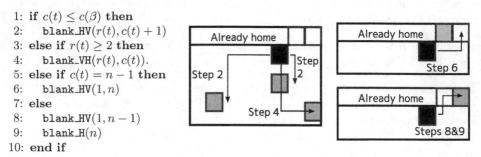

Fig. 15. blank_E_tile(t) moving β to East of t, which is only used for $t \in \{target(1, n − 1), target(1, n)\}$, where t is shaded and β is in black

Remark 7. *We guess that the algorithm in [5] also has the same problem without this kind of procedure when arranging corners. However, [5] has not argued on it, although we will see in the later in Proposition 16,* rescue_corner *does not enlarge the upper bound of the number of moves in this paper.*

The number #rescue_corner() is trivially obtained:

Proposition 11. *#rescue_corner() = 6.*

Next we present the procedure blank_E_tile moving β to East of a tile, which is used only for $target(1, n − 1)$ and $target(1, n)$. Recall that β was placed at South of the tiles $target(1, 1)$ through $target(1, n − 2)$. This procedure is similar to blank_S_tile for $target(1, k)$ through $target(1, n−2)$. Here we assume that β is located at $(2, n − 2)$ and $target(1, 1)$ through $target(1, n − 2)$ are home, when we start blank_E_tile. The number #blank_E_tile is estimated as follows.

Proposition 12. *For $t \in \{target(1, n − 1), target(1, n)\}$, #blank_E_tile(t) is*

$$
\begin{cases}
|c(t) − n + 3| + r(t) − 2 & \text{if } c(t) \le c(\beta)(= n − 2), \\
r(t) + c(t) − n & \text{if } c(t) \ge n − 1 \text{ and } r(t) \ge 2, \text{ and} \\
3 & \text{otherwise}
\end{cases}
$$

The procedure move_tile_Eblank in Fig. 16 is to move $target(1, n − 1)$ to $(2, n−1)$, or $target(1, n)$ to $(2, n−1)$ assuming that β is at East of the tile initially. This procedure utilizes NE-move, N-move, and E-move in [5], respectively shown in Figs. 17, 18, and 19. The number #move_tile_Eblank is given as follows.

Proposition 13. *#move_tile_Eblank(k) ≤ 8n − 18 for $k \in \{n − 1, n\}$.*

```
1: t ← target(1, k)
2: blank_E_tile(t)
3: repeat
4:     tile_NE_blankE(t)
5: until r(t) = 2 or c(t) = n − 1 holds
6: if row(t) = 2 then
7:     repeat
8:         tile_E_blankE(t)
9:     until c(t) = n − 1 holds
10: else
11:     repeat
12:         tile_N_blankE(t)
13:     until r(t) = 2 holds
14: end if
```

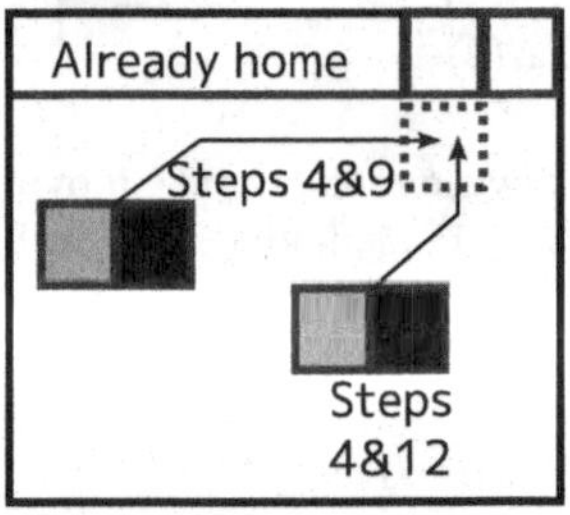

Fig. 16. move_tile_Eblank(k) moving $target(1, k)$ to $(2, n − 1)$, which is used only for $k \in \{n − 1, n\}$, where t is shaded and β is in black

```
1: blank_H(c(β) − 1)
2: blank_V(r(β) − 1)
3: blank_H(c(β) + 1)
4: blank_V(r(β) + 1)
5: blank_H(c(β) + 1)
6: blank_V(r(β) − 1)
```

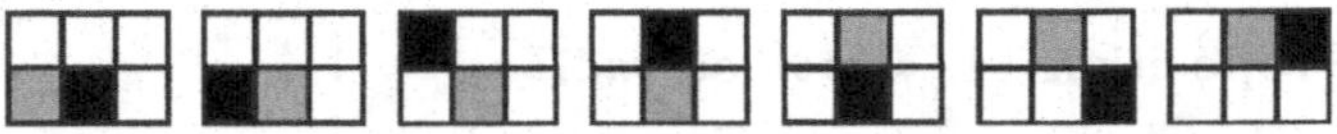

Fig. 17. tile_NE_blankE(t), moving the shaded tile t, where β is in black [5]

```
1: blank_V(r(β) − 1)
2: blank_H(c(β) − 1)
3: blank_V(r(β) + 1)
4: blank_H(c(β) + 1)
5: blank_V(r(β) − 1)
```

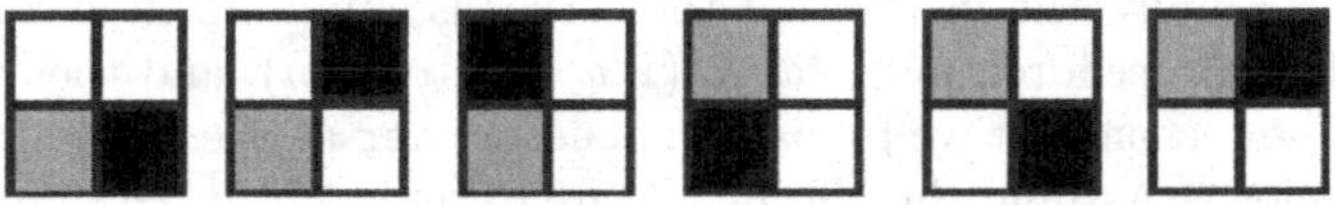

Fig. 18. tile_N_blankE(t), moving the shaded tile t, where β is in black [5]

```
1: blank_H(c(β) − 1)
2: blank_V(r(β) + 1)
3: blank_H(c(β) + 1)
4: blank_H(c(β) + 1)
5: blank_V(r(β) − 1)
```

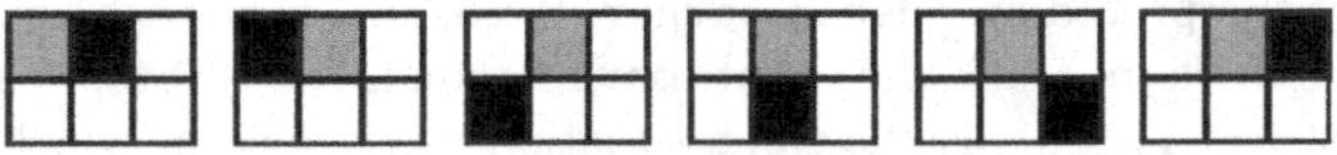

Fig. 19. tile_E_blankE(t), moving the shaded tile t, where β is in black [5]

Proof. The estimation is similar to those in the proof of Propositions 7, 8, and 9. The number #blank_E_tile(t) in Step 2 is at most $2n − 6$, which is obtained by the situation that t is at $(1, n)$. This also gives the maximum number of moves spent by the rest of the procedure: $(n − 2)$ NE-moves, i.e., $6(n − 2) = 6n − 12$ moves. In total, #move_tile_Eblank(k) $\leq (2n − 6) + (6n − 12) = 8n − 18$.   □

1: `blank_H(n − 1)`
2: `blank_V(1)`
3: `blank_H(n)`
4: `blank_V(2)`

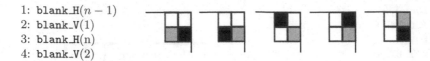

Fig. 20. `move_secondlast()` moving $target(1, n − 1)$ from $(2, n − 1)$ to $(1, n)$, where $target(1, n − 1)$ is shaded, β is in black, and the position of β is $(2, n)$ at the end

1: `blank_H(n − 1)`

Fig. 21. `move_last()` moving $target(1, n)$ from $(2, n − 1)$ to $(2, n)$, where $target(1, n)$ is shaded, β is in black, and the position of β is $(2, n − 1)$ at the end

The procedure `move_secondlast` in Fig. 20 is to move $target(1, n − 1)$ to $(1, n)$ assuming that $target(1, n − 1)$ is at $(2, n − 1)$ and β is at $(2, n)$ initially:

Proposition 14. $\#move\_secondlast() = 4$.

The next procedure `move_last` in Fig. 21 moves $target(1, n)$ to $(2, n)$, assuming that $target(1, n)$ is at $(2, n − 1)$ and β is at $(2, n)$ initially:

Proposition 15. $\#move\_last() = 1$.

Finally, the process for the corner is summarized as `row_corner` in Fig. 22. This procedure moves $target(1, n − 1)$ to $(1, n)$, and then $target(1, n)$ to $(2, n)$. The reason that we prepare `rescue_corner` at the beginning of this section is as follows. Assume that Steps 1-3 are excluded from `row_corner` and $target(1, n)$ is at its home $(1, n)$ when this procedure starts. Right after Step 4 is finished, $target(1, n)$ still resides at $(1, n)$. The followed procedure `move_secondlast` moves $target(1, n)$ to $(1, n − 1)$ while moving $target(1, n − 1)$ to $(1, n)$. Thus, we are not be able to process Steps 7-10. Avoiding this situation is the purpose of the procedure `rescue_corner`. If `rescue_corner` is applied, $target(1, n)$ will be at $(1, n − 1)$. After that, `move_secondlast` in Step 5 moves $target(1, n)$ to $(2, n − 1)$. This is the same situation as the one in which $target(1, n)$ was not at $(1, n)$ (in this case `rescue_corner` has not been applied) and then Step 7 is processed. Step 6 is to meet the condition on the position of β in Proposition 12: it moves β from $(2, n)$ to $(2, n − 2)$, which spends 2 moves. Steps 10 to 12 move $target(1, n−1)$ and $target(1, n)$ to their homes $(1, n−1)$ and $(1, n)$, respectively. One important property is that this procedure `row_corner` does not move tiles at $(1, 1)$ through $(1, n − 2)$. The number $\#row\_corner()$ is estimated as follows.

Proposition 16. $\#row\_corner() \leq 16n − 26$.

Proof. Suppose that $target(1, n)$ is at $(1, n)$. In this case, `rescue_corner()` in Step 2 is processed and then `move_tile_Eblank(n)` in Step 7 is not processed.

```
1: if tile(1, n) = target(1, n) then
2:    rescue_corner()
3: end if
4: move_tile_Eblank(n − 1)
5: move_secondlast()
6: blank_H(n − 2)
7: if the condition in Step 1 was false then
8:    move_tile_Eblank(n)
9:    end if
10: move_last()
11: blank_V(1)
12: blank_H(n)
13: blank_V(2)
```

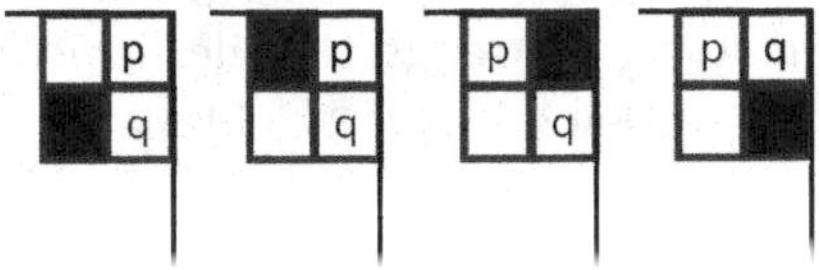

Fig. 22. `row_corner()`: The right figure only depicts the moves in Steps 11-13, where p and q respectively represent the tiles $target(1, n-1)$ and $target(1, n)$

Thus the upper bound of the total number of moves is estimated by the upper bounds of the four procedures, `#rescue_corner()`, `#move_tile_Eblank(n − 1)`, `#move_secondlast()`, and `#move_last()` based on Propositions 11, 13, 14 and 15 with the number of moves spent in Steps 6 and 11-13. That is, `#row_corner()` is at most $6 + (8n - 18) + 4 + 2 + 1 + 3 = 8n - 2$, where the last term 3 for Steps 11-13 comes from the observation that β is at $(2, n - 1)$ when `move_last` finished in Step 10.

Consider the next case that `rescue_corner()` in Step 2 is not processed and then `move_tile_Eblank(n)` in Step 7 is processed. The number of moves is obtained from the above value $8n - 2$ by subtracting 6 of `#rescue_corner()` (Proposition 11) and adding `#move_tile_Eblank(n)`. As a result, this case requires at most $(8n - 2) - 6 + (8n - 18) = 16n - 26$ moves, where the term $(8n - 18)$ is based on Proposition 13. This is larger than $8n - 2$ in the first case. □

3.6   Arranging the First Column

We arranged the first row in the previous subsections. Similar idea works for the first column, where the first tile $target(1, 1)$ is already at its final position by the arrangement of the first row. After arranging the first row, we can view the remaining part of the board is of size $(n - 1) \times n$. Based on this observation, first we swap the row-index and column-index of positions, tiles, and β. Then we apply Steps 2-4 of `1st_row` to this new board excluding the process for the first tile $target(1, 1)$. This corresponds to arranging the first column of the original board except for the first tile $target(1, 1)$. We call this procedure `1st_column`.

The number `#1st_column()` is estimated analogously to `#1st_row()`. There are two differences in the estimation: at the beginning of `1st_column`, (i) the blank β is at $(n, 2)$[4], and (ii) all the tiles in the first column are at their homes.

---

[4] This corresponds to $(2, n)$ in the original board.

Due to the space limitation, we omit the proofs of the corollaries and the propositions in this section, which are shown by similar discussions to the ones in the previous sections and are deferred to the full version of the paper.

The difference (i) affects Proposition 8. Here the location of β is $(n, 2)$ when we start to arrange the tile $target(1, 2)$, although β was $(2, n)$ in the proof of Proposition 8. Then we obtain the following corollary based on Proposition 8.

Corollary 17. $\#move\_tile\_Sblank(2) \le 7n - 12$ in $1st\_column()$. *(The proof is omitted.)*

For $3 \le k \le \lfloor (n+1)/2 \rfloor$, the estimation is the same as in Proposition 8:

Corollary 18. For $3 \le k \le \lfloor (n+1)/2 \rfloor$, it holds that $\#move\_tile\_Sblank(k) \le 8n - 2k - 11$ in $1st\_column()$.

For $k \ge \lceil (n+1)/2 \rceil$, we need to slightly change the statement, in which the upper bound is decreased by two compared to the one in Proposition 9. This decrease is based on the above mentioned difference (ii).

Corollary 19. For $\lceil (n+1)/2 \rceil \le k \le n-2$, it holds $\#move\_tile\_Sblank(k) \le 6n + 2k - 17$ in $1st\_column()$. *(The proof is omitted.)*

Based on the above Corollaries 17, 18, and 19, $\#1st\_column()$ is estimated:

Proposition 20. $\#1st\_column() \le 15n^2/2 - 39n + 52 + 3p(n)$. *(The proof is omitted.)*

As for the treatment of the corner, `row_corner()` can be also applied (We call it `column_corner()` for the purpose). Its analysis is slightly changed as in the proof of Corollary 19, based on the difference (ii):

Corollary 21. $\#column\_corner() \le 16n - 30$ in $1st\_column()$. *(The proof is omitted.)*

4   Conclusion

We designed an algorithm by modifying the known one[5]. The number of moves by our algorithm is at most $5n^3 - 21n^2/2 + 35n/2 - 141$ and the running time of the algorithm is $O(1)$ per move. The lower bounds on the number of moves shown in [5] is also applied to our algorithm, e.g., there is a case requiring $\Omega(n^3)$ moves to complete. Since the algorithms in this paper and in [5] spend $5n^3 + O(n^2)$ moves, an interesting research topic must be designing a simple and fast algorithm which spends $cn^3 + O(n^2)$ moves for some constant $c < 5$.

References

1. Bhasin, H., Singla, N.: Genetic based algorithm for N-puzzle problem. Int. J. Computer Applications **51**(22), 44–50 (2012)
2. Brüngger, A., Marzetta, A., Fukuda, K., Nievegelt, J.: The parallel search bench ZRAM and its applications. Annals of Operations Research **90**, 45–63 (1999)
3. Johnson, W., Story, W.E.: Notes on the "15" puzzle. American Journal of Mathematics 2(4), 397–404 (1879)
4. Levitin A., Papalaskari, M.-A.: Using puzzles in teaching algorithms. In: Proc. SIGCSE 2002, pp. 292–296 (2002)
5. Parberry, I.: A real-time algorithm for the $(n^2 - 1)$-puzzle. Information Processing Letters **56**, 23–28 (1995)
6. Ratner, D., Warmuth, M.: The $(n^2 - 1)$-puzzle and related relocation problems. J. Symbolic Computation **10**, 111–137 (1990)
7. Reinefeld, A.: Complete solution of the eight-puzzle and the benefit of node ordering in IDA*. In: Proc. Int. Joint Conf. Artificial Intelligence, pp. 248–253 (1993)
8. Schofield, P.D.A.: Complete solution of the eight puzzle. Machine Intelligence **1**, 125–133 (1967)

Variable Neighborhood Search for the Elementary Shortest Path Problem with Loading Constraints

Telmo Pinto, Cláudio Alves$^{(\boxtimes)}$, and José Valério de Carvalho

Centro Algoritmi, Escola de Engenharia, Universidade do Minho,
4710-057 Braga, Portugal
{telmo,claudio,vc}@dps.uminho.pt

Abstract. In this paper, we address the elementary shortest path problem with 2-dimensional loading constraints. The aim is to find the path with the smallest cost on a graph where the nodes represent clients whose items may have different heights and widths. Beyond its practical relevance, this problem appears as a subproblem in vehicle routing problems with loading constraints where feasible routes have to be generated dynamically. To the best of our knowledge, there are no results reported in the literature related to this problem. Here, we explore a variable neighborhood search approach for this problem. The method relies on constructive heuristics to generate feasible paths, while improved incumbents are sought in different neighborhoods of a given solution through a variable neighborhood search procedure. The resulting variants of the algorithm were tested extensively on benchmark instances from the literature. The results are reported and discussed at the end of the paper.

Keywords: Shortest path problem · Loading constraints · Variable neighborhood search · Computational study

1 Introduction

The last years have seen an increased interest for rich routing problems that consider practical constraints arising in real settings. A good example is the vehicle routing problem with loading constraints for which many different contributions have been described recently [2,5,10]. The problem merges two well-known problems in the field of combinatorial optimization, namely the classical vehicle routing problem and the packing problem that results from the need of building feasible routes when the items involved are 2- or 3-dimensional objects. Clearly, the resulting problem is NP-hard since it combines two problems that are already NP-hard. A comprehensive survey related to this family of problems can be found in [4].

The elementary shortest path problem with loading constraints arises in particular in the context of vehicle routing problems where the items have at least

© Springer International Publishing Switzerland 2015
O. Gervasi et al. (Eds.): ICCSA 2015, Part II, LNCS 9156, pp. 474–489, 2015.
DOI: 10.1007/978-3-319-21407-8_34

two dimensions, and the capacity of the vehicles is a strong constraint. The problem occurs typically when feasible routes have to be generated dynamically as happens for example in column generation based approaches. Given this strong connection between the two problems, we will briefly review in the sequel the main aspects and contributions related to vehicle routing problems with loading constraints.

Some versions of the capacitated vehicle routing problem with loading constraints (L-CVRP) consider that the loading and unloading of the items must be done without moving the other items that are in the vehicle. Therefore, the position of the items inside the vehicles depends directly on the sequence by which the clients are visited. This constraint is known as a sequential or LIFO (*Last-In, First-Out*) constraint, and the corresponding routing problems are referred to as sequential L-CVRP. The problems that do not enforce this constraint are called unrestricted L-CVRP. Moreover, if the items can be rotated inside the vehicles, the resulting problem is denominated by rotated L-CVRP.

Due to the inherent complexity of the L-CVRP, the vast majority of the approaches described in the literature are based on heuristic approaches. The only exact method reported so far is due to Iori *et al.* [5], who described a branch-and-cut method to solve the problem with 2-dimensional items (2L-CVRP). Gendreau *et al.* [2] proposed the first heuristic method for the 2L-CVRP for both the unrestricted and sequential case. Their approach is based on tabu search, and it allowed the visit of infeasible solutions by penalizing them in the objective function. In [10], the authors describe a guided tabu search approach for the 2L-CVRP, which is enhanced with diversification procedures that penalize long arcs in the solution. The feasibility of the routes is checked by means of five different heuristics.

Some attempts in solving the 2L-CVRP exactly through column generation are reported in [7], while this approach is seen by Iori and Martello in [4] as worthwhile of investigation in order to determine effectively exact solutions for this problem. The Dantzig-Wolfe decomposition principle on which column generation approaches are based has been extensively applied to vehicle routing problems. The standard reformulation that results from this decomposition is a set partitioning problem which is solved typically by dynamic column generation. The related pricing subproblem is an elementary shortest path problem with additional resource constraints. This problem has received much attention in the literature [1,8,9]. Applying the Dantzig-Wolfe decomposition to the L-CVRP yields also a set partitioning problem which can be solved through column generation. The pricing subproblem remains an elementary shortest path problem, but now the resource constraints become 2- or 3-dimensional packing constraints which are much more difficult to handle than other standard constraints as the capacity constraints, for example.

To the best of our knowledge, the elementary shortest path problem with loading constraints has never been addressed in the literature. In this paper, we describe and analyse a solution approach based on variable neighborhood search for the problem with 2-dimensional items and sequential constraints. To generate feasible routes, we use different constructive methods that handle the packing part of the problem through alternative strategies based on bottom-left and level packing placement rules. Local search is supported on several neighborhoods defined from both the routing and packing definition of the solutions. The combination of the different strategies described in this paper leads to different variants of the variable neighborhood search algorithm. The performance of these variants is evaluated and compared through extensive computational experiments on benchmark instances from the literature for the 2L-CVRP.

The paper is organized as follows. In Section 2, we define formally the elementary shortest path problem with loading constraints addressed in this paper, and we introduce the corresponding notation. In Section 3, we describe the constructive heuristics used to generate feasible solutions for the problem. The neighborhood structures and the details of our variable neighborhood search algorithm are described in Section 4. In Section 5, we report on the computational experiments performed to evaluate and compare the performance of our approach. Some final conclusions are drawn in Section 6.

2   The Elementary Shortest Path Problem with Loading Constraints

The elementary shortest path problem with 2-dimensional loading constraints (2L-ESPP) is defined on a graph $G = (V, E)$ with the set of nodes V representing the n clients of the problem plus the depot 0 from which the vehicle leaves initially and to which it should come back at the end of the visits, and E representing the set of edges of the graph. The travelling cost associated to the edges $(i, j) \in E$ will be denoted by c_{ij}. The loading area of the vehicle has a total width denoted by W and a maximum height of H units. Each client $i \in V \setminus \{0\}$ has b_i 2-dimensional items of width and height respectively equal to w_i and h_i. These dimensions are general, *i.e.* no particular constraints apply to the width and height of the items. The visit of a client implies that all his items are loaded on the vehicle. Hence, we assume that the load associated to any client $i \in V \setminus \{0\}$ fits in the vehicle. The 2L-ESPP consists in finding the minimum cost route for the vehicle that starts and ends at the depot 0 and that visits at most once the clients in V. Note that here we assume that there are negative costs for some edges of the graph, which happens typically when the problem is defined as a pricing subproblem of a column generation model for the 2L-CVRP.

In this paper, we address the case where the items of the clients have a fixed rotation, and where the sequential or LIFO constraint applies. The latter implies that, during the loading and unloading of the items of a given client, the items of all the other clients that are already in the vehicle cannot be moved. Furthermore, lateral movements of the items inside the vehicle are forbidden. Hence, the loading and unloading operations can only be done in a direction that is parallel to the left and right sides of the vehicle. Figure 2 illustrates the case of a feasible and an infeasible loading pattern for the instance of Example 1 according to this sequential constraint. In the example represented in this figure, the sequence of visits is $(0, 2, 1, 0)$. However, in the pattern (b) of Figure 2, one of the items of client 2 cannot be unloaded without moving first the items of client 1.

A solution for the 2L-ESPP is defined as a sequence of clients that starts and ends at the depot 0, together with a placement position for each item of the clients that are visited. Alternatively, the latter can be replaced by defining the sequence by which the items of each client should be loaded on the vehicle, and by defining the placement rule that is used. Let S denote the sequence of clients visited in a solution of the 2L-ESPP. We have

$$S = (s_1, s_2, \ldots, s_{|S|}),$$

with $s_1 = s_{|S|} = 0$, while $s_2, \ldots, s_{|S|-1}$ represent the clients visited by the corresponding route. The cost of the solution associated to S will be denoted by $z(S)$, i.e. $z(S) = \sum_{i=1}^{|S|-1} c_{s_i s_{i+1}}$. Moreover, let P define the order by which the items of the clients of S are placed in the vehicles, such that

$$P = (p_2, \ldots, p_{|S|-1}),$$

with $p_i = (p_i^1, p_i^2, \ldots, p_i^{b_{s_i}})$, $i = 2, \ldots, |S| - 1$, and p_i^j representing the index of the jth item of client s_i to be placed in the vehicle.

The following example illustrates through a small instance the details related to the definition of the 2L-ESPP and its solutions.

Example 1. Consider the instance of the 2L-ESPP represented in Figure 1. The set of nodes V is composed by $n = 4$ clients and the depot 0, i.e. $V = \{0, 1, 2, 3, 4\}$. The costs c_{ij}, $(i, j) \in E$, are shown beside the edges of the graph. The items of each client are identified through the tuple (i, k), with i representing the client and k the index of the item $(k = 1, \ldots, b_i)$. A feasible solution for this instance is the following:

$$S = (0, 2, 1, 0) \qquad \text{and} \qquad P = ((1, 3, 2), (1, 2)).$$

assuming that a standard bottom-left rule is used to place the items. The corresponding loading pattern is illustrated in Figure 2-(a). The cost of this solution is $z(S) = -5$. It is easy to see that all the items of the clients can be unloaded from the vehicle without lateral movements or moving the items of the other clients. Figure 2-(b) shows an alternative loading pattern that violates this sequential constraint.

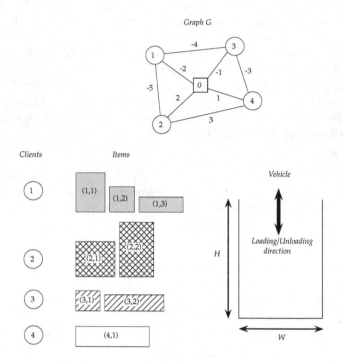

Fig. 1. Instance of the 2L-ESPP (Example 1)

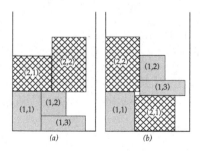

Fig. 2. Feasible (a) and infeasible loading pattern (b) (Example 1)

3   Building Feasible Solutions with Constructive Heuristics

To build an initial solution for the problem, we adopted a constructive approach in which the clients are added one by one to the route. The next client to be evaluated is inserted in the current route only if all his items can be loaded on the vehicle according to the constraints that apply to the loading patterns. The clients are evaluated following a nearest neighbor approach. Starting from the depot, two different strategies were considered to select the first client:

(FN) the first client to be evaluated is the one that is nearest to the depot 0
     $(i.e.\ argmin\{c_{0i} : i = 1, \ldots, n\})$;
(FR) the first client to be evaluated is selected randomly.

The acronyms (FN) and (FR) used above will be used later to distinguish between the variants that we obtain by using one of these two strategies. After selecting the first client, the remaining ones are evaluated by non-decreasing order of the cost of the edge that connects them to the last client in the current route. The client that is evaluated is inserted in the route if all his items can be loaded on the vehicle. If no more clients can be inserted, the route is closed by connecting the last client to the depot.

The hardest part when building a solution for the 2L-ESPP is to find (if it exists) a feasible arrangement of the items on the vehicle such that there are no overlaps, all the items are put inside the boundaries of the vehicle without rotation, and the sequential constraint is satisfied. To address this issue, we considered three different strategies:

(BL) a standard bottom-left placement rule;
(RBL) a revised bottom-left procedure;
(LP) a level packing approach.

The strategy (BL) consists in placing the next item in the bottom and leftmost free position of the vehicle that ensures that all the loading constraints that apply are satisfied. Each time an item is placed on the vehicle, at most four orthogonal free spaces are generated identifying the different areas of the vehicle where the next items can be placed. In turn, after placing an item in a free space, this free space (and all the others that are intersected by the item) is removed, and replaced by an updated set of free spaces. Placing an item according to the strategy (BL) is equivalent to finding the free space whose width and height are equal to or larger than the size of the item, and whose bottom and leftmost position is the smallest among all the free spaces provided that it leads to a feasible placement. After finding this free space, the item is placed in its bottom and leftmost position. Figure 3 illustrates the concept of free space. Free space 1 and 2 (black dotted lines) are the free spaces respectively at the left and right of the item generated after its placement on the vehicle. Free spaces 3 and 4 (grey dashed lines) are the free spaces respectively above and below the item. The revised bottom-left procedure (RBL) consists in selecting the free space where

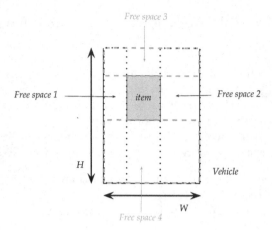

Fig. 3. Free spaces

the item produces the best fit, and then in placing the item in the bottom and leftmost position of this free space. The free space with the best fit is the one whose area is the nearest to the area of the item. The idea is to place the items in the areas of the vehicle where the packing generates the least possible waste, thus favouring the filling of holes. Again, only the free spaces whose bottom and leftmost position yields a feasible placement are considered for selection.

The last strategy (LP) that we explored consists in placing the items of the selected client in horizontal levels, and then in placing the levels on the vehicle one above another. The first item to be placed in a level determines the maximum height of the items that can be placed after it in that level. To select the level where the next item will be placed, we considered the following two strategies:

($LP.FF$) the next item is placed in the first open level where it fits;
($LP.BF$) the next item is placed in the open level where it best fits.

The level that was placed in the upmost position of the vehicle for the previous client is considered as an open level when placing the items of the next client. After placing all the items on the levels, the levels are placed on the vehicle so that the one with the largest remaining space is placed in the upmost position. One of the advantages of level packing approaches is that they ensure that the patterns satisfy the sequential constraint, thus avoiding the necessity of checking the placement positions before placing the items. The loading patterns resulting from this level packing approach are similar to the cutting patterns that arise in 2-dimensional guillotine cutting stock problems.

After choosing a client to insert in the route, his items are selected one by one, and placed in the vehicle according to the strategies described above. The next item to be placed is selected according to two different orderings based on the following criteria:

(OH) height of the items;

(OA) area of the items.

In both cases, the items are ordered in non-increasing order of their height and area, respectively.

4   Variable Neighborhood Search Algorithm

The strategies described in the previous section yield different variants of a constructive method for building feasible solutions for the 2L-ESPP. In order to improve the solutions generated by these algorithms, we developed a local search approach embedded into a variable neighborhood search (VNS) algorithm. The VNS metaheuristic was described first by Mladenovic and Hansen in [6], and since then, it has been applied with success to solve different combinatorial optimization problems [3]. The aim of VNS is to explore systematically different neighborhoods of the solutions to diversify the search and escape from local optima. Here, VNS is used to drive the search into 7 alternative neighborhoods of the solutions generated through the constructive algorithms arising from the combination of the different strategies described above. In the following section, we define the neighborhood structures that we used to support the local search procedures. The details of our VNS algorithm are given in Section 4.2.

4.1   Neighborhood Structures

The representation of the solutions of the 2L-ESPP defined in Section 2 relies on two main elements: the sequence by which the clients are visited in a given route, and the characterization of the loading pattern used to arrange the corresponding items in the vehicle so that all the loading constraints that apply (no overlaps, fixed orientation and the sequential constraint) are satisfied. To explore the search spaces defined through these two aspects of the solutions, we defined 7 neighborhood structures that can be divided into routing neighborhoods and packing neighborhoods. The definition of the neighborhood structures relies on the constructive algorithms defined above, and covers a broad range of possible movements. Let CH denote the constructive heuristic used to build the initial solution for the instance, and which is obtained by combining the strategies described in Section 3. In our implementation, we assumed that the constructive heuristic CH used to generate the initial solution is also the one that is used to define the neighbors of a given solution. When generating the neighbors, the difference is that part of the sequences of clients (and the corresponding sequence by which the items are placed in the vehicle) is fixed. The neighborhood structures are defined in the sequel.

Routing neighborhoods

NS_1 **Swapping two clients in the route**

Given a solution whose sequence of visited clients is S (as defined in Section 2), the neighbors of this solution consist in all the feasible solutions obtained by swapping two clients in this route, by applying CH with the resulting sequence of clients (keeping the sequence of items for each client), and by adding more clients from the last one forward in the route using CH. Let S_c denote the sequence of clients in the route associated to the current solution, such that:

$$S_c = (s_1, s_2, s_3, \ldots, s_{|S_c|-1}, s_{|S_c|}),$$

One of the neighbors of this solution obtained by swapping s_2 and $s_{|S_c|-1}$ is the following:

$$S'_c = (s_1, s_{|S_c|-1}, s_3, \ldots, s_2, s_{|S_c|}),$$

provided that the items of the clients in S'_c can be put in the vehicle using CH, and no more clients can be added at the end of S'_c. Moreover, if P_c defines the sequences by which the items of the clients of S_c are placed in the vehicle (again as defined in Section 2), $i.e.$

$$P_c = (p_2, p_3, \ldots, p_{|S_c|-1}),$$

then the corresponding sequences of items associated to S'_c will be

$$P'_c = (p_{|S_c|-1}, p_3, \ldots, p_2).$$

NS_2 **Shifting a client in the route**
The neighbors of a solution whose sequence of clients is S are obtained by choosing a client and placing it in a different position of the sequence, by applying CH with the resulting sequence of clients (keeping the sequence of items for each client), and by adding more clients from the last one forward in the route using CH. The following solution is a neighbor of the solution S_c defined above obtained by selecting the client s_2 and by placing it in the third position of the sequence, $i.e.$

$$S'_c = (s_1, s_3, s_2 \ldots, s_{|S_c|-1}, s_{|S_c|}).$$

Again, in this case, we are assuming that no more clients can be added at the end of the route by applying CH.

NS_3 **Removing a client from the route**
The neighbors of a solution are obtained by removing a client from the route, by applying CH with the resulting sequence of clients in the same conditions as in the previous neighborhood structures, and by inserting clients at the end of the sequence (before the depot and if they fit in the vehicle) using again CH.

NS_4 **Removing a client and all its successors from the route**
The neighbors of a solution are obtained by removing all the clients from a given position of the sequence up to the end, by adding a selected client

at the end of the sequence and by placing his items using CH. The vehicle is filled by applying strictly the heuristic CH starting from the last client that was inserted.

NS_5 **Exchanging a client by another in the route**
The neighbors of a solution are obtained by exchanging a client by another that is not in the sequence, by placing the items of the clients in the resulting sequence using CH, and by adding other clients at the end of the sequence (before the depot) using again CH. Note that the sequence by which the items of the clients are placed in the vehicle remains unchanged for all the clients that were already in the route.

Packing neighborhoods

NS_6 **Swapping two items**
The neighbors of a solution are obtained by selecting a client in the route and by swapping two items in the sequence that defines the order by which his items are placed in the vehicle. Then, the heuristic CH is used to build the solution that corresponds to these sequences of clients and items (if possible), and to add other clients at the end of the sequence (before the depot) if they fit in the vehicle. Let S^c be the sequence of visited clients in the current solution, and let P^c denote the corresponding sequences by which the items are placed in the vehicle. As an example, let S^c and P^c be defined respectively as follows:

$$S_c = (s_1, s_2, s_3, s_4, s_5),$$

with $s_1 = s_5 = 0$, and

$$P_c = ((p_2^1, p_2^2), (p_3^1, p_3^2, p_3^3), (p_4^1)).$$

A possible neighbor S_c' of this solution is defined as follows:

$$S_c' = S_c \qquad \text{and} \qquad P_c' = ((p_2^1, p_2^2), (p_3^2, p_3^1, p_3^3), (p_4^1)).$$

It is obtained by swapping the first and second item of s_3 in P_c, provided that all the items can be placed according to P_c' in the vehicle by applying CH, and that no more clients can be added at the end of the sequence.

NS_7 **Shifting an item**
The neighbors of a solution are obtained by selecting a client in the route and by placing one of his items in a different position in the sequence that defines the order by which the items of this client were inserted in the vehicle. As in the previous neighborhood structure, the heuristic CH is used to build the solution associated to these sequences of clients and items. The same heuristic is used to add other clients at the end of the sequence and before the depot, if possible.

4.2   Variable Neighborhood Search

To explore the search spaces defined through the neighborhood structures described in the previous section, we developed a variable neighborhood search algorithm that applies local search on these 7 neighborhoods. The initial solution is generated by applying one of the constructive heuristics that result from the combination of the different strategies described in Section 3, namely $\{(FN), (FR)\}$, $\{(BL), (RBL), (LP)\}$, $\{(LP.FF), (LP.BF)\}$ (if (LP) has been selected), and $\{(OH), (OA)\}$. Then, the 7 neighborhoods are explored in cycle until a maximum computing time limit is reached. A solution is generated in a shaking phase from the current incumbent solution in the neighborhood that is being explored, and a local search procedure is applied right after in the same neighborhood to determine an improved solution. In our implementation, we resorted to a first improvement local search procedure that stops when it finds a solution that is better than the solution generated in the shaking phase, or if no better solution exists in this neighborhood. Note that all the solutions that are explored are necessarily feasible solutions for the problem. Our variable neighborhood search algorithm is described in Algorithm 1. The constructive heuristic is denoted by `findInitialSolution()`, while the shaking and local search procedures are represented respectively by `shaking((S, P), NS_k)` and `firstImprovement((S', P'), NS_k)`, with (S, P) denoting the current incumbent solution, (S', P') the solution generated in the shaking phase, and NS_k the neighborhood that is being explored.

5   Computational Experiments

To evaluate and compare the performance of the different variants of our variable neighborhood search algorithm, we conducted an extensive set of computational experiments on 180 benchmark instances of the 2L-CVRP used by Iori *et al.* in [5] and by Gendreau *et al.* in [2]. Note that, in the former, the largest instances were not used due to their complexity. The number n of clients of these instances ranges from 15 up to 255, while the total number of items varies between 15 and 786. A complete description of the instances can be found in [2]. For our experiments, we multiplied all the costs (distances) associated to the edges by -1. The algorithms were coded in C++, and the tests were run on a PC with an i7 CPU with 2.9 GHz and 8 GB of RAM.

We tested the 16 variants of our algorithm resulting from the combination of the strategies described in Section 3, namely $\{(FN), (FR)\}$, $\{(BL), (RBL), (LP)\}$, $\{(LP.FF), (LP.BF)\}$ (if (LP) has been selected), and $\{(OH), (OA)\}$. The instances were divided in 22 groups according to the number of clients. The average results for each group are reported in Tables 1-3. In Table 1, we report on the results obtained with the standard bottom-left placement rule (BL) for all the possible combinations of strategies involving (FN), (FR), (OH) and (OA). Table 2 provides the results achieved with the revised bottom-left placement rule (RBL), while Table 3 gives the results when the level packing procedure (LP) is used with $(LP.FF)$. Although we tested the

---

Algorithm 1. Variable Neighborhood Search Algorithm

Input:
  I: instance of the 2L-ESPP;
  CH: constructive heuristic defined from the combination of the different
  strategies $\{(FN), (FR)\}, \{(BL), (RBL), (LP)\}, \{(LP.FF), (LP.BF)\}$
  (if (LP) has been selected), and $\{(OH), (OA)\}$;
  Set of neighborhood structures $NS = \{NS_1, NS_2, \ldots, NS_7\}$;
  Limit t_{max} on the total computing time;
Output:
  Feasible solution (S, P) of value $z(S)$;

$(S, P) := \texttt{findInitialSolution}();$
repeat
  $\quad k := 1;$
  $\quad$**while** $k \leq 7$ **do**
  $\quad\quad (S', P') := \texttt{shaking}((S, P), NS_k);$
  $\quad\quad (S'', P'') := \texttt{firstImprovement}((S', P'), NS_k);$
  $\quad\quad$**if** $z(S'') \leq z(S)$ **then**
  $\quad\quad\quad (S, P) := (S'', P'');$
  $\quad\quad\quad k := 1;$
  $\quad\quad$**end**
  $\quad\quad$**else**
  $\quad\quad\quad$k:=k+1;
  $\quad\quad$**end**
  $\quad$**end**
until $\texttt{cpuTime}() \leq t_{max};$
return (S, P) ;

---

level packing procedure (LP) with the best-fit rule $(LP.BF)$, we do not report the results here due to the lack of space. This latter strategy led to results that are very near from those obtained with the approach based on $(LP.FF)$ for these instances. All the tests were run with a maximum computing time limit of 3 seconds. The purpose was to test and compare the ability of each strategy in finding efficiently good incumbents for the problem. The meaning of the columns in these tables is the following:

$\quad n$: number of clients;
$\quad M$: average number of items;
$\quad inst$: number of instances in the corresponding group;
$\quad ord$: criterion used to order the items of a client $((OH)$ or $(OA))$;
$\quad z_{CH}$: average value of the initial solution generated using the constructive heuristic resulting from the combination of the strategies described in Section 3;
$\quad \%^{CH}_{fill}$: average percentage of space used in the vehicle by the initial solution generated using the constructive heuristic;
$\quad z_{VNS}$: average value of the best solution obtained with the variable neighborhood search algorithm described in Algorithm 1;
$\quad \%^{VNS}_{fill}$: average percentage of space used in the vehicle by the best solution obtained with the variable neighborhood search algorithm;

imp: percentage of improvement achieved with the variable neighborhood search algorithm, *i.e.* $imp = (z_{VNS} - z_{CH})/z_{CH}$.

Additionally, in Tables 1-3, the average results for all the instances are reported in the line avg.

Table 1. Computational results with (BL)

				(FN)					(FR)				
n	M	inst	ord	z_{CH}	$\%_{fill}^{CH}$	z_{VNS}	$\%_{fill}^{VNS}$	imp	z_{CH}	$\%_{fill}^{CH}$	z_{VNS}	$\%_{fill}^{VNS}$	imp
15	31,1	10	(OH)	-335,30	61,26	-349,30	65,07	4,18	-318,90	61,21	-341,30	63,87	7,02
20	39,5	10		-423,80	60,82	-483,00	67,33	13,97	-435,30	63,02	-475,70	66,56	9,28
21	39,4	10		-561,90	63,93	-621,70	69,91	10,64	-542,70	65,63	-594,00	66,81	9,45
22	39,4	10		-940,80	62,23	-1076,70	68,42	14,45	-966,50	64,32	-1049,40	68,26	8,58
25	56,0	5		-532,20	60,12	-561,80	66,56	5,56	-452,00	61,12	-550,60	67,96	21,81
29	57,8	10		-984,70	63,54	-1137,00	69,39	15,47	-969,90	62,10	-1149,50	65,53	18,52
30	63,8	5		-512,40	62,72	-561,00	69,22	9,48	-497,60	65,70	-556,20	69,92	11,78
32	62,5	15		-1718,87	61,84	-1831,33	69,02	6,54	-1619,07	60,55	-1858,00	69,52	14,76
35	74,4	5		-576,20	60,74	-648,80	68,46	12,60	-554,40	62,68	-639,80	63,40	15,40
40	79,2	5		-611,60	57,78	-754,20	71,58	23,32	-643,60	63,26	-766,80	69,46	19,14
44	86,2	5		-1250,40	58,76	-1432,00	70,46	14,52	-1262,80	63,70	-1431,00	70,26	13,32
50	105,2	5		-770,80	62,14	-853,60	70,30	10,74	-717,20	63,42	-826,20	71,18	15,20
71	146,0	5		-542,00	61,40	-607,00	70,20	11,99	-548,60	68,76	-602,20	72,96	9,77
75	150,3	20		-1068,45	64,47	-1215,80	70,75	13,79	-1038,85	63,16	-1201,70	70,69	15,68
100	204,3	15		-1399,80	65,99	-1561,00	72,73	11,52	-1376,00	65,90	-1554,40	72,47	12,97
120	245,6	5		-2552,20	70,72	-2659,00	73,26	4,18	-2483,40	68,16	-2711,20	73,06	9,17
134	271,4	5		-2545,80	70,42	-2805,00	73,52	10,18	-2545,80	69,58	-2748,20	76,90	7,95
150	294,4	5		-1880,40	70,14	-2026,40	74,20	7,76	-1838,20	65,48	-2025,80	71,90	10,21
199	399,6	15		-2308,40	67,25	-2490,47	74,90	7,89	-2225,07	68,77	-2442,20	72,68	9,76
240	484,8	5		-1133,00	64,64	-1247,80	74,98	10,13	-1138,20	73,66	-1228,40	76,54	7,92
252	504,4	5		-1439,60	65,14	-1521,00	77,68	5,65	-1416,60	62,96	-1516,80	77,80	7,07
255	509,0	5		-1022,00	67,26	-1124,80	77,54	10,06	-1030,80	71,20	-1092,00	77,42	5,94
			avg.	**-1141,39**	**63,79**	**-1253,12**	**71,16**	**10,67**	**-1119,16**	**65,20**	**-1243,70**	**70,69**	**11,85**
15	31,1	10	(OA)	-331,40	60,03	-351,40	62,74	6,04	-288,30	57,59	-348,00	65,21	20,71
20	39,5	10		-412,10	60,50	-477,10	66,77	15,77	-398,90	63,40	-468,90	66,70	17,55
21	39,4	10		-546,50	61,69	-612,10	67,42	12,00	-554,40	62,60	-619,30	69,18	11,71
22	39,4	10		-958,20	64,97	-1042,20	68,73	8,77	-914,90	60,91	-1052,10	67,20	15,00
25	56,0	5		-539,80	64,60	-576,20	67,60	6,74	-498,20	63,44	-581,40	70,00	16,70
29	57,8	10		-947,70	62,79	-1146,50	69,05	20,98	-980,80	62,16	-1145,10	67,06	16,75
30	63,8	5		-519,00	64,88	-541,60	66,46	4,35	-505,40	62,72	-542,20	67,98	7,28
32	62,5	15		-1718,93	60,99	-1852,87	67,44	7,79	-1617,80	61,50	-1891,47	69,35	16,92
35	74,4	5		-614,40	61,28	-651,00	66,64	5,96	-622,00	65,72	-666,80	69,14	7,20
40	79,2	5		-669,40	64,90	-727,60	70,74	8,69	-690,20	63,78	-740,20	69,06	7,24
44	86,2	5		-1250,40	58,76	-1468,00	68,66	17,40	-1226,40	64,44	-1411,40	67,84	15,08
50	105,2	5		-777,60	66,76	-826,60	66,28	6,30	-763,60	67,10	-865,40	70,86	13,33
71	146,0	5		-542,80	59,98	-587,60	72,30	8,25	-550,40	61,26	-614,60	71,64	11,66
75	150,3	20		-1098,30	67,02	-1211,10	73,04	10,27	-1032,45	63,61	-1207,85	70,57	16,99
100	204,3	15		-1411,00	67,97	-1575,00	72,86	11,62	-1414,87	70,09	-1573,73	72,59	11,23
120	245,6	5		-2548,60	71,56	-2706,60	73,10	6,20	-2389,20	69,76	-2680,60	72,44	12,20
134	271,4	5		-2562,40	71,24	-2724,00	71,46	6,31	-2365,40	71,04	-2710,40	72,84	14,59
150	294,4	5		-1884,80	70,00	-2003,80	74,18	6,31	-1788,00	66,22	-1965,20	75,50	9,91
199	399,6	15		-2286,27	69,96	-2448,40	76,00	7,09	-2252,60	70,74	-2477,67	74,36	9,99
240	484,8	5		-1157,00	71,94	-1227,00	77,50	6,05	-1153,20	73,22	-1249,60	76,54	8,36
252	504,4	5		-1427,60	60,56	-1529,60	76,60	7,14	-1443,40	73,22	-1551,00	77,56	7,45
255	509,0	5		-1044,40	75,38	-1093,60	78,94	4,71	-1018,20	73,76	-1084,20	76,78	6,48
			avg.	**-1147,66**	**65,35**	**-1244,54**	**70,66**	**8,85**	**-1112,21**	**65,86**	**-1247,60**	**70,93**	**12,47**

The constructive heuristic runs typically during a very few milliseconds, and hence most of the computing time is spent in the local search phase of the algorithm. Despite the small value used in our experiments for the maximum computing time, the results show that the variable neighborhood search algorithm can improve significantly the value of the solution. Depending on the quality of the initial solution the percentage of improvement goes up to nearly 43%. This percentage tends to be larger when the level packing procedure is used to build the loading patterns.

Table 2. Computational results with (RBL)

n	M	inst	ord	z_{CH}	$\%^{CH}_{fill}$	z_{VNS}	$\%^{VNS}_{fill}$	imp	z_{CH}	$\%^{CH}_{fill}$	z_{VNS}	$\%^{VNS}_{fill}$	imp
						(FN)					(FR)		
15	31,1	10	(OH)	-269,80	48,18	-339,30	61,98	25,76	-278,90	52,02	-339,60	61,92	21,76
20	39,5	10		-385,90	50,06	-460,50	61,63	19,33	-392,00	53,38	-452,70	60,74	15,48
21	39,4	10		-538,00	57,77	-595,70	64,20	10,72	-519,50	53,51	-607,70	67,10	16,98
22	39,4	10		-800,50	55,00	-1040,00	66,07	10,05	-800,50	60,76	-1034,00	63,75	16,22
25	56,0	5		-610,80	58,20	-571,80	67,92	10,05	-586,00	42,70	-582,20	68,40	43,00
29	57,8	10		-880,70	53,33	-1115,20	65,33	26,63	-893,10	56,23	-1087,90	66,37	21,81
30	63,8	5		-472,20	57,54	-552,80	71,52	17,07	-467,80	50,16	-553,00	64,96	18,21
32	62,5	15		-1564,73	54,04	-1828,80	65,40	16,88	-1561,93	55,82	-1881,33	67,51	20,45
35	74,4	5		-570,40	51,24	-648,20	65,74	13,64	-518,20	58,04	-653,40	65,90	26,09
40	79,2	5		-603,20	49,22	-703,00	68,38	16,55	-642,20	53,62	-712,80	67,58	10,99
44	86,2	5		-1243,00	53,34	-1409,40	67,56	13,39	-1119,20	50,90	-1378,20	67,04	23,14
50	105,2	5		-752,80	59,20	-790,80	66,22	5,05	-688,00	56,20	-797,20	67,94	15,87
71	146,0	5		-516,20	49,46	-609,20	68,18	18,02	-545,40	60,78	-622,20	73,04	14,08
75	150,3	20		-1026,05	59,16	-1175,25	67,61	14,54	-986,55	57,35	-1161,60	68,79	17,74
100	204,3	15		-1359,07	56,15	-1553,07	70,36	14,27	-1306,93	57,04	-1540,87	71,47	17,90
120	245,6	5		-2390,20	64,56	-2683,60	71,18	12,28	-2354,00	65,14	-2736,60	69,92	16,25
134	271,4	5		-2449,20	64,56	-2809,20	72,54	14,70	-2310,40	55,76	-2745,00	73,86	18,81
150	294,4	5		-1791,60	60,40	-1965,20	72,54	9,69	-1771,00	57,00	-1955,60	72,36	10,42
199	399,6	15		-2240,87	62,76	-2425,13	72,35	8,22	-2181,33	59,96	-2434,13	73,62	11,59
240	484,8	5		-1124,40	56,52	-1196,40	71,02	6,40	-1135,80	66,26	-1232,00	76,32	8,47
252	504,4	5		-1400,60	56,58	-1497,40	74,96	6,91	-1380,60	64,90	-1507,20	73,92	9,17
255	509,0	5		-1012,20	58,48	-1069,00	76,54	5,61	-974,20	55,86	-1098,40	74,52	12,75
			avg.	**-1092,96**	**56,21**	**-1229,45**	**68,61**	**13,54**	**-1059,25**	**56,52**	**-1231,12**	**68,96**	**17,60**
15	31,1	10	(OA)	-277,20	47,84	-335,20	64,64	20,92	-302,30	56,79	-341,90	64,97	13,10
20	39,5	10		-419,90	57,21	-460,60	66,86	9,69	-405,90	62,27	-452,80	63,39	11,55
21	39,4	10		-530,50	56,60	-601,90	65,96	13,46	-478,90	55,61	-603,90	65,52	26,10
22	39,4	10		-920,00	57,42	-1026,50	64,77	11,58	-805,80	50,90	-1003,20	64,28	24,50
25	56,0	5		-542,40	59,90	-553,20	65,62	1,99	-520,00	60,04	-588,00	67,96	13,08
29	57,8	10		-949,80	61,93	-1136,20	68,34	19,63	-883,30	56,59	-1134,20	66,34	28,40
30	63,8	5		-501,40	61,24	-555,40	68,74	10,77	-428,20	52,26	-546,40	65,58	27,60
32	62,5	15		-1623,93	55,68	-1873,07	67,30	15,34	-1545,73	57,83	-1821,27	67,06	17,83
35	74,4	5		-582,60	55,40	-625,00	69,32	7,28	-518,40	52,38	-646,80	67,54	24,77
40	79,2	5		-576,20	46,54	-712,80	68,70	23,71	-646,40	57,82	-715,20	67,90	10,64
44	86,2	5		-1272,20	58,42	-1452,60	65,58	14,18	-1149,80	51,56	-1451,80	69,58	26,27
50	105,2	5		-727,80	53,86	-806,80	69,18	10,85	-723,80	60,16	-797,80	66,12	10,22
71	146,0	5		-544,80	57,90	-601,40	72,34	10,39	-536,00	60,50	-604,40	70,12	12,76
75	150,3	20		-1038,00	59,69	-1206,15	69,42	16,20	-1042,10	61,34	-1185,75	69,85	13,78
100	204,3	15		-1380,80	60,09	-1561,20	71,57	13,06	-1366,20	61,53	-1470,87	69,83	7,66
120	245,6	5		-2351,40	59,96	-2648,40	72,50	12,63	-2339,00	54,56	-2750,60	72,30	17,66
134	271,4	5		-2582,00	67,64	-2723,40	72,74	5,48	-2280,00	66,50	-2669,20	73,94	17,07
150	294,4	5		-1867,20	64,46	-1990,40	71,82	6,60	-1785,60	68,32	-1951,40	74,38	9,29
199	399,6	15		-2244,80	62,98	-2419,67	74,59	7,79	-2196,80	61,68	-2468,33	74,16	12,36
240	484,8	5		-1161,40	70,04	-1233,80	74,10	6,23	-1102,80	58,10	-1232,40	79,26	11,75
252	504,4	5		-1403,60	58,66	-1500,20	72,10	6,88	-1411,00	62,80	-1532,00	72,40	8,50
255	509,0	5		-1033,20	64,10	-1082,00	76,36	4,72	-994,60	64,56	-1078,20	70,84	8,41
			avg.	**-1115,05**	**58,98**	**-1232,09**	**69,66**	**11,34**	**-1066,48**	**58,82**	**-1229,34**	**69,24**	**16,06**

The best average results are obtained using the strategies (BL), (FN) and (OH). The strategy (FN) that consists in inserting first the client that is nearest to the depot generates usually the best initial solutions when compared to (FR). In some cases, choosing randomly the first client to insert in the route yields better initial solutions, but even in these cases, the local search procedure tends to reach better solutions at the end of the computing time with the strategy (FN) than it does with the strategy (FR). Note that the results remain nearly the same for different runs of the strategy (FR). Ordering the items by height (OH) or by area (OA) has a more significant impact when the level packing procedure is used to place the items in the vehicle. When the bottom-left based strategies (BL) and (RBL) are used, these two orderings yield results that are not significantly different for these instances.

The variants of the algorithm based on the bottom-left placement procedures find solutions with an high percentage of used space in the vehicles. This percentage

Table 3. Computational results with (LP) and $(LP.FF)$

n	M	inst	ord	z_{CH}	$\%^{CH}_{fill}$	z_{VNS}	$\%^{VNS}_{fill}$	imp	z_{CH}	$\%^{CH}_{fill}$	z_{VNS}	$\%^{VNS}_{fill}$	imp
					(FN)					(FR)			
15	31,1	10	(OH)	-246,40	39,03	-293,80	51,74	19,24	-256,10	39,35	-293,30	51,00	14,53
20	39,5	10		-332,40	38,31	-394,40	54,39	18,65	-331,50	42,05	-399,70	50,76	20,57
21	39,4	10		-487,50	48,08	-526,70	51,37	8,04	-445,30	42,42	-539,90	54,67	21,24
22	39,4	10		-772,70	42,13	-901,80	50,89	16,71	-773,70	42,51	-897,70	52,67	16,03
25	56,0	5		-437,80	45,26	-475,00	51,22	8,50	-400,60	47,74	-477,20	49,16	19,12
29	57,8	10		-840,70	45,14	-1018,40	52,99	21,14	-816,40	40,43	-973,30	50,08	19,22
30	63,8	5		-431,40	46,90	-501,00	56,42	16,13	-419,00	46,62	-464,40	54,48	10,84
32	62,5	15		-1462,47	46,86	-1554,40	51,45	6,29	-1452,93	46,20	-1631,53	51,47	12,29
35	74,4	5		-532,40	43,68	-587,00	55,40	10,26	-468,20	37,52	-563,60	53,64	20,38
40	79,2	5		-556,60	39,88	-619,40	50,78	11,28	-541,60	46,42	-645,40	54,32	19,17
44	86,2	5		-1113,80	40,56	-1217,60	52,68	9,32	-1018,60	46,16	-1311,20	56,12	28,73
50	105,2	5		-666,60	41,30	-740,40	52,68	11,07	-667,40	41,84	-759,80	54,34	13,84
71	146,0	5		-519,20	49,80	-563,40	58,78	8,51	-477,40	43,30	-555,80	53,96	16,42
75	150,3	20		-984,20	46,16	-1095,85	54,78	11,34	-927,70	43,20	-1087,70	57,22	17,25
100	204,3	15		-1313,40	49,90	-1410,00	58,85	7,35	-1260,87	48,48	-1384,33	55,43	9,79
120	245,6	5		-2302,40	50,54	-2635,80	62,06	14,48	-2223,00	48,96	-2558,60	59,52	15,10
134	271,4	5		-2258,20	55,56	-2510,40	61,70	11,17	-2194,60	52,60	-2392,20	62,02	9,00
150	294,4	5		-1724,00	49,64	-1935,40	62,02	12,26	-1743,60	51,02	-1806,20	64,92	3,59
199	399,6	15		-2180,60	48,18	-2374,33	62,55	8,88	-2179,53	53,97	-2371,60	66,35	8,81
240	484,8	5		-1116,20	48,50	-1192,80	68,30	6,86	-1074,80	48,44	-1188,00	62,98	10,53
252	504,4	5		-1365,80	46,74	-1468,60	70,78	7,53	-1335,60	51,02	-1464,40	63,90	9,64
255	509,0	5		-986,20	50,70	-1039,80	65,70	5,44	-981,40	54,02	-1038,20	63,24	5,79
			avg.	**-1028,68**	**46,35**	**-1138,92**	**57,16**	**11,38**	**-999,54**	**46,10**	**-1127,46**	**56,47**	**14,63**
15	31,1	10	(OA)	-250,22	36,92	-295,78	50,67	18,21	-252,11	36,67	-287,78	46,26	14,15
20	39,5	10		-346,50	37,75	-395,10	50,59	14,03	-339,20	42,76	-412,80	51,44	21,70
21	39,4	10		-486,60	46,81	-538,10	51,53	10,58	-450,80	45,31	-549,70	51,89	21,94
22	39,4	10		-756,20	40,47	-895,30	51,72	18,39	-769,20	43,04	-879,10	48,35	14,29
25	56,0	5		-493,75	41,73	-525,75	51,18	6,48	-423,20	44,20	-476,40	54,86	12,57
29	57,8	10		-965,00	40,43	-1054,75	45,96	9,30	-815,00	45,36	-1026,20	50,45	25,91
30	63,8	5		-430,00	45,14	-490,80	55,38	14,14	-410,40	40,04	-483,20	55,86	17,74
32	62,5	15		-1427,80	42,59	-1581,13	51,43	10,74	-1412,64	43,27	-1683,79	50,77	19,19
35	74,4	5		-532,40	43,68	-591,00	51,36	11,01	-492,20	41,90	-563,00	51,52	14,38
40	79,2	5		-638,75	37,30	-692,25	43,70	8,38	-556,80	42,84	-642,40	55,06	15,37
44	86,2	5		-1121,80	41,66	-1253,60	57,18	11,75	-1067,20	38,30	-1312,00	58,38	22,94
50	105,2	5		-658,80	40,36	-754,20	54,00	14,48	-653,40	47,52	-748,40	55,02	14,54
71	146,0	5		-502,20	46,98	-554,40	54,72	10,39	-497,00	41,76	-555,00	49,10	11,67
75	150,3	20		-1032,58	46,18	-1119,79	53,54	8,45	-924,65	40,84	-1090,60	57,05	17,95
100	204,3	15		-1291,27	48,62	-1463,00	60,69	13,30	-1282,93	47,11	-1422,40	58,85	10,87
120	245,6	5		-2288,00	47,74	-2524,40	61,88	10,33	-2236,20	45,16	-2555,00	61,74	14,26
134	271,4	5		-2309,40	49,20	-2579,20	58,62	11,68	-2555,75	43,00	-2847,50	56,80	11,42
150	294,4	5		-1708,80	49,54	-1871,40	59,36	9,52	-1673,00	44,42	-1898,20	62,82	13,46
199	399,6	15		-2178,00	47,87	-2355,53	61,90	8,15	-2149,33	49,75	-2353,27	62,86	9,49
240	484,8	5		-1378,25	59,38	-1423,75	64,83	3,30	-1086,40	45,42	-1194,40	62,14	9,94
252	504,4	5		-1373,40	48,76	-1442,80	61,60	5,05	-1359,40	51,14	-1477,80	63,08	8,71
255	509,0	5		-974,00	43,44	-1031,00	57,68	5,85	-1194,75	53,03	-1262,75	66,70	5,69
			avg.	**-1051,99**	**44,66**	**-1156,05**	**54,98**	**10,61**	**-1027,34**	**44,22**	**-1169,17**	**55,95**	**14,92**

is typically higher for the largest instances. It goes up to 78.94% when the strategies (BL), (FN) and (OA) are used on the instances with 255 clients and an average of 509 items per instance. The percentage of used space tends to decrease with the level packing procedures. This trend was expectable given that the loading patterns generated through the level packing procedure are more constrained (guillotinable patterns) than those generated with the approaches relying on the bottom-left rules. In general, the results obtained with level packing procedures are outperformed by the bottom-left based approaches.

6   Conclusions

In this paper, we explored the first solution algorithm for the 2L-ESPP. The approach is based on constructive heuristics to generate initial feasible solutions for the problem, and on variable neighborhood search to look for improved incumbents. We described different alternative neighborhood structures based on the routing and packing characteristics of the solution. We provided also the first results concerning the resolution of this problem for a large set of benchmark instances of the 2L-CVRP. The results illustrate the effectiveness of the variable neighborhood search procedure in improving the solutions of the constructive heuristics. These results allowed the comparison between the different strategies described in this paper. Besides the practical relevance of the problem, these results may contribute for the resolution of the 2L-CVRP through column generation algorithms since the 2L-ESPP is the pricing subproblem that results from the corresponding Dantzig-Wolfe decomposition of this problem.

Acknowledgments. This work was supported by FEDER funding through the Programa Operacional Factores de Competitividade - COMPETE and by national funding through the Portuguese Science and Technology Foundation (FCT) in the scope of the project PTDC/EGE-GES/116676/2010 (Ref. COMPETE: FCOMP-01-0124-FEDER-020430), by FCT within the project scope: UID/CEC/00319/2013, and by FCT through the grant SFRH/BD/73584/ 2010 for Telmo Pinto (funded by QREN - POPH - Typology 4.1 - co-funded by MEC National Funding and the European Social Fund), and by FEDER funds through the Competitiveness Factors Operational Programme - COMPETE.

References

1. Feillet, D., Dejax, P., Gendreau, M., Gueguen, C.: An exact algorithm for the elementary shortest path problem with resource constraints: Application to some vehicle routing problems. Networks **44**(3), 216–229 (2004)
2. Gendreau, M., Iori, M., Laporte, G., Martello, S.: A Tabu search heuristic for the vehicle routing problem with two-dimensional loading constraints. Networks **51**(1), 4–18 (2008)
3. Hansen, P., Mladenovic, N., Pérez, J.: Variable neighborhood search: methods and applications. Annals of Operations Research **175**, 367–407 (2010)
4. Iori, M., Martello, S.: Routing problems with loading constraints. Top **18**(1), 4–27 (2010)
5. Iori, M., Salazar-Gonzalez, J.J., Vigo, D.: An exact approach for the vehicle routing problem with two-dimensional loading constraints. Transportation Science **41**(2), 253–264 (2007)
6. Mladenovic, N., Hansen, P.: Variable neighborhood search. Computers and Operations Research **24**, 1097–1100 (1997)
7. Pinto, T., Alves, C., de Carvalho, J.V.: Column generation based heuristic for a vehicle routing problem with 2-dimensional loading constraints: a prototype. In: XI Congresso Galego de Estatística e Investigación de Operacións, Spain (2013)
8. Righini, G., Salani, M.: Symmetry helps: bounded bi-directional dynamic programming for the elementary shortest path problem with resource constraints. Discrete Optimization **3**(3), 255–273 (2006)
9. Righini, G., Salani, M.: New dynamic programming algorithms for the resource constrained elementary shortest path problem. Networks **51**(3), 155–170 (2008)
10. Zachariadis, E.E., Tarantilis, C.D., Kiranoudis, C.T.: A guided tabu search for the vehicle routing problem with two-dimensional loading constraints. European Journal of Operational Research **195**(3), 729–743 (2009)

A Model-Based Heuristic for the Combined Cutting Stock and Scheduling Problem

Nuno Braga[1], Cláudio Alves[1](✉),
Rita Macedo[2], and José Valério de Carvalho[1]

[1] Centro Algoritmi, Escola de Engenharia, Universidade do Minho,
4710-057 Braga, Portugal
{nuno.braga,claudio,vc}@dps.uminho.pt
[2] Institut de Recherche Technologique Railenium, F-59300 Famars, France
Rita.SantosDeMacedo@univ-valenciennes.fr

Abstract. In this paper, we address a variant of the cutting stock problem that considers the scheduling of the cutting operations over time. The problem combines the standard objective of cutting stock problems, which is the minimization of the raw material usage, together with a scheduling term penalizing tardiness. There is tardiness whenever the last instance of an item is cut after its given due date. This problem has been analyzed recently considering one and two-dimensional items. In this paper, we describe a new pseudo-polynomial network flow model for this combined cutting stock and scheduling problem. A revised version of this model in which consecutive time instants are aggregated is used to define an heuristic solution procedure for the problem. This revised formulation is complemented by a time assignment procedure leading to good feasible solutions for the problem. Computational results on benchmark instances are provided illustrating the potential of the approach.

Keywords: Cutting stock · Scheduling · Integer programming · Heuristics

1 Introduction

The cutting stock problem is a well-known combinatorial optimization problem from the field of cutting and packing with applications in different industries and other contexts such as computer science and telecommunications. The standard problem consists in finding how to arrange a given set of items out of available stock rolls of a given size such that the number of used rolls is minimized. While the problem is NP-hard, many contributions have been provided in the literature in the last years, including exact solution approaches, lower bounding procedures and heuristic approaches (some examples can be found in [3–6,9]). Given the progresses made toward the resolution of large scale instances of the standard problem, many variants incorporating constraints arising in practical settings are being explored. In this paper, we explore one of these variants that combines the standard objective of minimizing the number of rolls used with a

© Springer International Publishing Switzerland 2015
O. Gervasi et al. (Eds.): ICCSA 2015, Part II, LNCS 9156, pp. 490–505, 2015.
DOI: 10.1007/978-3-319-21407-8_35

scheduling term penalizing the tardiness of the cutting operations. In particular, tardiness arises whenever the last instance of an item is cut after its pre-defined due date. This problem will be referred to as the combined Cutting Stock and Scheduling Problem (CSSP). Therefore, the CSSP consists in determining the set of feasible cutting patterns that covers the demand of the items, and the time instants when each of these patterns must be cut such that the total number of rolls and tardiness is minimized. Here, we will consider the case where the items are all one-dimensional objects characterized by a single size.

The first contribution related to the CSSP is due to Li [8]. In [7], the authors considered a related problem arising in the real context of the copper industry, but they did not solve the problem in an integrated way, considering instead a two-stage solution procedure. In [8], the author addressed the two-dimensional with stock rolls that may have different lengths and widths. A job is defined as a set of items to be cut whose sizes may be different, and that has both release and due dates. The author proposes different integer programming models for the problem where cutting patterns are associated to time periods, and describes different Linear Programming (LP) based heuristics and non LP-based heuristics to generate feasible cutting patterns and schedules. The quality of the heuristics is evaluated using small sized instances for which an optimal solution can be computed within a reasonable computing time. As it happens with other models proposed in the literature, the formulation proposed by Li [8] is generally not exact depending on the length of the time periods in the sense that an optimal solution of the model may not be optimal for the global problem.

The one-dimensional CSSP was addressed recently in [10] for the case where no release dates are enforced. The authors explore a column generation based formulation for the problem based on the assignment of cutting patterns to time periods, and they discuss on its extension to special cases. The planning horizon is divided into a fixed number of time periods, and an item i can only be assigned up to the ith of them. The authors solve it using a column generation procedure, where the pricing subproblem is a shortest path problem on a modified graph. While the computational results reported in [10] show that the approach of Reinersten and Vossen [10] can generate high quality solutions, it was shown by Arbib and Marinelli [1] that it may fail the optimal solution of the problem in the cases where the latter is not achieved through an *early due date first* rule.

In [1], Arbib and Marinelli describe the first exact approach for the CSSP, which is based on the solution of an integer programming model where, as in [10], cutting patterns are assigned to time periods. In their formulation, tardiness is modeled using inventory variables and equilibrium constraints. The authors consider a version of the CSSP that is similar to the one addressed in [10]. A simplified version of their model is solved through a refinement procedure that iteratively updates the length of the time periods whenever there is more than a single pattern in a late period. The authors describe a partial enumeration branch-and-price scheme to generate good incumbents for the problem, and report on computational experiments for instances with 20 item sizes. The optimality gaps achieved by their approaches go up to 35%.

In [2], Braga *et al.* explore an exact and compact assignment formulation for a version of the CSSP similar to the one discussed in [10] and [1]. The model is strengthened using different valid inequalities. Some of these inequalities are generated using dual-feasible functions which are applied on the knapsack constraints of the original model. A cutting plane procedure relying on these dual-feasible functions is described, and its performance is analyzed through computational experiments on benchmark instances for different sets of parameters.

In this paper, we explore a new pseudo-polynomial network flow model for the CSSP. The model is exact in the sense that it leads to an optimal solution for the problem even in cases where for example the *early due date first* rule does not yield the best solutions. Moreover, we describe a revised version of this model in which the time instants are aggregated, and the total tardiness of the cutting operations is overestimated. Using this revised model, we define an heuristic solution procedure for the CSSP where the precise scheduling of the cutting patterns is achieved through a time assignment model. The performance of the approach was evaluated using benchmark instances form the literature. The computational results show that this model-based heuristic generates good quality solutions with small optimality gaps on average.

In Section 2, we define formally the CSSP and its notation. The integer programming models described so far in the literature are reviewed in Section 3. Our alternative network flow model for the CSSP is described in Section 4. The components of our heuristic solution approach are presented in Section 5, including the revised network flow model and the time assignment formulation. The results of our computational experiments are presented and analyzed in Section 6. Finally, some conclusions are drawn in Section 7.

2   The Combined Cutting Stock and Scheduling Problem

The combined cutting stock and scheduling problem is characterized by n one-dimensional items with sizes w_i, $i = 1, \ldots, n$, and stock rolls of length W, whose supply is assumed to be unlimited. Each item i has a given demand b_i and a due date d_i until which all the items i should be cut. An item may eventually be cut after its due date but, in this case, the corresponding tardiness is penalized in the objective function of the problem.

The problem consists in finding the best set of feasible cutting patterns and the precise time instants when each of these patterns must be cut. Hence, the objective function of the problem considers two different terms: the standard raw material usage and the total tardiness associated to the cutting operations. According to the relative importance of each of these terms, different weights may be considered. In this paper, we will assume that they are equally important, and thus their weights are both equal to 1. In [1], the authors showed that a solution that minimizes the total material usage does not necessarily ensure at the same time the minimization of the total tardiness.

Here, jobs are directly related to item sizes. The (eventual) tardiness of a job is determined by the time instant in which the last instance of an item size is cut.

We will assume that it takes exactly one unit of time to cut a stock roll. As a consequence, the number of stock rolls cut up to the last instant on which an item i is cut corresponds to the time required to complete the job related to this item i.

3   Integer Programming Formulations

3.1   The Model of Reinersten and Vossen [10]

The integer programming model proposed by Reinersten and Vossen in [10] is an extension of the column generation model of Gilmore and Gomory [6] for the standard cutting stock problem. The model assigns cutting patterns to time periods. The planning horizon is partitioned into exactly n time periods, with n being the number of different item sizes (jobs) available. The model imposes the additional requirement that an item i can only be cut up to the ith period. Note that the items are assumed to be ordered in non-decreasing order of their due dates. As shown in [1], because of this specific requirement, the resulting model is not exact in general, but only in the cases where the *early due date first* scheduling rule yields the optimal solution of the problem. The model of Reinersten and Vossen [10] stands as follows.

$$\text{min.} \quad \sum_{p\in P}\sum_{k=1}^{n} x_{pk} + \sum_{i=1}^{n} M_i y_i \tag{1}$$

$$\text{s.t.} \quad \sum_{p\in P}\sum_{k=1}^{i} x_{pk} \leq d_i + y_i, \quad i=1,\ldots,n, \tag{2}$$

$$\sum_{p\in P}\sum_{k=1}^{i} a_{ip} x_{pk} \geq b_i, \quad i=1,\ldots,n, \tag{3}$$

$$x_{pk} \geq 0 \text{ and integer}, \quad p\in P, \ k=1,\ldots,n, \tag{4}$$

$$y_i \geq 0 \text{ and integer}, \quad i=1,\ldots,n. \tag{5}$$

The variables x_{pk} denote the run length of a cutting pattern p (from the set of feasible cutting patterns P) in the time period k, with $k=1,\ldots,n$, and a_{ip} the number of times an item i is considered in pattern p. The tardiness associated to an item i is represented through the variables y_i. These variables take the value 0 if the corresponding item is cut within its due date. If an item i is cut after its due date, the variable y_i will take a value that is equal to the number of rolls cut after d_i up to the last stock roll from which an item i is cut. Constraints (2) define the variables y_i. The demand constraints of the problem are modelled through constraints (3). The objective function (1) considers both a material usage term and a scheduling term penalizing the total tardiness. In (1), the weight related to the tardiness of a given item i is denoted by M_i.

As referred to in [10], if the due dates are the same for all the items and if these due dates are not binding, then model (1)-(5) is equivalent to the model of Gilmore and Gomory [6] for the cutting stock problem.

3.2 The Model of Arbib and Marinelli [1]

In [1], Arbib and Marinelli describe an alternative integer programming model for the CSSP. As in [10], their model assigns cutting patterns to time periods of a given length. Let D denote the length of the planning horizon. The authors partition this planning horizon into $n+1$ time periods whose latest time instant coincide with the due date of an item (items are again assumed to be ordered in non-decreasing order of their due dates): $[d_0, d_1], [d_1, d_2], \ldots, [d_n, d_{n+1}]$, with $d_0 = 0$ and $d_{n+1} = D$. Their model is defined as follows.

$$\text{min.} \quad \sum_{p \in P} \sum_{k=1}^{n+1} x_p^k + \sum_{i=1}^{n} \sum_{k>i} (d_k - d_{k-1}) y_i^k \tag{6}$$

$$\text{s.t.} \quad \sum_{p \in P} x_p^k \leq d_k - d_{k-1}, \quad k = 1, \ldots, n+1, \tag{7}$$

$$s_i^{k-1} + \sum_{p \in P} a_{ip} x_p^k = s_i^k + b_i^k, \quad i = 1, \ldots, n, \ k = 1, \ldots, n+1, \tag{8}$$

$$u_i^k y_i^k + s_i^{k-1} \geq 0, \quad i = 1, \ldots, n, \ k = i+1, \ldots, n+1, \tag{9}$$

$$x_p^k \geq 0 \text{ and integer}, \quad p \in P, \ k = 1, \ldots, n+1, \tag{10}$$

$$y_i^k \in \{0, 1\}, \quad i = 1, \ldots, n, \ k = i+1, \ldots, n+1 \tag{11}$$

$$s_i^k \text{ integer}, \quad i = 1, \ldots, n, \ k = i, \ldots, n+1. \tag{12}$$

As in (1)-(5), the set of feasible cutting patterns is denoted by P. The x_p^k variables represent the run length of a pattern p of P on the time period k, while the coefficients a_{ip} have the same meaning as in (1)-(5). The variables s_i^k denote the inventory level of an item i on the time period k, while the variables y_i^k, defined only for the time periods later than the due date d_i of i take the value 1 if the last instance of item i is cut in time period k or later. The objective function (6) is composed by a term related to the total material usage and by a scheduling term. Note that the latter overestimates the tardiness of an item i since it consistently considers that the cutting patterns cut in the time period k are scheduled for the last time instant of this period. The total run length of the cutting patterns scheduled for the time period k cannot exceed the length of this time period, given that it takes one unit of time to cut a stock roll whatever the pattern that is considered. This constraint is defined through (7). From a general standpoint, the parameter b_i^k denotes the demand of an item i which is expected to be delivered on the time period k. In our case, we have that $b_i^k = b_i$ if $k = i$, and 0 otherwise. Constraints (8) force the demands of the items to be satisfied, as they allow for an item i to be cut later than its due date by setting the corresponding s_i^k variable to a negative value. In this case, the corresponding y_i^k variables are set to 1 through the constraints (9), where u_i^k is an upper bound on the number of items i that may be delayed on the time period k.

Given that the number of feasible cutting patterns and corresponding assignments to time periods is exponential, this model can only be solved through dynamic column generation procedures for any medium or large scale instance of the CSSP. Furthermore, it requires an additional refinement procedure as described in [1] that iteratively updates the length of the time periods to achieve an optimal solution of the problem.

3.3   An Assignment Formulation

In [2], Braga et al. explore an alternative assignment formulation for the CSSP. The model considers time periods whose length is equal to 1, and it assigns items directly to these time instants instead of patterns as in the previous models. Let x_i^t, $i = 1, \ldots, n$, $t = 1, \ldots, D$ (D being the length of the planning horizon), be the assignment variables representing the number of items i cut at the time instant k. If a stock roll is cut at a time instant k, then a variable z^k is set to the value 1, and to 0 otherwise. To measure the tardiness of an item, the model resorts to binary variables y_i^k, $k = d_i + 1, \ldots, D$, which take the value 1 if the item i is cut on the time instant k, and 0 otherwise. The assignment model is defined formally as follows.

$$\text{min.} \quad \sum_{k=1}^{D} z^k + \sum_{i=1}^{n} \sum_{k=d_i+1}^{D} y_i^k \tag{13}$$

$$\text{s.t.} \quad \sum_{k=1}^{D} x_i^k = b_i, \quad i = 1, \ldots, n, \tag{14}$$

$$\sum_{i=1}^{n} w_i x_i^k \leq W z^k, \quad i = 1, \ldots, n, \ k = 1, \ldots, D, \tag{15}$$

$$x_i^k + y_i^{k+1} \leq L_i^{min} y_i^k, \quad i = 1, \ldots, n, \ k = d_i + 1, \ldots, D, \tag{16}$$

$$z^k \in \{0, 1\}, \quad k = 1, \ldots, D, \tag{17}$$

$$x_i^k \geq 0 \text{ and integer}, \quad i = 1, \ldots, n, \ k = 1, \ldots, D, \tag{18}$$

$$y_i^k \in \{0, 1\}, \ i = 1, \ldots, n, \quad k = d_i + 1, \ldots, D. \tag{19}$$

The objective function is modelled through (13). The demand of each item is guaranteed through the constraints (14), while (15) ensure the feasibility of the assignments by forcing the total size of the items cut from a stock roll to be smaller than or equal to its length W. Constraints (16) are the defining constraints of the variables y_i^k, $i = 1, \ldots, n$, $k = 1, \ldots, D$, ensuring that a variable y_i^{k+1} is equal to one only if the variable y_i^k is set to one. In (16), L_i^{min} are upper bounds on the number of items i that may be cut from a stock roll.

Note that although the model (13)-(19) is compact, its continuous lower bound is typically weak. In [2], the authors show how to strengthen it by using different families of cutting planes. In particular, they use the knapsack constraints (15) to derive valid inequalities that proved to have a non-negligible impact on the quality of the continuous lower bound of the model.

4   A New Exact Pseudo-Polynomial Network Flow Model

The CSSP can be formulated using an alternative pseudo-polynomial network flow model defined on a graph $G = (V, A)$ where V denotes the set of vertices and A the set of arcs. The vertices in V represent discrete positions within the stock rolls from the leftmost position 0 up to the rightmost vertex W, $i.e.$ $V = \{0, 1, 2, \ldots, W\}$. The arc set A is divided into D sets A^t, $t = 1, \ldots, D$, with D being the length of the planning horizon. An arc (i, j) in A^t represent an item of size $j - i$ cut from the position i of a stock roll at the time instant t. The waste in the stock rolls is represented through arcs of unit length. The CSSP can be modelled as a minimum cost flow problem as follows.

$$\text{min.} \quad \sum_{t=1}^{D} z^t + \sum_{i=1}^{n} \sum_{t=d_i+1}^{D} y_i^t \tag{20}$$

sub. to

$$-\sum_{(r,s)\in A^t} x_{rs}^t + \sum_{(s,u)\in A^t} x_{su}^t = \begin{cases} z^t, & \text{if } s = 0, \\ -z^t, & \text{if } s = W, s = 0, \ldots, W, t = 1, \ldots, D, \\ 0, & \text{otherwise,} \end{cases} \tag{21}$$

$$\sum_{(r,r+w_i)\in A^t} \sum_{t=1}^{D} x_{r,r+w_i}^t = b_i, \quad i = 1, \ldots, n, \tag{22}$$

$$\sum_{(r,r+w_i)\in A^t} x_{rs}^t \leq L_i^{min} y_i^t, \quad \forall (r, s) \in A^t, t = d_i + 1, \ldots, D, \tag{23}$$

$$y_i^{t+1} \leq y_i^t, \quad i = 1, \ldots, n, \quad t = d_i + 1, \ldots, D, \tag{24}$$

$$z^{t+1} \leq z^t, \quad t = 1, \ldots, D - 1, \tag{25}$$

$$x_{rs}^t \in \{0, 1\}, \quad \forall (r, s) \in A^t, \ t = 1, \ldots, D, \tag{26}$$

$$y_i^t \in \{0, 1\}, \quad i = d_i + 1, \ldots, n, \ t = 1, \ldots, D, \tag{27}$$

$$z^t \in \{0, 1\}, \quad t = 1, \ldots, D. \tag{28}$$

The flow variables x_{rs}^t indicate whether an item of size $s - r$ is cut from the position r of a stock roll at the time instant t, while the y_i^t variables take the value 1 if at least one item i is cut at time instant t or later for $t > d_i$, and 0 otherwise. The tardiness of an item i is represented as the sum of these y_i^t variables from d_i+1 up to D, $i.e.$ $\sum_{t=(d_i+1)}^{D} y_i^t$. The variables z^t, $t = 1, \ldots, D$, indicate whether a stock roll is being cut at time instant t, and hence, $\sum_{t=1}^{D} z^t$ correspond to the total number of rolls used. The objective function (20) considers both the total material usage and the total tardiness related to the selected schedule. Constraints (21) ensure the conservation of flow along the graph G. The demand constraints are represented through (22). Constraints (23) force the y_i^t variables to take the value 1 if an item i is cut at time instant t. In this latter case, constraints (24) allow the activation of all the previous $y_i^{t'}$ variables from $t' = t$ down to $d_i + 1$. In constraints (23), L_i^{min} stands again for an upper bound on the number of items i that may be cut from a stock roll. As referred to above, it takes exactly one unit of time to cut a stock roll. Hence, cutting a stock roll at a time instant t means that other rolls have been cut previously up to t. Constraints (25) support this definition by forbidding a stock roll to be cut at time instant $t + 1$ if none has been cut at time t.

Model (20)-(28) is different from (1)-(5) and (6)-(12) in the sense that it is not explicitly based on the scheduling of cutting patterns but on the direct assignment of cuts to time periods. The model is exact since it imposes no restrictions on the time periods when each item can be cut, as happens in [10]. Furthermore, its size is pseudo-polynomial, and hence it is more tractable than the two other integer programming models described in [10] and [1]. The motivation behind the development of this model is essentially based on having a strong model that leads to exact optimal solutions for the CSSP but at the expense of a computational effort that is lower than the one required to solve (6)-(12). The latter relies on the dynamic generation of columns through the solution of a sequence of pricing problems. Although (20)-(28) requires also a dynamic management of its variables, we argue that it is much more efficient to do than for (6)-(12). The other models from the literature were included in this paper for the sake of completeness and also to make easier the identification of the main differences and similarities between the different modeling approaches.

5   A Model-Based Heuristic Approach

5.1   Overview

In this section, we describe a solution procedure for the CSSP based on (20)-(28) that takes advantage of its structure and size to generate feasible solutions of good quality. The approach relies on a revised version of this network flow model that is smaller than (20)-(28) and which is solved to optimality using an optimization solver. This revised version aggregates the time instants used' in (20)-(28) into time periods of a given length. It does not define a precise schedule for the cutting operations since the cuts are scheduled by period, and the order by which these cuts should be performed within the time period is free. As a consequence, the tardiness associated to each item is evaluated by overestimating its value within each time period and for each item. Hence, the solution of the revised network flow model provides an upper bound for the CSSP whose value is in fact greater than any feasible schedule of the cuts within the given time periods. For ease of presentation, we will refer to this upper bound as an overestimated upper bound. At a second stage, we compute the precise schedules through a procedure that consists in solving different time assignment models related to the time periods in order to define the time instants when the cutting patterns generated by the revised network flow model should be performed. The solution for the CSSP achieved through this time assignment procedure allows for a correct estimation of the corresponding tardiness, thus providing an upper bound for the problem that is smaller than the one obtained with the revised network flow model.

Figure 1 provides an outline of this model-based heuristic approach, where P stands for an instance of the CSSP. In the next subsections, we introduce formally both the revised network flow model and the time assignment formulation referred to above.

5.2 A Revised Network Flow Model

The network flow model (20)-(28) relies on time periods of unit length. To reduce its size, longer time periods may be used by aggregating different consecutive time instants. Let T denote the number of time periods that are considered. Note that T may take any value, and hence it does not have to be restricted to the number of different items, nor the last time instant of a time period has to coincide with the due date of a given item. The flow variables of the revised network flow model are similar to those of (20)-(28), except that they are now defined as general integer variables. Let $[a_{t-1}, a_t]$ represent the time period t, with a_t being the last time instant of period t. We have that $a_0 = 0$ and $a_T = D$, where D stands for the length of the planning horizon. The revised network flow model is defined formally as follows.

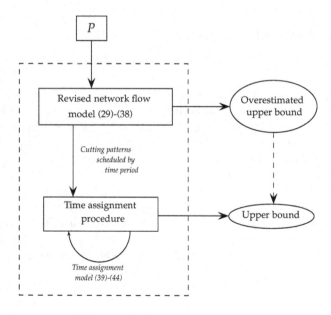

Fig. 1. Outline of the heuristic

$$\text{min.} \sum_{t=1}^{T} z^t + \sum_{i=1}^{n} \sum_{t=t_i}^{T} (a_t - a_{t-1}) y_i^t \tag{29}$$

sub. to

$$\sum_{(r,s)\in A^t} x_{rs}^t - \sum_{(s,u)\in A^t} r_{su}^t = \begin{cases} z^t, & \text{if } s = 0, \\ -z^t, & \text{if } s = W, s = 0, ..., W, t = 1, ..., T, \\ 0, & \text{otherwise.} \end{cases} \tag{30}$$

$$\sum_{(r,r+w_i)\in A^t} \sum_{t=1}^{T} x_{r,r+w_i}^t = b_i, \quad i = 1, \ldots, n, \tag{31}$$

$$\sum_{(r,r+w_i)\in A^t} x_{rs}^t \leq L_i^{min} y_i^t, \quad \forall (r,s) \in A^t, t = t_i, \ldots, T, \tag{32}$$

$$y_i^{t+1} \leq y_i^t, \quad i = 1, \ldots, n, \quad t = t_i, \ldots, T, \tag{33}$$

$$z^t \leq a_t - a_{t-1}, \quad t = 1, \ldots, T, \tag{34}$$

$$z^t \geq (a_t - a_{t-1}) y_i^{t+1}, \quad i = 1, \ldots, n, t = t_i, \ldots, T-1, \tag{35}$$

$$x_{rs}^t \geq 0 \text{ and integer}, \quad \forall (r,s) \in A^t, \ t = 1, \ldots, T, \tag{36}$$

$$y_i^t \in \{0,1\}, \quad i = 1, \ldots, n, \ t = t_i, \ldots, T, \tag{37}$$

$$z^t \geq 0 \text{ and integer}, \quad t = 1, \ldots, T. \tag{38}$$

The definition of (29)-(38) is similar in many aspects to the definition of (20)-(28). One of the key differences lies in the definition of the tardiness associated to an item i. In the objective function (29), the contribution to the total tardiness of an item i that is cut in a given late time period t is equal to the length of t. This value may overestimate the total tardiness of an item i by considering implicitly that the last instance of this item is always cut at the last time instant of a time period. The value t_i represents the time period that includes the due date d_i of item i, and it can be defined as follows: $t_i = \min\{t' : a_{t'} \geq d_i, t' = 1, \ldots, T\}$. The z^t variables become also general integer variables representing the number of stock rolls cut at time period t. Constraints (34) ensure that no more than $a_t - a_{t-1}$ rolls are used at time period t, given that it takes exactly one unit of time to cut a stock roll. Constraints (35) are an extension of (25) to the case where more than one stock roll can be cut at a given time period, and time periods may not have all the same length. These constraints state that a stock roll can be cut at a time period $t + 1$ only if all the available rolls at the time period t have been effectively used.

The drawback of model (29)-(38) is that, while the cutting patterns are clearly determined, the schedule of the corresponding cutting operations may not be completely defined since it may not be obvious how to determine the order by which the cutting patterns should be cut within a time period. To overcome this difficulty, we resort to a time assignment procedure that relies on the integer programming model described in the next section and that takes as input the cutting patterns generated by (29)-(38). This procedure determines the best schedule of these cutting patterns within each time period.

5.3 A Time Assignment Model

Let P' denote the set of cutting patterns generated through the resolution of model (29)-(38) for a given time period t' of length T' defined through $[a_{t'-1}, a_{t'}]$, and let α_{ip} represent the number of items i in pattern p. Note that, if $|P'| < T'$, the value of T' is set to $|P'|$ such that the definition of the resulting time period becomes $[a_{t'-1}, a_{t'-1} + |P'|]$. To schedule the cutting patterns of P' within the time period t' so as to minimize the total tardiness associated to the items, we consider the following assignment formulation with additional constraints.

$$\min \sum_{i=1}^{n} y_i \tag{39}$$

sub. to

$$\sum_{t=a_{t'-1}}^{a_{t'}} \lambda_{pt} = 1, \quad p \in P', \tag{40}$$

$$\sum_{p \in P'} \lambda_{pt} = 1, \quad t = a_{t'-1}, a_{t'-1} + 1, \ldots, a_{t'}, \tag{41}$$

$$y_i \geq (t - d_i)\lambda_{pt}, \ i = 1, \ldots, n, p \in P' \text{ such that } \alpha_{ip} > 0,$$
$$\qquad t = a_{t'-1}, a_{t'-1} + 1, \ldots, a_{t'}, \text{ such that } t > d_i, \tag{42}$$

$$\lambda_{pt} \in \{0, 1\}, \quad p \in P', t = a_{t'-1}, a_{t'-1} + 1, \ldots, a_{t'}, \tag{43}$$

$$y_i \geq 0 \text{ and integer}, \quad i = 1, \ldots, n. \tag{44}$$

The assignment variables λ_{pt} determine whether the cutting pattern p is assigned or not to the time period t (of unit length), while the variables y_i represent the tardiness of item i. The objective function (39) consists in the minimization of the total tardiness associated to the items. Constraints (40) and (41) are standard assignment constraints that ensure respectively that all the cutting patterns are assigned to a time instant in t', and all the time instants have exactly one cutting pattern assigned to it. Constraints (42) force the y_i variables to take at least a value equal to the largest tardiness of the different instances of the item i cut in this time period t'. Together with (39), they ensure that the tardiness of each item included in the patterns of P' is correctly evaluated.

6 Computational Experiments

To evaluate the performance of our model-based heuristic, we conducted a set of computational experiments on benchmark instances from the literature [1]. The instances are characterized by 20 items whose sizes w_i vary in the interval $[100, 7000]$, $i = 1, \ldots, n$, and by large stock rolls of length $W = 10000$. The demands b_i of the items are in the set $[1, 100]$, and hence these are cutting stock instances. The due dates of the items are defined as follows: $d_1 = \ldots = d_t = \frac{\overline{C}_{max}}{2}$ and $d_{t+1} = \ldots = d_n$, with $t = \lfloor \frac{n}{2} \rfloor$, and $\overline{C}_{max}$ being the makespan of a solution that minimizes the number of used stock rolls. Twenty different instances were

Table 1. Computational results on benchmark instances with $D = 2\overline{C}_{max} + 1$

Inst.	T	best lb	Revised network flow model (29)-(38)				Time assign. model (39)-(44)		final gap
			ub	t_{ub}	gap	t	ub	t	
1	20	240	240	38,80	0,00	38,80	240	0,00	0,00
2		400	548	1305,75	0,00	2948,68	538	0,10	10,25
3		4??	4??	?1,??	11,48	3304,74	407	0,06	6,27
4		475	518	?7,91	0,00	??0,??	498	0,09	4,84
5		309	309	22,04	0,00	??,??	309	0,09	0,00
6		421	436	64,43	0,00	395,63	436	0,04	?,??
7		348	364	598,73	0,00	2485,34	349	0,02	0,29
8		451	451	36,18	0,00	100,32	451	0,03	0,00
9		514	580	2812,27	4,47	3304,74	570	0,11	10,89
10		384	416	416,77	4,39	3307,80	403	0,04	4,95
11		257	280	1957,64	10,37	3301,70	280	0,04	8,95
12		503	518	53,73	0,00	235,30	506	0,02	0,60
13		399	424	36,32	0,00	420,39	419	0,06	5,01
14		416	462	689,43	11,95	3302,98	441	0,06	6,01
15		391	405	40,37	0,00	782,03	391	0,01	0,00
16		362	382	30,55	0,00	185,29	382	0,08	5,52
17		435	483	1684,70	0,00	2987,75	479	0,09	10,11
18		267	267	1615,56	0,00	1910,44	267	0,00	0,00
19		332	351	1993,93	5,90	3302,20	332	0,00	0,00
20		497	542	494,00	0,00	745,31	530	0,05	6,64
1	40	240	240	289,55	0,00	290,75	240	0,00	0,00
2		488	544	696,35	15,44	3502,77	543	0,00	11,27
3		383	404	1669,50	8,88	3501,47	402	0,02	4,96
4		475	505	630,00	0,00	630,15	498	0,05	4,84
5		309	309	1170,56	0,00	1170,56	309	0,00	0,00
6		421	436	182,50	0,00	1460,40	431	0,03	2,38
7		348	353	448,38	1,58	3503,98	349	0,02	0,29
8		451	451	173,00	0,00	2007,54	451	0,01	0,00
9		514	582	2537,88	24,59	3503,00	581	0,05	13,04
10		384	406	2651,07	6,14	3502,38	404	0,03	5,21
11		257	337	3500,00	24,53	3502,39	333	0,05	29,57
12		503	504	705,12	0,00	985,02	504	0,02	0,20
13		399	416	923,80	0,00	1892,92	415	0,03	4,01
14		416	463	2411,17	11,97	3501,66	445	0,05	6,97
15		391	393	331,08	0,00	331,08	391	0,00	0,00
16		362	372	197,81	0,00	574,92	372	0,03	2,76
17		435	496	3500,00	13,54	3505,28	494	0,05	13,56
18		267	-	-	-	-	-	-	-
19		332	351	3336,00	5,88	3504,08	332	0,01	0,00
20		497	554	1368,25	11,26	3501,75	554	0,03	11,47

generated in that way. The tests were run on a PC with an i7 CPU with 3.5 GHz and 32 GB of RAM. Furthermore, we used CPLEX 12.5 callable library to implement some of the optimization subroutines.

One of the aims of our computational experiments is to analyse the impact of some important parameters on the quality of the solutions provided by our approach. We compare in particular the performance of our heuristic for different numbers T of time periods in model (29)-(38), and for different lengths D of the planning horizon. In Tables 1-4, we report on the results for two values of T, i.e. $T \in \{20, 40\}$, and four values for D with $D = \overline{C}_{max} + \alpha\overline{C}_{max} + 1$ and $\alpha \in \{0.05, 0.1, 0.5, 1\}$. Both parameters determine the length of model (29)-(38). The larger their value is, the larger will be the size of this model. For $T = 20$, a time limit of 3300 seconds was used to solve (29)-(38), while the time assignment model (39)-(44) was solved within a maximum time limit of 300 seconds. For $T = 40$, these time limits were respectively equal to 3500 and 100 seconds. The reason behind these differences in the computing times is based on the

Table 2. Computational results on benchmark instances with $D = 1.5\overline{C}_{max} + 1$

Inst.	T	best lb	Revised network flow model (29)-(38)				Time assign. model (39)-(44)		final
			ub	t_{ub}	gap	t	ub	t	gap
1	20	240	240	97,63	0,00	97,63	240	0,00	0,00
2		488	548	1010,83	0,00	3286,98	535	0,09	9,63
3		383	412	268,45	1,17	3303,79	407	0,06	6,27
4		475	505	35,46	0,00	123,70	498	0,09	4,84
5		309	309	21,11	0,00	21,54	309	0,00	0,00
6		421	436	49,18	0,00	948,90	432	0,06	2,61
7		348	353	246,75	0,00	2971,29	349	0,01	0,29
8		451	451	150,00	0,00	703,05	451	0,03	0,00
9		514	569	1753,09	0,91	3304,38	564	0,11	9,73
10		384	405	750,18	4,17	3304,61	403	0,03	4,95
11		257	280	3300,00	9,08	3301,46	270	0,04	5,06
12		503	516	819,64	5,43	3308,27	516	0,09	2,58
13		399	424	578,45	0,00	741,89	420	0,05	5,26
14		416	451	2050,00	4,66	3305,00	440	0,07	5,77
15		391	393	302,45	0,00	587,69	391	0,01	0,00
16		362	377	39,54	0,00	639,21	377	0,06	4,14
17		435	483	330,06	1,04	1487,23	478	0,08	9,89
18		267	267	1153,00	0,00	1262,22	267	0,00	0,00
19		332	341	1178,36	3,12	3303,21	332	0,00	0,00
20		497	530	485,50	0,00	512,46	530	0,04	6,64
1	40	240	240	223,96	0,00	223,96	240	0,00	0,00
2		488	552	2870,10	15,65	3502,97	551	0,05	12,91
3		383	405	1723,26	6,23	3504,86	405	0,04	5,74
4		475	499	393,73	0,00	613,31	498	0,05	4,84
5		309	309	51,14	0,00	52,01	309	0,00	0,00
6		421	436	2408,82	4,77	3507,28	431	0,03	2,38
7		348	353	600,93	1,55	3509,89	349	0,01	0,29
8		451	451	420,00	3,15	3503,87	451	0,01	0,00
9		514	613	1065,82	21,80	3502,70	610	0,06	18,68
10		384	395	2784,60	2,73	3502,47	391	0,02	1,82
11		257	330	3500,00	22,93	3502,41	326	0,01	26,85
12		503	504	768,55	6,85	3508,00	504	0,02	0,20
13		399	416	2496,82	0,00	2923,29	415	0,03	4,01
14		416	435	2851,00	6,33	3504,26	429	0,03	3,13
15		391	393	440,63	0,00	1913,24	391	0,01	0,00
16		362	372	212,60	0,00	1473,37	372	0,02	2,76
17		435	482	2955,72	8,14	3502,28	478	0,04	9,89
18		267	267	3335,56	0,02	3503,22	267	0,00	0,00
19		332	351	3500,00	5,71	3502,38	332	0,01	0,00
20		497	554	2996,00	11,63	3501,40	554	0,03	11,47

fact that (39)-(44) is easier to solve to optimality as T grows. Typically, for all the instances used, we were able to find the optimal solution within these time limits. In all the cases, the total computing time is restricted to 3600 seconds. The meaning of the columns in Tables 1-4 is the following:

Inst.: instance number;

T: number of time periods in the revised network flow model (29)-(38);

best lb: best known lower bound for the instance;

ub: upper bound provided by the corresponding model within the time limit;

t_{ub}: time spent to reach the best upper bound provided by the revised network flow model (29)-(38);

gap: optimality gap associated to the solution provided by the revised network flow model (29)-(38);

t: time (in seconds) required to obtain the upper bound of the corresponding model;

Table 3. Computational results on benchmark instances with $D = 1.1\overline{C}_{max} + 1$

Inst.	T	best lb	Revised network flow model (29)-(38) ub	t_{ub}	gap	t	Time assign. model (39)-(44) ub	t	final gap
1	20	240	240	109,90	0,00	109,90	240	0,45	0,00
2		488	538	659,60	0,00	2308,64	537	0,09	10,04
3		308	414	1106,02	0,00	1080,52	407	0,03	6,27
4		475	408	21,10	0,00	00,09	498	0,56	4,84
5		309	309	22,22	0,00	22,00	309	0,02	0,00
6		421	431	52,93	0,00	117,18	131	0,0	0,0
7		348	349	71,08	0,00	102,51	349	0,00	0,00
8		451	451	69,20	0,00	637,35	451	0,03	0,00
9		514	581	3265,18	4,57	3301,02	579	0,09	12,65
10		384	403	1204,99	4,32	3303,21	403	0,03	4,95
11		257	270	3300,00	5,75	3301,40	268	0,03	4,28
12		503	516	176,73	0,00	1659,84	515	0,05	2,39
13		399	420	910,95	0,00	936,44	420	0,05	5,26
14		416	465	692,26	12,04	3300,97	465	0,06	11,78
15		391	391	49,51	0,00	49,51	391	0,02	0,00
16		362	373	18,94	0,00	48,81	372	0,03	2,76
17		435	478	3054,19	3,60	3300,92	478	0,06	9,89
18		267	267	1601,07	0,00	1490,50	267	0,00	0,00
19		332	333	3543,27	0,79	3302,92	332	0,00	0,00
20		497	533	893,21	0,00	1131,80	532	0,03	7,04
1	40	240	240	367,44	0,00	295,82	240	0,00	0,00
2		488	553	2834,88	15,21	3502,69	552	0,04	13,11
3		383	419	2028,25	9,36	3501,59	409	0,04	6,79
4		475	498	773,36	0,00	1466,09	498	0,05	4,84
5		309	309	67,02	0,00	67,02	309	0,00	0,00
6		421	431	194,48	0,00	961,68	431	0,03	2,38
7		348	349	859,28	0,00	1358,65	349	0,02	0,29
8		451	451	1593,36	3,46	3505,67	451	0,01	0,00
9		514	632	2153,93	24,03	3501,74	627	0,07	21,98
10		384	394	3500,00	2,45	3502,50	393	0,02	2,34
11		257	282	3500,00	9,81	3502,60	282	0,03	9,73
12		503	504	367,67	0,00	1774,61	503	0,01	0,00
13		399	416	1823,73	0,90	3500,64	415	0,03	4,01
14		416	432	3082,00	5,67	3501,94	429	0,03	3,13
15		391	391	772,95	0,00	774,01	391	0,00	0,00
16		362	372	126,00	0,00	2344,11	372	0,02	2,76
17		435	488	3040,86	9,48	3502,11	484	0,02	11,26
18		267	-	-	-	-	-	-	-
19		332	332	560,67	0,50	3508,38	332	0,01	0,00
20		497	598	2278,94	21,65	3501,56	598	0,03	20,32

final gap: optimality gap between the solution provided by the time assignment model (39)-(44) and the best lower bound for the problem, *i.e.* *final gap* $= (ub - best\ lb)/best\ lb$, with ub being the value of the best solution given by the time assignment model.

A $-$ entry means that CPLEX failed in finding an integer solution for the revised network flow model (29)-(38).

The average gaps between the upper bounds provided by our approach and the best known lower bounds are typically low, usually less than 6%. The best average gap is obtained with $T = 20$ and $D = 1.5\overline{C}_{max} + 1$. In this case, the average gap is equal to 3,88%. In a non negligible number of cases, our model-based heuristic yields the proven optimal solution for the problem. It is important to note that our heuristic does not start with a feasible solution computed using a different method. Hence, the results reported here illustrate solely the capacity of our approach to generate good incumbents for this problem.

Table 4. Computational results on benchmark instances with $D = 1.05\overline{C}_{max} + 1$

Inst.	T	best lb	Revised network flow model (29)-(38)				Time assign. model (39)-(44)		final gap
			ub	t_{ub}	gap	t	ub	t	
1	20	240	240	41,25	0,00	41,25	240	0,00	0,00
2		488	538	858,00	0,00	2629,48	535	0,09	9,63
3		383	407	826,00	0,00	2331,42	406	0,05	6,01
4		475	498	204,82	0,00	253,09	498	0,08	4,84
5		309	309	149,37	0,00	149,37	309	0,00	0,00
6		421	431	24,00	0,00	299,18	431	0,04	2,38
7		348	349	460,09	0,00	1024,76	349	0,01	0,29
8		451	451	287,50	0,00	524,09	451	0,03	0,00
9		514	565	2018,29	0,00	2831,39	561	0,10	9,14
10		384	403	2250,55	4,32	3301,79	403	0,03	4,95
11		257	268	3300,00	5,04	3301,39	268	0,03	4,28
12		503	516	496,10	0,00	2183,01	515	0,06	2,39
13		399	421	125,50	0,00	714,25	420	0,05	5,26
14		416	441	2117,27	7,26	3301,07	435	0,07	4,57
15		391	391	93,26	0,00	93,26	391	0,00	0,00
16		362	372	21,64	0,00	23,17	372	0,03	2,76
17		435	478	869,83	0,00	1032,35	478	0,04	9,89
18		267	267	2983,05	0,00	2983,05	267	0,00	0,00
19		332	332	147,57	0,00	2037,29	332	0,01	0,00
20		497	563	1275,33	0,00	3353,01	562	0,08	13,08
1	40	240	240	189,07	0,00	189,07	240	0,00	0,00
2		488	535	3289,82	11,62	3501,30	535	0,03	9,63
3		383	405	2485,00	6,34	3502,41	404	0,04	5,48
4		475	498	888,80	0,00	942,96	498	0,05	4,84
5		309	309	1611,05	0,00	1612,18	309	0,00	0,00
6		421	431	512,80	3,17	3503,66	431	0,01	2,38
7		348	349	1130,50	0,69	3501,30	349	0,00	0,29
8		451	451	381,18	3,12	3502,69	451	0,01	0,00
9		514	639	1838,57	24,62	3504,59	639	0,05	24,32
10		384	392	3065,47	1,86	3502,17	391	0,02	1,82
11		257	1194	3502,35	78,70	3502,35	1192	0,06	363,81
12		503	504	515,25	0,00	936,44	504	0,02	0,20
13		399	416	1736,50	8,45	3502,55	413	0,01	3,51
14		416	442	3432,80	8,14	3501,62	442	0,03	6,25
15		391	391	661,00	0,00	662,64	391	0,00	0,00
16		362	371	99,00	0,00	2830,68	371	0,02	2,49
17		435	483	2790,00	7,88	3502,44	483	0,03	11,03
18		267	-	-	-	-	-	-	-
19		332	332	315,00	0,50	3502,27	332	0,00	0,00
20		497	635	2559,93	24,45	3501,08	627	0,05	26,16

The revised network flow model (29)-(38) becomes naturally more difficult as the value of T increases. As a consequence, the optimality gap of the solutions obtained by solving this model with CPLEX tends to increase with T. This result appears to have an impact on the quality of the final solution. The average optimality gaps between the upper bound and the best known lower bound for the problem are smaller for $T = 20$ than they are for $T = 40$. The same happens with the value of D: the optimality gap of the solutions of the revised network flow model (29)-(38) are usually smaller as the value of D decreases.

The number of time assignment models (39)-(44) that are solved increases with T, while the size of these models decreases as T gets larger. However, the time assignment model is solved to optimality in a few milliseconds for both $T = 20$ and $T = 40$. As the results in Tables 1-4 show, this model helps in improving the solution obtained with the revised network flow model by computing a precise schedule of the patterns over time, and hence by reducing the overestimation of the total tardiness made by the revised network flow model.

7   Conclusions

We described a new pseudo-polynomial integer programming model for the combined cutting stock and scheduling problem, from which we derived an heuristic solution approach. We described a revised and more compact version of this model which is paired with a time assignment model that allows for a clear definition of the scheduling plan. A computational study on benchmark instances of the literature was conducted to evaluate the impact of some parameters on the performance of the corresponding heuristic. The results show that the heuristic could generate good quality solutions with a small optimality gap for most of these instances.

Acknowledgments. This work was supported by FEDER funding through the Programa Operacional Factores de Competitividade - COMPETE and by national funding through the Portuguese FCT in the scope of the project PTDC/EGE-GES/116676/2010 (Ref. COMPETE: FCOMP-01-0124-FEDER-020430), and by FCT within the project scope: UID/CEC/00319/2013.

References

1. Arbib, C., Marinelli, F.: On cutting stock with due dates. Omega **46**, 11–20 (2014)
2. Braga, N., Alves, C., de Carvalho, J.V.: Exact solution of combined cutting stock and scheduling problems. Accepted for publication in the CMS2014 Special Volume of Lecture Notes in Economics and Mathematical Systems, 11th International Conference in Computational Management Science. Springer (2014)
3. Degraeve, Z., Peeters, M.: Optimal integer solutions to industrial cutting-stock problems: Part 2, benchmark results. INFORMS Journal on Computing **15**(1), 58–81 (2003)
4. Dyckhoff, H.: A new linear programming approach to the cutting stock problem. Operations Research **29**, 145–159 (1981)
5. Fekete, S., Schepers, J.: New classes of fast lower bounds for bin packing problems. Mathematical Programming **91**, 11–31 (2001)
6. Gilmore, P., Gomory, R.: A linear programming approach to the cutting stock problem. Operations Research **9**, 849–859 (1961)
7. Hendry, L., Fok, K., Shek, K.: A cutting stock and scheduling problem in the copper industry. Journal of the Operational Research Society **47**(1), 38–47 (1996)
8. Li, S.: Multi-job cutting stock problem with due dates and release dates. Journal of the Operational Research Society **47**, 490–510 (1996)
9. Perrot, N.: Integer Programming Column Generation Strategies for the Cutting Stock Problem and its Variants. Ph.D. thesis, Université Bordeaux 1 (2005)
10. Reinertsen, H., Vossen, T.: The one-dimensional cutting stock problem with due dates. European Journal of Operational Research **201**, 701–711 (2010)

A New Competitive Implementation of the Electromagnetism-Like Algorithm for Global Optimization

Ana Maria A.C. Rocha$^{(\boxtimes)}$, Andreia Silva, and Jorge Gustavo Rocha

Algoritmi Research Centre, University of Minho, 4710-057 Braga, Portugal
arocha@dps.uminho.pt, andreia56837@gmail.pt, jgr@di.uminho.pt

Abstract. The Electromagnetism-like (EM) algorithm is a population-based stochastic global optimization algorithm that uses an attraction-repulsion mechanism to move sample points towards the optimal. In this paper, an implementation of the EM algorithm in the Matlab environment as a useful function for practitioners and for those who want to experiment a new global optimization solver is proposed. A set of benchmark problems are solved in order to evaluate the performance of the implemented method when compared with other stochastic methods available in the Matlab environment. The results confirm that our implementation is a competitive alternative both in term of numerical results and performance.Finally, a case study based on a parameter estimation problem of a biology system shows that the EM implementation could be applied with promising results in the control optimization area.

Keywords: Global optimization · Unconstrained minimization · Matlab environment · Electromagnetism-like algorithm · Derivative-free algorithm

1 Introduction

Many real life global optimization problems that arise in areas such as physics, chemistry, and molecular biology, involve multimodal and non-differentiable nonlinear functions of many variables that are difficult to handle by gradient-based algorithms. As a result, many researchers have devoted themselves in finding reliable stochastic global optimization methods that do not require any derivative computation.

The Electromagnetism-like (EM) algorithm, developed by Birbil and Fang [3,4], is a population-based stochastic search method for global optimization that mimics the behavior of electrically charged particles. The method uses an attraction-repulsion mechanism to move a population of points towards optimality. The EM algorithm is specifically designed for solving bound constrained problems in the form:

$$\min \ f(x)$$
$$\text{s. t. } \ x \in \Omega, \tag{1}$$

© Springer International Publishing Switzerland 2015
O. Gervasi et al. (Eds.): ICCSA 2015, Part II, LNCS 9156, pp. 506–521, 2015.
DOI: 10.1007/978-3-319-21407-8_36

where $f : \mathbb{R}^n \to \mathbb{R}$ is a nonlinear continuous function and $\Omega = \{x \in \mathbb{R}^n : lb \leq x \leq ub\}$ is a bounded feasible region. We do not assume that the objective function f is convex and may possess many local minima in the feasible region.

The goal of global optimization is to find the globally best solution of problems, in the presence of multiple local optima. There are several stochastic search methods that were developed to find the global minimum such as the Simulated Annealing, Genetic Algorithm, Particle Swarm Optimization, Ant Colony Optimization and Evolutionary Methods [6,10,11,13]. Some of these methods may combine the search process with local refinements like random search methods or gradient-based methods.

Systems biology has been responsible for a revolution in the ability of understanding biologic events and organisms. It made possible to have an comprehensive, quantitative and temporal analysis of the interaction between the components of a biological system. Through a mathematical model it is possible to summarize the knowledge of a biological system allowing to make experimentally verifiable predictions. In systems biology, the determination of parameters is a central challenge, because the majority of the mathematical models present some characteristics that make the problem difficult to solve, as the large number of parameters to be estimated as the highly non-linearity of the problems [14,16,28–30]. Hence, there is a need for global optimization methods capable of solving this kind of problems efficiently.

In the Matlab environment, the Global Optimization Toolbox [19] provides methods that search for global solutions. It includes global search, multistart, pattern search, genetic algorithm, and simulated annealing solvers. These methods use interactive tools for defining and solving optimization problems and monitoring the solution progress.

The aim of this paper is to present a new implementation of the EM algorithm in the Matlab® environment. To accomplish this assessment, we started by porting the algorithm to Matlab programming language and a set of functions were defined in order to develop the EM with a similar syntax to the methods available in the Matlab Global Optimization Toolbox. Tests were performed against a well known set of problems, and some benchmarks were performed to validate this implementation. With a stable implementation in Matlab, we select a well known problem to assess the application of this algorithm to estimate biological parameters.

This paper is organized as follows. Section 2 gives a general overview of the EM algorithm. The implementation details, functionalities and how to use EM in the Matlab environment are described in Section 3. Section 4 reports the results of the numerical experiments carried out with a benchmark set of unconstrained problems, as well as a comparison with other stochastic based methods. The parameter estimation problem, called α-pinene, is briefly described and solved by EM in Section 4.2. It is compared against other stochastic methods. Finally, the paper is concluded in Section 5 and some remarks for future developments are presented.

2   Electromagnetism-Like Algorithm

The EM algorithm simulates the electromagnetism theory of physics by considering each point in the population as an electrical charge that is released to the space. The charge of each point is related to the objective function value and determines the magnitude of attraction of the point over the others in the population. The better the objective function value, the higher the magnitude of attraction. The charges are used to find a direction for each point to move in subsequent iterations. The regions that have higher attraction will signal other points to move towards them. In addition, a repulsion mechanism is also introduced to explore new regions for even better solutions [3,4].

The EM algorithm is described in Algorithm 1 and comprises four main procedures: initialization of the algorithm, computation of the total force, movement along the direction of the force and a simple random line search algorithm [22,23].

Algorithm 1. EM algorithm

1: Initialization;
2: **while** stopping criteria are not met **do**
3:     Compute Force
4:     Move Points
5:     Local Search
6: **end while**

The "Initialization" procedure starts by randomly generating a sample of m points. Each point is uniformly distributed between the lower and upper bounds. Then, for each point of the population the objective function value is calculated. Finally the point which yields the least objective function value, denoted by the best point of the population, x^{best}, is identified as well as its corresponding objective function value f^{best}.

In the "Compute Force" procedure, each particle charge is calculated by

$$q^i = \exp\left(-n\frac{f(x^i) - f(x^{best})}{\sum_{k=1}^{m}(f(x^k) - f(x^{best}))}\right), \quad i = 1, \ldots, m. \tag{2}$$

that determines the power of attraction or repulsion for the point x^i. In this way the points that have better objective function values possess higher charges. After the charge calculation, the total force exerted on each point x^i is calculated by adding the individual component forces, F_j^i, between any pair of points x^i and x^j. According to the electromagnetism theory, the total force, F^i, is inversely proportional to the square of the distance between the points and directly proportional to the product of their charges:

$$F^i = \sum_{j\neq i}^{m} F_j^i = \begin{cases} (x^j - x^i)\frac{q^i q^j}{\|x^j - x^i\|^2} & \text{if } f(x^j) < f(x^i) \text{ (attraction)} \\ (x^i - x^j)\frac{q^i q^j}{\|x^j - x^i\|^2} & \text{if } f(x^j) \geq f(x^i) \text{ (repulsion)} \end{cases}, i = 1, \ldots, m. \tag{3}$$

The "Move Points" procedure uses the total force vector, F^i, to move the point x^i in the direction of the force by a random step length λ. The best point, x^{best}, is not moved. To maintain feasibility, the force exerted on each point is normalized and scaled by the allowed range of movement towards the lower bound or the upper bound of the set Ω, for each coordinate. Thus, for $i = 1, \ldots, m$ and $i \neq best$

$$x^i = \begin{cases} x^i + \lambda \frac{F^i}{\|F^i\|}(ub - x^i) & \text{if } F^i > 0 \\ x^i + \lambda \frac{F^i}{\|F^i\|}(x^i - lb) & \text{otherwise} \end{cases} \tag{4}$$

The random step length λ is assumed to be uniformly distributed in $[0,1]$.

Finally, the "Local search" procedure performs a local refinement around a point or around each point of the population [3,4]. In this implementation, this procedure is only applied to the best point of the population, since previous works showed it was sufficient and efficient in the improvement of the accuracy of EM [22–24]. This procedure implements a simple random line search algorithm, using the maximum feasible step length

$$\Delta = \delta_{local}(\max[ub - lb]) \tag{5}$$

with $\delta_{local} > 0$, to guarantee that the local search always generates feasible points [1,3,22–24]. A random movement of length Δ is carried out and if a better position is obtained within a maximum number of local iterations, x^{best} is updated.

3   MATLAB Implementation

Matlab® is a technical computing environment that integrates numerical and matrix analysis which has a simplistic environment to the user. Nowadays, Matlab is the standard tool that supports introductory and advanced mathematical based courses which is the choice tool used in industrial research.

3.1   Implementation Details

The original version of the Electromagnetism-like algorithm discussed on Section 2, was implemented as a Matlab global optimization solver. The implementation respects the structure of other optimization solvers where each module defines a specific routine of the algorithm.

The user can see the details of the `emalgorithm` using the following command.

```
>> help emalgorithm
```

The basic call of EM algorithm is:

```
>> [x,fval,exitflag,output]=emalgorithm(Fun,lb,ub,options)
```

The input and output arguments are described next.

Output Arguments. The output variables are x, as the minimizer, fval is the value of the objective function at the solution x, exitflag is a flag describing the exit condition and output is a structure that shows some values about the iterative process and other information. The possible values of exitflag and the corresponding exit conditions are detailed in Table 1.

Table 1. Exitflag conditions

Flag	Condition
1	Average change in value of the objective function over options.StallIterLimit iterations less than options.TolFun
5	options.ObjectiveLimit limit reached
0	Maximum number of function evaluations or iteration exceeded
-1	Optimization terminated by the output or plot function
-2	No feasible point found
-5	Time limit exceeded

The description of the items about the iterative process concerning the output argument are presented in Table 2.

Table 2. Output structure

Item	Description
problemtype	Type of problem: bound constrained
iterations	Total number of iterations
funccount	Total number of function evaluations
message	Termination message of the solver
totaltime	Time taken by the solver

Input Arguments. The input variables are Fun, which is the objective function, lb and ub are the lower and upper bounds on the variables, respectively, and options (optional) define parameters used to control the algorithm.

To know the list of the fields in the options structure as well as the valid parameters and the default parameters use:

```
>> help emoptimset
```

The available options are presented in Table 3 and can be defined with:

```
>> emoptimset(option, value)
```

where option should be a valid term and value is the assigned value of the option. To check the default option values of the EM algorithm, use:

```
>> emoptimset
```

Table 3. EM algorithm optional parameters

Option	Type	Default value	Description
PopulationSize	positive scalar	10*numberOfVariables	Defines the population size
TolFun	non-negative scalar	1e-6	Termination tolerance on function value
StallIterLimit	positive integer	500*numberOfVariables	Number of iterations over which average change in objective function value at current point is less than TolFun
MaxFunEvals	positive integer	3000* numberOfVariables	Maximum number of function evaluations
MaxIter	positive integer	Inf	Maximum number of allowed iterations
MaxLocalIterations	positive integer	5	Maximum number of iterations allowed in the local search procedure
TimeLimit	positive scalar	Inf	Total time (in seconds) allowed for optimization
ObjectiveLimit	scalar	-Inf	Minimum objective function value desired
Delta	positive scalar	1e-3	Positive scalar that defines the local refinement (δ_{local})
PrecisionTolerance	non-negative scalar	1e-3	
Display		final	Controls the level of display results, valid values, final, iter, diagnose, and off
DisplayInterval	positive integer	10	Interval for iterative display
PlotFcns	function_handle		Plot function(s) to be called during iterations (@emplotbestf, @emplotbestx, @emplotstopping, @emplotmeanf)
PlotInterval	positive integer	1	Interval at which PlotFcns are called

3.2   Example Usage

Suppose we want to run the EM to solve the Branin function [12]:

$$\min \left(x_2 - \frac{5.1}{4\pi^2}x_1^2 + 5\frac{x_1}{\pi} - 6 \right)^2 + 10 \left(1 - \frac{1}{8\pi} \right) \cos(x_1) + 10$$

s. t. $-5 \leq x_1 \leq 0,$

$$10 \leq x_2 \leq 15$$

Firstly, the function to be optimized should be created:

```
function y = branin(x)
% Branin function
y = (x(2)-(5.1/(4*pi^2))*x(1)^2+5*x(1)/pi-6)^2+...
    10*(1-1/(8*pi))*cos(x(1))+10;
end
```

Then, the following commands should be done in order to define the lower and upper bounds of the problem, define some options to control the algorithm, if desired, and finally call the EM algorithm:

```
lb = [-5 0];
ub = [10 15];
options=emoptimset('Display','iter');
[x,f,exitflag,output]=emalgorithm('branin',lb,ub,options)
```

The obtained solution is:

```
Iteration    f-count       f(x)
       10      256         0.397899
       20      469         0.397887
       30      679          0.397887
       40             889       0.397887
             ...
      270     5720         0.397887
      280     5930         0.397887
Maximum number of function evaluations exceeded: increase
options.MaxFunEvals.

x = 3.1415     2.2750
f =   0.3979
exitflag =  0
output =
      iterations: 284
       funccount: 6014
         message: 'Maximum number of function evaluations exceeded'
       totaltime: 1.3600
     problemtype: 'boundconstraints'
```

4 Numerical Results

Computational tests were performed on a PC with a 2.94GHz Pentium IV micro-processor and 4Gb of memory. At first, we compare the performance of EM algorithm with Genetic Algorithm (GA) and Simulated Annealing (SA) when applied to a benchmark set of nine functions. They are all stochastic and derivative-free methods available in the Global optimization Toolbox of Matlab. The GA is a method based on natural selection, selecting, at each step, individuals at random from the current population to be parents and uses them to produce the children for the next generation [10]. The SA mimics the physical process of heating a material and then slowly lowering the temperature to decrease defects in order to minimize the energy optimization problem [11]. GA and EM are population-based methods, while SA is a single point method. Secondly, we compare the EM method with other stochastic methods when solving a parameter estimation problem, the isomerization of α-pinene problem.

4.1 Application to a Test Set of Problems

In order to evaluate the efficiency of the implemented algorithm we considered nine test functions [12], summarized in Table 4, as benchmarks for comparing global search methods. The columns of the table refer to the name of the problem, 'Prob.'; information about the abbreviation usually used, 'Abbr.'; the dimension of the problem, 'n'; the known global minimum available in the literature, 'f^*'; the default box constraints, 'Box constraints'; the number of known local minimizers, 'loc', and the number of known global minimizers, 'glob'.

Table 4. Test functions

Prob.	Abbr.	f^*	n	Box constraints	loc	glob
Branin	BR	0.3978874	2	$[-5,0] \times [10,15]$	3	3
Golstein-Price	GP	3.0000000	2	$[-2,2]^2$	4	1
Hartman 3	H3	-3.8627821	3	$[0,1]^3$	4	1
Hartman 6	H6	-3.3223680	6	$[0,1]^6$	4	1
Six-hump camel	C6	-1.0316285	2	$[-3,3] \times [-2,2]$	6	2
Shekel 5	S5	-10.1531997	4	$[0,10]^4$	5	1
Shekel 7	S7	-10.4029406	4	$[0,10]^4$	7	1
Shekel 10	S10	-10.5364098	4	$[0,10]^4$	10	1
Shubert	SHU	-186.7309088	2	$[-10,10]^2$	760	18

In order to assess the performance of the implemented EM, it is compared with the GA and SA packages available in the Matlab Global Optimization Toolbox. For more details on the methods, see [19]. The justification for the choice of these solvers is due to the fact that they are implemented in Matlab and are derivative-free methods, such as EM. Besides, other global stochastic methods in Matlab environment, like the implemented in functions `GloabSearch` and `Multistart`, they are not used for comparison since they use gradient-based methods in the local search phase.

The EM algorithm used the default parameters, set by $PopulationSize = 20$, $\delta_{local} = $ 1e-3 and $MaxLocalIterations = 5$. For a fair comparison, GA and SA also used the default parameter values. The value of f^* was set to the parameter $ObjectiveLimit = f^*$. If the optimal solution is not reached within an absolute tolerance of 1e-5, the algorithm stops for a maximum of 10000 function evaluations, $MaxFunEvals = 10000$. In order to obtain statistically significant results, we run each method 30 times.

Table 5. Comparison of EM results with GA and SA

Prob.	EM			GA			SA		
	f_{best}	StdDev	N_{avg}	f_{best}	StdDev	Navg	f_{best}	StdDev	N_{avg}
BR	0.39789	7.77e-07	1086	0.39789	8.96e-07	825	0.39789	3.03e-06	6077
GP	3.00000	2.89e-06	1187	3.00000	2.09e+01	2048	3.00000	3.02e-06	485
H3	-3.86278	2.99e-06	1662	-3.86278	1.41e-01	1875	-3.86278	1.76e-04	9815
H6	-3.32237	3.66e-02	2682	-3.32237	5.84e-02	5584	-3.32075	1.49e-02	10001
C6	-1.03163	2.22e-06	491	-1.03163	2.28e-06	579	-1.03163	1.93e-06	1794
S5	-10.15320	3.30e+00	6175	-5.05520	5.93e-09	10020	-10.15308	3.50e-01	10001
S7	-10.40294	3.16e-06	3010	-5.08767	7.20e-09	10020	-10.40288	1.54e-01	10001
S10	-10.53641	5.16e-06	2728	-5.12848	5.22e-09	10020	-10.53638	1.36e+00	10001
SHU	-186.73091	1.30e-05	6155	-186.73091	5.83e+01	7599	-186.73091	5.11e-06	492

Table 5 reports the name of the problem, 'Prob.'; the best function value obtained by each algorithm, 'f_{best}'; the standard deviation between the best function values obtained by the algorithm, '$StdDev$'; the average number of function evaluations, 'N_{avg}'. Generally, EM reaches the global optimal solutions with less function evaluations than GA and SA. From $StdDev$ column we can conclude EM obtained more consistent and effective solutions except for problem S5, where EM converged to attractive locals, for some runs (see Fig. 1). For problems S5, S7 and S10, GA didn't reach the global optimum (always converged to local solutions), and EM has more accurate solutions than SA.

4.2   Parameter Estimation Problem

Most biological processes are nonlinear dynamic systems, which are usually modeled as a nonlinear programming problem subject to dynamic (usually differential-algebraic) constraints that describe the evolution over time of certain quantities of interest. In the context of bioprocess engineering optimization, the parameter estimation problem goal is to find the set of parameters of a mathematical model to obtain the best possible fit to the existing experimental data [17,26,30]. Since these problems can be non-convex, some of the main optimization methods based on the gradient, such as, Levenberg-Marquardt or Gauss-Newton [21], may not work well, because if there is a local minimum very close to the global minimum, the method may fail, failing to reach the optimal solution. Thus, there is a need for using global optimization methods to ensure convergence to a global optimal solution. The literature presents some advances in global optimization for problems of parameter estimation in dynamic systems in deterministic methods [8,18] and in stochastic methods [15,20,25].

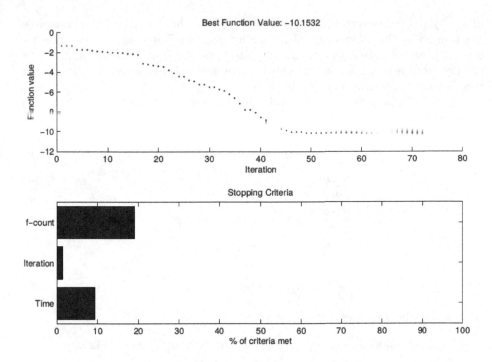

Fig. 1. Example of a run of EM algorithm with Shekel 5 problem

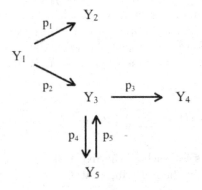

Fig. 2. Mechanism for thermal isomerization of α-pinene

In this paper, we are interested in solving the isomerization of α-pinene problem, that is a parameter estimation problem that arises from the modeling of the chemical phenomena of an isomerization reaction. The α-pinene estimation problem was firstly discussed by Box et al. [5] in 1973 and by Bates and Watts [2] in 1988 and by Seber and Wild [27] in 1989. Fuguitt and Hawkins [9] in 1947 studied the reaction of α-pinene where (y_1) is thermally isomerized to dipentene (y_2) and allo-ocimene (y_3). Allo-ocimene in turn yields pyronene (y_4) and a dimer (y_5). The converting of allo-ocimene (y_3) to the dimer (y_5) is a reversible reaction while the other conversions are irreversible. The goal of α-pinene estimation problem is to estimate the five rate constants ($p_1, \ldots, p_5$). Fig. 2 shows the proposed reaction scheme for this homogeneous chemical reaction describing the thermal isomerization of α-pinene.

The model is defined by the following system of five ordinary differential equations:

$$\frac{d[y_1]}{dt} = -(p_1 + p_2)y_1 \tag{6}$$

$$\frac{d[y_2]}{dt} = p_1 y_1 \tag{7}$$

$$\frac{d[y_3]}{dt} = p_2 y_1 - (p_3 + p_4)y_3 + p_5 y_5 \tag{8}$$

$$\frac{d[y_4]}{dt} = p_3 y_3 \tag{9}$$

$$\frac{d[y_5]}{dt} = p_4 y_3 - p_5 y_5 \tag{10}$$

considering the given set of time intervals:

$$t = \{1230.0, 3060.0, 4920.0, 7800.0, 10680.0, 15030.0, 22620.0, 36420.0\} \tag{11}$$

It is assumed that the initial values (i.e. at time $t \approx 0$) of one reactant and four products are $y_1(0) = 100$, $y_2(0) = 0$, $y_3(0) = 0$, $y_4(0) = 0$, $y_5(0) = 0$.

The goal of the α-pinene estimation problem is to estimate (predict) the unknown coefficients, $p_1, \ldots, p_5$, minimizing the objective function that corresponds to a weighted distance measure between the experimental values, corresponding to the measured variables, and the predicted values for those same variables, formulated as:

$$f(p) = \sum_{j=1}^{5} \sum_{i=1}^{8} (y_j(p, t_i) - \widetilde{y}_{ji})^2 \tag{12}$$

The experimental values for each of the five responses ($y_1, \ldots, y_5$) in the eight interval times are given in the following matrix,

$$\widetilde{y}_{ji} = \begin{bmatrix} 88.35 & 7.3 & 2.3 & 0.4 & 1.75 \\ 76.4 & 15.6 & 4.5 & 0.7 & 2.8 \\ 65.1 & 23.1 & 5.3 & 1.1 & 5.8 \\ 50.4 & 32.9 & 6.0 & 1.5 & 9.3 \\ 37.5 & 42.7 & 6.0 & 1.9 & 12.0 \\ 25.9 & 49.1 & 5.9 & 2.2 & 17.0 \\ 14.0 & 57.4 & 5.1 & 2.6 & 21.0 \\ 4.5 & 63.1 & 3.8 & 2.9 & 25.7 \end{bmatrix}$$

The best known solution is $p_1^* = 5.9256\text{e-}5$, $p_2^* = 2.9632\text{e-}5$, $p_3^* = 2.0450\text{e-}5$, $p_4^* = 2.7473\text{e-}4$, $p_5^* = 4.0073\text{e-}5$ with optimal value $f^*(p) = 19.872$.

Hereafter, the performance of EM algorithm will be tested when solving the hard optimization problem from the chemical and bio-process engineering area - the isomerization of α-pinene problem.

We used EM as described in Section 3 with the default parameters: population size of 20 points, $m = 20$, $\delta_{local} = 1e - 3$ and MaxLocalIterations=5. The lower bounds considered for the five parameters arise from physical considerations, $p_i \geq 0$, and we took the upper bounds to be $p_i \leq$ 5e-4, for $i = 1, \ldots, 5$.

Table 6 shows the average results obtained with EM, among 10 independent runs, when running the code with a time limit of 50 seconds.

Table 6. EM solutions after 50 seconds

	f_{best}	f_{median}	f_{worst}	N_{feavg}	It_{avg}
EM	19.874010	20.325178	21.805897	4070	147

The best solution found by EM gives p_1 =5.9260e-05, p_2 =2.9632e-05, p_3 =2.0470e-05, $p_4 = 2.7425$e-04, p_5 =3.9906e-05 and was found after 3824 function evaluations and 147 EM iterations. The variable values of the best solution obtained are represented in Fig. 3. Fig. 4 depicts the convergence curve of the objective function value along the iterations and the stopping criterion levels of the optimization process. The relation between experimental data and the predicted values of the α-pinene using the EM best solution is illustrated in Fig. 5.

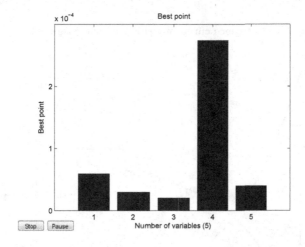

Fig. 3. Best solution found by EM for thermal isomerization of α-pinene

In the following, we intend to compare the results obtained with the implemented EM algorithm when solving the problem of isomerization of α-pinene, with the ones obtained with GA and SA packages available in the Matlab Global Optimization Toolbox, using the default parameter values. Table 7 reports the best, f_{best}, the average, f_{avg}, the worst, f_{worst}, solution values and the average number of function evaluations, N_{avg}, of the 10 runs. Each run was limited to 50 seconds, as the stopping condition for all the stochastic solvers.

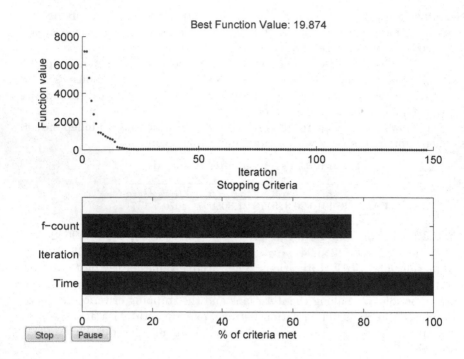

Fig. 4. Convergence curves for the α-pinene case study

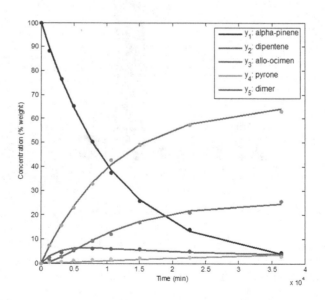

Fig. 5. Experimental data versus model prediction for the α-pinene case study

Table 7. Comparison results for the isomerization of α-pinene problem

Solver	f_{best}	f_{avg}	f_{worst}	N_{avg}
EM	19.874010	20.549177	21.805897	4070
GA	20.414430	167.137782	817.643129	4350
SA	24.718882	54.241137	108.635052	3982

From Table 7 we may conclude that EM algorithm achieved more accurate solutions with less computational effort, when compared with GA and SA algorithms.

In [7], we can find the solution obtained with the SSm method, a Scatter Search procedure with an improvement method, when solving the α-pinene: $f_{best} = 19.872000$, $f_{avg} = 24.747000$, $f_{worst} = 68.613000$ and $N_{avg} = 1144$. In fact, SSm reached the known global minimum but we remark that the improvement method used a gradient-based method implemented in the fmincon command of Matlab. However, the average results obtained with EM are more closer to the global optimal solution, with a standard deviation of 0.73279 of the ten best solutions found. Moreover, the relative error of the best solution found by EM algorithm was 1e-4 .Thus, EM gives competitive results when comparing with the SSm method, and was able to get good solution to the isomerization of α-pinene problem.

5   Conclusions and Future Work

In this paper, we implemented the EM algorithm in the Matlab® environment as a useful function for the scientific community with similar syntax to other optimization functions there already available. The implementation was considered successfully when tested under a set of benchmark problems that were solved and compared with other stochastic and derivative-free methods available in the Global Optimization Toolbox of Matlab. The implementation is freely available at http://www.norg.uminho.pt/arocha/code.htm.

The application of the EM algorithm for solving the hard isomerization of α-pinene problem showed good and competitive results when comparing with the other solvers in the Global Optimization Toolbox. This example shows how important it is to optimize mathematical models. The optimization plays a key role on systems biology. This parameter estimation problem was taken form the real world, and it could bring many advantages, to reduce costs of experimental measurements, for example.

In the future, we intend to integrate EM in the Matlab Optimization Graphical Interface, creating a user-friendly interface. As other global optimization functions available in the global optimization toolbox, we want to include hybrid functions to improve the quality of the solutions. The final purpose is to include the EM code as a Matlab built-in function in the Global Optimization Toolbox for unconstrained optimization problems.

Acknowledgments. This work has been supported by FCT (Fundação para a Ciência e Tecnologia, Portugal) in the scope of the project PEst-UID/CEC/00319/2013.

References

1. Ali, M.M., Golalikhani, M., Zhuang, J.: A computational study on different penalty approaches for solving constrained global optimization problems with the electromagnetism-like method. Optimization **63**(3), 403–419 (2014)
2. Bates, D., Watts, D.: Nonlinear Regression Analysis and Its Applications. John Wiley & Sons, Inc. (2008)
3. Birbil, S.I., Fang, S.-C.: An electromagnetism-like mechanism for global optimization. J. Global Optim. **25**, 263–282 (2003)
4. Birbil, S.I., Fang, S.-C., Sheu, R.L.: On the convergence of a population-based global optimization algorithm. J. Global Optim. **30**, 301–318 (2004)
5. Box, G.E.P., Hunter, W.G., MacGregor, J.F., Erjavec, J.: Some problems associated with the analysis of multiresponse data. Technometrics **15**, 33–51 (1973)
6. Colorni, A., Dorigo, M., Maniezzo, V.: Distributed Optimization by Ant Colonies. In: European Conference on Artificial Life, pp. 134–142. Elsevier Publishing (1991)
7. Egea, J.A., Rodríguez-Fernández, M., Banga, J.R., Martí, R.: Scatter search for chemical and bio-process optimization. J. of Global Optimization **37**(3), 481–503 (2007)
8. Esposito, W., Floudas, C.A.: Global optimization for the parameter estimation of differential-algebraic systems. Ind. Eng. Chem. Res. **39**(5), 1291–1310 (2000)
9. Fuguitt, R.E., Hawkins, J.E.: Rate of the thermal isomerization of α-pinene in the liquid phase. J. Am. Chem. Soc. **69**, 19–322 (1947)
10. Goldberg, D.E.: Genetic Algorithms in Search. Optimization & Machine Learning. Addison-Wesley (1989)
11. Ingber, L.: Simulated annealing: practice versus theory. Mathl. Comput. Modelling **18**, 29–57 (1993)
12. Jones, D.R., Perttunen, C.D., Stuckman, B.E.: Lipschitzian optimization without the Lipschitz constant. J. Optimiz. Theory App. **79**, 157–181 (1993)
13. Kennedy, J., Eberhart, R.: Particle Swarm Optimization. Proceedings of IEEE International Conference on Neural Networks **4**, 1942–1948 (1995)
14. Kohl, P., Noble, D.: Systems biology and the virtual physiological human. Mol Syst Biol. **5**, 292 (2009)
15. Larrosa, E.: New heuristics for global optimization of complex bioprocesses. Master Thesis, University of Vigo, Spain (2008)
16. Lieu, C., Elliston K.: Applying a causal framework to system modeling. In: Bringmann, P., Butcher, E., Parry, G., Weiss, B. (eds.) Ernst Schering Research Foundation Workshop. Systems Biology, ser., vol. 61, pp. 139–152. Springer, Heidelberg (2007)
17. Lillacci, G., Khammash, M.: Parameter estimation and model selection in computational biology. PLoS Computational Biology **6**(3), e1000696 (2010)
18. Lin, Y., Stadtherr, M.A.: Deterministic global optimization for parameter estimation of dynamic systems. Ind. Eng. Chem. Res. **45**, 8438–8448 (2006)
19. MATLAB and Global Optimization Toolbox Release 2013, The MathWorks, Inc., Natick, Massachusetts, United States (2013)
20. Moles, C.G., Mendes, P., Banga, J.R.: Parameter estimation in biochemical pathways: a comparison of global optimization methods. Genome Research **13**(11), 2467–2474 (2003)

21. Nocedal, J., Wright, S.: Numerical Optimization. Springer Series in OperationsResearch. Springer (1999)
22. Rocha, A.M.A.C., Fernandes, E.M.G.P.: Modified movement force vector in an electromagnetism-like mechanism for global optimization. Optim. Method. Softw. **24**, 253–270 (2009)
23. Rocha, A.M.A.C., Fernandes, E.M.G.P.: Performance profile assessment of electromagnetism-like algorithms for global optimization. AIP Conf. Proc. **1060**, 15–18 (2008)
24. Rocha, A.M.A.C., Fernandes, E.M.G.P.: Numerical study of augmented Lagrangian algorithms for constrained global optimization. Optimization **60**, 10–11 (2011)
25. Rodríguez-Fernández, M., Egea, J.A., Banga, J.R.: Novel metaheuristic for parameter estimation in nonlinear dynamic biological systems. BMC Bioinformatics **7**, 483 (2006)
26. Schittkowski, K.: Parameter estimation in systems of nonlinear equations. Numerische Mathematik **68**(1), 129–142 (1994)
27. Seber, G., Wild, C.: Nonlinear Regression. John Wiley & Sons, Inc. (1989)
28. Sun, J., Garibaldi, J.M., Hodgman, C.: Parameter estimation using metaheuristics in systems biology: A comprehensive review. IEEE/ACM Trans. Comput. Biol. Bioinformatics **9**(1), 185–202 (2012)
29. Villaverde, A., Egea, J., Banga, J.: A cooperative strategy for parameter estimation in large scale systems biology models. BMC Syst. Biol. **6**, 75 (2012)
30. Zhan, C., Yeung, L.F.: Parameter estimation in systems biology models using spline approximation. BMC Syst. Biol. **5**, 14 (2011)

Does Beam Angle Optimization Really Matter for Intensity-Modulated Radiation Therapy?

Humberto Rocha[1]([✉]) , Joana M. Dias[1,2] , Brigida C. Ferreira[3,4] ,
and Maria do Carmo Lopes[3,4]

[1] INESC-Coimbra, Rua Antero de Quental, 199, 3000-033 Coimbra, Portugal
hrocha@mat.uc.pt
[2] Faculdade de Economia, Universidade de Coimbra, 3004-512 Coimbra, Portugal
joana@fe.uc.pt
[3] School for Allied Health Technologies, 4400-330 Porto, Portugal
bcf@estsp.ipp.pt
[4] Serviço de Física Médica, IPOC-FG, EPE, 3000-075 Coimbra, Portugal
mclopes@ipocoimbra.min-saude.pt

Abstract. The beam angle optimization (BAO) in intensity-modulated radiation therapy (IMRT) consists on the selection of appropriate radiation incidence directions and can theoretically influence the quality of the IMRT plans, both by improving tumor coverage and by obtaining better organs sparing.However, in clinical practice, the importance of BAO is yet to be acknowledged and, most of the time, beam directions continue to be either equispaced or manually selected by the treatment planner, not making integral part of the optimization loop that is typically devoted to the fluences optimization.During this optimization loop, the treatment planner needs to tune different parameters, in a time consuming process, including, most of the time, objective function weights associated to the different structures included in the optimization procedure. Traditionally, the treatment planning is seen as a sequential process that starts with the selection of the irradiation beam directions.When irradiation beam directions are selected through a BAO procedure that uses the optimal values of the fluence optimization (for a given set of initial objective function weight parameters), does the posterior tuning of the objective function parameters completely jeopardize the BAO effort? The goal of this paper is to contribute to clarify this question, to perceive the importance of BAO in IMRT and its timing within the treatment planning optimization loop. For a study set of ten clinical cases of head-and-neck tumors treated at the Portuguese Institute of Oncology of Coimbra we showed that optimized beam angle sets obtained statistical significant better results ($p - value < 0.001$) than the traditional equispaced configuration. Furthermore, in our tests, despite optimal beam angle sets being always preferable to equispaced configurations, the choice of the weight parameters for angle selection showed influence on the correspondent plan quality.

Keywords: Radiotherapy · IMRT · Beam angle optimization · Direct Search

© Springer International Publishing Switzerland 2015
O. Gervasi et al. (Eds.): ICCSA 2015, Part II, LNCS 9156, pp. 522–533, 2015.
DOI: 10.1007/978-3-319-21407-8_37

1 Introduction

Intensity-modulated radiation therapy (IMRT) is a modern technique with a treatment planning that is usually a sequential process where initially a given number of radiation incidence directions are selected followed by the optimization of the radiation intensities considering those irradiation directions. The selection of appropriate irradiation directions in IMRT treatment planning – beam angle optimization (BAO) problem – is the first problem that arises in treatment planning, but its optimal solution is highly dependent on the optimal solution of the fluence map optimization (FMO) problem – the problem of deciding what are the optimal radiation intensities associated with each set of beam angles. When the BAO problem is not based on the optimal FMO solutions, the resulting beam angle set has no guarantee of optimality and has questionable reliability since it has been extensively reported that optimal beam angles for IMRT are often non-intuitive [20]. Nevertheless, in current clinical practice, most of the time, beam directions are considered equispaced or continue to be manually selected which requires many trial and error iterations between selecting beam angles and computing fluence patterns until a suitable treatment is achieved. Therefore, regardless the evidence presented in the literature that appropriate irradiation directions can lead to a plan's quality improvement [2,10,21], BAO continue to be either ignored or not making integral part of the treatment planning optimization loop that is typically dominated by the FMO problem.

IMRT treatment planning generally has to balance a multitude of risks and goals that should be, ideally, addressed by multiobjective approaches to achieve acceptable compromises defined *a priori* or selected *a posteriori* from a database of Pareto-optimal generated plans. Recently, *a posteriori* [6] and *a priori* [4] multi-criteria approaches have been proposed for IMRT. However, the inclusion of a full BAO integrated into a fluence map multiobjective optimization framework has still many underlying issues including clinically acceptable computational times. Moreover, the approaches proposed in most of the literature and offered by the vast majority of the treatment planning systems (TPS) – treatment plan dedicated commercial software applications – consider weighted mathematical formulations to model the FMO problem and balance the multiple conflicting objectives (e.g., achieve a high tumor dose while giving a low dose to neighboring organs). In current clinical practice, this balance is attempted by a treatment planner trying to steer the TPS by interactively tuning plan parameters towards a better solution. For each patient, the planner makes a first choice for the different parameters (including the weights assigned to each structure incorporated in the optimization loop) based on local protocols or experience. The TPS then generates a treatment plan with corresponding dose distribution. If the dose distribution is not satisfactory, then the planner tune the parameters (based on experience) for a new run of the algorithm until the dose distribution is considered satisfactory or the time to further improve it runs out. Typically, this parameter tuning (often called "optimization") is done for a fixed set of beam radiation incidence directions that are either chosen by the planner or are obtained after an initial BAO procedure. As referred above, the optimal solution of the BAO should depend on the optimal solution of the FMO problem. However, the optimal incidence radiation directions are obtained using a given set of parameters, in particular a given set of objective function weights associated to the different structures included in the optimization procedure. Afterwards, does the tuning of the fluence map objective function parameters completely jeopardize the initial BAO effort? Is there a timing to perform BAO in order to diminish deterioration of BAO results? The goal of this paper is to contribute to clarify this questions, to perceive the importance of BAO in IMRT and its timing within the treatment planning optimization loop.

2 Methods and Materials

2.1 Study Patients

Ten clinical examples of head-and-neck tumors treated at the Portuguese Institute of Oncology of Coimbra (IPOC), signalized as complex cases where proper target coverage and organ sparing, in particular parotid sparing, proved to be difficult to obtain with the typical 7-beam equispaced coplanar treatment plans, were selected. The patients' computed tomography (CT) sets and delineated structures were exported via Dicom RT to a freeware computational environment for radiotherapy research (CERR).

In general, the head-and-neck region is a complex area to treat with radiotherapy due to the large number of sensitive organs in this region (e.g., eyes, mandible, larynx, oral cavity, etc.). The spinal cord and the brainstem are some of the most critical organs at risk (OARs) in the head-and-neck tumor cases. These are serial type organs, i.e., organs such that if only one functional subunit is damaged, the whole organ functionality is compromised. Therefore, if the tolerance dose is exceeded, it may result in functional damage to the whole organ. Thus, it is extremely important not to exceed the tolerance dose assigned for these type of organs. Other than the spinal cord and the brainstem, the parotid glands are also important OARs. The parotid gland is the largest of the three salivary glands. A common complication due to parotid glands irradiation is xerostomia. This secondary radiation effect decreases the quality of life of patients undergoing radiation therapy of head-and-neck, causing difficulties to swallow. The parotids are parallel organs, i.e., if a small volume of the organ is damaged, the organ functionality may not be affected. Their tolerance dose depends strongly on the fraction of the volume irradiated. Hence, if only a small fraction of the organ is irradiated the tolerance dose is much higher than if a larger fraction is irradiated. Thus, for these parallel type structures, the organ mean dose is generally used as an objective for inverse planning optimization. In this retrospective study, the OARs used for treatment optimization were limited to the spinal cord, the brainstem and the parotid glands.

For the head-and-neck cases in study the planning target volume (PTV) consisted of PTV_{70} and $PTV_{59.4}$ corresponding to different prescribed doses.

Table 1. Tolerance and prescribed doses for all the structures considered for IMRT optimization

Structure	Mean dose	Max dose	Prescribed dose
Spinal cord	–	45 Gy	–
Brainstem	–	54 Gy	–
Left parotid	26 Gy	–	–
Right parotid	26 Gy	–	–
PTV$_{70}$	–	–	70.0 Gy
PTV$_{59.4}$	–	–	59.4 Gy

The prescription dose for the target volumes and tolerance doses for the OARs considered in the optimization are presented in Table 1.

2.2 FMO Model

For optimization purposes, radiation dose distribution deposited in the patient, measured in Gray (Gy), needs to be assessed accurately. Each structure's volume is discretized in voxels (small volume elements) and the dose is computed for each voxel using the superposition principle, i.e., considering the contribution of each beamlet. Typically, a dose matrix D is constructed from the collection of all beamlet weights, by indexing the rows of D to each voxel and the columns to each beamlet, i.e., the number of rows of matrix D equals the number of voxels (V) and the number of columns equals the number of beamlets (N) from all beam directions considered. Therefore, using matrix format, we can say that the total dose received by the voxel i is given by $\sum_{j=1}^{N} D_{ij}w_j$, with w_j the weight of beamlet j. Usually, the total number of voxels considered reaches the tens of thousands, thus the row dimension of the dose matrix is of that magnitude. The size of D originates large-scale problems being one of the main reasons for the difficulty of solving the FMO problem.

Many mathematical optimization models and algorithms have been proposed for the FMO problem, including linear models [18], mixed integer linear models [9], nonlinear models [5], and multi-criteria models [3]. Most of the FMO models in the literature belong to a class of constrained optimization models such that an objective function is optimized while meeting dose requirements. A variety of criteria may be considered to be included in the objective function, leading to many different objective functions. It is beyond the scope of this study to discuss which formulation of the FMO problem is preferable. Romeijn *et al.* [19] demonstrated that most of the treatment plan evaluation criteria proposed in the medical physics literature are equivalent to convex penalty function criteria when viewed as a multicriteria optimization problem. Here, we will use a convex penalty function voxel-based nonlinear model [1]. The conclusions drawn regarding this particular model are valid also if different weighted mathematical formulations to model the FMO problem are considered. In this model, each voxel

• is penalized according to the square difference of the amount of dose received by the voxel and the amount of dose desired/allowed for the voxel. This formulation yields a quadratic programming problem with only linear non-negativity constraints on the fluence values [18]:

$$\min_w \left[\underline{\lambda}_i \left(T_i - \sum_{j=1}^{N} D_{ij} w_j \right)_+^2 + \overline{\lambda}_i \left(\sum_{j=1}^{N} D_{ij} w_j - T_i \right)_+^2 \right]$$

$$s.t. \quad 0 \leq w_j \leq w^{max}, \; j = 1, \ldots, N,$$

where T_i is the desired dose for voxel i, $\underline{\lambda}_i$ and $\overline{\lambda}_i$ are the penalty weights of underdose and overdose of voxel i, w^{max} is the maximum beamlet intensity allowed and $(\cdot)_+ = \max\{0, \cdot\}$. This nonlinear formulation implies that a very small amount of underdose or overdose may be accepted in clinical decision making, but larger deviations from the desired/allowed doses are decreasingly tolerated [1].

The optimal solutions obtained ensure that the resulting treatment is the best possible with respect to the weighting parameters (λ) used. Since it is impossible to attribute effective clinical meaning to the weight parameters, the 'optimal' weighting scheme is unknown and the choice of the weights is typically a long trial-and-error process until a satisfactory solution is achieved. Furthermore, for beam angle optimization it is not clear how traditional trial-and-error parameter tuning should be incorporated or managed.

2.3   BAO Approach

In order to model the BAO problem as a mathematical programming problem, a quantitative measure to compare the quality of different sets of beam angles is required. For the reasons presented before, our approach for modelling the BAO problem uses the optimal solution value of the FMO problem as measure of the quality of a given beam angle set. Many authors consider non-coplanar angles [2,4,7,11–13] which result in potentially improved treatment plans [14, 22]. However, despite the fact that almost every angle is possible for radiation delivery, the use of coplanar angles is predominant. For simplicity, only coplanar angles will be considered.

Let us consider n to be the fixed number of (coplanar) beam directions, i.e., n beam angles are chosen on a circle around the CT-slice of the body that contains the isocenter (usually the center of mass of the tumor). In our formulation we consider all continuous $[0°, 360°]$ gantry angles instead of a discretized sample. Since for $\alpha, \beta \in [0°, 360°]$, the angle $360° + \alpha$ is the same as the angle $\alpha \in [0°, 360°]$ and the angle $-\beta$ is equivalent to the angle $360° - \beta \in [0°, 360°]$, we can avoid a bounded formulation. A basic formulation for the BAO problem is obtained by selecting an objective function such that the best set of beam angles is obtained for the function's minimum:

$$\min f(\theta_1, \ldots, \theta_n)$$

$$s.t. \ \theta_1, \ldots, \theta_n \in \mathbb{R}^n.$$

Here, the objective $f(\theta_1, \ldots, \theta_n)$ that measures the quality of the set of beam directions $\theta_1, \ldots, \theta_n$ is the optimal value of the FMO problem for each fixed set of beam directions.

This BAO formulation facilitates the use of pattern search methods (PSM). PSM are derivative-free optimization algorithms that require few function evaluations to progress and converge and have the ability to better avoid local entrapment making them a suitable approach for the resolution of the highly non-convex BAO problem [15–17]. PSM are directional search methods that use positive bases to move in a direction that produces a decrease in the objective function. The main feature of a positive basis, that motivates PSM, is that for any given vector, in particular for the gradient vector, there is a vector of the positive basis that forms an acute angle with the gradient vector which means that it is a descent direction. PSM are organized around two phases at every iteration: one that assures convergence to a local minimum (poll), and the other (search) where flexibility is conferred to the method allowing searches away from the neighborhood of the current iterate. PSM are used to address the BAO problem.

2.4 Computational Tests

Our tests were performed on a 2.66Ghz Intel Core Duo PC with 3 GB RAM. The computational tools developed within MATLAB and CERR – computational environment for radiotherapy research [8] – were used to obtain the dosimetric data input for treatment plan optimization and also to facilitate convenient access, visualization and analysis of patient treatment planning data. To address the convex nonlinear formulation of the FMO problem we used a trust-region-reflective algorithm (*fmincon*) of MATLAB 7.4.0 (R2007a) Optimization Toolbox. A tailored version of pattern search methods that include beam s-eye-view dose metrics in the search step so that directions with larger dose metric scores are tested first improving results and computational time [16] was used to tackle the BAO problem.

The FMO problem is inherently a multicriteria optimization problem with conflicting objectives. Despite the convex nonlinear formulation being commonly used for FMO and BAO, it requires the subjective decision of assigning penalty weights to be used, which is an handicap. A set of parameter that produces acceptable treatment plans for the equispaced beam angle configuration [16] corresponds to assign $\underline{\lambda}_i = \overline{\lambda}_i = 4$ to the target volumes and $\overline{\lambda}_i = 2$ to the OARs and remaining tissue. For this set of penalty weight parameters, an optimal beam angle set was obtained for each patient using our BAO approach and denoted *BAO*. For a unitary set of penalty weight parameters, i.e. considering the previous parameters equal to one, an optimal beam angle set was also obtained for each patient using our BAO approach and denoted *BAO1*. For each of the 10 patients, three different 7-beam angle configurations – *BAO*, *BAO1* and the equispaced configuration denoted *Equi* – were tested considering 100 different fluence map objective function weight parameters selected randomly in an automated way in the interval [2,8] for tumors and [1,4] for OARS and remaining tissue.

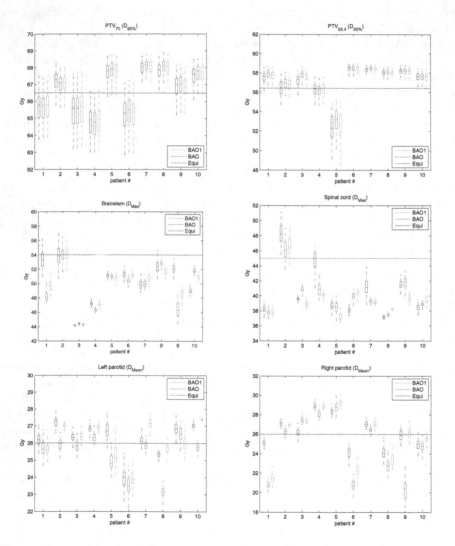

Fig. 1. Target coverage and organ sparing metric values of 100 treatment plans for each patient and for the three beam angle configurations tested: *Equi*, *BAO* and *BAO1*

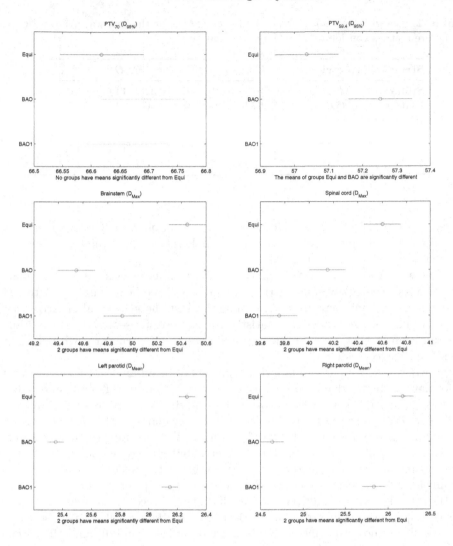

Fig. 2. Multiple comparison interactive graph of the overall target coverage and organ sparing metrics mean and confidence interval differences between the three beam angle configurations tested: *Equi, BAO* and *BAO1*

3   Results

Treatment plans are typically compared by judging their cumulative dose-volume histograms (DVHs) and/or by comparing different metrics that usually include the dose delivered to 95% of the tumor ($D_{95\%}$) and the mean (D_{mean}) or maximum doses (D_{max}) of OARs. Boxplots of the metric values considering the 100 treatment plans corresponding to the different penalty weight parameters for each of the 10 patients are displayed in Fig. 1 comparing the performance of the

Table 2. Mean values and corresponding $p-value$ of the one-way Anova for the structure metrics considered in IMRT optimization.

Structure (metric)	Equi	BAO	BAO1	p − value
Spinal cord (D_{max})	40.6015 Gy	40.1501 Gy	39.7543 Gy	< 0.001
Left parotid (D_{mean})	26.2670 Gy	25.3533 Gy	26.1470 Gy	< 0.001
Right parotid (D_{mean})	26.1721 Gy	24.6396 Gy	25.8298 Gy	< 0.001
Brainstem (D_{max})	50.4473 Gy	49.5464 Gy	49.9163 Gy	< 0.001
PTV$_{70}$ $(D_{95\%})$	66.6182 Gy	66.6899 Gy	66.6643 Gy	0.5017
PTV$_{59.4}$ $(D_{95\%})$	57.0359 Gy	57.2516 Gy	57.1582 Gy	0.0242

three beam angle sets. By simple inspection we can observe that organ sparing metrics are better for the optimized beam angle sets, in particular for BAO, for similar target coverage metric values.

One-way Anova (using MATLAB Statistics toolbox) was used to compare the means of the aggregated metric results of the three beam angle sets, to test the hypothesis that they are all the same, against the general alternative that they are not all the same. The statistical results are presented in Table 2 and we can confirm that metric means are different for all structures with statistic significance except for PTV70.

Since we are comparing more than two sets of beam angles, we need further information about which pairs of means are significantly different, and which are not. A multiple comparison procedure using the Tukey's honestly significant difference criterion was performed (post-hoc test available in MATLAB Statistics toolbox) and the output is displayed in Fig. 2. We can observe that there are no significant differences for the PTV70 metric but for the remaining structures, the optimized beam angle set obtained always better results with statistical significance $(p-value < 0.001)$. Furthermore, the optimized beam angle set BAO, obtained with weight parameters that lead to acceptable treatment plans for the equispaced configuration, clearly presented the overall better results.

We should note that mean results must be judged carefully and their main purpose is to compare overall results. For instance, mean results for the left parotid displayed in Table 2 show that BAO obtained an average sparing of 1.5 Gy compared to equispaced configurations for the 10 patients. However, for some patients the improvement is residual while for others is important (up to 6 Gy) – see Fig. 2. In an ideal forthcoming automated treatment planing, where computational time and effort are less limitative, results should be evaluated case by case and if BAO makes a strong impact in a single patient being better in overall average, that should prove its importance.

4   Discussion and Conclusions

The benefits of BAO are yet to be fully acknowledged in clinical practice. Regardless the evidence presented in the literature that appropriate radiation beam incidence directions can theoretically lead to a plan's quality improvement, in practice, the choice of beam angles is, most of the time, simply regarded as just another set of parameters that need to be selected and whose update can lead to the exact same unpredictable result as changing the weight parameters of many TPS weighted mathematical formulation to model the FMO problem. Updating beam incidence directions is even more risky for IMRT since it has been extensively reported that optimal beam angles for IMRT are often non-intuitive.

This study aims to be a contribute for the perception of the interest of BAO in IMRT by learning if the update of the fluence map objective function parameters jeopardize an initial BAO effort and if there is a timing to perform BAO in order to diminish potential result's deterioration. For our study set of ten clinical cases of head-and-neck tumors treated at the Portuguese Institute of Oncology of Coimbra we showed that optimized beam angle sets obtained statistical significant better results than the traditional equispaced configuration in terms of target coverage and organ sparing metrics. Furthermore, in our tests, despite optimal beam angle sets being always preferable to equispaced configurations, the choice of the weight parameters for angle selection showed influence on the correspondent plan quality. Therefore, the first result of this study lead to the conclusion that optimized beam angles can lead to better treatment plans than the traditional equispaced beam angles. The second result indicate that BAO should be performed only when a set of parameters that lead to an acceptable treatment plan for the equispaced beam angle configuration is found. These conclusions fully answer the questions raised in the introduction of the paper/study.

Finally, this study further validates our BAO approach. To the best of our knowledge, this is the first study that compares different beam angle sets using many treatment plans corresponding to different FMO weight parameters. When comparing optimized beam angle sets with the equispaced beam angle configuration using a single set of FMO weight parameters, the conclusions can be biased by the fact that larger weight parameters may lead the BAO procedure to obtain better results for the corresponding structures. Therefore, this study also contributes to a fair comparison of beam angle sets when FMO problem is a weighted mathematical formulation.

Acknowledgments. This work was supported by FEDER. COMPETE. iCIS (CENTRO-07-ST24-FEDER-002003). Portuguese Foundation for Science and Technology under project grants UID/MULTI/00308/2013 and PTDC/EIA-CCO/ 121450/2010.

References

1. Aleman, D.M., Kumar, A., Ahuja, R.K., Romeijn, H.E., Dempsey, J.F.: Neighborhood search approaches to beam orientation optimization in intensity modulated radiation therapy treatment planning. J. Global Optim. **42**, 587–607 (2008)
2. Aleman, D.M., Romeijn, H.E., Dempsey, J.F.: A response surface approach to beam orientation optimization in intensity modulated radiation therapy treatment planning. INFORMS J. Comput.: Computat. Biol. Med. Appl. 21, 62–76 (2009)

3. Breedveld, S., Storchi, P.R.M., Keijzer, M., Heemink, A.W., Heijmen, B.J.M.: A novel approach to multi-criteria inverse planning for IMRT. Phys. Med. Biol. **52**, 6339–6353 (2007)

4. Breedveld, S., Storchi, P., Voet, P., Heijmen, B.: iCycle: integrated, multicriterial beam angle, and profile optimization for generation of coplanar and noncoplanar IMRT plans. Med. Phys. **39**, 951–963 (2012)

5. Cheong, K., Suh, T., Romeijn, H., Li, J., Dempsey, J.: Fast Nonlinear Optimization with Simple Bounds for IMRT Planning. Med. Phys. **32**, 1975 (2005)

6. Craft, D., Monz, M.: Simultaneous navigation of multiple Pareto surfaces, with an application to multicriteria IMRT planning with multiple beam angle configurations. Med. Phys. **37**, 736–741 (2010)

7. Das, S.K., Marks, L.B.: Selection of coplanar or non coplanar beams using three-dimensional optimization based on maximum beam separation and minimized non-target irradiation. Int. J. Radiat. Oncol. Biol. Phys. **38**, 643–655 (1997)

8. Deasy, J.O., Blanco, A.I., Clark, V.H.: CERR: A Computational Environment for Radiotherapy Research. Med. Phys. **30**, 979–985 (2003)

9. Lee, E.K., Fox, T., Crocker, I.: Integer programming applied to intensity-modulated radiation therapy treatment planning. Ann. Oper. Res. **119**, 165–181 (2003)

10. Liu, H.H., Jauregui, M., Zhang, X., Wang, X., Dongand, L., Mohan, R.: Beam angle optimization and reduction for intensity-modulated radiation therapy of non-small-cell lung cancers. Int. J. Radiat. Oncol. Biol. Phys. **65**, 561–572 (2006)

11. Lu, H.M., Kooy, H.M., Leber, Z.H., Ledoux, R.J.: Optimized beam planning for linear accelerator-based stereotactic radiosurgery. Int. J. Radiat. Oncol. Biol. Phys. **39**, 1183–1189 (1997)

12. Meedt, G., Alber, M., Nüsslin, F.: Non-coplanar beam direction optimization for intensity-modulated radiotherapy. Phys. Med. Biol. **48**, 2999–3019 (2003)

13. Mišić, V.V., Aleman, D.M., Sharpe, M.B.: Neighborhood search approaches to non-coplanar beam orientation optimization for total marrow irradiation using IMRT. Eur. J. Oper. Res. **205**, 522–527 (2010)

14. Pooter, J.A., Romero, A.M., Jansen, W.P.A., Storchi, P.R.M., Woudstra, E., Levendag, P.C., Heijmen, B.J.M.: Computer optimization of noncoplanar beam setups improves stereotactic treatment of liver tumors. Int. J. Radiat. Oncol. Biol. Phys. **66**, 913–922 (2006)

15. Rocha, H., Dias, J.M., Ferreira, B.C., Lopes, M.C.: Selection of intensity modulated radiation therapy treatment beam directions using radial basis functions within a pattern search methods framework. J. Glob. Optim. **57**, 1065–1089 (2013)

16. Rocha, H., Dias, J.M., Ferreira, B.C., Lopes, M.C.: Beam angle optimization for intensity-modulated radiation therapy using a guided pattern search method. Phys. Med. Biol. **58**, 2939–2953 (2013)

17. Rocha, H., Dias, J.M., Ferreira, B.C., Lopes, M.C.: Pattern search methods framework for beam angle optimization in radiotherapy design. Appl. Math. Comput. **219**, 10853–10865 (2013)

18. Romeijn, H.E., Ahuja, R.K., Dempsey, J.F., Kumar, A., Li, J.: A novel linear programming approach to fluence map optimization for intensity modulated radiation therapy treatment planing. Phys. Med. Biol. **48**, 3521–3542 (2003)

19. Romeijn, H.E., Dempsey, J.F., Li, J.: A unifying framework for multi-criteria fluence map optimization models. Phys. Med. Biol. **49**, 1991–2013 (2004)

20. Stein, J., Mohan, R., Wang, X.H., Bortfeld, T., Wu, Q., Preiser, K., Ling, C.C., Schlegel, W.: Number and orientation of beams in intensity-modulated radiation treatments. Med. Phys. **24**, 149–160 (1997)
21. Voet, P., Breedveld, S., Dirkx, M., Levendag, P., Heijmen, B.: Integrated multi-criterial optimization of beam angles and intensity profiles for coplanar and non-coplanar head and neck IMRT and implications for VMAT. Med. Phys. **39**, 4858 (2012)
22. Wang, X., Zhang, X., Dong, L., Liu, H., Gillin, M., Ahamad, A., Ang, K., Mohan, R.: Effectiveness of noncoplanar IMRT planning using a parallelized multiresolution beam angle optimization method for paranasal sinus carcinoma. Int. J. Radiat. Oncol. Biol. Phys. **63**, 594–601 (2005)

Solving Constrained Multilocal Optimization Problems with Parallel Stretched Simulated Annealing

Ana I. Pereira[1,2($\boxtimes$)] and José Rufino[1,3]

[1] Polytechnic Institute of Bragança, 5301-857 Bragança, Portugal
{apereira,rufino}@ipb.pt
[2] Algoritmi R&D Centre, University of Minho, Campus de Gualtar,
4710-057 Braga, Portugal
[3] Laboratory of Instrumentation and Experimental Particle Physics,
University of Minho, Campus de Gualtar, 4710-057 Braga, Portugal

Abstract. Constrained multilocal programming optimization problems may be solved by solving a sequence of unconstrained problems. In turn, those unconstrained problems may be solved using techniques like the Stretched Simulated Annealing (SSA) method. In order to increase the solving performance and make possible the discovery of new optima, parallel approaches to SSA have been devised, like Parallel Stretched Simulated Annealing (PSSA). Recently, Constrained PSSA (coPSSA) was also proposed, coupling the penalty method with PSSA, in order to solve constrained problems. In this work, coPSSA is explored to solve four test problems using the l_1 penalty function. The effect of the variation of the reduction factor parameter of the l_1 penalty function is also studied.

Keywords: Unconstrained optimization · Parallel computing

1 Introduction

Multilocal programming aims to identify all local maximizers of unconstrained or constrained nonlinear optimization problems. More formally, a constrained multilocal programming problem may be defined by the following formulation:

$$
\begin{aligned}
\max \ & f(x) \\
\text{s.t. } & h_k(x) = 0, \ k \in E \\
& g_j(x) \le 0, \ j \in I \\
& -l_i \le x_i \le l_i, \ i = 1, \dots, n
\end{aligned}
\tag{1}
$$

where at least one of the n-dimensional functions $f, h_k, g_j : \mathbb{R}^n \to \mathbb{R}$ is nonlinear, and E and I are index sets of equality and inequality constraints, respectively. Since concavity is not assumed, the nonlinear optimization problem can have many global and local (non-global) maxima. Consider the feasible region (search space) defined by $\mathcal{R} = \{x \in \mathbb{R}^n : -l_i \le x_i \le l_i, i = 1, \dots, n; \ h_k(x) = 0, k \in$

© Springer International Publishing Switzerland 2015
O. Gervasi et al. (Eds.): ICCSA 2015, Part II, LNCS 9156, pp. 534–548, 2015.
DOI: 10.1007/978-3-319-21407-8_38

E; $g_j(x) \leq 0, j \in I$}. Thus, the purpose of the maximization problem (1) is to find all local maximizers, i.e., all points $x^* \in \mathcal{R}$ such that condition (2) holds:

$$\forall x \in V_\epsilon(x^*) \cap \mathcal{R}, \ f(x^*) \geq f(x). \tag{2}$$

where $V_\epsilon(x^*)$ is a neighborhood of x^*, with a positive ray ϵ.

It is also assumed that problem (1) has a finite number of isolated global and local maximizers. The existence of multi solutions (local and global) makes this problem a great challenge that may be tackled with parallel solving techniques.

Methods for solving multilocal optimization problems include evolutionary algorithms, such as genetic [1] and particle swarm [13] algorithms, and additional contributions, like [6,15,20,23,24]. Stretched Simulated Annealing (SSA) was also proposed [14], combining simulated annealing and a stretching function technique, to solve unconstrained multilocal programming problems.

In previous work [16,18], Parallel Stretched Simulated Annealing (PSSA) was introduced as a parallel version of SSA, based on the decomposition of the feasible region in several subregions to which SSA is independently applied by a set of processors. Several domain decomposition and distribution approaches were explored, leading to successively increasing levels of numerical efficiency.

More recently, the parallel solving of constrained multilocal programming problems was also proposed, through coPSSA [19] (constrained PSSA), that couples the penalty method with PSSA. Basically, coPSSA creates a homogeneous partition of the iteration set of the l_1 penalty function [11]; each iteration subset is run in parallel, by different processors of a shared memory system, and each specific iteration invokes PSSA; this, in turn, involves additional processors; these processors are usually from a distributed memory cluster, but may also be from the same shared memory host, once PSSA is a MPI-based application [8].

In this paper, coPSSA is explored, with the l_1 penalty function, to solve four well known test problems [4], in order to analyze the kind of performance gains that may be expected with a reasonable set of parallel configurations. Moreover, the effect of the variation of the reduction factor parameter τ of the l_1 function is analyzed to investigate the existence of values leading to faster convergence.

The rest of the paper is organized as follows. Section 2 revises the basic ideas behind SSA and PSSA. Section 3 covers the basics of penalty method with l_1 penalty function, and provides some details on coPSSA design and implementation. Section 4 presents performance and numerical results from the evaluation of coPSSA. Finally, Section 5 concludes and defines directions for future work.

2   Unconstrained Optimization

2.1   Stretched Simulated Annealing

Stretched Simulated Annealing (SSA) is a multilocal programming method that solves bound constrained optimization problems. These may be described as:

$$\max_{x \in X} \varphi(x), \tag{3}$$

where $\varphi : \mathbb{R}^n \to \mathbb{R}$ is a given n-dimensional multimodal objective function and X is the feasible region defined by $X = \{x \in \mathbb{R}^n : -l_i \leq x_i \leq l_i, i = 1, ..., n\}$.

SSA solves a sequence of global optimization problems in order to compute the local solutions of the maximization problem (3). The objective function of each global problem is generated using a stretching function technique [12].

Let x^* be a solution of problem (3). The mathematical formulation of the global optimization problem is as follows:

$$\max_{x \in X} \Phi_l(x) \equiv \begin{cases} \hat{\phi}(x) \text{ if } x \in V_\varepsilon(x^*) \\ \varphi(x) \text{ otherwise} \end{cases} \quad (4)$$

where $V_\varepsilon(x^*)$ is the neighborhood of solution x^* with a ray $\varepsilon > 0$.

The $\hat{\phi}(x)$ function is defined as

$$\hat{\phi}(x) = \bar{\phi}(x) - \frac{\delta_2[\text{sign}(\varphi(x^*) - \varphi(x)) + 1]}{2\tanh(\kappa(\bar{\phi}(x^*) - \bar{\phi}(x)))} \quad (5)$$

where δ_1, δ_2 and κ are positive constants, and $\bar{\phi}(x)$ is

$$\bar{\phi}(x) = \varphi(x) - \frac{\delta_1}{2}\|x - x^*\|[\text{sign}(\varphi(x^*) - \varphi(x)) + 1]. \quad (6)$$

To solve the global optimization problem (4) the Simulated Annealing (SA) method is used [5]. The Stretched Simulated Annealing algorithm stops when no new optimum is identified after r consecutive runs. [15, 16] provide more details.

2.2   Parallel Stretched Simulated Annealing

As a parallel implementation of the SSA method, PSSA was thoroughly described in previous work [16]. In this section, only a brief description is provided.

SSA applies a stochastic algorithm, ι successive times, in the feasible region of the bound constrained problem. The same algorithm may be applied to the subregions of a partition of the feasible region, with the aim of improving the number of optima found. Moreover, each subregion may be processed independently, once there are no data or functional dependencies involved. SSA is thus an embarrassingly parallel problem, calling for a Domain Decomposition approach.

In this regard, PSSA supports both homogeneous and heterogeneous decomposition of the feasible region, as well as static or dynamic assignment of subregions to processors. The various combinations of these possibilities are materialized in three PSSA variants: i) PSSA-HoS (Homogeneous decomposition, Static assignment), where subregions are defined only once, have equal size and processors self assign the same number of subregions; ii) PSSA-HoD (Homogeneous decomposition, Dynamic assignment), different from PSSA-HoS only in subregions being assigned to processors on-request (thus possibly in varying number); iii) PSSA-HeD (Heterogeneous decomposition, Dynamic assignment), based on an adaptive recursive refinement of an initial homogeneous partition of the feasible region, leading to an unpredictable number of subregions, of variable size, dynamically generated and processed on-demand, until certain stop criteria are met. PSSA-IIoS and PSSA-HoD have the same numerical efficiency (*i.e.*, both find the same number of optima), but PSSA-HoD is faster due to its workload auto-balancing. PSSA-HeS usually finds more optima, but is also the slowest of the PSSA variants, once it typically searches many more subregions.

PSSA was written in C, runs on Linux systems, and it builds on MPI [8] (thus following the message passing paradigm). It is a SPMD (Single Program, Multiple Data) application that operates in a *master-slaves* configuration (*slave* tasks run SSA in subregions, under coordination of a *master* task), and may be deployed in shared-memory systems and/or in distributed-memory clusters.

3   Constrained Optimization

3.1 l_1 Penalty Method

There are three main classes of methods to solve constrained optimization problems [9,25]: 1) methods that use penalty functions, 2) methods based on biasing feasible over infeasible solutions, and 3) methods that rely on multi-objective optimization concepts. In this work constraints are handled using a class 1 method with the l_1 penalty function. This function is a classic penalty [11] defined by

$$\varphi(x,\mu) = f(x) - \frac{1}{\mu}\left[\sum_{k \in E}|h_k(x)| + \sum_{j \in I}[g_j(x)]^+\right]$$

where μ is a positive penalty parameter that progressively decreases to zero, along k_{max} iterations. A lower bound $\mu_{\min}$ is defined and μ is updated as follows:

$$\mu^{k+1} = \max\left\{\tau\mu^k, \mu_{\min}\right\} \tag{7}$$

where $k \in \{1, ..., k_{max}\}$ represents the iteration, $\mu_{\min} \approx 0$ and $0 < \tau < 1$.

To solve the constrained problem (1), the penalty method solves a sequence of bound constrained problems, based on the l_1 penalty function, as defined by

$$\max_{x \in X} \varphi(x, \mu^k). \tag{8}$$

Problem (8) is solvable using PSSA. It is possible to prove that the solutions sequence $\{x^*(\mu^k)\}$ from (8) converges to the solution x^* of problem (1) [11,15].

The penalty method stops when a maximum number of iterations (k_{max}) is reached, or successive solutions are similar, accordingly with the next criteria:

$$\left|f(x^k) - f(x^{k-1})\right| \leq \epsilon_1 \ \wedge \ \left\|x^k - x^{k-1}\right\| \leq \epsilon_2 \tag{9}$$

where k is a given iteration of the penalty method.

The optimum value for the reduction factor τ of the update expression (7) is an open research issue in the optimization field [10,21]. This paper presents an analysis of the effect of the variation of τ in the l_1 penalty function method, when solving multilocal programming problems by PSSA. The main goal is to identify the best value τ in order to obtain the optima set in the shortest time.

3.2 Constrained PSSA

Constrained PSSA (coPSSA) was already introduced in [19], albeit in the context of a simpler evaluation scenario than the one explored in this paper. Here coPSSA is revised, and some clarification is provided on its design and implementation.

As stated in the previous section, problem (8) may be solved using SSA, including its parallel implementations, like PSSA. However, coupling PSSA with a serial implementation of the penalty method is of limited benefit performance-wise: in the end, some extra optima may be detected (due to the extra efforts of PSSA), but the penalty method will not run faster. A possible approach to increase the performance of the penalty method is to parallelize its execution.

As it happens, the parallelization of the penalty method is trivial: this method will execute a certain maximum number of times or iterations, as defined by the parameter k_{max}; each iteration uses successively lower values of the penalty parameter μ; these values are completely deterministic, as given by (7); therefore, a partition of the iteration space $\{1, ..., k_{max}\}$ may be defined, a priori, such that mutually disjoint subintervals of this space are assigned to different processors or search tasks (with one task per processor); each task will then run the penalty method along its iteration subspace; as soon as a task reaches convergence (in accordance to criteria (9)), the others will stop the search (before moving on to its next iteration, a task checks if other has already converged to the solutions).

Having different processors/tasks starting the optima search with different values of μ, scattered along the interval of possible values $\{\mu^1, ..., \mu^{k_{max}}\}$, may allow to reach convergence sooner, as compared to a single full sequential scan of that interval. However, it was noted that this strategy does not necessarily pay off; it must be tested for each and different constrained optimization problem, once the value(s) of μ that lead to convergence are unpredictable by nature.

coPSSA adopts the strategy above, for the parallelization of the penalty method, by performing a homogeneous decomposition of the iteration space: given P search tasks and k_{max} iterations, any search task t_p (with $p = 1, ..., P$) will iterate through approximately the same number of iterations, $w = \lceil \frac{k_{max}}{P} \rceil$, where $\lceil x \rceil$ is the smallest integer value not less than x; the iteration subinterval for a task t_p is then $\{k_{left}^p, ..., k_{right}^p\}$, where $k_{left}^p = ((p-1) \times w) + 1$ and $k_{right}^p = p \times w$; when a uniform width w is not possible ($k_{max} \bmod P \neq 0$), coPSSA makes the necessary adjustments in the last iteration subinterval (the one for task p); for instance, with $k_{max} = 100$ and $P = 2$, it is obtained $w = \lceil 50.0 \rceil = 50$ and so t_1 will iterate through $\{1, ..., 50\}$ and t_2 will iterate through $\{51, ..., 100\}$; with $P = 3$ it is obtained $w = \lceil 33.3(3) \rceil = 34$, and so t_1, t_2 and t_3 will iterate through $\{1, ..., 34\}$, $\{35, ..., 68\}$ and $\{69, ..., 100\}$, respectively.

Like PSSA, coPSSA was also coded in C, for performance reasons, and also to reuse previous code from one of the authors. In coPSSA, search tasks are conventional UNIX processes, that synchronize and exchange data using classical System V IPC mechanisms, like semaphores and shared memory[1]. When coPSSA

---

[1] Alternatives like Pthreads, OpenMP or even MPI, may be explored in the future.

starts, it forks P child processes (one per search task), where P is usually defined to be the number of CPU-cores of the host running coPSSA. Each child (search task) will then run the penalty method through a specific iteration subinterval; for each particular iteration, PSSA must be invoked; this is simply accomplished using the `system` primitive, that spawns second-level childs to run the `mpiexec` command[2]; in turn, this command runs PSSA, in a MPI master-slaves configuration, in a set of hosts / CPU-cores that are defined by coPSSA; basically, coPSSA is supplied with a bare MPI "machinefile", from which extracts the computing resources to be assigned to each PSSA execution; the results produced by PSSA executions are written in specific files, which are then checked by the search tasks, in order to detect a possible convergence.

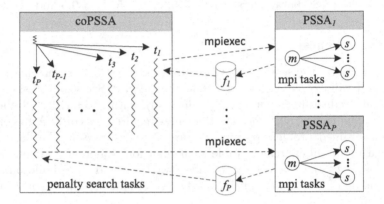

Fig. 1. The coPSSA application and its interactions with PSSA (t_p are coPSSA search tasks; m and s are PSSA tasks (master and slaves); f_p are result files from PSSA)

Figure 1 is a representation of coPSSA, including its interactions with PSSA. coPSSA may thus be viewed as a hybrid application, in the sense that combines its internal usage of shared-memory based parallelism, with an external component (PSSA) specially suited to exploit distributed-memory parallelism.

4   Evaluation

4.1   Setup

The experimental evaluation performed in the context of this work was carried in a commodity cluster of 9 hosts (1 frontend, and 8 worker nodes), with one Intel Core-i7 4790K 4.0GHz quad-core CPU per node, under Linux ROCKS version 6.1.1, with the Gnu C Compiler (GCC) version 4.4.7 and OpenMPI version 1.5.4.

coPSSA was always executed in the cluster frontend, with the number of search tasks (P) ranging from 1 to 8 (no overload was observed for $P > 4$,

---

[2] As a side note, we had to switch the MPI implementation, from MPICH2 to Open-MPI, because a bug in MPICH2 corrupts *stdio* in coPSSA after `mpiexec` returns.

despite the frontend having only a 4-core CPU). PSSA executions took place in the 8 worker nodes; these offer a total of 32 CPU-cores, fully specified in the base MPI hostfile supplied to coPSSA, that are used, four at a time, to service the PSSA execution requests of each coPSSA task; thus, each PSSA execution always consumed 4 cores, with 1 core for the master process, and 3 cores for slave processes; in order to fully utilize these 3 cores, the number of subregions for each PSSA execution was set to be as close as possible (in excess) of 3; for 2-dimensional problems, like the ones tested, this implies 4 subregions [16]. The PSSA variant used throughout the tests was always PSSA-HoD: not only it is appropriate for a fixed number of subregions, but is also the fastest variant.

All tests shared the penalty algorithm parameters $k_{max}{=}100$, $\mu^0{=}1.0$, $\mu_{min} = 10^{-6}$ and $\tau \in \{0.1, ..., 0.9\}$, and the convergence parameters $\epsilon_1 = \epsilon_2 = 10^{-4}$. The PSSA numerical parameters were $r = 5$, $\delta_1 = 1.5$, $\delta_2 = 0.5$ and $\kappa = 0.05$.

4.2   Search Times

Tables 4 to 7 show the search times measured for the selected benchmark problems, for all valid combinations of P (number of coPSSA search tasks) and τ (the penalty reduction factor under evaluation), with $P = 1, 2, ..., 8$ and $\tau = 0.1, 0.2, ..., 0.9$. The search times are the times required by coPSSA to converge to the constrained problem optima. Figures 2 to 5 represent the data of the tables, with four subfigures per table, that offer four different and complementary perspectives on search times (the first two – (a) and (b) – build on a horizontal reading of the table, and the last two – (c) and (d) – build on a vertical reading):

(a) "Search times for each τ": one curve per τ value, based on the search times obtained with a fixed τ, when varying the number P of search tasks; allows to verify the influence of the variation of the degree of search parallelism (P) on the search times produced by a specific τ; allows also to identify the τ value that ensures the lowest (absolute minimum) search times;

(b) "Average search times for $P \geq 1$ and $P \geq 2$": one curve, where each point is the average of the search times obtained with a fixed τ and all possible values of P (*i.e.*, $P \geq 1$); another curve, where each point is the average of the search times obtained with a fixed τ and all values of $P \geq 2$; the first curve allows to deduce the best τ, irregardless of the use of sequential ($P = 1$) or parallel ($P \geq 2$) searches; the second curve allows to conclude which τ value is the best when only parallel searches are made;

(c) "Search times for each P": one curve per P value, based on the search times obtained with a fixed P, when varying τ; allows to verify the influence of the variation of the value of τ on the search times produced by a specific P; allows also to identify the P value that ensures the lowest search times;

(d) "Average search times for all τ values": a single curve, where each point is the average of the search times obtained with a fixed P and all possible values of τ; allows to deduce the best P, irregardless of the value of τ.

For each of the above perspectives, these are the main conclusions that may be derived from the experimental data[3]:

(a) in all problems, when increasing the number of search tasks, the trend followed by the search times for each τ is mostly downwards, eventually followed by stabilization; this trend is more regular for Problems G6 and G8 (see Figures 3a and 4a), slightly less regular in Problem G12 (see Figure 5a), and much more irregular in Problem G11 (see Figure 4b). It is also possible to identify the values of τ that attain the absolute minimum search times; these times are boxed (within a tolerance of 5%), in Tables 4 to 7; the corresponding values of τ are gathered in the following table:

Table 1. Values of τ that ensure the absolute minimum search times

Problem	$P = 1$	$P \geq 2$
G6	0.1	0.5 , 0.6
G8	0.1	0.7
G11	0.3	0.9
G12	0.1	0.6 to 0.9

the previous table allows to conclude that when only sequential searches are used ($P = 1$), the absolute minimum values for search times are attained with small values of τ; however, if only parallel searches are conducted ($P \geq 2$), the lowest search times are achieved with higher (mid range to maximum) values of τ; one should note, though, that these observations cannot be generalized, because the τ values that produce the absolute minimum search times may not be the ones that produces the lowest average search times;

(b) the values of τ that ensure the lowest average search times depend on the specific problem, and also depend on whether both sequential or parallel searches are admitted ($P \geq 1$), or only parallel searches ($P \geq 2$) are allowed; in Tables 4 to 7, the columns "avg. $P \geq 1$" and "avg. $P \geq 2$" show the average search times attained by each τ, when $P \geq 1$ and $P \geq 2$, respectively; in each column, the lowest average search time is in bold (within a tolerance of 5%); the related values of τ are gathered in the following table:

Table 2. Values of τ that ensure the lowest average search times

Problem	$P \geq 1$	$P \geq 2$
G6	0.1	0.1 to 0.6
G8	0.1	0.8 , 0.9
G11	0.3	0.3
G12	0.1	0.1 to 0.6

the previous table allows to conclude that if using a sequential or a parallel search is indifferent ($P \geq 1$), then single small values of τ are best in order

---

[3] Note: for Problem G6, no values are shown for $\tau = 0.9$, because no convergence was reached within the maximum of $k_{max} = 100$ iterations.

to attain the lowest search times, on average (coincidentally, such values of τ match the values in Table 1 that ensure the absolute minimum search times for $P = 1$); however, if only parallel searches are admissible ($P \geq 2$), there may be several best values of τ to chose from, like in problems G6 and G12, that share the same range of values (0.1 to 0.6), or like in problem G8, with another range (0.8 to 0.9); problem G11, though, is an exception, because the best τ value is always the same (0.3) for sequential and parallel searches;

(c) when increasing τ, problem G6 (Figure 2c) and problem G12 (Figure 5c) exhibit a similar trend on the search times for each P (these times progressively increase for $P = 1$, but for $P \geq 2$ they are substantially lower and stable, and only start to increase for higher values of τ); for problem G8 (Figure 3c), the trend is also upwards for $P = 1$, but now is reversal for $P \geq 2$ (as τ values increase, the search times will decay); finally, for problem G11 (Figure 4c), the pattern is very irregular, independently of P (increasing τ may lead to an unpredictable surge or decline of search times);

(d) irregardless of τ, the use of parallel searches typically pays off for all problems (in general, increasing the level of parallelism (P) will decrease the search times); this downwards trend is more regular in problems G8 (Figure 3d), G11 (Figure 4d), and G12 (Figure 5d), where the lowest average search times are achieved with the higher values of P (values 7 and/or 8); in problem G6 (Figure 2d), both $P = 4$ and $P = 7$ provide the best average search times; all these times are shown encircled in the tables (within a tolerance of 5%).

4.3   Number of Optima

Another important issue is the number of optima found. Table 3 shows that number as an average: for each problem, sequential only ($P = 1$) and parallel only ($P \geq 2$) searches are considered; for each of these categories, it is presented the mean of the number of optima found with all values of τ tested (0.1,...,0.9); along with the mean, the coefficient of variation is also presented, in parenthesis.

Table 3. Average number of optima found (and coefficient of variation)

Problem	$P = 1$	$P \geq 2$
G6	4.9 (20.3%)	5.7 (12.7%)
G8	6.0 (0%)	6.0 (0%)
G11	11.4 (21.5%)	11.9 (18.4%)
G12	10.8 (7.7%)	10.2 (6.9%)

The table allows to conclude that: i) for problem G8, the number of optima found is the same, independently of P and τ; ii) for the other problems, there are indeed differences between using a single sequential search and using parallel searches; but these differences are small, and they may either favor parallel searches (problems G6 and G11) or sequential searches (problem G12); it seems, however, that parallel searches exhibit a smaller dispersion of the optima number.

Table 4. Search times values for Problem G6 (seconds)

	P								avg.	avg.
τ	1	2	3	4	5	6	7	8	$P \geq 1$	$P \geq 2$
0.1	33.7	7.7	7.7	7.7	7.7	7.7	7.7	7.7	**10.9**	**7.7**
0.2	41.8	7.7	7.7	7.7	7.7	7.7	7.7	7.7	11.9	**7.7**
0.3	78.7	7.7	7.7	7.7	7.7	7.7	7.7	7.7	15.6	**7.7**
0.4	95.8	7.7	7.7	7.7	7.7	7.7	7.7	7.6	18.7	**7.7**
0.5	85.8	7.7	7.7	7.7	7.7	7.7	7.1	7.8	17.4	7.8
0.6	142.4	7.7	7.7	7.7	7.9	8.2	7.2	8.7	24.7	**7.8**
0.7	159.9	7.8	7.7	7.8	9.0	7.6	8.8	8.7	27.1	8.2
0.8	263.2	91.3	31.7	8.7	14.8	17.6	8.7	12.7	56.10	26.5
0.9										
avg.	111.7	18.1	10.7	7.8	8.8	9.0	7.8	8.6		

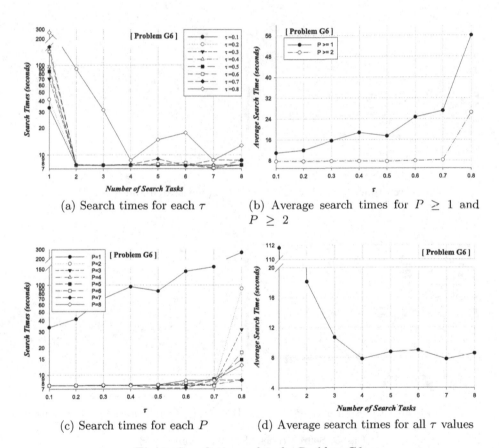

(a) Search times for each τ

(b) Average search times for $P \geq 1$ and $P \geq 2$

(c) Search times for each P

(d) Average search times for all τ values

Fig. 2. Search times plots for Problem G6

Table 5. Search times values for Problem G8 (seconds)

τ	1	2	3	4	P 5	6	7	8	avg. $P \geq 1$	avg. $P \geq 2$
0.1	14.9	15.0	15.2	15.0	15.0	14.9	15.0	14.9	**15.0**	15.0
0.2	19.3	19.5	19.5	19.3	19.5	19.5	19.5	19.3	19.4	19.4
0.3	22.8	22.6	22.3	22.6	22.7	22.6	22.3	20.1	22.3	22.2
0.4	25.8	23.1	22.8	22.8	22.7	18.8	9.8	9.9	19.5	18.6
0.5	35.0	23.0	22.8	22.8	10.5	9.9	9.9	10.4	18.0	15.6
0.6	41.2	22.9	22.7	9.8	9.7	8.5	9.9	10.0	16.8	13.4
0.7	51.2	22.9	9.7	10.0	9.7	9.7	9.6	6.8	16.2	11.2
0.8	77.6	9.6	10.0	9.3	9.0	9.8	9.5	8.4	17.9	**9.4**
0.9	157.3	9.7	10.0	9.7	7.4	9.4	11.2	11.8	28.3	**9.9**
avg.	49.5	18.7	17.2	15.7	14.0	13.7	12.9	13.0		

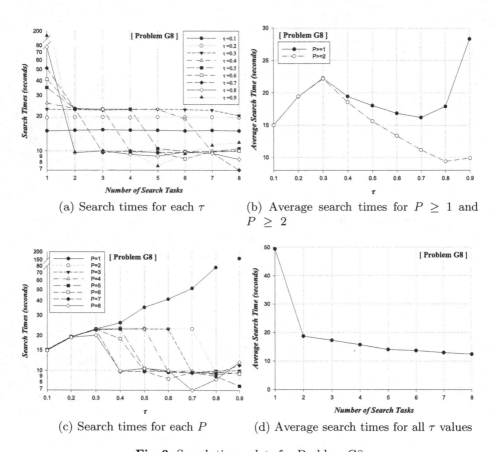

(a) Search times for each τ

(b) Average search times for $P \geq 1$ and $P \geq 2$

(c) Search times for each P

(d) Average search times for all τ values

Fig. 3. Search times plots for Problem G8

Table 6. Search times values for Problem G11 (seconds)

τ	P 1	2	3	4	5	6	7	8	avg. P ≥ 1	avg. P ≥ 2
0.1	274.8	24.0	23.8	24.0	28.3	23.9	23.8	28.4	56.4	25.2
0.2	127.4	88.1	24.1	23.6	23.6	23.0	23.7	28.3	46.4	34.5
0.3	15.2	14.9	16.0	14.9	11.0	15.0	14.9	15.1	15.0	15.0
0.4	264.1	48.3	23.4	24.9	56.2	23.6	41.8	28.9	63.4	35.0
0.5	100.2	94.1	84.1	64.8	88.9	12.7	19.8	28.1	61.6	56.1
0.6	89.9	90.2	89.9	24.5	37.7	34.4	27.5	11.3	50.7	45.1
0.7	318.3	56.2	103.0	56.2	54.5	53.8	27.6	48.8	89.8	57.1
0.8	184.2	41.9	69.5	28.9	16.2	44.3	25.7	23.1	54.2	35.7
0.9	38.1	37.6	37.6	32.8	9.5	37.6	11.9	19.0	28.0	26.6
avg.	157.1	55.9	52.3	32.7	36.6	29.8	24.1	25.6		

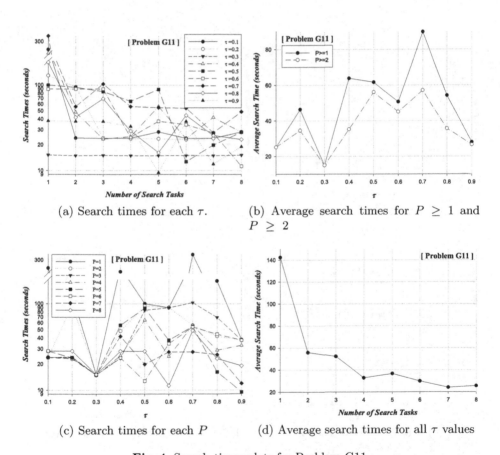

(a) Search times for each τ.

(b) Average search times for P ≥ 1 and P ≥ 2

(c) Search times for each P

(d) Average search times for all τ values

Fig. 4. Search times plots for Problem G11

Table 7. Search times values for Problem G12 (seconds)

τ	1	2	3	4	5	6	7	8	avg. $P \geq 1$	avg. $P \geq 2$
0.1	67.6	14.1	14.1	14.1	14.1	14.1	14.1	14.1	**20.8**	**14.1**
0.2	79.5	14.1	14.1	14.1	14.1	14.1	14.1	14.1	22.3	**14.1**
0.3	87.3	14.1	14.1	14.1	14.1	14.1	14.1	13.5	23.2	**14.0**
0.4	116.9	14.2	14.1	14.1	14.1	13.5	14.1	14.1	26.9	**14.0**
0.5	154.9	14.1	14.1	14.1	14.1	14.3	14.1	15.2	31.8	**14.3**
0.6	142.4	14.1	14.1	14.1	14.1	16.3	13.1	16.4	30.6	**14.6**
0.7	202.1	21.4	14.2	15.2	16.3	16.3	13.3	13.1	39.0	15.7
0.8	92.4	90.1	13.1	36.7	16.3	15.1	16.3	15.0	36.9	29.0
0.9	124.7	125.6	83.6	19.1	45.1	12.9	13.3	12.8	54.6	44.6
avg.	118.7	35.8	21.7	17.3	18.0	14.5	14.1	14.3		

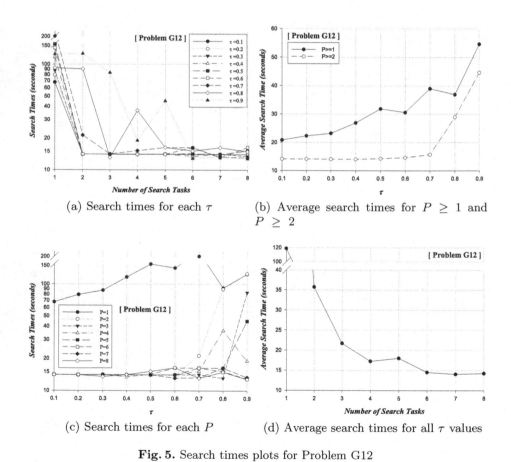

(a) Search times for each τ

(b) Average search times for $P \geq 1$ and $P \geq 2$

(c) Search times for each P

(d) Average search times for all τ values

Fig. 5. Search times plots for Problem G12

5   Conclusions and Future Work

In this paper coPSSA is explored, as a hybrid application that solves constrained optimization problems, by integrating a numerical l_1 penalty method with a parallel solver of unconstrained (bound constrained) problems (PSSA).

The effect of the variation of the number of search tasks, and of the penalty parameter reduction factor (τ) was studied, in the context of the l_1 penalty function. With base on the analysis of the results obtained with the four tested problems, it is possible to conclude: i) increasing the number of search tasks typically decreases the search times; ii) for the tested problems, smaller values of τ typically imply lower average search times; iii) for some problems, the number of optima found does not depend on the number of search tasks neither on the value of τ, while other problems are sensitive to the variation of those factors.

In the future, the research team intends to refine this work, by solving more constrained problems (including problems with more than 2 dimensions), and exploring higher levels of parallelism (*i.e.*, by running coPSSA with $P \gg 8$).

Acknowledgments. This work was been supported by FCT (Fundação para a Ciência e Tecnologia) in the scope of the project UID/CEC/00319/2013.

References

1. Chelouah, R., Siarry, P.: A continuous genetic algorithm designed for the global optimization of multimodal functions. Journal of Heuristics **6**, 191–213 (2000)
2. Eriksson, P., Arora, J.: A comparison of global optimization algorithms applied to a ride comfort optimization problem. Structural and Multidisciplinary Optimization **24**, 157–167 (2002)
3. Floudas, C.: Recent advances in global optimization for process synthesis, design and control: enclosure of all solutions. Computers and Chemical Engineering, 963–973 (1999)
4. Hedar, A.-R.: Global Optimization Test Problems. http://www-optima.amp.i.kyoto-u.ac.jp/member/student/hedar/Hedar_files/TestGO.htm
5. Ingber, L.: Very fast simulated re-annealing. Mathematical and Computer Modelling **12**, 967–973 (1989)
6. Kiseleva, E., Stepanchuk, T.: On the efficiency of a global non-differentiable optimization algorithm based on the method of optimal set partitioning. Journal of Global Optimization **25**, 209–235 (2003)
7. León, T., Sanmatías, S., Vercher, H.: A multi-local optimization algorithm. Top **6**(1), 1–18 (1998)
8. Message Passing Interface Forum. http://www.mpi-forum.org/
9. Michalewicz, Z.: A survey of constraint handling techniques in evolutionary computation methods. In: Proceedings of the 4th Annual Conference on Evolutionary Programming, pp. 135–155 (1995)
10. Mongeau, M., Sartenaer, A.: Automatic decrease of the penalty parameter in exact penalty function methods. European Journal of Operational Research **83**, 686–699 (1995)
11. Nocedal, J., Wright, S.: Numerical Optimization. Springer Series in Operations Research. Springer (1999)

12. Parsopoulos, K., Plagianakos, V., Magoulas, G., Vrahatis, M.: Objective function stretching to alleviate convergence to local minima. Nonlinear Analysis **47**, 3419–3424 (2001)

13. Parsopoulos, K., Vrahatis, M.: Recent approaches to global optimization problems through particle swarm optimization. Natural Computing **1**, 235–306 (2002)

14. Pereira, A.I., Fernandes, E.M.G.P.: Constrained multi-global optimization using a penalty stretched simulated annealing framework. In: AIP Conference Proceedings Numerical Analysis and Applied Mathematics, vol. 1168, pp. 1354–1357 (2009)

15. Pereira, A.I., Ferreira, O., Pinho, S.P., Fernandes, E.M.G.P.: Multilocal programming and applications. In: Zelinka, I., Snasel, V., Abraham, A. (eds.) Handbook of Optimization. ISRL, vol. 38, pp. 157–186. Springer, Heidelberg (2013)

16. Pereira, A.I., Rufino, J.: Solving multilocal optimization problems with a recursive parallel search of the feasible region. In: Murgante, B., Misra, S., Rocha, A.M.A.C., Torre, C., Rocha, J.G., Falcão, M.I., Taniar, D., Apduhan, B.O., Gervasi, O. (eds.) ICCSA 2014, Part II. LNCS, vol. 8580, pp. 154–168. Springer, Heidelberg (2014)

17. Price, C.J., Coope, I.D.: Numerical experiments in semi-infinite programming. Computational Optimization and Applications **6**, 169–189 (1996)

18. Ribeiro, T., Rufino, J., Pereira, A.I.: PSSA: parallel stretched simulated annealing. In: AIP Conference Proceedings, Numerical Analysis and Applied Mathematics, vol. 1389, pp. 783–786 (2011)

19. Rufino, J., Pereira, A.I., Pidanic, J.: coPSSA - Constrained parallel stretched simulated annealing. In: Proceedings of the 25th Int. Conference Radioelektronika 2015, pp. 435–439 (2015)

20. Salhi, S., Queen, N.: A Hybrid Algorithm for Identifying Global and Local Minima When Optimizing Functions with Many Minima. European Journal of Operations Research **155**, 51–67 (2004)

21. Shandiz, R.A., Tohidi, E.: Decrease of the Penalty Parameter in Differentiable Penalty Function Methods. Theoretical Economics Letters **1**, 8–14 (2011)

22. Surjanovic, S., Bingham, D.: Virtual Library of Simulation Experiments: Test Functions and Datasets. http://www.sfu.ca/ssurjano

23. Tsoulos, I., Lagaris, I.: Gradient-controlled, typical-distance clustering for global optimization. http://www.optimization.org (2004)

24. Tu, W., Mayne, R.: Studies of multi-start clustering for global optimization. International Journal Numerical Methods in Engineering **53**, 2239–2252 (2002)

25. Wang, Y., Cai, Z., Zhou, Y., Fan, Z.: Constrained optimization based on hybrid evolutionary algorithm and adaptive constraint-handling technique. Structural and Multidisciplinary Optimization **37**, 395–413 (2008)

Probabilistic Clustering of Wind Energy Conversion Systems Using Classification Models

Paula Odete Fernandes[1,2] (✉) and Ângela Paula Ferreira[1,3]

[1] Polytechnic Institute of Bragança, Campus de Santa Apolónia,
Apartado 1134 5301-857, Bragança, Portugal
{pof,apf}@ipb.pt
[2] UNIAG and NECE, Covilhã, Portugal
[3] CISE - Electromechatronic Systems Research Centre,
University of Beira Interior, Covilhã, Portugal

Abstract. This research intends to give insights on the pattern aggregation of wind energy conversion systems technologies through identification of homogeneous groups within a set of wind farms installed in Portugal. Pattern aggregation is performed using Hierarchical Cluster Analysis followed by Discriminant Analysis, in order to validate the results produced by the first one. The clustering support matrix uses three independent variables: installed capacity, net production and capacity factor, in a per year basis. Cluster labelling allows the identification of two homogenous groups of wind farms, whose main attributes are based on the technological conversion system trend: (1) asynchronous generator based technology and (2) direct driven synchronous generator based technology, with higher capacity factors.

Keywords: Wind farms · Wind turbine generators · Cluster analysis · Discriminant analysis

1 Introduction

The integrated use of renewable energy sources and mature technologies in power systems contributes to strengthen domestic economies, by reducing dependency on imported fossil fuels and dramatically reduce greenhouse-gas emissions, over time.

In this scenario, the exploitation of wind energy resources play an important key role, capable of decarbonising the power sector. Wind energy is a clean source and enfolds an environmental friendly technology. Its renewable character and the fact it does not pollute during the operational phase makes it one of the most promising energy source in reducing environmental problems at both global and local levels.

The technological maturity of wind energy conversion systems contributes to a small economic differential cost with regard to conventional technologies. In fact, wind energy systems are more competitive than other renewable energies, apart from for hydro energy systems. This scenario suggests that, in the following decades, wind energy will remain the main commitment to new electrical generation capacity.

© Springer International Publishing Switzerland 2015
O. Gervasi et al. (Eds.): ICCSA 2015, Part II, LNCS 9156, pp. 549–560, 2015.
DOI: 10.1007/978-3-319-21407-8_39

According to the Global Wind Energy Council (GWEC), installed wind power capacity has grown to cumulative worldwide installation level of 369,5 GW, with 51,5 GW alone installed in 2014. Europe's total installed capacity has been surpassed for the first time by Asian markets (India, China). Nevertheless, by the end of 2014, the wind power capacity installed would produce energy enough to cover approximately 10% of the Europe's electricity consumption, in a normal wind year [1]. Portugal accounts for about 5% of the wind energy capacity installed in European Union, with approximately 4,9 GW of accumulated installed capacity in 2014 which is capable to generate about 15% of the electrical energy consumption [2].

Despite the advantages inherent to a renewable energy source, wind energy has also some unfavourable conditions. The small power density of the wind leads to wide and material extensive turbines, thereby hindering the on-site assembly and the electrical infrastructure. Technological developments are being made in order to increase unit power wind generators to optimize the impact on the ground for onshore wind farms or for offshore applications [3].

Concerning the prime source, wind is stochastic in nature and essentially ruled by random meteorological changes. Due to its intermittent and unpredictable behaviour, wind energy conversion systems are not dispatchable, i.e., they do not have the ability to produce electrical energy following load requirements. The inherent variability of wind power is also raising concerns regarding the reliability and cost-effectiveness of the transmission and distribution power systems while supporting large wind farms [4]. High penetration levels of wind energy implies structural changes in power systems as, for instance, the usage of storage systems and/or coupling hydro and wind systems to smooth the output pattern [5]. Progress and improvements are being made in renewable energy integration into power systems. Development of energy storage provided, for instance, by bi-directional Vehicle-to-Grid technologies and intelligent networking will allow a greater penetration potential of wind energy.

The life span of the first wind farms is coming to an end, which implies repowering processes aiming an augmented efficiency and reliability of the wind turbines. From the available technologies of the conversion systems, it is far from clear which of them is the optimal.

The motivation for this work is supported by the absence of a deterministic certainty in allocating outputs of wind farms, regarding the technological conversion system trends [6,7]. Therefore, this study aims at giving some insights into technological approaches for wind turbines, using probabilistic clustering to identify homogeneous groups.

Clustering wind farms allocate different units into a group which contains some common characteristics, which may be used to reduce the size and the order of mathematical models and also to perform pattern classification into extensive multi-dimensional data set [8-11].

Previous studies had been performed on pattern aggregation of wind farms using probabilistic clustering [12,13]. This work addresses the problem through a methodology based on two multivariate analyses: Hierarchical Cluster Analysis and Discriminant Analysis. In order to identify the clusters characteristics, it is also performed an exploratory descriptive analysis. In addition to previous publications, this work considers an increased reference data set of wind farms and extended time span in order to establish a comparative analysis with previous results.

The paper is organized as follows: next section presents technological trends on wind energy conversion systems, Section 3 overviews the clustering and validation methodologies, Section 4 applies the proposed approach to a case study, presents the main results and discussion and, finally, Section 5 rounds up the paper with the main conclusions.

2   Wind Energy Conversion System Technologies

Despite the fact first development of commercial wind energy technology began in the late 1930s, only after the oil crises of the 1970s, there have begun economical incentives to develop the technology further [14]. Since the 1980s, there has been a significant consolidation of the design of wind turbines. This section describes main design styles in wind energy conversions systems and points out the technological trends of the conversion system drive train.

2.1   Design Styles

The mainstream commercial market uses horizontal axis wind turbines, meaning the rotating axis is parallel to the ground. This option is inherently more efficient than vertical axis. Concerning the number of blades, the aerodynamic efficiency and reduced acoustic noise emission establish three-bladed rotor design.

Other important issue related with the design of a wind energy conversion systems is the mean of limiting rotor power in high operational wind speeds. There are two main approaches: stall and pitch control.

In stall regulated machines, speed regulation is intrinsic to the aerodynamic design, without any change of the rotor geometry. Under this control approach, wind turbine runs at approximately constant speed even when the wind speed is high, without producing excessive power. The constant speed is achieved through the connection of the electric generator to the grid. Regarding this aspect, the grid behaves like a large flywheel, holding the speed of the turbine nearly constant irrespective of changes in wind speed.

Pitch control involves pitching the blades (i.e., turning the wind blades about their main axis) in order to regulate the power the rotor extracts from wind. This control involves an active control system, which should sense the blade position and defines appropriate changes of blade pitch, according to the measured output power.

Another important and decisive design issue of the wind turbines is the use of variable rotational speed versus fixed speed, with consequences on the overall performance of the system [15,16].

The constant speed turbine designs consist on generators operating at fixed speed when producing power, directly connected to the utility grid which, through the generator, holds the speed constant. This concept makes use of Squirrel Cage Induction Generators (SCIG) with a geared drive train to adapt the rotational speed to the frequency of the grid. With this design, the wind energy capture and also the power quality in the utility grid are reduced.

Variable speed wind energy systems allows operation below rated power, enabling increased energy capture, and also above rated power, even over a small speed range, which can substantially ease pitch system duty and reduce output power variability. This exploitation mode of wind energy systems improves the power quality when compared with constant speed systems. Variable speed wind energy systems may be implemented using synchronous or asynchronous generators, allowing wider or narrower wind speed ranges, respectively.

Solutions based on asynchronous generators, the so called Doubly Fed Induction Generators (DFIG), with the stator windings directly connected to the grid and a partial scaled electronic converter between the rotor and the grid, allow a low to moderate variation of the rotor speed. Since the power converter is partially scaled, typically one third of the rated power of the system [17], this solution is somewhat cost effective but, on the other hand, there are limitations to control effectively the grid variables, which translates in a deficient quality power system [18]. It should be pointed out that this concept uses a geared drive train to match the low rotational speed promoted by wind velocities to the higher efficient rotational speed of this generator type.

Solutions based on Synchronous Generators (SG) use full scaled electronic converters. The electrical energy is generated at variable frequency (strictly related to the rotational speed of the rotor) and then converted to the frequency of the grid. This concept takes advantage of the wide speed range operation allowed by the full scale converter between the generator and the grid, which also allows boosting the grid stability and performance. Additionally and when compared with DFIG, this type of generators requires lower ratio gearboxes, or even its omission, which translates in higher reliability and lower maintenance costs [18]. Gearboxes are one of the most expensive components of the wind turbine system and require significant repair or overhaul before the intended life span of the entire system is reached. Thus, the simplification introduced in the drive train by the absence of these components improves significantly the reliability of the wind turbine system and helps bring the cost of wind energy back to a decreasing trajectory [19].

2.2   Technological Trends

Under the premise of high variability and intermittency of wind speed, the actual demand on power quality issues compels for generators featuring variable speed, which is the dominant trend in the actual market.

Comparing partial speed range systems, promoted by DFIG, and full-range variable speed drives based on SG, the later bring some attractions, specially on operational flexibility and power quality issues, but also have some drawbacks related with the higher power of the electronic converter, with the same rating of the generator [20]. In fact, there was never a clear case for full variable speed range on economic grounds, with small energy gains being offset by extra costs and also additional losses in the power converter.

Another technological trend is related with direct driven generators, i.e., gearless systems. The direct drive systems of Enercon [21] are long established, and gearless systems or with low ratio gearboxes, using Synchronous Permanent Magnet Generator (SPMG) technology have emerged in recent years [22]. In fact, some manufacturers that in past had based their technology on asynchronous generators are now moving to SPMG with full sealed converter [23].

Permanent magnet technology allows a higher power-to-volume ratio and fully rated power converter based systems can be applied without design hardware modifications in both 50 Hz or 60 Hz power systems, which increases flexibility for international developers operating in multiple wind markets [14].

Concerning the power control in high operational wind speeds, the design issues of pitch versus stall and degree of rotor speed variation are evidently connected. The stall-regulated design remains viable, but pitch control offers potentially better output power quality, while overall costs of both systems remain similar [14].

3 Clustering and Validation Methodology

Clustering, i.e., partitioning objects/cases into similar groups, is a problem with several alternatives in mathematics as well as in applied sciences. Cluster analysis aims at recognizing groups of similar records and, therefore, helps to discover distribution of patterns and interesting correlations in data sets [24, 25].

To attain the main goal of this research study, a hierarchical cluster analysis has been applied, using the methodology proposed by Ward [24], which is the most commonly used for problems similar to the one under analysis, providing a more consistent solution and it is also recommended for quantitative variables measured on a ratio scale. In this methodology, an objective function, defined as the sum of squares of deviations of the individual observations compared with the average of the group, is minimized, aiming at creating groups which have maximum internal cohesion and maximum separate external distance [24]. This method uses the variance to evaluate distances between clusters, which results in an efficient approach when compared with other hierarchical methods (for instance, nearest neighbour, furthest neighbour and median clustering).

The Ward's distance between clusters C_i and C_j, $D_w(C_i, C_j)$, is the difference between the total within cluster sum of squares for the two clusters separately, and within cluster sum of squares, which results from merging the two clusters in cluster C_{ij} [24, 25], i.e.,

$$D_w(C_i, C_j) = \sum_{x \in C_i} (x - r_i)^2 + \sum_{x \in C_j} (x - r_j)^2 - \sum_{x \in C_{ij}} (x - r_{ij})^2 \qquad (1)$$

where r_i is the centroid of C_i, r_j is the centroid of C_j and r_{ij} is the centroid of C_{ij}.

To implement a dissimilarity measure between subjects, it is selected the Euclidean Distance Squared. The distance is defined as the square root of the sum of the squared differences between the values of i and j for all the selected variables $(1, 2,..., p)$, [26]:

$$D_{ij} = \sqrt{\sum_{k=1}^{p} |x_{ik} - x_{jk}|^2} \tag{2}$$

where x_{ik} is the value of the variable k for cases i and x_{jk} is the value of the variable k for cases j.

An agglomerative algorithm is used in order to produce a sequence of clustering schemes of decreasing number of clusters at each step. The clustering scheme produced at each step results from the previous one by merging the two closest clusters into one.

One of the most important issue in clustering analysis is the evaluation of the produced results to find the partitioning which best performs the underlying data. To accomplish this requirement, the performed methodology applies a Discriminant Analysis (DA). The basic purpose of DA is to estimate the relationship between a single categorical dependent variable and a set of quantitative independent variables [26]. Discriminant Analysis involves the determination of a linear equation like regression that will predict which group the case belongs to, minimizing the within-group distance and, simultaneously, maximizing the between-group distance, thus achieving maximum discrimination [26]. The basic idea underlying Discriminant Function Analysis is to determine whether groups differ with regard to the mean of a variable, and then to use that variable to predict group membership. Given p variables and g groups, it is possible to establish $m = \min(g-1, p)$ discriminant functions, DF_i, in the form given by [27]

$$DF_i = a + \sum_{k=1}^{p} b_{ik} x_k \quad (i = 1 \ldots m) \tag{3}$$

where a is a constant, b_{ik} is the discriminant coefficient and x_k is the independent variable.

The discriminant model has the following assumptions [28]:

1. Multivariate normality, given that data values are from a normal distribution;
2. Equality of variance-covariance within group, i.e., the covariance matrix within each group should be equal;
3. Low multicollinearity of the variables; when high multicollinearity among two or more variables is present, the discriminant function coefficients will not reliably predict group membership.

4   Wind Farms Clustering

Wind farms clustering is addressed here by applying the previous described methodology to a case study in order to identify possible clusters and their main attributes.

4.1   Case Study

The chosen support variables used to cluster wind farms are the installed capacity, net production and capacity factor, given in a per-year basis.

Installed capacity of a wind farm is the rated power of each wind turbine multiplied by the number of wind turbines within each farm. The rated power of each wind turbine translates the power each array train is able to convert from mechanical to electrical energy, under given test wind speed conditions. The data set comprises wind farms with installed capacities higher than 10 MW. It should be noted that some wind farms have been submitted to overpowering processes under the considered time span, *i.e.*, installed capacity of each farm in the per-year basis is not constant.

The net production per year measures the output of each farm, in terms of electrical energy delivered to the grid, considering programmed and random unavailabilities (failures) of wind turbines. At present, this output is not constrained by load demand or wholesale markets, as currently regulated.

The capacity factor is the ratio of actual productivity in a year to its theoretical maximum. Higher capacity factors indicate a better utilization of the installed capacity, which helps to reduce investment costs. In fact, capacity factors are particularly important on evaluating the overall economics of wind farms. Typically, capacity factors need to be elevated to values about 50% (or better) to make a modern wind farm commercially viable [18]. The capacity factor relates with the other two variables under analysis, nevertheless its inclusion aims at sensing the wind availability in each farm. Considering a particular farm, with a given number of wind turbines using a specific technological drive train, if the capacity factor increases from one year to another, it means that the meteorological conditions in the later favoured the wind source. On the other hand, comparing wind farms with similar wind availability profile, if they have different capacity factors, it implies that the design technology and layout of the infrastructure are performing differently, assuming that maintenance schemes are similar. From the original data set, two wind farms were excluded from the analysis, given that they performed as outliers regarding this variable, deviating from regular wind energy profiles.

Final data set comprises 32 wind farms from two promoters acting in the wind energy sector in Portugal, EDP and GENERG. The information was collected from institutional and technical Annual Reports available from [29,30], including a time span of 4 years (from 2010 till 2013).

4.2   Results and Discussion

The analysis of a dendrogram, using the Ward linkage method, is a common way to anticipate the hypothetical optimal number of wind farms' clusters as well as their composition. This graph also allows observing the distance level at which there is a combination of wind farms and clusters. From the obtained dendrogram, shown in Fig. 1, at the rescaled distance of 15, it is straightforward the definition of two notable groups.

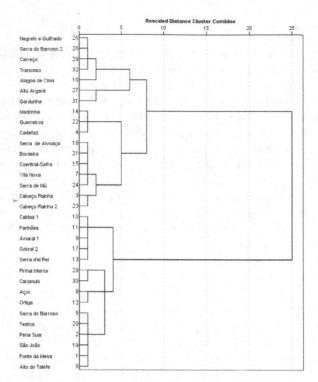

Fig. 1. Dendrogram graph for the data set under analysis

In order to strengthen the identification of the optimal number of clusters, it is also used the test R-Squared (R-Sq) based on the analysis of variance (ANOVA) information, namely the ratio between Sum of Squares Between Groups or Clusters and the Total Sum of Squares. The results of the relativized distance between clusters is shown in Fig. 2, from which a solution of two clusters has been chosen, explaining 28% of the total variance. Table 1 outlines wind farms that pose similar features allocated in the 2 identified clusters. The first cluster comprehends 15 wind farms, labelled cluster A, and the second cluster has 17 wind farms, with the label S.

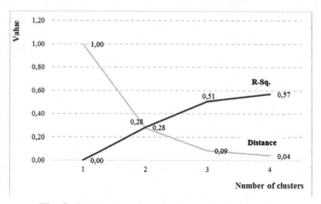

Fig. 2. Optimal number of clusters for the data set

Table 1. Clustering components based on variables installed capacity, net production and capacity factor

Cluster	Wind Farms		
A	Açor	Fanhões	São João
	Alto do Talefe	Ortiga	Serra d'el Rei
	Amaral 1	Pena Suar	Sobral 2
	Caldas 1	Pinhal Interior	Serra do Barroso
	Caramulo	Pinhal da Murta	Tocton
S	Alagoá de Cima	Madrinha	Serra de Alvoaçã
	Bordeira	Negrelo e Guilhado	Serra de Mú
	Cabeço Rainha	Guerreiros	Serra do Barroso 2
	Cadafaz	Cabeço Rainha 2	Alto Arganil
	Coentral-Safra	Gardunha	Vila Nova
	Carreço	Trancoso	

The following step consists in application of the Discriminant Analysis to validate the results produced by applying the Cluster Analysis, as previously described. According to this analysis, it is possible to observe significant mean differences for all predictors - independent variables - installed capacity, net production and capacity factor in the clusters A and S - dependent variables. The discriminant function reveals a significant association between groups and all predictors. Regarding the DA results, with only two groups, one Discriminant Function has been produced with an eigenvalue of 3,002. The canonical correlation is given by the multiple correlation between the predictors and the discriminant function. With only one function, the DA provides an index of overall model fit, which is interpreted as being the proportion of variance explained (R-Sq). The relation between the canonical discriminant function and the clusters reveals a greater correlation, approximately 87%. The significance of the discriminant function, evaluated by the Wilks' lambda, is considerably high (p value < 0,001), which means that the model has significant discriminatory power, and provides the proportion of total variability not explained, *i.e.,* the converse of the squared canonical correlation, evaluated in 25%. The classification results reveal that 93,8% of wind farms are classified correctly into 'Cluster A' or 'Cluster S' groups, which can be considered excellent. The wind farms in Cluster S were classified with slightly better accuracy (94,1%) than wind farms in Cluster A (93,3 %).

Table 2 shows the results for the variables into the different clusters, Capacity Factor (C_F , in %), Net Production (W , in GWh) and Installed Capacity (P , in MW) for the time span under analysis.

Table 2. Summary of descriptive statistics by cluster and variables

Cluster	n.°	Variable	Minimum	Maximum	Mean	Std. Deviation
A	15	P (MW)	10,00	144,00	29,80	37,134
		C_F (%)	23,42	28,36	25,80	1,293
		W (GWh)	22,35	328,80	66,86	83,418
S	17	P (MW)	10,00	114,00	28,98	23,843
		C_F (%)	27,86	32,77	29,690	1,548
		W (GWh)	28,20	280,00	74,14	58,012

From the descriptive statistics it is possible to observe that both clusters have a high dispersion of the installed capacity, with similar mean values. Main difference between clusters is the mean value of the capacity factor, with a low dispersion: cluster A has lower capacity factors than cluster S, meaning that the overall performance of wind farms in the later cluster had a better performance, for similar installed capacities. In consequence, the mean net production of cluster S is higher than the one of cluster A, whereas the high standard deviation observed for this variable in both clusters follows obviously the one observed for the installed capacity.

After the definition of the clusters and their components by the proposed methodology, it is now possible to explore dominant attributes in each cluster, in order to find particular patterns. The attribute looked into each cluster is the technological trend of the energy conversion system utilized in the different farms. It is possible to observe that Cluster A has a predominance of technology based on Asynchronous generators (73,3% of the farms), while in Cluster S the dominant technology is based on direct driven Synchronous generators (76,5% of the wind farms). As previously stated, this technological concept, using full variable speed range, improves substantially the efficiency of the system which corroborates the higher capacity factors observed within this clusters. Moreover, the absence of the gearbox component increases the reliability and allows reduced maintenance schemes which, together with the wide speed range operation, results in an increased capture of the disposable wind energy.

5   Conclusions

The main objective of this research was to identify homogeneous groups within a data set of wind farms of two promoters acting in the energy sector in Portugal, based on two multivariate analyses using a support matrix with three independent variables: installed capacity, net production and capacity factor, in a per year basis. In a first stage it is used Hierarchical Cluster Analysis followed by a Discriminant Analysis, in order to validate the results produced by the first one. Based on both methodologies, from the obtained results, it has been possible to identify two clusters, explaining 28% of the total variance.

Regarding the DA outcomes, one discriminant function has been produced with an eigenvalue of 3,002. The relationship between the canonical discriminant function and the clusters reveals a satisfactory positive correlation, about 87%. Also, the significance of the discriminant function, evaluated by the Wilks' lambda test, is significantly high which means that the model has significant discriminatory power, and provides the proportion of total variability not explained, *i.e.,* the converse of the squared canonical correlation is evaluated in 25%. The classification results shows that 93,8% of original grouped cases are correctly classified.

Following the clustering and validation methodology, it has been possible to identify the technological trend based on the wind turbine generator type in each cluster: Cluster A, with asynchronous generator based technology and cluster S, mainly using direct driven synchronous generator based technology.

From the descriptive statistics regarding data set of both clusters, it is possible to infer that cluster S presents higher mean value of capacity factor than cluster A, which is a good indicator that wind farms using the technological trend based on direct driven synchronous generators have a better performance than the ones based on geared asynchronous generators. The latter, may reduce the initial costs of the drive trains, by using partially scaled electronic converters, but as a counterpart, the increase in the converted energy over the lifespan may offset the higher initial cost

References

1. EWEA: Wind in power, 2014 European statistics. http://www.ewea.org/ (access date: February 2015)
2. ENEOP: Eólicas de Portugal, S.A. http://www.eneop.pt/ (access date: January, 2014)
3. Jonkman, J., Butterfield, S., Musial, W., Scott, G.: Definition of a 5-MW Reference Wind Turbine for Offshore System Development. National Renewable Energy Laboratory, Technical Report NREL/TP-500-38060 (2009)
4. Karki, R., Billinton, R.: Cost-Effective Wind Energy Utilization for Reliable Power Supply. IEEE Transactions on Energy Conversion **19**, 435–440 (2004)
5. Denault, M., Dupuis, D., Couture-Cardinal, S.: Complementarity of Hydro and Wind Power: Improving the Risk Profile of Energy Inflows. Energy Policy **37**, 5376–5384 (2009)
6. Arabian-Hoseynabadi, H., Tavner, P.J., Oraee, H.: Reliability Comparison of Direct Drive and Geared Drive Wind Turbine Concepts. Wind Energy, Wiley Online Library **13**, 62–73 (2010)
7. Polinder, H., van der Pijl, F.F.A., de Vilder, G.J., Tavner, P.J.: Comparison of Direct-Drive and Geared Generator Concepts for Wind Turbines. IEEE Transactions on Energy Conversion **21**, 725–733 (2006)
8. Chicco, G., Ilie, I.S.: Support Vector Clustering of Electrical Load Pattern Data. IEEE Transactions on Power Systems **24**, 1619–1628 (2009)
9. Ali, M., Ilie, I.S., Milanovic, J.V., Chicco, G.: Wind Farm Model Aggregation Using Probabilistic Clustering. IEEE Transactions on Power Systems **28**, 309–316 (2013)
10. Andrada, M.F., Vega-Hissi, E.G., Estrada, M.R., Garro Martinez, J.C.: Application of k-means clustering, linear discriminant analysis and multivariate linear regression for the development of a predictive QSAR model on 5-lipoxygenase inhibitors. Chemometrics and Intelligent Laboratory Systems **143**, 122–129 (2015)
11. Bouguettaya, A., Yu, Q., Liu, X., Zhou, X., Song, A.: Efficient agglomerative hierarchical clustering. Expert Systems with Applications **42**, 2785–2797 (2015)
12. Fernandes, P.O., Ferreira, A.P.: Wind farms model aggregation using probabilistic clustering. In: 11th International Conference of Numerical Analysis and Applied Mathematics 2013, ICNAAM 2013, Rhodes, Greece, pp. 618–621. AIP Conference Proceedings (2013)
13. Fernandes, P.O., Ferreira, A.P.: Pattern aggregation of wind energy conversion technologies using clustering analysis. In: 2014 14th International Conference on Computational Science and its Applications (ICCSA), pp. 105–110 (2014)
14. EWEA: Wind Energy – The Facts. Routledge (2009)
15. Hansen, A.D., Iov, F., Blaabjerg, F., Hansen, L.H.: Review of Contemporary Wind Turbine Concepts and their Market Penetration. Wind Engineering **28**, 247–263 (2004)
16. Hansen, A.D., Hansen, L.H.: Wind Turbine Concept Market Penetration over 10 Years (1995–2004). Wind Energy, Wiley Online Library **10**, 81–97 (2007)

17. Blaabjerg, F., Iov, F., Kerekes, T., Teodorescu, R.: Trends in power electronics and control of renewable energy systems. In: 14th International Power Electronics and Motion Control Conference (EPE-PEMC 2010), Ohrid, Republic of Macedonia, pp. k1–k19 (2010)
18. Mathew, S., Philip, G.S.: Advances in Wind Energy and Conversion Technology. Springer (2011)
19. Musial, W., Butterfield, S.: Improving wind turbine gearbox reliability. In: 2007 European Wind Energy Conference, Milan, Italy (2007)
20. Baroudi, J.A., Dinavahi, V., Knight, A.M.: A Review of Power Converter Topologies for Wind Generators. Renewable Energy, Elsevier **32**, 2369–2385 (2007)
21. Enercon: Enercon – Technology. http://www.enercon.de/en-en/21.htm (access date: June 2011)
22. Conroy, J.F., Watson, R.: Low-Voltage Ride-Through of a Full Converter Wind Turbine with Permanent Magnet Generator. Renewable Power Generation, IET **1**, 182–189 (2007)
23. GE_Energy: Wind Turbines. http://www.ge-energy.com/products_and_services/products/wind_turbines/ (access date: June 2011)
24. Ward, J.: Hierarchical Grouping to Optimize an Objective Function. Journal of the American Statistical Association **58**, 236–244 (1963)
25. Hardle, W., Simar, L.: Applied Multivariate Statistical Analysis, 3rd edn. Springer, Heidelberg (2012)
26. Johnson, R.A., Wichern, D.W.: Applied Multivariate Statistical Analysis, 6th edn. Pearson (2007)
27. Klecka, W.R.: Discriminant Analysis. Sage Publication, Inc. (1980)
28. Garson, G.D.: Discriminant Function Analysis. Statistical Associates Publishers (2012)
29. EDP: Energias de Portugal. http://www.edp.pt/ (access date: January 2015)
30. GENERG: Energia Natural. http://www.generg.pt/ (access date: January 2015)

Workshop on Cities, Technologies and Planning (CTP 2015)

A Smart Planning for Smart City: The Concept of Smart City as an Opportunity to Re-think the Planning Models of the Contemporary City

Ilaria Greco[✉] and Angela Cresta

Department of Low, Economic, Management and Quantitative Methods, University of Sannio, Via delle Puglie, 1 82100, Benevento, Italy
{ilagreco,cresta}@unisannio.it

Abstract. A "smart city" is generally meant as a technologically advanced cities, capable of joining "competitiveness" and "sustainability", by integrating different dimensions of development (economic, mobility, environment, people, living and governance), becoming self-sufficient. A broad definition that suggested different and varied visions, from which are derived identification systems and different classifications of Smart cities.

The paper part from a reflection, already started by the time the authors, on the theme of "Smart City" as "Senseable city", which means to focus the discussion no more on "how cities can be smarter", but on "how intelligent technologies can lead us to rethink the patterns of urban development by making them fair and inclusive, as well as efficient and sustainable". About, the paper tries to explore a new field of research related to the relationship between urban planning and smart city. The Smart city is certainly an opportunity to re-think the contemporary city in an innovative way, with systems of resource management in the city, aimed at improving the quality of life with respect to which the processing of integrated strategic planning is a key requirement for sustainable development of the city.

The reference is timely to the European initiative FP7-ENERGY-2012-SMARTCITIES which provides specific actions dedicated to urban planning as a basis for the construction of the Smart Cities. The initiative includes a long-term vision and refers to an urban planning in able to manage, direct and govern energy policies.

Keywords: Smart city · Urban planning · Urban development · Energy policies

1 A Critical Review of the Concept of "Smart City". The Different Approaches[1]

At the beginning of the Nineties, David V. Gibson, George Kometsky, Raymond W. Smilor published the Technopolis Phenomenon (1992), in which the term *Smart City*

---

[1] The paper is the result of a common reflection of the authors; however, the single sections can thus be attributed to: Ilaria Greco paragraphs 1, 4 and Angela Cresta paragraphs 2 and 3.

O. Gervasi et al. (Eds.): ICCSA 2015, Part II, LNCS 9156, pp. 563–576, 2015.
DOI: 10.1007/978-3-319-21407-8_40

was used for the first time connected to an urban development more and more dependent on technology and on innovation and globalization phenomena, mainly by an economic vision [23].

Despite numerous studies, the wide, literature, the various contributions on this topic and the many applications in models, actions and policies, we still do not have a shared definition of the term, or rather the concept of Smart City [24], [11].

Smart city continues to be a "label Urban" [29], a fuzzy concept, often used improperly[38] declined with many different meanings. In fact, as highlighted by Giffinger et al. (2007), the term is not used in a holistic way, but it is used for various aspects, which range from Smart City as an IT-district to a Smart City regarding the education (or smartness) of its inhabitants.

As pointed out by Hollands (2008), this vagueness of terminology is not only a problem of a uniform framework for benchmarking, but rather a deliberate choice to hide behind a general artificial all the contradictions that characterize the new urban forms [29].

However, over the last decade we have multiplied the attempts semantic and operational to "bring order" among disparate definitions of the concept and to achieve at least a shared vision of "smart city".

Without going into details of the various attempts to arrive at an univocal definition of a smart city, we can summarize the different ways in which it has been interpreted the concept of smart city into three types of approaches [41]: (1) a *techno-centered approach* characterized by a strong emphasis on new technologies and infrastructure that ITC would be the key to the smart city [8], [54]; (2) a *human-centered approach* where there is a large weight of social and human capital in defining the smart city[6]; (3) an *integrated approach* that defines a smart city from the possession of both the foregoing qualities, because the intelligent city has to ensure integration between technology and human and social capital to create the suitable condition for a continuous and ongoing process of growth and innovation [30], [10].

According to the *techno-centered approach*, there is who defines the Smart Cities of the technologically advanced city, where the sector's most important appears to be that of ICT, which would play an indispensable role in the realization of an intelligent city, technologically advanced and able to be competitive in a world scene [39]. This vision focuses on infrastructural innovation looking at citizens as end-consumers. This concept is to bring the vision of the first scholars, who spoke of the city as smart as those cities are able to leverage technology to improve the overall efficiency of the urban environment. New technologies applied to the city are an element that facilitates and sets new horizons to build and remodel an urban environment where data and information are shared, retrieved and processed to give real-time solutions. The challenge of the Smart City is that of a city that was relaunched with new technology but with the basic human intelligence that manages an intelligent city by nature. Hollands, in this regard, underlines the risk that the smart city is only a variation of the high-tech entrepreneurial city [43], [28].

According to the *human-centered approach,* largely widespread in the second half of the 2000s, the social capital represents the crucial element for building up a Smart City. The technologies, more and more widely available, are intended as "enabling tools", but insufficient to make "smart" an urban context, only by themselves. The scholars who support such a vision focus, consider, human and social capital as a starting lever for a "smart" development, recognizing a direct relationship between human capital and urban development. For this to happen it is important the presence of a "creative class", in terms both of an entrepreneurial class capable of innovating products and processes, both of a highly skilled labor force. Berry and Glaeser (2005) and Glaeser and Berry (2006) show, for example, that the most rapid urban growth rates have been achieved in cities where a high share of educated labour force is available [6], [11]. Smartness has to do with intelligence, so smart cities can simply be considered as the contrary of stupid cities [53]. Smart cities may be the city that uses its resources wisely, efficiently and effectively, to become economically advanced and self-sufficient.

The *integrated approach*, at present the most widely shared, combines the previous visions, looking at smart city as a city capable of use ICT in an extensive and intelligent way, in order to improve the overall urban performances and, above all, the quality of life of citizens. In this sense, the Smart city is that is able to manage the resources available in an intelligent, connected to a process of improving the *quality of life*, pivotal factor and common goal for the very definition. The capacity of a city to be intelligent to be assessed as a whole: the city should be viewed as an organic whole as a network, as a linked system [30]. The idea is that a smart city represents the final goal of a virtuous path – along which investments are addressed to achieve a sustainable growth, in economic and environmental terms – aimed at improving the quality of life of citizens and based on the involvement of settled communities - is currently more and more widespread [11]. After the models of urban sustainability represented by the "green city" and the "creative city", taking shape new paradigm for the modern city, the *"Smart city"* as *"intelligent city"*, in response to problems of congestion, pollution and physical degradation of modern city [26].

This approach is also what in Europe has influenced a number of studies and research on the subject, and also led several institutional initiatives aimed at building up at European Smart Cities. To arrive at a definition that brings together different criteria of analysis and the previously mentioned aspects, a few years ago the Vienna University of Technology - in collaboration with the University of Ljubljana and the Delft University of Technology - gave birth to a research on European medium-sized cities (with population less than 500,000 inhabitants). Later, this research became the ranking instrument of approximately 1600 city of EU27, plus Iceland, Liechtenstein, Norway and Switzerland.

This project, called "European smart cities", was born as part of a wider project ESPON 2013 (ESPON Project 1.1.1) and showed not only a final ranking of 70 cities, but it has remained a reference model to identify factors that make cities "smart" [13]. In this context, smart cities can be identified and ranked along six main axes or dimensions, that are: *a smart economy; smart mobility; a smart environment; smart people; smart living; and, finally, smart governance*. These six axes connect the traditional theories of urban growth and development, with the modern aspect of sustainable development of a city. Then, a middle city can be defined as "smart" when investments in human and social capital and traditional (transport) and modern (ICT) communication infrastructure fuel sustainable economic development and a high quality of life, with a wise management of natural resources, through participatory governance [11].

"A Smart City is a city well performing in six characteristics, built on the 'smart' combination of endowments and activities of self-decisive, independent and aware citizens: mobility, environment, people, living, governance, economy": this is the most complete definition dictated by the Report of the European Smart Cities, in line with the new European vision for the future development of global cities (Horizon 2020 Urban Forum, Digital Agenda - Strategy 20.20.20, Decree Digitalia, etc.) [24].

Starting from these dimensions of analysis, several other studies have been done (Boyd Cohen, City Protocol, Smart City in Europe, MIT Senseable City Lab, The European House - Ambrosetti , iCity Lab PA Forum, etc.).

But even this interpretation seems still limited. In fact, if a smart city is a city that knows how to exploit their human capital so that there is a creative and qualified context for economic development, other very important factors that are not exclusively linked to economic growth seem to be neglected. First of all the "equity": a city should not just be smart, but its smartness must cover all the inhabitants!

The *equity* is the new dimension that completes the process of smartness ensuring the development of a city in terms not only di smart city, but of *SENSEable City*. If the "*smart city*" is a city where investments in human and social capital, in the participation processes and in the technology infrastructure, are directed to sustainable and competitive economic development, ensuring a high quality of life and providing for the responsible management of natural and social resources from a shared governance, "the *SENSEable city*" is the which encourages dialogue between the different elements that make up the urban life, and that encourages more informed and fair decisions about their urban environment, with a new approach to the urban planning and an efficient use of networks. Therefore *SENSEable* city is a city that is smartness and fairness at the same time[2].

In this regard, it is interesting to take the warning issued by the sociologist and economist Sassen (2011), who believes that the new challenge is the attempt to "urbanize the technologies", that make them actually useful to new urban needs [49].

2   How to Redesign the Contemporary City: The New Challenges

The phenomenon of Smart Cities is born from the need to set forward-looking policies for the sustainable development of cities, reinforcing some very important issues for the contemporary city: technological innovation, the production of renewable energy, the Information Communication Technologies (ICT), the need for a sustainable use of resources. In this regard, the nascent concept of Smart Cities seems to indicate the main routes and needed to be taken in order for a city to become sustainable, competitive and self-sufficient [37]).

---

[2] For a reinterpretation of the concept of smartness according to the values of the *Equity City Index and a* possible evolution of the *smart city* into four different models of *Smart-Equitable Cities* and policies see Greco I., Bencardino M. (2014), The paradigm of the modern city: SMART and SENSEable Cities for smart, inclusive and sustainable growth. In B. Murgante et al., Lecture Notes in Computer Science and Its Applications, Pag.579-597, Springer.

The city today is considered the artificial ecosystem for excellence where you can deal with the problems related to climate change and the urgency of the effects of years of over-consumption of resources, which have led the city to have to find appropriate tools for the least possible impact on the territory. It is a living organism, which consumes food, water, energy and produces waste. This system to feed and regenerate needs to glass, plastic concrete, brick [34].

At the end of the twentieth century, the crisis of the economic model based on Fordism and the development of the tertiary sector impress a further acceleration to the city's transformation from an industrial city to a city of services, resisting the loss of its functions redesigning itself, individually and as part of the overall urban system.

The recent and rapid changes that have transformed the economy, especially in the last century, inevitably have started processes of urban transformation which, however, have succeeded and sometimes superimposed in a convulsed and often unregulated drawing of the cities, both in developed countries and in the other ones. In a few years, we have seen the transformation of the industrial cities in cities of services up to the current forms of conurbation and urban agglomeration, that have had as a single common factor to extend and subtract more and more space to a more rarefied urban space.

During the nineteenth century, the economic production function of the city takes on a prominent role on all other and the city became central to the development of national economies. This is the moment in which cities are beginning to grow and to expand in the suburbs. Thus, the economic and industrial development, the increase of population and the increase in urban population are linked, throughout the century, with a double thread in a cumulative growth. This was much more evident in the cities of the northern hemisphere (especially in North American and European ones) where the possibility of work related to the localization of mineral deposits formed a precise condition for regional development and industrial cities, which have designed the space, according to precise geographic forms, opposing workers' quarters to bourgeois ones, in an urban pattern of well distinguished areas [4].

But the industrial city has its beginning and its end, leaving its inheritance. The twentieth century was especially the century of the city and the suburbs. Far from growing and multiplying according to an idealized model of rationalist "modern city" rethinking existing cities, in most cases the suburbs have come to be arranged "like wildfire" around antique and nineteenth centers, laying the foundations for the development of the current shapeless metropolis and megalopolis [15].

The current urban transformations require us a further reflection. In fact, since 2009 for the first time in human history, the urban population has been surpassed the one established in the rural area and, according to a report from the Worldwatch Institute, by 2050 the 70% of the Earth's population will live in the city, a percentage than in industria-lized countries will rise up to 84%, by the middle of the twenty-first century [45].

The contemporary city now occupies 2% of the earth's surface, is home to 50% of the world population, consumes 75% of the total energy and is the cause of '80% of CO_2 emissions. The uniqueness of this stage is that these changes have not resulted in the reduction or containment of urban development of the city, which continue to grow, mutate and diversify its forms. The cities that have developed and grown in the industrialization have dilated and have become territorial systems of various types in which the flows of the industry are only one element of the complexity and, through agglomeration and conurbation, they have developed over more and more vast areas [47].

Based on the numbers above, the cities are seen in perspective, as the places will have to find solutions to problems that plague the society and will have to accept challenges regarding climate change, globalization and sustainability. The biggest challenge is to maintain and improve the standard of life of the growing numbers of the population at a rate of 1/10 of CO_2 emissions we produce today.

Therefore, it becomes necessary to solve these problems and search a solution to the current uncontrolled spatial development of the city. In particular, environmental issues have become central to economic development and their solution must start from urban areas. In fact, the cities are responsible for the majority of pollution, producing up to 70% of the total emissions of carbon dioxide, even occupying a residual portion of the earth surface. These problems appear more acute in developing countries, where the increase of the urban population are higher than in industrialized countries and it is expected that in the coming decades there will be 95% of the world's urban population growth .

In Europe, the threats to sustainable urban development are taken into particular consideration and are addressed in an integrated approach, taking into account both of environmental issues and of those social, economic, cultural and political ones. In fact, since 1999, through ESDP European Spatial Development Perspective (ESDP), the EU has been beginning to lead to programs focused to an integrated territorial development, primarily oriented to territorial balance and cooperation between the cities of the local territorial systems.

Today, the end of a period of steady economic growth causes stagnation and economic decline in many cities, particularly in those ones that are not European capitals and in the old industrial cities of Western Europe. That is leading to the gradual withdrawal of the welfare state in most European countries. So, in the framework program for research and innovation "Horizon 2020" the aging population, the low-density urban sprawl, that threatens sustainable development and that makes the service more expensive difficult to secure, the over-exploitation of resources, the lack of public transport networks, the risks to biodiversity and, finally, the issues related to the protection and maintenance of the land, threatened by the widespread hydrogeological, are all considered central problems [19].

It is clear that the "human overhead" can not be supported by the urban centers as well as designed until now and the need for smart cities and smart communities can not remain much longer concepts that are prerogative of a restricted array of experts, but will instead become a shared concept for improving the quality of life and for adapting it to the future needs of urban housing.

Today, the city must be able to prove to be self-sufficient for success. Must be able to create an agenda with common objectives, to grow in an integrated way and share the same vision; new public-private cooperation and new strategic directions that allow integration between Hard and Soft infrastructure, to create the development of a sustainable urban environment. Therefore, the development of tangible and intangible infrastructure can not be addressed only to the economic, environmental and political efficiency but rather they should promote social inclusion, quality food and good life.

So, a smart city is not a project but the beginning of an overall process of *sensing* and *actuating* for the transformation of the city, where there are particular needs of citizens, active and passive actors in the process.

3   The Energy Challenges for Cities. The European Initiative FP7-ENERGY 2012 SMARTCITIES for an Integrated Planning Sustainable of Resource

Although the integrated approach is currently the most widespread in scientific literature, the institutional initiatives aimed at building up Smart Cities are still characterized by a sectorial approach. Policies, plans and actions are, in fact, mainly oriented to engineering of selective interventions.

The implemented measures concerned essentially the "high impact" sector as the energy, the transport of goods, the mobility, the waste management, and based mainly on high-tech solutions. Also the sustainability has been seen so far strictly in energy and environment key, through choices and technologies that save energy, or from a functional point of view, through integration of e-participation techniques such as online consultation and deliberation over proposed service changes to support the participation of users-citizens in the democratisation of decisions taken about future levels of provision.

Also at European level, the smartness is mainly read in the environmental and energy key, so much so that both in the Strategic Plan for the Energy Technologies of 2007 (European Commission, 2007), and in the resulting Technology Roadmap (European Commission, 2009), there is precise and explicit reference to the smart city and a specific budget dedicated to this axis [22].

Moreover, the concept of Smart cities is in the Annual Programme of Work of 2012, drawn up by the European Commission; it focuses on the theme of cooperation between countries and EU Member States and is part of the *Seventh Framework Programme* (FP7) *for Research and Technological Development* (2007-2013). The program has a budget of 50 billion euro and aims on the enormous potential existing in the field of research and innovation, seen as key factors for competitiveness, employment, sustainable growth and social progress. The Initiative of Smart Cities is inserted inside the theme 5, relative to the energy policies, with the primary objective to address the European challenge towards more sustainable energy systems, less dependent on imported oil and based on a mix of different energy sources, in particular renewable energy.

The energy theme of the Seventh Framework Programme of 2012 focuses on the implementation of the *Strategy Energy Technology Plan* (SET-Plan), one of the pillars of European energy and climate policies. Most of the topics supports the SET-Plan European Industrial Initiatives (EII) Technology Roadmaps and implemetation Plans. The Roadmaps specify the research and indicate the needs for the next ten years, with the aim to accelerate the growing market of the most promising technologies in the area of solar, wind, bio-energy, smart grids, the Carbon Capture and Storage and smart cities. The focus of the implementation dell'EIIs was the result of a concentration of activities, objectives and priorities decided by mutual agreement, by industry, Member States and the European Commission. An important element of the initiative is to encourage the integration of FP7 projects with national and regional projects more relevant, to establish models of interventions that facilitate the sharing of knowledge and the dissemination of results.

In FP7 was reinforced research on energy efficiency to achieve results. In this section you enter the policies related to the Smart Cities and Communities, as the issues of energy efficiency and climate change objectives are cornerstones for FP7. The new area dedicated to the Smart Cities and Communities, is part of the Activity 8 (Energy Efficiency and Savings), whose objectives can only be achieved through a holistic approach and with special attention to new technologies. In addition, the innovative parts of the work program also strengthen the cohesion of the non-technological aspects, such as the involvement of users, through the active participation of citizens, the development of methodologies already tested and industrial leadership. It is, therefore, of a cross-sectoral approach between different public and private sectors in different fields of action.

The Initiative FP7-ENERGY-2012-Smart Cities and Communities is part of SETPlan and includes a number of topics related to energy, such as energy efficiency, power networks, the production of renewable energy and other issues related to urban electricity, heating and cooling, transport, waste and water management. The arguments challenged focus mainly on the energy dimension, as is expected from Energy-efficient Buildings (EeB) Public Private Partnership, which is clearly aimed at identifying technical solutions, economic and financial implications for improving the energy efficiency of city and neighborhoods; in this way it supports the initiative Smart Cities.

In this context the European call spurs, in fact, the city to experience innovative measures to accelerate the implementation of those technologies which allow a considerable reduction of emission of carbon. To get a potential replication of interventions, at least three cities in the United States and or Associated Nations must team up to propose a project under the call FP7-ENERGY-2012-SMARTCITIES.

The issues related to the call FP7-ENERGY-SMARTCITIES refer to *urban planning and the role of energy strategies* as a basic requirement for the success of the project. It speaks specifically of "Strategic sustainable planning and screening of city plans"[3]. It is the first action dedicated to a urban planning as a basis for the construction of the Smart Cities; it aims to create strategic models of sustainable planning to direct the flow of energy efficiency of different sectors and in different cities throughout Europe and intend to support the city with innovative and ambitious projects, which provide integrated urban planning.

All the interventions for the city must be directed towards energy efficiency through redevelopment of a major share of the housing stock, the energy systems,

---

[3] The other two themes are: i) Large scale systems for urban area heating and/or cooling supply and ii) Demostration of nearly Zero Energy Building Renovation for cities and districts.

heating and cooling, using smart grids, climate adaptation and mitigation, efficiency of the aqueducts, efficient waste collection, special treatment of them recycling and energy use; and yet, transport systems and adequate and modern mobility. Successful projects will gather Proposals of those cities, who will present, with credible evidence, topics ambitious and innovative planning, those, namely, who will find an ideal mix of all these elements, and will indicate the timeline, costs and the period of return on investment.

The outcomes expected from the planning will have to demonstrate that the integrated approach follows an economically better result than the individual project; and this will allow the exchange and dissemination of the results of the Indicators of Key Factors.

The initiative embraces a long-term and refers to an urban planning can manage, direct and govern energy policies. For the first time you refer to a sustainable urban planning in energy policies that also collects policies initiated earlier. The call smart Comminuties Cities lets to start a new urban experimentation, able to give new perspectives to the cities that decide to adopt energy policies aimed at sustainable integrated planning of resources.

4   Smart Strategies for Innovative Methods and Planning Practices: Some Reflections

Today, the urban planning is called to consider a new theme, that of the Smart City as a concept to study and analyze to develop as much as possible objectives and practices for a smart urban development. There is much talk of sustainable city, town to zero CO_2 emissions; it speaks of a city capable of exploiting the resources it has available in an intelligent and sustainable, to be able to win his biggest challenge: reducing the level of pollution.

We are facing a very important transition, which enormously takes into account the sustainable development, which in recent years has been the center of the project dimension and sees the transaction from traditional urban planning based on the mechanistic principle in solving problems related to the city, to an organic vision of the urban and metropolitan.

The city is certainly a complex system and, therefore, difficult to govern, but this is the challenge that today the Smart Cities offer, that is be able to give a new vision of the city system, through improved resource management and a systemic vision of choices of urban development. Focus on Smart City concept means using the word itself as a tool to share the same vision of the future. Means sharing perspectives development and ideas that lead to success. It is an essential tool to achieve concrete actions from different perspectives [53].

To realize a malleable concept, elusive, unpredictable and often rhetorical which is to Smart city, the Cities must find the right balance between local resources and policies aimed at the implementation of smart strategies for re-thinking methods and innovative planning practices.

The Smart City concept can be defined as a concept "opportunistic mobility": opportunistic because it may serve to give a jolt to the city, the way to lead their expansion, their development and resource management; mobility as potentially able to mobilize resources, both economic and financial, creative and human capital. The opportunities are often obvious, though still little concrete.

Still, according to Jan Vogelij the Smart City concept can be defined with two main verbs: *advantage* of the available resources of the city to enhance the strengths and work on weaknesses to *solve* problems that relate to the specific case.

An eco approach can help to balance the inputs and outputs. The efficiency is not only the creation of the effects resulting from the minimum cost, but it is also possible to search results and benefits from synergies. A good structure and urban planning helps to optimize efficiencies

The concept of Smart City can generate success precisely because it is able to maximize the technological applications to improve the functional efficiency of the social, economic and physical. A smart city is relatively dense, with a mixed of uses that connects different activities. In a smart city it comes to safety, health and social inclusiveness as characteristics essential for a well-designed urban environment. It is important to be able to stimulate the minds of the community for future innovation in the city. It is also essential to initiate processes of cultural regeneration, but at the same time we need to enhance the story of a city, to be considered as a value. The past of each area offers possibilities of natural and cultural assets that can be used as future urban development. "A smart city connects the past through the present with the future" [53].

The concept of Smart city, therefore, embodies the holistic view of things, according to which the functional sum of the parts is greater/different from the sum of the performance of the parts taken individually. The success of the Smart Cities is to have an integrated planning, cooperation between different sectors that are the key elements to contrast the problems of planning today, such as the fragmentation of the different administrative parts of the city in the first place and the city physical as a result of policies and projects after. It can be, therefore, a tool that is part of the urban planning and helping to address sustainable development strategies and long-term technological innovation, in order to improve the quality of life of the inhabitants. A tool to re-evaluate the central role of public service to the city, to stimulate synergies between public and private actors, and is an indispensable tool for re-put the issues of planning and management "intelligent" city in the foreground.

Amsterdam is an essential focus that acts as the glue between definitions, sectoral experiences on the one hand and policy and practice dictated by the European Union on the other, where he finds himself concretely action, policies and long-term vision of a transition process called Smart city.

The inter-scalar approach adopted by the Structure Plan 2040, from the local to the strategic vision of Amsterdam Smart City, in fact, part from the exploitation of the resources that the city and its metropolitan area have. Point to the existing city as the central hub of the future development, identify in areas already built the potential to attract population and working class, introducing sustainability in the parks to give breath to a dense city, converting the brown fields in areas of new construction and be able to connect this with the surrounding metropolitan area seems certainly be a winning strategy. But, the real novelty of the new structural plan 2040 are the energy issues: strategies to prepare the city to the new era after the fossil. The strategic plan focuses on strong themes of energy conservation and use of alternative energy sources as keys of metropolitan development [44].

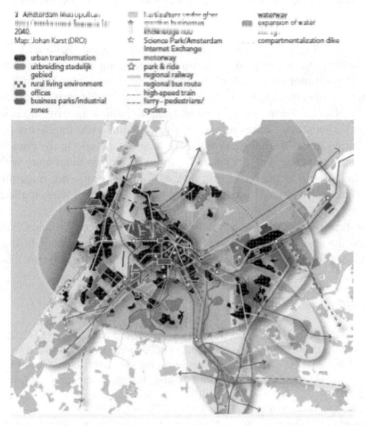

Fig. 1. Amsterdam Metropolitan Area Development Scenario for 2040. (https://www. amsterdam. nl/... /plan-01-2011-eng.pdf, p. 7).

In Italy, the Smart City concept is beginning to be acknowledged and made official during the agreement SMAU (International Exhibition of Information & Communication) - ANCI (National Association of Italian Municipalities), establishing the "Italian Observatory on Smart cities", and above all with the introduction by the Ministry of Education of the strategy Smart City as a strategy for the entire country.

The main objective of the observatory founded by the SMAU-ANCI is to sensitize the public administration on theme of Smart Cities and try to facilitate a meeting between government and the private sector. The Observatory aims to understand how, through the support of ICT, we can improve the quality of life in urban spaces, in terms of mobility, energy policies, waste and services to citizens.

Many cities, among which Bologna, Genova, Milano, Piacenza, Pisa, Venezia e Torino are working on a follow-up to the EU directive, which, through the Communication to the European Parliament, the Council, the European Economic and Social Committee and the Committee of the Regions COM (2009) 519 final "Investing in the development of Low Carbon Technologies (SET - Plan)", stimulates Smart cities to set up projects in order to improve the quality of life of citizens, engaging in a decisive way and on all fronts on the problems of climate change.

However, in Italy the pilot projects presented to the call of the European Smart cities and Communities refer to certain neighborhoods or areas, believed to be particularly interesting to accommodate advanced technologies, specific actions and measures of innovation, aimed at achieving the objectives established in Europe: CO_2 reduction, development of new integrated systems, new public-private relationship.

Cities like Genoa, Turin, Bologna are working to build long-term strategies: Genoa in a vision perhaps more infrastructure, but always linked to improving the lives of citizens. Turin aims at a strategic vision to reintegrate the city in the competitive circuit. Bologna is opening up to a smart planning of metropolitan area, with a social approach to the term in the broad overview of the Smart city, not forgetting the importance of making a city of Bologna to more vocations, which looks to the complexity of the city.

Still far, therefore, a systemic and integrated vision of the city would develop all the resources in the area. The Italian cities seem chasing the new concept of Smart City to not be excluded from the dynamics of European urban, rather than do own this.

References

1. AA.VV. Smart Cities nel mondo, CITTALIA-fondazione ANCI Ricerche (2012)
2. AA.VV. Smart Cities in Italia: un'opportunità nello spirito del Rinascimento per una nuova qualità della vita, ABB e The European House-Ambrosetti (2012). http://www.abb.it/
3. Annunziato, M.: La roadmap delle Smart Cities. Energia, Ambiente, Innovazione **4-5**(1), 33–42 (2012)
4. Batty, B.: The New Science of Cities. The MIT Press (2013)
5. Bencardino, M., Greco, I.: Smart Communities. Social Innovation at the Service of the Smart Cities. TeMA Journal of Land Use Mobility and Environment, Smart City Planning for Energy, Transportation and Sustainability of the Urban System, Special Issue, giugno 2014, 39–51 (2014)
6. Berry, C.R., Glaeser, E.L.: The divergence of human capital levels across cities. Papers in Regional Science **84**, 407–444 (2005)
7. Berthon, B., Guittat, P.: Ascesa delle città intelligent. Outlook, 2-2011. Accenture (2011). http://www.accenture.com
8. Cairney, T., Speak, G.: Developing a "smart City": Understanding Information Technology Capacity and Establishing an Agenda for Change (2000). http://trevorcairney.com/wp-content/uploads/2012/11/IT_Audit.pdf
9. Camagni, R.: On the concept of territorial competitiveness: sound or misleading?. Urban Studie **39**(13), 2395–2411 (2002)
10. Campbell, T.: Beyond smart cities. How cities network, learn and innovate. Earthscan, Londra-New York (2012)

11. Caragliu, A., Del Bo, C., Njkamp, P.: Smart cities in Europe. Paper presented at the Conference III Central European Conference in Regional Science, CERS (2009)
12. Cardone, M.: La rivoluzione delle smart city è in corso, QualEnergia.it (2012). http://www.qualenergia.it/articoli/20120802-la-rivoluzione-delle-smart-city-in-europa-e-negli-usa
13. Cierra, M · Città Creativa 3.0. Rigenerazione urbana e politiche di valorizzazione delle armature culturali In; Cammelli, M., Valentino, P.A. (eds,) Citymorphosis. Politiche Culturali Per Città Che Cambiano, Giunti, Firenze, pp. 213–221 (2011)
14. Centro Regionale di Informazione Nazioni Unite (Unric), Città, 18 giugno 2012 (2012). http://www.unric.org/it/rio20/28161-citta-18-giugno-2012
15. Ciorra, P.: La fine delle Periferie. Nascita e morte della periferia moderna (2010). http://www.treccani.it/enciclopedia/lafinedelleperiferie(XXI-Secolo)/
16. Cittalia, Smart cities nel mondo. Cittalia e Fondazione Anci Ricerche, Roma (2011)
17. Cohen, B.: The Top 10 Smart Cities on the Planet (2012). www.fastcoexist.com/1679127/the-top-10-smart-cities-on-the-planet
18. Cohen, B.: The Top 10 Smartest European Cities. Fonte (2013). http://wwwfastcoexist.com
19. Comunicazione COM 808 della Commissione al Consiglio, al Parlamento Europeo. Programma quadro di ricerca e innovazione "Orizzonte 2020" (2011)
20. Crivello, S.: Competitive city and sustainable city: some reflections on the relationship between the two concepts. Sociologia Urbana e Rurale 97, 52–67 (2012)
21. Deakin, M.: From city of bits to e-topia: taking the thesis on digitally-inclusive regeneration full circle. Journal of Urban Technology 14(3), 131–143 (2007)
22. De Luca, A.: Come (ri)pensare la smart city. EyesReg Giornale di Scienze Regionali 2(6), 143–146 (2012)
23. Gibson, D.V., Kozmetsky, G., Smilor, R.W. (eds.): The Technopolis Phenomenon: Smart Cities, Fast Systems. Global Networks. Rowman & Littlefield, New York (1992)
24. Giffinger, R., Kraman, H., Fertner, C., Kalasek, R., Pichler-Milanovic, N., Meijers, E.: Smart Cities - Ranking of European medium-sized cities. Centre of Regional Science, Vienna (2007) http://www.smart-cities.eu
25. Greiner, A., Dematteis, G.: Geografia umana. Un approccio visuale, Torino, UTET (2012)
26. Greco, I., Bencardino, M.: The Paradigm of the Modern City: SMART and SENSEable Cities for Smart, Inclusive and Sustainable Growth. In: Murgante, B., Misra, S., Rocha, A.M.A., Torre, C., Rocha, J.G., Falcão, M.I., Taniar, D., Apduhan, B.O., Gervasi, O. (eds.) ICCSA 2014, Part II. LNCS, vol. 8580, pp. 579–597. Springer, Heidelberg (2014)
27. Glaeser, E.: Triumph of the City: How Our Greatest Invention Makes Us Richer, Smarter, Greener. Penguin Books, Healthier and Happier (2011)
28. Harvey, D.: From Managerialism to Entrepreneurialism: The Transformation in Urban Governance in Late Capitalism. Geografiska Annaler 71(1), 3–17 (1989)
29. Hollands, R.G.: Will the real smart city please stand up? Intelligent, progressive or entrepreneurial? City 12(3), 303–320 (2008)
30. Kanter, R.M., Litow, S.S.: Informed and Interconnected: A Manifesto for Smarter Cities, Working Paper 09-141, Harvard Business School. http://www.hbs.edu/faculty/Publication%20Files/09-141.pdf
31. Komninos, N.: Intelligent cities: innovation, knowledge systems and digital spaces. Spon Press, London (2002)
32. Komninos N. (2011), Smart Cities are more competitive, sustainable and inclusive, Cities. Brief, n.2.
33. Kotkin J. (2009), The World's Smartest Cities. http://www.forbes.com

34. Landry, C.: The Creative City, A Toolkit for Urban Innovators, London. Earthscan, Sterling (2000)
35. Murgante, B., Borruso, G.: Cities and Smartness: A Critical Analysis of Opportunities and Risks. In: Murgante, B., Misra, S., Carlini, M., Torre, C.M., Nguyen, H.-Q., Taniar, D., Apduhan, B.O., Gervasi, O. (eds.) ICCSA 2013, Part III. LNCS, vol. 7973, pp. 630–642. Springer, Heidelberg (2013)
36. Murgante, B., Borruso, G.: Smart cities or dumb cities? Città, riqualificazione urbana e pioggia di dispositivi elettronici. Rivista GEOmedia **17**(4) (2013)
37. Murgante, B., Borruso, G.: Smart city or smurfs city. In: Murgante, B., Misra, S., Rocha, A.M.A., Torre, C., Rocha, J.G., Falcão, M.I., Taniar, D., Apduhan, B.O., Gervasi, O. (eds.) ICCSA 2014, Part II. LNCS, vol. 8580, pp. 738–749. Springer, Heidelberg (2014)
38. Nam, T., Pardo, T.A.: Conceptualizing smart city with dimensions of technology, people, and institutions. In: Proceedings of the 12th Annual International Conference on Digital Government Research (2009)
39. Niger, S.: La città del futuro: smart city, smart community, sentient city. (2012) www.astrid-online.it
40. Papa, R.: Smart Cities: Researches, Projects and Good Practices for the City. TeMA Journal of Land Use Mobility and Environment **6**(1) (2013)
41. Papa, R., Gargiulo, C., Galderisi, A.: Towards an Urban Planners' Perspective on Smart Cities. TeMA Journal of Land Use Mobility and Environment **6**(1) (2013)
42. Partridge, H.: Developing a human perspective to the digital divide in the Smart City (2014). http://eprints.qut.edu.au/1299/1/partridge.h.2.paper.pdf
43. Peck, J., Tickell, A.: Neoliberalizing space. Antipode **34**(3), 380-404 (2004)
44. Plan of Economically strong and sustainable Structural Vision: Amsterdam 2040. https://www.amsterdam.nl/.../plan-01-2011-eng.pdf
45. Potter, G.: Urbanizing the Developing World (2012). http://vitalsigns.worldwatch.org/vs-trend/urbanizing-developing-world
46. Ratti, C.: Smart city, l'onda può partire da noi, Avoicomunicare (Telecom) Interview. https://www.avoicomunicare.it/blogpost/ambiente/smart-city-l-onda-puo-partire-da-noi
47. Sen, A.K.: Commodities and Capabilities. Oxford University Press, Oxford (1985)
48. Soja, E.W.: Seeking Spatial Justice. University of Minnesota Press, Minneapolis (2010)
49. Sassen, S.: Who needs to become 'smart' in tomorrow's cities'. In: Keynote Speech at the LIFT Conference "The Future of Smart Cities'' (2011)
50. Schaffers, H., Komninos, K., Pallot, M., (eds.): Smart Cities As Innovation Ecosystem Sustained by the Future. Internet, Fireball White paper (2009). http://www.fireball4smartcities.eu/
51. Shapiro Jesse, M.: Smart Cities: qualità della vita, produttività, e gli effetti di crescita del capitale umano **88**(2), 324–335 (2006)
52. Townsend, A.: SMART CITIES: Big Data, Civic Hackers, and the Quest for a New Utopia. W.W. Norton and Company, New York (2013)
53. Vogelij, J.: Some thought about Smart Cities. Soest, NL (2011)
54. Washburn, D., Sindhu, U.: Helping CIOs Understand "Smart City" Initiatives, Forrester Research Inc. (2010). http://www.uwforum.org/upload/board/forrester_help_cios_smart_city.pdf
55. Woods, E., Bloom, E.: Smart Cities, Intelligent Information and Communication Technology Infrastructure in the Government, Buildings, Transport, and Utility Domains, Executive Summary, Pike Research, Cleantech Market Intelligence (2011)

Better Decisions for a Better Quality of Life: The Potential of Rural Districts Supported by e-governance Tools

Arnaldo Cecchini and Alessandro Plaisant[✉]

Department of Architecture and Planning, University of Sassari,
Palazzo Pou Salit, Piazza Duomo 6 07041, Alghero, Italy
{cecchini,plaisant}@uniss.it

Abstract. E-participation in decision making is gaining ground. An increasing number of administrational structures currently try to foster innovation processes at all levels of government, supported by EU cooperation strategies and funding. The aim of the paper is to test the usefulness and opportunities deriving from a mix of e-participation and "proactive" decision support tools by defining a policymaking process,. The authors present an overview of the work-in-progress of policymaking through a series of methods and tools to encourage participation and identifying consistent and low cost policy choices in PA. We conclude that this bottom-up approach, which is very different from the technology-based smart city paradigm, is extremely useful. It provides institutions with a simple, immediate and free use of a pool of ideas. It also supports administrational structures by organizing a number of basic services in low density rural areas affected by structural limitations in terms of accessibility, loss of population and public services in a better manner and at lower cost.

Keywords: e-participation · Strategic urban planning · Decision support · Self-sustainability

1   Constructing a Shared Framework for Policymaking

Better organizational planning needs new forms, less top down, more participation, more flexibility and more environmental awareness. The implementation of strategic policymaking can no longer be developed only in one level of government, as was often the case previously. Some means for renewal must be found.

Furthermore, we need to explore innovative itineraries for an interactive policymaking, open to different perspectives, supported by "proactive" decision support tools to encourage participation and identifying consistent and low cost policy choices for PA. This approach can be elaborated in at least three perspectives. Firstly, by re-thinking the role of public action in urban governance and involving the citizens through public participation, to manage processes at all levels in a process of sensitivity and mutual learning. This means adjusting tasks and competencies of the techno-structures of PA to new generations of tools and strategies, in order to deal with complex situations. Secondly, strategic policymaking can take root only in societies capable of self-organizing with support from institutions that provide the conditions

© Springer International Publishing Switzerland 2015
O. Gervasi et al. (Eds.): ICCSA 2015, Part II, LNCS 9156, pp. 577–592, 2015.
DOI: 10.1007/978-3-319-21407-8_41

for these organizational skills to develop, where weak or absent. For example, the role of the conventional structures for the implementation of policies could be different in the future: it could be assumed by unforeseen subjects, with new forms of structuring, which may be different for different policies or projects, and, in some cases, they may consolidate on an institutional basis. Last but not least, alternative approaches could be tested in building plans as well as support administrations to better and cheaply organize a number of basic services and facilities.

Since internet has become a place of content production, there has been increased focus on the interaction and active production of content, including co-production as a leisure ([14, 15], [11]). Furthermore, the presence of tools and applications for easy access have increased the number of daily active users on the Internet. This cultural revolution is important from several points of view. New opportunities to choose modes of communication affect our choice of media (The New York Times or Twitter for instance), science (peer-to-peer research) and work and innovation (models of social innovation). The ocean of free time hours, that we spend (the so-called Shirky's "cognitive surplus" [15]) to create Wikipedia-size collectively edited projects is another . This paradigm puts focus on Public Administration. One example is the Open-Government Initiative, introduced by President Obama in 2009 [9] strengthens democratic participation in the USA by supporting suggestions and opened up for better guidance in making better and more informed choices.

The Web 2.0 environment has enabled evident advantages in public interaction. For instance social interactions aren't necessarily conditioned by constraints in terms of space, distance, residence, or time, which was the case in the past. Furthermore, the asynchronicity of the response allows us to take part in processes as informed persons, and then participate by expressing our opinion. Also, there is no limit to the number of consultations. The latter would, of course, be impossible without the Internet [11]. Furthermore, the Web opens new forms of participatory governance, that can be summarized in the concept of "online or wiki-policymaking". Cottica [4] points out some features that help us understand these new tools and modes for policymaking. The acceptance of new tools to support public opinion- citizen journalism - for instance, is a novelty. As these mechanisms started to occur "[...] public administrations lose, in part, control of their actions, but, in exchange they receive exchange analysis capacity and decision-making by collective intelligence. Hence, PA becomes more effective" (*Ibid.*). The community organization is another novelty. Government authorities can support the creation of a community, but cannot control it. The "social experience" of Kublai is explanatory: a community for creative people that - in different roles - can identify, plan, communicate and carry out initiatives to produce economic feedbacks. The already quoted principle of transparency enables shared access to Open Data among administrational organisations. First of all, the rules for interaction must be established and enforced though, even if the result will be a bulletin board of complaints. The last two principles suggest the ease of understanding in communication, by speaking for themselves through single ideas and not on behalf of their organization, and the award of merit and promotions through evaluation systems related to criteria such as popularity and quality of ideas. On the contrary, wiki policymaking is affected by two types of inaccessibility: the technological and cognitive digital divide, especially in low density residential areas, affected by structural limitations in terms of accessibility, loss of population, and public services consequently. We think that open government and wiki poli-cymaking should be used in parallel with more traditional strategies for policymaking and that traditional strategies should not be entirely abandoned. Various group-management techniques can operate together with computerized tools to represent knowledge and better explore interactively the space of opportunities. Our emphasis is not so much on informational support or on methodological aspects characterizing all the steps of the plan, as it is on developing a mix of techniques, tools and services more or less innovative, measured directly in the contexts.

2   The Self-sustainable Territorial Management: An Interactive and Participatory Policymaking Process

We propose an interactive and participatory self-sustainability-oriented policymaking process, to create an interactive dimension, with the dual intent of feeding the territo-rial debate and exercising urban functions in low density residential rural areas. Various levels of government, different parts of the territory and different territorial subjects join together to harmonise rules for planning and management of the forms and processes that concern fields of collective discussion, such as the environment, services, leisure, etc., depending on their role, relevance, competences and the activi-ties they carry out, and according to their own point of view or the interests they rep-resent. In this form of planning the value represented is not that of the single groups, single hamlets or boards, but that of efficacy of public policies on problems of collec-tive interest and acknowledgement of the strategies and procedures agreed both in municipal and supra-municipal urban planning instruments and in regulation instruments of the boards or groups involved in the agreement.

2.1   Building an Organizational Space for Interaction

There is a vast literature about initiatives to be pursued to increase the range and the quality of projects that can be generated in a space of interaction. A space of interac-tion is configured as an open organisational space (for citizens but also visitors), in which relations, interpersonal exchanges and inter-subjectivity are condensed. In this space, the policy-making process grants proper importance to what ordinary participa-tory pathways leave suspended or sometimes ignore: the process of construction and continuous reconstruction of sense and significance of the decisional situations in-volving territories, together with the patrimony of successes and failures, and projects and stories connected with them. The space of interaction, with the aid of instruments that favour participation in the territorial debate (e.g. telematics access and open mapping, but also instruments aiding decisions on and evaluation of policies), must guarantee that all subjective rationalities that concur altogether in constructing inter-subjective rationality, those representing institutional ownership, technical knowledge (the experts and the techno-structures) and the context (the citizens and all subjects involved), are represented and organised according to a system of common rules [16]. On the one hand, it can act and become valid through appropriate administrative instruments that sanction their procedures, on the other it can define shared rules of self-organization, reinforcing the role of the administration for the governance of the territory. Thus, an interactive space supported by specific decision support instruments and tools, linked to real services, can help exercise urban functions in a shared way.

Some guidelines allow to clearly draw the outlines of the space of interaction in the case of low density residential rural areas: a continuous sensitivity-making process, together with procedures and tools sustainability-oriented (environmental, social, economic and institutional sustainability) about present and future actions; an accessibility, not only physical, but also informational and cognitive, to services and facilities for everybody; the implementation of community development projects; the identification and qualification of public spaces and disused buildings; the diversity enhancement; a synergistic cooperation between communities and administrative techno-structures; information and tourism promotion. The first step consists in identifying who will participate in the space of interaction. The participatory process, organized in 4 circular concentric levels, is the core of the project success. Each level is expected to activate several tools for communication, information and participation for each phase of the process, supported by a Decision Support System (DSS). The 1st level is composed by citizens, organizers, a local support group and planners. Actors are directly involved to promote the use of e-participation tools and their involvement must be participative and formal (newspaper's articles, public meetings, focus groups, DS and e-participation tools). The 2nd level is composed by the municipalities, operators, associations, ancillary institutions, whose involvement in the process must be linked to the construction of agreements and joint programmes and, therefore, to create consensus on choices (newspaper's articles, questionnaires, public meetings, focus groups, DS and e-participation tools). The 3rd level is the supra-local level of governance, whose involvement in the process must be reached on consultation (specific questionnaire, DS and e-participation tools). The 4th level is composed by the "wiki Community" and, potentially, all visitors, whose involvement in the process should be reached by a multi-way approach (e-participation tools). Among them, we distinguish 4 categories of actors and stakeholders: wage earners and the techno-structures of the municipalities, which manage the DSS platform and implement the tools; volunteers, which use their free-time to develop the system; interactive users (city users, visitors and tourists), which express their ideas, opinions and evaluations and, at the bottom, passive users, which use the services without implementing them. In the space of interaction, everyone can work at he/she prefers, sharing skills and knowledge, and possibly changing the range of interaction over time according to their needs.

2.2   Which Instrument for Policymaking? Operating Modes of Interaction

A participatory policymaking process doesn't require high investment in terms of money and time. The self-sustainability policymaking process is composed by three main phases or steps: 1. the problem definition; 2. the discussion; 3. the implementation and evaluation (see Fig. 1).

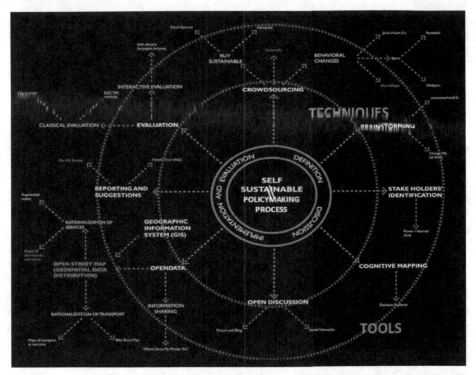

Fig. 1. The policymaking process

Each level of the process highlights a series of techniques and methods with their own specific interactive tools. The combination of techniques or their different use according to the different phases or functions, rather than the selection by existing techniques or the creation of hybrids, are particularly important for the construction of government policies. The aim is the flexibility of the process, so that we can modify the choices based on the needs of the problem addressed. Note that there is a massive number of e-participation tools, but for the purpose of this paper we will only point out two families of instruments: a) those involved on the conventional aspects of the administration – enforcing its action – and b) those that work on strategic aspects. We define here as conventional aspects, those for which authorities are already required to operate and provide services to citizens. In this sense, several tools allow citizens to report the daily problems of their neighbourhood. This type of instrument has been developed in countless variations, clones and applications in different fields. The system basically works well and is based on the principle that "many eyes see more than few" ([8; 11]). Then there are the tools that are customized to foster strategic processes in policy-making. Among them we recall platforms for collecting ideas and opinions (borrowed from the commercial logic of crowdsourcing), consultative instruments for co-drafting and suggesting changes in documents, speeches, behaviours (*Ibid.*).

The problem definition is the recognition of the decision, the process by which situations that require a response can be recognized. It concerns fields of collective discussion, such as the environment, services, etc. As Meltsner [7] argues, the definition of a policy problem requires the identification of the actors, the motivations and the beliefs, the resources and the places.

2.3   Tools for Representing Knowledge and Encouraging Interaction

Several techniques and tools support decision makers for representing knowledge upon specific problems. The technique of "crowdsourcing" combines the efforts of numerous self-identified volunteers or part-time workers, where each contributor of their own initiative adds a small portion to the greater result. For our purpose, crowdsourcing technique can be supported by the use of some specific tools, such as platforms for collecting ideas and opinions: *Ideastorm*, *UserVoice*, *Ideascale*, or some web applications to "buy sustainable", e.g. *Food sprout* (Naber & Chang, 2011), to manage "behavioural changes", e.g. *Mom maps* (Miller, 2009), *Junk.Hunt.Ca* (Thumb Genius Software, 2010), *Runtastic* (GmbH, 2009) or *We-Sport*. Brainstorming technique supports to bring out ideas through users' creativity. Some customized tools can be useful to manage this step, for example, *ChangeByUs*[1], a website run by the New York City's Office of the Mayor showing a public noticeboard in the city of New York to allow citizens to share ideas, create projects, discover resources, and make the administration to make the city more liveable. Each user can imagine to participate in a huge on-line "Metaplan", and there is an important section named "who is listening", managed by experts and technostructures in PA. Third technique to be considered in the first step is "stakeholder analysis", to identify potential networks of actors and stakeholders, such as new coalitions, formal or otherwise, that can be formed to assess the attitudes regarding the potential changes due to a plan or a project. Tools for SA in form of grid, network maps, stakeholder rolethink etc. [5] support decision makers to consider a "list" of stakeholders, broken down to an appropriate level, with the aim to establish a framework of shared strategies to be undertaken.

To be participatory, a planning process must include interaction, mediation and negotiation in each phase of the process and a continuous accompaniment of these procedures by the institutions. For example, in the deliberative approaches assume considerable importance the discussion and the debate among participants within the path that will lead to a decision. More correctly, public discussion, rational arguments and constructive debates represent the backbone of every process configured as a deliberative. In this way, the general goal is the integration of technical, administrative and political capacities of the institutions with capacities, knowledge, practices and experience rooted in the context, with the aim to trace a map of the process together. Various group management techniques for representing knowledge and discussion are used together with different analytical techniques, within successive steps of the planning processes, so as to better manage wicked and ill-defined problems. In this way the computerized techniques used move further away from acting simply as "decision support" tools and move closer towards becoming more proactive, "decision aiding" instruments [17]. There is a vast literature about techniques supporting decision makers in this step of the process. We consider the testing of the cognitive mapping technique through

---

[1] http://nyc.changeby.us/.

software program[2], that allows to prompt the development of a model by highlighting values, beliefs, and assumptions an individual has about a particular issue [5]. By mapping and visualizing the debate, it captures not only knowledge about the background context but also an understanding of the relationships that underlie such background knowledge through the recognition of goals, strategies/key issues, actions and opportunities (Fig. 2).

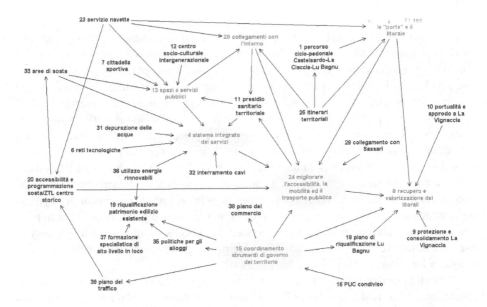

Fig. 2. DS tool for representing and analyse knowledge and highlighting key issues: cognitive mapping and Decision Explorer®

A number of analyses support the highlighting key issues in the model, by using, for example, cluster analysis, domain analysis and centrality analysis. The latter give an indication, on the one hand, of the importance of some fundamental themes within the model and, on the other hand, about the necessity of further tests on less represented constructs. Therefore, through investigation about clusters we are able to define which areas of interest and main strategic options are arisen in the interactive space, in addition to the issues with the higher score both in central and in domain analysis. On another point of view, "open discussion" techniques strengthen networking capacities offered by the internet to support discussion and orient points of view. Just think about the role of social media (*Youtube, Facebook, Twitter*) in the Northern African revolutions or protest movements in Spain ("Indignados"), US ("Occupy Wall Street") and Greece ("Aganaktismenoi") for exchanging information, organizing protest events, mobilizing participants.

---

[2] *Banxia Decision Explorer*, 3.3.0 v. (Banxia Software ltd.). The authors have tested cognitive mapping software in different public participation workshops with PA and three strategic plans for small Italian towns.

Some tools for online communities support to empower them through collective intelligence: *Assembl* (Imagination for People), helps communities to focus on moving an unstructured debate toward a co-built set of innovative ideas, through a process of "rough consensus". *DebateHub* (Open University's Knowledge Management Institute) allows to debate ideas and prioritise contributions in order to collectively organise and progress best argued ideas. *Edgesense* (Wikitalia) adds social network analytics to *Drupal* forum and community sites, by representing online conversations as a network of comments to foster collective intelligence processes.

The last step of the process is the implementation and evaluation phase, where we focus on techniques of "data collecting", mainly Open data, matched with the use of crowdsourcing tools on a geographical basis. At this step, group of ideas are structured and make it usable directly for designing services and facilities. To enable consistent policies choices and actions we need different forms of implementation (and evaluation consequently). We take into account new associations with variable geometry according to the possible scenario put in place. They can involve not only formal institutions, but also public and private entities of various kinds, which can be leaded to identify themselves in a project proposal. This approach cannot be embedded in a background of conventional techniques and tools for policymaking. From the implementation side of the process, we focus mainly on several aspects of policymaking and related techniques and tools: representation; sharing information; reporting and suggestion and rationalization of services. New tools should be intuitive, easy to use, and attractive. They must take advantage of the game dimension, as in the Italian platform for urban gaming *CriticalCity.org* [10].

Current model of representation of the city are digital, three-dimensional, multi-resolution and in real time: this is *City Model* [3]. *Citysensing* (Normal projects) is a platform for collecting, processing, and visualizing data about the city and its hidden and secret systems. It hosts collaborative data collection events, collect and process data from sensors, mobile phones, research, and observation, and make it available for anyone to explore. City Model and Sensing grant reciprocal meaning and efficacy, supporting the multi-agent interaction and processes of governance.

GeoMedia (Intergraph®) and several free software/open source GIS application allow users to work with and conduct analysis on geographic information as well as produce maps from that analysis assembled in different formats and for different uses e.g. *QGis* (qgis.org) an intuitive graphical interface cross-platform application and *Grass* (GNU GPL), mostly used for modelling and analysis.

Where Does My Money Go? is part of *OpenSpending* (4iP), enables UK citizens to know how public funds and taxes get spent, starting from the salary: thus, people can find information about government finance from countries across the world. It enables UK citizens to know how public funds and taxes get spent, starting from the salary. We define here an exemplar e-participation tool that support administrations for conventional aspects, those for which authorities are already required to operate and provide services to citizens. *FixMyStreet* (MySociety) allows citizens to report the daily problems of their neighbourhood, e.g. water leaking, holes and street drainage, traffic light bent, etc. (Fig. 3).

Fig. 3. Output from FixMyStreet (MySociety)

This type of instrument has had a large following and countless variations, clones and applications in different fields. *OpenStreetMap* (Coast, 2004) is a collaborative project, inspired by the success of *Wikipedia*, to create free editable maps of the world, by experiences of the users. Ground surveys are performed by mappers on foot, bicycles, car, boat. They can collect data using GPS devices, aerial photography, and other free sources. These crowdsourced data are then made available under the Open Database License. About the rationalization of public transport, *Moovit* (Tranzmate, 2011) is a community-driven application for real time public transport information and GPS navigation about plan trips across buses, trams, trains, rapid transit (metro, etc.), along with nearby stops and stations. It integrates and share real time data, crowdsourced by the people, with more static public transport schedules. From the point of view of the accessibility to places and services, to facilitate the opportunities to move and use different forms of urban mobility, *BikeShareMap* (UCL Casa) (Fig. 4) is a GPS web app that shows the network of thousands of bicycles available at self-services docking stations around the city.

There are also e-tools that support the "walkability" in communities, that is the ability to walk through places, and provide support point-to-point for reaching destination by foot. In addition to the well-known *Mapquest* and *Google maps*, *Walkscore* and *Walkanomics* assess the pedestrian friendliness of an area and then calculate this information into a score. Foursquare is a social media app that helps to find perfect places to go when looking for food, services, entertainment, and other amenities. *TransitScreen* provides just-in-time information to support walkers to find interconnection with alternative forms of transportation from screens displayed in transit stations and in other locations such as in residential and governmental buildings [13].

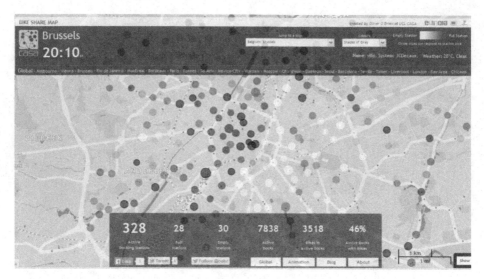

Fig. 4. Output from BikeShareMap of Brussels, Belgium (UCL Casa)

According to the idea of *Streetscape*'s president Shaffer, a Denver-based firm providing pedestrian amenities for public spaces, administrators "must increase their awareness of mobile solutions that exist (or are in development)" and, in the same time, consider the subset of people who "couldn't care less about technology", by giving information in traditional landmarks "that pedestrian already know to use to get around a city" (*Ibid.*).

As regards the last step of public policy assessment, apart from the traditional families of multi-criteria evaluation methods and techniques such as: multi-objective/goal programming; multi-attribute utility theory methods (Analytical Hierarchy Process - AHP, Multi-Attribute Utility Theory (MAUT); outranking methods (ELECTRE); preference disaggregation methods (UTA); rough set theory methods [18], our emphasis is on interactive multi-objective evaluation models for policy selection and prioritization. The software package known as ASA^3 (actors-strategies-actions), for example, allows to compare and prioritize policies depending on three criteria: the interest and the relevance of the options and actions for actors involved, especially synergies among options and actions, and necessary resources. These three criteria are essential for our purpose, as they represent the three fundamental dimensions of public policy decision-making, namely (i) the effectiveness and efficacy of actions (the relevance criterion), (ii) the complexity of and the effects on the political arena (the interest criterion), and finally (iii) the efficiency and resource constraints (the resources criterion) ([2]).

---

[3] ASA Software, LAMP, University of Sassari. ASA is developed and by one of the authors and tested by both of them in several strategic plans for Italian small towns. For more technical details see [2]).

3   From "Top Down" Smart Cities to "Bottom up" Digital Hamlets

To come full circle, an organizational space for interaction is made possible by a web portal with the DSS platform, that provides a valuable interface between citizens, authorities and the private sector and represents solution-oriented innovative approaches in a public and participatory dimension. Furthermore, without forgetting to stress the importance of finding ways to communicate with people who are not technically adept, this model of development has an eminently innovative and self-organisational dimension as well, that is, with regards to "autopoietic" systems [6], it refers to a system as a network of processes (relations) of internal elaboration of input coming from outside, capable of regenerating and transforming them in stable development energy and social growth. In this way, the computerized techniques and e-tools used move closer towards becoming more proactive, "decision aiding" instruments and the perspectives of each hamlets or districts will depend on the capacities of each one to offer different opportunities based on their own environmental conditions and options.

3.1   Low Density Rural Areas as a Test Bench: A Case Study in North-western Sardinia

To make the point about policymaking in a low density residential rural areas, we highlight on a work-in-progress about an experience [12] in a geographical region in the north western Sardinia (Italy), named Nurra. It is the second largest plain of the island, that covers a surface of about 800 km², located between the city of Sassari, and the towns of Porto Torres and Alghero. Specifically, this area is an explanatory example about changeable relationship between city and countryside. In fact, the region was interested by a huge work of soil drainage during Fascism, that continued also after the II World War by the ETFAS Regional agency4, firstly made by convicts from the Alghero jail, emigrants and, after the war, by refugees. Despite being next to one of the most populated areas of Sardinia, it has one of the lowest population densities in Italy, approximately 5 inhabitants per km², with all related problems, such as: lack of primary and neighbourhood services; lack of people involvement in policymaking, despite a strong social cohesion; lack of an efficient transport system; etc. Close to Alghero, in the countryside marked by canals organized in an orthogonal matrix, there are several rural hamlets: Santa Maria la Palma (1059 inhabitants), Guardia Grande (270), Maristella (424), Tramariglio (8), Sa Segada (395) and Villassunta (20). Fertilia (1703) surely plays a key role in the area and it is also considered a hamlet, despite the fact that it was a Fascist "new town", built in 1936. At the same time, these hamlets are poorly managed and marginal to the historic town of Alghero, that is a tourist destination thanks to low cost routes. Thus, in 2014 the new administrations of the municipalities of Alghero and Sassari are going to move toward a new

---

4 ETFAS, Ente Trasformazione Fondiaria Agricola Sarda.

management, aimed to support the accessibility and the facilities for the inhabitants of the hamlets.

SWOT analysis can describe a brief overview of some potential and critical issues to represent the territory, such as: the environmental and social issues, the transport system, the services, the economic structure and the tourism related activities.

Starting with the environmental structure, the natural area of the Calich Lagoon is afflicted by hydro-geological risk and water pollution. Then, the irrigated agricultural areas of the Nurra, that are loosing the settlements values because of unsustainable forms of land-use. Finally, the cliffs of Capo Caccia, part of the Regional Park of Porto Conte and its Marine Protect Area, that show poor interactions between communities and the institution.

As far as the social structure is concerned, note that resident population of the area has grown about 2,000 people in ten years (44.000 in 2007). Definitely, these are not large numbers, but, in this region, it is a substantial growth. In this context, the rejuvenation of the population is only related to the possibility of constructing policies to attract and maintain the younger population with a more affordable housing market. The areas relating to neighbourhoods like the historic centre of Alghero and the more rural areas, are those with the highest rate of older people. Yet, as for public transport system, the bus routes are dispersive, unclear and badly managed for lack of information and communication. Although there are a good number of rides, buses travel often empty and do not provide the basic services, so people prefers to move by car. This poor organization of public transport is associated with a total absence of alternative transport systems e.g. bike routes (except Alghero-Fertilia). Note that the main county roads link Alghero and individual hamlets (mainly the greater ones), leaving the smaller almost completely isolated, like Guardia Grande.

Any direct link is provided between the hamlets, just a dense capillary network of connections among hamlets that refers to the mesh of the land reclamation. A more complete transport analysis concerns timing and distance from one hamlet to another. Nevertheless, the lack of primary and district services in hamlets is crucial: these low density residential areas are detached from the town of Alghero and the few services have unpractical locations, any link with public spaces and poorly managed. For example, the bank and the post office are available only in Fertilia and S. Maria la Palma and the only service offered everywhere is about religious purpose (one church is present in each hamlet). This is the element that characterize each hamlet by an individual sense of identity. Fertilia presents some features more similar to a town than a hamlet, with a direct connection to Alghero, and hosts a broad range of services that in most of the hamlets are not present (nursery, primary and secondary schools, police station and emergency medical service, soccer field).

Despite a bit more relevant population (and tourist arrivals in the summer season), primary services are not enough and public spaces are limited. S. Maria la Palma hosts different levels schools, public areas and equipped areas for sport, public services (included the library) and commercial activities. It is the largest hamlet, which plays the pin role in the organizational structure of the territory, despite the inefficiency of the links. The other smaller hamlets hosts mainly basic services, e.g. general store and bars, with playgrounds and soccer fields.

As far as the economic structure is concerned, the municipality of Alghero has a level of commercial and industrial activities much greater than the rest of the territory, with a successful agricultural industry, where stands out three most important alimentary companies: S. Giuliano olive oil mill, the wineries of S. Maria la Palma and Sella & Mosca, with the related vineyards; a number of small farmers and a significant tourism industry with 870.000 international tourist arrivals, within a 3,6 millions as floating population in 2007 (VI Alghero Tourism Report, DADU, Uniss).

3.2   Towards "Digital" Hamlets

Under these circumstances, a web portal with the DSS platform can support decision makers to manage a mix between interactive tools and real services, working according to an alternative work schedules among low density residential areas, e.g. rural hamlets and districts. A DSS increases the effectiveness of the policymaking, as it provides support to the decision-makers coping with problems that cannot be managed by ordinary planning. In our case, the main task of the DSS is to extract useful information for decision-making from a significant amount of data in a short time. Thus, it makes different techniques and tools available, which are adaptable for specific purposes to the needs of the different users (both residents and visitors) in a "bottom-up" self-sustainable and participatory management system.

First step is the survey of the public spaces in the hamlets with flexible and multipurpose features along with the survey of the services. The spaces can be chosen because of their strategic location and size. Usually it is better to redevelop unused spaces in a central location, surrounded by green spaces or close to significant element for the hamlet. Thus, each hamlet will have spaces (both indoors and open air) that will represent the node in the network and host services and facilities. The primary services are picked out where weak or even absent: the pharmacy, the register office (with a hypothetical election service for referendum), medical services (immunizations, general practitioner, counselling service), heavy waste disposal, information campaigns (political, social and solidarity) and public information more clear and transparent on water and air quality. A real action plan will allow to identify the alternative services located in various hamlets depending on the day of the week. By contrast, as regards the community services, we choose activities covering different age groups, some will be stable (but not open everyday), while others will be itinerant in the hamlets in alternative days: e.g. creative workshops, aggregation centers for elderly persons, post-school activities and sports. By contrast, as regards the itinerant activities, we take into consideration the mobile library, the *PlayBus* and *CineBus* so as to provide cultural services in the rural areas to engage young people in different activities, by stimulating creativity and sense of aggregation and to reach people who do not usually have the opportunity to attend libraries. The entire network of libraries (fixed and mobile) will be connected into the library system, where people can reserve and get the required books through the *LibraryBus*. In the same way, *PlayBus* is a vehicle that weekly transports an entire collection of games in different places of rural or semi-rural areas, such as squares, courtyards, gardens and streets, where groups of volunteers carry out different kinds of animation programs of activities surrounding entertainment and education for children. Hence, the *CineBus* is an example of itinerant cinema for about 30 people that daily move around the hamlets and an itinerant farmer's market, that is present only in town. The design of some spaces by micro-interventions (public lighting, green spaces, pedestrian amenities), where these itinerant activities take place, can make them recognizable and usable even without organized activities. Finally, a bike sharing service and bike paths, linking the hamlets to the places of interest are organized. The web portal and the DSS platform support the management system and the promotion of the network of services as well. In this way, users can book, inquire about events, places and services in advance and perform different activities by linking with the platform. It also will suggest to the users the most useful and available e-tools to match with a specific service (Fig. 5).

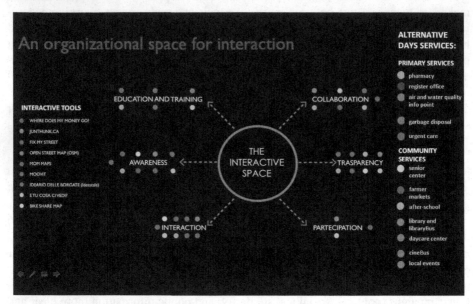

Fig. 5. The scheme of DSS platform to organize the space of interaction

To cope with and bridge the technological and cognitive divides, a crucial role is play at the government level. In fact, even if programs does not explicitly fund broadband infrastructure, subsidies are used to upgrade existing telephone networks, so that they are capable of delivering high-speed services, to buy laptops to children residing in poor and isolated regions (OLPC, 2009), to incorporate the community members into the process of creating the services. All these actions have to be supported by learning path to gain the basic internet-oriented skills. Thus, for the elderly especially, easy access modes and cultural preparations are very important, supported by informal learning and peer group support.

In closing, ICT oriented EU programming must be taken in account to fund the research project. The EAFRD 2014-2020 (European Agricultural Fund for Rural Development), in line with the Europe 2020 strategy, agree about some broad objectives of rural development support for 2014-2020, through six EU-wide priorities, including fostering knowledge transfer and innovation in agriculture, forestry and rural areas. Supported actions are related to a package of measures, and Member States may include within their rural development programmes thematic sub-programmes, contributing to the Union priorities for rural development, aimed to address specific needs identified, in particular in relation to: on-line health care services; small and medium enterprises support; local public services; ICT centers for young people; internet points/public communication, broadband infrastructures.

4   Conclusion

We conclude that this bottom-up approach to innovative e-participation practices is extremely useful for three reasons. First, it provides administrational organizations with a simple, immediate and free use of a pool of ideas, suggestions and proposals previously unconceivable. Second, it can help local administrations organize a number of basic services and facilities in a better way and at a lower price, hence urban presence can be perceived also in rural areas. A web portal with a DSS platform manages a mix between interactive tools and real services, working according to an alternative work schedule in hamlets and districts. Third, it encourages and enables citizens and visitors, planning communities and decision making practitioners to successfully implement community and development projects.

Through this framework we can observe the interaction between analysis and strategy and between processes of social interaction and technology as well as the integration between technical, administrative and political skills of institutions and techno-structures with knowledge, practices, experiences and needs of local context.

References

1. Auer, S.R., Bizer, C., Kobilarov, G., Lehmann, J., Cyganiak, R., Ives, Z.: DBpedia: a nucleus for a web of open data. In: Aberer, K., et al (eds.) The Semantic Web. LNCS, pp. 722–735. Springer, Heidelberg (2007)
2. Blecic, I., Cecchini, A., Plaisant, A.: Constructing strategies in strategic urban planning: a case study of a decision support and evaluation model. In: Murgante, B., Gervasi, O., Iglesias, A., Taniar, D., Apduhan, B.O. (eds.) ICCSA 2011, Part II. LNCS, vol. 6783, pp. 277–292. Springer, Heidelberg (2011)
3. Borga, G.: City Sensing. Approcci, metodi e tecnologie innovative per la Città Intelligente, Franco Angeli, Milano (2013)
4. Cottica, A.: Wikicrazia. Governare ai tempi della rete, Navarra Editore, Palermo (2010)
5. Eden, C., Ackermann, F.: Making strategy: the journey of strategic management. Sage Publications, London (1998)
6. Maturana, H., Varela, F.: Autopoiesis and Cognition: the Realization of the Living, 1st edn., 1973, 2nd 1980, p. 78 (1980)
7. Meltsner, A.J.: Policy Analysis in the Bureaucracy. UC Press, Berkeley (1976)
8. Noveck, B.S.: Peer-to-patent: collective intelligence, open review and patent reform. Harvard Journal of Law and Technology 20(1), 123–162 (2006)

9. Obama, B.: Memorandum for the heads of executive departments and agencies on transparency and open government, February 16, 2011 (2009). http://www.whitehouse.gov/the_press_office/TransparencyandOpenGovernment
10. Osimo, D.: "8 strumenti gratuiti per migliorare la politica", Chefuturo! Lunario dell'innovazione, 20 maggio (2012)
11. Plaisant, A., Verona M.M.: "Ideas for a better place: e-participation tools supporting decision making process at the local level. In: 7th National Conference on Informatics and Planning – Input 2012 Proceedings (INPUT 2012), Cagliari, May 10–12 (2012)
12. Plaisant, A., Pinna, M.T., Talu, V.: Digital hamlets: innovative methods and e-participation tools supporting policymaking at the local level. In: 2nd International Conference of Urban Planning and Sustainable Development (UPSD 2013), Harbin May 15, 2013, p. 70–94. Harbin Institute of Technology press, Harbin (2014)
13. Scott, M.: Smart Cities and the Technology of Walking, UrbanWebcity (March 24)
14. Shirky, C.: Here comes everybody: the power of organizing without organizations. Penguin Press, New York (2008)
15. Shirky, C.: Cognitive surplus: creativity and generosity in a connected age. Penguin Press, New York (2010)
16. Weick, K.: Senso e significato nell'organizzazione. Cortina, Milano (1997)
17. Wyatt, R.G.: Computer-Aided Policymaking. E & FN Spon, London (1999)
18. Zopounidis, C., Doumpos, M.: Multi-criteria Decision Aid in Financial DecisionMaking: Methodologies and Literature Review. Journal of Multi-Criteria Decision Analysis **11**, 167–186 (2002)

Measuring the Standardized Definition of "smart city": A Proposal on Global Metrics to Set the Terms of Reference for Urban "smartness"

Maria-Lluïsa Marsal-Llacuna[✉]

University of Girona, Girona Smart City Chair Campus Montilivi,
Politècnic Building III, Room 121 17003, Girona, Spain
luisa.marsal@udg.edu

Abstract. Since the end of 2014 a definition on "smart cities" agreed by international standardization bodies exists. As commonly said in the business community, *'measurement is the first step leading to control and eventually to improvement. If you can't measure something, yo can't understand it. If you can't understand it, you can't control it. If you can't control it, you can't improve it' (H. James Harrington).* Therefore, if we want to understand what the smart cities initiative is all about we have to start measuring its brand new definition. After that we'll be able to monitor and control the performance of cities in terms of smartness, and this will lead to the possibility of improvement. In this research we present a set of indicators that specifically serve to measure the newly agreed and acknowledged international definition on smart cities. This set of indicators will be now tested to measure the smartness of the city of Girona, a small-medium sized city in Spain. At the completion of this pilot we'll be able to design an index summarizing the set (or subsets) of indicators, as a future steps of this research. A summarizing index will help to get an overview and to understand the overall performance of a city in terms of smartness. Then, targeted actions should be undertaken according to the results revealed by the specific indicators so that the overall smartness can be improved.

Keywords: Smart cities · Sustainable city indicators · Urban resilience · Smart cities standardization

1 Introduction: What's Smartness in the Smart City?

The ISO/TMB SAG (International Standards Organization Technical Management Board, Strategy Advisory Group), on January 2015 launched the following definition of "smart city". This ISO definition is being automatically adopted by the other international, more specific standardization bodies (International Electrotechnical Commission IEC, International Telecommunications Union ITU, Comite Européen de Normalisation CEN):

© Springer International Publishing Switzerland 2015
O. Gervasi et al. (Eds.): ICCSA 2015, Part II, LNCS 9156, pp. 593–611, 2015.
DOI: 10.1007/978-3-319-21407-8_42

A smart city is one that
...dramatically increases the pace at which it improves its **social economic and environmental (sustainability) outcomes,** *responding to challenges such as climate change, rapid population growth, and political and economic instability...*
...by fundamentally improving how it **engages society,** *how it applies collaborative leadership methods, how it works across disciplines and city systems, and how it* **uses data information and modern technologies....**
... in order to provide better services and quality of life to those in and involved with the city (residents, businesses, visitors), now and for the foreseeable future, without unfair disadvantage of others or degradation of the natural environment.

The ISO/CD standard 37101, now under public consultation (12/02/2015) as last step before its final approval, defines the term "smartness" as it follows: *Smartness means to contribute to* **sustainable development and resilience**, *through soundly based decision making and the adoption of a long and short term perspective.*

In an attempt of putting together both definitions to provide an aggregated and simplified definition of what is "smartness in the smart city", we collected the key messages and concepts (highlighted in bold in the above original definitions), and came up with the following summary: **Smartness in the smart city is when the three pillars of sustainability (environmental, economic and social) are safeguarded while urban resilience is being improved by making use of technologies of information**

At the light of the existence of this internationally standardized consensus of what is smartness in the smart city, we elaborated a pioneering measurement exercise to provide metrics to the definition of smartness, as a first step for the monitoring and control of the performance of cities, and subsequent improvement of their smartness.

2   Global City Indicators as Metrics Defining the Terms of Reference of Smartness in the Smart City

The smart cities initiative has been underway for some time, in Spain and in the world. The prospects are that the movement will continue its propagation, with an increasing number of cities, companies, and other involved stakeholders interested in adopting the "smart city" label. Nowadays the use that these participating parties are doing of the "smart city" brand is quite random and has no good grounds since a commonly and widely accepted basis for the elaboration of terms of reference of what a smart city is didn't exist before. The newly international definition on smart cities launched by standardization bodies is that basis for the elaboration of those terms of reference. This recently approved definition of smart cities is therefore a necessary first step but not sufficient since it will be essential to elaborate indicators to objectively measure the performance of cities in terms of smartness according to that definition. Equally necessary will be the elaboration of a standard, in accordance with the definition and developing the indicators. The existence of a standard is the way to regulate the labelling of cities and their related smart services and products, otherwise the smart cities movement might fall in the risk of banalization and uncertain self-labelling. Both factors would discredit the initiative itself as well as damage all formal efforts that international and national communities are doing to provide the relevance and solvency that the smart cities movement deserves.

The set of indicators we propose here are aimed to be a contribution to that eventual standard of definition of smartness in the smart city. Accordingly, we presented the results of this work to the Spanish National Standardization Body (AENOR, Asociación Española de Normalización). Now this work is being reviewed by AENOR and it will be internationalized soon by being presented to ISO's next summer smart cities summit for their consideration. An eventual standard on the smartness of cities would be adopted by cities themselves for the marketing and labelling of their smart city strategies.

The smart cities label would be achieved if results (performance) obtained in the indicators measuring the definition would be above the thresholds set by the standard. Hence, the proposed indicators are aimed to become Key Performance Indicators (KPI) of the ISO's definition of smart city. As said, since the adoption of the terms of reference defining a smart city is made by cities the proposed KPI do not apply to products, solutions and services serving cities to become smarter. All these industrial and technological tools dedicated to help cities to become smarter would need specialized standards regulating their specific purposes. In summary, because smart cities definition:

1) has to be adopted by cities,
2) has international acknowledgement and applicability,
3) is based in the safeguarding and the strengthening of the resilience of the three pillars of sustainability, and
4) understands Information Technologies as a tool to support the initiative but in no case an end in itself;

we studied the instruments, charters, treaties, and other declarations internationally acknowledged and adopted that would fulfill premises 1, 2, and 3, to help us develop and set more holistic and systemic grounds to the definition. Moreover, we understand that proposed KPI indicators measuring smart cities definition will add the fourth component, for being themselves Information Technologies instruments. The results of the State of the Art of this search on international declarations revealed two documents accomplishing with premises 1 to 3, which are:

The European Charter for the Safeguarding of Human Rights in the City [1]
and,
The Global Charter-Agenda for Human Rights in the City [2]

Both documents were acknowledged by the United Nations at the 50th and 60th anniversaries of the Universal Declaration of Human Rights of 1948 [3]. Both Charters seek to ensure the right to the city for all citizens, in the broader sense of social sustainability and are widely acknowledged by cities around the world and supported by many other global institutions. The protection of human rights on an international scale was advanced greatly by the Universal Declaration of Human Rights in 1948. Many other conventions and treaties followed this seminal document, endorsing, expanding, or simply acknowledging its fundamental principles. Both Charters contribute to this positive trend, by adding specific urban rights, and linking human rights and citizens' rights. Moreover, cities which sign the Global Charter-Agenda also endorse instruments which follow the Declaration, such as the International Covenant on Civil and Political Rights (1966) [4], the International Covenant on Economic, Social and Cultural Rights (1966) [5], the Millennium Declaration (2000) [6], the Declaration for the 60th Anniversary of the United Nations (2005) [7], and the Vienna Declaration (1993) [8]. For the European Charter, the following instruments are also endorsed by signatory cities: the European Convention for the Protection of Human Rights and Fundamental Freedoms (1950 and 2010) [9], and the European Social Charter (1961 and 1996) [10].

3   Ongoing Standardization Works on Smart Cities. Analysis of Covered and Uncovered Aspects

As previously mentioned, according to international standardization bodies, smart cities are designed to promote urban development, while safeguarding environmental, social and economic sustainability, making use of intelligent technologies [11], [12]. Accordingly, ISO entitles its committee for the standardization on smart cities ISO/TC 268 "Sustainable Development in Communities", and only refers to "smartness" when referring to metrics. This is because IT-based measurements make a process or an action smart[1]. At the European level, CEN recently created the "Smart and Sustainable Cities and Communities Coordination Group (SSCC-CG)", for the purpose of standardizing smart cities in Europe. In comparison with ISO, we note the addition of *smart* to the *sustainable* terminology, and the addition of *cities* to *communities*. Accordingly, the more general definition of smart cities given by CEN is *"A Smart City is a city seeking to address sustainability issues via ICT-based solutions on the basis of a multi-stakeholder, municipally based partnership"*[2,3]. National standardization institutions follow the steps of international standardization bodies, and therefore, for smart cities standardization, they also understand that sustainable urban development, supported by IT-based solutions, is the purpose of the smart cities initiative. According to the Brundtland report (1987) [13], sustainable development is *"development that meets the needs of the present without compromising the ability of future generations to meet their own needs"*. Hence, the smart cities initiative, thanks to its IT-based approach, is in a unique position to achieve not only sustainable, but also objectively measurable urban development.

Putting all the above pieces together, the target for standardization bodies will be to elaborate standards measuring the performance of cities in the three pillars of sustainability; economic, environmental, and social. ISO has initiated standardization

---

[1] ISO/TC 268/SC 01 "smart urban infrastructure metrics". Project PWI37151. General principles and requirements.

[2] Directorate General for Internal Policies in the EU. Mapping smart cities in the EU., IP/A/ITRE/ST/2013-02. Jan. 2014.

[3] European Innovation Partnership on Smart Cities and Communities Operational Implementation Plan: First Public Draft.

works, mostly covering the environmental[4] and economic[5] aspects of sustainability. Although some aspects of social sustainability are also covered in the recently approved standard ISO 37120[6] (the first standard on the smart cities field launched by ISO), in the opinion of the author, social sustainability requires wider standardization coverage.

This is for two main reasons:

1) The citizen must be at the core of any city strategy (including smart city strategies). Therefore, social sustainability has to be carefully studied, taken fully into account, and safeguarded through standards.

2) Aspects of social sustainability which are already covered by existing standards (ISO 37120) correspond only to issues regarding tangible aspects. Meanwhile, intangible aspects of social sustainability are not taken into consideration. Intangibles need to be considered, and standards supporting the corresponding social policies need to be created.

This first international standard which is already approved in the domain of standardization of smart cities is *ISO 37120:2014 on Sustainable Development and Resilience of Communities. Indicators for City Services and Quality of Life* [11]. This standard was developed by Working Group 2, on Urban Indicators, within the Technical Committee ISO/TC 268 for the Sustainable Development in Communities. The indicators for City Services (CS) and Quality of Life (QoL), presented in ISO standard 37120, are designed to be global city indicators. In other words, these indicators will measure the performance of any city in the world in the areas of CS and QoL. These Global City Indicators are structured around 16 "themes". They include Economy, Education, Energy, Environment, Finance, Fire and emergency response, Governance, Health, Recreation, Safety, Shelter, Solid Waste, Telecommunication and innovation, Transportation, Wastewater, and Water and Sanitation. City performance relative to each of these themes is measured by a set of indicators, which collectively tell a "story" [11]. Overall, 100 indicators have been developed. Recognizing the differences in resources and capabilities between developed and developing cities, the overall set of 100 indicators has been divided into two categories. There are 46 "core" indicators, included as requirements in the standard, regarding which all cities are expected to report on, and 54 "supporting" indicators, regarding which all user cities are encouraged (but not expected) to report on. In addition, Annex A of the standard includes 39 "profile" indicators, which provide basic statistics and background information that help determine which cities are of interest for comparison purposes.

Below is the list of 100 indicators, distinguishing the core [c] and supporting [s] indicators, which are included in clauses 5 to 21 of the ISO 37120:

---

[4] ISO/TC 268/SC 01 "smart urban infrastructure metrics". Project PWI37151. General principles and requirements.

[5] ISO/TC 268/WG 01 "Management Systems standards for sustainable development in communities" Project ISO 37101 Internal Design Specifications.

[6] ISO/TC 268/WG 02 "Urban indicators" ISO 37120 Sustainable Development and Resilience. Indicators for City Services and Quality of Life.

5. Economy
[c] 5.1 City's unemployment rate
[c] 5.2 Assessed value of commercial and industrial properties as a percentage of total assessed value of all properties
[c] 5.3 Percentage of city population living in poverty
[s] 5.4 Percentage of persons in full-time employment
[s] 5.5 Youth unemployment rate
[s] 5.6 Number of businesses per 100 000 population
[s] 5.7 Number of new patents per 100 000 population per year.
  6. Education
[c] 6.1 Percentage of female school-aged population enrolled in schools
[c] 6.2 Percentage of students completing primary education: survival rate
[c] 6.3 Percentage of students completing secondary education: survival rate
[c] 6.4 Primary education Student/teacher ratio
[s] 6.5 Percentage of male school-aged population enrolled in schools
[s] 6.6 Percentage of school-aged population enrolled in schools
[s] 6.7 Number of higher education degrees per 100 000 population
  7. Energy
[c] 7.1 Total residential electrical use per capita (kWh/year)
[c] 7.2 Percentage of city population with authorized electrical service
[c] 7.3 Energy consumption of public buildings per year (KWh/m2)
[c] 7.4 The percentage of total energy derived from renewable sources, as a share of the city's total energy consumption
[s] 7.5 Total electrical use per capita (kWh/year)
[s] 7.6 Average number of electrical interruptions per customer per year
[s] 7.7 Average length of electrical interruptions (in hours)
  8. Environment
[c] 8.1 Fine Particulate Matter (PM 2.5) Concentration
[c] 8.2 Particulate Matter (PM10) Concentration
[c] 8.3 Greenhouse gas emissions measured in tons per capita
[s] 8.4 NO2 (nitrogen dioxide) concentration
[s] 8.5 SO2 (sulphur dioxide) concentration
[s] 8.6 O3 (Ozone) concentration
[s] 8.7 Noise Pollution
[s] 8.8 Percentage change in number of native species
  9. Finance
[c] 9.1 Debt service ratio (debt service expenditure as a per cent of a municipality's own-source revenue)
[s] 9.2 Capital spending as a percentage of total expenditures
[s] 9.3 Own-source revenue as a percentage of total revenues
[s] 9.4 Tax collected as a percentage of tax billed
  10. Fire and emergency response
[c] 10.1 Number of firefighters per 100 000 population
[c] 10.2 Number of fire related deaths per 100 000 population
[c] 10.3 Number of natural disaster –related deaths per 100 000 population
[s] 10.4 Number of volunteer and part-time firefighters per 100 000 population
[s] 10.5 Response time for emergency response services from initial call
[s] 10.6 Response time for fire department from initial call
  11. Governance
[c] 11.1 Voter participation in last municipal election (as a percentage of eligible voters)
[c] 11.2 Women as a percentage of total elected to city-level office

[s] 11.3 Percentage of women employed in the city government workforce
[s] 11.4 Number of convictions for corruption/bribery by city officials per 100 000 population
[s] 11.5 Citizens' representation: number of local officials elected to office per 100 000 population
[s] 11.6 Number of registered voters as a percentage of the voting age population
    12. Health
[c] 12.1 Average life expectancy
[c] 12.2 Number of In-patient hospital beds per 100 000 population
[c] 12.3 Number of physicians per 100 000 population
[c] 12.4 Under age five mortality per 1 000 live births
[s] 12.5 Number of nursing and midwifery personnel per 100 000 population
[s] 12.6 Number of mental health practitioners per 100 000 population
[s] 12.7 Suicide rate per 100 000 population
    13. Recreation
[s] 13.1 Square meters of public indoor recreation space per capita
[s] 13.2 Square meters of public outdoor recreation space per capita
    14. Safety
[c] 14.1 Number of police officers per 100 000 population
[c] 14.2 Number of homicides per 100 000 population
[s] 14.3 Crimes against property per 100 000
[s] 14.4 Response time for police department from initial call
[s] 14.5 Violent crime rate per 100 000 population
    15. Shelter
[c] 15.1 Percentage of city population living in slums
[s] 15.2 Number of homeless per 100 000 population
[s] 15.3 Percentage of households that exist without registered legal titles
    16. Solid waste
[c] 16.1 Percentage of city population with regular solid waste collection (residential)
[c] 16.2 Total collected municipal solid waste per capita
[c] 16.3 Percentage of city's solid waste that is recycled
[s] 16.4 Percentage of the city's solid waste that is disposed of in a sanitary landfill
[s] 16.5 Percentage of the city's solid waste that is disposed of in an incinerator
[s] 16.6 Percentage of the city's solid waste that is burned openly
[s] 16.7 Percentage of the city's solid waste that is disposed of in an open dump
[s] 16.8 Percentage of the city's solid waste that is disposed of by other means
[s] 16.9 Hazardous Waste Generation per capita (tons)
[s] 16.10 Percentage of the city's hazardous waste that is recycled
    17. Telecommunication and innovation
[c] 17.1 Number of internet connections per 100 000 population
[c] 17.2 Number of cell phone connections per 100 000 population
[s] 17.3 Number of landline phone connections per 100 000 population
    18. Transportation
[c] 18.1 Kilometers of high capacity public transport system per 100 000 population
[c] 18.2 Kilometers of light passenger transport system per 100 000 population
[c] 18.3 Annual number of public transport trips per capita
[c] 18.4 Number of personal automobiles per capita
[s] 18.5 Modal split (percentage of commuters using a travel mode to work other than a personal Vehicle)
[s] 18.6 Number of two-wheel motorized vehicles per capita
[s] 18.7 Kilometers of bicycle paths and lanes per 100 000 population
[s] 18.8 Transportation fatalities per 100 000 population
[s] 18.9 Commercial air connectivity (number of non-stop commercial air destinations)

19. Urban planning
[c] 19.1 Green area (hectares) per 100 000 population
[s] 19.2 Annual number of trees planted per 100 000 population
[s] 19.3 Areal size of informal settlements as a percentage of city area
[s] 19.4 Jobs/housing ratio
20. Wastewater
[c] 20.1 Percentage of city population served by wastewater collection
[c] 20.2 Percentage of the city's wastewater that has received no treatment
[c] 20.3 Percentage of the city's wastewater receiving primary treatment
[c] 20.4 Percentage of the city's wastewater receiving secondary treatment
[c] 20.5 Percentage of the city's wastewater receiving tertiary treatment
21. Water and Sanitation
[c] 21.1 Percentage of city population with potable water supply service
[c] 21.2 Percentage of city population with sustainable access to an improved water source
[c] 21.3 Percentage of population with access to improved sanitation
[c] 21.4 Total domestic water consumption per capita (liters/day)
[s] 21.5 Total water consumption per capita (liters/day)
[s] 21.6 Average annual hours of water service interruption per household
[s] 21.7 Percentage of water loss (unaccounted for water)

4   Convergences and Divergences Between Charters Regarding Citizenship Rights. Shortcomings that Should be Addressed and Ambitions that Should be Realized

The main differences between the European Charter and the Global Charter-Agenda lie in their geographic scope, and the agenda or local action plan that only the Charter-Agenda contains. Specifically, in the Global Charter-Agenda for Human Rights in the City, each right is accompanied by an action plan to be used by local governments when tackling its implementation. The "Agenda" character of the Global Charter can reduce its theoretical appeal, compared to the European Charter, since while it makes the content more concrete, it also creates major challenges [14]. Finally, it is important to mention that the Global Charter-Agenda presents more basic rights (such as the provision of water and electricity), because it has to respect the diversity of circumstances in different cities around the World. On the other hand, the European Charter presents more sophisticated rights, which reflect the European level of development.

We elaborated an exercise to identify equivalent and divergent rights between the two Charters. We found more convergences than divergences. Although a detailed analysis of the divergences found follows, the main divergence between the Charters lies in the absence in the Global Charter-Agenda of the following rights, which are included in the European Charter:

- Article VI: International municipal cooperation;
- Article VII: the Principle of subsidiarity;
- Article XIV: the Right to work;
- Article XVII: the Right to health;
- Article XXII: Consumers' rights;

- Article XXV: Local administration of justice
- Article XXVII: Preventive measures, and
- Article XXVIII: Taxation and budgetary mechanisms

The absence of an article on measures for International Municipal Cooperation in the Global Charter Agenda (Art. VI, EU Charter) can be easily understood. What is less clear is the non-inclusion of subsidiarity principles, since the Global Charter Agenda is committed to universal access to domestic utilities (Art. XII, GL Charter).

In less developed countries, the excessive cost of basic services when they are transferred from cities to central state bodies (and/or externalized) is well known, and well documented, and has been studied since the rise of the concept of sustainable urban development in the nineties (e.g. by Drakakis [15], [16], [17]). Therefore, and considering the call for universal access to public utilities in the Global Charter-Agenda, the author believes that this should be covered by the Global Charter-Agenda. The author also believes that elements of the right to work should be introduced into the Global-Charter, since this has been proven to directly affect poverty rates [16]. It is clear that municipal administrations cannot directly create labor markets, and it is not their responsibility, but they can act indirectly through policies and initiatives. And, considering that measures included in the Global Charter Agenda have a greater impact, since it is an Agenda too, the inclusion of labor market promotion could significantly reduce poverty. The relationship between poverty and unemployment has become a more pressing issue since it was included in the United Nations Millennium Development Goals (Goal 1, target 1B), with a due date for overall completion of the MDG strategy in 2015.

It is problematic to include the Right to Health, which is only in the European Charter, since it is not a municipal responsibility. This is why it was previously stated that the Charters do not fully succeed in limiting their scope to the rights and duties of cities and citizens, and sometimes proclaim rights which are within the remit of regional and national governments. In the case of the right to health, public health providers are not local bodies, but are constituted by national authorities. How can cities safeguard a right of citizens which is not within their remit? A similar issue arises with respect to the rights of consumers (Art. XXII). Of course, cities can establish control mechanisms for weights and measures, and conduct quality checks and inspections regarding sanitary standards, zoning regulations, and compliance with municipal ordinances/regulations. However, many business activities of economic operators (such as for example those located in private malls) are completely under the jurisdiction of national authorities. Regarding Article XXV of the European Charter on the Local Administration of Justice, although law and justice are primarily the responsibility of national governments, in the opinion of the author, efficiency of the judiciary and access to justice could be enhanced if this were covered in the Global Charter-Agenda. At least with respect to municipal courts and public mechanisms for extra-judicial solutions, such as mediation and arbitration (alternative dispute resolution). Similar comments apply to Article XXVII on Preventive Measures and Article XXVIII on Taxation and Budgetary Mechanisms of the European Charter, which include the ombudsman as a local mediator, and participatory budgets, respectively. The author maintains that similar articles would be desirable in the Global Charter-

Agenda, in order to promote access to justice, enhance citizen participation, and increase transparency, all of which are municipal responsibilities.

Other differences, perhaps of less importance, are rights included in the European Charter but not in the Global Charter-Agenda, such as:

- Article II: Non-discrimination for reason of color, age, sexual orientation, language, religion, political opinion, ethnic, national or social origin, or level of income
- Article III: Linguistic and cultural freedom including the protection of minority linguistic group
- Article IV: Accessible public transport
- Article VIII: Right of municipal suffrage
- Article VIII: Right to demonstration
- Article X: Right to receive social familiar protection, including subsidies and social housing
- Article XVII: Right to social personal services
- Article XIII: Social awareness rising through education
- Article XVI: Housing for women being victims of violence and/or prostitution
- Article XVIII: Promotion of energy saving measures and recycling
- Article XX: Promotion of environmentally friendly vehicles& pedestrian areas
- Article XXI: Quality and accessibility in leisure areas, and sustainable tourism
- Article XXIII: Evaluation of the efficiency of public services
- Article XXIV: Publication of municipal regulations& contractual transparency

Finally, minor differences regarding rights which are included in the Global Charter-Agenda but not in the European Charter include:

- Article III: Combating of any form of physical or mental violence and the fostering of coexistence end mediation between social groups
- Article III: Availability of imprisonment facilities and right to safety against any type of violence, including enforcement agencies
- Article IV: Fight discrimination against women and promote full development in equality with men
- Article V: Decent conditions and secondary education for children
- Article VI: Decentralized distribution of public services
- Article VII: Freedom of opinion and expression
- Article VII: Advocacy of free and pluralist sources of information
- Article X: Security on legal ownership titles and titles of occupancy, and regulation of the housing market with the promotion of features of urban centrality
- Article XII: Right to equal access to in quantity and quality domestic utilities
- Article XII: Drinking water, sanitation and food, ensuring that no one is deprived form those basics for the lack of economic means

After this analysis of differences and commonalities between both Charters, we can summarize the scope of the Global Charter-Agenda by saying that it safeguards the right of citizens to basic sustainable urbanization, including rights to civil liberties, socio-economic rights, woman's and children's' rights, and the right to a clean environment. Recapping the missing points identified by the author, we find the rights to civil liberties on social and judicial advocacy (corresponding to Articles XXV and

XXVII of the EU-CH), participatory budgets (corresponding to Article XXVIII of the EU-CH), and the right to work (corresponding to Article XIV of the EU-CH), which should be adapted and included in the Global Charter-Agenda.

5   Allocation of ISO 37120 Indicators in the Rights of the Charters. Identification of Missing Rights and/or Missing Indicators

We can observe from the international standard ISO 37120 on city services and quality of life that clause 5 on Economy includes some indicators on labor market (5.1 to 5.6). This aligns with the author's position that the "Right to Work" is missing from the Global Charter-Agenda (but not the European Charter). As previously discussed, local administrations cannot directly create labor markets but only act indirectly through the implementation of policies and actions.

Another subject included in the standard which is not considered in the Global-Charter Agenda, but addressed by the European Charter, is the "Right to Health". As previously mentioned, this is outside the remit of municipalities, since public health providers are the responsibility of national governments. Therefore, from Clause 12 of the ISO 37120 on Health, the author suggests removing the following indicators: 12.2 Number of in-patient hospital beds per 100 000 population, 12.3 Number of physicians per 100 000 population, 12.5 Number of nursing and midwifery personnel per 100 000 population, and 12.6 Number of mental health practitioners per 100 000 population. On the other hand, but still regarding health issues, the author agrees with the inclusion of the following indicators: 12.1 Average life expectancy, 12.4 Under age five mortality per 1 000 live births, and 12.7 Suicide rate per 100 000 population, since they have less to do with the provision of health care and can be influenced by policies and actions which promote a healthy city.

The ISO standard also has a clause on Finance, which includes indicators to measure the debt service ratio, capital spending as a percentage of total expenditures, own-source revenue as a percentage of total revenues, and the tax collected as a percentage of tax billed. Indicator 11.4, Number of convictions for corruption/bribery by city officials per 100 000 population (Clause 11 on Governance), is also noteworthy. These indicators prove that it is possible to measure the sustainability of public finances. This reinforces the suggestion made by the author in the previous section, to include an article in the Global Charter-Agenda on Preventive Measures and Taxation and Budgetary mechanisms, in line with the European Charter.

Indicator 18.9 Commercial air connectivity (number of non-stop commercial air destinations), included in Clause 18 on Transportation, raises a subject which is missing from both Charters, and is perhaps more relevant for the European Charter, which includes a Principle on International Cooperation. In the opinion of the author, both Charters should include a "Right to Internationalization", since labor markets, education, family life, economic/business activities, and social networking are increasingly global and transnational in nature. Citizens should have the right to be included in the global aspects of all of these life domains, to take advantage of all opportunities to improve their quality of life.

The last theme missing from both Charters, according to the opinion of the author, after analyzing the ISO standard, is the right to "Efficient response to emergencies". In Clause 10 on Fire and Emergency response, the ISO standard includes a set of indicators measuring the capacity of cities to respond to fire, natural disasters, and emergencies. Today, as the debate on resilience gains momentum (and it is being included in the latest ISO definition), Charters should consider including a right to be protected from externally induced anthropogenic risk, such as war, emergencies, and natural disasters. After all, most socio-economic rights are compromised if citizens do not enjoy a resilient city that takes measures against preventable or foreseeable major risks.

We elaborated a second exercise consisting of distributing the ISO 37120 indicators amongst the different rights contained in the Global Charter-Agenda. The allocation of indicators in the rights is an exercise to weigh and reveal what is important and acknowledged by international standardization bodies in terms of rights of citizens in the city. In the next and final section, we will present the indicators included in standard ISO 37120 together with the indicators we developed to cover all of the articles of the Global Charter-Agenda.

6 Results and Next Steps: Elaboration of Indicators to Measure Smartness According to the Smart Cities Definition Based on the Global Charter Agenda for Human Rights in the City. Future Research on the Elaboration of a Summarizing Index.

As discussed in the first section of this paper, the only way for cities to prove their accomplishments and asses their performance regarding a given theme is through indicators and monitoring. In conformity with the internationally acknowledged definition of smart cities, cities wishing to be considered smart will have to prove their performance in the areas of social, economic, and environmental sustainability. In order to help cities assess and prove their environmental and economic sustainability, they can use a considerable number of existing instruments (frameworks, sets of indicators, indexes, monitoring schemas, recommendations, etc.). The same cannot be said for social sustainability, for which there are only instruments addressing public participation (which is necessary but not sufficient).

From the analysis on how existing smart city standards are covering what is being proposed by internationally acknowledged instruments safeguarding urban aspects of sustainability, namely both Charters on citizenship rights in the city, we learned that there are still many uncovered aspects. Although some of the existing indicators proposed by the ISO 37120 standard could be used as metrics to define the smartness of cities, a large set of new ones needs to be developed. In this research we created indicators measuring the rights included in the Global Charter Agenda for Human Rights in the City, a citizen-centered instrument addressing all aspects of urban sustainability, which provides a perfect coverage of the conceptual aspects of the definition of smartness in the smart city. And, by providing the indicators measuring those rights we are covering the technological aspect of the definition since indicators are themselves information technologies instruments.

Therefore, this research can be considered the first of a kind in providing a systemic and holistic measurement instrument to comprehensively assess the overall performance of cities in meeting smartness, through the evaluation of the safeguarding of urban sustainability and the improvement of resilience of cities. In this context, the safeguarding of human rights in the city promoted by the Charters covers the whole spectrum of *sustainabilities*, and specially the social one, the less addressed by existing indicators. Hence, a city with good performance in the indicators measuring the safeguarding of human rights in the city can be considered smart.

The construction of indicators for The Global Charter-Agenda of Human Rights in the City has been based on the following criteria:

- MEASURABLE AND ASSESSABLE: indicators of a quantitative nature. Only in isolated cases, where no quantitative indicator was possible (not even binary), have qualitative indicators been provided. In the latter case, indicators have to be understood as attributes.
- REPLICABLE AND REUSABLE: indicators have been designed respecting global reality, and are therefore applicable and valid in any urban area in the world.
- SCALABLE AND SIMPLE: indicators serve any urban settlements of any size, whether town or city. Some indicators lose their meaning in towns, because they simply do not have to provide certain services. In such cases, some indicators are not applicable.
- REALISTIC AND FAIR: indicators respond to the actual availability of data. Sometimes the information to feed a given indicator is obtained from non-municipal sources. Therefore, it is advisable to organize a "call for data", which brings all stakeholders together to participate in the assessment of the safeguarding of human rights in the city. The "call for data" should reach parties such as municipal agencies, utilities and private service providers, and NGOs. To be fair, the indicators should only address municipal responsibilities.

Figure 1 shows the indicators we elaborated for the Global Charter-Agenda for the Safeguarding of Human Rights in the City, in combination with the distribution of ISO 37120 existing indicators on city services and quality of life. For the sake of consistency we utilized the structure of the chapters and the order of the articles of the European Charter, to present the articles of the Global Charter-Agenda. The central column of the figure shows these articles. On both sides of this central column of articles, we present the indicators. Indicators on the left side of the central column correspond to articles with common content between the two charters, while indicators on the right side of the central column are for articles exclusively in the Global Charter-Agenda. ISO indicators are presented in boxes with a black thin profile, and our proposed indicators are in boxes underlined in bold grey. Articles outside of this central column in light grey boxes are not originally in the Global Charter-Agenda, but rather are imported from the European Charter to present some indicators proposed by ISO. Finally, there is one box of indicators not present in any Charter, corresponding to clause 10 of ISO 37120 on Fire and Emergency response, located within Article XXIII Efficiency of public services.

Part I : General Provisions

I. Right to the City		I. Right to the City	(n° of NGO's dedicated to solidarity per 100 000 inhab.)
			(n° of associations with a political, social and ecological background per 100 000 inhab.)
		...1 c) All city inhabitants have the **right to participate** in the **configuration and coordination of territory as a basic space** and foundation for peaceful life and coexistence. *[I.1.b), d), I.2, I.3]*	(n° of channels to measure quality of life of citizens (e.g. surveys, questionnaires, etc.) per 100 000 inhab.)
		...City inhabitants have the right to **respect the rights and dignity of others.** *[I.1.a), b), c), d), I.2, I.3]*	general statement, no indicators required
II. Principle of Equality of Rights and Non-Discrimination		I. Right to the City	
		IV.1. Every man and woman benefits from equal rights to become fully-fledged citizens in the life of the city. *[I.1.a), c), I.2, I.3]*	general statement, no indicators required
		IV. Right of Women and Men to Equality	
		IV.1. All city inhabitants have the right **not to be treated in a discriminatory manner by reason of their gender.**	general statement, no indicators required
		...regulations, prohibiting **discrimination against women...**	11.2 Women as a % of total elected to city-level office
			11.3 % of women employed in the city government workforce.
			6.1 % of female school-aged population enrolled in schools
			6.5 % of male school-aged population enrolled in schools
		IV.3. All city inhabitants **refrain from engaging in any act or practice that may be detrimental to women's rights.**	(number of complaints for violating womens' rights per 100 000 inhab.)
III. Right to Cultural, Linguistic and Religious Freedom		**VII. Freedom of Conscience and Religion, Opinion and Information**	
		VII.1. All city inhabitants have the right to freedom of **religion, conscience and thought,** and agree with the **freedom to change their religion or belief,** and freedom —alone or in community with others and in public or private— **to manifest their religion or belief** in teaching, practice, worship and observance.	(% of municipal budget dedicated to support and provide means to the different religious groups and beliefs present in the city)
		b) All city inhabitants have the right to **freedom of opinion and expression.** This right includes freedom **to hold opinions** without interference and to seek, **receive and deliver information and ideas through any media.** These rights may be subject only to such limitations that are necessary for the protection of public safety, order, health or morals, or for the protection of others' rights and freedoms, in the framework of national legislation.	(% of municipal budget dedicated to learn children's' own religion in schools)
		VII.2. Belonging minorities of all city inhabitants have the **freedom to manifest their religion or beliefs,** including the right of members to profess this right. It is up to this to profess it.	
		The city ensures that **everyone is able to hold opinions without interference.** *[...] [VII.2.], [VII.3]*	(% of municipal budget devoted to the safeguard against the creation of ghetto communities)
		VII.3. City inhabitants have the duty and the responsibility to **respect everyone else's religion, beliefs and opinions.** *[VII.1, VII.2]*	
IV. Protection of the most Vulnerable Groups and Citizens		**VI. Right to Accessible Public Services**	
		VI.2.1. Bearing in mind the difficulties with regard to this growth, the city takes urgent measures to improve the quality of life and opportunities of its inhabitants, especially those of lesser means as well as people with disabilities. The city in conjunction with the protection of the rights of the elderly and vulnerable groups among communities. *[...]*	(% of municipal budget devoted to the protection of vulnerable groups including people with lesser means, disabled, children and elderly)
		V. Rights of Children	
		V.1. It is the prime responsibility of all inhabitants, respect the **dignity and rights of children including those of disabled children.** *[V.1.V.2]*	
		III. Right to Civic Peace and Safety in the City	
		III.2. The city ensures the **security and physical and mental safety** of all its inhabitants, and takes measures to **combat acts of violence,** regardless of who the perpetrators may be. *[...]* The city adopts measures to **combat school and domestic violence** and, in particular, **violence against women and vulnerable groups, such as children, the elderly and the disabled.** The city assumes its role in the **management of social tensions,** in order to **prevent friction between the different groups** that live in the city from turning into actual conflict. To this end, it **fosters coexistence, social mediation and dialogue among those groups.** *[III.1, III.2, I.1, III.3]*	(% of municipal budget dedicated to the mediation with social groups, conduct civic actions against violence, the promotion of good coexistence and personal safety) (number of complaints for personal violent actions and affectation of safety)
		II. Right to Participatory Democracy	
		II.1.1. The city particularly encourages the participation of women in full exercise of their citizenship at each stage of the urban life as well the **participation of minority groups.** It promotes the **participation of children** in matters directly relevant to them. This it promotes the assessment of the inhabitants in collective and community issues. To this end, it **facilitates the participation of civil society, including human rights protection associations,** in the resolution of problems and the improvement of measures which serve the demands of all the citizens. *[II.1.a), b), II.2, II.3]*	(% of municipal budget devoted to promote the participation and integration of vulnerable groups in the civic life of the city)
		XII. Right to Sustainable Urban Development	
		XII.1. All city inhabitants have the right to enjoy urban development, with a local, **social integration.** *[XII.1.a), b), XII.2, XII.3]*	
V. Duty of Solidarity		I. Right to the City	
		I.1.d) All city inhabitants have the **right to available spaces** and resources allowing them to be **active citizens.** The working and common spaces shall be **respectful of everyone's values and of the value of pluralism.** *[I.1.a), b), c), I.2, I.3]*	(surface in municipal buildings available to citizens to perform civic and citizenship activities)
		VI. Right to Accessible Public Services	
		VI.3. City inhabitants use social services responsibly *[VI.1, VI.2]*	general statement, no indicators required
		XII. Right to Sustainable Urban Development	
		XII.3. In fulfilling their responsibility, city inhabitants act in a manner that **respects the environment and promotes energy saving** and the **good use of public installations, including public transportation.** *[XII.1.a), b), XII.2]*	general statement, no indicators required
		II. Right to Participatory Democracy	
		II.3. City inhabitants participate in local affairs, contact other persons and institutions. They **take part in decisions** and contact them, and **express their opinions** towards other individuals and adopts in a **spirit of tolerance and pluralism.** City inhabitants take part in local policy matters of the common interest, for the benefit of the community. *[II.1, II.2, II.3]*	general statement, no indicators required
		XII. Right to Sustainable Urban Development	
		XII.2. The inhabitants develop solidarity in an active community which to promote sustainable planning and a sustainable urban context, in the benefit of current and future generations. *[XII.1.a), b), XII.2, XII.3]*	general statement, no indicators required
VI. International Municipal Cooperation		I. Right to the City	
		I.2.1. [The signatories of the Charter are encouraged to **develop contact with neighbouring cities and territories** with the aim of **building caring communities and regional capitals.** As a framework and summary of all rights provided for in this Charter-Agenda, the above right will be satisfied to the degree inasmuch each and every one of the rights described therein are fully effective and **guaranteed domestically.** *[I.1.a), b), c), d), I.2, I.3]*	(n° of agreements with neighbouring cities to cooperate whit the aim of building caring communities and regional capitals)
VII. Principle of Subsidiarity			

Part II : Civil and Political Rights in the City

VIII. Right of Political Participation	11.1 Voter participation in last municipal selection (as a percentage of eligible voters) 11.6 Number of registered voters as a percentage of the voting age population	**II. Right to Participatory Democracy**	
		II.1.All city inhabitants have the right to **participate in political and city management processes,** in particular, with to participate in the **decision-making process of local public policies.**	(existence of municipal regulation or legislation to rule local referendums)
	11.5 Citizens' representation: number of local officials elected to office per 100 000 population	II.1.b. **question local authorities** regarding their public policies, and to assess them *[...]* The city promotes a **quality participation of its inhabitants in local affairs,** and to assess them, their **ability to impact on local decisions** *[...] [II.1.a), b), II.2, II.3]*	(n° of municipal referendums per year) (n° of public meetings per year) (n° of public debates per year)
		VIII. Right to Peaceful Meeting, Association and to Form a Trade Union	
		VIII.1. All city inhabitants have the right to **freedom of peaceful assembly and association,** which includes the **right of individuals to associate together.**	(n° of public demonstrations and collective gatherings which have been repressed over the total of public meetings per year)

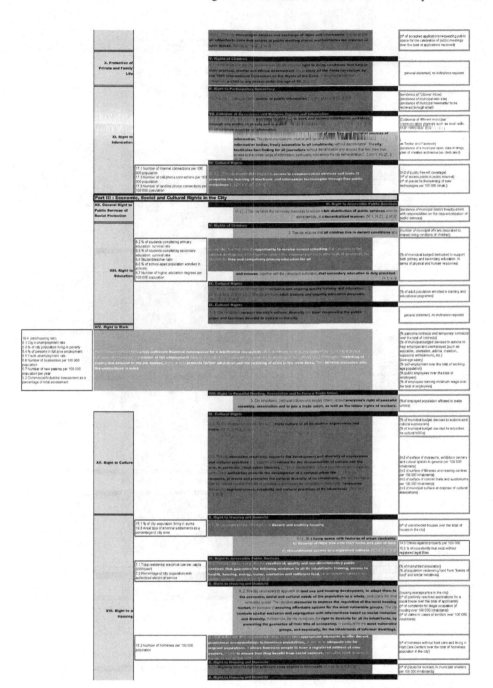

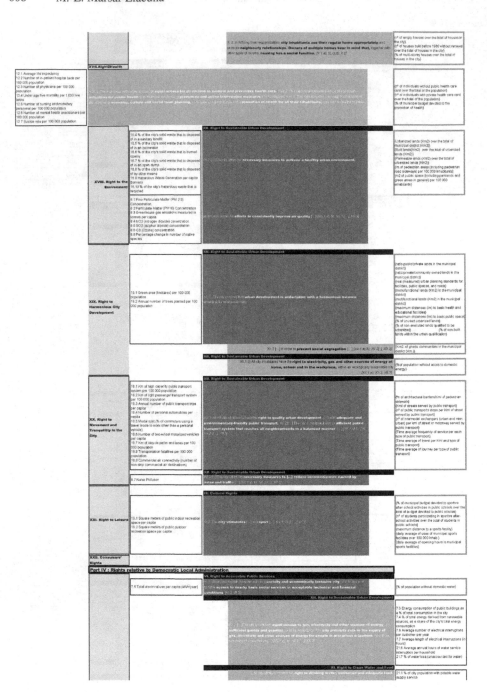

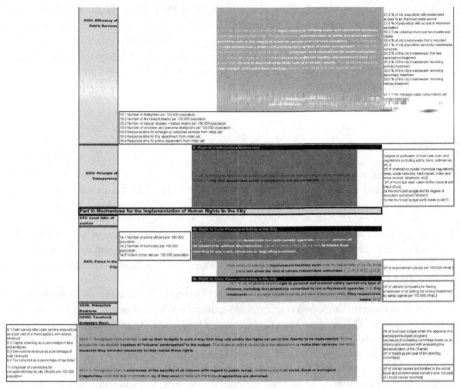

Fig. 1. Proposed indicators measuring the rights included in the Global Charter-Agenda for Human Rights in the City, in combination with ISO 37120 indicators

This research is a work in progress and the next efforts will go in the direction of elaborating one or more than one summarizing indices to visualize in a more synthetic way the performance of cities in terms of smartness. It is still too early to know but we believe we'll elaborate three indexes, one of each summarizing indicators measuring the three pillars of resilience and sustainability (economic, social and environmental). We already started the previous exercise which is to distribute the different rights of the Charters in these three pillars. We also started the elaboration of indicators for the European Charter for the Safeguarding of Human Rights in the City and we are using the guidance of the different thematic chapters in which the European Charter is presented to distribute the rights in the three pillars of resilience and urban sustainability.

Acknowledgments. This research is consuming a huge amount of work and dedication. The elaboration of indicators would not have been possible without the valuable contributions and help of this team, who disinterestedly met with me once a week during months, always active, helpful, and bringing lots of great ideas: Dr. Pere Soler (University of Girona, director the Director of the Joint Master's Program in Youth and Society (MIJS); Dr. Imma Boada (University of Girona, director of the Institute of Informatics and its Applications); Dr. Joaquim

Meléndez (University of Girona, director of the Doctoral Program in Technology); Ms. Anna Serra (Lawyer at Red Cross Girona); Mr. Fran Quirós (Responsible of Cooperation Programs at Charity Girona); Mr. Lluís Puigdemont (Responsible of the Rights Department at Charity Girona); Dr. Montse Aulinas (Project Manager at Grup Fundació Ramon Noguera); Ms. Yolanda García (Responsible of social programs at Grup Fundació Ramon Noguera)

Special thanks to Mr. Mark Segal, international consultant on democratization issues, for his valuable comments and general editing support during the elaboration of this research

References

1. United Cities and Local Governments, Committee on Social Inclusion, Participatory Democracy and Human Rights UCLG- CISDP (1998) The European Charter for the Safeguarding of the Human Rights in the City. (WWW document). http://www.uclg-cisdp.org/en/right-to-the-city/european-charter (accessed December 18, 2014)
2. United Cities and Local Governments, Committee on Social Inclusion, Participatory Democracy and Human Rights (2006) Global Charter-Agenda for human Rights in the City. (WWW document). http://www.uclg-cisdp.org/en/right-to-the-city/world-charter-agenda (accessed December 18, 2014)
3. United Nations, General Assembly (1948) Universal Declaration of Human Rights. (WWW document). http://www.un.org/en/documents/udhr/ (accessed December 18, 2014)
4. United Nations, Office of the High Commissioner for Human Rights (1966) International Covenant on Civil and Political Rights (WWW document). http://www.ohchr.org/en/professionalinterest/pages/ccpr.aspx (accessed December 18, 2014)
5. United Nations, Office of the High Commissioner for Human Rights (1966) International Covenant on Economic, Social and Cultural Rights (WWW document). http://www.ohchr.org/EN/ProfessionalInterest/Pages/CESCR.aspx (accessed December 18, 2014)
6. United Nations, General Assembly (2000) United Nations Millennium Declaration (WWW document). http://www.un.org/millennium/declaration/ares552e.htm (accessed December 18, 2014)
7. United Nations, Office of the High Commissioner for Human Rights (2005) Declaration for the 60th Anniversary of the United Nations (WWW document). http://www.ohchr.org/en/udhr/pages/60udhr.aspx (accessed December 18, 2014)
8. United Nations, Office of the High Commissioner for Human Rights (1993) Vienna Declaration and Programme of Action (WWW document). http://www.ohchr.org/EN/ProfessionalInterest/Pages/Vienna.aspx (accessed December 18, 2014)
9. Council of Europe (1950 and 2010) Convention for the Protection of Human Rights and Fundamental Freedoms. (WWW document). http://conventions.coe.int/treaty/en/treaties/html/005.htm (accessed December 18, 2014)
10. Council of Europe (1961 and 1996) European Social Charter. (WWW document) http://www.coe.int/T/DGHL/Monitoring/SocialCharter/ (accessed December 18, 2014)
11. International Standards Organization (2014) 37120 Sustainable Development and Resilience. Indicators for City Services and Quality of Life. Standard briefing note and outline (available on-line and standard on purchase) (WWW document). http://www.iso.org/iso/catalogue_detail?csnumber=62436 (accessed December 18, 2014)

12. International Electrotechnical Commission (2014) White paper. Orchestrating Infrastructure Smart Cities (WWW document). http://www.iec.ch/whitepaper/smartcities/?ref=extfooter (accessed December 18, 2014)
13. United Nations, World Commission on Environment and Development: Our Common Future. Oxford University Press, Oxford (1987)
14. Drown, A.: The Right to the City: Road to Rio 2010. International Journal of Urban and Regional Research 37(3), 957–71 (2013)
15. Drakakis-Smith, D.: Third World cities: sustainable urban development, part 1. Urban Studies 32(4-5), 659–77 (1995)
16. Drakakis-Smith, D.: Third world cities: sustainable urban development, part 2. Population, labour and poverty. Urban Studies 33(4–5), 673–701 (1996)
17. Drakakis-Smith, D.: Third World cities: Sustainable urban development, part 3. Basic needs and human rights. Urban Studies 34(5–6), 797–823 (1997)

Benchmarking Smart Urban Mobility:
A Study on Italian Cities

Chiara Garau[✉], Francesca Masala, and Francesco Pinna

DICAAR – Department of Civil and Environmental Engineering and Architecture,
University of Cagliari, Cagliari, Italy
{cgarau,fpinna}@unica.it, fmasala@outlook.it

Abstract. The sustainable development of transport systems has generated particular interest within the initiatives in the field of 'smart cities'. This paper is part of current approaches to smart cities benchmarking and it focuses on the definition of a quantitative methodology, capable to evaluate urban mobility through a synthetic indicator of mobility. In this work, we propose a valid method for any city in the world, although its application focuses on the Italian context. This is because this paper shows a first phase of a broader research.

Keywords: Smart cities · Sustainable urban development · Smart mobility index · Quantitative analysis

1 Introduction

Recent years have seen the emergence of a new vision of the city in which the single term 'Smart City' has enclosed concepts such as new technologies, sustainability, intelligence, competitiveness, inclusiveness, place-based promotion and efficiency [1,2,3]. However, in this context, mobility and transport planning play a key role in economic, and social development [4]. Indeed, in according to Ortolani (2014), 'between 1990 and 2007 the EU-27 road transport is increased by 29% and the car ownership, in the same time, is increased by 34%'. Moreover, the current transport model 'is accountable for 23% of energy consumed in Europe, and about three quarters of which depends on road transport. It is estimated that energy consumption in this sector will increase around 80% for 2030' [5]. Problems caused by the increase and by the indiscriminate use of private cars are widely recognized in the literature, and today there are numerous studies that discuss the issue of sustainable mobility [6,7,8,9,10,11,12] and of transport planning. Below, some studies on this topic are discussed.

Giffinger et al. (2007) [1], analyze six key aspects of smart cities in European cities, including mobility, using a methodology of comparison. This takes into account

---

C. Garau, F. Masala, and F. Pinna—This paper is the result of the joint work of the authors. In particular, paragraphs 2.1; 4 and 2.2.3 have been jointly written by the authors. Chiara Garau has written paragraphs 1 and 3 Francesco Pinna has written paragraphs 2; 2.2 and 2.2.1 and Franacesca Masala has written paragraph 2.2.2. Chiara Garau and Francesco Pinna have revised the whole paper and checked for its comprehensive consistency.

© Springer International Publishing Switzerland 2015
O. Gervasi et al. (Eds.): ICCSA 2015, Part II, LNCS 9156, pp. 612–623, 2015.
DOI: 10.1007/978-3-319-21407-8_43

the presence of certain factors that affect mobility, without checking how each element contributes to the smartness of the analyzed cities.

Moeinaddini et al. (2014), use the indicators proposed by UITP (the international organisation for public transport authorities, operators, and policy decision-makers) for "estimating the relationship between urban structure variables and daily private motorized trips in various cities around the world and [they] utilize this relationship to introduce a mobility sustainability index for evaluating transportation in cities at the macro-level" [13]. In addition, they argue that studies proposing a quantitative assessment of sustainable urban mobility are limited.

In relation to the literature considered, the innovativeness of the paper is not simply linked to the application of a similar methodology, but it is aimed to highlight the efficiency and effectiveness of the actions undertaken in the field of smart mobility. This is done through the use of (specific and synthetic) indicators capable of quantitatively analyze the problem of mobility. Therefore, the innovation consists in having identified quantitative indicators in relation to physical and infrastructural capital that characterizes every single city, in order to apply the studied methodology in those contexts that have demonstrated their smartness at least in terms of public sharing of the data.

Based on these premises, the purpose of this work is to define a set of indicators that evaluate the smartness in the field of mobility, through their aggregation. The Italian context has been chosen as a field of study, not only to assess how the Italian cities are approaching the development of sustainable mobility, but also to the greater ease by the authors to find the necessary data compared to other countries. To this end, the paper is organized as follows: in section 1, the methodology of aggregation of a system of indicators is presented. In section 2, these are combined to define a synthetic indicator of mobility. In section 3, the methodology, defined in the previous sections, is applied to Italian case studies. Finally, the paper concludes by analysing the obtained results.

2   Construction of a Synthetic Indicator of Urban Mobility

The authors felt the need to classify the topic from a quantitative point of view in order to operate with measurable data and to provide an objective assessment of smart mobility. This allowed to address the issue by checking place-based processes, without falling back on subjectivity, typical of the qualitative assessment. The analysis is also proposed as a starting point to provide a benchmarking of the urban mobility.

The study is also linked to the state of the art on the quantitative analysis available in the literature [14, 15, 16], also paying attention to experiences carried out on the topic [17].

The selected indicators are of two types:

— *Measurable indicators*, determined by a ratio of two or more different quantities;
— *on/off indicators*, a score was attributed to the yes/no answer.

The collected and then reworked data derived from several sources among which the following:

— *Dati ambientali nelle città. Qualità dell'ambiente urbano* (Environmental data in cities. Quality of the urban environment)*;* and *Ambiente urbano: gestione eco-compatibile e smartness* (Urban environment: eco-management and smartness) [18, 19];

— *Official web sites* [20] of the public transport companies (such as: Tper, Ataf, CTM, GTT, etc.), of bike and car sharing [21] (such as: Bicincittà, ICS - Iniziativa Car Sharing),

— *institutional web sites of the municipalities* [22] tested as case studies.

2.1 The Selection Criteria of the Cities

In the preliminary phase, a general analysis of the Italian situation was made through which 78 cities, with an urban extension of between 50 and 250 km^2, have been identified and studied. Among all of these, the attention was focused on those cities that, in relation to mobility, put at disposal the necessary data to calculate specific indicators. We had difficulty on the availability of data, so that the initial number of 78 cities was reduced to 17. Therefore, the cities analyzed in this paper were careful to the monitoring and to the publication of what emerged in their territory in the field of sustainable mobility and thus of the smartness.

In conclusion, the choice has been based on two basic principles: 1) the availability of the greater number of data so as to identify a sufficient number of specific indicators. 2) the dimension of the city so as to compare data among the cities.

2.2 Methodology

The proposed methodology for evaluating smart mobility can be summarized in the following three phases:

1. Data collection and selection of indicators;
2. Standardization of indicators;
3. Aggregation of indicators.

The three phases are better explained below.

2.2.1 Data Collection and Selection of Indicators

In this phase, the data were grouped according to six main variables:

— *Public transport:* (the territorial coverage and its use are evaluated through the analysis of the annual users and of the physical and management characteristics of the transport network);

— *Cycle lanes:* (the territorial coverage and the availability per inhabitant are evaluated through the analysis of the extension of the transport network and its availability per capita);

— *Bike sharing*: (the territorial coverage and the availability are evaluated by analyzing the service from the place-based perspective and per capita);

— *Car sharing*: (with the same criteria of bike sharing, as well as structured by Burlando et al., 2007 [23]);

— *Private mobility support system:* (operational solutions based on the application of new technologies in order to expand the range of services of private mobility and improve efficiency. Among these were considered: Variable message sign, SMS service for traffic alerts, Electronic payment park systems, Applications for mobile devices);

— *Public transport support system:* (services for users of urban mobility using information technology. They allow the dissemination of information in real time, play a decisive role in preventing or decongest critical situations in traffic flows and facilitate some payment transactions. Electronic bus stop signs, Electronic ticket payment system, Information on routes, schedules and waiting times, Travel planner for the route calculation, Travel tickets online).

In literature, the latter two indicators are considered very important. For example, Mobile Smart City Benchmarking Report (2014) reports: "[they] are playing a pivotal role in enabling smart cities around the world. Real-time data from mobile phone networks can underpin key smart city services, such as intelligent traffic management systems and disaster response solutions. Mobile networks can provide very detailed information about where individuals, devices and machines are located, how fast they move from one location to another, how clustered devices are together, and how this data changes over time and in relation to historical patterns" [24].

Each of these variables summarises specific indicators. The first four variables are determined from the ratio of two or more different quantities, while the last two use a binary measurement system (yes/no). This allowed to determine specific indicators and consequently the main variables (Table 1).

Table 1. Variables and indicators to evaluate smart mobility

Variables	Indicators	Specific Indicators		Unit
Public transport	I_{PT}	I_{BND}	Bus network density	km/100km^2
		I_{DPT}	Demand for public transport	passengers per year/inhabitants
		I_{TLC}	Traffic lights centralized	n°/total
Cycle lanes	I_{CL}	I_{CLD}	Cycle lanes density	km/km^2
		I_{CLI}	Cycle lanes for ten thousand inhabitants	km/10.000 inhabitants
Bike sharing	I_{BS}	I_{BSD}	Bicycle station density	n°/km^2
		I_{BPI}	Bicycle per thousand inhabitants	n°/1.000 inhabitants
Car sharing	I_{CS}	I_{CI}	Car for ten thousand inhabitants	n°/10.000 inhabitants
		I_{SI}	Station for ten thousand inhabitants	n°/10.000 inhabitants
Private mobility support system	I_{PMSS}	I_{VMS}	Variable message sign	yes=1,00; no=0,00
		I_{STA}	SMS service for traffic alerts	yes=1,00; no=0,00
		I_{EPPS}	Electronic payment park systems	yes=1,00; no=0,00
		I_{AMD}	Applications for mobile devices	yes=1,00; no=0,00
Public transport support system	I_{PTSS}	I_{EBSS}	Electronic bus stop signs	yes=1,00; no=0,00
		I_{ETPS}	Electronic ticket payment system	yes=1,00; no=0,00
		I_{RSWT}	Information on routes, schedules and waiting times	yes=1,00; no=0,00
		I_{TPC}	Travel planner for the route calculation	yes=1,00; no=0,00
		I_{TTO}	Travel tickets online	yes=1,00; no=0,00

2.2.2 Standardization of Indicators

The chosen indicators are not comparable, because expressed in different measurement units (such as: passengers/inhabitants, bicycle per thousand inhabitants, etc.). At this stage it is therefore necessary to obtain indicators that are independent of their specific unit of measurement: in according to Mazziotta et al. (2008) [16] one of the most used methods is to create indicators compared to a common range, in order to obtain measurement scales of the same amplitude (for instance, between 0 and 1 or between 0 and 100). In this way the immediate comparability between the considered indicators is made possible.

The scale of values is between 0.01 to 10, where the minimum value corresponds to 0.01 and the maximum value is 10, in order to reduce the number of decimal places. The formula used to create these indicators is the following:

$$x_{ir} = \{[x - \min(x_i)] / [\max(x_i) - \min(x_i)]\} *10 \tag{1}$$

where:
x_{ir} = standardized indicator
x = indicator
$\min(x_i)$ = minimum value of the indicator;
$\max(x_i)$ = maximum value of the indicator.

Table 2. Example of the standardized indicator I_{PT}

Cities	Public transport $I_{PT}=[(I_{BND}+I_{DPT}+I_{TLC})/3]$			
	I_{BND}	I_{DPT}	I_{TLC}	I_{PT}
Ancona	2,24	4,09	0,01	2,11
Bari	3,92	2,05	0,01	1,99
Bologna	3,40	10,00	7,50	6,97
Bolzano	1,72	5,16	8,80	5,23
Brescia	5,63	7,75	10,00	7,79
Cagliari	1,74	4,26	10,00	5,33
Catanzaro	1,25	1,94	0,01	1,07
Florence	9,18	8,78	7,70	8,55
Frosinone	3,03	0,57	0,01	1,20
Lecce	2,45	0,45	4,30	2,40
Palermo	3,22	1,27	0,01	1,50
Piacenza	1,37	2,94	8,60	4,30
Prato	3,21	0,01	0,01	1,08
Salerno	5,52	1,44	0,80	2,59
Teramo	0,01	1,24	0,01	0,42
Turin	10,00	8,19	4,90	7,70
Treviso	3,28	3,65	4,10	3,68

The values of the standardized indicators will be reported in the following way: each variable consists of a series of columns (one for each specific indicator concerning this variable) and one column that indicates in bold the variable of the indicator.

Table 2 shows the example of the variable Pubblic transport, composed of the following specific indicators:

Bus network density (I_{BND}),
— Demand for public transport (I_{DPT});
— Traffic lights centralized (I_{TLC})

All of them generate the indicator Public transport (I_{PT}).
All other variables are calculated similarly.

2.2.3 Aggregation of Indicators

The aggregation of indicators described in Table 1 was made in this way:

— the indicators within the same variable (see Table 2) were obtained through the arithmetic mean, whose general formula is:

$$x = (x_1 + x_2 + ... + x_n)/ n \qquad (2)$$

— the synthetic indicator of smart mobility is obtained through the geometric mean of the six indicators (one for each variable). The generic formula is:

$$x = (x_1 * x_2 * ... * x_n)^{\wedge(1/n)} \qquad (3)$$

In according to Mazziotta et al. (2008), the use of both different arithmetic and geometric mean assumes for implicitly that there is a certain interchangeability between the indicators belonging to the same variable and, instead, a lower substitutability between the variables that determine the synthetic indicator of smart mobility.

The following formula shows how the synthetic indicator of smart mobility is obtained:

$$I_{SM} = (I_{PT} * I_{CL} * I_{BS} * I_{CS} * I_{PMSS} * I_{PTSS})^{\wedge (1/6)} \qquad (4)$$

In conclusion, it is obtained a set of 18 specific indicators according to six variables in order to define the synthetic indicator of smart mobility.

This synthetic indicator can be calculated for each city worldwide, allowing comparison as described in the next paragraph.

3   Results: Ranking of the Italian Case Studies

The application of the described methodology has allowed to evaluate smart mobility in 17 Italian cities similar for place-based extension (figure 2). In fact, as previously declared, the choice has fallen on urban realities characterized by urban extension of not less than 50 km^2, and not more than 250 km^2.

Fig. 1. Spatial distribution of the Italian cities selected for the ranking

Table 3 shows the results of the specific measurable indicators, standardized through the application of the formula (1) and the indicator for each variable determined through the arithmetic mean, with the application of the formula (2).

Table 3. Measurable indicators and aggregation for variable

Cities	Public transport $I_{PT}=[(I_{BND}+I_{DPT}+I_{TLC})/3]$				Cycle lanes $I_{CL}=[(I_{CLD}+I_{CLI})/2]$			Bike sharing $I_{BS}=[(I_{BSD}+I_{BPI})/2]$			Car sharing $I_{CS}=[(I_{CI}+I_{SI})/2]$		
	I_{BND}	I_{DPT}	I_{TLC}	I_{PT}	I_{CLD}	I_{CLI}	I_{CL}	I_{BSD}	I_{BPI}	I_{BS}	I_{CI}	I_{SI}	I_{CS}
Ancona	2,24	4,09	0,01	2,11	0,01	0,07	0,04	0,01	0,01	0,01	0,01	0,01	0,01
Bari	1,92	7,05	0,01	1,00	0,38	0,01	0,19	2,70	3,86	3,28	0,01	0,01	0,01
Bologna	3,40	10,00	7,50	8,97	0,47	1,17	6,82	1,70	2,20	1,99	7,61	8,69	8,15
Bolzano	1,72	5,16	8,80	5,23	7,36	6,54	6,95	0,20	5,72	2,96	0,91	1,60	7,05
Brescia	5,63	7,75	10,00	7,79	9,89	8,34	9,12	10,00	10,00	10,00	2,31	3,10	2,70
Cagliari	1,74	4,26	10,00	5,33	6,02	6,05	6,03	1,20	3,16	2,18	4,85	3,81	4,33
Catanzaro	1,25	1,94	0,01	1,07	0,30	0,76	0,53	0,01	0,01	0,01	0,01	0,01	0,01
Florence	9,18	8,78	7,70	8,55	6,40	2,94	4,67	0,60	2,79	1,70	3,13	5,00	4,07
Frosinone	3,03	0,57	0,01	1,20	0,85	1,58	1,21	1,10	5,44	3,27	0,01	0,01	0,01
Lecce	2,45	0,45	4,30	2,40	1,23	6,40	3,82	0,40	4,84	2,62	6,42	1,31	3,86
Palermo	3,22	1,27	0,01	1,50	0,80	0,06	0,43	0,01	0,01	0,01	5,07	7,74	6,41
Piacenza	1,37	2,94	8,60	4,30	4,65	10,00	7,32	0,30	2,84	1,57	0,01	0,01	0,01
Prato	3,21	0,01	0,01	1,08	4,23	3,80	4,02	0,50	1,40	0,95	0,01	0,01	0,01
Salerno	5,52	1,44	0,80	2,59	0,37	0,10	0,23	0,50	0,84	0,67	0,01	0,01	0,01
Teramo	0,01	1,24	0,01	0,42	0,30	2,19	1,25	0,40	2,05	1,22	0,01	0,01	0,01
Turin	10,00	8,19	4,90	7,70	10,00	2,35	6,17	8,30	8,37	8,34	10,00	10,00	10,00
Treviso	3,28	3,65	4,10	3,68	7,93	9,69	8,81	3,10	8,56	5,83	0,01	0,01	0,01

The analyzed variables show how the emerging cities are those of northern Italy. An evident factor relates to the variable of car sharing; cities that do not yet have this service are 9, most of them identified in the south of Italy.

Table 4 shows the standardized specific binary indicators on/off and its related indicators for the variables Private mobility support system and Public transport support system.

Table 4. Binary indicators and aggregation for variables

Cities	Private mobility support system $I_{PMSS}=[(I_{VMS}+I_{STA}+I_{EPPS}+I_{AMD})/4]$					Public transport support system $I_{PTS}=[(I_{EBSS}+I_{ETPS}+I_{RSWT}+I_{TPC}+I_{TTO})/5]$					
	I_{VMS}	I_{STA}	I_{EPPS}	I_{AMD}	I_{PMSS}	I_{EBSS}	I_{ETPS}	I_{RSWT}	I_{TPC}	I_{TTO}	I_{PTSS}
Ancona	0,01	0,01	10,00	0,01	2,51	10,00	0,01	10,00	0,01	10,00	6,00
Bari	0,01	0,01	10,00	0,01	2,51	10,00	10,00	10,00	10,00	0,01	8,00
Bologna	10,00	10,00	10,00	10,00	10,00	10,00	10,00	10,00	10,00	10,00	10,00
Bolzano	10,00	0,01	10,00	0,01	5,01	10,00	10,00	10,00	0,01	0,01	6,00
Brescia	10,00	0,01	10,00	10,00	7,50	10,00	10,00	10,00	0,01	0,01	6,00
Cagliari	10,00	0,01	0,01	0,01	2,51	10,00	10,00	10,00	0,01	10,00	8,00
Catanzaro	0,01	0,01	0,01	0,01	0,01	0,01	0,01	0,01	0,01	0,01	0,01
Florence	10,00	0,01	10,00	0,01	5,01	10,00	10,00	10,00	10,00	10,00	10,00
Frosinone	0,01	0,01	0,01	0,01	0,01	0,01	0,01	0,01	0,01	0,01	0,01
Lecce	10,00	0,01	10,00	0,01	5,01	10,00	10,00	10,00	0,01	10,00	8,00
Palermo	0,01	0,01	10,00	0,01	2,51	0,01	10,00	10,00	0,01	10,00	6,00
Piacenza	10,00	0,01	0,01	0,01	2,51	0,01	10,00	0,01	0,01	10,00	4,01
Prato	0,01	0,01	0,01	0,01	0,01	0,01	10,00	0,01	0,01	10,00	4,01
Salerno	0,01	0,01	0,01	0,01	0,01	0,01	0,01	0,01	0,01	0,01	0,01
Teramo	0,01	0,01	0,01	0,01	0,01	0,01	0,01	0,01	0,01	0,01	0,01
Turin	10,00	0,01	10,00	10,00	7,50	10,00	10,00	10,00	10,00	10,00	10,00
Treviso	10,00	0,01	10,00	10,00	7,50	10,00	10,00	10,00	10,00	10,00	10,00

The data on binary indicators on /off always show that cities of northern Italy are the most equipped with private mobility support system, in contrast to other cities (such as: Frosinone, Prato, Salerno, Catanzaro, and Teramo) that do not have any of these facilities.

The situation improves for the equipment of public transport support system; excellent equipment for Bologna, Turin, Treviso and Florence, followed by Cagliari, Bari, and Lecce. In contrast, Frosinone, Salerno, Catanzaro, and Teramo do not have this type of service.

Table 5 presents the results obtained for each individual variable and it is important to consider how the synthetic indicator of smart mobility is obtained through the application of the formula (4).

Table 5. Synthetic indicator of Smart Mobility

Cities	Smart mobility $I_{SM}= (I_{PT}*I_{CL}*I_{BS}*I_{CS}*I_{PMSS}*I_{PTSS})^{\wedge(1/6)}$						
	I_{PT}	I_{CL}	I_{BS}	I_{CS}	I_{PMSS}	I_{PTSS}	I_{SM}
Ancona	2,11	0,04	0,01	0,01	2,51	6,00	0,22
Bari	1,99	0,19	3,28	0,01	2,51	8,00	0,79
Bologna	6,97	6,82	1,99	8,15	10,00	10,00	6,52
Bolzano	5,23	6,95	2,96	7,05	5,01	6,00	5,32
Brescia	7,79	9,12	10,00	2,70	7,50	6,00	6,65
Cagliari	5,33	6,03	2,18	4,33	2,51	8,00	4,27
Catanzaro	1,07	0,53	0,01	0,01	0,01	0,01	0,04
Florence	8,55	4,67	1,70	4,07	5,01	10,00	4,90
Frosinone	1,20	1,21	3,27	0,01	0,01	0,01	0,13
Lecce	2,40	3,82	2,62	3,86	5,01	8,00	3,93
Palermo	1,50	0,43	0,01	6,41	2,51	6,00	0,94
Piacenza	4,30	7,32	1,57	0,01	2,51	4,01	1,30
Prato	1,08	4,02	0,95	0,01	0,01	4,01	0,34
Salerno	2,59	0,23	0,67	0,01	0,01	0,01	0,09
Teramo	0,42	1,25	1,22	0,01	0,01	0,01	0,09
Turin	7,70	6,17	8,34	10,00	7,50	10,00	8,17
Treviso	3,68	8,81	5,83	0,01	7,50	10,00	2,27

The results show that cities of northern Italy achieve an enough high score in almost all variables analyzed, while cities of south Italy obtain the lowest values, excepting for some cities that succeed in emerging with discrete scores, such as Cagliari and Lecce.

Aggregating the 18 indicators in a synthetic indicator of smart mobility has finally allowed not only to evaluate the smart mobility according to the equipment measured in the cities under study, but also to generate a city ranking (Figure 2).

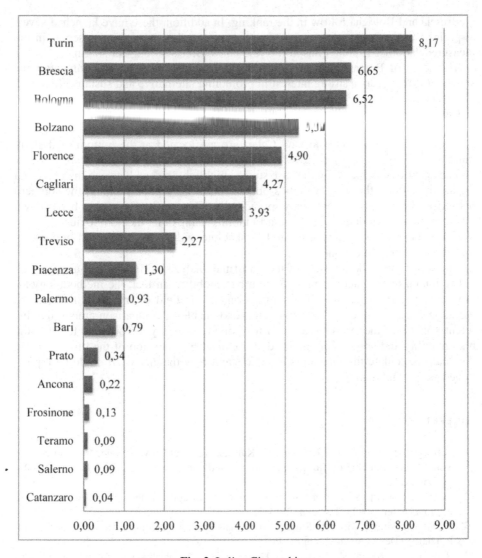

Fig. 2. Italian City ranking

4   Conclusions

Through the application of the presented methodology, it has been possible to generate a cities ranking (Figure 2) in which Turin is the most advanced in terms of smart mobility. Surely, this is due to the special interest that this city has shown in wanting to be a leader of the Italian Smart Cities, activating a number of projects and widening the process of transformation as a smart city in the entire metropolitan area [25].

Brescia and Bologna follow in the ranking. In addition, these have launched several projects in the smart context, and for that reason, they have the top scores in the measured variables (Public transport; Cycle lanes; Bike sharing; Car sharing).

The score of Bolzano and Florence is also good. In fact, its governments have shown in recent years a strong interest in promoting alternative and sustainable modes of transport.

Cagliari is in sixth position, and is the first city of the south Italy that appears in the ranking. It is followed by Lecce, Treviso and Piacenza.

Frosinone, Teramo, Salerno, and Catanzaro are located at the bottom of the city ranking.

The analyses conducted on the case studies highlight the gap between the north and the south of Italy: the northern cities obtain almost all a good ranking, the southern ones instead manage to emerge only sporadically. This confirms what has been shown by other studies that determine the smart mobility using other methodologies.

The used methodology is still valid in absolute and can be replicated in contexts different from the Italian ones.

The results describe the state of the art of the analyzed cities but can also give an indication on how to act to improve the urban mobility. In fact, the methodology allows to verify the outcomes of further actions in the field of smartness through the variation of each interested indicator. This also enables a comparison among the different policies of the cities. Namely, if the policies of a city are oriented towards the bike mobility and administrators decide to double the extension of the bike network, they can recalculate the indicators to understand how the city itself could be repositioned within the ranking.

References

1. Giffinger, R., Fertner, C., Kramar, H., Kalasek, R., Pichler-Milanović, N., Meijers, E.: Smart Cities: Ranking of European Medium-Sized Cities. Vienna University of Technology, Vienna (2007)
2. Garau, C.: Smart paths for advanced management of cultural heritage. Regional Studies, Regional Science 1(1), 286–293 (2014)
3. Garau, C.: From Territory to Smartphone: Smart Fruition of Cultural Heritage for Dynamic Tourism Development. Planning Practice and Research 29(3), 238–255 (2014)
4. Redman, L., Friman, M., Gärling, T., Hartig, T.: Quality attributes of public transport that attract car users: A research review. Transport Policy 25, 119–127 (2013)
5. Ortolani, C.: The cycling as a driver of a renewed design and use of public space within the neighborhoods. CSE-City Safety Energy (1), 51 (2014)
6. Banister, D.: The sustainable mobility paradigm. Transport Policy 15(2), 73–80 (2008)
7. Bertolini, L.: Integrating Mobility and Urban Development Agendas: a Manifesto. disP - The Planning Review 48(1), 16–26 (2012)
8. Litman, T., Burwell, D.: Issues in sustainable transportation. Int. J. Global Environ. 6(4), 331–347 (2006)
9. Lopez-Lambas, M.E., Corazza, M.V., Monzon, A., Musso, A.: Rebalancing urban mobility: a tale of four cities. Urban Design and Planning 166(DP5), 274–287 (2013)

10. Papa, E., Lauwers, D.: Smart Mobility: Opportunity or Threat to Innovate Places and Cities? In: 20th International Conference on Urban Planning and regional Development in the Information Society, REAL CORP 2015, pp. 543–550 (2015)
11. Litman, T., Burwell, D.: Issues in sustainable transportation. Int. J. Global Environ. **6**(4), 331–347 (2006)
12. Litman, T.: The new transportation planning paradigm. ITE Journal 83(6) (2013)
13. Mooinaddini, M. Asadi-Jlickuri, Zu Zaly Shah M: An urban mobility index for evaluating and reducing private motorized trips. Measurement 03, 02 (2015)
14. Ciccarelli, A.: Una metodologia statistica per l'analisi di competitività delle province. Working Paper dell'Istituto Guglielmo Tagliacarne (2003)
15. Gismondi, R., Russo, M.A.: Definizione e calcolo di un indice territoriale di turisticità: Un approccio statistico multivariato. STATISTICA n.3 (2004)
16. Mazziotta, C., Mazziotta, M., Pareto, A., Vidoli, F.: La costruzione di un indicatore sintetico di dotazione infrastrutturale: metodi e applicazioni a confronto. XXIX Conferenza Italiana di Scienze Regionali (2008)
17. Debnath, A.K., Chin, H.C., Haque, M.M., Yuen, B.: A methodological framework for benchmarking smart transport cities. Cities **37**, 47–56 (2014)
18. ISTAT, Dati ambientali nelle città - Qualità dell'ambiente urbano (2014). http://www.istat.it/it/archivio/129010
19. ISTAT, Ambiente urbano: gestione ecocompatibile e smartness (2014). http://www.istat.it/it/archivio/141296
20. Tper Bologna: http://www.tper.it/; CTM Cagliari: http://www.ctmcagliari.it/; GTT Torino: http://www.gtt.to.it/; Ataf Firenze: http://www.ataf.net/it/ataf.aspx?idC=2&LN=it-IT; CAP Autolinee Prato: http://www.capautolinee.it/; SETA Piacenza: http://www.setaweb.it/; Brescia Mobilità: http://www.bresciamobilita.it/
21. Bicincittà: http://www.bicincitta.com/default.asp; ICS - Iniziativa Car Sharing. http://www.icscarsharing.it/main/i-gestori
22. Comune di Bolzano. http://www.comune.bolzano.it/homepage.jsp; Comune di Firenze. http://www.comune.fi.it/export/sites/retecivica/index.html; Comune di Palermo. http://www.comune.palermo.it/; Comune di Cagliari. http://www.comune.cagliari.it/portale/
23. Burlando, C., Arduino, G., Nobile, D.: Il car sharing come business development area: analisi del settore, strategie d'impresa e ricadute socio economiche, No. 0703 (2007)
24. GSMA, Connected Living, Mobile Smart City Benchmarking Report, (2014). http://www.gsma.com/connectedliving/wp-content/uploads/2014/02/2649_GSMA_benchmarking_Report_Web.pdf
25. TorinoSmartCity. http://www.torinosmartcity.it/.

The Role of Public-Private Partnerships
in School Building Projects

Critical Elements in the Italian Model for Implementing Project Financing

Gianluigi De Mare[1(✉)] and Federica Di Piazza[2]

[1] University of Salerno, Via Giovanni Paolo II, 132, Fisciano (SA), Italy
gdemare@unisa.it
[2] Technical Unit Project Finance - Italian Government, Viale Della Mercede, 9, Rome, Italy
f.dipiazza@palazzochigi.it

Abstract. Given the problems of a legislative, procedural, financial and strategic nature that the application of project financing has encountered in Italy over the last 20 years, a thorough review in this paper is conducted of the prospects offered by this method in the specific case of completing works for primarily social purposes, including educational facilities. Following a comparison with the performance records of other countries in the European Community and in relation to recent legislative initiatives by the Italian Government, it is possible to define the contractual, organisational and operational methods. These methods appear the best suited to attracting the interest of a private investor and to increasing the quality of the projects to be realised, together with all the benefits of the level of educational service provided. The combined experience of the two authors in the fields of academic study and government research has been put to use in realising projects, which are generally considered unattractive by private lenders.

Keywords: Public-private partnership · School building · Financing of public works

1   Introduction

With the launch of the first phase of its School Building Plan and the forthcoming ratification of Legislative Decree no. 104/2013[1], the Italian Government has recently made the broad theme of welfare infrastructure one of its central priorities. The general framework that includes these measures relates to a widely accepted belief that an efficient education system is a key element for the socio-economic development of a country over the long-term.

Sections 1, 3 and 4 are developed by Di Piazza F.; Section 2 by De Mare G.; and Section 5 by both authors.
[1] Decree implementing Article 10 of Education Law no. 104/2013, signed by the Ministry of Infrastructure and Transport and the Ministry of Economics and Finance.

© Springer International Publishing Switzerland 2015
O. Gervasi et al. (Eds.): ICCSA 2015, Part II, LNCS 9156, pp. 624–634, 2015.
DOI: 10.1007/978-3-319-21407-8_44

Social building projects are one component of a wider and more diverse category of infrastructure programmes that influence and determine the quality of life of the population. The sector incorporates various types of construction, including schools and hospitals, judicial buildings (courts and prisons), residential complexes - notably social housing [1,2] - and sports, recreational and cultural facilities. In a context of economic crisis, the goal of an achieving sustainable growth and of an increasing the competitiveness of an economic system based on innovative services [3] is assigned to this type of structures: over the period 2007 -2013, against a 1.5% contraction in European GDP, public spending on social projects increased by 10.7% [4].

At national level, during the same period, one can point out a progressive reduction of investment in the education system[2] (See Figure 1), although overall the level of spending per student in Italy is just above the OECD average in both pre-primary and primary schools[3] (See Figure 2).

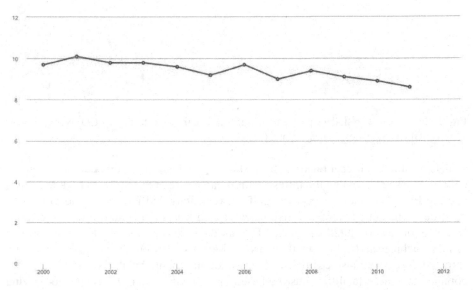

Fig. 1. Investment in education compared to public spending in Italy (Source: OECD Statistics, Education at a Glance)

---

[2] From 1995 to 2008, public and private spending increased by 8% in real terms, before recording a significant decrease of 11% between 2008 and 2011. In particular, there was an investment of around 75.6 billion Euros in 2008, which gradually decreased to 71.8 billion in 2011.

[3] Compared with an OECD average total annual expenditure on primary schooling of US $ 8,290, spending in Italy amounts to $ 8,440; conversely, expenditure per secondary student is 7% lower than the OECD average (US $ 9,280).

Pre-primary education   Primary education

Secondary education   Tertiary education

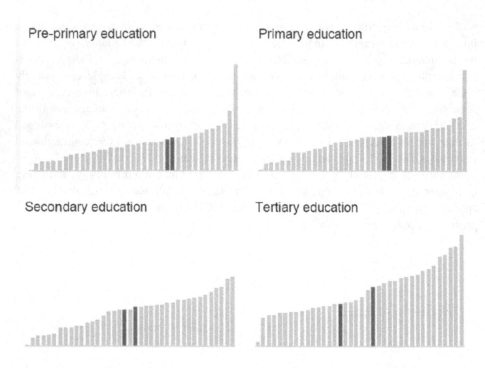

Fig. 2. Investment per student per year in Italy (red) compared to the OECD average (blue) (Source: OECD Statistics, Education at a Glance)

As recent international rankings have shown [5], the loss of competitiveness in the country is also affected by the infrastructural gap between the Italian school buildings and the European ones. Indeed, the gulf between Italy and Europe is due to insufficient development of more effective teaching and learning methods and to the obsolescence of school building stock. The gradual disappearance of the traditional school/teaching model requires the construction of school buildings that are not only safe and decent, but also ecological, accessible, and usable by all the members of the community. These facilities must be based on a flexible plan, to respond to varying educational needs in a more adaptable and effective manner, with outdoor spaces and multipurpose places for learning.

Therefore, the investment in school buildings is seen as a long-term investment, aimed at increasing the physical, human and intangible assets of the country. It also involves a wide variety of different interventions and of users, since it ranges from pre-school to secondary and tertiary education, to university institutions and to residential colleges.

This study, after highlighting the critical issues involved in applying forms of PPP (public private partnership) in Italy, will provide an overview of the role and of the implementation of the school building programme as part of the broader and strategic education sector. In the study there are four parts:

1. the first is a summary of the problems observed in Italy when applying a financial model based on project financing;
2. the second, again from the Italian perspective, shows the salient features of the school building programme launched by the Italian government;
3. the third part adopts an international perspective and it focuses on the role of private subjects in the creation of school infrastructure by defining whether, in relation to limits and to opportunities the instruments of partnership can prove a useful tool in promoting effectiveness and efficient public spending in the education sector;
4. the conclusions describe lines of actions to follow in Italy in order to determine PPP school building schemes which are able to maximise the efficiency of state and community funding.

2   The Critical Issues Involved in the Application of PPP in Italy

Studies conducted at the Department of Civil Engineering, University of Salerno, within the context of the Genovesi Award (with over 50 project financing schemes analysed and monitored over the last four years), have revealed various problems and critical issues in relation to this instrument.

These can be classified into three main areas of interest.

Regulations and procedures
There arises in the first place the issue of fragmentation, involving the contracting authorities and the markets of reference.

As is known, the recent EC Directives (24 and 25/2014) are concerned with the simplification of procurement procedures, by increasing recourse to the most economically advantageous offer, by encouraging use of negotiated procedures without calls for tender and by introducing the possibility of issuing joint calls between administrations (even transnational) interested in developing similar projects (like in the case of underground systems and strategic infrastructure projects). The issue of involving wider markets in project financing initiatives was examined by this perspective and it has provided interesting prospects for the many small local authorities (with fewer than 5 thousand inhabitants) that are currently subject to the Law 135/12 for the incorporation of basic public functions.

The conducted studies also show that performances of an acceptable standard can only result from a combination of markets at supra-municipal level, because of the require of a substantial critical mass for financial investments which are not particularly profitable.

There is also evidence of the continuing deregulation to which the norms in question have been subjected, as well as the host of different opinions and interpretations that rarely refer to specific bodies within the sector. This leads to problems in updating the various operators, with repercussions both on the reliability of calls for tender and on the legitimate execution of the bidding processes. In addition, there is the litigation, both during and after the process, which often accompanies the work at every stage of its useful life.

Financial aspects

A second theme, that also emerges frequently, concerns the bankable aspects of the initiatives. The time that the plans are confirmed should represent the moment of direct engagement for banking institutions, in order to involve them directly in the development phase of the project.

Another factor with negative impact is the attempt to force into the area of project financing those works which in terms of the amounts involved and of (limited) interest should continue to pursue the natural path of public financing and traditional contracting.

Strategic profiles and policy management

We need to consider the impact on the feasibility of the work, or even on the simple practicality of the project idea, resulting from the limited disclosure of preliminary design documents prepared by public authorities. The series of problems that can lead to unsuitable documentation is not sufficiently clear to policy makers. They are too ready to entrust an important role, like the role of roughing out the activity that triggers the dynamics of the investments, to individual technicians or teams unspecialised in the field, who lack the necessary interdisciplinary approach to tackle an issue as complex as project finance.

We should also mention the notable problems caused by excessive bureaucratic delays and by changes in scenario resulting from shifts in policy direction by government bodies (local or national).

Another salient element is the lack of independent technical skill and judgment displayed by the public bodies. This causes an incongruous sharing of risk, to the extent that the private investor is often covered against any forecasting error and that is too readily granted them to make changes during the course of construction. These issues tend to distort both the intrinsic reasoning behind the project initiative (i.e. the assumption that the costs of the project will be paid off over time through an income stream and not through a fast and calculated return on the capital invested) and the principles of competitiveness governing the award of contracts.

3   The Italian Government Plan for School Construction

The launch of the first phase of the School Building Plan involves funding for three areas of activity: the first, known as "Beautiful Schools", includes all minor maintenance operations; the second, known as "New Schools" concerns refurbishment work or the construction of new complexes; finally, "Safe Schools" involves work to bring buildings into line with regulations and safety standards.

Currently, the three schemes have different levels of funding and types of implementation, which can be summarised as follows [6].

By means of Resolution no. 22 of 30.06.2014 by the Committee of Economic Planning (CIPE), 400 million Euros have been allocated to the "Safe Schools" programme, designed to cover 1,639 interventions at an average cost of approximately 160,000 Euros.

These operations will be facilitated by simplified procedural measures, which will enable a rapid employment of funds. Within the same area of activity, co-financing is envisaged (in a proportion not exceeding 50%) for a further 1,226 interventions, using resources derived from bidding discounts and other economies resulting from initial interventions. The co-financing quotas for the individual operations will be allocated directly by the Ministry of Education, University and Research (MIUR).

For the "Beautiful Schools" category is allocated a budget of 150 million Euros. Intended for minor maintenance work, redecoration and functional improvements. The financing involves 7,751 school complexes for the year 2014 and about 10,000 operations during 2015. The procedural model envisages that funds are delivered directly by the Ministry of Education to schools, for which a review into funding allocation is still ongoing.

The third area of operation, known as "New Schools", has been provided with total funding of 800 million Euros (400 million for 2014 and the same amount for 2015), to be spent on refurbishment work or on the construction of new school buildings. Financing is being achieved through the release of funds under the Stability and Growth Pact. The first 404 operations were identified following applications by more than 4,000 municipalities, which gave notice that such work could begin immediately. For those other municipalities which applied for funding or the release of the operations agreement initiating in 2015, a new possibility now arises with the forthcoming Economic and Financial Planning Document, and with subsidised loans currently being activated at the expense of the State.

In spite of the many procedures for granting loans and of different bodies involved in implementing the Plan, a key component for the effectiveness and efficiency of spending is related to the methods employed for completing the projects.

4   School Building and PPP

The operation of the Plan just described and its implementation in line with principles of effectiveness and efficiency in public spending may be based on working models that envisage the involvement of private capital.

Before examining the technical elements involved in a PPP school building programme, it is appropriate to refer to the general perspective, questioning the extent to which the involvement of private capital improves the overall performance of the education system. From this perspective, and acknowledging the key role played by the public sector in the field of education to compensate for market failure and ensuring fairness of service, international experience has shown that the impact of private capital in the educational system represents an opportunity to introduce elements of positive innovation [7]. There are various models of private involvement in the education system (see Figure 3) and, in systems where educational funding is by public bodies (lower left quadrant), they range from contract outsourcing to the provision of school services or other forms of subsidy (vouchers and credit systems).

Providing with services

		Private	Public
Financing	**Private**	• Private schools • Private universities • Student accommodation • Tutoring	• Rates for users • Student loans
	Public	• Vouchers • School credit • Contractual outsourcing	• Public schools • Public universities

Fig. 3. Private-public partnership and education: Explanatory matrix (Source: Edited by Word Bank 2009)

Moreover, simply by examining the role of private investment in the construction/renovation of school buildings, we can see that the results of some studies show a positive correlation between the construction/restoration of buildings under PPPs and the educational results achieved. In particular, if we compare academic performance achieved in schools constructed under PFIs in the UK [8,9] with those built using traditional contracts, it emerges that where private capital was involved there was a 44% improvement in school performance and educational motivation - described in terms of truancy - also decreased in comparison with conventional schools.

At international level, one of the most popular models of PPP in the field of education is based on the public financing of services, even when they are provided by private entities as part of contract outsourcing aimed at the construction and management of the school. In this instance, as already outlined above, private financing seems able to ensure better value for money than traditional forms of contract. The form of the partnership arrangement envisages that the contracting entity (government or local authorities) agrees to the construction/refurbishment and maintenance of school buildings over the long-term (typically 20-30 years). This operational scheme, widely used in Britain and occasionally applied in Italy[4], is based on the undertaking of risk by the private entity in relation to the efficient management of the project facilities (core and collateral).

Together with the positive elements, which support the use of PPP schemes in school building programmes, it should also be noted that, in the economic-financial context, the private provider supplies services directly to the public administration

---

[4] In Italy, in the face of increasing interest in the PPP market, there are only limited examples of PPPs promoted by the public bodies [10,11]. There are some signs of a growing use of fund management companies (SGRs) in school building schemes, and especially in student housing and facilities (see the example of Tor Vergata, Rome). These include contractual arrangements whereby the public body/university concedes use of the area in front of the building and management of the facilities.

and so the operation represents substantially a market risk borne by the public admin-istration (a "cold" operation). As a result, school building can be compared to other sectors (such as traditional hospital building) where the private entity that constructs and manages the asset receives profits from payments (in various forms) made by the public administration. Given this structural limitation, the widespread use of the mod-el allows to the public body to defer payments over the term of the concession. On the private front, the main point of weakness is connected to the high risk still perceived by the market in the service and social infrastructure sectors.

To maximise the attractiveness of the investment, it is possible to identify three forms of incentive involving the legal, financial and technical aspects of the opera-tion.

1. On the legal front, at the planning stage it is opportune to devote special attention to the various contract options for formalising the involvement of the private part-ner, taking into account the level of responsibility that the private and public sec-tors wish to assume, and the economic and operational requirements of the public entity concerned (see Figure 4).

Types of public and private partnership	Main features
Conventional agreement	The public entity outsources the design and the construction to a private partner on the basis of specific funding requirements
Build – Own – Transfer (BOT) agreements	The private partner designs, builds, finances and manages the economics of the construction for the period of the concession, at the end of which he transfers the asset to the public entity
Build – Operate – Own – Transfer (BOOT) agreements	The private partner obtains a contract for the building, for the managing of the maintenance and for the providing with services, with the possibility of generating revenue over a period of time such to guarantee the return on investment. At the end of the concession, the private partner transfers the asset to the public entity
Build – Operate – Own – Subside – Transfer (BOOST) agreements	The private partner designs, builds and manages the economics of the construction also through the transfer of public resources
Entrusting of maintenance services	The public entity outsources the building maintenance to the private partner
Build – Own – Operate (BOO) agreements	The public entity transfers both the property and responsibility for the building and for the services to the private partner for the entire economic life of the asset

Fig. 4. Models of PPP in Europe (Source: Edited by Word Bank 2009)

2. On the financial side, it is advisable to increase the size of return on investment by extending the services provided by the private partner in the area of management (involving payment or fees). This is in order to reduce the level of public funding needed to support the economic and financial equilibrium of the operation and transfer the risk to the private sector. From this perspective, the model of implementation known as comprehensive contracting (commonly used in modern hospital building) enables the private entity to achieve management economies through the provision of services associated both with infrastructure (maintenance of the building and also efficient energy management), and with non-educational core facilities (e.g. infrastructure for sports, associations, the service industry). These can guarantee ancillary earnings from, for example, operational services (canteen, bar, cleaning) or educational facilities (lessons, tutoring, transportation, summer activities, etc.). In this way, the public entity can promote, simultaneously and working with a single agent, a dual structure of activities provided by the private partner, while encouraging the latter to organise and deliver the services in an efficient manner.

3. From the viewpoint of design, given that one of the critical factors determining the success of the education service is related to the quality of buildings, special attention must be paid to minimising the element of risk in construction and maximising economies of scale. The concentration of educational levels in single complexes associated with different but interrelated functions, and the standardisation of layout and building techniques, can ensure less risk of uncertainty with regard to construction times and to investment and operating costs (energy and maintenance).

5   Conclusions

By an examination of the general picture, this article has attempted to evaluate whether, and to what extent, PPP models used in the educational construction sector can provide a solution to ensure efficient spending in relation to the school building plan and, more generally, for public investment in this sector. Although by no means an exhaustive critique, this contribution aims to define the extent of the opportunities and problems - as well as the prospects for future study - to be found in the relationship between public and private investment in education and, more specifically, in school infrastructure.

To this end, we have cited the results of recent studies conducted at international level that indicate considerable positive potential in the involvement of private partners, both in the context of the more general education system and in the construction/maintenance and management (also at a profit) of its relative housing stock. With regard to this second case, certain studies have shown that, confronted by a structural requirement for the transfer of funding by a public entity in the context of a school building project (due to the inherently public nature of the investment), a deferral of payments and more effective management of the building and non-educational core services can offer a credible prospect to update these models.

The use of these instruments in the Italian context is not without problems. In general terms, this is because of the following reasons:

- excessive fragmentation of contracting, and a continuous and opaque mutation of sector regulations;
- limited involvement of banks in the planning process;
- insufficient technical provision on the part of the public administration;
- undue bureaucracy in decision-making mechanisms and authorisation procedure.

More specifically:

- for the operators, there remains a perception of high risk in committing to long-term investments in social projects, even though interest is increasing among operators and institutional lenders with regard to this sector;
- for the implementing body, the Italian school system is characterised by a fragmentation of authority, with the creation of school infrastructure delegated to various local entities which operate in the same area (so raising transaction costs and limiting opportunities to achieve economies of scale by combining different levels of schooling).

Nevertheless, increasing the effectiveness and efficiency of PPP operations with respect to the current framework is a challenge that is presently supported by both state and community investment. To encourage the use and increase of private capital involvement in PPP contracts, it seems desirable to promote:

- the standardisation of processes by defining standard forms of agreement for school construction projects (based on the experience of the Technical Project Financing Unit   linked to the *Agreement Model for hospital building*);
- the definition of common procedural methods and the spread of *best practices* which enable market operators and public entities to structure activities in accordance with models of innovative educational management.

References

1. De Mare, G., Nesticò, A., Tajani, F.: The rational quantification of social housing. In: Murgante, B., Gervasi, O., Misra, S., Nedjah, N., Rocha, A.M.A., Taniar, D., Apduhan, B.O. (eds.) ICCSA 2012, Part II. LNCS, vol. 7334, pp. 27–43. Springer, Heidelberg (2012)
2. Calabrò F., Della Spina L.: The public-private partnerships in buildings regeneration: a model appraisal of the benefits and for land value capture. In: 5th International Engineering Conference 2014 (KKU-IENC 2014). Advanced Materials Research, Vol. 931–932, pp. 555–559 © (2014). Trans. Tech. Publications, Switzerland doi:10.4028/www.scientific.net/AMR.931-932.555 (2014)
3. Madera R., Pisano A.: Risparmio, Investimenti a lungo termine e crescita sostenibile in Long-term investing in Europe: re-launching fixed, network and social infrastructure, Rome (2014)
4. OECD: Society at a Glance. In: OECD Social Indicators. OECD Publishing (2014)

5. OECD: Centre for Educational Research and Innovation, Innovative Learning Environments. OECD Publishing (2013)
6. http://www.istruzione.it/edilizia_scolastica
7. Word Bank: The Role and Impact of Public - Private Partnership in Education (2009)
8. Hodge, G.A., Greve, C.: Public-private Partneships: an International Performance Review. Public Administration Review **67**(3), 545–558 (2007)
9. Hodge G.A., Greve C., Boardman A.E.: International Handbook on Public-private partnership. Edward Elgar Pubblishing (2010)
10. XII Rapporto Finlombarda, Asili Nido in Partenariato Pubblico Privato. Un Manuale Operativo, Cresme (2011)
11. Tajani, F., Morano, P.: Concession and lease or sale? A model for the enhancement of public properties in disuse or underutilized. Wseas Transactions on Business and Economics **11**, 787–800 (2014)

Contribution of Geomatics Engineering and VGI Within the Landslide Risk Assessment Procedures

Francesco Mancini[1(✉)], Alessandro Capra[1], Cristina Castagnetti[1], Claudia Ceppi[2], Eleonora Bertacchini[1], and Riccardo Rivola[1]

[1] DIEF, University of Modena and Reggio Emilia, Modena, Italy
{francesco.mancini,alessandro.capra,cristina.castagnetti,
eleonora.bertacchini,riccardo.rivola}@unimore.it
[2] DICATECh, Technical University of Bari, Bari, Italy
claudia.ceppi@poliba.it

Abstract. This paper presents a literature review on the methodology called Volunteered Geographic Information (VGI) and its use for Landslide Risk Assessment (LRA). General risk assessment procedures are discussed and the potential contributions of VGI are identified, in particular when quantitative characterization of factors such as Hazard, Vulnerability and Exposure is required. The review shows that the standard LRA procedures may benefit from input given by surveyors when performing hazard assessments, while crowdsourced data would be a valuable support in vulnerability/damage assessment studies. The review also highlights several limitations related to the role of VGI and crowdsourcing in LRA.

Keywords: Landslide risk assessment · Geomatics engineering · VGI

1 Introduction

In the last decade, the Volunteered Geographic Information (VGI) tools were indicated as a valuable resource in the evaluation and assessments of risks arising from natural hazards and for a rapid and comprehensive inventory of exposed assets [1,2,3,4,5].

Among the range of possible natural or man-induced disasters, this paper provides a literature review on the possible roles played by VGI and geomatics engineering as support to the Landslide Risk Assessment (LRA) procedure. Initially, possible approaches to the quantitative LRA will be introduced with a short insight to the existing international framework. Successively, a review of useful geomatics engineering techniques, ranging from terrestrial to satellite-based, used in the monitoring of slope failure phenomena will be introduced. Finally, a discussion on the role of VGI and crowdsourcing in the field of LRA will be provided by illustrating issues arisen after the relevant literature.

2 Landslide Risk Assessment: Quantitative Approaches

As presented by the Centre for Research on the Epidemiology of Disasters (CRED) in its annual statistical review, a total of 330 natural triggered disasters were reported in

© Springer International Publishing Switzerland 2015
O. Gervasi et al. (Eds.): ICCSA 2015, Part II, LNCS 9156, pp. 635–647, 2015.
DOI: 10.1007/978-3-319-21407-8_45

2013 [6]. Worldwide, all the monitored phenomena caused 96.5 million of victims (21,610 killed) with a very high percentage (88%) coming from low income economies. The recorded economic damages decreased in comparison to the last decade. Within the wide range of possible natural disasters, the CRED provided a classification that can be found in [1]. Landslides are listed under the geophysical disasters, which caused costs the 82% below their 2003-2012 annual average and mostly due to the Sichuan, China, earthquake. In particular, the geophysical disasters accounted worldwide for 32 episodes (9.7% of total; 7.1 million victims; 1,166 deaths).

As said, many of the geophysical natural disaster were reported over region belonging to developing countries. Here, deficiencies in existing digital maps and assets inventories could represent a limiting factor whenever a quantitative risk assessment procedure is sought.

In the quantitative analysis of risks related to landslide hazards and investigations on slope failure phenomena, an increasing interest has been recently showed by the scientific community and stakeholders. In this field, the assessment of direct and indirect damages to properties and assets take on an increasing importance in addition to the development of reliable procedures and methodologies able to predict potential hazards to landslide. Beside this, increasing attention is now placed in the mitigation procedures able to reduce losses due to landslides by means of effective planning and management processes.

However, in spite of improvements in hazard recognition, prediction, mitigation measure, and effectiveness of early warning systems, worldwide landslide activity is widely reported. For countries affected by landslide risks an improvement in the effectiveness of funds allocation procedures is a requirement in addition to a careful vulnerability assessment of exposed assets. Hazard, risk, vulnerability and exposure are some keywords in the LRA procedure. A detailed list of keywords and definitions was provided in [7] by the United Nations International Strategy for Disaster Reduction.

According to Crozier and Glade [8], the concept of risk refers to a dual component: the likelihood of an adverse happening and its consequences. However, the adverse event has to be recognized and defined as occurrence and consequences triggered by this adverse event. A widely accepted definition of risk is the following: "the exposure or the chance of loss due to a particular hazard for a given area and reference period" [9]. Mathematically, it could be expressed by the multiplication between the probability that a hazard impact will occur and the consequences of such an impact. The Varnes' formula defines $R = H \times V \times E$, being H the hazard, E the exposure and V the vulnerability components.

In 2009 the UNISDR defines an hazard as "a dangerous phenomenon ... that may cause loss of life, injury or other health impacts, property damage, loss of livelihoods and services, social and economic disruption, or environmental damage". Moreover, a particular hazard is quantitatively described by the frequency of occurrence of different intensities for different areas. Here, the contributions of surveyors play a fundamental role because of the ability by traditional and novel methodologies to detect and represent the magnitude and spatial pattern of an investigated phenomenon. Scientific studies/maps, long-term monitoring, historic reports on past incidence of hazards (in particular the location), frequency and severity of the events constitute the wide range of useful products for a hazard assessment procedure. In addition, the UNISDR defines the exposures as "people, property, systems, or other elements present in hazard zones that are thereby subject to potential losses". Exposure is very often referred as "elements at risk". It is strongly connected with the concept of vulnerability which represents the degree of loss to a given element, or set of elements at risk, resulting from the occurrence of natural phenomena with defined magnitude. The degree of vulnerability could be expressed over a scale ranging from 0 (no damage) to 1 (total loss).

In the quantitative risk assessment, hazard, vulnerability, damage and exposure have to be carefully evaluated with respect to a geographical extent and spatial detail. The relevant literature introduced several approaches to quantitative risk assessment, as summarized in Table 1.

Table 1. Approaches to quantitative risk analysis as found in literature. The table follows an increasing level of complexity of the methodology from top to bottom. In the definition column, common values are introduced only once (after [10], with modifications).

Risk formulation	Definition	Source
Risk = H x C	C: Consequence (potential worth of loss) H: Hazard	Einstein (1988)
Rs = H x V	Rs: Specific Risk V: Vulnerability	Varnes (1984)
Rt = Rs x E = (H x V) x E	Rt: Total Risk E: element at risk	Varnes (1984)
Rt =∑(Rs x E) = ∑ (H x V x E)	V: Vulnerability	Fell (1994)
Rs = P(Hi) x ∑(E x V x Ex) Rt = ∑ Rs (landslide event 1,...n)	P(Hi): Hazard for a particular magnitude of landslide (Hi) E: total value of elements at risk, Ex: Exposure	Lee et Jones (2004)
R(DI) = = P(H) x P(S/H) x P(T/S) x P(L/T)	R(DI): individual risk P(S/H): Probability of spatial impact P(T/S): Probability of temporal impact P(L/T): Probability of loss of life for an individual hazard	Morgan et al. (1992)
R(PD) = = P (H) x P (S/H) x V(P/S) x E	R(PD): Specific risk property P(H): Hazard P(S/H): Probability that landslide impact the property V(P/S):Vulnerability E: Value of Property	Dai et al. (2002)

During the last 10 years, the increasing availability of geographical data, from authoritative sources or crowdsourcing processes, has encouraged the use of statistical and multivariate approaches in the task of hazards/susceptibility prediction to landslide [11]. Investigations related to the landslide susceptibility assessment could be based on qualitative and quantitative approaches (see Figure 1 for an overview).

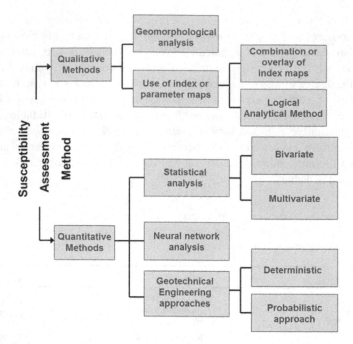

Fig. 1. Classification of landslide susceptibility assessment into qualitative and quantitative approaches

For instance, in the geographical assessment of landslide susceptibility a GIS multivariate analysis could be adopted. In this analysis, possible causal factors have to be related to landslide occurrences.

Causal factors and a reference landslide inventory, where location and description of past occurrences are reported, have to be previously designed within the GIS environment. Causal factors derived from the elevation model are resumed under the term "morphometric". Others are relevant to lithology, drainage system, existing infrastructures and anthropogenic sources.

In this analysis causal factors have to be connected to landslide occurrences by particular functions whose parameters have to be defined. It follows the idea that "the past are the key of the future".

The acquisition, storage and management of such data greatly benefits of methodologies provided by geomatics engineering.

3   Approaches by Geomatics Engineering in the Delineation of Landslide Hazard

A comprehensive review of possible methodologies adopted by surveyors in the investigation of slope failure phenomena goes beyond the scope of this work. However, in this sections a gallery of some application of geomatics engineering performed in the past by authors to landslide monitoring and hazard assessment well be provided.

In the LRA, methodologies belonging to the geomatics engineering are mainly focused on the task of detecting hazards due to slow and very slow movements. Among the variety of available techniques the following will be briefly introduced: *real time monitoring by multi sensor approach, GNSS (Global Navigation Satellite System), UAV (Unmanned Aerial Vehicle) proximity survey, GB-SAR (ground-based radar interferometry), TLS (Terrestrial Laser Scanning) and satellite radar interferometry (DInSAR and Permanent Scatterers Interferometry©).* Geomatic could contributes by providing geographic data on which *statistical and multivariate approaches to landslide hazard prediction* are based on. Hereafter, the above mentioned approaches are very briefly discussed with a pros and cons balance.

3.1   Real Time Monitoring by Multi-sensor Approach

The integration of various techniques is nowadays accessible, allowing to identify possible hazard to slope failures at increasing reliability. By the multi-sensor approach, Automated Total Stations (ATS), GNSS receivers and clinometers represents some of the used technologies.

ATS requires the availability of a suitable site, located outside the affected area, and several reflectors within the monitored zone. Additional reflectors need to be installed over stable positions, serving as control points for data correction. When periodic surveys are required, a forced centering device is often used to assure repeatability of ATS-based positioning. The inter-visibility between ATS and peripheral prisms could represent a drawback in addition to the stability of both the ATS and control prisms [12]. The latter, in particular, if small displacements are sought. To this purpose monuments are checked by GNSS surveys.

The stability of monuments hosting the reference ATS is mandatory and could be achieved by bi-directional clinometer able to measure tilting movements. It contributes to the monitoring of the reference consistency among subsequent observations.

The stability of control prisms is of great concern because their coordinates are used to compute geometric corrections which are subsequently applied to all raw measurements in order to correct refractions effects due the atmospheric influence on the electronic distance measurements. Such errors can achieve some centimeters of magnitude if no correction is introduced. The monitoring station on Figure 2 was designed by authors to detect potential slope failure phenomena in the northern Italian Apennines.

3.2   GNSS (Global Navigation Satellite System)

In the detection of slow and very slow displacements, the GNSS methodology based on the relative-static positioning was globally used by episodic or continuous monitoring. A careful designing of a network, composed of reference and monitoring points at useful locations, is a requirements to understand and model kinematic phenomena. Displacements could be detected at monitoring points (constituting nodes of the network) only and possible instabilities of reference stations established within a 3D reference frame. In the GNSS relative positioning the precision is very high and the error model accurately defined but the number of monitoring points is rather low

Fig. 2. Integrated monitoring system for unstable slopes: master unit at the top; below a GNSS remote site for continuous monitoring, reflectors and reference prism/GNSS

(depending on the spatial point density) and field efforts by surveyors significant. Anyhow, GNSS measurements are able to draw the superficial displacement field at variable (but very often reduced) geographical resolution.

In Figure 3 some results provided by the GNSS monitoring over a small village located in southern Apennines (Italy) are depicted with a delineation of landslides bodies as detected by the geomorphological surveys.

3.3   UAV (Unmanned Aerial Vehicle) Proximity Survey

More recently, multi-rotors UAV systems have proven to be a very useful tool for very high resolution DSM (Digital Surface Model) and orthophotos generation within geomorphological investigations [13,14]. Due to the initial stage of such application to unstable slope, only few of them could be retrieved from literature. See for example [15] for a cutting–edge investigation of sliding phenomena by UAV systems. These UAV-based methodologies use collections of unordered, non-metric, aerial images and data analysis based on classical computer vision approaches.

In particular, the flexible 3D surface reconstruction based on the Structure from Motion (SfM) approach is widely used as rapid, inexpensive and highly automated method. Besides the good quality of elevation model produced, orthophoto at unprecedented spatial resolution can be produced over hazardous area.

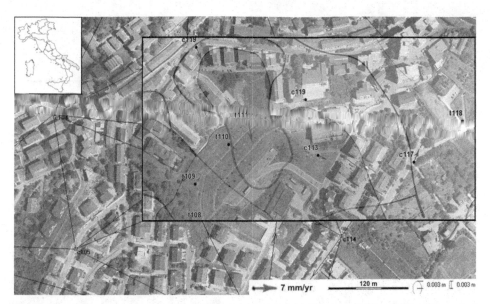

Fig. 3. Displacements over a small portion of the Bovino's (Foggia, Italy) landslide as revealed by the GNSS monitoring. Annual velocities (mm/yr) detected during the year 2009 at nodes and a delineation of the landslide body are superimposed to the GNSS network geometry. Error ellipses are depicted in green (see the reference ellipse in the lower right side).

3.4   GB-InSAR (Ground-Based Interferometric Synthetic Aperture Radar)

Spot monitoring campaigns with GB-InSAR allow the rapid assessment of landslide activity [16] even in radar-hostile, partially vegetated slopes and with high contents of humidity (ground and atmospheric). However, solutions could be affected by the processing strategy due to the parameters used (for instance number/timing of raw scenes, coherence of the images over time and space; shape/extent of the area and number or sampling rate of processed scenes). At very low displacement rates (i.e. few mm during the survey period) and with predominantly vegetated grounds, the processing strategy can affect the outcomes significantly and the detection of small displacements very hard.

Under favorable conditions the installations of GB-InSAR sensors in a suitable place allows the monitoring of slow slope movements in near-real time, being also possible to operate at distances of up to few km from the radar sensor. Results can be visualized "on site" through a 2D/3D displacement map thanks to a GIS interface. See Figure 4 for a displacement map from GB-InSAR data collected at Romanoro (Modena, Italy) with displacement (along the Line Of Sight, LOS) of relevant points (PS) as detected from surveys.

3.5   TLS (Terrestrial Laser Scanning)

With respect to other geomatics techniques, the main advantage in the use of TLS lies in providing a continuous geometric description of surfaces and changes by a multi-temporal

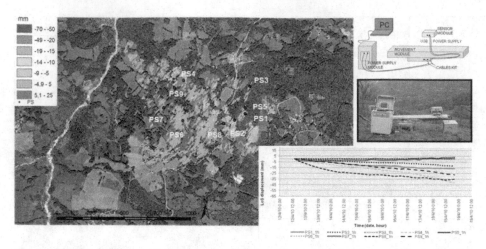

Fig. 4. 2D LOS displacement at Romanoro (Modena, Italy) landslide from processing of GB-InSAR image acquired at 1 hour rate. Time series have been reported for some representative points. In the upper-right insets a depiction of the radar sensors and its components.

approach. The main drawback is the impossibility to identify punctual features at desired locations, resulting in troubles to determine displacements. Despite that, terrestrial laser scanning is widely used to support landslides monitoring and some attempts have been carried out to detect geomorphological changes over time [17].

The main difficulty in comparing successive laser scanning surveys concerns the alignment process. Indeed, the reliability of final results is dependent on the accuracy of alignment process of multiple point clouds. An efficient solution would be the direct alignment, which requests a stable fixed position for the TLS placement at each campaign as well as to fix the orientation by acquiring specific markers during successive surveys. In the indirect approach, a manual or automatic recognition of homologous points on point cloud pairs is required and 3D transformation computed to align point clouds. The vegetation filtering is often required while surveying unstable slopes in order to represent the ground surface only. Once the alignment has been achieved, several strategies are available for surfaces reconstruction: the multi-resolution meshing approach, based on the Delaunay 2.5D triangulation from each scanning position, proved to be more successful in describing complex local morphologies than grid approaches [18]. See Figure 5 for results of TLS surveys to the Collagna (Modena, Italy) rockslide.

3.6   Satellite Radar Interferometry (DInSAR and Permanent Scatterers Interferometry[©])

Since 20 years, the satellite sensing based on SAR (*Synthetic Aperture Radar*) technology has been providing valuable information in the LRA. Thanks to methodologies such as the Differential SAR Interferometry (DInSAR) and, more recently, the Permanent Scatterers Interferometry (PSI[©]), several radar satellites were used to provide impressive information about slow superficial movements. Depending on the

Fig. 5. Results from the Collagna (Modena, Italy) rockslide monitoring (see photograph in the upper left image); the laser scanner during surveying (upper right); Digital Terrain Model obtained by integrating airborne and terrestrial laser scanning (bottom left) and morphological changes over the period 2010-2013 obtained by multi-temporal TLS surveys (bottom right)

geometry of satellite acquisition, elevation maps and deformation maps could be processed by Differential Interferometry. Fringes represent differences in elevations or displacements at large geographical extent.

The alternative PSI[©] method is based on the statistical analysis of radar response from permanent scatterers with suitable geometry at the ground. A variation in the slant range from satellite to targets among repeated passes is likely due to displacements towards or away from the sensor. Displacements can be solely detected along the LOS and a decomposition of displacements along the vertical and West-Est directions is only possible by the combined analysis of ascending and descending orbits. A potential displacement along a slope will be detected with an opposite sign by the ascending and descending orbits. The methodology is not sensitive to displacements in the north-south direction and over vegetated areas.

4   On the Potential Role of Geomatics Engineering and VGI in the Landslide Risk Assessment Procedure

As stated in section 2, a LRA procedure is a complex task and needs for an integrated approach. According to [19], risk assessment "takes the output from risk analysis and assesses these against values judgements, and risk acceptance criteria". As introduced in section 3, the monitoring of landslide by the geomatics engineering is able to address the complex issues of hazard evaluation and support the census of element at risk. However, an exhaustive LRA could also benefit from community based knowledge. There are several stages at which values and judgments enter in the decision-making process by underpinning consideration about the relevance of risks and the associated consequences. It happens when the identification of a range of possible alternatives for managing risks are formulated. These types of judgment are relevant to the risk evaluation procedure, for instance where three categories of risks could be identified: acceptable, tolerable and intolerable [20].

Such judgments are strongly influenced by psychological, cultural and social perspectives. Hence, a multitude of factors contributes to risk perception, and it may vary greatly among individuals belonging to a community. Therefore, the role played by the "communication of risk" and the "understanding of risk" could be complex.

Despite adversarial attitude and widespread skepticism about the reliability and involvement of the volunteered information, in the framework of the LRA the role of VGI is unquestionably useful under particular conditions. It is the case of rare events, such as those induced or exacerbated by climate change, in which the potential role of individuals may be similar to that played by the early warning system. Under some conditions, the landslide phenomenon may assume an evolution from slow to very fast. Only few slopes could be instrumentally monitored and, in the case of sudden development of the sliding phenomenon, there are no terrestrial or satellite-based methodologies able to provide information at the required temporal rates.

In such situations, information collected from citizens living within areas subjected to landslide risk could help in identifying possible precursory phenomena and constitute a potential early warning system for authorities. These kinds of Community-Based Early Warning Systems (CBEWS) could contribute towards a reduction of economic losses after a natural phenomenon occurs and in the mitigation of direct and indirect effects on goods, people and properties. The CBEWSs are supposed to be an ideal tool, being able to provide the communities and disaster risk manager with anticipatory information on a potential impending phenomenon and improve the preparedness against adverse phenomena. Detractors of such an approach drive the attention on possible false positive responses from CBEWS and the needs of a reliable procedure able to provide a judgment about the credibility of information from the users.

The VGIs philosophy could support EWS especially in developing countries where inventories, existing data infrastructures and available equipment are not able to cope with a rapid and widespread monitoring of emergency situations. Even though a risk evaluation can be conducted with data from instrumental survey and monitoring procedures, a complete LRA needs the implementation of intangible data. The latter could be based on the knowledge by communities about the specific risk. Obviously, an integration between expert and community based knowledge could be also an opportunity. Risk maps developed through collaboration between researchers and communities are the simplest way to represent and inform about a specific risk. Beside this, a detailed description of the whole process would be useful in addition to guidelines to support any decisional phase.

Due to thsese motivations, the LRA could greatly benefits from massive information coming from crowdsourcing, technical and/or scientific knowledge and VGI. In the hazard assessment procedures surveyors coming from professional or scientific communities can be a primary source of knowledge by providing the extremely wide variety of data and results on the magnitude and extent of monitored phenomena. The problems are related to the way surveyors can disseminate data, results and knowledge about surveyed hazards. A common point about dissemination of data would be required by taking also into account issues related to the data heterogeneity (arising from different methods, production stage, etc.) and varying level of uncertainty of observation and results.

In view of expected implementations of VGI systems as a tool for risk assessment to such phenomena, some open questions have to be faced. A first one is related to the minimum level of skillfulness and knowledge required by contributing people while a second relies with the amount and reliability of available information, especially over highly vulnerable areas with poor dataset or within regions where geographical database are not in use. Several other task have to be faced thoroughly: the willingness by users to contribute, difficulties in the access to knowledge by potential contributors (critical for poor qualified group of people), the reliability of contributions and to the need of a long-term maintenance of initiatives. Nevertheless, the introduction of the VGI concepts could be a solution for some of the issues arisen in this paper. For instance, it is a shared opinion that the conceptual match between elements at risk and VGI is an applicable framework.

5   Conclusion

In the scientific community involved in the field of risk assessment related to natural disasters is a common thought the VGI could be a solution for some of the tasks. In particular, the Landslide Risk Assessment procedure may benefit from the use of VGI and crowdsourcing in the strengthening of existing Spatial Data Infrastructures and "authoritative" or "conventional" data and whenever data are missed.

Nevertheless, limitations to the use of VGI in the natural disaster can be found in recent literature. Firstly, some gaps in the use of VGI for natural hazard assessment must be filled as well as the need for more robust case studies and experimental research to support this promising field [1]. Manfré et al. [2] introduced the needs of training for involved volunteers and minimum number of volunteers. Another key aspect was introduced in [21] by Camponovo and Freundschuh who discussed the need for more research on the quality of the categorization (i.e., attribute data) of volunteered emergency data. Coleman [22] stated that the VGI is not the ultimate solution to all geospatial data updating and maintenance challenges now faced by mapping organizations. The contribution of the scientific community in this field could be placed in the establishment of a rigorous framework and workflow able to provide reliable results and reduce the uncertainty of basic information used in the Landslide Risk Assessment procedures.

References

1. Horita, F.E.A., Degrossi, L.C., de Assis, L.F.G., Zipf. A., de Albuquerque. J.P.: The Use of Volunteered Geographic Information (VGI) and Crowdsourcing in Disaster Management: a Systematic Literature Review. AIS Electronic Library (AISeL) (2013)
2. Manfré, L.A., Hirata, E., Silva, J.B., Shinohara, E.J., Giannotti, M.A., Larocca, A.P.C., Quintanilha, J.A.: An Analysis of Geospatial Technologies for Risk and Natural Disaster Management. ISPRS Int. J. Geo-Inf. 1(2), 166–185 (2012)
3. Schelhorn, S.J., Herfort, B., Leiner, R., Zipf, A., de Albuquerque, J.P.: Identifying elements at risk from openstreetmap: the case of flooding. In: Proceedings of the 11th International ISCRAM Conference. University Park, Pennsylvania (2014)
4. Horita, F.E., de Albuquerque, J.P.: An approach to support decision-making in disaster management based on volunteer geographic information (VGI) and spatial decision support systems (SDSS). In: Proceedings of the 10th International Conference on Information Systems for Crisis Response and Management, Baden-Baden, Germany, pp. 12–15 (2013)
5. Poser, K., Dransch, D.: Volunteered Geographic Information for Disaster Management with Application to Rapid Flood Damage Estimation. Geomatica 64(1), 89–98 (2010)
6. Guha-Sapir, D., Hoyois, Ph, Below, R.: Annual Disaster Statistical Review (2013). The numbers and trends. Centre for Research on the Epidemiology of Disasters, Brussels (2014)
7. United Nations International Strategy for Disaster Reduction (UNISDR): Terminology on Disaster Risk Reduction. United Nations (2009)
8. Crozier, M.J., Glade, T.: Landslide hazard and risk: issues, concept and approach. In: Glade, T., Anderson, M., Crozier, M.J. (eds.) Landslide Hazard and Risk, pp. 1–40. John Wiley and Sons (2005)
9. Varnes, D.J.: IAEG. Landslide Hazard Zonation - A Review of Principles and Practice. Commission on Landslides, p. 60. UNESCO, Paris (1984)
10. Duzgun, H.S.B., Lacasse, S.: Vulnerability and acceptable risk in integrated risk assessment framework. In: International Conference on Landslide Risk Management and 18th Vancouver Geotechnical Society Symposium (2005)
11. Mancini, F., Ceppi, C., Ritrovato, G.: GIS and Statistical Analysis for Landslide Susceptibility Mapping in the Daunia Area (Italy). Nat. Hazards Earth Syst. Sci. 10(9), 1851–1864 (2010)
12. Castagnetti, C., Bertacchini, E., Corsini, A., Capra, A.: Multi-sensors Integrated System for Landslide Monitoring: Critical Issues in System Setup and Data Management. Eur. J. Remote Sens. 46, 104–124 (2013)
13. Mancini, F., Dubbini, M., Gattelli, M., Stecchi, F., Fabbri, S., Gabbianelli, G.: Using Unmanned Aerial Vehicles (UAV) for High-resolution Reconstruction of Topography: The Structure from Motion Approach on Coastal Environments. Remote Sens. 5(12), 6880–6898 (2013)
14. Bertacchini, E., Castagnetti, C., Corsini, A., De Cono, S.: Remotely piloted aircraft systems (RPAS) for high resolution topography and monitoring: civil protection purposes on hydrogeological contexts. In: SPIE Remote Sensing, pp. 924515–924515. International Society for Optics and Photonics (2014)
15. Turner, D., Lucieer, A., de Jong, S.M.: Time Series Analysis of Landslide Dynamics Using an Unmanned Aerial Vehicle (UAV). Remote Sens. 7(2), 1736–1757 (2015)

16. Corsini, A., Berti, M., Monni, A., Pizziolo, M., Bonacini, F., Cervi, F., Ciccarese, G., Ronchetti, F., Bertacchini, E., Capra, A., Gallucci, A., Generali, M., Gozza, G., Pancioli, V., Pignone, S., Truffelli, G.: Rapid assessment of landslide activity in emilia romagna using gb-insar short surveys. In: Landslide Science and Practice, vol. 2, pp. 391–399. Springer, Berlin (2013). ISBN: 9783642314452, Rome, 3-7, October 2011

17 Corsini, A., Castagnetti, C., Bertacchini, E., Rivola, R., Ronchetti, F., Capra, A.: Integrating Airborne and Multi-temporal Long-range Terrestrial Laser Scanning with Total Station Measurements for Mapping and Monitoring a Compound Slow Moving Rock Slide. Earth Surf. Proc. Land. **38**, 1330–1338 (2013)

18. Castagnetti, C., Bertacchini, E., Corsini, A., Rivola, R.: A reliable methodology for monitoring unstable slopes: the multi-platform and multi-sensor approach. In: Proceedings of the SPIE, Earth Resources and Environmental Remote Sensing/GIS Applications V (2014)

19. Fell, R., Ho, K.K.S., Lacasse, S., Leroi, E.: State of the art paper 1-A framework for landslide risk assessment and management. In: Proceedings of the International Conference on Landslide Risk Management, Vancouver, BC, Canada (2005)

20. Fell, R.: Landslide Risk Assessment and Acceptable Risk. Canadian Geotechnical Journal **31**, 261–272 (1994)

21. Camponovo, M.E., Freundschuh, S.M.: Assessing Uncertainty in VGI for Emergency Response. Cartogr. Geogr. Inf. Sci., 1–16 (2014)

22. Coleman, D.J.: Potential Contributions and Challenges of VGI for Conventional Topographic Base-mapping Programs. In: Sui, D., Elwood, S., Goodchild, M. (eds.) Crowdsourcing Geographic Knowledge: Volunteered Geographic Information (VGI) in Theory and Practice. Springer Science+Business Media, Dordrecht (2013)

Planning with Nature: Green Areas Configuration and Natural Cooling in Metropolitan Areas

Marialuce Stanganelli[✉] and Carlo Gerundo[✉]

Department of Civil, Architectural and Environmental Engineering,
University of Naples Federico II, Naples, Italy
{stangane,carlo.gerundo}@unina.it

Abstract. This paper focuses on urban planning strategies to adapt cities to the increasing rising of temperatures during summer heat waves. The main target is to investigate which configuration and distribution pattern of green spaces could effectively improve natural cooling of urban environments. Although the benefit that green areas give to natural cooling is well known, this kind of studies has hardly been carried out, especially at an urban scale where it is crucial to define quantities and density of green areas to address open spaces design. To reach this goal, a methodology based on the interpretation of the statistical correlation among temperature, urban parameters and green areas configurational indicators was implemented and applied to the case study of Naples metropolitan area, performing all the analysis in a GIS. Results provide guidelines to improve natural cooling in urban areas adopting the most effective configuration and distribution of green areas within a densely built context.

Keywords: Climate-change · Natural cooling · Urban heat island · Green infrastructures

1   Urban Adaptation to Heat Waves

Climate change (CC) is one of the defining problems of the 21st century and its impact on cities will be severe, as foreseen by the Fifth Report of the Intergovernmental Panel on Climate Change [1]. CC most serious effect on urban areas is the increase in frequency, intensity, and/or duration of extreme weather events such as heavy rainfall, warm spells and heat events, with consequential dramatic risk to cities comfort and liveability, human health and anthropic activities [1].

With regard to adaptation to extreme heat events, many examples of environmental stewardship can be found in high-income countries. In locations with large daily variations in temperature, an incisive response can include upgrading homes with passive cooling system in order to reduce solar and internal heat gains, while enhancing natural ventilation or improving insulation [2], [3]. Air conditioning and other forms of mechanical cooling could be no more the solution to urban high temperature since electricity generation contributes to Greenhouse gases emissions. Usually interventions to promote natural cooling do not go beyond architectural scale.

© Springer International Publishing Switzerland 2015
O. Gervasi et al. (Eds.): ICCSA 2015, Part II, LNCS 9156, pp. 648–661, 2015.
DOI: 10.1007/978-3-319-21407-8_46

Urban planning could play a crucial role in developing urban resilience to CC effects and contribute to adaptive efforts by promoting natural cooling through a, sustainable urban design including Ecosystem services supply [4], [5]. A rethinking of urban planning and a reassessment of the discipline itself is necessary. Planners and researchers should focus their efforts on the development of new tools to improve cities climatic performances and new ways to use Ecosystem services as active urban structural materials'.

Due to CC, in next decades, heat waves are expected to increase in number and intensity [1]. This phenomenon is extremely dangerous for human health: data on the number of deaths due to the European heat wave of 2003 (about 35000) make it one of the more devastating natural disasters of last decade [6]. Projections suggest that 1-in-20 year hottest day is likely to become a 1-in-2 year event by the end of the 21st century in most regions [1]. In big cities, heat waves further increase their intensity and dangerousness because of the phenomenon of Urban Heat Island (UHI). UHI is a thermal anomaly affecting large urban settlements where temperatures are higher than surrounding rural areas. The intensity of this phenomenon can be quantified as the maximum difference between the average temperature of urban air and the one of surrounding rural environment. The difference appears to be more pronounced at night than during the day. For instance, during summer periods the temperature difference between urban and suburban areas can range from +1 °C to +3 °C in daylight, while at night it can reach values ranging from +7° to +12° C [7].

The high vulnerability of Metropolitan areas to heat waves effect is due to the widespread overbuilding, the prevalence of paved surfaces on green areas, the use of building materials with low ability to dissipate heat, the morphology of some urban tissues which obstruct natural ventilation, the huge amount of emissions caused by human activities (traffic, industrial plants, heating and air conditioning systems for household use).

During heat waves, UHI can produce peak demands for energy consumption with consequent power blackouts in metropolitan areas; increasing of air pollution and health problems and mortality [8].

Some scholars estimated that every degree increase (K) adds a significant supply to the air conditioning load evaluated between 5% and 10% of total consumption [9]. As an indirect secondary effect, air conditioning increase causes a proportional growth of several pollutants emissions in the atmosphere, as well as the rise of Greenhouse gases production (CO_2 mostly). Moreover, high temperatures intensify photochemical reactions of pollutants in the air: for every degree of temperature over 22°C, accident by smog increases by 5% [10].

UHI could be particularly harmful for human health because the temperatures increase also during the night, threatening human body resilience to extreme heat without giving night relief [11].

Until now, UHI has been studied in four different research areas, which almost never have converged in a global and systemic study. Studies have been developed concerning the following single problems:

1. identification of areas interested by UHI phenomenon using temperature measurement techniques (remote sensing, direct survey, temperature sensors) or processing meteorological statistical data;

2. identification of cool building materials which may mitigate heat increase in urban areas, (i.e. roof covering, street pavements, green roofs and walls);
3. evaluation of urban form influence on temperature increase;
4. analysis on influence of UHI on energy consumption and Greenhouse gases emissions.

According to literature, acting on urban shape it is possible to reduce the flow of heat stored in urban structure and the anthropogenic heat flux: the geometry, spacing and orientation of buildings and outdoor spaces strongly influence the microclimate in the city [12].

In general, UHI are influenced by many factors concerning urban form as:

— formation of Urban Canyons, where heat is hardly dissipated due to the absence of ventilation [13];
— urban tissues exposure considering solar and main winds orientation [14]; wind can cushion major temperature fluctuations, reducing the occurrence of hot spots [15]; some studies pointed out that network grids having street canyons parallel to the prevailing wind flow minimize the shelter, increase the wind speed and the dissipation of stored heat, and therefore contribute to mitigate UHI [16];
— abundance and localization of green spaces, since greatly increasing the total surface of green areas within a city may significantly reduce temperatures at the city scale [17], but several studies hint that different distributions of green areas, provide significantly different effects on urban microclimate[18], [19].

Although many studies have been already developed on urban planning and UHI issues there is an overall lack of strategies and applicative tools to measure the effectiveness of actions developed.

With regard to urban green areas, even though the benefit that they give to natural cooling is well known, there are still no specific evidence-based recommendations on how best to incorporate them into an urban area, in terms of abundance, distribution and type of greening [20].

This paper focuses on urban planning strategies to adapt cities to the increasing rising of high temperatures during summer heat waves. The main target is to investigate which configuration and distribution of green spaces could significantly improve natural cooling of urban environments.

2   A Methodology to Investigate How Urban Structure Influences Temperature

In order to analyze the links between Ecosystem Services provision (green areas, water basins) and high temperatures in a metropolitan area, a methodology based on the evaluation of the correlation curves "temperature - urban parameters" was developed. This methodology was applied to the case study of Naples metropolitan area.

Climate could depend on many different natural parameters: morphology, altitude, orientation, ventilation, soil permeability, green areas and water basins. The combination of these parameters allows the identification of different microclimate zones. Moreover, within a homogeneous microclimate zone, other antrophic parameters could affect temperatures, above all building materials and urban tissue age.

The Metropolitan area of Naples is an interesting field of application because of its several different microclimatic conditions, from coastal zones to hill areas, to inner plans. The first step developed in this study was the identification, within the case study area, of different Microclimate Zones having homogeneous conditions of natural microclimate (i.e. distance from the sea, land morphology) and consistent conditions of urban environment. Within each Microclimate Zone, Test Areas (TAs) with similar features of building materials and urban tissue age were identified.

To measure main urban structure features that could contribute to increase UHI phenomenon, the Check Indicators described in Table 1 were selected and calculated in each TA.

Table 1. Check indicators calculated to analyze TAs urban features

Name	Calculation scheme	Units
Land Cover Ratio	(built surfaces)/(area of TA)	m^2/m^2
Building Density	(building volume)/(area of TA)	m^2/m^2
Percentage of Green Areas	(green areas)/(area of TA)	m^2/m^2
Non Permeable Surface Index	(non permeable open areas)/(area of TA)	m^2/m^2
Mean Building Height	(building volume in the TA)/(built surfaces in the TA)	m

Specific influence of water basins was taken into account evaluating, for each TA, its distance from the sea shore line and/or lakesides.

CI and landscape metrics calculation in TAs required the set up of a geographical database for Naples metropolitan area including: land use map, topographical database map, built environment, road maps, orthophoto.

Geographic Information System (GIS) routines and spatial analysis were developed to calculate the majority of Check Indicators.

The link among urban structure features and temperature increase was investigated performing a correlation analysis between Check Indicators value in any TA and its average temperature.

A more in-depth study was carried out for green areas through a spatial pattern analysis aiming to identify which should be the optimal shape and distribution of green areas in an urban environment. The connection among these metrics and average temperature in each TA was also examined, executing a correlation analysis.

3   Geographical Database Implementation and Data Mining in the Case Study Area

The Metropolitan Area of Naples is one of the most densely populated in Italy. It has an area of 1,171.13 km², and a total population of about 3 million inhabitants.

Its geomorphological configuration is the result of a complex intertwining of volcanic systems, limestone massifs and pyroclastic-alluvial plains.

The climate is typically Mediterranean except for few inner zones where it is more continental. The inland plain, where urbanization has been spreading out, is flanked by two volcanic systems: Vesuvius, to the east, and *Campi Flegrei*, to the west. Between them the most densely populated centers are concentrated. The urbanized area has an average population density of 8000 inhabitants per square kilometer.

The widespread and speculative overbuilding has transformed much of the rural inland plain into a huge periphery. Moreover, this transformation erased the pre-existing rural green areas without providing a suitable supply of urban parks or green spaces (Fig. 1). These factors produced a substantial climate alteration, in particular in the inner areas where extremely high values in temperature are detectable (in some cases over 50 °C).

With regard to high urban temperatures, it is necessary to specify that it is possible to identify two types of UHI: surface and atmospheric UHI.

Surface UHI refers to the temperature of urban surfaces exposed to the sun's rays that are hotter than air. In summertime urban surfaces temperature could reach more than 50°C during the day, while difference with air temperature is smaller during nighttime. Atmospheric UHI refers to warmer air in urban areas compared to cooler air in rural surroundings. This phenomenon is not particularly intense throughout the

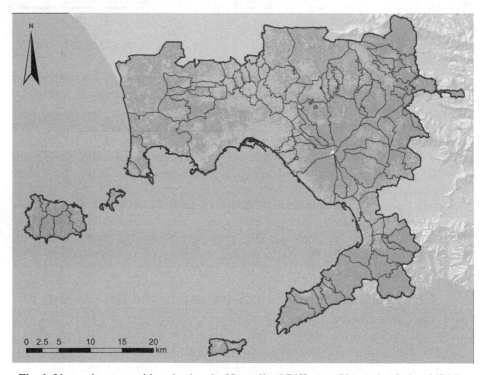

Fig. 1. Vegetation map achieved using the Normalized Difference Vegetation Index (NDVI)

day, while it becomes more pronounced during nighttime due to the slow heat release from urban surfaces.

To identify UHI, scientists use direct and indirect methods. Direct methods, such as fixed weather stations and mobile traverses, are used to identify atmospheric UHI [7], [11]. In this research, remote sensing, an indirect measurement technique, was used to estimate surface UHI. Thermal values were detected by a temperature map, produced by the GIS Service of the Province of Naples. This map was carried out processing data acquired by the air transported sensor MIVIS (Multispectral Infrared and Visible Imaging Spectrometer) during the summer 2005.

The entire detection consisted of 116 strips which cover 1170 km^2 of the whole Metropolitan Area. The high resolution images obtained (3 × 3 meters space grid) allowed a precise and detailed analysis of phenomena, so that it was possible to identify the temperature of each single building.

To perform the analysis carried out in this study, georeferenced data (UTM-WGS84), made available by the GIS Services of the Province of Naples, were used. Two informative layers called "permeable surface" and "non-permeable surface" were extracted by the Topographic Database of Campania Region, derived from the Regional Digital Map (1:5000 scale, edition 2004/2005). Building heights were calculated subtracting the heights of the Digital Surface Model (DSM) to the ones of the Digital Terrain Model (DTM). Both DSM and DTM has a grid of 1 × 1 meter and were processed from data detected using LIDAR (Light Detection and Ranging) sensor air transported (flights occurred between 2009 and 2012).

The Orthophoto raster images employed were extracted from the Orthophoto of Campania Region (scale 1:5000, detected in 2004/2005)

To calculate Check Indicators ArcGis 9.3 was used, while urban green areas landscape metrics analysis was carried out employing Fragstats 4.2.

3.1   Check Indicators Calculation

To examine linkages between temperature, urban morphology, water and green areas, 33 same sized squared areas (1 km^2) were extracted from the different Microclimate Zones, in the west side of the Province of Naples. In order to take into account the several different urban morphologies recognizable, each area was split into 16 same sized square homogenous TA, to a total of 553 TAs.

TAs were classified according to urban tissue and building features mainly comprised in. Urban tissue classification was derived from the analyses performed to draw up the Province of Naples Regional Plan, which identifies three principal residential classes:

— Historical centers, made up by historical urban fabrics located in the central core of cities and towns, characterized by high density, irregular texture of urban block, traditional masonry buildings constructed before 1936;

— Modern high density areas made up by urban tissues built between 1936 and 1965, generally characterized by high density, regular texture, regular urban block, concrete buildings, and a satisfying public space provision;

— Modern low density areas made up by urban tissues recently built, generally characterized by irregular grids, concrete buildings, presence of several rural areas and/or non-built lots, and a scarce public space provision;
— Mixed areas, including TA where more than one of the above three classes were recognizable; Rural areas, where none of the four classes above was present.

To take into account the role of water basin in UHI mitigation, TAs were further classified considering their distance from sea and lakes. This classification was achieved performing a GIS routine with ArcGis 9.3 to create multiple buffers (1000, 2000, 5000 meters) from the shore line and a single buffer (1000 meters) from lakeside. The value of the buffer where TA centroid falls into was attributed to each one of them (Fig. 2).

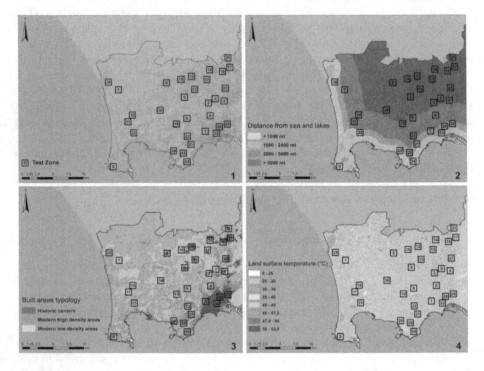

Fig. 2. TAs position *(1)*; buffer distances from sea and lakes *(2)*; built areas typology *(3)*; MIVIS temperature map used in the case study *(4)*

Another GIS routine was developed to assess temperatures of TAs: a specific tool was created using Model Builder to calculate the mean values reported in the temperature map for each TA (Fig. 3).

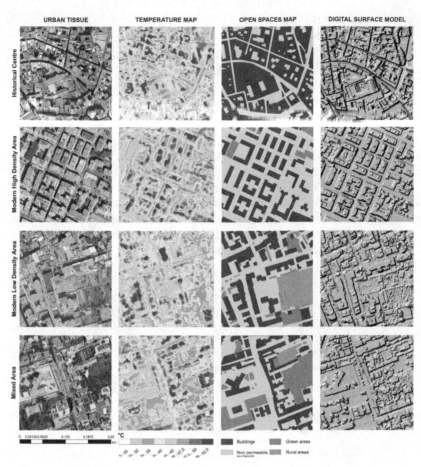

Fig. 3. Examples of urban tissues analyzed compared with the corresponding temperature map, open spaces map and DSM

Five Check Indicators were selected to assess and describe urban structure and morphology:

— Non-permeable Surfaces Index: expressed by the ratio between paved open areas (i.e. streets, parking areas, courtyard) and the total amount of open areas;
— Percentage of Green Areas: i.e. the area occupied by vegetation, public parks, pitches;
— Land cover ratio: expressed by the ratio between the built surface and the surface of TA. The built surface is the horizontal plane projections of the buildings shapes;
— Building density: expressed by the ratio between buildings volumes and the surface of the TA;
— Mean height of buildings.

CIs were calculated combining different informative layers:

— permeable surfaces, corresponding to the sum of green areas, pitches, fallow non-built lots, rural areas;

— non permeable surfaces;
— built surfaces (containing the height of each building).

The degree of association between each Check Indicators and mean temperature were estimated, for each TA, calculating the Pearson linear correlation coefficient.

3.2   Configurational Analysis of Green Areas

Configurational analysis of Green Areas was carried out by calculating a set of metrics belonging to the theoretical framework of landscape ecology using Fragstats 4.2 [21].

The selected metrics describe the configuration of Green Areas at the class level providing the most information about green areas shape, position and spatial distribution within each single TA, with a low degree of redundancy.

Furthermore, some patch metrics were selected and summarized at the class level using a variety of distribution statistics that provide first and second-order summary statistics of the patch metrics for the focal class such as: mean, area-weighted mean, standard deviation, and coefficient of variation (Fig. 4).

Metrics group	Name	Description [units]	Relevant correlation with mean temperature (*)
Area & edge	Patch area	[ha]	•
	Patch radius of gyration	It measure how far across the landscape a patch extends its reach [m]	×
	Largest patch index	It expresses the percentage of the TA comprised of the single largest patch of class; as such, it's a simple measure of dominance [%]	•
	Total edge	It expresses an absolute measure of total edge length of patches belonging to the same class [m]	•
Shape	Perimeter-area ratio	[none]	×
	Fractal dimension index	It reflects shape complexity across a range of spatial scales (patch sizes) [none]	×
	Perimeter-area fractal dimension	It's similar to FRAC, but it is applied to a collection of patches at the class level [none]	×
	Related circumscribing circle	It expresses the ratio of patch area to the area of the smallest circumscribing circle [none]	×
Core area	Core area	It represents the area in the patch greater than a specified depth-of-edge distance from the perimeter, set equal to 10 m [ha]	•
Isolation	Euclidean nearest neighbor distance	It expresses the shortest straight-line distance between a focal patch and its nearest neighbor of the same class [m]	×
	Connectance index	It represents the percentage of all possible inter-patch connections less then a specified threshold distance, set equal to 20 m [%]	×
	Proximity index	It quantifies the spatial context of a patch, in relation to its neighbors of the same patch class [none]	•
Aggregation	Clumpiness index	It measures the aggregation degree of a patch class [none]	×
	Patch cohesion index	It measures the physical connectedness of the corresponding patch type [none]	×
	normalized Landscape shape index	It provides a simple measure of patch class aggregation or clumpiness [none]	×
	Interspersion and juxtaposition index	It measures interspersion or intermixing of a patch class [%]	×
Subdivision	Effective mesh size	It measures the subdivision degree of a patch class [ha]	•

(*) • relevant indirect correlation detected; × no correlation detected

Fig. 4. Features of the used landscape metrics in Fragstats 4.2

The selected metrics were grouped as follows: area & edge metrics; shape metrics; core area metrics; isolation metrics; aggregation metrics; subdivision metrics.

Finally, the degree of correlation between surface temperature and Green Areas metrics in each TA were estimated by correlation coefficient calculation.

Correlation coefficient between Percentage of Green Areas and mean temperature is the only one having always a significant value for each urban tissue and each buffer distance from sea and lakes. In addition, the more distance from water basins increases, the more the correlation between mean temperature and other CI calculations decreases (Fig. 5).

Interesting results were achieved from Green Areas configurational analysis. In particular, we observed a beneficial mean temperature decrease in the cases described below (Fig. 6).

With regard to green areas dimension results show that:

— the longer green areas frontline is (total amount of Green Areas perimeters in a TA), the more mean temperature of TA decreases (Total Edge);
— the higher percentage of TA surface is comprised in the largest green area, the less the temperature is (Largest Patch Index);
— the more heterogeneous Green Areas extensions and shapes are, the higher is the contribution to mean temperature reduction (standard deviation of Green Areas surfaces and standard deviation of Core Area).

With regard to green areas distribution results show that:

— concentration of smaller and closer Green Areas have a positive effect on mean temperature reduction (Area-weighted mean Proximity Index);
— the less Green Areas are subdivided, the more mean temperature tends to decrease (Effective mesh size).

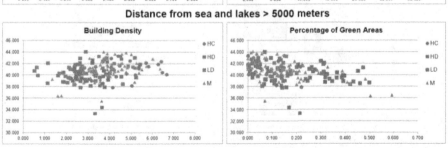

Fig. 5. Relationship between main temperature and some CI, calculated for TAs belonging to different urban tissues and located at growing distance from shore line and lakesides

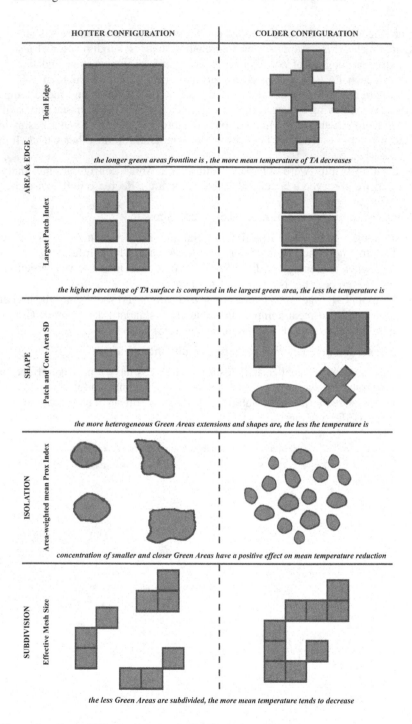

Fig. 6. Example of how different green areas distribution could produce a temperature decrease

4   Results and Urban Planning Involvements

The insane and deplorable decay of contemporary cities is even due to a thoughtless use of natural resources in urbanized areas. Until recent years, natural elements had never been considered as effective 'essential building materials' of cities, with a structural role in city performances. Moreover, urban policies adopted in the last century have allowed cities to erase natural features from their tissue. "Modern cities have literally dispersed and camouflaged the natural substrate of their sites. Many of these substrates have been altered beyond recognition: waters have been covered or diverted, topographies erased or manipulated, forests shredded or fragmented - the list is without end" [22]. The key concept of this kind of urban development was that cities could go ahead without nature or even despite nature and this concept has dramatically affected climatic comfort of cities and buildings. Air conditioning introduced the belief that technology could do better than nature and replace natural cooling, warming and air change mechanism with technological provisions. The speedy spread of building entirely depending from air conditioning has lead cities to increase their energy needs and Greenhouse gases emissions, worsening the overall urban climate conditions.

That's why nowadays there is a call for a new ecologically informed urbanism. It is necessary to reconsider the importance of the ecological role of natural features within cities and to develop new tools to improve urban ecosystem performances through an accurate provision of green spaces, water and wind.

Research results showed that urban parameters play a different role in UHI arising in a very densely built up metropolitan area with a Mediterranean climate. The results of correlation analysis between Check Indicators and mean temperature showed that in TAs farther from sea and lakes building parameters have a weaker correlation with mean temperature. On the contrary, TAs closer than 5000 meters are more influenced by Check Indicators as Building Density, Land Cover Ratio or Mean Height than the farther TAs.

In other words, mean temperature values in TA very far from the sea seem to be independent from urban morphology. It can be attributed to the absence in the inland plans of wind flows able to mitigate UHI. Urban tissues close to the seaside can benefit from sea breeze that is able to neutralise the negative influence that high urban density has on temperature. Then, urban morphology has a crucial role in easing natural flow of air and it is an essential resource for natural cooling in areas provided by natural breezes.

Percentage of Green Areas shows a strong inverse correlation with temperatures in every kind of areas, independently from their location (near or far from sea and lakes) and urban morphology (historical or contemporary tissue).

Our research verified the crucial role that green areas has in mitigating the dangerous effect of heat waves and highlighted some configurations and allocations of green areas within a building context that better respond to this aim. In the metropolitan area investigated the presence of green areas guarantee a temperature decreasing of about -0,5 °C for every increase of 10% of green areas: a good configuration could provide a further temperature decrease up to 2 °C.

Substantial differences in temperature are observable in TAs, belonging to the same Microclimate Zone and having a similar Percentage of Green Areas. Here, results showed that a more efficient distribution and configuration of green areas respond to the following key concept:

— *Synergy*: of course, the bigger a green area is, the best is its climatic performance, anyway if a big green area is isolated in urban fabric, its benefic effect has a low range of action. Without increasing the overall extension of green areas, coupling together a big areas with smaller ones at a few distance could guarantee a better result in temperature decreasing;

— *Pervasiveness*: for equal areas, better performances are observed when green frontline is more extended. This imply that shapes having an higher level of pervasiveness in building fabric are more efficacious then regular shapes presenting a compact frontlines;

— *Heterogeneousness*: the presence of many green areas with differences in shape and extension performs better;

— *Multiplicity*: a multiplicity of small green areas at a small distance performs better than few green areas with a greater extension but more distant from each other. It is better Small areas but regularly distributed in the urban tissue;

— *Continuity*: continuity among different green areas interspersed in a urban tissue guarantee a better climatic performance.

This study suggests that Urban Green is not only a matter of quantity and extension (i.e. how many square meters for each inhabitant), but also an unavoidable mean to improve ecological performance of urban environment. Then, beyond quantity, distribution and configuration have their weight too.

The design reference framework could be clearly defined for new urban development but the huge problem is how to improve existing urban fabric to enhance their resilience to heat waves frequency increasing. Results of this study clearly suggest a way: to increase density of small green areas; to augment the pervasiveness into building fabric of green spaces and whether it is possible to connect existing green areas in order to create a green urban infrastructure. This last could be the back-up of a requalification strategy and support other elements such as bike lanes, walking paths and recreational areas.

Acknowledgements. Paragraphs 1 and 5 are due to Marialuce Stanganelli; paragraphs 2, 3 and 4 are due to Carlo Gerundo. Special thanks have to be done to the GIS Service of the Province of Naples for the provision of geographic database and land use data.

References

1. IPCC Fifth Assessment Report, Mitigation of Climate Change (2013)
2. Holmes, M.J., Hacker, J.N.: Climate change, thermal comfort and energy: Meeting the design challenges of the 21st century. Energy and Buildings **39**, 802–814 (2007)
3. Roberts, S.: Effects of climate change on the built environment. Energy Policy **36**, 4552–4557 (2008)
4. ESPACE, Climate Change Impacts and Spatial Planning Decision Support Guidance (2008) www.espace-project.org
5. Matthews, T.: Climate Change Adaptation in Urban Systems: Strategies for Planning Regimes. Urban Research Program, Research Paper 32. Griffith University, Brisbane (2011)
6. EM DAT – The International Disaster Database – Centre for Research on Epidemiology from Disasters (CRED) http://www.emdat.be/

7. Bonafè, G.: Microclima urbano: impatto dell'urbanizzazione sulle condizioni climatiche locali e fattori di mitigazione. Area Meteorologica Ambientale, Servizio IdroMeteorologico, ARPA, Emilia Romagna (2006)
8. Zauli Sajani, S., Tibaldi, S., Scotto, F., Lauriola, P.: Bioclimatic characterization of an urban area: a case study in Bologna (Italy). Int. J. Biometeorol. **52**, 779–785 (2008)
9. Akbari, H., Levinson, R.M., Rainer, L.: Monitoring the Energy-Use Effects of Cool Roofs on California Commercial Buildings, Energy and Buildings **37**, 1007 1016 (2005)
10. Taha, H., Douglas, S., and Honey, J.: The UAM sensitivity analysis, The August 26-28 1987 Oxidant Episode. In: Taha H. et al. (eds) Analysis of Energy Efficiency and Air Quality in the South Coast Air Basin - Phase II. Lawrence Berkeley Laboratory Report LBL-35728, Berkeley (1994)
11. Di Cristo, R., Mazzarella, A., Viola, R.: Hourly discomfort conditions in the city of Naples (Southern Italy) estimated by the heat index. Nat. Hazards. **40**, 373–379 (2007)
12. Kleerekoper, L., van Esch, M., Baldi, T.: How to make a city climate-proof, addressing the urban heat island effect. Resources, Conservation and Recycling **64**, 30–38 (2012)
13. Oke, T.R., Nunez, M.: The Energy Balance of an Urban Canyon. Journal of Applied Meteorology **16**, 11–19 (1977)
14. Esch, M., van Bruin-Hordijk, T., de Duijvestein, K.: The influence of building geometry on the physical urban climate: a revival of 'light, air and space'. In: PLEA2007 – 24th Conference on Passive and Low Energy Architecture (2007)
15. Taleb, D., Abu-Hijleh, B.: Urban heat islands: Potential effect of organic and structured urban configurations on temperature variations in Dubai, UAE. Renewable Energy **50**, 747–762 (2013)
16. Radhi, H., Fikry, F., Sharples, S.: Impacts of urbanisation on the thermal behaviour of new built up environments: A scoping study of the urban heat island in Bahrain. Landscape and Urban Planning **113**, 47–61 (2013)
17. Bowler, D.E., Buyung-Ali, L., Knight, T.M., Pullin, A.S.: How effective is 'greening' of urban areas in reducing human exposure to ground level ozone concentrations, UV exposure and the 'urban heat island effect'? (Systematic Review No. 41). Retrieved from Collaboration for Environmental Evidence website: http://www.environmentalevidence.org/Documents/SR41.pdf (2010)
18. Li, J., Song, C., Cao, L., Zhu, F., Meng, X., Wu, J.: Impacts of landscape structure on surface urban heat islands: A case study of Shanghai, China. Remote Sensing of Environment **115**, 3249–3263 (2011)
19. Connors, J., Galletti, C., Chow, W.L.: Landscape configuration and urban heat island effects: Assessing the relationship between landscape characteris-tics and land surface temperature in Phoenix, Arizona. Landscape Ecology **28**(2), 271–283 (2013)
20. Bowler, D.E., Buyung-Ali, L., Knight, T.M., Pullin, A.S.: Urban greening to cool towns and cities: A systematic review of the empirical evidence. Landscape and Urban Planning. **97**, 147–155 (2010)
21. McGarigal, K., Marks, B.J.: FRAGSTATS: spatial pattern analysis program for quantifying landscape structure. In: General Technical Report. PNW-GTR–351, Portland (1995)
22. Girot, C.: Vision in motion: representing landscape in time. In: Waldheim, C. (ed.) Landscape urbanism reader. Princeton Architectural Press, New York (2006)

Recreation Tourist Areas. An Exam on Recreational Business Districts in Olbia (Sardinia)

Silvia Battino[1], Giuseppe Borruso[2(✉)], and Carlo Donato[1]

[1] DiSEA– Department of Economic and Business Sciences, University of Sassari,
Via Muroni, 25 01700, Sassari, Italy
{sbattino,cadonato}@uniss.it
[2] DEAMS – Department of Economic, Business, Mathematic and Statistical Sciences,
University of Trieste, Via a. Valerio, 4/1 34127, Trieste, Italy
giuseppe.borruso@econ.units.it

Abstract[1]. Research concerned about cities paid a particular attention, particularly in the past, related to the central places of a city, or the areas where central activities characterizing cities are located. That led to the definition of the so called Central Business District - also known as "Downtown" or "the City". Some related concepts have been also developed for addressing tourism and recreation. That led to defining the Recreational Business District, a – not necessarily geographically – central part of a city generally hosting free time and leisure, lived by locals, visitors and tourists. Here a research is carried on, in order to evaluate the ways in which recreational areas tend to distribute and concentrate on urban environments. The analysis is carried on considering the municipality of Olbia in Sardinia (Italy). The idea is that of comparing the spatial distribution and extent of such Recreational areas in parts of a territory quite different in shapes, population and, in general geographical characters. The analysis was also performed over recreational activities considering their level of connectedness by means of social networks ad media, examining the pattern they draw in a tourist area. Different techniques are used, including a point pattern analysis based on Kernel Density Estimation.

Keywords: City · Urban centre · Density estimation · Olbia · Sardinia · Italy · Tourism · Recreational business district · Central business district

1   Introduction

Cities have been for decades the object of study of scholars from a variety of disciplines, from sociology, to urban planning to geography. Particularly from the early decades of the past century different scholars focused on the characteristics of cities in

---

[1] The paper derives from the joint reflections of the three authors. Silvia Battino realized paragraphs 1 and 3, while Giuseppe Borruso wrote paragraphs 2 and 4. Carlo Donato wrote paragraph 5. The geographical visualization and analysis, where not otherwise specified, have been realized using QGIS 2.6.2.

© Springer International Publishing Switzerland 2015
O. Gervasi et al. (Eds.): ICCSA 2015, Part II, LNCS 9156, pp. 662–674, 2015.
DOI: 10.1007/978-3-319-21407-8_47

terms of land use and land occupation by different activities, as well as on their evolution in space and time ([1] [2] [3] [4] [5]). The studies highlighted the presence of a central area, easily definable particularly in North American cities, where some kinds of activities could concentrate, as the Central Business District (CBD), where a low population density could be reordered and a majority of activities dedicated to services and the tertiary sector.

Variations to the general 'model' of a monocentric city with a well defined CDD in its centre, included the possibility of having polycentric cities and the gemmation of the CBD in other areas dedicated to services and financial activities. These are characterized by tertiary functions and activities. Tourism is part of these activities and particularly during the second half of the Twentieth century, played an important role in the changes intervened in urban systems, converting them often in mono-functional space and leading to different zoning division [6].

A Recreational Business District (RBD) can be therefore hypothesized and represented together with other functional zoning of the city. Such areas are characterized by the presence of activities that can be enjoyed by tourists as well as locals. As reported in another research [29] connected to the RBD at urban level in cities, such areas have been studied by Stansfield and Rickert in the Seventies of the Twentieth century [6], definable as an area of concentration of goods and services, used by visitors and tourists and clustered around natural phenomena or architectural, cultural and historical attractions. Zanini and Lando pointed out as [7] in time the sprading of consumption habits pushed the RBD to focus also on retail activities, accommodation and leisure, activities highly specialized and often organized in order to provide a general flavor of an area with its own polarization role ([8] [9]). Getz on the contrary [10] highlighted how in most European cities the Central Business District share with the Tourism Business District common space, and also we can highlight a same concept in identifying a common set of spaces between the CBD and the tourist district or RBD. This is quite acceptable as the metropolitan services serving the CBD satisfy generally also the tourists and locals' needs in free time.

Here in this paper we tackle a first analysis to highlight the RBD of the city of Olbia (Sardinia Island, Italy).

The rest of the paper is organized as follows. In Paragraph 2 the methods adopted and the type of analysis carried out is presented, while Paragraph 3 is focused on a short description of the study area and on the data used. Results and discussion are presented in Paragraph 4 and the Conclusions are dedicated to some remarks concerning the recreational and tourist aspects of the city are presented, together with suggestions for future research activities and directions.

2   Methods Adopted. Analysis of Recreational Activities as a Point Pattern

The research performed in this paper is based on a point pattern analysis, in which the events of the point pattern are the geographical locations of recreational activities, attributed to a set of qualitative and quantitative information useful for an analysis over tourist settlements and organization. The analysis foresaw a collection of urban activities and their classification in terms of free time and recreation, moving then to georeference them, following similar research based on the Central Business Districts and Recreational Business Districts of cities as Sassari and Trieste ([11] [12] [13] [29]). A list of activities at urban level particularly dedicated to recreation was therefore collected from the Italian Yellow Pages and improved by fieldwork, and also verifying the presence of the activities on the Internet and on social networks and media.

A *Kernel Density Estimation* (KDE) was used as a consolidated technique to represent point events in space as a 3D function, continuously representing density and its variation over the study region, in order to visualize the phenomenon as a kind of 'heat map' of areas of concentration of point features in a given area. ([11] [12] [13] [14] [15] [16] [17] [18] [19] [20] [21])

As reported in many research papers on the topic, the function is characterized by a general formula in which a three dimensional window spans over the study region and visits all the events in a point pattern, assigning weights according to a distance function and using weights stored in the point event database [22].

$$\hat{\lambda}(s) = \sum_{i=1}^{n} \frac{1}{\tau^2} k\left(\frac{s - s_i}{\tau} \right) \tag{1}$$

$\hat{\lambda}(s)$ is the density estimation of the point pattern measured at location s, while s_i represents the observed i^{th} event. $k(\ )$ is the kernel weighting function and the parameter τ is the radius of research of the function, or bandwidth, to be centered in location s, and searching for events s_i to be computed into the density function [23]. The continuous density function is represented, in a GIS environment, by means of grid cells whose values represent either a density or a probability function. The variation of values between neighboring cells is smooth so that their distribution approximates a 3D distribution.

3   The Study Region and the Data

The Municipality of Olbia covers a surface of 376,10 km along the North Eastern coast of Sardinia region, occupying an area surrounded by hills and low altitude peaks while the coastal region is characterized by a set of beaches and promontories cut by geological concretions.

57,889 people live in Olbia (1 January 2014 – www.demo.istat.it) and it represents the 4th municipality of Sardinia Island after Cagliari, Sassari and Quartu Sant'Elena and it represents an area with a certain economic dynamism, attracting people from other farther locations in Sardinia, being this demonstrated by the fact that it is the only Municipality having tripled its dimension since the Fifties of the past century.

The polarizing capacity of the city were highlighted since the Seventies and continued in the following decades with the increase of the urban functions hosted by the city, thanks also to an economic development related to the growing tourism industry.

With the growth of the tourist flows towards "Costa Smeralda" since Seventies of the past century, Olbia was nearly forced to become a gateway for such region and to host services dedicated to tourists, developing passengers' airport and port serving a wide part of the region ([31] [32]). Only recently the city has discovered and started to strengthen its own tourist capacity in order to become a tourist urban system capable of distributing substantial, and high value tourist services, changing its nature from a pass-by location to a touristic destination able to compete with other destinations of the island.

The same Municipality of Olbia is characterized by the presence of the main homonymous urban centre (Olbia), where 80% of the population of the municipality is concentrated, as well as the industries, port and services are located, and a set of coastal centres (Cugnana Verde, Portisco, Porto Istana, Porto Rotondo), fitted with marinas and tourist ports, mainly dedicated to holidays and characterized by houses, lodges and structures occupied only during the summer vacation period.

In order to perform the analysis, we collected several data concerning the recreational activities located into the Municipality of Olbia and in the neighbouring areas. As a source for the activities, we used the Yellow Pages service ([24] accessed March 2015) to collect the activities of the City of Olbia. Data were integrated with on-field surveys and categorized onto 4 main classes, these representing some of the activities providing a sense of recreation. Data were georeferenced at address point level using the GIS data provided by the Sardinia Region Open Data portal [26] and checked via on-line geocoding services as the platform GPSVisualizer [27]. As many recreational activities – as agritourism – are located out of town where georeferenced address points are scarcer, direct check and on-line geocoding using Google and Bing Maps APIs where used.

The activities selected are dedicated to recreation and count for a total of 281 activities as reported in Table 1.

Table 1. The activities determining the RBD in the city of Sassari
Source: Our elaboration from Yellow Pages [24] (2014)

Sub Category	Hotels and Agritourism	Restaurants	Discos and night clubs	Bars and candies	Total
N.	148	96	6	31	281

The georeferenced recreational were visualized on a digital map of the Municipality of Olbia in order to understand how they distribute over the territory, both at urban and extra-urban level.

We aimed at analyzing the spatial distribution of the recreational activities in the different areas of the study region, i.e., considering Olbia as an urban centre, its neighborhood and also the coastal areas – as Porto Rotondo – that are specialized and dedicated to tourist activities. Our aim was also the analysis of such structures in terms of their presence on social networks and, therefore analyzing the dotation in terms of webpages and profiles over the most popular social networks and media, as Facebook, Twitter, Google+ or Instagram (Figure 1).

Olbia maintains a quite high percentage of presence on the web, both in terms of a webpage for each activity as well as in terms of the presence on social networks and media, particularly if compared to other similar research carried on [29]. Of the 281 activities considered, nearly 60% has a website and a higher 66% is profiled over a social network or media, while little less than 50% both hosts a website and a social profile (Table 2).

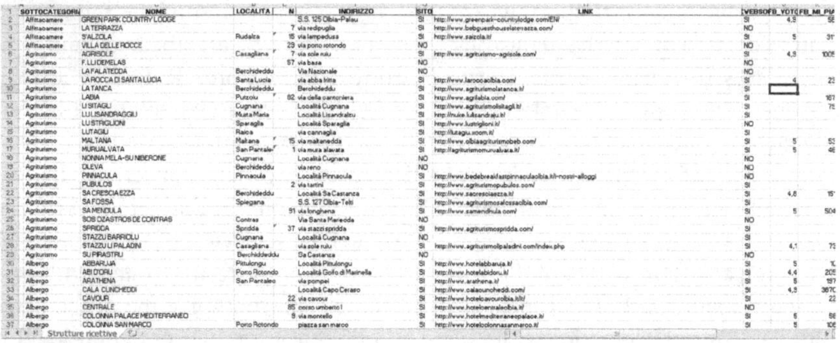

Fig. 1. Part of the recreational activities and their presence on the Web and Social Networks (2015) Source: Our elaboration on GIS data from Yellow Pages [24] and Social networks and media

Discos and restaurants are the categories with higher presence over the social networks (100% and 90.63% respectively), while discos (66.67%), hotels and agritourism (65.54%) lead in terms of percentage of structures hosting website, and they also lead in terms of being both present on the Internet and on social networks and media (66.67% and 60.42% respectively) Among the social networks considered, a set of attributes has been chosen for the most popular applications. Facebook, Twitter, Google+ and Instagram were considered as the social media and network to check for each economic activity. Of such activities, characteristics as the 'Likes', 'Followers', 'Following', etc. were considered for investigation. As a general element, however, being Facebook the most widely used medium of social interaction, its presence in terms of 'likes' was then considered as a weighting factor for the most refined analyses.

Table 2. Recreational activities grouped by their presence on the web and on social networks Source: Our elaboration from Yellow Pages; direct field data collection [24] (2015)

	Activities		Activities (%)	
	Yes	No	Yes	No
Web site.	168	113	59.79	40.21
Social network / media	186	95	66.19	33.81
Web site + Social network / media	137	144	48.75	51.25
Total	281		100	

4   Results and Discussion

A first visual analysis can be done on the scatterplot of point features referred to recreational activities.

An initial observation of the point pattern given by the recreational activities shows some concentration in the urban area of Olbia (centre of the map) as well as on coastal settlements, those dedicated mainly to vacation periods. An initial kind of analysis was performed over the pure distribution of recreational activities, considering also a first their degree of connection via the Internet and social networks and media.

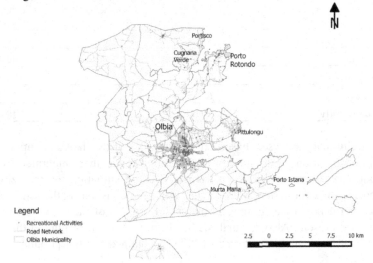

Fig. 2. Recreational activities in Olbia and neighboring centres
Source: Our elaboration on GIS data from Yellow Pages [24]

The recreational activities considered are mainly located in the urban centre of Olbia, where nearly 72% of the elements analysed are concentrated. Porto Rotondo follows with nearly 9% of activities are located. Lower figures characterize the other coastal areas and the inner Olbia region.

The aggregation of data collected and analysed observing their presence on the web and social profiling offers some suggestion over the characters of the places and of their recreational supply. Porto Rotondo remains dominant in terms of activities connected via the web and social tools with over 70% of presence. Lower figures characterizes the simultaneous presence of social networks and websites, that displaying that the different activities can be interested more by one medium than the other one. Olbia on the other hand is characterized by the lower absolute value of activities with an Internet website but with the higher figures in terms of the combination of places hosting both a website and a social profile.

This kind of results is consistent with the different characters of the locations and of the activities located there. Porto Rotondo and the other coastal locations are – as said – characterized as places where tourism is the main activity, often also concentrated in a limited period of time of the year. So it is quite understandable that the activities hold a strong presence on the social networks and media and the Internet, in order to reach non local customers. On the other hand Olbia with its important urban component presents a mix of activities more open to innovation and a presence on the ICT tools and a more 'traditional' component not so interested in that (Table 3).

Table 2. Percentage of Recreational Activities Grouped by their presence on the web and on social networks
Source: Our elaboration from Yellow Pages; direct field data collection [24] (2015)

Area	Website	Social network / media	Web site + Social network / media
Olbia	55.94	69.31	59.79
Inland region	66.67	44.44	44.44
Porto Rotondo	70.83	75.00	58.33
Coastal region	68.18	59.09	54.55
Olbia Municipality	59.79	66.19	48.75

These preliminary analyses, based on a scatterplot of the data and supported by a first set of observation on the quality of distribution, was than integrated by a more refined point pattern analysis using a Kernel Density Estimation for analyzing possible hotspots in the overall distribution. That was done considering the total amount of activities, than using some subsets, as the social profiles of the activities, as well as those not at all present. Also, an analysis has been performed weighting the activities by their social 'reputation' (in terms of "likes" obtained via the popular social network Facebook).

Different bandwidth were tested, while here we present the results related to a bandwidth of a 500 m one, proofed in other research of being a good compromise, for urban level analysis, in terms of effectiveness of the message, smoothing of the 3D estimation surface and aggregation of the data.

We decided to maintain such a bandwidth also for the recreational activities and for the social recreational activities, as such a distance is compatible with an average 5 to 10 minutes walking distance, therefore approximating a level of accessibility to services at urban scale.

The results from the density analysis on central activities are portrayed in Figure 3a, where an overview of the concentration of activities in the municipality is presented. We can notice the main hot spots in the urban centre of Olbia, as well as in Porto Rotondo. In Figure 4a we provide a more refined zoomed representation of such density estimation and here it is evident where activities concentrate, namely in the historical centre of the city and along the major high streets and connecting roads, as Corso Umberto I and Viale Aldo Moro (North of the historical city centre). The more concentrated areas of recreational activities follow mainly a North-South orientation, culminating into the historical city centre located in proximity of the city's port. Other hot spots can be noticed moving out Westwards from the city centre.

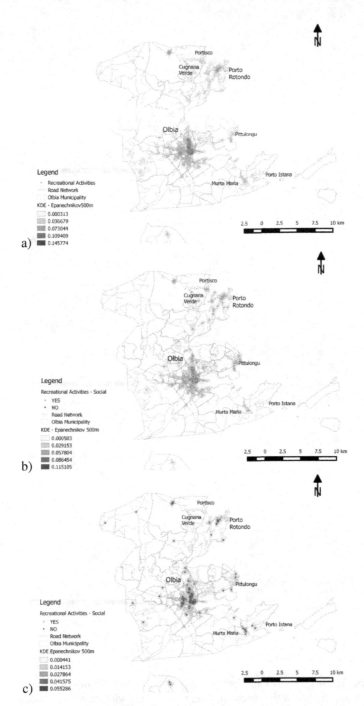

Fig. 3. Kernel Density Estimation over Recreational business activities in the city of Olbia – a) all activities; b) activities with a social profile; c) activities without a social profile
Source: Our elaboration on GIS data from Yellow Pages [24]

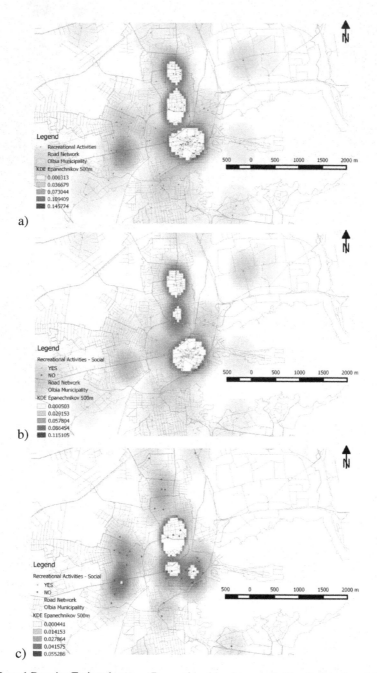

Fig. 4. Kernel Density Estimation over Recreational business activities in the city of Olbia – a) all activities; b) activities with a social profile; c) activities without a social profile [particular; inner white areas = hot spots].

Source: Our elaboration on GIS data from Yellow Pages [24]

The analysis continues with the discrimination of the presence of the activities on social network platforms, therefore dividing the original set of recreational activities in two subsets, as the one having and those not having a social profile.

Figure 3b shows a density analysis performed over the recreational business activities with a social profile. Lower density values are expected, also because the number of the events is smaller than in the previous case, being the 'social' ones a subset of the entire database. The hotspots are located in similar positions than in the previous analysis and although with different shapes they present similar characteristics. Figure 4b helps in moving more easily in the city of Olbia and here also the central area pops up, as well as the main radial segments where activities are clustered.

Some considerations can still be done when performing the density analysis over the non-social activities (Figure 3c and Figure 4c). Here some differences with respect to the general pattern appears, as well as with the distribution of the social activities. On one side hot spots can be always detected in the same areas. That means that 'central' places (as Olbia) host both social and non-social activities in a quite interlinked urban mix. On the other side if you focus on the second subset, as that of the non-social activities, these can also be found in visible clusters outside the city centres in the Western and Northeastern part of the city. Also, as highlighted also by the first set of observations, the inner regions of the municipality present similar characters in terms of a scarcity of presence over social networks.

The analysis is completed by a weighted density estimation, in which not only the 'presence/non presence' is evaluated, but we try to provide a rough indicator of such presence providing a measure of its importance.

As mentioned in the first part of this work, during the collection of the data we gathered for each activity its presence on the most popular social networks and media [28], inserting also the figures related to such presence (i.e., following and followers for Twitter platforms; number of friends and 'likes' on the Facebook page, etc.).

Of such an amount of data we focused on this paper just on the most popular one, as Facebook, as the presence on other media is surely less frequent and therefore the analysis on just other networks and media could be misleading in terms of the pattern arising. So from the Facebook profile we derived some other information concerning the friendship relations, the number of visitors on the profile and the likes expressed by people accessing the profile or asking for friendship.

So a first indicator was selected, as the number of 'likes' on the page and that used as a weight for the density estimation. In Figure 5 we portray the results of such an analysis, related to the entire territory of the Municipality of Olbia.

Of course the general pattern already observed tends to persist; however such pattern is reinforced by some sites peaking over the territory with some local 'champions'. Going to observe more in detail the data, it can be noticed that two discos, located in Olbia and Porto Rotondo, as well as a hotel resort in the Northeastern outskirts of Olbia present very high value of 'likeness' therefore highlighting single peaks in correspondence of such locations. On the contrary in area characterized by quite high number of activities and 'likes', a more homogeneous, central hotspot seems to appear.

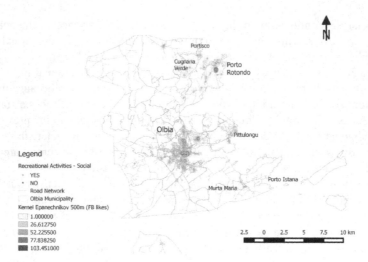

Fig. 5. Kernel Density Estimation on Recreational business activities with social profiles - weighted by n. of Facebook "Likes".

Source: Elaboration on GIS data from Yellow Pages [24]

5   Conclusions

As in previous research works on Central and Recreational Business Districts, the research produced some results concerning the spatial distribution of activities in space, leading to the representation of clusters and areas of concentration, highlighting areas where tourists and other people in search of leisure and amusement concentrate. In the analysis we considered the Municipality of Olbia from different points of view: as an urban centre, whose importance in touristic terms is growing, and as an area characterized by a set of 'new towns' or 'new settlements', developed in time and particularly dedicated to tourist activities, often occupied just during the summer period by non-residents.

Another level of the analysis considered the presence of social networks and media in the recreational activities. Here some points observed in other contexts [29] are confirmed, where the more touristic is a location, being that a settlement or a part of a city, the more 'social' it is in terms of its activities. The research carried on shows also a preference – at urban level – for recreational activities for being located in the central parts of the city and this is valid also for those more linked to the world of social networks and media. As concluded in a previous paper "The presence on popular social networks and media enhances an activity's own attractive capacity and therefore allows playing a real touristic role" [29].

The research needs to be further explored, on one side to consider "the activities dedicated explicitly to tourism and their relationship with the wider ones targeted on recreation – these latter including actually also locals as consumers and not just tourists" [29], while on the other side there is the need to focus on the role of social networks and media. On this point the interest is twofold. We need to examine how the different social networks and media contribute to the 'social success' of a location and there is the need to integrate this analysis with evaluations in space and time, of quantity and quality of interaction of people in recreational areas.

References

1. Burgess, E.W.: The growth of the city. In: Park, R.E., Burgess, E.W., Mc Kenzie, R.D. (eds.) The City, pp. 47–62. University of Chicago Press, Chicago (1925)
2. Christaller, W.: Die Zentralen Orte in Suddeutschland. Fischer, Jena (1933)
3. Hoyt, H.: The structure and Growth of Residential Neighborhoods in American Cities. U.S. Government Printing office, Washington D.C. (1939)
4. Harris, C.D., Ullman, E.L.: The nature of cities. Annals of the American Academy of Political and Social Science **242**, 7–17 (1945)
5. Alonso, W.: Location and land use. Toward a general theory of land rent. Harvard University Press, Cambridge (1965)
6. Stansfield, C.A., Rickert, J.E.: The recretional business district. Journal of Leisure Reseach **2**, 213–225 (1970)
7. Zanini F., Lando F.: Impatto del turismo sulla struttura terziaria urbana. Note di Lavoro. DSE, Università Ca' Foscari di Venezia vol. 5, pp. 1–25 (2008)
8. Timothy, D.J., Butler, R.W.: Cross-border shopping. A North American perspective. Annals of Tourism Research **22**(1), 16–34 (1995)
9. Jasen-Verbeke, M.C.: Leisure shopping. A magic concept for the tourism industry. Annals of Tourism Research **12**(1), 9–14 (1991)
10. Getz, D.: Planning for tourism business districts. Annals of Tourism Research **20**(3), 583–600 (1993)
11. Battino, S.: Estensione e delimitazione dei core urbani della città di Sassari. Bollettino A.I.C. **143**, 29–48 (2011)
12. Battino, S., Borruso, G., Donato, C.: Analyzing the central business district: the case of sassari in the Sardinia Island. In: Murgante, B., Gervasi, O., Misra, S., Nedjah, N., Rocha, A.M.A., Taniar, D., Apduhan, B.O. (eds.) ICCSA 2012, Part II. LNCS, vol. 7334, pp. 624–639. Springer, Heidelberg (2012)
13. Battino, S., Borruso, G.: Analisi GIS del Central Business District di Sassari. Visualizzazioni cartografiche. Atti 16 Conferenza ASITA (Vicenza 6-9 novembre), pp. 183–190 (2012)
14. Thurstain-Goodwin, M., Unwin, D.J.: Defining and Delimiting the Central Areas of Towns for Statistical Modelling Using Continuous Surface Representations. Transactions in GIS **4**, 305–317 (2000)
15. Borruso, G.: Il ruolo della cartografia nella definizione del Central Business District. Prime note per un approccio metodologico. Bollettino dell'Associazione Italiana di Cartografia **126-127-128**, 255–269 (2006)
16. Borruso, G., Porceddu, A.: A tale of two cities. density analysis of CBD on two midsize Urban Areas in Northeastern Italy. In: Borruso, G., Lapucci, A., Murgante, B. (eds.) Geocomputational Analysis for Urban Planning. Studies in Computational Intelligence, vol. 176, pp. 37–56 (2009)
17. Borruso, G.: Network Density Estimation: a GIS Approach for Analysing Point Patterns in a Network Space. Transactions in GIS **12**, 377–402 (2008)

18. Danese, M., Lazzari, M., Murgante, B.: Kernel density estimation methods for a geostatistical approach in seismic risk analysis: the case study of potenza hilltop town (Southern Italy). In: Gervasi, O., Murgante, B., Laganà, A., Taniar, D., Mun, Y., Gavrilova, M.L. (eds.) ICCSA 2008, Part I. LNCS, vol. 5072, pp. 415–429. Springer, Heidelberg (2008)

19. Danese, M., Lazzari, M., Murgante, B.: Geostatistics in historical macroseismic data analysis. Transactions on Computational Sciences 6(5730), 324–341 (2009)

20. Murgante, B., Danese, M.: Urban versus Rural: the decrease of agricultural areas and the development of urban zones analyzed with spatial statistics. International Journal of Agricultural and Environmental Information Systems (IJAEIS) 2(2), 16–28 (2011). Special Issue on Environmental and agricultural data processing for water and territory management. IGI Global

21. Gatrell, A.: Density estimation and the visualisation of point patterns. In: Hearnshaw, H.M., Unwin, D.J. (eds.) Visualisation in Geographical Information Systems. Wiley, Chichester (1994)

22. Levine, N.: CrimeStat III: A Spatial Statistics Program for the Analysis of Crime Incident Locations. Ned Levine & Associates, the National Institute of Justice, Houston, Washington, DC (2004)

23. Battino S.: Turismo sostenibile in Gallura: prospettiva vincente o modello illusorio? I principali caratteri distintivi del cuore turistico della Sardegna, Bologna, Patron (2014)

24. Italian Yellow Pages. http://www.paginegialle.it

25. SardegnaOpendata. http://dati.regione.sardegna.it/dataset?tags=toponimi+e+numeri+civici

26. GPSVisualizer. http://www.gpsvisualizer.com/

27. Chainey, S., Reid, S., Stuart, N.: When is a hotspot a hotspot? A procedure for creating statistically robust hotspot maps of crime. In: Kidner, D., Higgs, G., White, S. (eds.) Socio-Economic Applications of Geographic Information Science, Innovations in GIS. 9. Taylor and Francis, London (2002)

28. Cosenza, V.: Social Media Statistics. http://vincos.it/social-media-statistics/

29. Battino, S., Borruso, G., Donato, C.: Some preliminary remarks on the recreational business district in the city of sassari: a social network approach. In: Murgante, B., Misra, S., Rocha, A.M.A., Torre, C., Rocha, J.G., Falcão, M.I., Taniar, D., Apduhan, B.O., Gervasi, O. (eds.) ICCSA 2014, Part II. LNCS, vol. 8580, pp. 629–641. Springer, Heidelberg (2014)

30. RAS (REGIONE AUTONOMA DELLA SARDEGNA), Piano Paesaggistico Regionale – Ambiti di Paesaggio - Scheda Ambito n 18 Golfo di Olbia, Cagliari (2006)

31. Battino, S., Donato, C.: Olbia gateway passeggeri: infrastrutture, problemi e prospettive. In: Borruso, G., Danielis, R., Musso, E. (a cura di) (eds.) Trasporti, logistica e reti di imprese. Competitività del sistema e ricadute sul territorio, pp. 154–159. Milano, Franco Angeli (2010)

32. Battino, S.: Stime quantitative ed un primo studio cartografico delle seconde residenze nel processo di litoralizzazione della fascia costiera del Nord-Est della Sardegna. Bollettino A.I.C. 150, 4–19 (2014)

GeoLapse. A Digital Space-Based and Memory-Related Time-Capsule App

Letizia Bollini[✉], Giulia Busdon, and Annalisa Mazzola

Department of Psychology, University of Milano-Bicocca,
Piazza Dell'Ateneo Nuovo 1 20126, Milan, Italy
letizia.bollini@unimib.it

Abstract. The paper explores the relationship between space and time through the memory based experience of serendipity. GeoLapse – the name of the prototypal mobile app – is based on the idea of allowing users to send messages located in space and – simultaneously or asynchronously – in time aimed to create a sort of *digital time-capsule*.

Keywords: Space-based interaction design · Geo-based experience design · Time-based interaction design · Digital time capsule · Digital heritage · Emotional interface design

1   Memory: A Time Driven Experience

If space is the dimension physically experienced by our senses, time is the state of mind that marks our daily life. According to the philosophical point of view of a classical thinker such as Plato and Aristotle the experience of time – past, present and future – is deep connected with memory activities of our brain as opposed to physical life and on it are based consciousness and knowledge. Afterwards, Augustine of Hippo in the development of his reflections on the nature of time adopts a metaphor - the *stock* - to describe the activity of memory in spatial terms. His approach to intentionality, memory, and language as these phenomena are experienced within consciousness and time anticipated and inspired the insights of modern phenomenology and hermeneutics. Husserl writes: "The analysis of time-consciousness is an age-old crux of descriptive psychology and theory of knowledge. The first thinker to be deeply sensitive to the immense difficulties to be found here was Augustine, who labored almost to despair over this problem." [1]. Hegel elaborating the concept of memory in connection with the Western culture and history identifies in language the specific medium memory is related on. Nietzsche will definitely put in doubt the relationship between long-memory recalling and *oblivion* – the right to forget – as an indispensable existential condition. [2] The memory lives therefore in connection with its opposite, forgetfulness, a sort of duality that permeates the human experience in its relationship with the everyday life, in his relationship with others and with the environment which these experiences *phenomenally* take place.

© Springer International Publishing Switzerland 2015
O. Gervasi et al. (Eds.): ICCSA 2015, Part II, LNCS 9156, pp. 675–685, 2015.
DOI: 10.1007/978-3-319-21407-8_48

On the other hand the memory lives associated with the phases of time and the processing that our psychic world makes – sometimes mistakenly – of it. *Déjà vu*, *lapsus*, *serendipity* are some of the paradoxical phenomena that give rise to distorted experiences of time and that at the same time provide suggestions to play with the time parameter as story-telling drivers. Travel *through time* is one of the literary, science-fiction, cinematographic theme most fascinating and explored in the artistic culture.

If in human history writing, the records, portraiture, the diaries, the correspondence and photography were the tools to keep track of time and to transmit evidence of the past for posterity, in the contemporary world these tools have been superseded by social networks and mobile devices.

2   Time Capsule: A Spatial Located Memory

As initially mentioned, the experience of time is closely related to the space in which it occurs. Among the instruments intentionally used to keep track of memory and the past, the time capsules are the most specifically designed for this purpose. "A container used to store for posterity a selection of object thought to be representative of life of a particular time" is the definition given by Jarvis [3]: a sort of broadcast message addressed to unknown receivers to communicate in a one-way-only to the future. A time capsule can be opened only by a given moment on and in the place where it was originally located. If the selection of objects – that means the message to be sent – is one of the objectives of the time capsule its location is the crucial element for its future interaction and discoverability.

Once again the location of the memory - the spatial dimension of time - is the driver of the relationship between past, present and future and the key connecting link among generations able to evoke *another* space and time.

Currently, however, social networks have completely reversed and reinvented this relationship: a *tweet*, a change of *status*, a *post* generate a continuous time flow that allows you to find, even after many years, your thoughts and emotions. This virtual storytelling becomes social, but often removes the spatial dimension and relationship with the places of life, *depriving* the experience of an essential component that is rooted in the place where the events happen where space is a dimension definitely physical, but also symbolic, metaphorical and collective. On the other hand the relationships ground of digital networks creates a new form of conceptual space within which people build new experiential bonds with its own existential streaming and sharing offered to their connections.

3   GeoLapse: A Digital Space-Based Time Capsule

Geolaps is a mobile application– a prototype – designed as a sort of *digital time capsule* able to reconnect the temporal and spatial in the user experience both typical memory.

GeoLapse users don't know when or where a message is, because the application is based on the fact that an addressee could receive a *Glap* only when is in the right place at the right moment, decided and only known by the sender.

A *Glap* is the base unit of *GeoLapse* and it is the message itself. The name came from a contraction of GeoLapse itself. GeoLapse is born as a Serendipity Time Capsule which allow each user to discover sudden and unknown attraction points in their everyday spaces in order to awake old or new emotions linked to the space itself.

The name GeoLapse has been created melting two significant words: 'geo' and 'lapse'. The word 'geo' originates itself from the greek work γεο- and γεω- (from γῆ as "Earth") and it is historically used with the meaning of "dilated land as the terrestrial globe." The second one, 'lapse', is defined by the Oxford English Dictionary as "a period of time between the two things that happen." In addition exist a cinematic technique called "time-lapse" used for documenting visible or not natural events whose evolution in time are little perceptible by the human eye.

Naming in this way GeoLapse would suggest the importance of the space around us and across the passage of time. Users are totally involved into the space and time dimensions during the use of GeoLapse Themes like time, space and memory are close to the system. They are experienced by both senders and addressees in the interaction with GeoLapse: the first ones composing the Glap-s live a moment of reflection about past, present and future like those who created real Time Capsules, the others as they open the Glap-s live an experience in the real world in order to strengthen or create a bond with the places of their lives.

As Time Capsules capture proper objects of the time of production to allocate them to the distant future and communicate with it, in the same way GeoLapse wants that individuals reflect on their lives carrying messages to be deployed in the real so each recipients can experience a particular moment. In this way we can create a bridge between generations, people, emotions and spaces in order to communicate our present thoughts and culture without temporal limits. The key point of GeoLapse is surprising people by chance. In this way the bond with the spaces and people who lived in came out suddenly and let users think about their lives and relation in. A sudden emotion is awaken by both the reading of the Glap and the being in a specific place. In this way GeoLapse wants to be a mobile application which helps to reconnect people to each other and to the real world creating an interaction between real and digital.

3.1 Immaterial Needs: Designing Emotion

When the moment of analyzing user needs has come, it was immediately clear that GeoLapse responds needs that are immaterial and emotional. They are not practical or functional in the conventional way but surely they are still needs. Being able to reconnect people, memories and space as GeoLapse does, is a new concept in the mobile experience world.

GeoLapse has been designed with the purpose to offer a new emotional experience to users that are overwhelmed by everyday life routine such as work, school, family and all the variety of duties that occur in their lives. GeoLapse wants to make people smile while they are in a rush taking a train, and it does by delivering messages in time and space that are unexpected.

In this time, when relationships are often taken for granted, an application like GeoLapse wants to re-give people emotions they maybe forgot just to make them feel loved and happier one day or another

3.2 Possible Future Scenarios: A Co-design Approach

Designing the GeoLapse experience, the team imagined four different future scenarios in which the application could help people reconnect themselves with memories and special places. The scenarios design phase was based on the concept that each user, in different time of his life, could fit himself in more than one scenario: they reflect emotional situation more than concrete ones.

To my future self

The user can send Glap-s to his future self so he can set goals to achieve about his personal life and career.

When he will recive the Glap he will find out if he has fullfilled his expecta-

To all mankind

It's the public kind of Glap.

The user can tell a story about a personal experience or event happened in the past, in a particular location.

Every GeoLapse user will be able to receive the Glap when they'll be close to

To my beloved ones

The user can send Glap-s to his beloved ones, aiming to surprise them by awakening old memories or telling untold secrets, using GeoLapse with a more romantic

To my legacy

GeoLapse users who are consciuous of their future premature death can send Glap-s to their families and friends aminig to stay alive in their lives even if they're

Once the scenarios have been drafted, it has been necessary to define which kind of elements people considered fundamental in order to compose a Glap.

To make it stimulating for both the team and the users, we decided to pursue a Co-Design approach: by sorting a workshop out, this technique let users take active part in the designing phase of the application contents without working on temporary prototypes but making them.

The Co-Design experience has been used to discover the core and undeniable elements to awake and recreate memory in order to make it as intense as possible. These elements will then be employed to compose Glap-s to let GeoLapse users live an experience related to people, emotions and space.

The experience has been fulfilled by involving a total of 20 subjects divided into two workshop sessions.

The workshop has been structured into four sections:

1. Introducing the project explaining the main GeoLapse functions and the interaction dynamics to let subjects better understand the Co-Design experience

2. A first set of questions aimed to define the background experience of the subjects regarding their knowledge of the main messaging applications

3. The core of the workshop involved the using of a set of seven cards for each individual. Each card represented one of the typical elements for the preservation of the memory (Picture, Text, Video, Audio record, Object, Book, Perfume) plus and empty card which could be filled with an extra content according to the subjects. Each subject had to order the set of cards from the most important to the less, following their sensation. The result of this section were 20 different sequences that has been put together in a general one using a score system: the first two position obtained 2 points, the last one lost 1 point and the central positions gained each 1 point.

4. A second set of question asked the subjects to imagine the use of GeoLapse in their everyday life.

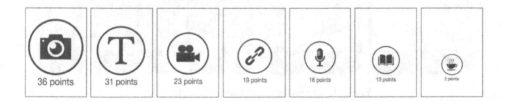

" I'd surely be surprised, moved and happy because another person wanted to make me remember a special shared moment. "   " I would be very glad thinking that someone considered me deserving a special message like that. "   " I would feel happy and surprise, thinking that someone is or was thinking of me in that time in that moment. "

Fig. 1. Co-design workshop results and highlights

3.3   Geolapse App: An Experimental Simulation

After the basic dynamics have been designed, it has been created a storyboard that shows the essential interactions acted in GeoLapse.

It will be shown both sides of the usage of GeoLapse: the application seen as sender and recipient. For the sender-side it will be displayed how a user can compose a Glap by putting together the different kinds of elements and how the will insert the data for the future delivery of the Glap. It will then be shown how the recipient-side get the notifications and how the Glap could be seen as the recipient gets it.

As it will be argued in the next section of this paper, the storyboard also helped the team to go through with the testing phases about the notification system.

3.4   Engaging the User: Alerting Messaging

Considering the fact that not all the users believe in fate or chance so not all the individuals like the idea of "not knowing if there is a message to be read" or where it is located. In addition it is difficult to be in the right place at the right moment to read the message and when we create a mobile application we have to deal with already existing systems.

So since a GeoLapse user don't know when or where a message is it was necessarily to introduce a notification system which inform people only about the presence of particular Glap-s for them in the real world.

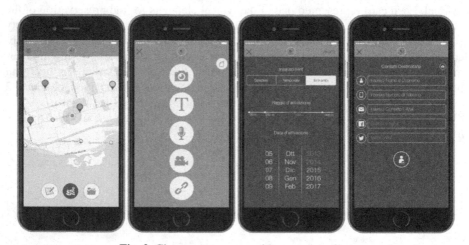

Fig. 2. Glap message composition user interface

https://www.youtube.com/watch?v=LFH-ozWwKG4&feature=youtu.be

The starting point of the GeoLapse notification system design was an analysis of the standard alerting message system. In a cross device view all the last generation smartphone allow the users to customize the visualization of their alert messages: it can be controlled by the OS (Operation System) or by the Application itself. In this way ours devices let us setting the main view of alert messages. It can be shown as

1.   Pop-Up message, which allows a block of text to appear in the middle of the screen and carries the information required. It is the most informative kind of visualization.

2.   Badge message, which show a number on the application icon and doesn't show any further information but only inform about the fact that a trigger occurs. It is the less informative kind of visualization.

3.   Status Bar message, which show a preview of the alert message in the superior part of the device's screen. It is the middle informative kind of visualization.

In any case every notification view let the user know which is the application who ⟨ʍʘʍ ɕ ʍ the ʍʘʍʘʍ ʘʘ ʍʘʍ ʘʘʍʘ ʘʘ ʍʘʍ ʘʘ ʍʘʍʘ ʘ in informed. The sender of the alert message is always recognizable but especially the users can respond directly to these events and can understand which is the information that they need to know.

Furthermore, to understand the real users' behavior about their alerting messages it was opened an online questionnaire to understand how they react on the notification on their devices.

Trying to figure out how and what users have active on their devices we realized that on 50 responses the trend has been to open the message in most cases which let us understand that the individual differences are the most relevant aspects on this theme.

In order to create more engagement between the users and GeoLapse and to get more involvement and curiosity the Geolapse alerting system misses these key elements. First of all in fact the sender does not reveal the actual content of the Glap nor his identity. In this way the receivers have to find the Glap-s in the real world to discover who is writing and which is the real content of the message. The whole purpose is to make the addressee curious about discovering the identity of the sender, the location and the content itself.

Fig. 3. Notification interface of a Glap

There have been designed three different kind of alerting message: space related, time related and space-time related.

All of them have the same content:

•   A 200-character-long text message considered the real hint

•   An optional multimedia element (Picture, Video, Audio Record, External Link), considered the memory acrtivator

•   A default text referring to the kind of notification message.

The differences between this three kind of alerting messages are the activating dynamic and the central element on which the contents are presented.

The space related one, is based on a map and its activation depends on the proximity to the message. The closeness of activation is decided by the sender choosing on a range (from 50 meters to 1 kilometers) suggested by the application itself.

The time related typology bases itself on a timeline showing the life span of the Glap and it is activated at the moment that the sender set up. This kind of message has no commitment with the closeness to the place.

The last one is a mixed version of the previous two. For this reason has a map and it is activated as the addressee is next to the message, but it has also a timeline showing the lifetime of the Glap.

The same formulation of three different types of notification will be considered useful for understanding every type of person and to provide an alternative: they can be placed along a continuum ranging from total randomness (notification spatial) in absolute accuracy (notification spatial and temporal) being warned of the presence of a Glap.

Fig. 4. Glap message notification user interface

3.5   Evaluating the Serendipity Experience

After the design phase we wanted to test and evaluate the experience that comes from the interaction with a system based on serendipity. The characteristic of asynchrony between time, place and the user is crucial to the alerting system of GeoLapse: in fact, the initial hypothesis was that users could see these missing as a negative element and unmotivated.

Testing the usability of the alerting messages we wanted to understand how the information stored in was helpful and how it works as a memory activator. Testing the experience which comes from the interaction with them we wanted to investigate especially the actual satisfaction and the emotions originate from GeoLapse.

We obtained a total sample of 21 respondents aged between 17 and 53 years. The sample was randomly divided into 3 groups, each consisting of 7 individuals in order to minimize errors resulting from the administration procedure, to assess whether there were differences in the perception of the types of alerting message and which could actually work better considering the purpose of the application .

The first group was asked to test the notification space and compare it with the space and time related kind of alert message. The second group instead has been asked to test the time related one and in the same way of the first group to compare it with the space and time related alert message. The third and last group of subjects finally has to test all three types of notification messages.

Before starting the testing phases the application has been described and the aim was illustrated to the participants for let them better understand how GeoLapse really works.

How far do you consider "close" a place important to you?

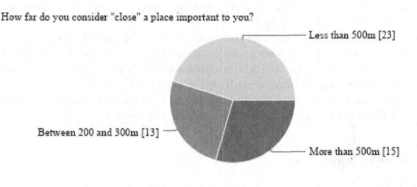

Less than 500m [23]

More than 500m [15]

Between 200 and 300m [13]

What do you do with something not urgent that has an expiration date?

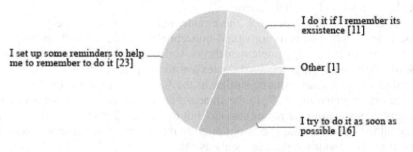

I do it if I remember its exsistence [11]

I set up some reminders to help me to remember to do it [23]

Other [1]

I try to do it as soon as possible [16]

❝ This kind of notification messages creates in me more curiosity than standard ones ❞

❝ I think this will get a step further than the traditional alerting systems. The fact that the recipient should understand a notification with clues and localization assumes -a different importance of the message [...] everything will arouse an emotion to the recipient, who will be involved more closely ❞

❝ The fact that the alerting messages atcivate themselves in a real position, next to me, in the world make them more real ❞

Fig. 5. User testing phase results and Highlights

The testing phase was divided into 6 step.

In the Step 1 an alert message in a Pop-up visualization was shown and were asked the following questions to evaluate an initial reaction: "What do you think? What are you going to do? What do you expect to find once you'll open it?

The Step 2 was about the real usability task. It concerned the exploration of the content of the alert message. Its purpose was to let the users to discover the content in the notification and for us to understand how them interact with this digital elements (the operation sequence consist in Open the notification, Scroll the content, Save the media and Set a reminder).

During the Step 3 a Score Usability System was introduced to the subject to understand their experience with the alerting system in a usability view. It is a questionnaire based on 5-point Likert scale which investigate in a quick and dirty way about the reception of the interface and their ability.

The Step 4 and 5 represent the comparative phases. Each group repeats the 2nd and 3rd phase using a different kind of alert message (Group 3 has done this phase for each alert message - total of 3 times).

The Step 6 represent the last and most important phase the one about the semistructured interview with questions on issues that require a response qualitative. During the interview we were able to reach the opinion about the perception of the system:

* How it was usefull or different from the already existent one
* How the users consider the alert messages: personal, real or digital element
* Which kind of alert message prefer and the reasons why

4   Conclusions

Although the application has its own complexity both for conceptual aspects and interaction dynamics, the most critical and significant issues for its future development are related to privacy and alerting system.

The lack of synchronicity between the events and the people in the time/space where the Glap is activated is an important concept to be explored both in terms of user experience that technology implementations. Unlike other mobile apps like Foursquare® where the message is received contextual to the place and conditions of co-presence, dynamics GeoLaps is more similar to the functions of the search service of your device implemented in iOS 7. In the standard condition the user is not aware of its proximity with a significant point/message, but the device that identifies it for him even in conditions in which the app is not directly turned on, or at list in notification mode. To meet this condition the user is always the track in its geolocation and trips from its own device with significant implications for what concerns the personal privacy. On a technical level, however, the implementation issue moves from the app itself at the level of the operating system in connection with the native services of the device (GPS etc.) to be able to constantly follow the user and allow them to interact with Glaps that have been scattered in the physical/virtual space addressed to him by another user, in another space/time.

Acknowledgments. Although the paper is a result of the joint work of all authors, Letizia Bollini is in particular author of parts 1, 2, 3 and 4, Giulia Busdon is author of paragraphs from 3.1 to 3.3 included and Annalisa Mazzola is author of paragraphs 3.4 to 3.5.

References

1. Husserl, E.: Phenomenology of Internal Time-Consciousness, p. 21. Indiana UP, Bloomington (1964)
2. Nietzsche, F.: Vom Nutzen und Nachteil der Historie für das Leben (1874)
3. Jarvis, W.E.: Time Capsules: A Cultural History (2002)
4. Phone, A., Mishra, A., Vengyal, R. John Phone: a study of industrial phone deformations for flexible thin-film smartphone notifications. In: Proceedings of the SIGCHI Conference on Human Factors in Computing Systems. ACM (2013)
5. Krug, S.: Don't make me think! A common sense approach to web usability. Pearson Education India (2000)
6. Lavazza, M.C.: Comunicare la User Experience. Apogeo Editore (2012)
7. Leonhardi, A., Kubach, U., Rothermel, K.: Virtual information towers. a metaphor for intuitive. In: Proceedings of the Third International Symposium on Universitat Stuttgart Digest of Papers. IEEE (1999)
8. Linderman, M., Fried, J.: Defensive Design for the Web: How to improve error messages, help, forms, and other crisis points. New Riders Publishing (2004)
9. Misti, R., Bagnara, S.: Psicologia ambientale. Il Mulino (1978)
10. Rizzo, F.: Strategie di co-design: teorie, metodi e strumenti per progettare con gli utenti. Franco Angeli (2009)
11. Xu, Z., Zhu, S.: Abusing Notification Services on Smartphones for Phishing and Spamming. WOOT (2012)

A SWE Architecture for Real Time Water Quality Monitoring Capabilities Within Smart Drinking Water and Wastewater Network Solutions

Grazia Fattoruso[1](✉), Carlo Tebano[2], Annalisa Agresta[3], Bruno Lanza[2],
Antonio Buonanno[1], Saverio De Vito[1], and Girolamo Di Francia[1]

[1] UTTP/Basic Materials and Devices Laboratory, ENEA Portici Research Centre,
P.Le E. Fermi 1 80055, Portici, NA, Italy
grazia.fattoruso@enea.it
[2] UTTP/CHIA Laboratory, ENEA Portici Research Centre,
P.Le E. Fermi 1 80055, Portici, NA, Italy
[3] Civil, Architectural and Environmental Engineering Department,
University of Naples, Via Claudio 21, Naples, Italy

Abstract. The world is facing a water quantity and quality crisis. These global concerns are addressing water sector operators to smart technological solutions that realize the so-called *smart* drinking water and wastewater networks. Water quality preservation is one of the essential services that smart water utilities have to guaranteed. The water quality monitoring systems include a variety of in situ sensors with several sensor protocols and interfaces. Sensor integration as well as real time sensor readings accessibility and interoperability across the interconnected layers of functionality needed for a comprehensive smart water network solution are the challenges should be tackled. The objective of this research work has been to develop a standardized OGC SWE (Sensor Web Enablement) architecture that enables the integration and real time access to the various continuous and networked sensors can be installed along drinking water and wastewater networks, and real time sensor data browsing, querying and analyzing capabilities across the components of a smart water network solution. Furthermore, a web based geo-console and a QGIS SOS client application have been developed ad hoc for supporting utilities to effectively manage their water treatment and optimize quality-testing processes.

Keywords: SWE framework · Sensor web · Smart water utilities (SWNs) · Water quality monitoring and modeling · Sensor networks · QGIS · SOS client · Map viewer

1 Introduction

Water is recognized essential to sustain life nevertheless the world is facing a water quantity and quality crisis, caused by continuous population growth, industrialization, food production practices, increased living standards and poor and unsustainable water use strategies as well as inadequate management systems and practices.

© Springer International Publishing Switzerland 2015
O. Gervasi et al. (Eds.): ICCSA 2015, Part II, LNCS 9156, pp. 686–697, 2015.
DOI: 10.1007/978-3-319-21407-8_49

These global concerns are addressing water sector operators to smart technological solutions that promise more efficient and sustainable water systems, realizing the so-called *smart drinking water and wastewater networks*.

A comprehensive smart drinking water/wastewater network solution involves several interconnected layers of functionality [1]. They are basically:

-   measurement *and sensing instrumentation* that collects data on water flow, quality, acoustics, supply and other critical parameters;
-   real-time two-way *communication channels* (e.g. two-way radios, cellular networks) that allow utilities to gather data from networked measurement and sensing devices automatically and continuously;
-   *basic data management software* that enables utilities to process the collected data and present an aggregated view via basic visualization tools and GIS, simple dashboards or even spreadsheets and graphs;
-   *real-time data analytics and modeling software* that enables utilities to derive actionable insights from network data. Dynamic dashboards allow utility operators to monitor their water/wastewater network in real time for hazards or anomalies. At the same time, hydraulic and water quality modeling software can help operators understand the potential impact of changes in the network and analyze different responses and contingencies. Pattern detection algorithms can draw on historical data to help distinguish between false alerts and genuine concerns, and predictive analytics allows operators to consider likely future scenarios and respond proactively and effectively
-   *automation and control tools* enable water utilities to conduct network management tasks remotely and automatically. This layer provides tools that interface with the real-time data analytics and modeling software, leveraging communication channels and the physical measurement and sensing devices within the network. Many utilities have existing SCADA systems that can be integrated with smart water networks to further enhance their control over the network.

Thus, the smart water network technology arms the control room operators with a comprehensive and more effective set of decision-making capabilities for a sustainable and optimized water utilities management. Especially, these capabilities enable the operators to quickly assess events as they occur, identify potential problems before they reach a critical level, respond to operational challenges, and minimize downstream effects [2].

Among services to be provided by smart drinking water and wastewater utilities, the water quality preservation can be considered essential. A drinking water supply has to be safe as well as satisfactory.

A smart water network solution for drinking water quality preserving includes networked sensors for chlorine, pH, biological indicators and other chemicals as well as heavy metals along several network locations[3].

Chlorine is widely used as a disinfectant in drinking water systems for its advantages such as high oxidation potential, long-term disinfection until the water reaches the consumer, excellent disinfection effectiveness and relatively low cost. However, there are some disadvantages such as the decay of its concentration along the water distribution network and the formation of undesirable DBPs (Disinfection by Products) that are characterized by recognized toxicity and carcinogenicity.

The water quality dynamics can continuously be monitored and assessed by real-time modeling based on the *live data* (e.g. free chlorine) gathered by networked probes along the drinking water system [4], [5]. Thus, water system operators can quickly compare water quality data against regulatory requirements (e.g., maximum contaminant levels) over time and along the water network as well as identify water quality changes and anomalous events, forecasting potential hazards and adverse public health impacts .

Wastewater - spent or used water from farms, communities, villages, homes, urban areas or industry, if un-managed, can be a source of pollution, a hazard for the health of human populations and the environment. Wastewater can be contaminated with a myriad of different components: pathogens, organic compounds, synthetic chemicals, nutrients, organic matter and heavy metals, carried along the water from different sources. Unregulated or anomalous/malicious discharge of wastewater along the sewage network may affect the wastewater treatment plants operation and consequently pollute the receiving water body [6].

A continuous and real time monitoring of several water quality parameters along a sewage network by a distributed heterogeneous sensor network coupled with predictive water quality modeling capabilities, can enable operators of the wastewater utility to quickly assess contamination events as they occur, due to unregulated or anomalous/malicious discharge into the network, alerting the wastewater treatment plants, eventually identifying the possible anomaly discharge points, and hence minimizing the environmental and social impacts [7].

Anyway, recent global smart water surveys still reveal a large gap between *need* and *reality*. Most of water utilities still rely entirely on manual collection of water quality samples, which can take several days [1], though they are strongly aware of the need of continuous real-time data on water quality integrated with predictive system dynamics scenarios for making informed decisions and not relying on experience and intuition.

Thus, a smart water network solution for water quality monitoring includes a variety of in situ sensors to be managed with several sensor protocols and interfaces. Sensor integration as well as real time sensor readings accessibility and interoperability become essential requirements to be meet within a such solution [8].

The recent OGC Sensor Web Enablement (SWE) framework could play a crucial role providing fundamental building blocks for developing the required capabilities [9], [10]. The focus of this research work has hence been to investigate and develop a SWE architecture for water quality monitoring capabilities in smart water/wastewater solutions which enables the integration and the real time access by the several smart components of the various continuous and networked sensors can be involved, as well as the retrieve of the events and alerts triggered through sensors. More specifically, the developed architecture realizes sensor related services and data delivery, enabling real time sensor data browsing, querying and analyzing capabilities across the components of data management as well as data analytics and modeling within a smart solution. A web based geo-console developed ad hoc enables the access to real time and time series observation data returned by SOS instances, visualizing the (near-)

real time sensor state through three hours graphs, popups, and hence the alert state and predictive water quality dynamics through thematic layers. Similarity, a QGIS SOS client application has been developed ad hoc for adding the same capabilities to the open source GIS desktop QGIS.

The developed architecture has been applied to two case studies that respectively involve the water quality monitoring and managing for real aqueduct of Santa Sofia (Southern Italy) and the wastewater network of the Marsa Lubrunlu city (Tannlo in Italy)

2   Developing the SWE Architecture

Within smart water/wastewater network solutions, the kinds of sensor resources to be managed can be stationary or in motion and gather data in-situ or remotely, while the sensor hardware, sensor protocols and interfaces are various. For implementing the sensor integration and interoperability requirements, we have exploited the OGC-SWE technological framework [9], [11].

2.1   The OGC SWE Framework

Sensor Web paradigm refers —Web accessible sensor networks and archived sensor data that can be discovered and accessed using standard protocols and application programming interfaces (APIs) [9]. Sensor Web can hence be seen as a huge internet based on sensor network and data archive[12], [13].

The SWE technological framework refers an infrastructure which enables an interoperable usage of sensor resources by their *discovery*, *access*, *tasking*, as well as *eventing* and *alerting* within the Sensor Web in a standardized way. It hides the underlying layers, the network communication details, and heterogeneous sensor hardware, from the applications built on top of it.

More specifically, the SWE framework makes available standard for encoding sensor data and metadata (OGC Observations & Measurements, OGC Sensor Model Language) as well as standards for service interfaces to access sensor data (OGC Sensor Observation Service); standards for subscribing to alerts/events (OGC Sensor Alert Service) and controlling sensors (Sensor Planning Service)[9].

Thus, within this framework, our objective have been to implement SWE services for water quality sensor observations gathered by *in situ* sensors virtually connected into a network infrastructure, based on web technologies, that enables us to remotely and real time access *live* sensor data using web. In this way, real time water quality data can be integrated into a spatial data infrastructure, an essential requirement for reliably water quality dynamics predicting and eventually dispatching time critical alerts.

2.2   The Architetture Design

The proposed SWE architecture provides functions ranging from integrating and real time accessing the various networked sensors and their observations, to retrieving events and alerts triggered through sensors, enabling observation data browsing, querying and analyzing capabilities within client application.

According to the functionality requirements to be implemented, this architecture (Figure 1) has been designed and developed including server components such as:

- a geodatabase for storing geospatial and sensor data;
- a map server for publishing on the web spatial and sensor data, thematic maps and predictive scenarios related to the hydraulic and water quality dynamics;
- SOS (Sensor Observation Service) Server for accessing descriptions of sensors and their collected observation data by standardized web service interfaces.

The client applications developed are both web that desktop. In particular, a web based geo-console (geospatial HCIs) has been developed ad hoc for accessing real time and time series observation data returned by SOS instances, visualizing them through graphs (time series) or popup (real time observations or predictive values at fixed time intervals), for analyzing the sensor state especially their alert state and the predictive water quality dynamics. Another client application is *QGIS SOS client application* developed ad hoc for adding the ability to the open source GIS desktop QGIS to access sensor to data served by SOS server and deliver them to the available simulation models.

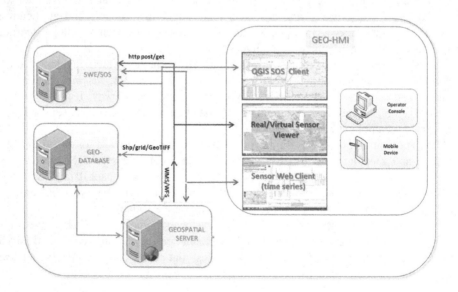

Fig. 1. The architecture design for water quality monitoring capabilities within smart water/ wastewater network solutions

As regards the SWE component of the proposed architecture, it consists of the services SOS (Sensor Observation Service), SES (Sensor Event Service) and WNS (Web Notification Service) (Figure 2) [9], [13], [14].

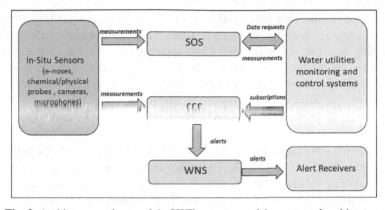

Fig. 2. Archietcture schema of theSWE componet of the proposed archietcture

The SOS component allows clients (e.g. QGIS SOS client developed ad-hoc by ENEA) to access descriptions of associated sensors and their collected observations by a standardized web service interface. SES provides notification services, with stream processing capabilities. Alerting and notification capabilities provides support for creating alarms and filter constructs by system users. Users hence use created alarm constructs to subscribe to live sensor feeds and continuously receive notifications once events are detected during live streams processing. The notifications are processed by WNS that notifies the user by sending an sms with the notification received by SES.

2.3   The Component Diagram

The software component diagram of the proposed architecture is showed in the Figure 3. In particular, the implementation of the proposed SWE component has been deployed by 52°North framework [15] version 3.x, customized using the standards OGC SOS 1.0 and 2.0. The deployed SOS endpoint uses a database PostgreSQL 9.3.1 with a spatial extension PostGIS 2.x to store observation values and sensor metadata.

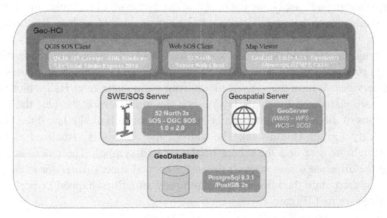

Fig. 3. Component diagram of the proposed SWE architectura

The *map server* has been developed by using GeoServer 2.6.x [16], an open source server for sharing geospatial data as web services that enables spatial data interoperability, publishing data from any major spatial data source using open standards. GeoServer is a OGC compliant implementation of a number of open standards such as Web Feature Service (WFS), Web Map Service (WMS), and Web Coverage Service (WCS). Additional formats and publication options are available including Web Map Tile Service (WMTS) and extensions for Catalogue Service (CSW) and Web Processing Service (WPS).

Geospatial data within the case studies has been published by WMS e WFS.

The implementation of the GeoDatabase component has been deployed by l'object-relational database PostgreSQL 9.3.1 with the spatial database extender PostGIS 2.x that adds support for geographic objects allowing location queries to be run in SQL.

The web SOS client application has been developed as a extending the open source Sensor Web Client while the QGIS SOS client has been developed by using Nokia QT libraries of the QT Creator [18], an integrated development environment which is part of the SDK for the Qt GUI Application development framework. The libraries have been compiled by using Microsoft SDK for Windows 7.1 and Visual Studio Express 2010, for Windows platform (32 bit and 64 bit) and Linux platform.

The map viewer has been developed by GeoExt2 [19], an Open Source JavaScript framework that enables building desktop-like GIS applications through the web. Through this framework GIS functionality of OpenLayers with the user interface of the ExtJS 4.2.x library provided by Sencha have been combined. ExtJS 4.2.x library anables the development of javascript desktop-like applications by using HTML5 e CSS3 standards.

3   The Application Layer

For accessing and browsing within maps the various continuous and networked sensors, and queering and analyzing the observation data both in real time and at prefixed time intervals, a web based geo-console has been developed ad hoc. By this console, the utility operators can visualize the measurements gathered by sensors really installed along the drinking water and wastewater networks as well as predictive measurements returned by on line hydraulic and water quality simulation models, served by SOS services. Within the viewer, monitoring parameters (e.g. H2S, chlorine, rainfall, etc.) are visualized by single layers. For the multi-sensor stations, the real time observations of all active sensors can be visualized as well as the last three hours observations by a graph (Figure 4a). The state of the sensors is visualized as thematic layer through the size and the colour of the related symbol. The colour scale associated to the different sensor states is *green* for normal state, *orange* for attention state and *red* for alert state then occurs when anomaly situations happen or prefixed threshold values are met (Figure 4b).

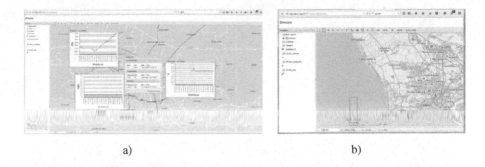

a)   b)

Fig. 4. The web based geo-console. a) View of all real time observations (clorine, temperature, pressure) gathered by a multi-sensors station installed along the real acqueduct of Santa Sofia (South Italy); b) View of a thematic clorine state layer

Through a toolbar, user can access two specific viewers that act as application layers. One viewer handles the rendering by graphs of queried time series (Figure 5). Similarity, the other viewer handles the rendering of the on line modeling simulations by graphs and thematic maps.

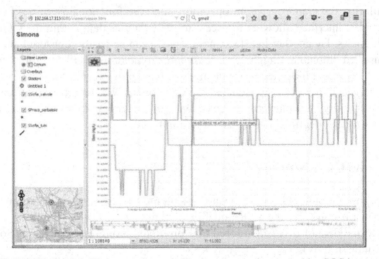

Fig. 5. The Web SOS Client: graphes of chlorine time series served by SOS istances

As a desktop client, a QGIS SOS client application (Figure 6) has been developed ad hoc for adding the ability to the open source GIS desktop QGIS to access sensor to data served by SOS and deliver them to the simulation models (i.e SWMM, Epanet/MSX), plugins of the QGIS software (e.g. GHydraulics) [20].

This client application offers easy to use interfaces within QGIS to run the *GetCapabilities* operation by SOS server, to request a service description containing the spatial and temporal extent of the offered observations as well as a list of the sensors and observed features.

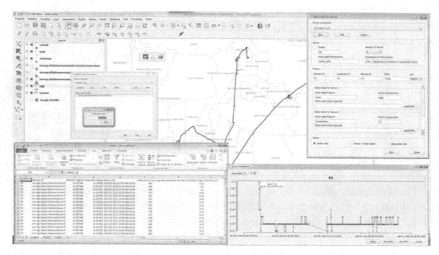

Fig. 6. The QGIS SOS client application: interface for adding sensors and editing observations into SOS server *(on the rigth)*; interface for browsing the observations data served by SOS services *(down in the window); interface for adding and quering SOS layes into QGIS view (on the left)*

Similarity, the *GetObservation* operation, the core functionality of the SOS, is run by a specific interface allowing to access to observations data within the QGIS viewer by table views. By another interface, new sensors can be easily registered and observations inserted, implementing the *Transactional profile* of the SOS specification.

Thus, this client hides technical details of SOS services and protocols, so that even non-experts can use all components of the developed SWE architecture, transparently, within a friendly GIS platform, coupled with the spatial analysis and geo-processing tools as well as integrated numerical modeling tools.

4   The Case Studies

The proposed architecture have been implemented within two application scenarios. One scenario concerns the water quality dynamics (i.e. DBPs formation as well as free chlorine decay) to be monitored and predicted along the real Santa Sofia aqueduct (Southern Italy).

The case study of Santa Sofia aqueduct involves a wireless network of sensors installed along the system which monitors in real time and continuously physical and chemical water parameters (i.e. pressure, residual chlorine, conductivity, temperature and pH) and delivers *live* data to online hydraulic and water quality models [4], [5]. Additionally, on-demand deployable sensor platforms for measuring DBPs are installed into specific locations along the water network.

The SWE architecture implementation encapsulates the different sensor types into the SOS server, so that the gathered data can be accessed as SOS services by the developed web based geo-console as well as the QGIS SOS client in the way described in the previous sections.

The alerting functionalities have been implemented by using a SES instance in conjunction with a WNS one. Thus, a sms is sent when critical scenarios occur i.e. water contaminants levels do not meet regulatory requirements (e.g., a disinfection residual higher than 0.2 mg/l for free chlorine and DBPs concentrations higher than 30μg/l).

This platform arms the control room operators of the Santa Sofia aqueduct with a comprehensive set of decision making capabilities, enabling them to continuously monitor the drinking water quality integrity, confirm normal system performance, optimize emergency response. Just as an example, through the geo-console, the operators can visualize in real time the alert state of the chemical (e.g. chlorine) sensors by thematic layers as well as by the alert sms promptly sent. At the same time, they can assess the dynamics of the anomaly situation, observed in a specific location, along the network, visualizing the simulated observations as well as the impact of the emergence on the population, and so on.

The second scenario concerns the development of a monitoring and warning system for contaminations along the wastewater network of Massa Lubrense city (Souther Italy) [7]. Different sensor types are involved in this case study ranging from total immersion probes for monitoring qualitative and quantitative parameters such as PH, COD, NH3, and water level and conductivity to multi-sensor platforms composed by an e-nose for NH3, H2S, in conjunction with temperature and humidity sensors for adjusting the gas readings, and a microphone. The networked sensor platforms have been installed in strategic locations of the wastewater network.

By implementing the proposed SWE architecture, the control room of the Massa Lubrense sewage network is armed with real time and continuous monitoring capabilities of the state of the wastewater network that enable it to quickly assess contamination events as they occur, especially due to unregulated or anomalous/malicious discharge into the network, alert the wastewater treatment plants, and eventually identify the possible anomaly discharge points, and hence minimize the environmental and social impacts.

Through the developed SWE architecture, the displaying of real time and time series data gathered by the networked sensors installed along the sewage network is fulfilled by SOS instance. For the real-time notification, the sensor data are transferred to the SES instance which filters the incoming data with regard to alert criteria specified. Thus, if a matching alert condition is found by the SES, the according alert is dispatched, by sending the notification request to the WNS instance.

5   Conclusion

The water utilities integrity especially in the water quality preservation component is best assured by a real time and continuous monitoring of the physical and chemical parameters coupled with a real time modeling of hydraulic and water quality dynamics along water/wastewater networks.

While most of the water utilities have active automation and control systems, they still rely entirely on manual collection of water quality samples, which can take several days, for monitoring the water quality. Though they are moving to smart water solutions for water quality preservation and more generally for an more efficient and proactive management and control of the water utilities.

For enabling an interoperable usage of different sensor resources involved in smart water quality solutions, through their *access* as well as *eventing* and *alerting*, a SWE based architecture has been proposed. It has allowed hiding the underlying layers, the network communication details, and heterogeneous sensor hardware, to the applications (e.g. geo-console as well as system dynamics models) built on top of it. Thus, SWE services have been implemented within the two case studies for water quality sensor observations gathered by *in situ* sensors virtually connected into a network infrastructure, based on web technologies, that enables the application layers to remotely and real time access *live* sensor data using web. In this way, real time water quality data can be integrated into a spatial data infrastructure, enabling their access to water/wastewater networks dynamics models as well as systems for monitoring water quality and dispatching time critical alerts.

For the future, this architecture will developed for mobile devices (tablet, smartphone), enabling the operators to monitor the water quality integrity in situ.

Acknowledgments. This research work has been funded by POR FESR 2007-2013/SIMONA Project and by POR Campania 2000/2006 M. 3.17/ACQUARETI

References

1. White paper online: Water 20/20 Bringing Samrt Water Network in Focus by Sensus. https://sensus.com/smartwaternetworks
2. Boulos, P.F., Wiley, A.N., Colo, B.: Can We Make Water Systems Smarter? (2013). Online paer: http://dx.doi.org/10.5991/OPF.2013.39.0015
3. HallJeffrey, J.S., Szabo, G., Panguluri, S., Meiners, G.: Distribution System Water Quality Monitoring: Sensor Technology Evaluation Methodology and Results. A Guide for Sensor Manufacturers and Water Utilities. Water Infrastructure. Environmental Protection Agency - EPA 600/R-09/076 (2008)
4. Fattoruso, G., et al.: Simulation of Chlorine Decay in Drinking Water Distribution Systems: Case Study of Santa Sofia Network (Southern Italy). In: Di Natale, C., Ferrari, V., Ponzoni, A., Sberveglieri, G., Ferrari, M. (eds.) Sensors and Microsystems. LNEE, vol. 268, pp. 467–470. Springer, Heidelberg (2014)
5. Fattoruso, G., et al.: Use of Kinetic Models for Predicting DBP Formation in Water Supply Systems. In: Di Natale, C., Ferrari, V., Ponzoni, A., Sberveglieri, G., Ferrari, M. (eds.) Sensors and Microsystems. LNEE, vol. 268, pp. 471–474. Springer, Heidelberg (2014)
6. Corcoran, E., Nelleman, C., Bajer, E., Bos, R., Osborn, D., Savelli, H.: Sick Water? The central role of wastewater management in sustainable development. A Rapid Response Assessment. United Nations Environment Programme, UN-HABITAT, GRID-Arendal (2010). ISBN: 978-82-7701-075-5
7. De Vito, S., Fattoruso, G., et al.: An integrated infrastructure for distributed wastewater quality monitoring and decision support. In: Proceedings of the AISEM, Trento , February 3-5, 2015

8. Glasgowa, H.B., Burkholdera, J.M., Reeda, R.E., Lewitusb, A.J., Kleinmana, J.E.: Real-time remote monitoring of water quality: a review of current applications, and advancements in sensor, telemetry, and computing technologies. Journal of Experimental Marine Biology and Ecology **300**, 409–448 (2004)

9. Bröring, A., Echterhoff, J., Jirka, S., Simonis, I., Everding, T., Stasch, C., Liang, S., Lemmens, R.: New Generation Sensor Web Enablement. Sensors **11**, 2652–2699 (2011)

10. Gibbons, P., Karp, D., Ke, Y., Noth, S., Seshan, N.T., An Architecture for a Worldwide Sensor Web. IEEE Pervasive Comput. **2**, 22–33 (2003)

11. Delin, K., Jackson, S., Some, R.: Sensor Webs. NASA Tech. Briefs **23**, 90 (1999)

12. Delin, K.: The Sensor Web: A Macro-Instrument for Coordinated Sensing. Sensors **2**, 270–285 (2001)

13. Shafi, S., Reshi, A.A.: Real Time Sensor Web Enabled Water Quality Monitoring System Using Service Oriented Architecture (SOA). International Journal of Scientific & Engineering Research **3**(5) (2012)

14. Jirka, S., Bröring, A., Stasch, C. Applying OGC Sensor Web Enab lement to Risk Monitoring and Disaster Management. Online paper

15. 52° North Initiative for Geospatial Open Source Software GmbH

16. 52 North Sensor Web Community. http://52north.org/communities/sensorweb

17. GeoServer: http://geoserver.org/

18. PostgreSQL/PostGIS: http://postgis.net/

19. QTCreator: https://qtproject.org/wiki/Category:Tools::QtCreator

20. GeoExt 2: http://geoext.github.io/geoext2/

21. Fattoruso, G., Tebano, C., Agresta, A., Buonanno, A., De Rosa, L., De Vito, S., Di Francia, G.: Applying the SWE Framework in Smart Water Utilities Domain. In: Compagnone, D., Baldini, F., Di Natale, C., Betta, G., Siciliano, P. (eds.) Sensors. LNEE, vol. 319, pp. 321–325. Springer, Heidelberg (2015)

"Serpentone Reload" an Experience of Citizens Involvement in Regeneration of Peripheral Urban Spaces

Federico Amato[1], Sara Bellarosa[2], Giuseppe Biscaglia[3], Luca Catalano[4],
Antonio Graziadei[5], Annalisa Metta[6], Beniamino Murgante[1(✉)],
Maria Livia Olivetti[6], Pasquale Passannante[2], Annalisa Percoco[7],
Gerardo Sassano[5], and Francesco Scaringi[3]

[1] University of Basilicata, 10, viale dell'Ateneo Lucano 85100, Potenza, Italy
{federico.amato,beniamino.murgante}@unibas.it
[2] NUR Association, 6, Via Rocco Scotellaro 85100, Potenza, Italy
sbellarosa@gmail.com, passannantep@yahoo.it
[3] Basilicata 1799 Association, 1, Piazza Crispi 85100, Potenza, Italy
gvbiscaglia@gmail.com, f.scaringi@me.com
[4] Osa Architettura e Paesaggio, 175, Via Cristoforo Colombo 00174, Rome, Italy
lucacatalano@osaweb.it
[5] Paesaggi Meridiani Association, 24, Via Pienza 85100, Potenza, Italy
{antonio,gerardo}@paesaggimeridiani.it
[6] Department of Architecture, University of Roma Tre, 10, Largo Marzi, Rome, Italy
{annalisa.metta,marialivia.olivetti}@uniroma3.it
[7] Municipality of Potenza, Piazza Matteotti, Potenza, Italy
annalisa.percoco@gmail.com

Abstract. Suburbs are often very contradictory places. Despite great part of urban population live there, these parts of cities are mostly considered as degradation places. The topic of suburbs regeneration is relevant today. Nevertheless, often expensive interventions implemented by local authorities fail to regenerate their public spaces. This paper presents the experience of "Serpentone reload", a workshop based on participatory reactivation of abandoned or underused spaces and buildings in "Cocuzzo/Serpentone" neighbourhood in Potenza (Basilicata, Italy). The workshop particularly focused on the reuse of the "Ship", an underground building, completed in 2010, never used, because it has been perceived as an extraneous element, the result of an imposition and not the outcome of shared choices. The experience is particularly significant, because it shows how low cost interventions, realized with citizens involvement, could contribute to the regeneration of peripheral urban areas more than expensive and complex imposed interventions.

Keywords: Public space · Co-creation · Public housing neighbourhood · Temporary design

1 New Planning Paradigms "Lighter, Quicker, Cheaper"

Small actions to achieve great changes: at glance it can seems a pun, but it contains the key of a new approach to the city. Traditional approaches to planning define rules,

© Springer International Publishing Switzerland 2015
O. Gervasi et al. (Eds.): ICCSA 2015, Part II, LNCS 9156, pp. 698–713, 2015.
DOI: 10.1007/978-3-319-21407-8_50

parameters and indices, produce choices and great transformations, pursuing the ambitious research of permanent and definitive solutions, that most often do not take into account the changing nature and the extreme dynamism of the urban organism.

The scenarios imagined by urban plans in several cases are out-dated before being implemented, mainly because of the rapid changes of the initial conditions and the unpredictable and often innovative uses introduced in the meantime by the inhabitants. Consequently urban landscape become full of unfinished or abandoned "relics", unnecessary infrastructure, empty or underused, public spaces without meaning for the community, rural fragments trapped within the urban fabric only by chance, saved by buildings. Massive urban transformations are almost always characterized by significant public investments in interventions, spreading over several years, high soil consumption, scarce involvement of local communities, production of unused urban spaces at the margins of great interventions (the so-called "urbanscrap"), devastation of fragile suburban and rural landscapes. In many cases these "heavy" interventions do not provide significant and lasting improvements to inhabitants life quality.

An alternative way to these interventions, taking into account the poor produced results, is emerging, with increasing success and efficiency; a new approach to urban design based on the following three terms "Lighter, Quicker, Cheaper" [1] [11] [12] [13] [14].

This approach, based on the need to do less and better, using the few available economic resources in a more effective way, rethinks city starting from its built part.

The movement Everyday Urbanism [3] considers fragility and potential of built city, often hidden, as a strength, in order to give space to "fragments of happy cities that continually take shape and vanish, hidden in the unhappy cities" [2]. Possible and current strategies are then based on reactivation and reuse of fragile and marginal places and abandoned or underutilized buildings, refusing to find permanent solutions. This logic is based on micro-temporary interventions, a sort of "urban acupuncture", or tactical urbanism [5], producing sustainable benefits on the whole urban organism revitalizing parts of forgotten city, encouraging their re-appropriation by inhabitants.

Examples of successful experiences consist of temporary projects characterized by strong innovation and social creativity, related to the world of culture, associations, start-up projects of temporary strong innovation and social creativity, related to the world of culture and associations, start-up of small enterprises and handicrafts, etc.

The temporary reuse has the great advantage of maintaining the process open, ensuring adaptability of the project to urban dynamics changes and different needs and demands that might occur over time. Another important aspect that characterizes new scenarios of contemporary city is an active inhabitants involvement in transforming and managing the urban landscape.

There are a lot of examples of unstructured groups of citizens who undertake voluntary actions in taking care of portions of the city.

The success of such experiences is the evidence of efforts of local communities in improving the current city and in shaping the city of tomorrow.

From a planning point of view, errors made prove that the city must be thought by people who live in it and interventions must be the result of participatory processes, containing bottom-up instances, avoiding decisions imposed by few people. Shared decisions inevitably have greater strength and a better chance of success.

From the citizen point of view the achievement of participation spaces requires a great effort of responsibility and civic sense towards what is shared.

The road is imagining and building together, a city of space reuse, involvement and sharing, a community as a place for everyone as the result of a common effort.

2   Urban Gaps of Potenza Municipality

Considering the issues discussed in the previous section, the city of Potenza could be a laboratory where it is possible experimenting "urban acupuncture" interventions. The most urgent need of this city is to adopt a strategy based on patching urban mosaic, on regenerating urban space and on re-appropriation of urban places.

Thinking about urban mosaic or spaces to be regenerated by adopting a different approach to the city, looking especially to social relations, with the aim of re-imagining the city itself. The choice of the type of city is, therefore, strictly related to the social dimension, the so-called city of individuals, where people biographies, with their social relationships intertwined with the urban space, are relevant and can guide policy making and administration. The imperative that now arises is to design cities having people as a reference, to deal with wishes as well as with needs. In practice, beyond the city itself it is important to consider also "looking and feeling", in other words image and sentiment of a city. When approaching government of a city, it is fundamental to capture its new complementary dimensions, direct expression of the wishes of its inhabitants: actual city, city living, imagined city and dreamed city. The demand of cities, today, increasingly appears not only as the demand for new services or adequate functions or spreading of urban quality, but the demand of everything that can help making the city a unique and irreproducible good [19].

There is the need to produce urban projects based on the growing demand of beauty and attractiveness, able to reflect expectations of inhabitants. The action will be focuses on urban public spaces, real life scene, meeting and social interaction.

The physical dimension of a city is important, but it is equally important its density of social relationships. This aspect has to be understood as the place where processes of cohesion and social exclusion occur, the place of cultural norms regulating behaviour, of identity that is expressed in urban public space. If the city is a field of complex relationships it is also a place of policy and practice of change and transformation construction, based on visible (buildings, spaces and places) institutional and social aspects. The urban political process should be seen as a social construction result of social interactions. Facing and designing urban policy means, therefore, taking into account social problem solving issues, in an effort to find rational and useful solutions to the correlation among inhabitants, city users and spaces. Reinterpreting urban policy in this dimension means to contribute to the reduction of inequalities, to the management of expressed or latent social conflicts, to the building of a sense of community where it is missing or greatly weakened. The urban design for Potenza municipality has to be centred on these reflections, through a massive cognitive and participatory mobilization [7] [8], able to meet a new right to the city [6], based on the demand of beauty, variety, usability, security, sustainability, etc. The need of an individual and collective welfare is the new paradigm, more complex to satisfy, based more on immaterial aspects than on physical changes of the city. On the other hand, the same right to the city is not confined to individual liberty to access urban resources, but has the right to change us by changing the city. It is a collective right, rather than individual, which inevitably requires the exercise of a common power

3   The Experience of Serpentone Reload

3.1   A Ship in a Mountain Town

"Serpentone Reload" experience, started with a workshop held in September 2014, having the aim to study "Cocuzzo" neighbourhood in Potenza (Basilicata, Italy), is based on previously described principles. In 1998 the Italian Ministry of Public Works published a call, with the aim of low quality suburbs regeneration. The main aim was to improve the quality of neighbourhoods with lack of environmental quality and services, paying particular attention to the recovery of buildings constructed in modern periods and energy efficiency improvements [9].

The Ministry selected and funded forty-six projects. The project submitted by Potenza Municipality, concerning the regeneration of two neighbourhoods, "Poggio Tre Galli" and " Cocuzzo", has been included within this funding [10].

The first neighbourhoods the typical example of a dormitory suburb. The second one is characterized by high complexity in terms of housing with strong social conflicts [17]. In this neighbourhood, despite considerable accessibility problems and an enormous lack of public spaces [15], the hugest housing of the city is located [18]. This building, entirely devoted to social housing, is five hundred meters long and forty meters tall and has a sinuous shape which generated the name "Serpentone", literally big snake. This huge building based on "Unitéd'habitation" principles became a symbol of urban decay in a suburban area of the city.

The regeneration project, funded by the ministry in 1998,was based on the following elements:

- lowering of two floors of the building, realizing a multifunctional roof with a heating system;
- accessibility improvement;
- realization of underground parking;
- creation of a huge square.

In the adjacent area of Poggio Tre Galli neighbourhood the construction of new social housing, a parking area and a neighbourhood square was planned.

The whole project was funded for more than 10 million euro by the Ministry, 2 million euro by the local organization of Social Housing management (Azienda Territoriale per l'Edilizia Residenziale, ATER), 625,000 euro by Potenza Municipality and 4,8 million euro by private investors.

However once obtained funding, the project was radically changed.

Fig. 1. The "Serpentone" shortly after its completion, when it was disconnected from the central part of Potenza; the subsequent urban development has included the neighbourhood in the urban part of the city

Enric Miralles was commissioned to work on the design of a large public space in Cocuzzo neighbourhood. Unfortunately after few months a cancer was diagnosed to the architect, which caused his fast death. The project was then assigned to the architect Marco Casamonti. His idea of improving the urban space of Cocuzzo neighbourhood was based on the transformation of Via Tirreno, the only road of the neighbourhood, in a large park. The proposed project considered the strengthening of a road surrounding Serpentone and Serpentino, a building similar to the first with a smaller size and distant few meters from it. In this way the road that separates the two huge buildings could be destined to park [16].

The main idea was to transform Via Tirreno in a green corridor "animated by the presence of rocks" in the lower part of the neighbourhood and a "ship" with huge concrete sails ten meters tall in the upper part.

To date, the ship is the only part of the regeneration project realized. This impressive concrete object, since its creation, is at the centre of a strong controversy. Today the ship, made with significant public investments is basically not used.

Fig. 2. A view of the "ship" from the arcades of "Serpentone";the ship inhibits correct lighting to the first floors of "Serpentino"

The roof garden is totally abandoned, forgotten both by residents and municipality, which stopped all kind of maintenance. The huge "cargo", the space under the roof garden, was never completed.

The ship, therefore, caused a decline in quality of life for residents, who, as a protest against the municipality administration, unable to collect needs and requirements, have continued their daily life in perfect indifference as if it never existed. It is not uncommon to meet inhabitants declaring proudly that they had never stepped on the roof garden of the Ship.

In this context the workshop "Serpentone Reload" has developed, on September 14-21 2014, with four main objectives: project and participation, strategies, communication, implementation.

Project and participation means designing, through the active involvement of citizens, interventions able to define and characterize areas of aggregation, socialization and interaction within Cocuzzo neighbourhood.

These interventions have always been realized through the actual technical skills of participants, taking into account availability of resources and time.

Strategy means working on identifying initiatives and events designed to encourage sharing of workshop space.

Communication refers to the identification and definition of the best strategies of networking intervention and initiatives related to it.

Realization finally implies how project interventions have to be realized in self-construction by citizens. In this way the sense of belonging to a community becomes stronger. All citizens feel involved in a small and voluntary urban transformation, transmitting enthusiasm and pleasure of taking care of a small and collective garden.

The interior space looks like a construction site, making difficult to image specific uses for it.

The Municipality, despite the important amounts of resources invested in the construction of this huge space, usesit as a deposit. The impact of the ship on the quality of the urban space is bad.

The idea of the park, the metaphor of the ship and the rocks appear distant and elusive. This ship stranded between two large buildings, "Serpentone" and "Serpentino", caused the occurrence of significant problems. Partially obstructing the only way crossing the neighbourhood, compromising accessibility because the interventions of the road annular network have been realized only partially. Furthermore, the impressive body of the concrete ship inhibits correct lighting to the first floors of "Serpentino".

3.2   Story of a Method of a Territorial Re-Animation

Serpentone Reload was imagined as a workshop in which participants were at the same time actors of a path specifically created for them and interpreters of the reality submitted to them.It was not a traditional top down participatory process due to the lack of institutions, nor a result of a re-appropriation bottom-up process.

Serpentone Reload was an educational experience with the aim, more than working on space configuration, to suggest several new uses of abandoned or underutilized spaces. The awareness of places potentiality has been promoted, organizing meetings between citizens on one hand and practitioners, students and artists and the other hand.

"Territory is the use made of it" [4]; starting from this definition Cocuzzo neighbourhood could not be analysed or reinterpreted starting from spaces as architectural places, but from the use or non-use that the local community makes of it. Paying attention to everyday practices, the first part of the workshop attempted to define various neighbourhood identities, investigating recognizable practices in the neighbourhood and around it.

A first introduction of the place occurred through a series of urban ethnography tools, identified and tested by coordinators and then proposed to participants, in order to activate knowledge between participants and neighbourhood and to build tools for the analysis of the place and the possible needs expressed by citizens.

Cognitive walks in the neighbourhood have been conceived to explore a portion of everyday neighbourhood activities through clichés and citizens telling. Attention of participants and their objectives were aimed at identifying actions that consider existing daily practices.

Interviews with experts selected by coordinators on key issues concerning the neighbourhood, tried to investigate not only their personal visions, but have deepened the individual discourses as a synthesis of collective paths.

This sort of anthropological exercise, based on neighbourhood walks, informal meetings, experts interviews, observation forms of place life, has been the starting point of the working week.

Some key elements have been identified and submitted to participants attention: identity, clichés, conflicts, relationship with the ship.

Identity of Cocuzzo neighbourhood is characterized by a negative stigmatization diffuse in the whole city of Potenza, which inevitably is reflected in a negative self-perception by the neighbourhood.

The cliché related to this social representation describes Cocuzzo neighbourhood as a dangerous, degraded, peripheral and inhospitable place. The improvement of neighbourhood connection with the central part of the town changed completely the whole geography of the city.

But what are the "common places" of the neighbourhood in the true sense? Do they exist? Are they used as such or are they abandoned and empty? The ship is one of them, but not used, nor valorised; on the contrary, in the collective imagination of the whole city, is one of the best expressions of neighbourhood degradation. Another key element of the neighbourhood is conflict: social mixing realized in the neighbourhood (Serpentone and Serpentino, social housing, are clearly separate from the other buildings) produced a clear socioeconomic and spatial distinction. All the inhabitants are joined against the ship, which can reconstruct a social fracture catalysing a sort of hate against the intervention.

The negative assessment shared by everyone is not based on its use, but it is the result of a sort of political imposition of an expensive and not understood intervention.

This reaction of strong rejection made the ship invisible, and therefore unusable, for local residents. A form of passive resistance to decision-making approach imposed from the top that, obviously, did not take into account citizenship needs. This passive reaction accentuates the expropriation (or its perception) of public space, already occurred with the ship construction.

These are questions and issues resulting from analyses and surveys before the workshop.

Walking, observing, meet crossing the place were useful in creating a specific knowledge of the place, where each participant was able to compare their own knowledge of the place with the representation provided by neighbourhood inhabitants.

After a walk in the neighbourhood it was asked to each participant to immediately provide a question to motivate his activities within the workshop, immediately establishing what were priorities and interests of each participant. During the workshop they were asked to update this question according to the newly received information.

The methodology of this workshop was based on an analysis of needs defined with the objects of the research and on several measures of territorial re-animation based on local specific skills.

Outputs of the performed analyses were the basis of a common reflection developed in the second part of the workshop, mainly based on project and self-construction activities.

In the case of Serpentone, also temporary interventions, in the construction of small artefacts, that often characterize tactical urbanism approach, appeared to be incongruous and "out of scale". Geometric and symbolic-perceptive aspects have been considered. The first one compare small artefacts to the size of the ship and residential buildings, the second ones analyse in which context these interventions can be useful. In this case small artefacts are suitable only at proximity dimension, without having any impact on relations between the neighbourhood and other parts of the city.

It was therefore considered more useful to address workshop contribution towards forms of activation of ideas and proposals. A path of project-action-event has been activated, more incisive on awareness level than on space configuration and intentionally quite ephemeral.

Due to workshop shortness it would not be possible to build relationships with the residents, able to define a clear and comprehensive idea of shared expectations and projects.

The risk was to catapult in the neighbourhood one more unrequested "object" or "collection of objects", with a great probability to create the same rejection mechanisms that characterized the initial situation. Hence, the desire to work on "almost nothing" in physical terms and instead having a huge production in terms of imagination, open to new places interpretations which would generate persistent transformations in a slowest time. Two actions have been pursued: a concrete and immediate possibility of use the ship and its roof-garden, to build a new shared imagination. In the first case, a game mechanism has been adopted: the game of "It can". Engaging students attending neighbourhood schools, an attempt to reverse the prevailing attitude of complaint and dissatisfaction with the ship has been tried, in order to highlight all possible enjoyable things that it was possible to realize.

The result was a surprising, large and inventive list of opportunities, emerged at first with suspicion, gradually with a lively succession of proposals. Several opportunities have been immediately concretised, animating the park with games and meetings, which have demonstrated with simple and instantaneous concreteness, the possibility to live the ship and its garden, only apparently unfriendly.

Others have produced micro-places, made by workshop participants with the inhabitants, realizing basic gardening activities (selective cuttings of tall grass, pruning of existing trees and removal of dead ones, planting a precious Liriodendrum tulipifera, arranging seasonal flowering plants) and setting up spaces for games (a sandy track for bowls, a mini football, volleyball and basketball fields). The strength of these operations, although so minute, was to evoke a spontaneous mark of names: the Solarium, the Playground, the Ascent, the Well, the Mirador, indicated by signposts on original drawing, immediately become meeting places.

Fig. 3. Workshop participants guide students from school to the "ship"

Fig. 4. The Solarium, "space created during the workshop, attracts children and local residents

Fig. 5. Few games built by workshop participants were enough to bring the neighbourhood children on the ship

The second action has been working in the direction of a gradual removal of the suffering neighbourhood stigma. The ship was the emblem of "internal" stigmatization, generated by the same inhabitants. It was necessary then to try to exorcise the negative connotation, using game and irony. The ship appeared stranded. It was necessary to put it into a harbour, able to speak about effervescence, flurry of activity, relationships, coming and going, voices, departures and returns, loves, adventures, explorations. The port was supposed to be the whole neighbourhood, which consequently was imagined such as a swarm of ships and boats, arriving and departing from somewhere. Hence the idea of using a simple object came: the paper boat, too familiar and almost universal. Miniaturizing the ship was a way to tame it, have it friend, make it harmless, through a game. Thus, the cargo became a laborious and joyful place for paper boats production. Around the tables of work, a heterogeneous community of workers has alternated: workshop participants, neighbourhood children and their families, representatives of associations and local authorities.

The construction of the small paper boats became a ritual of the community: sitting around the tables, while their hands, more and more skilled, bent the paper, people talked about their stories, showing not only worries and discomforts, but also desires, concrete proposals, expertise to make available. During the workshop days, the cargo has hosted a permanent laboratory of small boats, stimulating, in the "distracted" repetitiveness actions, talks, secrets, stories and variegated forms of hospitality.

If you know how to make a paper boat it takes no more than a minute: with the workforce of participants, inhabitants, students, it has been possible to produce in a week ten thousand small boats with a really collective work.

Fig. 6. In the cargo of the "ship" workshop participants, neighbourhood residents and members of municipality administration building paper boats

Fig. 7. The small boats invade the roof garden of the "ship", defining paths that lead to the various activities undertaken

Thousands of small boats have animated Serpentone "harbor": in rows, more or less ordered, or in groups (as grouped in the race) invaded the garden, braving the strong winds of a September Sunday.

The small boats covered parked cars; simulated flower gardens; marked paths, indicating interesting trajectories to follow in the meadow; sailed in the fountain, where finally the water returned.

A further issue has been added: the ship cargo, a generous space for quantities of certain beauty never used except as improper deposit. At the port inauguration, the ship opened its cargo space and showed its shipment: an instant-exhibition of videos reporting the neighbourhood, realized by workshop participants, and wonderful images by the photographer Salvatore Laurenzana.

The exhibition marked the opening of the art centre N.Av.E. - New Expressive Adventures (coordinated graphic image within the same workshop has been defined).

4   Results

4.1   "Shape is Emptiness and Precisely the Emptiness is Shape"

The challenge of "Serpentone Reload" was to involve neighbourhood residents and to reverse the negative sentiment expressed in general by the city against the ship, which had, not without reason, a national importance for the inconsistency of the project. Hence the idea of the project to build a multidisciplinary path starting a participation process to build active citizenship for continuing the garden maintenance and the spaces below. Between September and November residents have organized a school of gardening for garden maintenance, with weekly appointments of shared work. An annual calendar, on the model of an horticultural calendar, has been produced and distributed among inhabitants, where every month is accompanied by useful recommendations on the activities to be carried out in the garden.

Workshop activities also allowed to focus the attention of the whole town on Cocuzzo neighbourhood. At the same time, inhabitants realize all the potential that a structure such as the ship would have, as a sort of point of interest for the neighbourhood and for the city.

Here, Sariputra, form is emptiness and emptiness is precisely form (Heart Sutra). This Buddhist quote is a good metaphor for the image of the implementation idea. For Buddhism vacuum means the absence of itself, not only from a spatial point of view, but also from a time point of view: this implies that each element and phenomenon of the internal or external reality, is not only interconnected with the other, but it is also provisional and interconnected with the provisional nature of the other elements and phenomena. It determines a structure that is relative, relational and, at the same time, transitory, impermanent.

In that "empty" context the idea of a space that should not have to tend to a fixity structure, but maintains its performative fluidity was experimented.

Continuing the Buddhist metaphor, the empty "quarry", was intended as figures background that show their contours only through the reciprocal action between them, interaction guaranteed and made possible by the same background. The potential of that "empty" space has been used not only during the workshop, but also as a result, some activities of the City Festival of Hundred Stairs, which deals with dance and performative arts, were made in N.Av.E..

Fig. 8. The ship cargo has become the stage for performances of "City Festival of Hundred Stairs"

The real meaning of this activity was to show some of the potential of N.Av.E and what could be the future: a performative place able to take many forms in relation to the dynamics of participation achieved. Several meetings and seminars have been held to compare this experience to important similar experiences in Italy and Europe. In order to continue the place narration, it was thought the creation of short-time workshops about cinematographic writing to collect the experience of stories made by inhabitants of Serpentone and transpose them into a movies script. A public discussion has been opened and several social or cultural associations, theatrical or musical groups, requested to help to navigate N.Av.E as a privileged place of their activities serving the neighbourhood and the city.

5 Conclusions

Serpentone Reload laboratory was an innovative experience made available to different involved actors:

- to decision-makers, because it offered the opportunity to experiment inclusive policy processes and has demonstrated the power of regeneration of minute actions, in short times with few resources;

- to participants, young designers that have directly verified the possibilities to build a project that actually has not built any artefact, nothing tangible, measurable and quantifiable, understanding that this way is not a waiver of the project, but on the contrary it is an intentional amplification;
- to inhabitants, who accepted to question their refusing positions and to trust in themselves and their landscapes.

Many projects of public space share the tension to affirm opportunities and destinies in a very wide repertoire of co-creation.

In the same way Serpentone Reload, calls the project to reflect on their responsibilities and their own instruments with a reflection, factual and active, on public participation, often devalued.

Participation is devalued when it is confused with the consensus. When decision makers and designers abdicate their duty of projection of the future assuming roles of mediators that collect and nourish the wish list of their voters or customers. Experiences such as Serpentone Reload show that participating means sharing skills and points of view, finding solutions that are not accomplished but that continually become, taking place through time, modelling itself on the living body of the city.

Acknowledgements. Authors are grateful to Salvatore Laurenzana for pictures and to Mimmo Nardozza for workshop organizational coordination.

Authors are grateful to all workshop participants, Sara Adesso, Federico Amato, Giorgia Botonico, Bernardo Bruno, Antonio Canosa, Simone " lovetu" Cortese, Giuseppe D'emilio, Simona Di Giovanni, Francesco Fazio, Michela Galletti, Serena Lorusso, Mariangela Meliante, Rocco Morrone, Carmen Peluso, Daniela Policriti, Lydia Postiglione, Mariangela Russo, Alberta Caterina Santagata, Chiara Terranova, Elenonora Vaccaro.

References

1. Bravo, L., Carmagnini, C., Matityhou, N.: Lighter, Quicker, cheaper: towards an Urban Activism Manifesto. XXVIII Congeresso nazionale Istituto Nazionale di Urbanisitica, Salerno (2013)
2. Calvino, I.: Le città invisibili. Mondadori, Italia (1996)
3. Chase, J.L., Crawford, M., Kaliski, J.: Everyday urbanism. Monacelli Press, New York (2008)
4. Crosta, P.L.: Il territorio è l'usoche se ne fa. Franco AngeliEdizioni, Italia (2012)
5. Lydon, M., Bartman, D., Garcia, T., Preston, R., Woudstra, R.: Tactical Urbanism, short termaction ‖ Long term change, vol. 2 (2012). www.issuu.com/streetplanscollaborative/docs (retrieved on October 3, 2013)
6. Harvey, D.: The right to the city. New Left Review **53**, September-October, 2008
7. Murgante, B.: Wiki-Planning: the experience of basento park in potenza (Italy). In: Borruso, G., Bertazzon, S., Favretto, A., Murgante, B., Torre, C. (eds.) Geographic Information Analysis for Sustainable Development and Economic Planning: New Technologies, pp. 345–359. Information Science Reference IGI Global, Hershey (2012). doi:10.4018/978-1-4666-1924-1. ch023

8. Lorusso, S., Scioscia, M., Sassano, G., Graziadei, A., Passannante, P., Bellarosa, S., Scaringi, F., Murgante, B.: Involving citizens in public space regeneration: the experience of "garden in motion". In: Murgante, B., et al. (eds.) ICCSA 2014, Part II. LNCS, vol. 8580, pp. 723–737. Springer, Heidelberg (2014). doi:10.1007/978-3-319-09129-7_52

9. Pontrandolfi, P., De Fino, C., Guida, A., Guida, F.: La riqualificazione possibile. Periferia, metodo, progetto. Il Contratto di Quartiere della zona occidentale di Potenza, pp. 4–368. Ermes Edizioni, Potenza (2000)

10. Murgante, B.: Le vicende urbanistiche di Potenza. EdituiLErmes, Potenza (2005) ISBN 88-87687-70-6

11. Maynard, M.: Light, Quick And Cheap: The Big Shift in Urban Planning. Forbes (2013). http://www.forbes.com/sites/michelinemaynard/2013/10/23/light-quick-and-cheap-the-big-shift-in-urban-planning/

12. Silberberg, S., Lorah, K., Disbrow, R., Muessig, A.: Places in the Making: How placemaking builds places and communities. Department of Urban Studies and Planning (DUSP). MIT (2013). http://dusp.mit.edu/cdd/project/placemaking

13. Rube, K.: Lighter, Quicker, Cheaper Remaking Streets for Better Outcomes. Project for Public Spaces (2013). http://www.pps.org/

14. Sassano, G., Graziadei, A.: Cambiare il volto della città a partire da piccole (e concrete) azioni. Il Quotidiano della Basilicata (2015). http://www.ilquotidianodellabasilicata.it/news/cronache/733877/Cambiare-il-volto-della-citta-.html

15. Las Casas, G., Pontrandolfi, P., Murgante, B.: "La riqualificazione del Vallone di S. Lucia a Potenza: l'efficacia di programmi ed interventi per il recupero di un'area marginale della città" atti della XII Conferenza internazionale. Vivere e Camminare in Città. Le periferie. Bergamo - Brescia, 9-10 giugno (2005)

16. Murgante, B.: La barca di via Tirreno va: Dal contratto di quartiere alle polemiche di questi giorni. Il Quotidiano della Basilicata (2007)

17. Las Casas, G., Argento, R., Pontrandolfi, P.: Un'esperienza di urbanistica partecipata-comunicativa lungo il fosso di S. Lucia a Potenza: i primi passi del laboratorio urbano. In: Contributo dell'INU al Rapporto italiano per la Conferenza ONU Habitat II. Istanbul, giugno, Presidenza Consiglio dei Ministri (1996)

18. Murgante, B., Danese, M.: Urban versus Rural: the decrease of agricultural areas and the development of urban zones analyzed with spatial statistics. International Journal of Agricultural and Environmental Information Systems (IJAEIS) 2(2), 16–28 (2011). doi:10.4018/jaeis.2011070102

19. Percoco, A.: Quale ben-essere per la nostra città? Il Quotidiano della Basilicata (2014). http://www.ilquotidianodellabasilicata.it/news/politica/732604/index.html

City Visions: Concepts, Conflicts and Participation Analysed from Digital Network Interactions

Maria Célia Furtado Rocha[1], Gilberto Corso Pereira[2], and Beniamino Murgante[3(✉)]

[1] Federal University of Bahia, CAPES (proc. n. 11527/13-7), Salvador, Brazil
mariacelia.rocha@prodeb.ba.gov.br
[2] Federal University of Bahia, Salvador, Brazil
corso@ufba.br
[3] University of Basilicata, Potenza, Italy
beniamino.murgante@unibas.it

Abstract. The present article investigates the possibility of understanding the different city visions and conflicts arising in digital interactions in network of movements that discuss the use of urban space. We assume "expanded participation" as a way to include new social arrangements, present in cultures that use networked communication as a means to share visions, values and produce meanings and actions collectively. Digital network interactions make possible a form of participation based on personal expression of individuals interacting in social networks. The study analyzed the case of the movement Parco del Basento in the city of Potenza, Italy. The movement grown using as the main platform for the dissemination and discussion of ideas one Facebook page. The analysis of posts and their interactions was based on Social Network Analysis and Text Analysis Tools to investigated convergent positions of members on topics such as Urban Space and the city visions that emerge from discussions. Finally we analyze possibilities and difficulties found in the usage of this kind of tool.

Keywords: Digital social networks · Semantic networks · Civic participation

1 Introduction

Cities are complex organizations, portions of territory that polarize the geographic space and concentrate wealth, culture and people. They enable innovation and the economic and technological development. We live in an increasingly urban world and the expansion of cities has been the subject of numerous conflicts of interest between residents, administrators, planners and / or real estate developers. To plan the growth or evolution of a city involves the ability to anticipate and avoid problems associated with its development. Urban planning is a sophisticated instrument of political power because in essence determines how city dwellers can occupy the urban space, decide what activities can happen (and which can not) in certain areas.

Any intervention in the urban space will transform the use and value of this space, which results in a differentiated appropriation of land value produced by the intervention or the worth alteration that will be caused by the change in its use. Any decision of Urban

© Springer International Publishing Switzerland 2015
O. Gervasi et al. (Eds.): ICCSA 2015, Part II, LNCS 9156, pp. 714–730, 2015.
DOI: 10.1007/978-3-319-21407-8_51

Planning determines ultimately, who among the inhabitants will have advantages or dis-advantages that are distributed asymmetrically affecting residents differently. Logan and Molotch [17] consider that local conflicts on urban growth are central to the organization of cities. The central conflict is between residents, who use urban space to satisfy essential needs (live and work in the city) and entrepreneurs seeking financial returns, which is achieved generally intensifying land use (more buildings, verticalization, etc). For this reason, it is not surprising that the participation of society in the municipal planning processes are centered on proposals and action plans seeking to influence all steps of the process of decision-making of any kind, requiring broad access to information about operations, plans or places. The different stages of technological development always reflected in the urban space and in the form of cities. Transport technologies and the increase in speed of displacement led to greater spatial dispersion, the appearance of the phone reformatted the city allowing, among other things, the separation of factory and offices. We can consider the social organizations and current socio-technical networks as systems and environments with formations of different scales and dependent on digital technologies, which generate a range of spatial practices, organizational and interaction not previously prosecuted [15]. Today citizens are able to register and publicize their impressions and these can be anchored in the city itself. Planners should recognize that now our urban experience is not only influenced by urban form, but in different ways and forms of communication with which we interact daily. Social interactions of everyday life are in fact permeated by digital technologies, since the interactions between social networks are enhanced and contribute to organizing everyday urban life. It is the role of a contemporary urbanism exercise the possibilities that arise with current technologies. This paper investigates the possibility of understanding the different city visions and conflicts arising through data collected in digital interactions in network of movements that discuss the use of urban space. In this work, we assume "expanded participation" as a way to include new social arrangements, present in cultures that use networked communication as a means to share visions, values and produce meanings and actions collectively. Digital network interactions make possible a form of participation based on personal expression of individuals interacting in social networks. The study analyzed the case of the movement Parco del Basento in the city of Potenza, Italy. The movement was framed and grown using as the main platform for the dissemination and discussion of ideas one Facebook page as described by Murgante [23]. The analysis of posts and their interactions used text analysis tools and investigated the convergent positions of members on topics such as Urban Space, Conflict Visions and Politics and the *Polis*. Data was collected and processed as part of the research developed for the Maria Célia Rocha doctoral thesis at UFBa - Federal University of Bahia and with stage in UNIBAS - University of Basilicata, with support from CAPES.

2   Digital Network Interactions

Interactivity is a feature of the constitution of the texts on the Internet. The digital conversation is a textual type but conceived as orality. His way of writing is more flexible and informal. It is an oral interaction with use of graphical means, a "text spoken in writing" [22]. According Recuero [28] the terms interaction and conversation are not always synonymous. The language is adapted, modified and conventionalized to give talks dimensions of orality. For the author, the interactive exchanges between the actors in these environments have many similarities with oral conversation - fast and informal. Social networking sites have created new forms of circulation of information, led to the advent of new forms of conversation: collective conversations, asynchronous or synchronous, public and able to involve a lot of actors, which the author calls "network Conversation" whose most important feature is the spread between social groups through the connections between individuals [28]. Social networking sites are those that comprise the category of systems focused on exposing and publish the social networks of players [27]. These sites promote social contact, forge personal, professional or geographic connections and encourage weak ties (Van Djick, 2013), ties between people who know each other but they are not close friends, which, however, are critical to maintaining the cohesion of the social system [14]. These conversations, even when facing aspects of daily life, are likely to influence actions of the political system, contributing to the expansion of the arguments spectrum to the resonance of matters of interest common to the network and even for scheduling issues in mass media and institutions of the political system [6] [18]. Many studies have been devoted to understanding the political behavior based on interactions in digital social networks. Some of them are based on interactions collected in the Twitter microblogging site, such as that performed by Small [31] on the possibility offered by the platform for participation and political conversation; the reflections made by Artieri [3] talks about how this and other social networking sites are integrated with the mass media forms and make the plural and potentially visible public sphere. According to him, these sites do not play the role of representing themes of society, but have a role in drawing attention to these issues, to communicate life experiences able to relate the uniqueness of the plan to the collective, a political action consisting essentially of "contagious ". Many of these studies refer to the concept of social network, which first appears in a study of the principles of social stratification in a Norwegian island held by the British anthropologist A. Barnes, published in 1954. The anthropologist has established the hypothesis that all the inhabitants of that island would be connected by shorter or longer chain knowledge. On the other hand, Mercklé [20] says, Milgram [21] may have been the first to make the effort to empirically demonstrate some of these intuitions, as evidenced by his experience of "small world". He created chains between persons elected randomly from a national population and demonstrated that the individual is not alienated and separated from the rest of society: all are, in a sense, delimited by a tightly woven social fabric [21]. The notion of social network has expanded to designate sets of relationships between people or between social groups. It involves a set of methods, concepts, theories, models used in sociology as well as in other social science disciplines (anthropology, social psychology, economics, etc.) and considers as an object of study not the attributes of individuals (age, profession) but relations between individuals and the regularities they present, to describe them, to realize their formation and their transformations, analyze their effects on individual behavior. The proposal is to restore the individual behavior of the complexity of social relations subsystems that give them meaning and to which they give meaning (Merckle, 2004). A social network, in this view, can be defined as consisting of a set of social units and relationships established between them, directly or indirectly, through variable length chains. The network is used here as a kind of contextual variable. The social units may be individuals, informal groups of individuals or more formal organizations such as associations, companies or countries. The relationships that establish the elements can also be of various kinds, including all sorts of verbal and gestural interactions. The analysis of social networks reassembles the individuals from the existence of strong links between them, ie in function of the observation of cohesion and density that they form sets. The study of these relationships and regularities that feature is developed based on models and methods supported by mathematical tools based on graph theory and linear algebra and it establishes and institutionalizes its own domain (Merckle, 2004). The graphs provide a graphical representation of social networks, facilitate viewing and enlighten some of its structural properties that impact on their behavior (Barabasi, 2012). On the other hand, the graph theory provides an arsenal of concepts, theorems and algorithms by means of which the graphical figurative goes beyond simple representation to allows a mathematical treatment of new knowledge (Merckle, 2004). Concepts of graph theory seek describe the properties of nodes - its centrality - the arrangements relating between them - distance and connection paths between nodes - or global network properties, in particular through the density and connectivity notions. However graphs with dozens of nodes and hundreds of arcs/edges may become unreadable. The linear algebra and matrix calculus, another area of mathematics, will help overcome this barrier. Expressions such as network relationships, density, click popularity, isolation, prestige could then receive an operative mathematical definition, which allows to build indicators, measure them and then empirically test the hypotheses or check the propositions that could be wrong because of its largely metaphorical content (Merckle, 2004). General contributions to the study of networks will be brought by Watts and Strogatz [35] and Barabási and Albert [5], among others. The first claim that many real-world networks, social or other, behave like networks "Small World"; the second will describe properties of real complex networks of large size, whose growth occurs according to preferred and non-random connections, as predicted by Erdős and Rényi. The theory of random networks from Erdős and Rényi, introduced in 1959, says that if the network is large almost all nodes have approximately the same number of links in spite of completely random placement of links. But for Barabási and Linked [4], in real networks there is not a type of feature node, there is not an intrinsic dimension of these networks. In many real-world networks, the degree distribution, ie the number of the probability distribution of connections of nodes throughout the network [25] [4], typically has a long tail: most nodes have less than half the average number of links while there is a small fraction of hubs that appear much better connected than the average [34]. The degree distribution follows the power law free scales. Its characteristic is that many small events coexist with some very large. This indicates that in real networks, there is not a type of feature node; there is not an intrinsic dimension of these networks. Combining growth and preferential binding, Barabási and Albert [5] explained the hubs, nodes more connected. Power laws that they formulated in mathematical terms refer the notion that few large events lead most of the action, and the scale-free topology shows the existence of organizing principles acting at each stage of the network formation process [4]. In turn, the findings of Watts and Strogatz [35] allowed the identification of a universal class networks, a family of networks that share some properties (small network and dense clusters) regardless of their individual characteristics. With this metric and idealized models to solve social networking issues can be used in other disciplines [34]. These and other studies in the sometimes called "new" science of networks contributed ideas, introduced techniques and metrics, either by borrowing or "rediscovering" what's already time mathematicians, economists and sociologists developed. They contributed to the interdisciplinary synthesis of new analytical techniques, which has a large computing power and unprecedented amount of empirical data [34].

3   Text Analysis

Following this trend, the use of these techniques and metrics will be incorporated into the study of abstract relations between discrete entities, such as extracted from the written words that make up what Watts [34] calls "symbolic" networks, as opposed to "interactive" networks, which links describe tangible interactions capable of transmitting information, influence or material. In the first case, he warns, often is not clear how network metrics such as grade, size and centrality of the way, should be interpreted from the point of view of its consequences for certain social, physical and biological processes. This observation Watts [34] serves as a warning to those who seek to obtain the meanings conveyed in texts with use of these techniques. Moreover, according Merckle (2004) characterization of a network by observation of their relations, by observation of its form, must include the definition of relations and therefore the attention to its content. The question of the content can not be removed from the analysis of structures without the risk of not understanding the complexity of individual situations. This work is based on networked word representations and metrics of graph theory to the discovery and clarification of meanings conveyed in digital conversations. Clues provided by the textual analysis and visual analysis of words networks serve for the capture of the main meanings attributed by the Parco del Basento group to urban space in whose defense engages. The operation of political and cultural context and expressed values and motivations in networked conversations help to reveal answers to questions like: what are the characteristics of urban space required? what are the preferred actions to get it? Text analysis refers to a set of techniques for the study of social communication that allows access to meanings, values, social norms through language and symbolic expressions conveyed in texts and images in their various media [26] [33]. Carley [8] and Alexa [1] cite numerous techniques of textual analysis and say that, in the social sciences, the content analysis predominates. Alexa [1] defines text analysis as a field of practice, and the content analysis as a set of techniques to provide text interpretation. To Popping [26] the terms text analysis and content analysis are interchangeable: This type of analysis allows the researcher, from text and its context, make explicit all aspects of social processes, to better understand and make valid inferences and reproducible over them. The text analysis flourished during the Second World War, being used to answer questions such as: how often a word or specific term occurs in a text? how many articles examine specific topics? Then focused on the valence analysis to observe the positive or negative connotation given to issues and, in a third time, between 50 and 60 of the last century, pointed up the weights themes and words to indicate the relative intensity assessments and assertions [26]. Remained descriptive method until around the late 1900. The available statistical techniques, especially association analysis and correlation, allowed to investigate associations (or "contingency") between textual features. In the late 1960s researchers began to use the computer to sort and count words and phrases in "categories of meaning" in order to codify the terms without human interference, according to a predefined coding scheme. According Popping [26], the fact that the words and phrases have been tabulated, categorized, correlated and interpreted, but taken out of context, led at that time, the loss of confidence in using the method. However Schorott and Lanoue (1994, cited in [26]) found, for the period 1977 to 1986, a major expansion in the number of articles published using the text analysis techniques, particularly in research in communication; in the 1990s there was strong growth in its use in Sociology.

4   Semantic Networks

The networked interaction was analyzed using complex semantic networks. We try to characterize relationship between words used in the posts on Facebook pages from both groups. This work analyzes words and their relations forming semantic networks. From this point, by considering pairs of elements, it is possible identifying words with higher centrality metrics – an indication to reach main ideas expressed in the content of virtual posts and discussions.

Many text analysis computer-assisted emerge around 1980, with a preponderance of traditional text analysis, called thematic analysis. The testing of hypotheses is through research the frequency of occurrence of words in phrases categorized in a number of subjects. Association analysis can also be done by a simple quantification of the occurrence and co-occurrence of words in the text, without restricting them to predetermined subjects [26]. From occurrence of measures, correlations between the concepts can be computed; from the correlation measures between concepts may be applied other techniques such as cluster analysis, factor analysis, path analysis, etc. [26]. In the analysis of thematic content, you cannot answer several research questions using the same data as a consequence of that van Atteveldt [33] calls the semantic gap. This gap is due to the level of abstraction required by the researcher confronted with the fact that the words in a text often refer to actors and concrete issues, which requires the encoder the need to make judgments as to whether the actor or question are included in the concept. The other gap is due to the fact that, since categories and variables unstructured - for example the frequency of specific patterns of relationships - are used to measure complex phenomena, the researcher is defied to convert complex texts in simple scores which, again, requires interpretation. These difficulties, he says, require data to remain as close as possible to the text of the semantic point of view, a necessary condition for reuse or combination of databases. Semantic analysis of texts, in turn, is a type of content analysis that expands the possibility of answers to research questions. There are also counting concepts, but unlike the thematic analysis, there is still counting the relationships between concepts. The coding of these relationships allows one to test hypotheses by statistical analysis of a data matrix, and the ambiguities of natural language are avoided by use of semantic grammars, indicating which relations can be found and allow to consider the context of the phrases [26]. Data of natural language suffer a translation process for structured isomorphic forms that serve the inference mechanisms. It are then transformed by processes of abstraction in order to preserve and reveal entities and relations explicit or implicitly represented in the input data [11]. Each sentence is then coded according to the intended meaning - be it a description, be it a evaluation. The construction of the computer aided analysis corpus therefore involves conversion, selection of groups of words considered analysis units and their classification according to a scheme that can be performed a priori or emerging process. In the first case it assumes a confirmatory perspective, while in the latter assumes an exploratory view [1] [26]. In the confirmatory approach, words are defined independently and prior to any textual coding, while in exploratory approach the words are taken from texts. The choice of terms and relationships to be extracted requires prior knowledge of the field and the research itself context, the prerequisites to infer answers to questions. This includes the way the words are categorized and, together with the choice thereof, provides a way of interpreting the text and, therefore, the content analysis context [33]. Once connections have been established between concepts, one can build networks of semantically related concepts and expand the analysis of the relations between concepts in addition to their co-occurrence in the sentence. In the semantic analysis, the relationship between pairs of concepts is exploited. In the network approach, the information contained in the declarations made up of two related concepts are amalgamated and analyzed in terms of links, chains and concepts organizations [26]. This is exactly what makes the Network Text Analysis (NTA) which has originated in traditional techniques to index relations between concepts, syntactic grouping of words and hierarchical bonds / non-hierarchical words. It is based on the assumption that the language and the knowledge can be modeled as words networks and their relationships.

Semantic networks are structured representations of knowledge aimed at understanding and development of inference [11]. Knowledge is represented in a graph structure in patterns of interconnected nodes and arcs [32], or, in mathematical language, patterns of vertices connected by edges (for undirected relationship) or arcs (in the case of directed relationships) [16]. In the language of computer science, triple linking object-connector-object together form a network of objects, consisting of nodes and their links [33]. In semantic network nodes are concepts - abstract representations of ideas, thoughts, knowledge and meaning units - and links established a significant association between pairs of concepts. Once built, the representation of the network is then asked to answer the research questions. Concepts and their connections can then be characterized according to their position in the network. In the 90 Carley [8] states that analysis that went beyond the visual inspection began to consider connectivity between concepts by examining its density (number of links that connects with other concepts) to make inferences about their communicative prominence. In his study of the connectivity of the symbols based on semantic networks, the author deals with two more dimensions: consensus - once symbols to operate as such, must be connected to widely shared historical inferences - and conductivity that considers, the ability from a concept to function as a passage for the many ideas that flow on network paths. At that moment, the way interpreted in the measurement of focal concepts in meaning networks - a representation of messages in defined contexts. The higher the density, the greater the extension concept, and the more individuals share links and concepts related to the focal concept, the greater the social consensus. In turn, the conductivity of a concept increases as grows the number of paths connecting it to other focal concepts, which places it at the crossroads of multiple network paths [8]. Thus, the author has implemented measures related to the topology of the semantic network to emphasize the connections between concepts such as representation of the meaning of messages. Text analysis will incorporate measures developed in social network analysis to study groups, particularly in the analysis of mental maps built from the perspective of its members. Differences in the distribution of concepts and relationships between them come insights about similarities and differences in the content and structure of the texts. For making it possible to examine the data graphically and statistically, map analysis would allow the researcher to get close to the text and use the inferential ability of quantitative techniques to increase qualitative techniques [8]. As it contemplates the content analysis, the results can also be analyzed from a traditional analytical content perspective. The addition of relational information, however, allows the researcher to consider not only changes or differences in the floor over time, but changes or differences in meaning. Carley [8] proposes a comparative approach by graphic depiction of mental models of individuals and social groups, represented as networking concepts. The union of all the statements of members of a group would comprise the knowledge of the group about a topic in particular, as the intersection of the statements of all members of a group show concepts that are consensus in the group. In this method, the statements made explicit by the sources are extended by the knowledge of the context, which are the basis for inferences about the conceptual relationships [26]. Carley and Palmquist [9] used Network Text Analysis to build mental models, with software aid. To represent these models, the statements (made up of concepts related to each other) are displayed as networks of shared concepts. After identifying the concepts that will be used in the encoding of texts and after defining the relationships that can exist between them, the information is extracted and encoded as a set of statements. Finally, here is your disposal in a mental representation of the model in the form of "map" for presentation and statistical analysis. Diesner and Carley [12] describe similar process to condense data into concepts and link them according to statement formation rules. The rules for the preprocessing of text and for the formation of declaration, together form the encoding scheme. To support this process AutoMap (http://www.casos.cs.cmu.edu/projects/automap/) are used, software developed by the Center for Computational Analysis of Social and Organizational System - CASOS, at Carnegie Mellon University. During the pre-processing, there is the elimination of words that do not convey content, for example, proper names, conjunctions, articles and prepositions, depending on the case, and the translation of specific words in more general concepts. The authors refer to these processes as "exclusion" and "generalization". Deleting words is implemented in AutoMap through a list of words; generalization is implemented with the use of thesauri. The declaration of formation is implemented in software using the window method proposed by Danowski [10]. Thus, in AutoMap the resulting statements of the relationship between concepts are established based on a series of connection between adjacent concepts, according to a defined window length. The researcher, therefore, should be aware that their choices will affect the results, even before getting involved in the coding and analysis of texts. She also prevents to the fact that, if you choose to locate implicit knowledge beyond those explained in the text, it will require more than generalization, the effort to determine a set of concepts which other concepts are missing, which makes it difficult the automation of the process, because such choices require semantic and cultural interpretations of the data [8]. In any case, the presence of these words does not guarantee having the same meaning, or because the definitions attributed to them are different, either because the same word is placed in syntactically different positions, or because the same word is used in several semantic contexts.

5   Concepts Network: Structure, Content and Meaning

Paranyushkin [24] also starts from the idea of the existence of a constant process of interaction between concepts, which come into relationship and produce meaning. Meanings may be totally different even with the use of the same vocabulary. The author assumes that the closer together the words appear in the text the more related they are. Therefore, to represent texts, he chooses to use words as nodes of the network and its proximity and relationship between them.

According to him, the visual representation of text and graph can help improve understanding of the text by providing references to the underlying information retrieval, represent the strength of association between words groups, contexts and dynamics of multiple discourses present there. The author uses graph analysis tools for the detection of interrelated concepts communities and to find the most influential concepts.

The study networks with tools of graph theory and theory of complex networks can be a powerful aid to analysis of text, on the other hand its use often moves understanding of their contents for analysis of structure and dynamics of growth. This bias coincides with the assessment made by Merckle (2004) about the trend assumed by those who incorporated theory of resource graphs in studies related to the field of social studies. Many years after Barnes made use for the first time the notion of "social network", Merckle (2004) found that the dominant approach in sociology of social networks, the structural analysis, was marked by a strong tendency to simulation experimental, the development of abstract models of relational systems and the use of a hypothetical-deductive logic. For him, there would be a strong theoretical and methodological opposition between a "comprehensive" approach, based on analysis of egocentric networks, inherited from the anthropological tradition, and an approach "explanatory" sometimes tempted by structuralism, matter of sociometric tradition embodied by the structural analysis Anglo-Saxon. This same approach is reflected here in many of the present authors in this brief review of semantic networks, especially in the application of the theory of complex networks, dedicated to the study and comparison of network topology. Based on this, you can identify different application scales of these studies. Possibly comparative studies of network topology and its growth dynamics are more suited to the identification of major emerging themes. In particular taking a large database, the exploration of hypotheses based on standardized terms. Studies coming closer of the contexts in which concepts are applied (which involves many returns to the original text to verify the direction in which they are applied) allow further progress towards the variety of views and the identification of semantic fields prevailing in the debate. These studies benefit greatly from the use of centrality indices associated with graph theory. The challenge is to deal with the large amount of digital data available bearing in mind the multiple meanings for which the semantic network provides clues. Data visualization emerges as a research field just to account for this challenge. In turn, the analysis of semantic networks seems to come now being taken up as a tool for clarification, evaluation and synthesis of views in the discussion of solutions to urban problems. Some recent studies illustrate this aspect, including which proposes Social Media Geographic Information / Analysis [7] in support of the planning methodology and uses what it calls textual analysis space-time; the proposal for a framework that includes capturing the cultural significance of the landscape based on values expressed by its inhabitants (Cerreta and others, 2014); the study of the opinions of people with respect to the heritage-based message posted on Twitter (Monteiro and others, 2014) and studies of social interactions in network groups engaged in degraded urban areas [18] [30]. Finally, explore and discover the content of conversations in online social networks can help lighten "a routing area of individual and collective actions", since it falls within the size of the production of meaning. From this perspective, the case of the Parco del Basento is analyzed here.

6   Analysis of Conversations Networks

Networks established were analyzed, in general, from the discussions in the Parco del Basento group and comments on the themes framed Urban Space, Views in Conflict and Politics and *polis*.

Data were collected using Netvizz application [29]. Semantic networks were generated using the AutoMap v3.0.10.36 based on the post's comments. For each of the posts, the comments were gathered in a single text, the authors' names were eliminated, as well as dates and symbols. End points were added to judgments when omitted, as well as spelling corrections were made. Articles, conjunctions and prepositions were excluded but always after checking whether their elimination could change the meaning of the sentence. Adverbs of manner, place and questioning were generally excluded, most particularly when it came to generate the theme networks Urban Space. Semantic networks were generated with the method of windows on the AutoMap and viewed with the ORA NetScenes 3.0.9.9j version, trial version. The coding choices for extracting all nets were as follows: Unit: sentence; Directionality: bidirectional; Window size: 2. The components of each network were visualized with the ORA. Component is a subset of nodes in a network, so that there is a path between any two of its nodes, but you can not add any other node that has the same property. If a network consists of two components, a single link placed correctly can connect them. This connection is called the bridge. In general, a bridge is any link that, when cut, disconnects the graph [4]. It was decided to focus the analysis on the largest component of the network and eliminate peripheral concepts (pending concepts, according to the terminology used by ORA). The algorithm Newman for detection of communities helped identify concepts clusters more strongly related to denote topics of interest in the conversation. Visual inspection of the networks and their communities gave rise to numerous returns to the original text and the new refinements by processing software, until we get a simple to understand network in visual terms. Those generated semantic networks were consolidated by theme. Because of space limitations this text will deal only with theme "Urban Space". In this case we generated a network in order to obtain a global view of the group on a bill creating the park itself, supported by a vision of the qualities attributed to the resulting space of urban evolution. The terms most frequently use in relation to urban space were selected. With the help of lists of concepts of each network, we sought to identify words, referring to similar aspects, there with higher frequencies on other networks. We tried to unify terms with similar meaning. New generalized networks were then obtained. From this development began to emerge categories of urban space and actions on it. Initially, the concepts were classified as actors, initiatives, actions, elements of space, space qualities, qualities of actions and values and motivations. It was found that often the term fits into more than one category, but for simplification effect was chosen by one of them. In the step of generating and revising the network, the concepts are organized into semantic domains, or in similar meanings fields. Then, we identified which keywords best represent each of these semantic fields, thus giving rise to the analysis categories. The Urban Space theme network was generated with the ORA and then exported to Gephi 0.8.2 beta for viewing and further treatment. To categorize concepts was taken as a starting point emerging categories on the theme Urban Space, which were widespread and defined in table 1. Based on this categorization took place a visual analysis of network partitions. This allowed to know the group and the representative vision of who does what, how and where, with what motivations and results. Figure 1 shows the network as a whole, formed in six (6) communities. The colors correspond to the communities and the node size depends on the degree of each concept. The main interests of the group are revealed by the most central concepts in the network, shown in Table 2. Centrality value was calculated as the centrality indices ranking. The most prominent is the community that has the largest number of concepts located between the ten (10) on the 1st ranking positions in the case, the community 4. This community has 54% of the best-positioned concepts in ranking and contains 23.68% the total number of concepts in the network. In fact, it appears that the "community 4" plays a central role in the network. The subgraph corresponding to it shows the group's motivations associated with the creation of the park ("riscattare", "area_ex_cip_zoo", "area_verde", "cittadini"), resources mobilized for this ("petizione", "cittadinanza_attiva", "partecipata" in addition to the characteristics of the movement ("bataglia", "portare_avanti"). the movement, therefore, focuses on the defense of the project of creation of the Parco del Basento a specific area of the city and aims to rescue this space for a diverse use of it was intended during the process of urban development in the mid of the previous century, it is opposed to the real estate speculation process and argues that the area should be used as a green area by the citizens.

Table 1. Analytical categories

Category	Identifier	Definition
Actor	AT	Every individual, group, professional category or role played by individuals, entity or institution, whether declared as agent or patient actions, be it an object of evaluation or judgment.
Action	AC	Every physical or communicative action that has occurred or may occur, with possible impact on actors, space, institutions of public administration and social institutions of any kind, including politics, understood as the activity aimed at establishing the rules and to make the decisions to make possible the coexistence between groups of people (Bobbio, 2010).
Event, Institution, Initiative or Resource	EIR	Any initiative undertaken or idealized, either in physical space or in forums set up under the social relations, and their used or available resources to achieve them, including spatial element (area, area, city, place) referred.
Value, Quality and Motivation	VQM	Judgment assigned to any element of other categories of analysis, including with respect to the time when the action takes place.

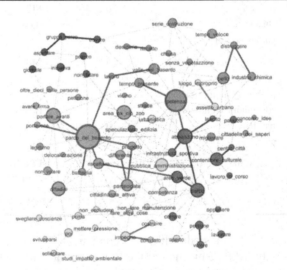

Fig. 1. Urban Space Network (Source: research data, Rocha, 2015)

The term spokesman ("portavoce") belongs to this community and appears on the large number of signatories of the petition in favor of the creation of the park, more than eleven thousand people at that time. Not connected to the mention of the group - which belongs to another community (identifier 0), which does not even have a term in the rankings above. The movement affirms itself as a form of active citizenship, participatory and different. The idea of participation is here associated with own architectural design idea of the park. The second most prominent community is number 5, which deals with the abandonment of both sports and cultural infrastructure as the existing parks and green areas, and shows the desire of people to use the green area to run. The word "abbandono", the third place in the ranking of centrality, has 10 degree and succeed "potenza" specifically in terms of this index. At this point, equals the expression "portare-avanti," which is the most prominent community network (the community 4) and deals with the willingness to carry forward the "battle" for the park. The mention of the city of Potenza is located in the community 1, which presents the characteristics of urban evolution and Basento Valley area. Here appears a protagonist of the changes in city characteristics: the chemical industry. According to group members, the city quickly would have turned into a city "closed" (referring to the large buildings that do not allow the view of the landscape), devoid of vegetation. An overview of the most important communities (4, 5 and 1), shown in Figure 2 below shows how the Parco del Basento group sees in his initiative a way to rescue the urban space from the mobilization of citizens of Potenza: the city for use by citizens.

Table 2. Urban Space - Core Concepts (Source: research data, Rocha, 2015)

Position	Concept	Centrality	Category	Community
1	parco_del_basento	40	EIR	4
2	potenza	36	EIR	1
3	abbandono	32	VQM	5
4	portare_avanti	25	AC	4
	area_ex_cip_zoo	25	EIR	4
5	cittadini	24	AT	4
6	pubblica_amministrazione	20	AT	2
7	contenitore_culturale	11	EIR	5
8	riscattare	10	AC	4
9	parco	8	EIR	5
	area_verde	8	EIR	5
10	partecipata	5	VQM	4
	bello	5	VQM	1

Potenza is represented by the node with the second highest number of relationships (degree 16); rivals only with the mention of the Parco del Basento (18). But while the concept occupies the second position in the ranking of centrality and join the second largest community (22.37% of the concepts of network), it is not one of the two most prominent communities in the sense used here, with the community 1 has only 15.38% of the concepts presented in the ranking of centrality. The exploration of the neighborhood (depth equal to 1) the main network node - "parco_del_basento" - reveals the distrust with the government, which is blamed for the lack of maintenance of urban infrastructure and is the target of group pressure. Taking the point of view of the categories, it appears that most - six (6) - the 13 concepts ranking of centrality refer to EIR (Event, Institution, or Resource initiative). In it are framed the initiative of the Parco ("parco_del_basento"), its deployment area ("area_ex_cip_zoo"), the city itself ("potenza") - which falls under subcategory space element (EE) - and equipment / cultural buildings ("contenitore_culturale") in that area. The EIR category appears in the centrality ranking with 46.15% of the concepts; on the network as a whole, it appears with 17.07% of the concepts.

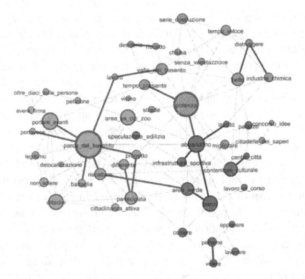

Fig. 2. Urban Space - Top communities (Source: research data, Rocha, 2015)

7   Final Considerations

The aim of this research was to capture the meaning assigned today to the urban space as a common good for groups present in digital social networks and the forms they assume to defend him. To identify the required space, with its main characteristics and uses, it was decided to apply social network analysis techniques at extracted semantic networks of conversations groups with Facebook profile that led forth movements for regeneration of degraded urban areas of the city of Potenza, southern Italy, particularly in 2012. Conversations and key terms used in them according to the ranking of the most central concepts were identified, according to a combination of centrality indices and neighborhoods of these nodes. A categorization of the concepts helped put highlighted actions, resources and qualities mentioned to initiatives and activities of the groups, as well as to the city and other aspects of space.

The assumption considered that social networking platforms have been used by groups to promote a collective understanding of the required quality of urban space common to all and on how they should act to defend it, regardless of the different views of democracy that may exist between its members. Despite their political beliefs and individual views, they want to participate and to influence the decisions that touch public life, particularly the public administration. They want to be actors, act materially and through words, publicly reveal their intentions. As if they were citizens of a polis, in the space where they share words and actions establish a kind of organized memory [2]. Network analysis, based on topology and graph theory, is attentive only to nodes and neglects the significance of the "holes" that ultimately make up the mesh [13]. For networks "symbolic", these holes represent silences, pauses, breaks, gaps own spoken language, full of expression and meaning, or are the product of the researcher's choices. One can not ignore the restriction arising from the need to simplify the network (and helping to word processing with computer assistance) through the construction of statements by the relationship between concepts from a distance fixed within the limits of a given text unit. All these strategies, so to speak, "flatten" the text and lead to impoverish its interpretation. It must be clear that semantic networks are not more than the modeling products by the researcher, which uses the simplification to unveil meanings, and so this should refer often to originally collected data. The decision to combine several network node centrality index also helps to put the concepts into line, concealing their functions in speech. This is shown particularly in the case of concepts with high centrality (betweenness centrality), that can connect communities with varying topics of interest and can assist in the underlying information retrieval, but that in the rankings prepared for this study often presented terms disparate in relation to most of the terms well positioned in each of the other indices used. This shows the need to deepen the method using a differential treatment of each of the centrality indices. The intention here was to reveal meanings through the discovery of the most important terms in accordance with the positions they take the statements made by members of the groups, certainly both coding choices as the combination method of centrality indices adopted were suitable, including regarding the level of detail necessary to obtain answers to research questions. The method was therefore effective to investigate and discover the meanings of speeches many of which remain hidden in a traditional content analysis.

References

1. Alexa, M.: Computer-assisted text analysis methodology in the social science (1997)
2. Arendt, H.: A condição humana, 11th edn. Forense Universitária, Rio de Janeiro (2010)
3. Artieri, G.B.: Connessi in pubblico: sfera pubblica e civic engagement tra mainstream media, blog e siti de social network. In: Bartoletti, R., Franca, F. (cura) (eds.) Comunicazione e civic engagement. Media, spazi pubblici e nuovi processi de partecipazione, pp. 97–116. Franco Angeli, Milano (2013)
4. Barabási, A.L.: Linked. How Everything Is Connected to Everything Else and What It Means for Business, Science, and Everyday Life. Plume, England (2003)
5. Barabási, A.L., Albert, R.: Emergence of Scaling in Random Networks. Science, v. 286 (1999)
6. Barros, S.A.R.: Deliberação Pública Online: Esferas Conversacionais ao Redor de Conteúdos de Tres Jornais Brasileiros. Dissertaçao de Mestrado, Programa de Pos-Graduação em Comunicação e Cultura Contemporaneas. Universidade Federal da Bahia, Salvador (2013)

7. Campagna, M.: The geographic turn in social media: opportunities for spatial planning and geodesign. In: Murgante, B., et al. (eds.) ICCSA 2014, Part II. LNCS, vol. 8580, pp. 598–610. Springer, Heidelberg (2014)
8. Carley, K.: Coding Choices for Textual Analysis: A Comparison of Content Analysis and Map Analysis. Sociological Methodology **23**, 75–126 (1993)
9. Carley, K., Palmquist, M.: Extracting, Representing and Analyzing Mental Models Institute for Software Research, Paper 40 (1992)
10. Danowski, J.A.: Computer Mediated Communication. A publication. Social Content Analysis Using a CBBS Conference. Communication yearbook, v. 6, pp. 905–924. Sage (1982)
11. Diesner, J., Carley, K.: Semantic networks. In: Barnett, G., Golson, J.G. (eds.), Encyclopedia of Social Networking (2011)
12. Diesner, J., Carley, K.: AutoMap1.2 – Extract, analyze, represent, and compare mental models from texts. Technical Report, Carnegie Mellon University, School of Computer Science, Institute for Software Research International (2004)
13. Galloway, A.R.: Rets et ré seaux dans la tragédie antique. In: Stiegler, Bernard. Réseaux sociaus. Culture politique et ingénerie des réseaux sociaux, pp. 37–66. FYP éditions, France (2012)
14. Granovetter, M.: The strenght of weak ties: A network theory revisited. Sociological Theorym **1**, 201–233 (1983)
15. Latham, R., Sassen, S.: Digital formations: constructing an object of study. In: Digital Formations: Information Technology and New Architectures in the Global Realm. Princeton University Press, New Jersey (2005)
16. Lemieux, V., Ouimet, M.: Análise Estrutural das Redes Sociais. Instituto Piaget, Lisboa (2008)
17. Logan, J.R., Molotch, H.L.: Urban fortunes: the political economy of place. California Press, Berkeley (1987)
18. Lorusso, S., Scioscia, M., Sassano, G., Graziadei, A., Passannante, P., Bellarosa, S., Scaringi, F., Murgante, B.: Involving citizens in public space regeneration: the experience of ``Garden in Motion''. In: Murgante, B., et al. (eds.) ICCSA 2014, Part II. LNCS, vol. 8580, pp. 723–737. Springer, Heidelberg (2014)
19. Maia, R.: Visibilidade Midiatica e Deliberação Publica. In: Gomes, W., Maia, R.C.M. (eds.) Comunicaçao e Democracia. Problemas & Perspectivas, pp. 165–194. Paulus, Sao Paulo (2008)
20. Mercklé, P.: Sociologie des réseaux sociaux. La Découverte, Paris (2004)
21. Milgram, S. (2003). El problema del mundo pequeño. Araucaria. Revista Iberoamericana de Filosofía, Política y Humanidades, v. 4, pp. 15–28 (2014). http://148.215.2.11/articulo.oa?id=28210402. Acesso em 17 dez
22. Modesto, A.T.T.: Processos interacionais na Internet: Análise da Conversação Digital. Tese de Doutorado, Faculdade de Filosofia, Letras e Ciencias Humanas da Universidade de Sao Paulo, Sao Paulo (2011)
23. Murgante, B.: Wiki-Planning: the experience of basento park in potenza (Italy). In: Boruso, G., et al. (eds.) Geographic Information Analysis for Sustainable Development and Economic Planning: New Technologies, pp. 345–359. GI Global, Hershey (2013). doi:10.4018/978-1-4666-1924-1. ch023
24. Paranyushkin, D.: Visualization of text's Polysingularity Using Nerwork Analysis. Nodus Lab, Berlin (2012)
25. Pereira, H.B.B., Fadigas, I.S., Senna, V., Moret, M.A.: Semantic networks based on titles of scientific papers. Physica A **390**(6), 1192–1197 (2011). Elsevier
26. Popping, R.: Computer-assisted Text Analysis. Sage, London (2000)

27. Recuero, R.A.: Redes Sociais na Internet. Porto Alegre: Ed. Sulina, (Coleção Cibercultura) (2009)
28. Recuero, R.A.: Conversação em Rede. Porto Alegre: Ed. Sulina (2012)
29. Rieder, B.: Studying Facebook via Data Extraction: The Netvizz Application, pp. 346–355. ACM WebSci (2013)
30. Rocha, M.C.F., Florentino, P.V., Pereira, G.C.: City as commons: study of shared visions by communities on facebook. In: Murgante, B., et al. (eds.) ICCSA 2014, Part II. LNCS, vol. 8580, pp. 486–501. Springer, Heidelberg (2014)
31. Small, T.A.: What the hashtag? A content analysis of Canadian politics on Twitter. In: Brian, D.L., Dan, M. (eds.) Social Media and Democracy. Innovations in participatory politics, pp. 109–127. Taylor & Francis, London (2012)
32. Sowa, J.F.: Semantic Networks. Encyclopedia of Artificial Intelligence, 2nd edn. Wiley, 1992 (1987). http://www.jfsowa.com/pubs/semnet.htm
33. Van Atteveldt, W.: Semantic Network Analysis. Techniques for Extracting, Representing, and Querying Media Content. Ph.D. thesis, Vrije Universiteit, Netherlands (2008)
34. Watts, D.J.: The "new" Science of Networks. Annua. Rev. Sociol
35. Watts, D.J., Strogatz, S.H.: Collective dynamics of 'small-world'networks. Nature 393(6684), 440–442 (1998)

Citizens Participation in Improving Rural Communities Quality of Life

Roberta Soligno, Francesco Scorza, Federico Amato,
Giuseppe Las Casas, and Beniamino Murgante

University of Basilicata 10, Viale Dell'Ateneo Lucano 85100, Potenza, Italy
soligno.roberta@libero.it, francescoscorza@gmail.com,
{federico.amato,giuseppe.lascasas,beniamino.murgante}@unibas.it

Abstract. The concept of participation is very difficult to clearly define, because of the complex framework and the experimental nature of any participatory process. These processes are more difficult to implement in small and isolated communities because of the particularities social structure. I several cases technologies can help citizens in participating in these activities, but the result is strongly related to the predisposition of a communities to use theses ICT tools.

"Hack my town" was an "hackathon" among Universities in order to find solutions to "smart villages and territories" problems. It took place in Glorenza, a small medieval village located in the North of Italy, in Val Venosta, near to Swiss boundary. The introduction of smart solutions in a small rural context could be not very easy: Glorenza is, in fact, characterized by a limited predisposition for both technological and social innovation. Participants experienced how a small mountain village can become a smart village through the synergy between students and their professors, stimulating from one side scientific discussion and from another side a creative environment, where new con-ceptual solutions can be found.

Keywords: Rural communities · S.W.O.T. analysis · Problems and objectives tree · Logical framework approach

1 Bottom-up Decision Making Process: A Community-Based Approach

The concept of participation is very difficult to clearly define, because of the complex framework and the experimental nature of any participatory process. Public participation to planning processes can be defined such as policies, techniques and strategies that are able to involve citizens and stakeholders in decision-making, and to support decision-makers in defining shareable scenarios for future development of cities.

Participation is, therefore, an approach which tries to obtain active involvement of different stakeholders in a planning process, in order to ensure a final, useful and usable result, closer to their needs. Considering this point of view, some important factors can be underlined. First of all participation involves people who using space, local actors, final beneficiaries that are determined depending on the case and the size of the

© Springer International Publishing Switzerland 2015
O. Gervasi et al. (Eds.): ICCSA 2015, Part II, LNCS 9156, pp. 731–746, 2015.
DOI: 10.1007/978-3-319-21407-8_52

project, according to appropriate considerations and analyses. There are no fixed rules, but specific features of the place interested by interventions allow to identify them. Then, participation should imply the existence of a final result: there is the need to get at a proposal that could be more or less accepted by institutions. Everything that does not produce a proposal is not participatory planning, but simple consultation. Arnstein [17] described the detailed aspects of participation, highlighting the possible manipulation that could concern participative processes. Through the metaphor of a ladder, Arnstein identified eight levels of participation, divided into three parts. Bottom rungs are "Manipulation" and "Therapy": these describe levels of "non-participation" that have been contrived to substitute for genuine participation. Their actual objective is not to enable people to participate in planning or conducting programs, but to enable power-holders to influence participants. The second band is "Tokenism", that includes "Informing", "Consultation" and "Placation": these levels describe conditions where citizens have not the power to insure that their views will be considered by decision makers; so only power-holders maintain the right to decide. The higher rungs are "Partnership", "Delegate Power" and "Citizen Control", levels of "citizen power" that describe increasing degrees of decision-making clout of citizens. However it should be clarified that 'participation' does not mean giving city government to people: for this task, citizens already elect institutional representatives, whose responsibility is to define and to implement policies. Participation has not to be the ultimate goal of governments, but it has to be considered as a valid support. Citizens ask for a level of participation allowing the construction of programs and public initiatives tuned with their priorities and their needs [16] and this means a democratic process, a collective construction of decisions.

1.1   Bottom-up and Top-down Approaches in Planning

In recent decades, removal of typical hierarchical relationships adopted in planning processes has been announced several times, theorizing a sort of convergence towards planning models that combine top-down policies, promoting the prescriptive feature of the plan, and bottom-up initiatives that increase value of local specificities introducing flexibility [1] [2].

It is usually supposed that "bottom-up policies are not adopted because of the lack of effective top-down policies" [3]. This is not a contradiction: even if bottom-up policies derive from social needs, they have to be included by local administrations, under the current top down framework. According to Murgante et al. [4], during 1960s the transition from a purely top-down approach to a 'reticular interactive' one interested strategic planning: knowledge and imagination of society began to play a fundamental role in order to discover desirable scenarios. If we consider Harvard [9] and Minnesota [10] models, SWOT analysis became a central instrument to examine internal and external environments, producing a stakeholder analysis taking into account organizations, groups, people and all citizens, who can have a key influence on strategic processes. Adopting the reticular approach, there is the widest possible involvement of all potential stakeholders in order to avoid possible conflicts which could stall the whole process and, above all, create a broad and shared planning vision. Visioning concerns not only actors, who can be represented by institutions, but it also considers the possibility that collective knowledge may stimulate the search for optimal solutions. According to research of convergence between bottom-up and top-down approaches, a plan sends and receives impulses from its community: on one side, it defines what are non-negotiable uses, defining constraints that should intercept safeguarding instances; on the other side, stakeholders propose possible transformations. A plan has to pursue the need of preservation and transformation, safeguarding collective interests and avoiding, at the same time, the possibility to lose any private investment [15]. In order to involve more people in planning process, it is fundamental to distinguish between simple citizens and organized stakeholders: stakeholders are influential subjects for an initiative, while in great part of cases citizens' opinions provide ideas or claims that often remain unheard. This distinction is important because, in our view, citizens could give impulses in terms of needs, claims, demands, imagination, ideas, projects, which can be accepted or not by a plan [1]. Visioning methods have been adopted in a lot of cases, in order to define fundamental and significant bottom-up contributions: one of these is the "workshop", a method which allows to identify the most significant requests and to choose priority ambits of intervention, through a careful analysis of needs and local problems. Unfortunately, this approach has been often applied in contexts where decision makers do not wish to share decisions with the community.

1.2   Community Involvement in Decision-Making Processes: From Advocacy to Wiki-Planning

In 1960s some experiences, developed in USA, underlined that a low level of quality of life is closely connected with the capacity of communities to define their own living conditions [8]. The strong rooted ness of residents to places where they live is a key factor in determining quality of towns and neighbourhoods, because it generates people desire to contribute to choices affecting their territory. A first attempt to involve a large number of citizens in decision-making process was done by "Advocacy Planning", a theory conceived and supported by Davidoff [14]. According to Davidoff's idea, a planner has to be pluralist and to represent different interests, especially the low-income people's ones: he argued that a planner not only has to analyse social problems in order to find solutions, but he also has to be a sort of lawyer for weakest social groups. "Advocacy Planning" is strictly connected with another theory called "Community Planning": adopted in 1960s, it represents a planning approach that carefully analyses citizens' interpretations of places where they live, giving less importance to technical and expert knowledge. In this view, planning is not only a technical activity: it is also a political activity in which the information produced by dialogues and comparisons with local communities is fundamental for governments. One of the most important exponents of this community-based idea is [7]: he encourages public sector to learn social and environmental values through participative processes, because a participated plan could be easily adopted without subsequent changes and its objectives reflect ideas supported by inhabitants.

Another form of citizen involvement is "Placemaking": rooted in community-based participation, it aims at creating liveable places in cities through the interaction between designers and citizens. The concept behind Placemaking was originated in 1960s, by several authors like Jacobs [8] and Whyte [18], who offered innovative ideas about design of cities. The approach considers community opinions as a fundamental instrument to improve urban spaces: a direct knowledge of a place can give significant information about its functioning, its problems and about people priorities. This information is then used to create a common vision for public places. This vision can quickly evolve into an implementation strategy, beginning with small-scale, doable improvements that can immediately bring benefits to public spaces and to people who use them. Since 2000, the diffusion of ICT has introduced important innovations about governance and democracy: the use of Internet can integrate traditional approaches to participation, as it can be a mean to inquire about citizens opinions concerning administration and decision making processes. Traditional participatory methods are often inadequate and inefficient: these types of interaction presuppose physical presence of citizens during organized meetings, but in a lot of cases people who can really help participation have not the possibility to attend such meetings. Unfortunately, economically active population (employers, professionals, entrepreneurs, etc.) does not have enough time to participate, consequently only children and elderly opinions are collected. Moreover, the participative phase begins when programs or projects have been already defined, and this fact causes a general mistrust towards public administration. These are not occasional situations in traditional participative processes: we could affirm that a useful solution come from electronic participation, based on an asynchronous interaction. In particular web platforms, expressly conceived to participated processes, can offer instruments encouraging a constructive dialog among citizens, technicians and PA, that helps the identification of specific objectives and that allows to inform a large number of people about the results of the process, obtaining a great reply about community's desire. In Wiki-planning approach, citizens unconsciously reach the higher levels of Arnstein's ladder, helping the typical steps of planning. The advent of Web 2.0, where people are voluntary sensors [12], allowed high levels of interactions thanks to the transition from a one-way to a two-way approach: in the first citizens are simply informed on what the contents of a plan are, while in the second one people can express their ideas, that could influence choices of the administration [1].

Social platforms can lead from a closed model of decision making based on professionals government and representative democracy where participation is mainly relegated to election [5], to an integration of representative democracy and weak forms of direct democracy, where a decision maker has the possibility of directly consulting citizens in order to take a particular decision. We could say that Wiki-planning theory derives from Advocacy Planning: Davidoff's idea made citizens aware of their own role and so they became able to balance the power of big public and private agencies. In Wiki-planning these actions take place through the help of virtual environment and cloud services and all people have the same position on the scale of responsibility.

2   Logical Framework Approach

Already used in the '70s, the LFA method can be now defined as an effective technique to analyse problems in order to identify objectives and activities that could solve them. It is therefore a mean that improves quality of projects, through an analytical approach to design and management of programs oriented to obtain specific objectives.

The use of the LFA allows:

- to clarify purposes and to justify the existence of a project;
- to clearly define key elements of a project;
- to analyze project formulation at an early stage;
- to facilitate communication among all parties involved;
- to measure success or failure of programs.

Moreover, the method has got a lot of vantages:

- it ensures the analysis of fundamental problems and local criticalities, in order to provide better information for decision makers;
- it is a guide for a systematic and logical analysis of key-elements that form a well-done project;
- it improves planning, underlining connections between project elements and external factors;
- it provides a better instrument for monitoring and analysis of project effects;
- management and administration benefit from standardized procedures that collect and assess information.

The use of this approach is strictly connected with the setting of development projects: these types of projects have the aim to bring desirable changes in the contest of implementation and in society in general. The definition of future expectations is a very important step because it makes possible to check at a later stage the measure of program success related to its objectives and to target groups [13].

With a synthetic operative definition, we can affirm that LFA is based on the design of the logical nexus framework in which a project is characterized by input of resources, implementation of certain activities and outputs that should contribute to desired future objectives.

Input, output and activities are therefore the base elements of the project. However they are not factors that influence its success: it depends not only on factors that can be controlled by project management, but also on a series of external assumptions. During design phase and implementation it is necessary to analyse and to control these external factors, because they could be the main cause of project failure [13], even if everything has been realized as expected.

So LFA can be considered useful not only in the early stage of project concept design, but also during the implementation of projects.

This approach is composed of two phases:

- analysis: this phase examines the existing situation in order to develop the desired future situation and to identify some strategies to achieve it; the analysis is done with the help of stakeholders, who contribute to the definition of main problems and objectives;

- synthesis: strategies are made clear in order to be applied; the Logical Framework Matrix allows to identify activities that have to be undertaken, available resources, resources that have to be found, and it allows to verify coherence and relevance of choices as regards context of implementation.

In the stage of analysis we can find:

- situation analysis, that is in turn composed by analysis of stakeholders, of problems and objectives;
- strategy analysis.

Instead, in the stage of synthesis we find:

- logical framework matrix;
- implementation.

In particular problems and objectives are analysed by means of a problems and objectives tree. First of all there is the need to find the focal problem derived from the available information about the existing situation. Then, the construction of the problems tree allows to organize problems considering the relationship between causes and effects. While the problems tree provides the negative image of reality from stakeholders' point of view, the objectives tree, which is its dual, outlines the desirable future, rewording all problems and making them into objectives (positive statements). The last tree allows to select the strategic axes of the project. These strategies became feasible projects through the help of the Logical Framework Matrix (LFM).

Matrix construction allows two levels of reading, referring to:

- overall and specific objectives of the program, expected results and activities which have to be undertaken to reach them (vertical axis);
- concreteness, relevance and measurability of each objective, result and activity, on the basis of objectively verifiable indicators and sources of verification.

In the framework proposed by Las Casas et al. [6], we can find two types of indicators: efficacy indicators, which measure the degree of objectives achievement, and effectiveness ones, which measure the relation between resources used and realized products.

Table 1. Logical Framework Matrix (Las Casas et al., 2009)

Intervention Logic	Objectively Verifiable Indicators			Source of Verification	Assumptions
Overall Objective	Context analysis 1) Objective pertinence 2) Objective relevance	Efficacy Indicators	Effectiveness Indicators		
Project Purposes					
Results/ Outcome					
Activities	Inputs				
Preconditions					

We can say that LFA is an operative tool to establish strategies and guidelines for project implementation and to understand the logic behind the project so that any changes are necessarily conformed to overall project design. Referring to this paper, the approach is presented as a support for a participative workshop experience, where it allowed to develop strategies, considering the needs of involved group and limiting the uncertainty that characterizes a development program.

3   The Experience of Glorenza

"Hack my town" workshop was presented as a challenge, a "hackathon" among Universities in order to find solutions to "smart villages and territories" problems, but it was more than a challenge: it was an occasion of cultural exchange and meeting among Universities coming from different places and fields of study (cfr. http://hackmytown.unibz.it). Sponsored by Free University of Bolzano, it took place in Glorenza, a small medieval village located in the North of Italy, in Val Venosta, near to Swiss boundary. Participants experienced how a small mountain village can become a smart village through the synergy between students and their professors, stimulating from one side scientific discussion and from another side a creative environment, where new conceptual solutions can be found. The workshop lasted three days: the work was developed during the second day, after a meeting with the local community that provided us a lot of information about the city and local problems. In spite of the short available time, the collected information allowed to develop significant solutions, optimized by a bottom-up approach that involved local people to highlight issues of the place: the involvement of citizens and local administrations may allow to avoid waste of public resources focusing the attention on local community motivations. The meeting with citizens showed that the most important problems perceived by the community were connected with the two pillars of local economy: agriculture and tourism. Local people underlined that agriculture is almost only based on apple growing: at first, this cultivation was situated in less extensive areas, but then global warming has led its presence also at higher altitude levels.

The use of pesticides in industrial apple growing has led problems to other cultivations and to livestock and this fact has represented a relevant concern for local community, which aims at introducing new agricultural practices based on different types of crops and at ensuring earnings for farmers. As regards the tourism, we can say that it represents one of the most important development axes. Stakeholders consider Glorenza such as a " tourism thermometer" thanks to its historical beauties. Even if the village is very attractive, tourism connected to historical and cultural riches seems to be not enough exploited: there is the need to attract more people for a longer period.

Problems might be caused by seasonal tourist traffic: people visit Glorenza above all during summer, while wintry tourism presents some difficulties. Summer holidays are favoured by high naturalness of the area and by a large number of activities such as hiking or climbing. Cycling has a great importance: there is an extensive network of bicycle paths and the rail network of Val Venosta allows transporting bicycles, making connections easy. At summer end, Glorenza begins to empty: during winter people prefer to stay out of the city, near to the ski lift. Moreover, tourist accommodation has problems too: there are some small hotels with few beds, and so they are not able to accommodate a large number of people. Stakeholders denounced also a weak cooperation among local administrations: for example, regarding Adige river, which is an important territorial resource, there are a lot of discussions but measures cannot be adopted because municipal districts aim at asserting their own interests. Finally the meeting underlined some considerations about the local community. According to stakeholders, people are characterized by a strongly traditionalist and conservative culture, which leads to lack of interest towards the village and some constraints towards innovative changes, such as the use of internet and technologies. Medieval beliefs persist among people: for example, property is generally inherited only by the eldest male. Many initiatives, as open-air markets or cycling tours, meet people dissent, because they bring noise in the city. As a main example of people behaviour towards innovation, stakeholders remembered that in the past Glorenza did not become the terminus of the railway line because of its inhabitants that opposed the action with a referendum. All collected information during this meeting have represented the base knowledge to understand what are positive and negative features of the place and to focalize the attention on particular resources that should be valorised. Through these information we have defined main investigation areas.

4   The Project

In this section of the work the project development is described. We point out elements connected to the application of the methodology in the specific "workshop" activity in order to demonstrate their usefulness. In Glorenza case, this approach has been used to find rational solutions for village problems reported by the local community and to elaborate bottom-up strategies that could be considered a valuable support in the development of the area. As previously described, LFA is composed of a phase of analysis and another phase of synthesis. The analysis phase includes:

- context analysis;
- evaluation of concerns emerged during the meeting with local community;
- S.W.O.T. analysis;
- problems and objectives tree;
- strategies identification.

The phase of Synthesis includes the Logical Framework Matrix.

Concerning the case study, it is important to underline that the implementation of workshop activities in Glorenza was based on only 3 working days. Such a short time allowed us to consider only some specific aspects in context analysis. So results discussed in this work have to be considered as a preliminary, not exhaustive attempt. Val Venosta territory is characterized by very high environmental values: there are many protected areas, as SIC (Site of Community Importance) and ZPS (Special Protection zone), and Parks. The area is served by mobility infrastructures of Autonomous Province of Bolzano and is crossed by one of the most important roads, linking Italy and Switzerland: SS 41 starts from Sluderno and ends in Tubre, at the mountain border post of

Müstair , passing through Glorenza. As regards the population, today Glorenza has 898 inhabitants: comparing official population data from National Statistical Institute between 2004 and 2014, there are not particular critical points in population structure, except a slight tendency to growing old. Geo-statistic assessment of population density, based on Kernel's method, shows a concentration of people along Val Venosta valleys and underlines a marginal position of the town compared with the nearby towns of Sluderno and Malles. About tourism, the local system called Mullou Venosta, that includes 5 municipalities (Curon Venosta, Glorenza, Malles Venosta, Prato allo Stelvio and Tubre), counts 1500 beds divided between hotels and other forms of tourist accommodation. This local system looks peripheral compared to the surrounding context regarding the accessibility to tourist services. Referring to Val Venosta system, web-mapping of facilities and services, developed through the reuse of web open data, underlines a fragmentary system, concentrated in villages that could be considered as the "doors" of the closely natural system. People concerns emerged from the participatory meeting allowed to identify the main problems according to the specific point of view of Glorenza local community. Summing up, it can be said that first of all Glorenza stakeholders denounced a risk linked to an agricultural practice based on a monoculture: the territory is interested almost exclusively by apple growing, an intensive cultivation that causes the alteration of the agricultural landscape and the excessive dependence of local economy on one seasonal production. Concerning tourism, people underlined a marked drop in wintry flow of visitors, caused by the relative distance of Glorenza from ski lift plants, closer to other towns in the valley. Moreover the village seems to have a supply of accommodation facilities limited to a small number of guests. Finally there were some considerations about cultural features of local community linked to traditional and conservative forms of economical and social organization. Glorenza people seem to be resistant to current forms of innovation (social, technological, economical, etc.) and not very inclined to contribute to village improvement, a typical behaviour of isolated mountain communities. The results of the analysis have been represented by S.W.O.T. matrix.

Strengths include:

- presence of a great environmental and cultural value;
- high architectural quality of the medieval town;
- good level in preservation and maintenance of the historical centre;
- balanced population structure;
- economical wellbeing of population;
- high quality of road infrastructures and mobility connection to provincial main centres for services.

Among the weaknesses we find:

- cultural features of local community, mainly based on traditional and conservative attitudes;
- scarce community involvement in decision-making processes and in territorial management;
- agricultural practice based on a monoculture;
- low level in the transformation of agricultural row material (milk, fruit and vegetables);

- widespread lack of interest among people about internet and technologies, or, in other terms, innovation;
- marked drop in wintry flow of visitors;
- lack of integrated prospects for tourism development.

Opportunities are:

- Glorenza's leading role in the territorial identity of Val Venosta;
- funding opportunities from European Union programs and policies;
- geographical proximity with strong economic systems (CH,AU);
- investments in internet high speed connection.

Finally, threats includes:

- lack of community collaboration to find innovative forms of development and cooperation;
- higher development of the nearby cities compared to Glorenza;
- lack of cooperation among tour operators and lack of trust towards new forms of tourist accommodation.

The problem tree, showed in following figure 1, shows problems identifying causes/effects logic: this elaboration has highlighted the inadequate exploitation of environmental and cultural resources as focal problem. Social innovation theme represents a transversal strategy, which can be considered as the fundamental basis (also the precondition) for territorial integrated development: main actors (individual citizens and associations) have to be part of a participative community, that considers itself able to contribute to territorial growth.

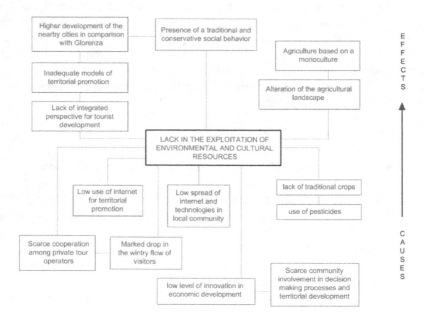

Fig. 1. Problem tree

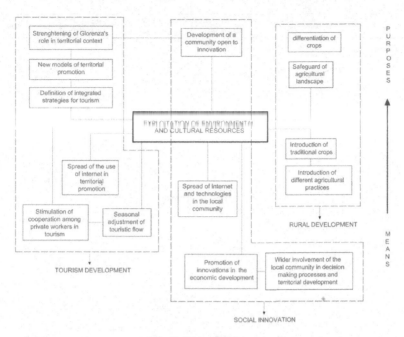

Fig. 2. Objectives tree

So, it becomes necessary to invest in "citizens' empowerment": in this way people realize they are able to improve the quality of their life and so they become aware of their central role. Glorenza's community will become a "smart community", if citizens are involved in decision making, reporting their needs and possible solutions for city problems. In a "smart" viewpoint, information spreading is relevant for community involvement: in Glorenza, investing in ICT means first of all bringing people near to web opportunities through education and knowledge. As regards tourism, public and private actors have to work for the safeguard of Glorenza's historical heritage because it represents a unique resource in the wider Val Venosta area, and they also have to promote innovative changes in tourist services and supply chain, in marketing and branding. Rural development aims at exploiting environmental resources and at protecting rural traditions. We can speak about "endogenous development" if the local community will cooperate in the selection and the promotion of territorial high value and resources as inputs for the construction of strategies. Territorial identity is a very important element because it makes a place different from another one: the reintroduction of typical products is a relevant strategic mean for rural valorisation and its contribute can improve the image of the area as a whole. For each strategy, Logical Framework Matrix has been created to clarify operational terms of implementation: the matrix identifies overall goals, project purposes, activities and results, highlighting principles of effectiveness and efficiency in public expense and verifying relevance and consistency of choices.

SOCIAL INNOVATION		OBJECTIVELY VERIFIABLE INDICATORS		SOURCE OF VERIFICATION	ASSUMPTION
		Efficacy Indicators	Effectiveness Indicators		
OVERAL OBJECTIVE	Development of a smart community open to innovation				
PROJECT PURPOSES	O.1 – People's involvement in decision making processes	N° involved people/ Tot	Shared projects and saving of resources	Questionnaires	Decisions based on the citizens'needs
	O.2 – Larger use of internet and technologies	Δ N° people that use internet and technologies		Questionnaires	Improvement in the spread of information
	O.3 - Development of a city that can be considered as a node for innovation networks	N° events; N° participants	Increased profitability of tourism in relation to the events	Tour operators' data; Departments for relations with research Institutes	
RESULTS/ OUTCOMES	P.1.1 – Participatory planning workshop as a support for the territorial management	N° stakeholders; N° meetings			Presence of participants that represent the widest number of stakeholders' needs (associations, institutions, technicians ...)
	P.1.2 – Space and equipment for meetings	Capacity	Management costs/ m²	Project data	Space that are always available and equipped
	P.1.3 – Open Data portal in order to increase government transparency and accountability	N° accesses	Available data/ Investment; Updated data/ Available data	web	Presence of citizens that choose to collaborate with the local administration
	P.2. - Courses that help the spread of digital technologies and social networks	N° qualified persons; Improvement in the knowledge of technology	N° Qualified persons/ Investment		Citizens' participation
	P.3.1 - Conference center	Capacity	Management costs/ m²	Project data	
	P.3.2 - Conference and workshop	N° conferences and workshop that are realized; N° participants; N° involved institutions	N° organized events/ Investment		Improvement in the spread of knowledge
	P.3.3 - Agreements with national and international research centers	N° involved research centres		Publications	Synergies among institutions and research centres
ACTIVITIES	A.1.1 – Individuation of stakeholders; Organization and implementation of workshops	INPUT			
	A.1.2 – Selection and adaptation of public places				
	A.1.3. 1 - Data research				
	A.1.3. 2 - Creation of suitable open datasets		Provincial Fund; European Regional Development Fund (ERDF) 2014/2020; Qualified staff for the implementation of the portal		Presence of a public authority able to favor transparence and cooperation
	A.1.3. 3 - Implementation of the portal				
	A.1.3. 4 - Data monitoring and data update				
	A.2. - Organization and implementation of training courses		European Social Fund (ESF) 2014/2020; Specialized agencies for the implementation of training courses		Citizens' participation
	A.3.1 – Selection and adjustment of public spaces				
	A.3.2 - Organization and implementation of meetings with Universities and organizations that support business growth		National and International Researchers		Cooperation with research centres and Universities
	A.3.3 - Networking actions		European Structural Fund 2014/2020		Cooperation with research centres

Fig. 3. "Social innovation" strategy Matrix

The figure 3 represents the first Matrix related to "Social innovation" strategy.

- As regards the social innovation, the overall objective is the development of a smart community open to innovation. The project expects:
  - people involvement in decision making processes;
  - larger use of internet and technologies;
  - development of a city that can be considered as a node for innovation networks.

TOURISM DEVELOPMENT		OBJECTIVELY VERIFIABLE INDICATORS		SOURCE OF VERIFICATION	ASSUMPTION
		Efficacy Indicators	Effectiveness Indicators		
OVERAL OBJECTIVE	Reorganization and exploitation of local resources				
PROJECT PURPOSES	O.1 – Strengthening of the summer tourism	Δ N° Tourists / year (June - September)	Revenues in the summer months/ Total revenues	Corporate data	
	O.2 – Increase of the wintry tourist flow	N° Tourists / years (except summer months)	Revenues (October - May)/ Total revenues	Corporate data	
	O.3 - Qualification and development of the tourist services and the supply chain	User satisfaction level	Increase in efficiency / Investment	Questionnaires	Coherence with the expectations of the target groups
RESULTS/ OUTCOMES	P.1.1 - Historical and artistic itinerary	N° points of interest of the (...)		?	Increase of the (...) resources
	P.1.2 - Programs and catalogues (also online)	N° distributed copies / N° online access	Compared with the investment	Web	
	P.2.1 - Public transport service that links Glorenza to the ski lift plants	N° users / month	Compared with the investment	Provincial Agency for Mobility	Presence of tourists that choose the public transport
	P.2.2. App for the monitoring of the transport service that can provide useful information about the lines, the waiting time and any critical weather condition	N° download		Provincial Agency for Mobility	
	P.3.1 - Efficient organizational models and qualified staff	N° qualified persons/ year	N° qualified persons that work in the tourism Industry		
	P.3.2 - Beds (B&B)	N° arrivals e presences / year; N° Beds	Presences/ available beds; Compared with the investment	Corporate data; Corporate data	Coherence with the needs of the target groups
	P.3.3 - Certification for the tourism businesses	N° certified businesses		Chamber of Commerce	Increase of the competitiveness of the tourist services
	P.3.4 - ICT services for the accommodation facilities and the restaurants	N° accomodation facilities and restaurants with ICT services/ Tot	Compared with the investment		Increase of the competitiveness of the tourist accomodation
ACTIVITIES	A.1.1 - Project and implementation of the itinerary	INPUT	Provincial funds for tourism development		
	A.1.2 - Project and implementation of the catalogue, also with multimedia devices				
	A.2.1. 1 - Selection of the routes		European Regional Development Fund (ERDF) 2014/2020; Provincial funds		
	A.2.1. 2 - Organization of the bus lines				
	A.2.1. 3 - Design and implementation of the ticket offices				
	A.2.2. 1 - Information gathering		European Regional Development Fund (ERDF) 2014/2020; Experts on data management		
	A.2.2. 2 - Creation of open datasets				
	A.3.1 - Organization and implementation of training courses		Provincial funds; European Social Fund (ESF) 2014/2020		Participation in training courses
	A.3.2 - Design and implementation				
	A.3.3 - Certification services for the tourism businesses				
	A.3.4 - Financial incentives for the ICT services		European Regional Development Fund (ERDF) 2014/2020		Presence of companies that take advantage of the financial incentives

Fig. 4. "Tourism development" Matrix

- The first purpose of the project is connected with the organization of participatory planning workshops and the creation of an open data portal in order to promote interactions among PA and local community and then to increase government transparency and accountability.
- According to the principles of openness and transparency, in this way, public administration redefines relations with citizens, that can have a continuous monitoring of the undertaken decisions.
- Project activities include also networking among Universities and research centres and the organization of conferences and meetings in order to introduce innovation and development into the area.
- European Structural Funds may be used for this purpose.
- The second strategy concerns "Tourism development" (figure 4).
  Reorganization and enhancement of local tourism development aims at:
  - strengthening of summer tourism;
  - increase of wintry tourist flow;
  - qualification and development of tourist services and supply chain.

RURAL DEVELOPMENT		OBJECTIVELY VERIFIABLE INDICATORS		SOURCE OF VERIFICATION	ASSUMPTION
		Efficacy Indicators	Effectiveness Indicators		
OVERALL OBJECTIVE	Protection of the rural landscape and exploitation of its peculiarities				
PROJECT PURPOSES	O.1 – Introduction of traditional crops and diversified farming practices	T/year products on the market	Compared with the investment	Rural Development Program; Provincial offices	Presence of farmers that trust the introduction of traditional crops
	O.2 – Development of organic agriculture	T/year products on the market	Compared with the investment	Rural Development Program; Provincial offices	Presence of farmers that choose organic farming
	O.3 - Development of the commercial chain of agricultural products	N° trading companies / N° products on the market / Δ local products on the market	Compared with the investment	Chamber of Commerce; Rural Development Program	Synergy with activities for tourism promotion
RESULTS/ OUTCOMES	P.1.1 - Pears and apricots	HA of apricot and pear plantations	Compared with the investment	Land registry office	
	P.1.2 - Product certifications	T/ years certified products		Chamber of Commerce; Product certification bodies	
	P.2.1 - Production rules for organic farming	N° farms that adopt organic standards		MIPAAF (Ministry of Agriculture and Forestry)	Presence of a large number of farmers that comply with the rules
	P.2.2 - Processing plants of organic products	N° processing plants	Compared with the investment	Chamber of Commerce; Provincial offices	Finished products are brought on the market
	P.2.3 - Organic certifications	T/ years certified organic products		Chamber of Commerce; Product certification bodies	Rise in value and in competitiveness of the local products
	P.3.1 - Shops and food tastings connected with tourism	N° shops		Chamber of Commerce; Web	
	P.3.2 - Marketing of local products	N° sales networks (also online)	Compared with the investment	Chamber of Commerce; Web	
	P.3.3 - Business networks	N° combined companies	Rationalization of business charges	Chamber of Commerce	Increase of the national and international competitiveness
ACTIVITIES	A.1.2. ; A.2.3 – Product certification services	Private funds			
	A.2.1 ; A.2.2. 1 – Financial incentives for farmers that choose organic agriculture	European agricultural fund for rural development (EAFRD); Provincial funding			
	A.2.2. 2 - Implementation of the processing plants of organic products				
	A.3.1 – Development of a local distribution system considering typical products as suitable elements to characterize the tourist supply	European agricultural fund for rural development (EAFRD); Provincial funding			Cooperation between agricultural sector and tourism
	A.3.2 - Participation of the producers in national and international marketing events	Incentives offered by the Chamber of Commerce; Private funds			
	A.3.3. 1 - Selection of companies for sharing the goals of innovation and competitiveness	Incentives offered by the Chamber of Commerce; Private funds; Provincial funding			
	A.3.3. 2 - Formulation of a common program				
	A.3.3. 3 - To draw up a contract				

Fig. 5. "Rural development" Matrix

In Glorenza, summer tourism is already well organized, then the project proposes the integration of hiking and outdoor activities, carried out during the summer, valuing the historical and architectural beauties in the city centre that differentiate the village from the neighbouring ones. The project aims at the creation of an itinerary that can valorise them, since they are tangible expressions of local culture, and the development of programs and catalogues for the visit. The problem of seasonal tourism is mainly caused by the relative distance of the city from ski lift plants: the project intends to propose the creation of appropriate public transport services, monitored through smart-phones which provide useful information about lines, waiting time and any critical weather condition. Accommodation facilities require a reorganization: there is the need to integrate the current offer with other forms of hospitality, such as "bed and breakfasts", which could be more suited to user needs. In order to qualify tourist offer, the plan provides training courses for staffs and the diffusion of ICT services for the effective promotion of tourist accommodation and restaurants. In order to preserve landscape and rural peculiarities of the area, the project provides the strategic line called "Rural development" (figure 5).

The specific purposes are:
- introduction of traditional crops such as pears and apricots;
- development of organic agriculture;
- development of a commercial chain of agricultural products.

The diversification of crops is based on the introduction of other typical cultivations, different from growing apples. Therefore it is necessary to awoken farmers towards the opportunities offered by the rediscovery of these typical crops and the possibility of a local production, in order to add more value to agricultural sector. The alteration of agricultural landscape, caused by pesticides, leads to consider the hypothesis of an organic agriculture development with the creation of new production rules and of new production facilities. As for the commercial chain of agricultural products, the project includes the development of a local distribution system, linked also to tourist and cultural exploitation, and the participation of producers in national and international marketing events.

Then, the development of business networks can also favour:
- the increase in competitiveness and productivity;
- the diffusion of know-how;
- the innovation development;
- the certification of production process;
- the internationalization of companies;
- the cost reduction of business management.

5   Conclusion

Referring to methodological aspects, the use of LFA proved appropriate to develop bottom up strategies during a participatory workshop. The steps of the method agree with the basic actions in the implementation of a workshop:
- identification and analysis of expressed problems;
- definition of objectives and activities to solve them;
- to foster a rational approach to the formulation of bottom up strategies.

The strict application of the method has allowed us to analyse problems expressed by the local community, acting as a guide to a systematic and logical analysis of connections among key elements of a well-structured project. The introduction of smart solutions in a small rural context could be not very easy: Glorenza is, in fact, characterized by a limited predisposition for both technological and social innovation. For this reason, there is the need to prepare citizens for necessary changes, making them aware about opportunities that ICT tools can give. The project experience has shown a particular value as it led to interact social groups, experts and representatives of public administration. If these activities became more widespread and systematic, they could lead to effective strategies and development projects based on people needs and on goals sharing. The described approach could represent a model for spreading participatory workshop applications, based on LFA method, in other peripheral areas characterized by similar social, environmental and economic features.

References

1. Murgante, B.,: Wiki-planning: the experience of basento park in potenza (italy). In: Borruso, G., Bertazzon, S., Favretto, A., Murgante, B., Torre, C. (eds.) Geographic InformationAnalysis for Sustainable Development and Economic Planning: New Technologies,Information Science Reference IGI Global, Hershey, PA (2012). doi: 10.4018/978-1-4666-1924-1.ch023
2. Lorusso, S., Scioscia, M., Sassano, G., Graziadei, A., Passannante, P., Bellarosa, S., Scaringi, F., Murgante, B.: Involving citizens in public space regeneration: the experience of ``garden in motion''. In: Murgante, B., Misra, S., Rocha, A.M.A., Torre, C., Rocha, J.G., Falcäo, M.I., Taniar, D., Apduhan, B.O., Gervasi, O. (eds.) ICCSA 2014, Part II. LNCS, vol. 8580, pp. 723–737. Springer, Heidelberg (2014)
3. Musco, F.: Rigenerazione urbana e sostenibilità. FrancoAngeli (2009)
4. Murgante, B., Tilio, L., Lanza, V., Scorza, F.: Using participative GIS and e-toolsfor involving citizens of marmoplatano – melandroarea in european programming activities. Journal of Balkans and Near Eastern Studies.Special Issue on E-Participation in SouthernEurope and the Balkans (2011). doi: 10.1080/19448953.2011.550809
5. Noveck, B.S.: Wiki government: How technology can make government better, democracy stronger and citizens more powerful. BrookingsInstitution Press, Harrisonburg (2009)
6. Las Casas, G., Scorza, F.: Un approccio 'contexbased' e 'valutazione integrata' per il futuro della programmazione operativa regionale in europa. In: Bramanti, A., Salone, C. (eds.) Lo Sviluppo Territoriale Nell'economia Della Conoscenza: Teorie, Attori Strategie Collana AISRe – Scienze Regionali, Vol. 41 (2009). ISBN: 978-88-568-1051-6
7. Forrester, J., Cambridge, H., Cinderby, S.: The value and role of GIS to planned urban management and development in cities in developing countries. City Development Strategies (1999)
8. Jacobs, J.: The Death and Life of Great American Cities. Random House, New York (1961)
9. Bryson J. M., Einsweiler R. C. (eds), Strategic planning: introduction. Journal of the American Planning Association (1987)
10. Bryson, J.M., Einsweiler, R.C. (eds.): Strategic Planning: Threats and Opportunities for Planners. Planners Press, American Planning Association, Chicago, Washington DC (1988)
11. Bryson, J.M.: Strategic planning for Public and Nonprofit Organization: A Guide to Strengthening and Sustaining Organizational Achievement. John Wiley, San Francisco (2004)
12. Gibelli, M.C.: Riflessioni sulla Pianificazione strategica. In: Rosini, R. (ed.) L'urbanistica delle aree metropolitane. Alinea, Firenze (1992)
13. Goodchild, M.F.: Citizen as sensors: the world of volunteered geography. GeoJournal **69** (2007)
14. The Logical Framework Approach (LFA). Handbook for objectives - oriented planning. Norad (1999)
15. Davidoff, P.: Advocacy and Pluralism in Planning. Journal of the American Institute of Planners (1965)
16. Pazienti, M.: Rapporti tra pianificazione strategica e sviluppo locale. In: Las Casas, G., Properzi, P. (eds.) Quadri Di Analisi Regionale Prospettive Di Interazione Multi-Settoriale. FrancoAngeli, Milano (2002)
17. Arnstein, S.R.: A ladder of citizen participation. Journal of the American Planning Association (1969)
18. Whyte, W.H.: The Social Life of Small Urban Spaces. The Conservation Foundation, Washington, D.C. (1980)

Citizen Participation and Technologies:
The C.A.S.T. Architecture

Francesco Scorza[1](✉) and Piergiuseppe Pontrandolfi[2]

[1] School of Engineering, Laboratory of Urban and Regional Systems Engineering,
University of Basilicata, 10 Viale Dell'Ateneo Lucano 85100, Potenza, Italy
francescoscorza@gmail.com
[2] DICEM, University of Basilicata, Via Lazazzera 75100, Matera, Italy
piergiuseppe.pontrandolfi@unibas.it

Abstract. The role of the participation has assumed a key dimension in all processes of physical planning and economic planning in the urban scale to the regional scale. Technological innovation, the spread of internet and mobile, have generated significant innovations compared to the models of management of participatory processes and interaction with communities and citizens. A critical element of strong within these processes is the ability to manage the information produced by the community (real and / or virtual) in order to develop guidelines and shared visions for the city and the territory. The paper analyses this issue in reference to a project proposal aimed at the development of participatory processes multi-scalar within the Project CAST (Active Citizenship for Sustainable Development of the Territory), selected by the Region Basilicata inside of a contract for the development of innovative and creative activities. The project aims to develop creative and innovative processes in the planning of the city and the territory they see the broad involvement of the population and local actors. The proposal is a first contribution to the operating institution of "Urban Center" in the two main urban centers of the Region Basilicata: Potenza and Matera "European Capital of Culture - 2019". This is attributable to a 2.0 approach the ability of citizens to define the project of the territory in which they live and work, contributing to the definition of a scenario shared for the promotion of local development projects and the development of urban regeneration policies.

Keywords: Participation · Urban regeneration · Tweetting maps · Virtual urban center

1 Introduction

Assuming that the role of participation has reached a key dimension in all physical and economic planning processes both at urban and territorial scale it is relevant to consider that technological innovation and the spread of internet and mobile technologies have produced significant changes in the participatory (we call it e-participation [1][2]) process management and in the models of interaction among/with citizens.

© Springer International Publishing Switzerland 2015
O. Gervasi et al. (Eds.): ICCSA 2015, Part II, LNCS 9156, pp. 747–755, 2015.
DOI: 10.1007/978-3-319-21407-8_53

The ability to manage the information (considering proper standards of interoperability [3]) produced by community (real and/or virtual information) in order to find shared visions and guidelines for cities and territories is a critical element in participative processes (cfr. [4]).

The work analyses this theme referring to a proposal for the development of multi-scalar participatory processes inside the project C.A.S.T. (Active Citizenship for Sustainable Development of the Territory) that has been selected by Region Basilicata through a call for application for the development of innovative and creative activities.

The project aims at developing innovative and creative processes in urban and territorial planning which lead to a wider involvement of population and local actors.

The proposal represents a first operative contribution for the foundation of "Urban Center" in the two main towns of Basilicata: Potenza and Matera. Such process fits also in the strategy of "MATERA European Capital of Culture – 2019". The general framework of the project looks at the citizens opportunity to determine the project of the territory in which they live and work through a 2.0 approach, contributing to the definition of a shared scenario for the promotion of local development projects and of urban regeneration policies.

In this paper we present briefly the project implementation context and we provide an analysis of the ICT infrastructure supporting project implementation.

2   CAST Project: An Overview of the Implementation Context

The project considers an interdisciplinary approach to the design and the implementation of participatory processes. Experts from different disciplines (in particular urban planning, architecture and sociology) will interact with local communities to experiment with innovative forms of e-governance also through the help of web-based tools.

CAST concerns the metropolitan area of Potenza and the town of Matera: in particular, the latter town will develop important initiatives regarding innovation and urban regeneration in consequence of its appointment as "European Capital of Culture 2019".

The project intends to promote social innovation that can be considered necessary for the growth of local contexts through actions aimed at increasing citizens' empowerment. This goal becomes very important also referring to the definition of strategies and intervention programs financed by the "new cohesion policy 2014-2020".

The activation of a workshop for urban and territorial planning derives from the idea of encouraging a wider participation in decision making processes and in the definition of territorial and economic programs of intervention based on an improved strategy of local development. The context in question is an internal area in Basilicata that is clearly characterized by economic and social weaknesses in spite of the presence of natural, cultural, economic and social resources (the area includes ten municipalities involved in the Metropolitan Structure Plan of Potenza).

These weaknesses are accentuated by depopulation: the abandonment of inner territories in favour of a rapid growth of Potenza leads to an excessive urbanization that causes social and economic imbalances according to the logic of competitiveness which has favoured the strongest areas.

The topics are territorial balance, social inclusion, organization of basic services following the principles of equity, efficiency and the adequate use of resources to promote sustainable changes. The implementation of these development interventions could rely on financial resources, above all EU funding, which will be available in the next years.

As far as the town of Matera, the project aims at clarifying the terms of a city development that includes the prospect of European Capital of Culture in 2019.

Inside the project C.A.S.T., we want to promote participatory activities that try to get cultural production innovations also through the strengthening of tourism development.

In particular, it intends to identify how the re-appropriation of some public spaces could be achieved through the involvement of inhabitants and favourite actors (associations, third sector operators…) and it tries also to exploit the great cultural and environmental heritage of the town.

Participation will involve the main interest groups and PVOs (Private Voluntary Organizations) that already work in these fields of intervention in order to reinforce "trust building" processes and to give to city users the opportunity to point out problems of public spaces for finding solutions of reuse and regeneration.

3   Technologies and Participation: An Integrated Open-Source Solution

The Project C.A.S.T. consists of actions aimed at the integration of open-source tools and frameworks in order to obtain a technological infrastructure which makes possible a high level of interaction among users.

The project idea derives from some experiences that has been recently developed [5]: "Cilentolabscape" can be considered as a significant example in which an integrated platform puts together traditional models for data display and advanced tools of social interaction. The system combines CMS features (Content Management System), a Geoportal to get territorial information, advanced systems for the management of online polls and votes, OGC services for data sharing according to OPEN DATA standards, the integration of social networks and the management of spatial social alerts for participation and collaborative mapping.

The system is formed by the following parts: a technological platform - CMS & SDI, a Geoportal, widgets, app POI builder, a mobile app for portal access.

4   The Integrated Technological Platform: Main Components

The figure below represents a schema of ICT components. Adopting an Open Source approach the system integrates several components for several information management: social streaming, geographic information, multimedia information. The conceptual design of the platform is mainly oriented to exploit opportunities coming from open data projected in a geographic dimension [6].

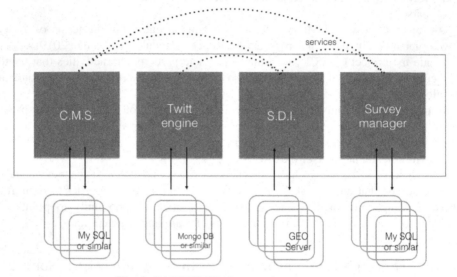

Fig. 1. CAST ICT Platform: main components

The design shows a Spatial Data Infrastructure (SDI), formed by a central database (PostgreSQL/PostGis), a geospatial application server (GeoServer) and a content management system (Joomla).

The project requires the installation of a central database, made through a RDBMS PostgreSQL with PostGIS Extension, that has the following tasks:

–   receiving, managing and recording spatial data, both raster and alphanumeric;

–   receiving, managing and recording data that comes from the content management system (Joomla).

The management and the display of portal contents depends on the CMS (Content Management System) which have traditional features and tools: the managing of static contents, News and Newsletter, Gallery, Download.

The SDI supplies mapping services and it will be implemented through a cluster with a load balancer; the Load Balancer Server will be configured by the Web Server Apache 2 with modjk.

Thanks to this load balancing method, the load of requests is split among different nodes (virtual and physical) and so requests are directed to less engaged ones.

The Spatial Data Infrastructure is the basis of the distribution and the use of mapping services.

It is composed by two different lines of reasoning, depending on data type (vector or raster): as far as vector data, a structure of geoDatabase cluster is implemented through a RDBMS PostgreSQL with PostGIS Extension.

The Geoportal

Spatial information, produced by participatory process, will be organized inside the Geoportal

In order to reach an adequate level of efficiency, this Geoportal must have some features such as:

- basemaps (Google, Bing, Open Street Map);
- Zoom and Pan;
- tools for the management of OCG server list;
- the display of map layers and dynamic overlay-mapping;
- tools for data filtering, user management and to save projects;
- Geo-coding (research of places and components);
- the management of Viewports (views of custom areas on maps).

These features are directed to different types of users that are potentially interested in the service: users could use them as nowadays they do with other available open-mapping tools or they could interact with the system through the production of layers and maps and the sharing of information with the community involved in the project.

Social Mapping and Social Allert

Nowadays social networks represent the most important data bank in the world.

As far as Twitter, a social network that provides a "short message" service with a localization tool, everyday there are about 400 million of tweets which are made available by the system through API.

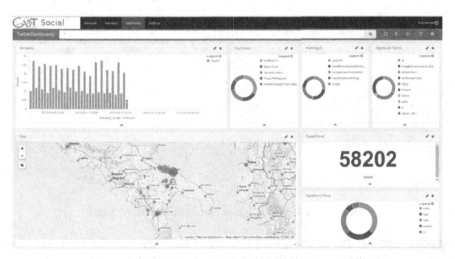

Fig. 2. Tweeting maps engine

A tweet can give geospatial information, that come from different mobile devices through GPS or GSM, and multimedia information, that organize images or texts in a synthetic and significant way.

Web-gis and social network integration improves the interaction between users and the system, increasing the quality of the participatory process thought for the project C.A.S.T.

Fig. 3. Tweeting maps engine: semantic search "Potenza"

Fig. 4. Tweeting maps engine: semantic search "Matera"

Social networks can be considered as a significant data bank for the project goal: according to its purpose, the interaction among users on the platform through social networks improves participatory processes.

The key element is the opportunity to indicate localized instances referring to punctual emergencies, which can be categorized on the basis of the discussion topics on the platform.

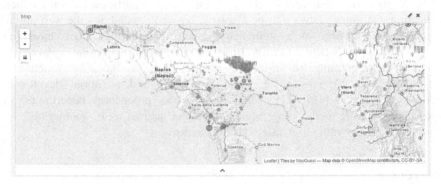

Fig. 5. Tweeting map: density of collected data on 3 months period

5   Creating and Experimenting "Virtual Urban Center"

The development of ICTs and the integration of traditional participative methods allow to find innovative means which might represent an important help for territorial government and urban regeneration, given the complex relationship among citizens and institutions.

In this sense the electronic participation can give a significant contribution, becoming a means to reach a higher level on the ladder of citizens' involvement.

The project C.A.S.T. does not want to replace traditional methods of participation, but it intends to support them through a widespread use of ICTs: there is the need to experiment with these new participative models and to take advantage of the low cost of ICTs, above all in situations where participation find it hard to become a firm procedure and where decision makers do not admit the usefulness of citizens' involvement.

The creation of "virtual urban centers" – that can favour the knowledge of contexts, documented and oriented information, an effective communication and a continuous interaction among different stakeholders – leads to a widespread participation and gives new possibilities that must be examined in a renewed approach to urban and territorial planning. Among implications of such approach we are interested in urban renovation and soil consumption [7] implications and also real estate consequences (see also [8] [9]).

6   Conclusions

This work presents the general framework for the implementation of a participatory process oriented at urban regeneration in a peripheral context. The main focus is to address the ICT platform for e-participation the main role of the project. As an ambitious goals but also as an effective choices the research group expect good performance from electronic participation more than traditional procedures.

The research, using the described ICT tolls will asses the process in order to highlight strengths, weaknesses and opportunities of e-participation in a concrete application. On going results will be available on www.castlab.it - the web portal of the project. We affirm, as a preliminary conclusion, that ICT tools based on a structured methodological framework is a mainstream operative approach for managing and exploiting the virtual information flows produced by community in order share strategic visions for urban renovation and territorial development. Previous experience [10] [11] demonstrated the procedural benefits of such approaches covering methodology applications and internet tools, gaining high level of participation according to Arnstein [12] classification.

References

1. Knapp, S., Coors, V.: The use of eParticipation systems in public participation: the VEPs example. In: Coors, V., et al. (eds.) Urban and Regional Data Management, pp. 93–104. Taylor and Francis, London (2008)
2. Lanza, V., Prosperi, D.: Collaborative E-Governance: Describing and Pre-Calibrating the Digital Milieu in Urban and Regional Planning. Taylor and Francis, London (2009)
3. Laurini, R., Murgante, B.: Interoperabilità semantica e geometrica nelle basi di dati geografiche nella pianificazione urbana. In: Murgante, B. (ed.) L'informazione geografica a supporto della pianificazione territoriale, pp. 229–244. FrancoAngeli, Milano (2008)
4. Pløger, J.: Public participation and the art of governance. Environment and Planning B: Planning and Design **28**, 219–241 (2001)
5. Attardi, R., Cerreta, M., Franciosa, A., Gravagnuolo, A.: Valuing cultural landscape services: a multidimensional and multi-group SDSS for scenario simulations. In: Murgante, B., Misra, S., Rocha, A.M.A., Torre, C., Rocha, J.G., Falcão, M.I., Taniar, D., Apduhan, B.O., Gervasi, O. (eds.) ICCSA 2014, Part III. LNCS, vol. 8581, pp. 398–413. Springer, Heidelberg (2014)
6. Izzi, F., La Scaleia, G., Dello Buono, D., Scorza, F., Las Casas, G.: Enhancing the spatial dimensions of open data: geocoding open PA information using geo platform fusion to support planning process. In: Murgante, B., Misra, S., Carlini, M., Torre, C.M., Nguyen, H.-Q., Taniar, D., Apduhan, B.O., Gervasi, O. (eds.) ICCSA 2013, Part III. LNCS, vol. 7973, pp. 622–629. Springer, Heidelberg (2013)
7. Amato, F., Pontrandolfi, P., Murgante, B.: Using spatiotemporal analysis in urban sprawl assessment and prediction. In: Murgante, B., Misra, S., Rocha, A.M.A., Torre, C., Rocha, J.G., Falcão, M.I., Taniar, D., Apduhan, B.O., Gervasi, O. (eds.) ICCSA 2014, Part II. LNCS, vol. 8580, pp. 758–773. Springer, Heidelberg (2014)
8. Morano, P., Tajani, F.: Urban renewal and real option analysis: a case study. In: Murgante, B., Misra, S., Rocha, A.M.A., Torre, C., Rocha, J.G., Falcão, M.I., Taniar, D., Apduhan, B.O., Gervasi, O. (eds.) ICCSA 2014, Part III. LNCS, vol. 8581, pp. 148–160. Springer, Heidelberg (2014)

9. Morano, P., Tajani, F.: Bare ownership evaluation. Hedonic price model vs. artificial neural network. International Journal of Business Intelligence and Data Mining **8**(4), 340–362 (2013)
10. Murgante, B., Tilio, L., Lanza, V., Scorza, F.: Using participative GIS and e-tools for involving citizens of Marmo Platano-Melandro area in European programming activities. Journal of Balkan and Near Eastern Studies **13**(1), 97–115 (2011)
11. Lanza, V., Tilio, L., Azzato, A., Casas, G.B., Pontrandolfi, P.: From urban labs in the city to urban labs on the web. In: Murgante, B., Gervasi, O., Misra, S., Nedjah, N., Rocha, A.M.A., Taniar, D., Apduhan, B.O. (eds.) ICCSA 2012, Part II. LNCS, vol. 7334, pp. 686–698. Springer, Heidelberg (2012)
12. Arnstein, S.R.: A Ladder Of Citizen Participation. Journal of the American Institute of Planner **35**(4), 216–224 (1969)

Innovations in Promoting Sustainable Development: The Local Implementation Plan Designed by the Province of Potenza

Francesco Scorza[1](✉) and Alessandro Attolico[2]

[1] Laboratory of Urban and Regional Systems Engineering, School of Engineering, University of Basilicata, 10, Viale dell'Ateneo Lucano 85100, Potenza, Italy
francescoscorza@gmail.com
[2] Province of Potenza, Planning and Civil Protection Office, P.zza M. Pagano, 85100, Potenza, Italy
alessandro.attolico@provinciapotenza.it

Abstract. The Province of Potenza engaged several international cooperation actions concerning local planning tools, sustainability, risk assessment and management, territorial resilience and renewable energy sources. Such effort contributed to the integrated strategy of boosting sustainable development at territorial scale by an action of coordinating municipalities in a common framework. This work describes the Local Implementation Plan developed in the framework of Renergy Project concerning the local implementation of best practices assessed through the cooperation activities developed in the INTERREG IV C project. The Province of Potenza leaded the partnership and stimulated a territorial strategy based on local issues and facing the Covenant of Majors and EU 2020 targets

Keywords: Energy efficiency · Covenant of majors · EU 2020 · New cohesion policy · Regional planning

1   Introduction

International cooperation projects represent an opportunities for local administrations to face common problems of the EU regional framework in a common way, exchanging experiences, case studies and discussing failures and inefficiency. In this case the RENERGY project (financed by INTERREG IV C EU Program) represent a success story of Best Practices exchange among European Regions. The project, tackling climate change and promoting a sustainable, secure and competitive energy supply, is based on the issue of reducing energy costs and consequently the dependence on primary energy sources. The overall aim of the project is to improve, by means of interregional cooperation, the effectiveness of regional development policies as well as to contribute to economic modernisation and increased competitiveness of European Regions and territories.

Apart form the technological innovation promoted by the project [1], one of the main output is the Local Implementation Plan: a strategic document affirming the

© Springer International Publishing Switzerland 2015
O. Gervasi et al. (Eds.): ICCSA 2015, Part II, LNCS 9156, pp. 756–766, 2015.
DOI: 10.1007/978-3-319-21407-8_54

route for transferring Learned Best Practices in the local policy framework, promoting investments and ensuring sustainable performance for the local communities [2].

In this work the main component of the Local Implementation Plan of the Province of Potenza are described. From the context analysis to the main pillars the research allow to enhance the operative contributions included in the documents for generating local development process based on an inclusive strategy for energy saving.

Conclusions regard the perspectives of operation implementation as a contribution for the territorial resilience scenario [3] [4] [5] as a mainstream component of the Provincial Structural Master Plan (PSP) recently adopted.

2   Energy Chain in Basilicata Region: Demand and Supply Analysis

In this section the energy chain in Basilicata Region is described in general figures. Such representation allows to understand the dimension of the system.

The supply chain is divided in: Fossil sources and Renewable Energy Sources (RES). Each component presents criticalities and local conflicts with identity values and resources.

Concerning the Fossil Sources Energy supply, the territory of the Basilicata Region is characterized by a considerable fossil sources availability, which is partly exploited.

There have been several oil and natural gas findings in the Basilicata Region between 1992 and 2006. More than 27% of the onshore findings in Italy over the last 15 years are situated on the Lucanian territory.

The Basilicata Region is top of the national ranking for its oil production thanks to the 4,3 million tons of oil extracted up to 2006, while as concerns the natural gas production, the Basilicata Region territory is the only one in Italy still steadily increasing (natural gas production from seabed minings: 10% of national production; natural gas production from onshore extractions : 47% of national production).

By analyzing the time series of the energy consumptions on the Basilicata Region territory, it can be said that most of the electric power produced in the Basilicata Region up to 2006 derived from the fossil fuel use, with a preponderance for the natural gas, which shows a use of 98% compared to other fossil sources n late 2006; this reflects the national trends.

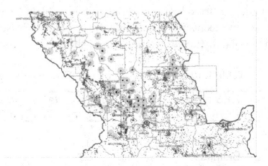

Fig. 1. Oil wells in the Province of Potenza

Ten thermal power plants fuelled by fossil fuels were in operation to 2005, for a total of 283 MWe and of 1.1185 GWht of gross production.

The Energy demand in Basilicata Region has not a very high amount for end-use. From the comparison with the national energy demand value, it is easy to see that the Lucanian consumptions represent less than 1% of the national consumptions.

The historical data on the regional electricity consumptions show a steadily increase of the demand during the twenty years from 1990 to 2010. This growth is being driven mostly by the industry development, but a significant increase, in relative terms, has occurred also in the other reference sectors. This puts the Basilicata Region in contrast with the average trends of the national consumptions, which are almost regular.

The Basilicata Region had an electricity gross production of 2.238 GWh in 2010; 2.171 GWh were intended for the consumptions. At the end of 2010 the gross consumptions was estimated at 3.174 GWh. On this basis, the Basilicata Region is not meeting the consumptions for a 30,1% share of the regional demand. This latter value was about 52,2% just two years earlier. The energy deficit reduction has been achieved thanks to the increase of the electricity production from renewable sources.

As concerns the future scenarios on the energy demand, the forecast in the RES report describe a tendential increase of 8,5% of the total consumptions (electricity and heat). This outcome confirms that the Basilicata Region is experiencing a period of growth and development in comparison to the other Italian regions.

The Regional Energy and Environmental Plan aim is to guarantee a support proper to the economic and social development needs thanks to the rationalisation of the whole energy sector and to a sustainable management of the land resources.

Ultimately, the whole energy sector programming is concerned with four overarching goals:

1) Energy consumptions and energy bill reduction;
2) Increase of the electricity production from renewable sources by fostering the supplied power generation;
3) Increase of thermal energy production from renewable sources by fostering the energy generation from biomass;
4) Creation of an energy district in Val d'Agri.

As concerns the first goal, the Region with Regional Law No 13/2006 constituted the Società Energetica Lucana (S.E.L.) whose role is the purchase for the public authorities which have to issue invitations to tender for the electricity and gas supply, in order to exploit the opportunities offered by the electricity and gas liberalized market.

The increase of the electricity production from renewable sources has been programmed within the plan by keeping to the limits shown in the table below:

Table 1. RES target according to PIEAR Basilicata

RES type	Installable power target by 2020 (MW)
Wind	981
Photovoltaic	359
Biomass	50
Hydroelectric	48

3   Local Constrains and Criticalities

Within the RENERGY project activities – in particular during the debate with the territorial stakeholders - the controversial issue of the energy-saving technologies application and of energy production from RES with the value of landscape and architecture protection and conservation has been touched upon.

It is a complex topic in which the theme of the territorial planning and the nature protection one intersect on different levels, as the technical disciplines of architectural restoration, the identity values system of a local community which identifying intangible functions and values in its historic settlement.

The large RES installations are the focus in relation to the landscape protection and conservation. The fact that the presence of large.wind power and photovoltaic plants alters the perception of the places is an evidence for the local population.

It is highlighted the need to develop a detailed analysis about this subject, also thanks to intervisibility evaluation tools, in order to define indirect impacts and appropriate forms of environmental offset.

In general, the public buildings in the Basilicata Region are a significant part of the towns and rural settlements historical heritage (it is however a typical case of the national trend). This is a historical inheritance which is often subject to preservation restrictions which do not foster the use of energy efficiency technologies and interventions.

There are no unambiguous opinions nor normative references on this profile. This is therefore a study and experimentation field where define intervention samples and procedures which might balance the energy efficiency targets with the architectural and functional conservation needs.

4   The Territorial Governance Tools

The energy policies have many interactions with all the territory governance and management tools. This is the case of the urban and territorial planning, the economic planning, the education and training programs for young people and communities, the technical and vocational training, the higher education and the research etc ...

The Implementation Plan of the Province of Potenza purposes are based on the following multi-level policy documents, which are directly associated with the Province of Potenza administrative functions and to the ongoing activities.

The main reference is the Provincial Structural Plan (PSP), which has been recently drawn up and approved in accordance with the Regional Law 23/99.

In particular, the PSP includes the energy policies theme as a dimension of the land planning according to a coordination and subsidiarity principle between the regional and the national levels.

Since administrative, territorial planning, complex material management tasks are given at the provincial administration level, the PSP functions as a complex planning tool in which the energy policies subject may generate positive integrations with the other territorial policies.

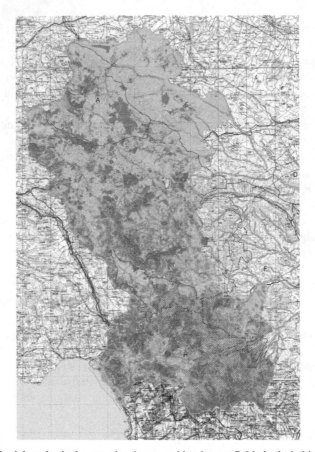

Fig. 2. Provincial ecological network scheme and landscape fields included in the PSP

The PSP is the planning act with which the Province plays an "active" policy coordination and connection role, by determining general provincial territory planning directions intended also to integrate the citizens working and mobility in the different life cycles, and to organize the equipment and the services, by ensuring the territory access and usability.

This PSP framework includes an energy planning dimension.

The main goals concerning energy and environment which shall be pursued within the territorial planning activities are summarised below.

- energy production, distribution and use systems optimisation for the major facilities (cogeneration plants projects heating installations, which are already under investigation within some projects in the Citadels of Knowledge Melfi and Potenza, etc.) and energy production, distribution and use systems verification in small plants (for example by fostering the replacement of inefficient heat generators, which is already in progress in the buildings owned by the Province);

- incentives programmes for the correct use of energy (targeted information campaigns and realisation of an "energy help desk" web);
- provincial field planning update (Transport Plan, Waste Management Plan, etc. , which are associated with energy and environmental aspects);
- special projects for the renewable energy sources exploitation;
- definition of guidelines and technical provisions within the planning tools (implementing technical standards, sustainable building code, etc.) which consider also energy rationalization projects (i.e. public lighting systems rationalization through installations review projects and programmed lamps replacement, buildings energy certification, etc.);
- information campaigns aimed at the proper use of energy through initiatives and demonstration projects.

5 Strategy Overview

The Province of Potenza, in the framework of territorial strategies defined at regional level and coherently with the objectives of the Covenant of Mayors, pursues a general objective of sustainability and energy self-sufficiency. That includes applications in all areas of demand/supply of energy and, in an operative view concerning the production of energy from RES.

It is a component of the political long-term strategy defined 'Abitare Futuro' that focuses the need to improve the quality of life for Basilicata citizens, in terms of equity, opportunity, development, employment, services.

The Implementation Plan includes three strategic lines that represent relevant areas of action that could contribute effectively to the current strategies framework and on going processes at territorial level.

The idea is to propose experimental projects implementing significant parts form experiences learned during the project Renergy by comparison Partners' Good Practices. In the following part of the document are described procedural aspects, visions and operative contributions contextualized in the territory of the Province of Potenza and coherent with local programming framework which also includes the perspective of the regional operational programs for 2014-2020.

Strategic line 1: Energy Villages

The role of small communities of Appennine Mountains and the development perspectives linked to a specialization in the areas of EE and RES

This strategic line includes 2 Pillars:

- P1 Energy Village Mainstream - Support for energy qualification of small local communities through integrated interventions (infrastructure, technologies and incentives) proposed through participatory planning.
- P2 Virtual LAB - ICT platform for the participatory design, communication and dissemination of results, education on energy efficiency, smart monitoring - in real time – of projects and local actions.

Table 2. Strategic line 1: Energy Villages (roles and responsibilities)

Role	Description	Entities
Promotor	The Province of Potenza, in the framework of the activities of institutional coordination between municipalities and the Basilicata Region, will develop an action of promoting strategic line in order to identify programmatic areas appropriate to carry out and finance activities	Province of Potenza in collaboration with Regional Administration for funding ad operative and procedural design
Owner	Public Administrations	*To be defined according with operative implementation*
Financing Entity / Resources	Local and National Funds, Regional Programs	*To be defined according with operative implementation*
Clients/Users	Local Municipality and local communities actively involved in the implementation	To be defined according with operative implementation among municipalities belonging to the peripheral areas of Basilicata Region
Key partnerships	Public – Private partnership will be encouraged in order to maximize the value of public funds and to promote market uptake	*To be defined according with operative implementation*
Consultants	Technical experts supporting the promoters and local communities in developing activities	*To be defined according with operative implementation*

Strategic line 2: ESCO - Engaged SMART Communities

From experience 'Futurenergy' to an integrated program of actions addressed to public schools.

This strategic line provides a direct action of the Province of Potenza in relation to the institutional competence on the management of a large portion of the school buildings on the territory. The proposed activities integrate ongoing actions and are intended to test and monitor successful actions to be extended to all the institutional holders of similar administrative functions in the management of public buildings.

Main activities to be developed for implementation:

-   Action 1: 50/50 – To stimulate energy saving actions in school communities in order to allocate saved financial resources to the direct needs of the institutions (new school activities, supports, benefits etc.) and to develop new investments for savings and energy retrofitting of school buildings

- Action 2: To share the optimization EE and RES through compensation measures and benefit sharing according to the model created by the project of Solar Community BOLOGNA
- Action 3: EE from behavioral styles awareness: you can save energy promoting sustainable behaviors in school communities
- Action 4: Private Sector cooperation: Evaluate tools and procedures to open the active participation of private sector in the interventions of EE on public property and on the networks and service infrastructure as envisaged in the PSP.

Table 3. Strategic line 2: ESCO - Engaged SMART Communities (roles and responsibilities)

Role	Description	Entities
Promotor	The Province of Potenza is the owner and the manager of the school real estate throughout the territory of competence	Province of Potenza
Owner	Province of Potenza	Province of Potenza
Financing Entity / Resources	Local and regional funds, Regional programs for education and infrastructures	*To be defined according with operative implementation*
Clients/Users	Schools and students communities	*To be defined according with operative implementation*
Key partnerships	PPPP (Public, Private, People Partnership) related to the identification of actions, behaviors, procedures to encourage sustainability and investment in RES	*To be defined according with operative implementation*
Consultants	Technical experts supporting the promoter	*To be defined according with operative implementation*

Strategic line 3: Stake Holders Involvement

From ENERGY LAB an Agenda for 'relevant stake holders' in EE and RES
Main activities to be developed for implementation:

- Action 1: Establish permanent consultation tables with the local actors of the public system, with private operators, with the associations and ONG engaged in the activities of planning and development of interventions in the field of 'EE and RES

Table 4. Strategic line 3: Stake Holders Involvement (roles and responsibilities)

Role	Description	Entities
Promotor	In the Project Renergy a significant part of activities was addressed to the involvement of local Stake Holders with the instrument of ENERGY LABS. This activity was coordinated by the Province of Potenza, which intends to maintain such actions in partnership with other relevant regional player constituting independent associative structures on the model of "network operators Slagelse" oriented to trust building and market development of EE and RES	Province of Potenza
Owner	--	--
Financing Entity / Resources	Associations and groups will be formed in a system of self-financing that does not require the intervention of public resources	*To be defined according with operative implementation*
Clients/Users	Local governments, businesses, associations	*To be defined according with operative implementation*
Key partnerships	PPPP (Public, Private, People Partnership)	*To be defined according with operative implementation*
Consultants	Technical experts supporting the promoter	*To be defined according with operative implementation*

6   Conclusions

This Local Implementation Plan, described synthetically in this work, holds two different level of operational contribution: on one hand it represents an effective example of framework implementation for lessons learned in a best practices exchange project; on the other fits within a territorial planning instrument (the PSP of the Province of Potenza) orienting general policies towards specific implementation projects.

Such contributions has an added values for coordinating the activities of local municipalities in a common strategy looking at local sustainable development. And this approach fits with Province institutional roles and functions.

The feasibility of the strategic lines implementation refers to financial sources identified among the intervention lines of the regional development programmes linked with the 2014-2020 New Cohesion Policy sources [6]. Also local funding sources and in particular those ones deriving from the Val d'Agri fossil sources exploitation. The main financial instruments for the LIP implementation are:

- *PO FESR 2014/2020 Basilicata Region* - The programme is the main tool to support the structural interventions in behalf of the territorial authorities, the companies and the citizens to achieve the development objectives within which the energy subject matter (in terms of EE and RES) represents a strategic component for the Basilicata Region. The integrated programming approach, which has produced efficient applications in the Basilicata Region (ex. PIT, PISUS, PIOT) represents the operating model which has to be developed in an experimental key for the Implementation Plan ENERGY VILLAGE.
- *PO Val D'Agri* - It is the operational tool for the Basilicata Region resources from oil extraction fields exploitation management. The program is about a specific territorial field which is also identified in the regional program as 'energy district' with specialisations in technological innovation and sustainability.

These conditions enhance to verify a territorial specialisation in order to include the Implementation Plan actions in an integrated actions and financial sources framework.

The perspective regards the integration of such strategic schema in the framework of Territorial Resilience strategy promoted by the Province of Potenza in cooperation with UNISDR [7] [8]. The Province developed a multi-stakeholders participatory process in order to include a number of stakeholders (crf. [9]). In fact energy efficiency and energy sustainability represents pieces of the jigsaw of territorial security. We refer to Disaster Risks Reduction, Prevention and Management and Climate change adaptation which are the up to date global discussion in terms of territorial and communities sustainable perspectives as discussed recently in the Sendai World Conference promoted by UNISDR.

References

1. Scorza, F., Attolico, A., Moretti, V., Smaldone, R., Donofrio, D., Laguardia, G.: Growing sustainable behaviors in local communities through smart monitoring systems for energy efficiency: RENERGY outcomes. In: Murgante, B., Misra, S., Rocha, A.M.A., Torre, C., Rocha, J.G., Falcão, M.I., Taniar, D., Apduhan, B.O., Gervasi, O. (eds.) ICCSA 2014, Part II. LNCS, vol. 8580, pp. 787–793. Springer, Heidelberg (2014). doi:10.1007/978-3-319-09129-7_57
2. Las Casas, G., Lombardo, S., Murgante, B., Pontrandolfi, P., Scorza, F.: Open Data for Territorial Specialization Assessment. Territorial Specialization in Attracting Local Development Funds: an Assessment. Procedure Based on Open Data and Open Tools, in Smart City. Planning for Energy, Transportation and Sustainability of the Urban System Special Issue, June 2014 Editor-in-chief: Rocco Papa, TeMA Journal of Land Use, Mobility and Environment, print ISSN 1970-9889 | on line ISSN 1970-9870 (2014)
3. Frazier, T.G., Thompson, C.M., Dezzani, R.J., Butsick, D.: Spatial and temporal quantification of resilience at the community scale. Applied Geography **42**, 95–107 (2013). doi:10.1016/j.apgeog.2013.05.004
4. Birkmann, J.: Risk and vulnerability indicators at different scales: Applicability, usefulness and policy implications. Environmental Hazards **7**(1), 20–31 (2007). doi:10.1016/j.envhaz.2007.04.002

5. Tilio, L., Murgante, B., Di Trani, F., Vona, M., Masi, A.: Resilient city and seismic risk: a spatial multicriteria approach. In: Murgante, B., Gervasi, O., Iglesias, A., Taniar, D., Apduhan, B.O. (eds.) Computational Science and Its Applications - ICCSA 2011. LNCS (LNAI), vol. 6782, pp. 410–422. Springer, Heidelberg (2011). doi:10.1007/978-3-642-21928-3_29
6. Scorza, F.: Improving EU cohesion policy: the spatial distribution analysis of regional development investments funded by EU structural funds 2007/2013 in Italy. In: Murgante, B., Misra, S., Carlini, M., Torre, C.M., Nguyen, H.-Q., Taniar, D., Apduhan, B.O., Gervasi, O. (eds.) ICCSA 2013, Part III. LNCS, vol. 7973, pp. 582–593. Springer, Heidelberg (2013). doi:10.1007/978-3-642-39646-5_42. ISBN: 978-3-642-39645-8, ISSN: 0302-9743
7. Johnson, C., Blackburn, S.: Advocacy for urban resilience: UNISDR's Making Cities Resilient Campaign. Environment and Urbanization 26(1), 29–52 (2014). doi:10.1177/0956247813518684
8. Djalante, R.: Review Article: Adaptive governance and resilience: The role of multi-stakeholder platforms in disaster risk reduction. Natural Hazards and Earth System Science 12(9), 2923–2942 (2012). doi:10.5194/nhess-12-2923-2012
9. Bäckstrand, K.: Multi-stakeholder partnerships for sustainable development: Rethinking legitimacy, accountability and effectiveness. European Environment 16(5), 290–306 (2006). doi:10.1002/eet.425

Smart Monitoring System for Energy Performance in Public Building

Francesco Scorza[(⊠)]

School of Engineering, Laboratory of Urban and Regional Systems Engineering,
University of Basilicata, 10, Viale dell'Ateneo Lucano 85100, Potenza, Italy
francescoscorza@gmail.com

Abstract. Real time information become an usual way for common citizen to access and use data coming from own information systems. This imply new issues for ICT application in main fields and domains such as 'energy efficiency', sustainability, energy management. According to the growing interest in energy saving as a relevant component of territorial sustainability we developed an application based on open-source technologies providing real-time open data of energy consumptions. This hw-sw system can be oriented to individual householder needs, such as to industrial purposes and public ones. The paper discuss the preliminary results of the application of such technologies on public schools building in an integrated project linking usage model of public spaces to citizens behaviours and consciousness concerning sustainability. Outcomes could influence territorial policies and projects in the framework of EU 2020 strategy and Covenant of Majors.

Keywords: Energy efficiency · Sustainability · Energy planning · Smart monitoring systems · Covenant of majors

1 Introduction

Climate change mitigation policy are linked to the carbon footprint idea of local communities and productive structures. Literature discusses widely about barriers to energy system changes, such as funding, financing and information, are well recognized,

The affirmed vision of this work is based on the definition of 'energy upgrades' as strategic investments motivated by environmental sustainability and corporate social responsibility [1].

How users can have information concerning their specific demand for energy and consequently their consumption? This represents a basic but complex question we try to face with this paper.

We define the users in the wider sense: it means each consumption point. In other words it is not the single citizen, or the industrial activities under investigation, but the 'reason way' an energy demand was determined by an anthropic activity.

© Springer International Publishing Switzerland 2015
O. Gervasi et al. (Eds.): ICCSA 2015, Part II, LNCS 9156, pp. 767–774, 2015.
DOI: 10.1007/978-3-319-21407-8_55

It is important to affirm that a common sustainable consciousness is growing due to increasing costs connected to energy and the effort in policy making at each level in promoting energy efficiency and sustainability toward common sense.

In this current situation the energy user express new need for information concerning his individual energy footprint. According to internet era principles, in order useful and successful, the information should be available on the net, accessible by multi-device system, accurate and in (near) real-time. In particular real time information becomes an usual way for common citizen to access and use data coming from information systems. This implies new issues for ICT application in main fields and domains such as 'energy efficiency', sustainability, and energy management.

The "individual" accounting of energy demand in household applications (heating, cooling, electricity) [2], but also for productive (industrial) and public applications, represents a tool for increasing energy efficiency. The individual assessment, in several cases, prevails on technical energy consulting and in particular it allow to identify the causes of any waste of energy and consequently to adopt strategies to improve energy efficiency.

According to the growing interest in energy saving as a relevant component of territorial sustainability we developed an application based on open-source technologies providing real-time open data of energy consumptions. This hw-sw system can be oriented to individual householder needs, such as to industrial purposes and public ones. The paper discuss the preliminary results of the application of such technologies on public schools building in an integrated project linking usage model of public spaces to citizens behaviours and consciousness concerning sustainability. Outcomes could influence territorial policies and projects in the framework of EU 2020 strategy and Covenant of Majors.

2   Energy Planning and Monitoring Issues

Sustainability in energy planning at territorial scale represents a new challenge speeded out in the framework of EU 2020 agenda. It means additional tasks for territorial administrations engaged with traditional planning issues and managing the territorial dynamics in a framework of strong community claims concerning development perspectives and strategic visions.

The complexity of a "context based" [3] approach (i.e. Place based approach [4]) managing together territorial development, environmental preservation, sustainable energy development, and socio economic growth appear more as a chimera than as a strategy. Especially in the "time of the crisis" which addressed on EU policy and regulations a share of negligence.

Several documents regard the long term EU policy vision is concerning sustainable development [5] [6] [7] and, more recently in the Europe 2020 strategy [8] the instances of knowledge and innovation, competitive economy, social and territorial cohesion were oriented to enrich the umbrella objective "to promote a smart, sustainable and inclusive EU growth".

The territorial planning faced the new strategic vision of sustainable Europe in several operative way. One of the most effective tools are represented by the "Covenant of Mayors" [9]. One of the results of such policy is the number of SEAP (sustainable energy action plan). A new operative planning tool addressing energy sustainability at local level (municipal level).

After the preliminary phase of dissemination and involvement of European Municipalities in the application of the Covenant of Major and the first generation of SEAP now the attention is shifted on the monitoring phase of implementation. In some case it means to measure investments but the target fulfilment depends also on communities involvement and individual behaviours.

This represents a background instance for this paper.

3   Energy, Complexity and Barriers

The challenge of moving to sustainable energy systems which provide secure, affordable and low-carbon energy services requires technological improvement, trust-building in consumers and professionals and information.

Energy systems consist of a complex structure including many actors, interacting through networks, leading to emergent properties and adaptive and learning processes. Although complexity science is not well understood by practitioners in the energy domain (and is often difficult to communicate), the research of complexity models in energy chain can be used to aid decision-making at multiple levels [10].

The application is based on an ICT system that allow user individual assessment of energy demand and the definition of an user-model of consumption that could drive towards sustainable behaviour producing relevant energy saving and energy costs reductions.

Common sense barriers toward sustainability and energy efficiency improvement depends on several factors. Among others:

- levels of satisfaction with energy (i.e. HVAC) systems,
- user confidence with measuring/distributing systems
- cultural barriers that prevent the delivery of low energy homes,
- people's acceptance of investments in energy efficiency plants or technology as a standard for sustainable homes.

It is interesting the analysis by Aldossary et al. [11] in Arabia Soudita based on a comprehensive survey was conducted across the country. The findings reveal limited public awareness as well as important socio-cultural barriers to the delivery of sustainable homes.

The ICT application looks at reducing such cultural obstacles pushing on the common interest in data exploitation through web and mobile interface combined with the general saving energy demand linked to on-going trend and economic crisis effects.

4   An ICT Infrastructure for Monitoring Energy Performance in Private and/or Public Building

One of the most relevant consumption point for energy users is represented by the building where they develop activities or they benefit services. Let's consider the house as the main consumption poitn for a citizen.

To define building energy performance imply an assesment model or matrix that, according to traditional practices connected the the complex system of rating and labelling within the framework of building energy certification schemes [12] (i.e. for italy CasaClima (cfr. [13]).

The issuer of smart monitoring system development already produced several example world-wide according to commercial or open prototyping and systems (among others: [14] [15] [16] [17]).

The individual assessment cannot respect such technical procedures. We look mainly at providing information intercepting the two main interest of the citizen: a measure of consumption; a generalized model to combine energy saving and usage of the building.

We developed a system based on an open hardware interfaces equipped with proper sensors allowing to produce real-time data concerning: temperature, humidity and lightning in rooms and portions of the building (indoor and outdoor) and a system for monitoring electricity consumption of main components of the building wiring.

Those equipment are connected with local wi-fi and produce data on real-time measure of the selected parameters.

The system store and manage data through a Energy Content Management System providing synthetic representation through a web interface.

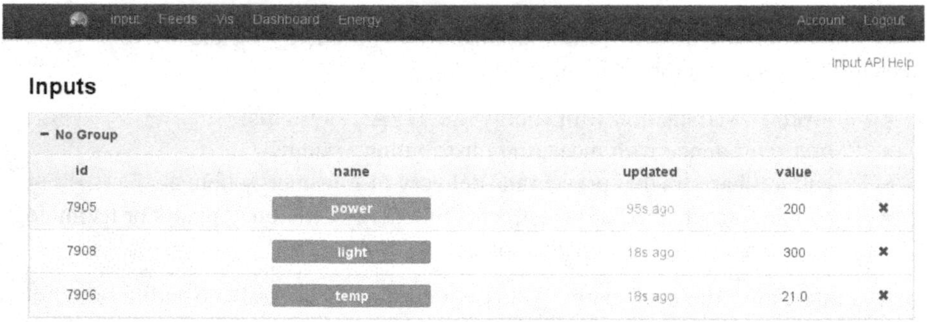

Fig. 1. Input interface for sensors

Time series graph, widgets and other graphic options allow to represent information and to generate an user friendly dashboard oriented to increase user understanding on energy consumption and building performance according to the usage scheme he adopted.

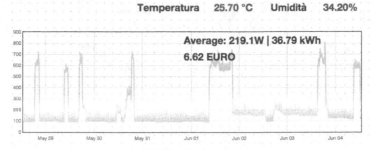

Fig. 2. Output information

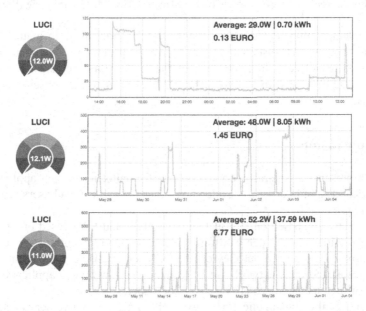

Fig. 3. Daily, weekly, monthly data representation of electrical consumption in a real monitoring point

An interesting application of this system was developed by the Province of Potenza (Italy) and by the Municipality of Avrig (Romania). They selected the system to realize a smart monitoring system for the assessment of energy performance in public school buildings [18].

Liceu - Valori in timp real

	Temperatura	Umiditate	Luminozitate		Consumul de curent pe cele 3 faze
Sala 2	14.5°C	50 %	12	F1	1268 W
Sala 14	19.3°C	33 %	3	F2	247 W
Sala 12	16.4°C	30 %	15	F3	12 W
Lab. Info	29.1°C	39 %	17		
Cancelarie	25.2°C	50 %	57		

Fig. 4. Real-time date from LICEUL TEHNOLOGIC MIRSA in Avrig (RO)

The implementation project aims at support education and information activities concerning energy and sustainability in education programs developed by public schools (www.ee-sms.eu). In a general view the applications are oriented both on the private and public case with increasing benefit according to the consciousness of the beneficiaries.

5   Conclusions

This work presents a general overview of smart monitoring system application and technological design in order to develop conscious behavior in end users according to sustainable principles.

The case study is based on ICT technologies (HW and SW) based on open source approach. The idea is that open hardware (such as arduino / genuino) are suitable not only for prototyping device but they can be used for extensive application providing excellent results. Sensors network can be realized in a low budget application and the diffusion of such technologies is expected to be wider.

A general goal of this application is to reach an active involvement of local communities. It means that a bottom up approach based on people commitment could contribute in trust-building and a positive competition on technological advance could be expected.

Planning implication depends on the scale of the analysis: at detailed scale we can expect implication in real estate analysis in public/private building management [19]; concerning territorial dimension such systems could generate information driving investments [20].

The implemented system has also the feature to provide open data concerning punctual consumption and it works according interoperability standards [21].

It is possible to affirm that conscious local communities could promote better and effective solution for sustainable cities and territories in short term view if supported by low-cost and accessible technologies.

References

1. Gliedt, T., Hoicka, C.E.: Energy upgrades as financial or strategic investment? Energy Star property owners and managers improving building energy performance. Applied Energy **147**, 430–443 (2015). doi:10.1016/j.apenergy.2015.02.028
2. Celenza, L., Dell'Isola, M., Ficco, G., Palella, B.I., Riccio, G.: Heat accounting in histori-cal buildings. Energy and Buildings **95**, 47–56 (2015). doi:10.1016/j.enbuild.2014.10.070
3. Las Casas, G., Scorza, F.: Un approccio "context-based" e "valutazione integrata" per il futuro della programmazione operativa regionale in Europa. In: Bramanti, A., Salone, C. (eds.) Lo sviluppo territoriale nell'economia della conoscenza: teorie, attori strategie. Collana Scienze Regionali, vol. 41. FrancoAngeli, Milano (2009)
4. Barca, F.: An agenda for a reformed cohesion policy: a place-based approach to meeting European union challenges and expectations. Independent report prepared at the request of the European Commissioner for Regional Policy, Danuta Hübner, European Commission, Brussels (2009)
5. COM: 264 final, European Commision, 2001 (2001)
6. COM: 658 final, European Commision, 2005 (2005)
7. COM: 772 final, European Commision, 2008 (2008)
8. COM: 639 final, European Commision, 2010 (2010)
9. Covenant of Mayors: European Commission (2013)
10. Bale, C.S.E., Varga, L., Foxon, T.J.: Energy and complexity: New ways forward. Applied Energy **138**, 150–159 (2015). doi:10.1016/j.apenergy.2014.10.057
11. Aldossary, N.A., Rezgui, Y., Kwan, A.: An investigation into factors influencing domestic energy consumption in an energy subsidized developing economy. Habitat International **47**, 41–51 (2015). doi:10.1016/j.habitatint.2015.01.002
12. Pérez-Lombard, L., Ortiz, J., González, R., Maestre, I.R.: A review of benchmarking, rating and labelling concepts within the framework of building energy certification schemes. Energy and Buildings **41**(3), 272–278 (2009). doi:10.1016/j.enbuild.2008.10.004
13. Schmitt, Y., Troi, A., Pichler, G., Sparber, W.: Klimahaus Casaclima - A regional energy certification system stimulates low energy architecture. In: Sun, Wind and Architecture - The Proceedings of the 24th International Conference on Passive and Low Energy Architecture, PLEA 2007, pp. 531–536 (2007). http://www.scopus.com/inward/record. url?eid=2-s2.0-84864138229&partnerID=40&md5=a8a5db6b9429f85182abffb20ae56618
14. Tristo, G., Bissacco, G., Lebar, A., Valentinčič, J.: Real time power consumption monitoring for energy efficiency analysis in micro EDM milling. International Journal of Advanced Manufacturing Technology **78**(9–12), 1511–1521 (2015). doi:10.1007/s00170-014-6725-3
15. Hwang, K.-I., Jeong, Y.-S., & Lee, D. G. (2015). Hierarchical multichannel-based integrated smart metering infrastructure. *Journal of Supercomputing*. doi:10.1007/s11227-015-1441-9
16. Jiao, W., Hagler, G.S.W., Williams, R.W., Sharpe, R.N., Weinstock, L., Rice, J.: Field assessment of the village green project: An autonomous community air quality monitoring system. Environmental Science and Technology **49**(10), 6085–6092 (2015). doi:10.1021/acs.est.5b01245
17. Celenza, L., Dell'Isola, M., Ficco, G., Palella, B.I., Riccio, G.: Heat accounting in historical buildings. Energy and Buildings **95**, 47–56 (2015). doi:10.1016/j.enbuild.2014.10.070

18. Scorza, F., Attolico, A., Moretti, V., Smaldone, R., Donofrio, D., Laguardia, G.: Growing Sustainable Behaviors in Local Communities through Smart Monitoring Systems for Energy Efficiency: RENERGY Outcomes. In: Murgante, B., Misra, S., Rocha, A.M.A., Torre, C., Rocha, J.G., Falcão, M.I., Taniar, D., Apduhan, B.O., Gervasi, O. (eds.) ICCSA 2014, Part II. LNCS, vol. 8580, pp. 787–793. Springer, Heidelberg (2014)
19. Morano, P., Tajani, F.: Least median of squares regression and minimum volume ellipsoid estimator for outliers detection in housing appraisal. International Journal of Business Intelligence and Data Mining **9**(2), 91–111 (2014)
20. Scorza, F.: Improving EU Cohesion Policy: The Spatial Distribution Analysis of Regional Development Investments Funded by EU Structural Funds 2007/2013 in Italy. In: Murgante, B., Misra, S., Carlini, M., Torre, C.M., Nguyen, H.-Q., Taniar, D., Apduhan, B.O., Gervasi, O. (eds.) ICCSA 2013, Part III. LNCS, vol. 7973, pp. 582–593. Springer, Heidelberg (2013)
21. Scorza, F., Casas, G.L., Murgante, B.: Overcoming Interoperability Weaknesses in e-Government Processes: Organizing and Sharing Knowledge in Regional Development Programs Using Ontologies. In: Lytras, M.D., Ordonez de Pablos, P., Ziderman, A., Roulstone, A., Maurer, H., Imber, J.B. (eds.) WSKS 2010. CCIS, vol. 112, pp. 243–253. Springer, Heidelberg (2010)

Author Index

Printed in the United States
By Bookmasters